TABLE OF CONTENTS
PART 1

PROCEEDINGS OF SYMPOSIA
IN PURE MATHEMATICS
Volume 38, Part 2

OPERATOR ALGEBRAS AND APPLICATIONS

AMERICAN MATHEMATICAL SOCIETY
PROVIDENCE, RHODE ISLAND
1982

PROCEEDINGS OF THE SYMPOSIUM IN PURE MATHEMATICS
OF THE AMERICAN MATHEMATICAL SOCIETY

HELD AT QUEENS UNIVERSITY
KINGSTON, ONTARIO
JULY 14–AUGUST 2, 1980

EDITED BY
RICHARD V. KADISON

Prepared by the American Mathematical Society
with partial support from National Science Foundation grant MCS 79-27061

Library of Congress Cataloging in Publication Data

Symposium in Pure Mathematics (1980: Queens University, Kingston, Ont.)
Operator algebras and applications.
(Proceedings of symposia in pure mathematics; v. 38)
Includes bibliographies and index.
1. Operator algebras–Congresses. I. Kadison, Richard V., 1925–
II. Title. III. Series.
QA326.S95 1982 512′.55 82–11561
ISBN 0-8218-1441-9 (v. 1)
ISBN 0-8218-1444-3 (v. 2)
ISBN 0-8218-1445-1 (set)

1980 Mathematics Subject Classification. Primary 46L05, 46L10;
Secondary 43A80, 81E05, 82A15.

TABLE OF CONTENTS
PART 2

AUTHOR INDEX

Proceedings of Symposia in Pure Mathematics
Volume **38** (1982), Part 2

CELEBRATION OF TOMITA'S THEOREM

Van Daele Alfons

ABSTRACT. We first state Tomita's famous theorem. Then we give a proof in a special case because we feel it helps a lot in understanding better the general case. Finally we indicate the main steps in the proof of the general case.

Let $\mathcal{H}$ be a Hilbert space and let M be a von Neumann algebra with a separating and cyclic vector ω.

1. NOTATION. Denote by S the closure of the conjugate linear map $x\omega \to x^*\omega$ where $x \in M$, and denote by F the adjoint of S.

2. PROPOSITION. *There exists a conjugate linear isometric involution* J *on* $\mathcal{H}$ *and a non-singular positive self-adjoint operator* Δ *such that*

i) $S = J\Delta^{1/2} = \Delta^{-1/2}J$
$F = J\Delta^{-1/2} = \Delta^{1/2}J$

ii) $\Delta = FS$ *and* $\Delta^{-1} = SF$

iii) $J\Delta J = \Delta^{-1}$.

Because $S\omega = \omega$ we also have $F\omega = \omega$ and $J\omega = \omega$ as well as $\Delta\omega = \omega$. All these results are obtained by considering the polar decomposition of S. Using spectral theory one also defines the one-parameter group of unitaries $\{\Delta^{it} | t \in \mathbb{R}\}$. We have $J\Delta^{it}J = \Delta^{it}$ and $\Delta^{it}\omega = \omega$ for all $t \in \mathbb{R}$.

We will now prove and discuss the following famous theorem of Tomita [4,5].

3. THEOREM. *We have* $JMJ = M'$ *and* $\Delta^{it}M\Delta^{-it} = M$ *for all* $t \in \mathbb{R}$.

We will first give a fairly complete proof in the case where S is bounded.

4. LEMMA. *We have* $SMS = M'$ *and* $FMF = M'$ *and also* $\Delta M\Delta^{-1} = M$.

Proof : For any x,y and z in M we have $SxSy\omega = Sxy^*\omega = yx^*\omega = ySx\omega$ and if we apply this result twice we have $SxSyz\omega = yzSx\omega = ySxSz\omega$. Since $M\omega$ is dense we also get $SxSy = ySxS$ and because this is true for all $x,y \in M$ we obtain $SMS \subseteq M'$. From $Fx'\omega = x'^*\omega$ when $x' \in M$ we get $FM'F \subseteq M$ by a similar calculation. Taking adjoints we obtain $SM'S \subseteq M$ and also $M' \subseteq SMS$. Therefore

1980 Mathematics Subject Classification 46 L 10

0082-0717/80/0000-0584/\$02.00

$SMS = M'$ and similarly $FM'F = M$. Finally since $\Delta = FS$ and $\Delta^{-1} = SF$ we get $\Delta M \Delta^{-1} = M$.

5. LEMMA. *For all* $z \in \mathbb{C}$ *we have* $\Delta^z M \Delta^{-z} = M$.

Proof : Define $H = \log \Delta$ and $\psi(x) = Hx - xH$ and $\phi(x) = \Delta x \Delta^{-1}$ when $x \in \mathcal{B}(\mathcal{H})$. Then $\phi = \exp \psi$ by a well known formula. We know that ϕ leaves M invariant. By the holomorphic calculus ψ can be expressed as an integral in terms of the resolvents of ϕ. Since $Sp\phi \subseteq Sp\Delta \,.\, Sp\Delta^{-1}$ we have that the resolvent set is connected in $\mathbb{C}$ so that also the resolvents leave M invariant. Then also ψ leaves M invariant. Finally because $\Delta^z x \Delta^{-z} = (e^{z\psi})(x)$ for all $z \in \mathbb{C}$, the lemma is proved.

It is clear how the theorem will follow in the bounded case. Indeed because $\Delta^z M \Delta^{-z} = M$ for all $z \in \mathbb{C}$, in particular $\Delta^{it} M \Delta^{-it} = M$ for all $t \in \mathbb{R}$. But also $\Delta^{1/2} M \Delta^{-1/2} = M$ which together with $S = J\Delta^{1/2} = \Delta^{-1/2}J$ and $SMS = M'$ implies $JMJ = M'$.

We now proceed to the general case. The proof we have given in the bounded case is a very natural one, and therefore we first tried to use similar arguments in the general case. Without to much difficulties we were able to find a suitable generalization for the map $\phi : x \to \Delta x \Delta^{-1}$ when Δ is unbounded. In fact a little later this map was extensively studied by Cioranescu and Zsido in their paper on the analytic generator [1]. We could not obtain an easy analogue of the result $SMS = M'$ however. Later Zsido in [7] and Kadison in [2] have obtained proofs of the Tomita theorem along these lines.

Our proof lies closer to the original one of Tomita, and in fact we would consider the following resolvent property as the central result both in our approach and in the one of Tomita.

6. LEMMA. *For all* $\lambda \in C \setminus [0,\infty]$ *we have* $(\Delta-\lambda)^{-1} M'\omega \subseteq M$.

When $\lambda = -1$ this result is essentially Sakai's linear Radon Nikodym theorem [3] :

Proof of the lemma for $\lambda = -1$: Take $x' \in M'$ and define $\varphi(x) = \langle x\omega, \omega \rangle$ and $\psi(x) = \langle x\omega, x'\omega \rangle$ where $x \in M$. If $0 \leqslant x' \leqslant 1$ then also $0 \leqslant \psi \leqslant \varphi$ and by Sakai's theorem there exists an $h \in M$ such that $0 \leqslant h \leqslant 1$ and $\psi(x) = \frac{1}{2}\varphi(xh+hx)$ for all $x \in M$. Using the definition of φ and ψ we get

$$\begin{aligned} \langle x\omega, x'\omega \rangle &= \frac{1}{2}(\langle xh\omega, \omega \rangle + \langle hx\omega, \omega \rangle) \\ &= \frac{1}{2}(\langle h\omega, x^*\omega \rangle + \langle x\omega, h\omega \rangle) \\ &= \frac{1}{2}(\langle Sh\omega, Sx\omega \rangle + \langle x\omega, h\omega \rangle) \end{aligned}$$

because $h = h^*$ and the definition of S. Since $\Delta = S^*S$ it follows that $h\omega \in \mathcal{D}(\Delta)$ and that $x'\omega = \frac{1}{2}(\Delta+1)h\omega$. This completes the proof.

Sakai's theorem can easily be modified to obtain the lemma for all λ of modulus 1. Using a 2 x 2 matrix trick, the general case can be obtained from the case with λ of modulus 1. This was pointed out to me by Haagerup. The resolvent property is also true in the more general setting of left Hilbert algebras. There the proof is completely different. For more details, see [6].

To get from the resolvent property to the main result, two intermediate steps are needed. The full proofs can also be found in [6].

7. LEMMA. *For any* $r > 0$ *we have*

$$\frac{\Delta^{1/2}}{\Delta + r} = \int_{-\infty}^{+\infty} \frac{r^{-it-1/2}}{e^{\pi t} + e^{-\pi t}} \Delta^{it}\, dt$$

8. PROPOSITION. *If* $r > 0$ *and* $x' \in M'$ *and* $h \in M$ *such that* $(\Delta + r)^{-1} x'\omega = h\omega$ *then*

$$h = \int_{-\infty}^{+\infty} \frac{r^{-it-1/2}}{e^{\pi t} + e^{-\pi t}} \Delta^{it}\, J\, x'^{*} J \Delta^{-it}\, dt.$$

The formula in the lemma is obtained first in scalar valued form by using complex integration, and then in operator valued form by using spectral theory. The proof of proposition 8 is rather technical. The following special case however clarifies the nature of this result.

Proof of the proposition for Δ bounded : The proof is obtained in two steps. First we show that

$$Jx'^{*}J = \Delta^{1/2} h\, \Delta^{-1/2} + r\, \Delta^{-1/2} h\, \Delta^{1/2}.$$

Take $y, z \in M$, then

$$\begin{aligned} \langle x'y\omega, z\omega\rangle &= \langle x'\omega, y^{*}z\omega\rangle \\ &= \langle \Delta h\omega, y^{*}z\omega\rangle + r\langle h\omega, y^{*}z\omega\rangle \\ &= \langle Sy^{*}z\omega, Sh\omega\rangle + r\langle h\omega, y^{*}z\omega\rangle \\ &= \langle z^{*}y\omega, h^{*}\omega\rangle + r\langle h\omega, y^{*}z\omega\rangle \\ &= \langle y\omega, zh^{*}\omega\rangle + r\langle yh\omega, z\omega\rangle \\ &= \langle y\omega, ShSz\omega\rangle + r\langle Sh^{*}Sy\omega, z\omega\rangle \end{aligned}$$

Hence $x'^{*} = ShS + r\, FhF$ and since $S = J\Delta^{1/2} = \Delta^{-1/2}J$ and $F = J\Delta^{-1/2} = \Delta^{1/2}J$ we get the desired formula.

The next step is to show that

$$x = \Delta^{1/2} h \Delta^{-1/2} + r\, \Delta^{-1/2} h \Delta^{1/2}$$

is equivalent with

$$h = \int_{-\infty}^{+\infty} \frac{r^{-it-1/2}}{e^{\pi t} + e^{-\pi t}} \Delta^{it}\, x\, \Delta^{-it}\, dt.$$

And this is infact nothing else but the formula of lemma 7 for the map $\Delta \cdot \Delta^{-1}$ in the place of Δ. The proof is essentially the same and again uses complex integration.

We now sketch the proof of theorem 3 in the general case :

Take $x' \in M'$, then by lemma 6,7 and proposition 8 we have that

$$\int_{-\infty}^{+\infty} \frac{r^{-it-1/2}}{e^{\pi t} + e^{-\pi t}} \Delta^{it} Jx'^{*}J\Delta^{-it} \, dt$$

belongs to M for all $r > 0$. Rewriting this formula with $r = \exp s$ and $s \in \mathbb{R}$, and using Fourier analysis it follows that $\Delta^{it} Jx'^{*}J\Delta^{-it} \in M$ for all $t \in \mathbb{R}$. Therefore also $\Delta^{it} JM'J\Delta^{-it} \subseteq M$ for all $t \in \mathbb{R}$. And then simple arguments show $JMJ = M'$ and $\Delta^{it} M\Delta^{-it} = M$.

BIBLIOGRAPHY

1. I. Cioranescu & L. Zsido : "Analytic generators for one-parameter groups". Tohoku Math. J. 28 (1976) 327-362.

2. R. Kadison : "Similarity of operator algebras" Acta. Math. 141 (1978) 147-163.

3. S. Sakai : "C*-algebras and W*-algebras", Ergebnisse der Mathematik und ihrer Grenzgebiete, Band 60, Berlin (1971).

4. M. Takesaki, "Tomita's theory of modular Hilbert algebras and its applications". Springer Lecture Notes in Mathemathics 128, Springer Verlag, 1970.

5. M. Tomita, "Quasi-Standard von Neumann algebras" Mimeographed note (1967).

6. A. Van Daele : "The Tomita-Takesaki theory for von Neumann algebras with a separating and cyclic vector". Proc. Int. School of Phys. Enrico Fermi, Ital. Phys. Soc. Publ., Bologna (1976).

7. L. Zsido : "A proof of Tomita's fundamental theorem in the theory of standard von Neumann algebras". Revue Comm. Pures et Appl. 20 (1975) 609-619.

DEPARTMENT OF MATHEMATICS
UNIVERSITY OF LEUVEN
LEUVEN, BELGIUM.

Proceedings of Symposia in Pure Mathematics
Volume **38** (1982), Part 2

POSITIVE CONES FOR VON NEUMANN ALGEBRAS

Huzihiro Araki

ABSTRACT. For the standard representation of a von Neumann algebra M on a Hilbert space H with a cyclic and separating vector η, a one parameter family of closed convex cones V_η^α is defined as the closure of $\Delta_\eta^\alpha M_+\eta$ where Δ_η is the associated modular operator, $0 \leqq \alpha \leqq 1/2$ and M_+ is the positive part of M. The properties of these cones are reviewed and the connection with the "positive part" of L_p-spaces for M is indicated.

§1. INTRODUCTION.

The motivation for considering the cone V_η^α, especially for a special value of α, dates back to the problem of finding a suitable representative vector ξ_φ for a given normal state (or normal positive linear functional) $\varphi \in M_*^+$, where a "representative" vector means the following relation for all $x \in M$:

$$(x\xi_\varphi, \xi_\varphi) = \varphi(x) \tag{1.1}$$

This problem is discussed in Section 2. It is now known that the vector representative exists and is unique for $0 \leq \alpha \leq 1/4$.

The operator A_φ^α affiliated with M such that $\xi_\varphi^\alpha = A_\varphi^\alpha \eta$ is referred to generally by the name "Radon-Nikodym derivative (of φ by ω_η)". In this connection, the unitary Radon-Nikodym cocycle and the relative modular operator are discussed in Section 3.

The cone V_η^α for $\alpha = 1/4$ does not depend on the vector η very much and is a natural object for M. Hence it is called the natural positive cone and denoted as $\mathcal{P}^\natural$. This cone will be discussed in Section 4. The main properties of this cone are that the map $\varphi \in M_*^+ \to \xi_\varphi^{1/4} \in \mathcal{P}^\natural$ is a bijective concave homeomorphism and that any Jordan automorphism (and hence any *-automorphism) of M induces and exhausts the automorphisms of the pair $(\mathcal{P}^\natural, H)$.

1980 Mathematics Subject Classification. 46L10, 46L50.

0082-0717/80/0000-0496/$03.75

The polar cone of V_η^α is $V_\eta^{\alpha'}$ where $\alpha'+\alpha = 1/2$. In particular, $V_\eta^{1/4}$ is selfdual (or selfpolar). They are interchanged by the operator J of the Tomita-Takesaki theory. This duality is actually related with the duality of L_p-spaces for von Neumann algebras. This will be discussed in Section 5. $V_\eta^{1/2p}$ (for $p \geqq 2$ and its L_p-completion for $1 \leqq p < 2$) is more or less interpreted as the positive elements of (the η-representation of) the L_p-space for M.

Some polar decomposition theorems holds with respect to V_η^α: Any ξ in the domain of $\Delta_\eta^{(1/2)-2\alpha}$ can be written uniquely as

$$\xi = u|\xi|_\alpha \tag{1.2}$$

where $|\xi|_\alpha \in V_\eta^\alpha$, u is a partial isometry in M and $u^*u = s^M(|\xi|_\alpha)$ (alternatively $uu^* = s^M(\xi)$). Actually arbitrary ξ has such a decomposition if $1/2 \geqq \alpha \geqq 1/4$, a result which is obtained from the polar decomposition theorem in L_p-spaces for von Neumann algebras (see (5.6) and (5.9)). Exactly the same statements holds if we replace M by M' and α by $(1/2)-\alpha$.

A vector ξ is in the domain of $\Delta_\eta^{(1/2)-2\alpha}$ if and only if

$$\xi = \xi_1-\xi_2+i(\xi_3-\xi_4) \tag{1.3}$$

with all ξ_i in V_η^α. The decomposition (5.10) is unique if we require either $s^M(\xi_1)\perp s^M(\xi_2)$ and $s^M(\xi_3)\perp s^M(\xi_4)$ or $s^{M'}(\xi_1)\perp s^{M'}(\xi_2)$ and $s^{M'}(\xi_3)\perp s^{M'}(\xi_4)$ in addition to $\xi_i \in V_\eta^\alpha$. If $\alpha = 1/4$ the two requirements are satisfied by one and the same decomposition.

§2. REPRESENTATIVE VECTORS.

If the vector state ω_η, defined by

$$\omega_\eta(x) = (x\eta,\eta), \tag{2.1}$$

dominates a $\varphi \in M_*$ in the sense that there exists $\lambda > 0$ satisfying

$$\lambda\omega_\eta(x) \geqq \varphi(x) \tag{2.2}$$

for all $x \in M_+$, then it is a classical (and easily verifiable) fact that there exists a unique positive operator $A'_\varphi \in M'$ such that

$$\varphi(x) = (x\xi_\varphi^{1/2},\xi_\varphi^{1/2}),\ \xi_\varphi^{1/2} = A'_\varphi\eta, \tag{2.3}$$

where M' denotes the commutant of M. It is known that the closure of $(M')_+\eta$ is $V_\eta^{1/2}$ and, in the above case, $\xi_\varphi^{1/2}$ is the unique vector representative of φ in $V_\eta^{1/2}$, with $\|A'_\varphi\|^2$ being the infimum of λ satisfying (2.2).

Under the same condition for φ, Sakai [1] proved the existance of $A_\varphi^0 \in M_+$ such that

$$\varphi(x) = (x\xi_\varphi^0, \xi_\varphi^0), \quad \xi_\varphi^0 = A_\varphi^0\eta. \tag{2.4}$$

In this case ξ_φ^0 is the unique vector representative of φ in V_η^0. Later, Takesaki [2] has shown that the unique representative ξ_φ^0 of any $\varphi \in M_*^+$ in V_η^0 exists and satisfies (2.4) for some positive selfadjoint operator A_φ^0 affiliated with M. The operator A_φ^0 is a kind of (the square root of) Radon-Nikodym derivative of φ by ω_η. (The terminology coincides with that of the measure theory when M is commutative, in which case $\Delta_\eta =$ 1 and all V_η^α are the same.) The vector ξ_φ^0 can be characterized [3] as the vector representative of φ nearest to the vector η with respect to the distance in H.

If ω_η dominates φ in the sense of (2.2), then $\varphi \in M_*^+$ has a vector representative ξ_η^α in V_η^α for each $\alpha \in [0,1/2]$. In this case, there exists an $A_\varphi^\alpha \in M$ $(0 \leqq \alpha \leqq 1/4)$ such that $\xi_{\eta_+}^\alpha =$ $A_\varphi^\alpha\eta$. [4] More recently, it has been proved that any $\varphi \in M_*^+$ has a unique vector representative ξ_φ^α in V_η^α for each $\alpha \in [0,1/4]$. (This is (1.2) with M replaced by M'.) [5,22]

If $\lambda\omega_\eta \geqq \varphi$, then there exists a unique $\xi_\alpha \in V_\eta^\alpha$ for each α $(0 \leqq \alpha \leqq 1/2)$ ("linear" Radon-Nikodym theorem) such that

$$\varphi(x) = (1/2)\{(x\eta, \xi_\alpha) + (x\xi_\alpha, \eta)\}$$

for all $x \in M$, where $\lambda\eta - \xi_\alpha \in V_\eta^\alpha$ (i.e. $0 \leqq \xi_\alpha \leqq \lambda\eta$ in the order structure given by V_η^α) and hence $\xi_\alpha = \Delta_\eta^\alpha h\eta$ for an $h \in M_+$ satisfying $0 \leqq h \leqq \lambda$. The case $\alpha = 0$ is due to Sakai and the generalization is due to Araki [13].

§3. RELATIVE MODULAR OPERATORS AND UNITARY RADON-NIKODYM COCYCLES.

For any vector $\xi \in H$, $s^M(\xi)$ denotes the smallest (orthogonal) projection operator $P \in M$ satisfying $P\xi = \xi$ and is called the M-support of ξ. A vector ξ is cyclic if and only if $s^{M'}(\xi) = 1$ and separating if and only if $s^M(\xi) = 1$. A conjugate linear operator $S_{\xi\eta}$ (i.e. $S_{\xi\eta}(c_1\zeta_1 + c_2\zeta_2) = \bar{c}_1 S_{\xi\eta}\zeta_1 + \bar{c}_2 S_{\xi\eta}\zeta_2$) with the kernel $(s^M(\xi)s^{M'}(\eta)H)^\perp$ and with the range dense in $s^{M'}(\xi)s^{M}(\eta)H$ is defined by

$$\left.\begin{array}{l} S_{\xi\eta}(\zeta+x\eta) = s^{M}(\eta)x^*\xi, \\ x\in M, \quad \zeta\in(1-s^{M'}(\eta))H. \end{array}\right\} \tag{3.1}$$

It is closable and the closure has the polar decomposition:

$$\bar{S}_{\xi\eta} = J_{\xi\eta}\Delta_{\xi\eta}^{1/2}, \quad \Delta_{\xi\eta} = S_{\xi\eta}^{*}\bar{S}_{\xi\eta}. \tag{3.2}$$

The operator $\Delta_{\xi\eta}$ is called the relative modular operator (of the pair ξ,η).

In particular, we denote

$$\Delta_{\xi\xi} = \Delta_{\xi}, \quad J_{\xi\xi} = J_{\xi}. \tag{3.3}$$

If η is cyclic and separating, Δ_η is the modular operator of the Tomita-Takesaki theory and is positive selfadjoint. J_η is an antiunitary involution (i.e. $(J_\eta\zeta_1,J_\eta\zeta_2) = (\zeta_2,\zeta_1)$ and $(J_\eta)^2 = 1$). The important result of Tomita is that for all $x\in M$

$$\sigma_t^{\omega_\eta}(x) \equiv \Delta_\eta^{it}x\Delta_\eta^{-it} \in M, \tag{3.4}$$

$$J_\eta x J_\eta \in M'. \tag{3.5}$$

The *-automorphisms σ_t^{ω} defined by (3.4) for $\omega = \omega_\eta$ is called the modular automorphism. If we write

$$\sigma_t^{\eta}(x) \equiv \Delta_\eta^{it}x\Delta_\eta^{-it} \tag{3.6}$$

for any x, we have $\sigma_t^{\eta}(M') = M'$ and, in fact, $\sigma_{-t}^{\eta}(x')$ for $x'\in M'$ is the modular automorphism for the vector state $\omega_\eta'(x') = (x'\eta,\eta) = (x'\in M')$ of M'.

The modular operator Δ_η for a given cyclic and separating vector η is characterized as a positive selfadjoint operator Δ such that (1) $\Delta\eta = \eta$, (2) $\Delta^{it}x\Delta^{-it} \equiv \sigma_t(x)\in M$ if $x\in M$ and (3) the automorphism σ_t of M satisfies the KMS-condition. The operator J_η for a given η is characterized as an antiunitary involutive operator J (i.e. $(J\xi,J\xi') = (\xi',\xi)$ for all ξ, $\xi'\in H$ and $J^2 = 1$) such that (1) $J\eta = \eta$, (2) $JxJ \equiv j(x)\in M'$ if $x\in M$ and (3) $(xj(x)\eta,\eta) \geqq 0$ for any $x\in M$. The last property implies

$$(x_1 j(x_1)\eta,\ x_2 j(x_2)\eta) \geqq 0 \tag{3.7}$$

for any $x_1 x_2\in M$ and motivates the consideration of the set $\{xj(x)\eta;\ x\in M\}$, whose closure turns out to be the natural positive cone $V_\eta^{1/4}$ to be discussed in the next section. The property (3.7) is then a half of the selfdual property of $V_\eta^{1/4}$, already mentioned in the preceding section.

The relative modular operators satisfy

$$\Delta_{\xi\eta}^{it} x \Delta_{\xi\eta}^{-it} = \sigma_t^{\omega_\xi}(x) \qquad \text{for } x \in M, \tag{3.8}$$

$$\Delta_{\xi\eta}^{it} x' \Delta_{\xi\eta}^{-it} = \sigma_t^{\eta}(x') \qquad \text{for } x' \in M, \tag{3.9}$$

where both ξ and η are assumed to be cyclic and separating. As a consequence,

$$(D\omega_\xi : D\omega_\eta)_t \equiv \Delta_{\xi\zeta}^{it} \Delta_{\eta\zeta}^{-it} \in M, \tag{3.10}$$

$$(D\omega'_\xi : D\omega'_\eta)_t \equiv \Delta_{\zeta\xi}^{it} \Delta_{\zeta\eta}^{-it} \in M', \tag{3.11}$$

and they turn out to be independent of the cyclic and separating vector ζ as well as independent of the vector representatives ξ and η of the faithful states ω_ξ, ω_η, ω'_ξ, ω'_η.

The one-parameter family of unitary operators $(D\omega_\xi : D\omega_\eta)_t$ $(-\infty < t < \infty)$ are called the unitary Radon-Nikodym cocycle. They are strongly continuous in t and have the following properties:

(1) The intertwining property:

$$(D\varphi : D\psi)_t \sigma_t^{\psi}(x) (D\varphi : D\psi)_t = \sigma_t^{\varphi}(x) \qquad (x \in M). \tag{3.12}$$

(2) The cocycle property:

$$(D\varphi : D\psi)_{s+t} = (D\varphi : D\psi)_s \sigma_s^{\psi}\{(D\varphi : D\psi)_t\} \tag{3.13}$$

$$= \sigma_t^{\varphi}\{(D\varphi : D\psi)\}_s (D\varphi : D\psi)_t, \tag{3.14}$$

$$(D\varphi : D\psi)_0 = 1. \tag{3.15}$$

(3) The chain rule:

$$(D\varphi : D\psi)_t (D\psi : D\omega)_t = (D\varphi : D\omega)_t. \tag{3.16}$$

Here all states (more precisely normal positive linear functionals) φ, ψ, ω are assumed to be faithful.

A. Connes [6,7,8] introduced the unitary Radon-Nikodym cocycle and found the above properties. In fact, the definition is for any faithful normal semifinite weight. Any continuous one-parameter group of unitaries in M satisfying the cocycle property (3.13) is of the form $(D\omega : D\psi)_t$ for some faithful normal semifinite weight ω. The relative modular operators are studied in [9].

§4. NATURAL POSITIVE CONES.

The cone $V_\eta^{1/4}$ is called the natural positive cone for M. If $\xi \in V_\eta^{1/4}$, then ξ is cyclic if and only if it is separating and, if so, $V_\xi^{1/4}$ coincides with $V_\eta^{1/4}$. For any state φ, there is the unique vector representative $\xi_\varphi^{1/4}$ in $V_\xi^{1/4}$. If ξ' is any cyclic and separating vector, then there is the vector representative ξ of the state $\omega_{\xi'}$ in $V_\eta^{1/4}$ and the two are related by a unitary operator u' in M': $\xi' = u'\xi$. The natural positive cone associated with ξ' is then related to $V_\eta^{1/4}$ by

$$V_{\xi'}^{1/4} = u'V_\xi^{1/4} = u'V_\eta^{1/4}. \tag{4.1}$$

The dependence of the operator $J_{\xi'\eta}$ on the cyclic and separating vector ξ' is given by

$$J_{\xi'\eta} = u'J_{\xi\eta} = u'J_{\eta\eta}. \tag{4.2}$$

In particular $J_{\xi\eta}$ is the same operator for all $\xi \in V_\eta^{1/4}$ and will be written simply by J. We also introduce the following notation:

$$j(x) = JxJ. \tag{4.3}$$

If $x \in M$, $j(x) \in M'$ and if $x' \in M'$, $j(x') \in M$.

The vector representative $\xi(\varphi) \equiv \xi_\varphi^{1/4}$ have the following properties:

$$\|\xi(\varphi_1)-\xi(\varphi_2)\|^2 \leqq \|\varphi_1-\varphi_2\| \leqq \|\xi(\varphi_1)-\xi(\varphi_2)\|\,\|\xi(\varphi_1)+\xi(\varphi_2)\|. \tag{4.4}$$

$$\xi(\lambda\varphi_1+(1-\lambda)\varphi_2)-\lambda\xi(\varphi_1)-(1-\lambda)\xi(\varphi_2) \in V_\eta^{1/4} \quad (0 \leqq \lambda \leqq 1). \tag{4.5}$$

$$\xi(\lambda\varphi) = \lambda^{1/2}\xi(\varphi) \qquad (\lambda \leqq 0) \tag{4.6}$$

The equation (4.4) implies that the bijective map $\varphi \in M_*^+ \to \xi(\varphi) \in V_\eta^{1/4}$ is a homeomorphism.

There exists the Radon-Nikodym derivative $A_{\varphi\omega}^{1/4} \in M$ with $\xi(\varphi) = A_{\varphi\omega}^{1/4}\xi(\omega)$ if and only if there exists $\lambda > 0$ such that

$\lambda\xi(\omega)-\xi(\varphi) \in V_\eta^{1/4}$. The infimum of such λ is in fact, $\|A_{\varphi\omega}^{1/4}\|$. This $A_{\varphi\omega}^{1/4}$ satisfies the chain rule: $A_{\varphi\psi}^{1/4}A_{\psi\omega}^{1/4} = A_{\varphi\omega}^{1/4}$. If $\lambda\xi(\omega)-\xi(\varphi) \in V_\eta^{1/4}$, then there exists a continuous family of $A(z) \in M$ for $0 \leq \text{Re } z \leq 1/2$ such that $A(z)$ is holomorphic for $0 < \text{Re } z < 1/2$, uniformly bounded for $0 \leq \text{Re } z \leq 1/2$ and $A(it) = (D\varphi, D\omega)_t$ and $A(1/2) = A_{\varphi\omega}^{1/4}$. In fact $A(z)$ is the closure of $\Delta_{\xi(\varphi)\xi(\omega)}^{z}\Delta_{\xi(\omega)}^{-z}$. A similar situation prevails if $A_{\varphi\omega}^{1/4}$ is unbounded but closable. In general, a closable $A_{\varphi\omega}^{1/4}$ affiliated with M might not exist.

A linear bijective mapping α of M preserving the adjoint $*$ and the Jordan product $x\circ y = (xy+yx)/2$ is called a Jordan automorphism of M. (A $*$-automorphism is a special case.) They induce a bijective mapping of φ:

$$(\alpha\varphi)(x) \equiv \varphi(\alpha x). \tag{4.7}$$

The induced mapping on $V_\eta^{1/4}$ can be extended to a unique unitary operator $U(\alpha)$ satisfying

$$U(\alpha)\xi(\varphi) = \xi(\alpha\varphi). \tag{4.8}$$

This provides a continuous representation of the group of all Jordan automorphisms of M. (A relation of a Jordan automorphism and a $*$-automorphism is in [10].) Conversely any unitary operator on H mapping $V_\eta^{1/4}$ into itself is $U(\alpha)$ for some α.

The natural positive cone has been introduced in [11], [12], [13], [6]. It has been geometrically characterized as a facially homogeneous, orientable, selfdual cone by A. Connes in [11]. Motivated by this work, Bellisard and Iochum geometrically characterized selfdual cones in a Hilbert space associated with infinite dimensional Jordan algebras called JBW-algebras, [14], [15]. On the other hand, characterizations of state spaces of Jordan algebras and C*-algebras have been given by Alfsen and Shultz. [16], [17]. Closely related discussion in connection with foundation of quantum mechanics is in [18]. We shall not go into these subjects.

§5. L_p-SPACES.

We define (the η-representation of) L_p-space for $\infty \geq p \geq 2$ by the following space and the norm [5].

$$L_p(M,\eta) = \{\zeta \in \bigcap_\xi D(\Delta_{\xi,\eta}^{(1/2)-1/p});\|\zeta\|_p^\eta < \infty\} \tag{5.1}$$

$$\|\zeta\|_p^\eta = \sup_{\|\xi\|\leq 1} \|\Delta_{\xi,\eta}^{(1/2)-(1/p)}\zeta\|. \tag{5.2}$$

This space is already complete. We define (the η-representation of) L_p-spaces $L_p(M,\eta)$ for $1 \leq p < 2$ as the completion of H with the following norm:

$$\|\zeta\|_p^\eta = \inf_{\|\xi\|\leq 1} \|\Delta_{\xi,\eta}^{(1/2)-(1/p)}\zeta\|. \tag{5.3}$$

For $p \geq 2$, each $L_p(M,\eta)$ is complete and mutually isomorphic where the following map induces an isomorphism of L_p^η onto $L_p^{\eta'}$:

$$\zeta \in L_p^\eta \to \Delta_{\eta',\eta'}^{(1/p)-(1/2)}\Delta_{\eta',\eta}^{(1/2)-(1/p)}\zeta \in L_p^{\eta'}. \tag{5.4}$$

For $p < 2$, the same mapping defined on a dense subset can be extended to the bijection of whole spaces for $\lambda\omega_{\eta'} > \omega_\eta$ and hence for general η and η' through $\omega_\eta + \omega_{\eta'}$.

The inner product (ζ,ζ') in H can be extended to all $\zeta \in L_p(M,\eta)$ and $\zeta' \in L_{p'}(M,\eta)$ with $(1/p)+(1/p') = 1$. We then have

$$\|\zeta\|_p^\eta = \sup\{\|(\zeta,\zeta')\|;\ \|\zeta'\|_{p'}^\eta \leq 1\}. \tag{5.5}$$

$L_p(M,\eta)$ can be identified with the dual $L_{p'}(M,\eta)^*$ of $L_{p'}(M,\eta)$.

$L_1(M,\eta)$ is isometrically isomorphic to M_* with the correspondence $\zeta \in L_1(M,\eta) \to \varphi(x) = (\zeta,x^*\eta)$. $L_2(M,\eta)$ is H itself. $L_\infty(M,\eta)$ is isometrically isomorphic to M with the correspondence $x\eta \in L_\infty(M,\eta) \to x \in M$.

Any $\zeta \in L_p(M,\eta)$ has a unique polar decomposition

$$\zeta = u|\zeta| \tag{5.6}$$

where u is a partial isometry in M such that $u^*u = s^M(|\zeta|)$ and $|\zeta|$ is in $L_p(M,\eta)_+$, which is defined to be the closure of $V_\eta^{(1/2p)} \cap L_p(M,\eta)$ (the closure unnecessary if $p \geq 2$, the intersection $\cap L_p(M,\eta)$ unnecessary if $p \leq 2$). Under the isomorphism (5.4) of $L_p(M,\eta)$ for different η's, this decomposition is independent of η. Any element $|\zeta|$ can be written as

$$|\zeta| = \Delta_{\xi\eta}^{1/p}\eta \tag{5.7}$$

for some $\xi \in V_\eta^{1/4}$ and we write $|\zeta| = \omega^{1/p}$ where $\omega = \omega_\xi \in M_*^+$. Then $\|\zeta\| = \omega(1)^{1/p}$. For $1 \leqq p < 2$ and $\zeta \notin H$, $|\zeta|$ is actually not in H. However the expression (5.7) makes sense for any $\xi \in V_\eta^{1/4}$ as an element in the dual of $L_{p'}(M,\eta)$:

$$(|\zeta|,\zeta') = (\Delta_{\xi,\eta}^{1/2}\eta,\ \Delta_{\xi,\eta}^{(1/p)-(1/2)}\zeta') \tag{5.8}$$

for all $\zeta' \in L_{p'}(M,\eta)$ where $(1/p)-(1/2) = (1/2)-(1/p')$ and $p' > 2$. Therefore, an arbitrary element of $L_p(M,\eta)$ is uniquely specified by a pair of $\xi \in V_\eta^{1/4}$ and a partial isometry u in M such that $u^*u = s^M(\xi)$. Such an element is denoted by $u\omega_\xi{}^{1/p}$. Its L_p norm is $\omega_\xi(1)^{1/p}$.

The operator J maps V_η^α onto $V_\eta^{\alpha'}$ where $\alpha'+\alpha = 1/2$. The polar decomposition (5.6) for $p \leqq 2$ (hence for any $\zeta \in H$) implies

$$J\zeta = j(u)J|\zeta| \tag{5.9}$$

where $J|\zeta| \in V_\eta^\alpha$ with $\alpha = (1/2)-(1/2p)$ and $\omega_{J\zeta} = \omega_{J|\zeta|}$. This is the existence (and uniqueness) of a vector representative of any $\varphi \in M_*^+$ in V_η^α for $0 \leqq \alpha \leqq 1/4$.

The unique decomposition (1.3) with $s^M(\xi_1) \perp s^M(\xi_2)$ and $s^M(\xi_3) \perp s^M(\xi_4)$ holds for $\xi \in L_p(M,\eta)$ and $\xi_i \in L_p(M,\eta)_+$. In particular, if $2 \leqq p \leqq \infty$, the decomposition becomes a special case of (1.3) with $\xi_i \in L_p(M,\eta)_+$ following from $\xi \in L_p(M,\eta)$.

The product of $\zeta_1 = u_1(\omega_{\xi_1})^{1/p}$ and $\zeta_2 = u_2(\omega_{\xi_2})^{1/q}$ can be defined by

$$\zeta_1\cdot\zeta_2 = u_1\Delta_{\xi_1\eta}^{1/p}u_2\Delta_{\xi_2\eta}^{1/q}\eta \in L_r(M,\eta) \tag{5.10}$$

if $1 \leqq r \leqq \infty$ and $p^{-1}+q^{-1} = r^{-1}$. This product is independent of the choice of η (under the identification by (5.4)) and for an appropriate choice of η, (5.9) makes sense as a vector in H (always makes sense if $r \geqq 2$). We have

$$\|\zeta_1\zeta_2\|_r \leqq \|\zeta_1\|_p\|\zeta_2\|_q. \tag{5.11}$$

The inequality (5.10) is essentially due to the multiple KMS condition [19], which says that the function

$$F(z) = (\Delta_{\xi_j,\eta}^{z_j'} x_j \Delta_{\xi_{j+1},\eta}^{z_{j+1}} x_{j+1} \cdots x_n \eta,\ \Delta_{\xi_j,\eta}^{\bar{z}_j''} x_{j-1}^* \Delta_{\xi_{j-1},\eta}^{\bar{z}_{j-1}} \cdots x_0^* \eta)$$

$$(\text{"="}\ (x_0 \Delta_{\xi_1,\eta}^{z_1} x_1 \Delta_{\xi_2,\eta}^{z_2} x_2 \cdots \Delta_{\xi_n,\eta}^{z_n} x_n \eta, \eta)) \qquad (5.12)$$

of n variables $z_1, \cdots, z_n$ with $z_j = z_j' + z_j''$ is bounded and continuous in the tube domain

$$\text{Re } z_j \geqq 0, \quad \text{Re } z_1 + \cdots + \text{Re } z_n \leqq 1,$$

and holomorphic in the interior of this tube domain with the following bound

$$|F(z)| \leqq (\prod_0^n \|x_j\|) \omega_\eta(1)^{1-\Sigma \text{Re } z_j} \prod_1^n \omega_{\xi_j}(1)^{\text{Re } z_j} . \qquad (5.13)$$

In the defining expression of $F(z)$, j, z_j' and z_j'' are to be chosen such that

$$\text{Re } z_1 + \cdots + \text{Re } z_{j-1} + \text{Re } z_j' \leqq 1/2,$$

$$\text{Re } z_n + \cdots + \text{Re } z_{j+1} + \text{Re } z_j'' \leqq 1/2.$$

L_p-spaces for von Neumann algebras have been studied by Haagerup [20], Hilsum [21], Kosaki [22, 23], Araki and Masuda [5].

BIBLIOGRAPHY

1. S. Sakai, "A Radon-Nikodym theorem in W*-algebras", Bull. Amer. Math. Soc. 71(1965), 149-151.

2. M. Takesaki, Tomita's theory of modular Hilbert algebras and its applications, Springer, 1970.

3. H. Araki, "Bures distance function and a generalization of Sakai's non-commutative Radon-Nikodym theorem", Publ. RIMS, Kyoto Univ. 8(1972), 335-362.

4. H. Araki, "One-parameter family of Radon-Nikodym theorems for states of a von Neumann algebra", Publ. RIMS, Kyoto Univ. 10 (1974), 1-10.

5. H. Araki and T. Masuda, to be published.

6. A. Connes, "Group modulaire d'une algébre de von Neumann", C. R. Acad. Sc. Paris, 274(1972), 1923-1926.

7. A. Connes, "Une classification des facteurs de type III", Ann. Sci. Ecole Norm. Sup. 4ème Sér. 6(1973), 133-252.

8. A. Connes, "Sur le théorème de Radon Nikodym pour les poids normaux fidèles semifinis", Bull. Sci. Math. 2ème Sér. 97 (1973), 253-258.

9. H. Araki, "Introduction to relative hamiltonian and relative entropy", Marseille preprint 75/p.782 (1975).

10. R. V. Kadison, "Isometries of operator algebras", Ann. Math. 54(1951) 325-338.

11. A. Connes, "Caracterisation des algèbres de von Neumann comme espaces vectoriels ordonés", Ann. Inst. Fourier (Grenoble) 26(1974), 121-155.

12. U. Haagerup, "The standard form of von Neumann algebras", Math. Scand. 37(1975), 271-283.

13. H. Araki, "Some properties of modular conjugation operator of von Neumann algebra and a non-commutative Radon-Nikodym theorem with a chain rule", Pacific J. Math. 50(1974),

14. J. Bellissard, B. Iochum, and R. Lima, "Cônes antopolaires homogènes et facialement homogènes", C. R. Acad. Sci. Paris, 282(1976), 1363-1365.

15. J. Bellissard and B. Iochum, "Homogeneous self dual cones versus Jordan algebras, The theory revisited", Ann. Inst. Fourier, Grenoble, 28,1(1978), 27-67.

16. E. M. Alfsen and F. W. Shultz, "State spaces of Jordan algebras", Acta Math. 140(1978), 155-190.

17. E. M. Alfsen, F. W. Shultz and H. Hauche-Olsen, "State spaces of C*-algebras", Acta Math. (to appear)

18. H. Araki, "On a characterization of the state space of quantum mechanics", Comm. Math. Phys. 75(1980), 1-24.

19. H. Araki, "Multiple time analyticity of a quantum statistical state satisfying the KMS boundary condition", Publ. RIMS, Kyoto Univ. Ser. A, 4(1968), 361-371.

20. U. Haagerup, "L^p-spaces associated with a von Neumann algebras", (Odense preprint)

21. M. Hilsum, Les espaces L^p d'une algèbre de von Neumann", (Université Pierre et Marie Curie preprint).

22. H. Kosaki, "Positive cones associated with a von Neumann algebra", to appear in Math. Scand.

23. H. Kosaki, "Positive cones and L^p-spaces associated with a von Neumann algebra", to appear in J. Operator Theory.

RESEARCH INSTITUTE FOR MATHEMATICAL SCIENCES
KYOTO UNIVERSITY
KYOTO 606, JAPAN

Proceedings of Symposia in Pure Mathematics
Volume **38** (1982), Part 2

GEOMETRIC ASPECTS OF THE TOMITA - TAKESAKI THEORY.

Christian F. Skau[1]

ABSTRACT. We show how the various objects of geometrical nature that is naturally encountered in the Tomita - Takesaki theory of von Neumann algebras characterize the faithful normal states they correspond to. We visualize the objects in question by using the reference frame of a standard form.

1. INTRODUCTION. Let M be a σ-finite von Neumann algebra acting standardly on the complex Hilbert space H with $(M,H,J,P^{\natural})$ a standard for for M (in the sense of [1]). Let φ be a faithful normal state on M. There exists a unique unit vector ξ_0 in $P^{\natural}$ such that $\varphi = \omega_{\xi_0}$. Let J_{ξ_0} and Δ_{ξ_0} be the conjugate linear isometric involution and modular operator, respectively, associated with (M,ξ_0) by the Tomita - Takesaki theory. Then $J_{\xi_0} = J$ and

$$P^{\natural} - P^{\natural} = \{\xi \in H \mid J\xi = \xi\}.$$

Set $H^{\natural} = P^{\natural} - P^{\natural}$. The restriction of the inner product on H to $H^{\natural}$ is <u>real</u> and J is the (real) symmetry with respect to $H^{\natural}$ (considering H in its real restriction). Set

$$P^{\#} = P^{\#}_{\xi_0} = \overline{M_+\xi_0}, \quad P^b = P^b_{\xi_0} = \overline{M'_+\xi_0}, \quad K = K_{\xi_0} = \overline{M_h\xi_0}, \quad \tilde{K} = \tilde{K}_{\xi_0} = \overline{M'_h\xi_0}.$$

Here $M_+(M'_+)$ denotes the positive part of M (M') and $M_h(M'_h)$ the hermitian part of M (M'), respectively. Let $C = C_{\xi_0}$ denote the centralizer of $\varphi = \omega_{\xi_0}$. We have $K = P^{\#} - P^{\#}$, $\tilde{K} = P^b - P^b$, $J(P^{\#}) = P^b$, $J(K) = \tilde{K}$.

2. REPRESENTATION OF K AND $\tilde{K}$ AS OPERATOR GRAPHS.

THEOREM 1. There exists an operator $A\colon H^{\natural} \to H^{\natural}$, $0 \leq A \leq I$, such that $(\text{graph}\,A, \text{graph}(-A)) \cong (K,\tilde{K})$, i.e. there exists an isometry from the real Hilbert space $H^{\natural} \oplus H^{\natural}$ onto H, in its real restriction, which maps graph A onto K and graph$(-A)$ onto $\tilde{K}$.

The operator A is unique up to isometric equivalence, i.e. if B is another operator with the same properties as A then there exists an isometry

---

1980 Mathematics Subject Classification. Primary 46L10; Secondary 46L30.
[1]Supported by the Norwegian Research Council for Sci. and the Humanities.

0082-0717/80/0000-0574/$02.00

W: $H^\natural \to H^\natural$ such that $B = WAW^*$. Specifically, we may choose

$$A = \frac{|I-\Delta^{\frac{1}{2}}|}{I+\Delta^{\frac{1}{2}}}\Bigg|_{H^\natural}, \quad \text{where} \quad \Delta = \Delta_{\xi_0}.$$

In order to visualize the content of Theorem 1 we present the following instructive, if simplified, figure.

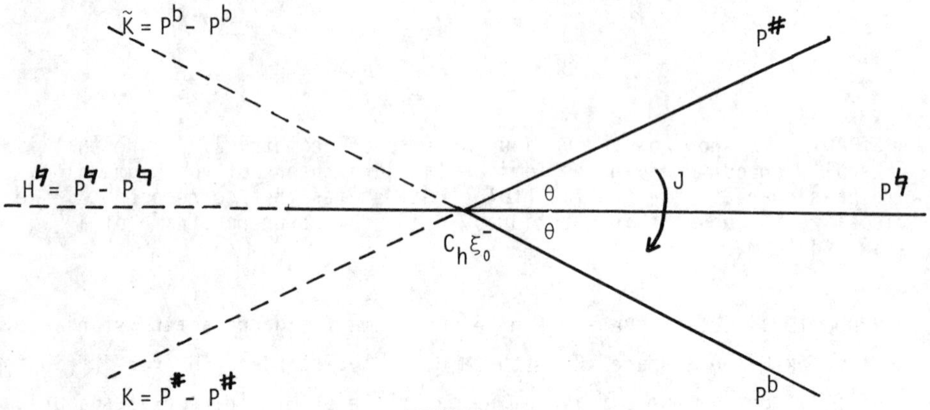

FIGURE 1.

As indicated in Figure 1 the intersection of $H^\natural$ with K (or $\tilde{K}$) is equal to $C_h\xi_0^-$. Also, by Theorem 1, we may give precise meaning to the notion of "angle" θ between $H^\natural$ and K (or $\tilde{K}$).

Now let ψ be another faithful normal state on M and let $\eta_0 \in P^\natural$ be the unique unit vector such that $\psi = \omega_{\eta_0}$. Using the subindex $_1$ to denote the various objects associated with (M, η_0) we get the following suggestive figure.

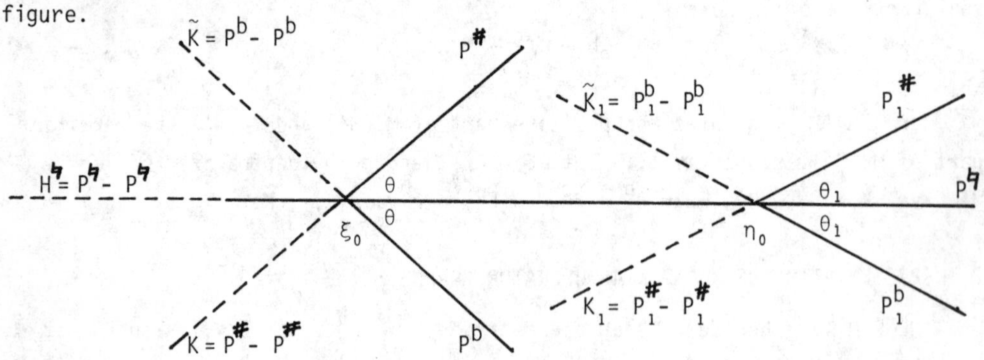

FIGURE 2.

We can show that if $\psi = \varphi \circ \alpha$ for some Jordan isomorphism α of M, then $\theta_1 = \theta$. However, the converse is not true in general. To get a satisfactory characterization of states up to Jordan isomorphisms we must bring the various cones into consideration.

3. THE CONES $P^{\#}$ AND P^{b}. Retaining the notation from Figure 2 we state the following theorem for factors only.

THEOREM 2. Let $(M,H,J,P^{\natural})$ be a standard form of the factor M. Let φ and ψ be two faithful normal states on M and let ξ_0 and η_0 be the two unique cyclic and separating unit vectors in $P^{\natural}$ such that $\varphi = \omega_{\xi_0}$, $\psi = \omega_{\eta_0}$. Then

(i) $\psi = \varphi \circ \alpha$ for some automorphism $\alpha: M \to M$ if and only if there exists a unitary operator u on H such that $u(P^{\#}) = P_1^{\#}$. (If M is finite we have to assume in addition that $u\xi_0 = \eta_0$).

(ii) $\psi = \varphi \circ \beta$ for some antiautomorphism $\beta: M \to M$ if and only if there exists a unitary operator v on H such that $v(P^{\#}) = P_1^{b}$. (If M is finite we have to assume in addition that $v\xi_0 = \eta_0$ and also that ξ_0 (or η_0) is not a trace vector for M).

The reason for the special assumptions in the finite case stems from the fact that only in this case may a $P^{\#}$-cone coincide with a P^{b}-cone. This fact can be rephrased in terms of modular groups and seems to be a result of independent interest:

LEMMA. Assume $\sigma_t^{\varphi} = \sigma_{-t}^{\psi}$, $\forall t \in \mathbb{R}$, where φ and ψ are faithful, normal, semifinite weights on M. Then M is semifinite.
If φ and ψ are finite, then M is finite.

4. THE "SHADOW" OF $P^{\#}$ IN $H^{\natural}$. Returning to Figure 1 it is tempting to consider how the cone $P^{\#}$ (or P^{b}) projects into $H^{\natural}$, in other words, where lies the "shadow" of $P^{\#}$ in $H^{\natural}$? Let $q = \frac{1}{2}(I+J)$ be the orthogonal (real) projection of H onto $H^{\natural}$. Set

$$Q_{\xi_0} = q(P_{\xi_0}^{\#}) \quad (= q(P_{\xi_0}^{b})).$$

It turns out that Q_{ξ_0} is a closed, proper cone in $H^{\natural}$ such that $P^{\natural} \subseteq Q_{\xi_0}$ for any $\xi_0 \in P^{\natural}$. Moreover, $Q_{\xi_0} = P^{\natural}$ if and only if ξ_0 is a trace vector for M. We also have the following theorem.

THEOREM 3. Let M be a factor and let $\xi_0, \eta_0 \in P^{\natural}$ be cyclic and separating unit vectors for M. Then $Q_{\xi_0} = Q_{\eta_0}$ if and only if

(i) $\eta_0 = \xi_0$

or

(ii) M is finite and $\eta_0 = \lambda\xi_0^{-1}$ for some $\lambda > 0$. (Recall that $P^{\natural} = L^2(M,\tau)_+$ in the finite case).

Proofs and further details can be found in [2] and [3].

BIBLIOGRAPHY

1. U. Haagerup, "The standard form of von Neumann algebras", Math. Scand., 37 (1975), 271-283.

2. U. Haagerup, C.F. Skau, "Geometric aspects of the Tomita-Takesaki theory II", To appear in Math. Scand.

3. C.F. Skau, "Geometric aspects of the Tomita-Takesaki theory I", To appear in Math. Scand.

DEPARTMENT OF MATHEMATICS
UNIVERSITY OF TRONDHEIM, NLHT
7000 TRONDHEIM, NORWAY

Proceedings of Symposia in Pure Mathematics
Volume 38 (1982), Part 2

LEFT HILBERT SYSTEMS

F. Combes

ABSTRACT. We show that a positive type measure on a locally compact group G cannot always be extended a s a regular weight on $C^*(G)$. More generally the formalism of the Gelfand-Segal representations is abstractly studied and related with Tomita's theory.

1. INTRODUCTION. Let G be a locally compact group, K(G) the convolution algebra of continous functions having a compact support.

1.1. Main problem. To any positive type measure μ on G can one associate a semi-finite and l.s.c. weight on $C^*(G)$ such that $K(G) \subset \mathfrak{N}_\phi$ and $\phi(g^* \ast f) = \langle \mu , g^* \ast f \rangle$ for f, $g \in K(G)$.

The answer is negative. Before looking at some example, this is some others questions closely related to the proceding one.

1.2. Let f be a state on a C^*-algebra A. It defines a representation π_f a canonical cyclie vector ξ. On the von Neumann algebra $M = \pi_f(A)''$ of the representation, the normal state $\tau = \omega_\xi | M$ is such that $f = \tau \circ \pi_f$, while on $M' = \pi_f(A)'$ the normal state $\tau' = \omega_\xi | M'$ is faithful.

Does there exist in any Gelfand-Naīmak-Segal construction of representations, canonical weights τ and τ' having analogous properties ?

1.3. When ϕ is a l.s.c., semi-finite weight on a C^*-algebra, such τ and τ' can be constructed, but not in very natural ways. For example, one can extend ϕ as a normal weight ϕ'' on the bidual A", reduce ϕ'' by its support and apply to the faithful, normal weight ϕ_o obtained the left Hilbert algebras theory. But the normal weight on A" extending ϕ is not unique and the relations between π_ϕ and π_{ϕ_o} are not clear.

0082-0717/80/0000-0513/$03.50

Even in this example, can one express Tomita's ideas (6) more naturally ?

1.4. On a C^*-algebra, J. Dixmier and A. Guichardet (1), (3) have shown that there is a good correspondance between semi-finite, l.s.c. traces and bitraces. As the definition of a bitrace can be formulated on any involutive algebra, bitraces are much more general than traces.

Is it possible to extend such a correspondance to weights ? What are "weighted" representations ?

1.5. M. Rieffel and J. Phillips have studied abstractly square integrable representations in the left-Hilbert algebras framework (5), (4).

What are exactly the connections between square integrable coefficients and bounded or closable vectors ? between square integrable representations and the G.N.S. representation ?

2. DEFINITIONS

2.1. We call left Hilbert system, an involutive algebra A, a left ideal L of A, a sesquilinear positive form s on L such that the system $\mathfrak{A} = (A, L, s)$ satisfies

1/ $s(ax, y) = s(x, a^* y)$ for all $x, y \in L$ and $a \in A$,

2/ $\forall a \in A, \exists k > o$; $s(ax, ax) \leqslant ks(x,x)$ for all $x \in L$,

3/ $(L \cap L^*)^2$ is dense in (L, s).

Right Hilbert systems are defined symetrically. To any left Hilbert system $\mathfrak{A} = (A, L, s)$ is associated the right Hilbert system $\mathfrak{A}^* = (A, L^*, s^*)$ where L^* is the right ideal adjoint of L and where $s^*(x, y) = s(y^*, x^*)$. One says that $\mathfrak{A}$ is unimodular when $\mathfrak{A} = \mathfrak{A}^*$ (one find again bitraces).

Examples. $(A, \mathfrak{N}_\phi, s_\phi)$ when ϕ is a l.s.c. semi-finite weight on a C^*-algebra A or $(K(G), K(G), s_\mu)$ when μ is a positive type measure on a locally compact group G.

2.2. The G.N.S. representation (Λ, H, μ) is constructed as usually : by separating and completing (L, s). Then A acts on the left...

More generally a representation ρ of $\mathfrak{A}$ is a representation ρ of A such that $\rho(L \cap L^*)$ is not degenerated (then $\rho(L \cap L^*)'' = \rho(A)''$).

A vector $\alpha \in H_\rho$ is bounded if there exists $k \in \mathbb{R}^+$, such that $\| \rho(x) \alpha \|^2 \leqslant k\, s(x, x)$ for all $x \in L$.

The vector $\alpha \in H_\rho$ is closable if for any sequence (x_n) in L such that $s(x_n, x_n) \to 0$ and $\rho(x_n - x_m)\alpha \to 0$ one has $\rho(x_n)\alpha \to 0$. There exist then a unique closed operator $\pi'(\alpha)$ admiting ΛL as a core such that $\pi'(\alpha)\Lambda x = \rho(x)$ for $x \in L$. This operator intertwines π and ρ ; it is bounded when α is bounded.

2.3. When $\alpha, \beta \varepsilon H_\rho$, the coefficient $c_{\alpha,\beta}: x \to (\rho(x)\alpha|\beta)$ form L to $\mathbb{C}$ is said square integrable when it is continous on (L,s). There is then a unique $t_\alpha \beta \varepsilon H$ such that

$c_{\alpha,\beta}(x) = (\rho(x)\alpha|\beta) = (\Lambda x \mid t_\alpha \beta)$ for all $x \varepsilon L$.

So we define an operator t_α whose domain is the set of $\beta \varepsilon H_\rho$ such that $c_{\alpha,\beta}$ is square integrable.

3. RELATIONS BETWEEN CLOSABLE, BOUNDED VECTORS AND SQUARE INTEGRABLE COEFFICIENTS.

Proposition. t_α is densely defined if and only if α is closable. Then $t_\alpha = \pi'(\alpha)$.

Corollary. If ρ has a square integrable coefficient $c_{\alpha,\beta} \neq 0$, then $\alpha_1 = P_K \alpha \neq 0$ is closable (here K is the closure of the domain of t_α).

4. THE CANONICAL RIGHT HILBERT ALGEBRA.

Let B be the set of bounded vectors in the G.N.S. representation (Λ, H, μ) of $\mathfrak{A} = (A, L, s)$. Then $N = \{\pi'(\alpha)\ ;\ \alpha \varepsilon B\}$ is a left ideal in M', where M is the von Neumann algebra generated by π, and $\pi' : \alpha \to \pi'(\alpha)$ is bijective from B on N.

Theorem. $\mathfrak{A}' = \pi'^{-1}(N \cap N)$ gifted with the involutive algebra structure making π' an anti-isomorphism is a right Hilbert algebra.

Example. if f is a state an a C^*-algebra A, and $\mathfrak{A} = (A, A, s_f)$ then $\mathfrak{A}' = M' \xi_f$, the support of $\tau = \omega_{\xi_f} \mid M$ is $\overline{\mathfrak{A}'}$.

Definition. A representation (not necessarely irreducible or factorial) will be called square integrable when the bounded vectors are a total set for ρ. (this definition, as well as the corollary are from M. Enock and J.M. Schwartz).

Corollary. The G.N.S. representation has the following universal property : for any square integrable representation ρ of $\mathfrak{A}$ there exists a normal homomorphism ϕ form $M = \pi(A)''$ on to $\rho(A)''$ such that $\rho = \phi \circ \pi$.

5. WEITHTED LEFT HILBERT SYSTEMS.

Does there always exist bounded vectors in the G.N.S. representation ? Is $\mathfrak{A}' \neq 0$? The answer is negative.

Definition. We say that $\mathfrak{A} = (A, L, s)$ is weighted when there exist a family (α_i) in H such that

$s(x, x) = \sup\, (\pi(x^* x)\alpha_i \mid \alpha_i)$ for all $x \varepsilon L$.

(Equivalent conditions are the density of $N = \pi'(B)$ in M' or the fact that $\mathfrak{A}'$ be a total set for π).

Theorem. $\mathfrak{A}'$, and $(\mathfrak{A}')'$ defined in $K = \overline{\mathfrak{A}'}$, give canonical weights τ' on M' faithful, and τ on M with support K, which are semi-finite and normal.

6. ARE THE REPRESENTATIONS DEFINED BY POSITIVE TYPE MEASURES WEIGHTED ?

Proposition. On $G = SL(2, \mathbb{R})$ there exist a positive type measure μ such that $\mathfrak{A}_\mu = (K(G), K(G), s_\mu)$ has no bounded vector different from zero.

In the cone P of positive type measures on $SL(2, \mathbb{R})$, R. Goodman has constructed a measure belonging to an extremal ray and such that π_μ is reducible, and contains two proper subrepresentation (2). If some vector $\alpha \neq 0$ in H_μ was bounded relatively to $\mathfrak{A}_\mu$, the coefficient $\phi = C_{\alpha,\beta}$ associated to α would be majorized in P by μ. By extremaltiy of μ, one would get $\mu = \lambda\phi$, and μ would be a pure continuous positive type function and π_μ would be irreducible.

Remarks. There always exist weighted positive type measures on a locally compact group. For example, the Dirac measure δ_e which defines the left regular representation of G, gives on K(G) a Tomita's algebra structure. This is some other examples.

Proposition. Let G be a locally compact group and μ a positive type measure on G.

(i) If μ is the Haar-measure of a compact subgroup H of G then μ is weighted.

(ii) Suppose G unimodular. If μ is bounded, or if $d\mu(x) = f(x)dx$ with $f \in L^2(G)$, then μ is weighted.

Corollary. On a compact group, all the positive type measures are weighted.

BIBLIOGRAPHY

1. J. DIXMIER, Traces sur les C^*-algebres, Ann, Inst, Fourier, 13 (1963), 219-262.

2. R. GOODMAN, Positive definite distributions and intertwining operators, Pacific J. Math, , 48 (1973), 83-91.

3. A. GUICHARDET, Caractères des algèbres de Banach involutives, Ann. Inst. Fourier, 13 (1962), 1-81.

4. J. PHILLIPS, Positive integrable elements relative to a left Hilbert algebra, J. Funct. Anal. , 43 (1973), 390-409.

5. M. RIEFFEL, Square integrable representations of Hilbert algebras, J. Funct. Anal., 3 (1969), 265-300.

6. M. TAKESAKI, Tomita's theory of modular Hilbert algebras and its applications (Lecture Notes, n° 128, Springer-Verlag, 1970).

UNIVERSITE D'ORLEANS
45046 ORLEANS CEDEX, FRANCE

Proceedings of Symposia in Pure Mathematics
Volume **38** (1982), Part 2

ITPFI FACTORS - A SURVEY

E. J. Woods

1. INTRODUCTION. This article surveys the role played by the ITPFI factors in the developments in the theory of von Neumann algebras during the last 14 years. As far as possible, the presentation is chronological.

Sec. 2 contains some preliminary material, including a brief description of the Connes-Takesaki flow of weights. Sec. 3 reviews briefly the early history. Sec. 4 is concerned with the period 1966-1972 when the major impact on the classification problem took place. Sec. 5 presents some later developments, especially those involving the flow of weights. Sec. 6 gives a brief assessment of the current situation and possible future developments, and contains a short list of open questions.

2. PRELIMINARY MATERIAL. All Hilbert spaces occuring in this article are separable.

DEFINITION 2.1. A von Neumann algebra M is said to be approximately type I (hereafter referred to as ATI) if it is of the form

$$M = (\bigcup_{n=1}^{\infty} M_n)''$$

where $M_n \subset M_{n+1}$ for each n and each M_n is a finite-dimensional matrix algebra.

This notion has been described by at least five different names in the literature. It was first introduced by Murray and von Neumann [35] for II_1 factors, where they used the name approximately finite. Since II_1 factors are finite, Dixmier [16] introduced the term hyperfinite in his description of the Murray-von Neumann results. But this name is not appropriate

1980 Mathematics Subject Classification. 46L35, 28D99.

when the factors are infinite. So Elliott and Woods [18] introduced the name approximately-finite dimensional, and Kadison and Ringrose [22] use matricial. The terminology ATI was suggested by Connes.

DEFINITION 2.2. A factor M is called ITPFI if it is of the form $M = \bigotimes_{\nu=1}^{\infty}(M_\nu,\phi_\nu)$ on the Hilbert space $H = \bigotimes_{\nu=1}^{\infty}(H_\nu,\Phi_\nu)$ where the M_ν are type I_{n_ν} factors acting on H_ν, $2 \leq n_\nu < \infty$, and $\phi_\nu(x) = (\Phi_\nu, x\Phi_\nu)$ is a faithful state on M_ν.

The name ITPFI stands for infinite tensor product of factors of finite type I, and was introduced by Araki and Woods [4].

DEFINITION 2.3. Let ϕ be a faithful state on the type I_n factor M. Then $Sp(\phi/M)$ denotes the ordered list $(\lambda_1, \ldots \lambda_n)$ where $\lambda_1 \geq \lambda_2 \geq \ldots \geq \lambda_n > 0$ and the λ_j are the eigenvalues of the operator ρ where $\phi(x) = \tau(\rho x)$, $x \in M$ and τ is the normalized trace on M. (If some λ has multiplicity m then it occurs m times in the list, the multiplicity being defined when M acts irreducibly on an n-dimensional Hilbert space.)

DEFINITION 2.4. Let M, M_ν, ϕ_ν be as in definition 2.2. By the eigenvalue list of M (relative to the given construction) we mean the sequence of lists $Sp(\phi_\nu/M_\nu) = (\lambda_{\nu 1}, \ldots , \lambda_{\nu n_\nu})$.

REMARK 2.5. Since $Sp(\phi/M)$ is a complete unitary invariant for the pair (M,ϕ), an ITPFI factor is determined up to unitary equivalence by its eigenvalue list. Hence ITPFI factors can be easily specified in precisely the same way that one specifies a measure which is an infinite product of atomic measures. In contrast to this, no purely spatial construction of an ATI non-ITPFI factor has yet been given. All the existence arguments use, in effect, the flow of weights at some point.

DEFINITION 2.6. Let T be an ergodic transformation on the Lebesgue measure space (X,μ). The factor $W^*(T,X,\mu) = L^\infty(X,\mu) \otimes_T Z$ given by the Murray-von Neumann group measure space construction is called a Krieger factor.

The name Krieger factor was first used by Connes [10], and is certainly appropriate in view of Krieger's substantial contributions in this area [25-31]. While it is true that (except for the

III_1 case) Connes has shown that the Krieger factors coincide with the injective factors [11], it nevertheless remains a useful terminology, especially for an injective factor which is given in the form $W^*(X,T,\mu)$.

It is elementary that ITPFI $\Rightarrow$ Krieger. The implication Krieger $\Rightarrow$ ATI is less obvious (see Krieger [28], proposition 2.2).

REMARK 2.7. The ATI factors are significant because of their "accessibility". For example, if G is a non-type I separable locally compact group, then <u>every</u> ATI von Neumann algebra occurs as $\Pi(G)''$ for some unitary representation Π [32]. And in some cases, Connes has shown that <u>only</u> ATI von Neumann algebras occur [11].

Finally, we present some technical material concerning the Connes-Takesaki flow of weights [12] which will be needed in Sec. 5. While the problem of <u>proving</u> the main properties is far from trivial, it is extremely easy to describe the flow of weights. And what is perhaps even more important, there is a straightforward prescription for constructing the flow for a given factor. (The term flow of weights as used here corresponds to what Connes and Takesaki [12] call the "smooth flow of weights".) To describe this flow one begins by noting that the Murray-von Neumann comparison of projections extends in an obvious way to the comparison of weights. The measure <u>algebra</u> of the flow has as elements the equivalence classes $[\phi]$ of integrable weights of infinite multiplicity. (A weight ϕ on a von Neumann algebra M has infinite multiplicity if $\phi \oplus \phi \sim \phi$, and it is integrable if $\{x: \int_{-\infty}^{\infty} dt\, \sigma_t^{\phi}(x^*x) < \infty\}$ is σ-weakly dense in M.) The flow is then defined by its action on the measure algebra as given by the equation $F_M(t)[\phi] = [e^t\phi]$, $t \in R$. While this description of the flow does suffice for some purposes (see for example [14]), it is not sufficiently concrete in general. An explicit construction of the flow of weights for a given type III_0 factor M_1 acting on the Hilbert space H_1 is obtained as follows. Let

$$H = L^2(R) \otimes H_1, \qquad (2.1)$$

$$M = \mathcal{L}(L^2(R)) \otimes M_1, \qquad (2.2)$$

and

$$\tilde{\omega} = \omega \otimes \phi \tag{2.3}$$

where ϕ is any state on M, and $\omega(x) = \tau(\rho x)$ where τ is the trace on $\mathcal{L}(L^2(R))$ and ρ is the diagonal operator given by $\rho(t) = e^{-t}$. Let

$$(V_s f)(t) = f(s+t), \quad f \in L^2(R) \tag{2.4}$$

and set

$$\theta_s = (\text{Ad } V_s) \otimes 1. \tag{2.5}$$

Since $V_s \rho V_s^{-1} = e^s \rho$, the centralizer $M_{\tilde{\omega}}$ is invariant under θ_s. It follows from [12] that

$$M_{\tilde{\omega}} = \int_X N(x)\, d\nu(x) \tag{2.6}$$

where N(x) is type II_∞ a.e., and that the flow is given by the action of θ_s on the centre of the centralizer $C_{\tilde{\omega}} = L^\infty(X,\nu)$. If $T(M) \neq \emptyset$ there is a discrete version of this which yields the base transformation for the flow built under the appropriate constant ceiling function. See [15] for a detailed application of these constructions.

3. THE EARLY HISTORY. In 1938 von Neumann introduced the notion of an infinite tensor product of Hilbert spaces [36]. For a more concise description of these spaces see Guichardet [19] or Woods [47]. von Neumann asserted in [36] that certain ITPFI factors were type III. It should be remembered that the existence of type III factors was a non-trivial problem in the beginning. Their existence was established in Rings of Operators III [37] which appeared four years after [34]. (The corresponding problem in ergodic theory, namely the existence of ergodic transformations which preserve no equivalent invariant measure, was not resolved until 1959.) An explicit proof of von Neumann's assertion, together with a partial type classification of ITPFI factors in terms of their eigenvalue lists, was given by Bures [6] in 1960. The type classification problem for ITPFI factors of bounded type (i.e. $M = \otimes(M_\nu,\phi_\nu)$ with M_ν type I_{n_ν} and $n_\nu \leq N$ for some $N < \infty$) was solved by Moore [33], and the general case was solved by Takenouchi [44] and Hill [21] (see also Pukanszky [40] and Araki [1]). Størmer used the Tomita-Takesaki theory to give a rather elementary solution of the type classification problem [43].

There have been some interesting applications of the infinite tensor product space constructions. The van Hove model of a quantum field interacting with fixed sources used this construction [45]. Perhaps the simplest mathematically rigorous construction of the Fock representation of the canonical commutation relations is as an infinite tensor product representation [2]. In 1964 Klauder, McKenna and Woods proved, by a very elementary argument, that two infinite tensor product representations of the canonical commutation relations (with respect to a fixed orthonormal basis in the test function space) were unitarily equivalent if and only if the reference product vectors were weakly equivalent in the sense of von Neumann [24]. From the viewpoint of providing an easily computable invariant, this result represented a significant advance. It was later generalized by Bures [7]. In 1965 Araki and Woods proved that an atomic complete Boolean algebra of type I factors necessarily arose from an infinite tensor product construction [3].

4. THE YEARS OF INFLUENCE. While the first 30 years produced many results they yielded, as far as the classification problem was concerned, only the existence of seven non-type I isomorphism classes and the uniqueness of the ATI II_1 factor.

In 1966 R. Powers proved that a certain one-parameter family of ITPFI factors R_x, $0 < x < 1$ were mutually non-isomorphic [41], and hence there are precisely c isomorphism classes of factors on a separable Hilbert space (the set of all von Neumann algebras on a separable Hilbert space has the cardinality of the continuum). The Powers' factors R_x are infinite tensor products of type I_2 factors with the constant eigenvalue list $((1+x)^{-1}, x(1+x)^{-1})$.

In the summer of 1967 H. Araki and J. Woods began to study the results of Powers. The crucial point in Powers' argument was his lemma 4.3 concerning the ratios of the eigenvalues in the eigenvalue list. A detailed study of this lemma led Araki and Woods [4] to introduce the <u>asymptotic ratio set</u> $r_\infty(M)$ for M an arbitrary ITPFI factor. This invariant can be defined by

$$r_\infty(M) = \{0 \leq x < \infty : M \sim M \otimes R_{f(x)}\}$$

where

$$f(x) = \begin{cases} x & \text{if } 0 \leq x \leq 1 \\ x^{-1} & \text{if } 1 < x < \infty, \end{cases}$$

R_0 is the type I factor, R_1 is the hyperfinite II_1, and R_x, $0 < x < 1$ are the Powers factors. Since in fact this definition corresponds to the end results of the paper, it is perhaps somewhat misleading from a historical point of view. We therefor present as well the original working definition in terms of the eigenvalue lists (see definitions 2.3 and 2.4). Given $M = \bigotimes_{\nu\varepsilon N}(M_\nu,\phi_\nu)$ and $I \subset N$, we write $M(I) = \bigotimes_{\nu\varepsilon I}(M_\nu,\phi_\nu)$ and $\phi(I) = \bigotimes_{\nu\varepsilon I}\phi_\nu$. The asymptotic ratio set of M is then the set of all $x \varepsilon [0,\infty)$ such that there exists a sequence of mutually disjoint finite index sets $I_n \subset N$, disjoint subsets K_n^1, K_n^2 of $Sp(\phi(I_n)/M(I_n))$ for each $n = 1,2,3 \ldots$ such that $\lambda \varepsilon K_n^1$ implies $\lambda \neq 0$, and a bijection ψ_n from K_n^1 to K_n^2 such that

$$\sum_{n=1}^{\infty} \sum_{\lambda\varepsilon K_n^1} \lambda = \infty$$

and

$$\lim_{n\to\infty} \sup_{\lambda\varepsilon K_n^1} |x - \psi_n\lambda/\lambda| = 0.$$

Araki and Woods proved that $r_\infty(M)$ must be one of the following sets $\{0\}$, $\{1\}$, $S_0 = \{0,1\}$, $S_x = \{0,1,x^n: n \varepsilon Z\}$, $0 < x < 1$, and $S_1 = [0,\infty)$. (The notation here has been changed so as to be consistent with that later introduced by Connes.) They proved that, except for the case S_0, $r_\infty(M) = r_\infty(N)$ implied that $M \sim N$. Thus for the first time since "Rings of Operators IV" [35] one had identified factors given by different constructions. They also introduced a second invariant $\rho(M)$ defined by

$$\rho(M) = \{0 \leqslant x < \infty: R_{f(x)} \sim R_{f(x)} \otimes M\}$$

where f(x) is defined above. They used the invariant ρ to exhibit a continuum of non-isomorphic factors in the class S_0. They gave some useful criteria for determining the asymptotic ratio set of a <u>given</u> ITPFI factor (see lemma 9.3 of [4]), and they showed that certain factors occuring in physics ([2],[5]) were the unique factor in the class S_1. It is perhaps worth remarking that, at this point in time, it would have taken a great deal of courage to believe that a general structure theory would soon appear. So while the above decomposition of the type III ITPFI factors corresponds precisely to that later obtained by Connes, there was no way of knowing its significance at the time. O. Nielsen's proof [38] that the central decomposition of a von Neumann algebra could be rearranged as an integral over von Neumann algebras of "pure" asymptotic ratio type was, of course, a comforting sign (a von Neumann algebra is of "pure" asymptotic ratio type if a.e. factor occuring in its central decomposition has the same asymptotic ratio set).

At this time Krieger had already begun his extensive investigation [25-31] of the factors $W^*(X,T,\mu)$ (see definition 2.6). This was a closely related problem and, in particular, he was able to translate the invariant $r_\infty(M)$ directly into an invariant for ergodic transformations (X,T,μ) [29].

While Powers' result fixed the number of isomorphism classes, it said nothing about the intricacy of the classification problem. In October 1971, Woods used the invariant $\rho(M)$ to exhibit a non-smooth family of ITPFI factors, thereby proving that the classification of factors was not smooth [46]. (If the equivalence classes can not be labelled by a countable number of Borel parameters then one says that the equivalence relation is not smooth. The Borel structure for von Neumann algebras was introduced by Effros [17].) The explicit construction which was used provided sufficient control to permit the later observation that the resulting ρ sets were uncountable. This was not a question at this time. But after Connes' identification of the ρ sets with his T sets via $T \in T(M) \Leftrightarrow \exp(-2\pi/T) \in \rho(M)$, together with his results that T(M) was a subgroup of R and that all countable subgroups of R occured as T(M) for some M acting on a separable Hilbert space, it certainly became a question. I first became aware of this issue during discussions with Connes in Paris in January 1973, and realized that the construction mentioned above settled the question. A few months later Krieger told me that he was hoping to prove his isomorphism theorem, and that in any case he had identified T(M) with the point spectrum of the associated flow of weights. Since the existence of ergodic flows with uncountable point spectra was an open question, he was quite interested to learn of the existence of uncountable T sets. Somewhat later, Osikawa gave a direct construction of an ergodic flow with uncountable point spectrum [39]. (Since the above developments are not recorded in the literature, I have taken the liberty of setting them forth here.)

At this point in the discussion there begins a shift in the pattern of results, which ultimately brought a remarkable amount of order into the somewhat unstructured picture of late 1971.

In September 1971 Connes, who had just attended the Battelle lectures of Takesaki on modular automorphisms, was preparing to expound the Araki-Woods results in Dixmier's seminar. He immediately observed, while returning home on the Paris metro, that the role played by the adjoint operation (see for example lemma 5.7 of [4]) allowed him to obtain these invariants from the modular automorphisms. More precisely, for ITPFI factors one has $S(M) = r_\infty(M)$ and $T \in T(M) \Leftrightarrow \exp(-2\pi/T) \in \rho(M)$ [8]. The latter relationship makes it understandable that Araki and Woods missed the fact that T(M) is a subgroup of R. All other general properties of $r_\infty(M)$ and $\rho(M)$ for M ITPFI were established in [4] (however Connes substantially simplified the proof of lemma 5.8 of [4], see [8], théorème 1). The observation of Connes discussed above was the starting point for his remarkable thesis "Une classification des facteurs de type III" [9]. It is undoubtedly the most important role played by the ITPFI factors in the heroic developments during the period 1971-1975.

The next problem we consider is the existence of non-ITPFI Krieger factors. While this question is now overshadowed by the beautiful structure theory which was developed, it should be remembered that it probably attracted more attention than any other single technical question from 1970 to 1972 (the full significance of the question of the "uniqueness of the hyperfinite II_∞" had yet to appear). This question was, and still is, of interest because of the fact that if non-type I factors occur in some situation, then usually all ATI factors, and in some cases only ATI factors, will occur (see Remark 2.7). Since much was known about ITPFI factors and they are easy to construct (see Remark 2.5), there was considerable interest in knowing whether or not it would become necessary to develop methods for dealing with ATI but non-ITPFI factors. And of course the existence of such factors would mean that the structural properties which distinguish them from the ITPFI factors could well be interesting. In 1970 Krieger constructed a $W^*(X,T,\mu)$ which would eventually be shown to be non-ITPFI. But at the time he could only prove that the pair $(W^*(X,T,\mu),\mathfrak{M})$ where $\mathfrak{M}$ is the maximal abelian subalgebra appearing in the group measure space construction, could not be obtained by a tensor product construction [30]. In July 1972 Connes exploited the Tomita-Takesaki theory of modular automorphisms to give the first proof that Krieger's example was not ITPFI [9].

5 LATER DEVELOPMENTS. There have been some specific applications. Connes and Woods proved that the Powers factors were asymptotically abelian [13], thus resolving the question of Sakai concerning the existence of infinite asymptotically abelian factors. Katznelson has investigated diffeomorphisms of the circle with Lebesgue measure [23]. He has shown that the Krieger factors obtained from a C^2-diffeomorphism with irrational rotation number are ITPFI, and that every ITPFI factor is the Krieger factor of some C^∞-diffeomorphism of the circle.

However most of the activity in the last few years has been concerned with the problem of understanding why an ATI factor can fail to be ITPFI. The original example of Krieger [30] obtained by his "tower construction" technique (in effect a construction based on the flow of weights), was sufficiently complicated that it shed little light on the ITPFI versus non-ITPFI phenomenon. In 1976, Connes and Woods obtained an example for which, in the context of the flow of weights, the argument was quite elementary [14]. Their starting point was the observation that if M is ITPFI then the Sakai flip σ on $M \otimes M$ belongs to $\overline{\text{Int}}(M \otimes M)$. Now the equation $\text{Mod}\ \alpha\ [\phi] = [\phi \circ \alpha]$ defines a continuous homomorphism from Aut M into the polish group of automorphisms of the measure space X_M of the flow. Clearly $\alpha \in \overline{\text{Int}\ M}$ implies that $\text{Mod}\ \alpha = 1$. The argument is then completed by showing the existence of a flow such that the image of the Sakai flip can not be the identity map. While the argument is elementary and involves a straightforward spatial property, the crucial existence step is carried out in the flow. In this sense, a direct spatial construction remains to be given.

In 1976 Connes and Woods began a general investigation of the ITPFI versus non-ITPFI phenomena. This question is best understood in the context of Krieger's theorem [31] which states, in part, that the mapping given by the Connes-Takesaki flow of weights from Krieger factors of type III_0 (with algebraic isomorphisms as the equivalence relation) to non-transitive ergodic flows (with the usual equivalence relation) is one-to-one and onto between equivalence classes. This is a most remarkable theorem. Taken together with Connes' results on injective factors [11], it implies that in certain situations in analysis (for example the representation theory of connected separable locally compact groups) the factors which occur and are not of type III_1, are known and classified. The question of what property characterizes the flows arising from ITPFI factors thus seems to be a rather natural one. It was precisely this question that Connes and Woods took up in 1976, and we now present a brief description of their results [15].

We begin with the definition of a new property in ergodic theory which resulted from this work. Let $\alpha : G \to \mathrm{Aut}\,(X,\nu)$ be a homomorphism from a locally compact group G to the group of automorphisms of a Lebesgue measure space (X,ν). α is said to be approximately transitive (hereafter referred to as AT) if given $\varepsilon > 0$ and $f_1, \ldots f_n \in L^1_+(X,\nu)$, there exist $f \in L^1_+(X,\nu)$ and $\lambda_1, \ldots \lambda_n \in L^1_+(G,dg)$ such that

$$||f_j - \int_G dg\, \lambda_j(g)\, f\circ\alpha_g\, d\nu\circ\alpha_g/d\nu||_1 \leq \varepsilon.$$

An AT action is necessarily ergodic. If G = Z and α is AT then the transformation $T = \alpha(1)$ is said to be AT. A transformation which admits a strong approximation by periodic transformations (i.e. there exists a sequence of Rohlin towers which generate the measure algebra) is AT. A flow built under a constant ceiling function is AT if and only if the base transformation is AT. The main results are

(1) If M is a Krieger factor of type III_0 then the flow of weights is AT if and only if M is ITPFI, and

(2) an AT transformation which preserves a finite measure has zero entropy.

The proof that ITPFI implies AT follows rather easily from the explicit construction of the flow given in Sec. 2, together with the observation that if M is an ITPFI factor then there exists an increasing sequence E_n of conditional expectations onto finite type I subfactors such that $E_n x \to x$ strongly as $n \to \infty$ for all

$x \in M$. The link between the spatial properties of M and the flow is provided by the observation that a finite projection e in the centralizer defines a measure $\mu_e \prec \nu$ by the equation $d\mu_e/d\nu =$ Trace $e(x)$ where $e = \int^{\oplus} e(x)\, d\nu$ (see Eq. (2.6)). To prove the converse that AT implies ITPFI, one starts with Connes' result that if M is a Krieger factor then there exists an increasing sequence E_n of conditional expectations onto finite-dimensional subalgebras [10], and the observation that a certain "product property" implies ITPFI. Now let M be a Krieger factor, ϕ a given state on M, $\varepsilon > 0$, V a strong neighbourhood of 0 in M, and $T_1, \dots T_n \in M$. Then there exists $K < \infty$ such that $T_j \in E_K M + V$, $j = 1, \dots n$. The problem is then to use the AT of the flow together with the above relation between projections and measures to imbed $E_K M$ into a finite type I subfactor N so that $\| \phi - \phi_N \otimes \phi_{N^c} \| \leq \varepsilon$ where $N^c = N' \cap M$ so that $M = N \otimes N^c$, and ϕ_N (resp. ϕ_{N^c}) denotes the restriction of ϕ to N (resp. N^c). The product property is then satisfied, so that M is ITPFI. Finally, we remark that the entropy result allows one to deduce the existence of non-ITPFI factors. Again it should be noted that the crucial existential step is carried out in the flow.

6. EPILOGUE. We have sketched some of the developments involving ITPFI factors during the period 1966-1980, especially the impact on the classification problem. While it is unlikely that the ITPFI factors will again play such a profound role in the development of operator algebras, it is likely that they will continue to be of interest. It seems probable that there will be increasingly many applications of operator algebras, and some of these may well ask questions about ITPFI factors. Although much is known about these factors, much more remains to be known. Our survey would be incomplete without a short list of open questions to indicate the unknown territory. It should be emphasized that the following list is far from exhaustive and is merely intended to illustrate the nature of some current questions.

(1) Give an "explicit computation" of the flow of weights for an ITPFI factor in terms of its eigenvalue list. Of course an ITPFI factor can be given in many ways, and hence has many diff-

erent eigenvalue lists. At present, any computations for any ITPFI factor would be a welcome result. A few very special cases have been computed by Connes (private communication). Because of Krieger's theorem, a good answer to this problem might be helpful in resolving some of the following ones.

(2) Which flows arise from ITPFI factors of bounded type? (See Sec. 3 for the definition of bounded type.) Krieger has given an example of an ITPFI factor which is not of bounded type [30]. There are a number of variations on this question. For example, the question of bounded versus unbounded type can be considered from the purely spatial point of view.

(3) Determine the ATI factors whose flow of weights preserves a measure (resp. a finite measure). This is perhaps the most natural question which proceeds from the flow to the factor, and is therefor in some sense the counterpart to the question of Connes and Woods [15] described in Sec. 5. Since the Connes-Takesaki construction is from the factor to the flow, questions in this direction may be more difficult. Of course the total number of questions of this nature is extremely large (it amounts to compiling a bilingual dictionary). A reasonable goal is to tackle a few of the more obvious ones in order to develop techniques for dealing with the transition from the factor to the flow and vice versa.

(4) Does there exist an ITPFI factor of bounded type which is not ITPFI_2? A factor is said to be ITPFI_2 if it can be expressed in the form $\bigotimes_{\nu=1}^{\infty} (M_\nu,\phi_\nu)$ with M_ν type I_2 for all ν.

(5) Does there exist a type III_0 ITPFI_2 (resp. ITPFI) factor M such that $M \otimes M \sim R_x$ for some $0 < x < 1$? Krieger has given a type III_0 ATI factor M with $M \otimes M \sim R_x$ [28]. The ITPFI_2 question is certainly easier if one proceeds from the eigenvalue list approach, but even this question seems difficult. This illustrates the fact that while there is an explicit formula for S(M) in terms of the eigenvalue list, in practice it may by very difficult to compute S(M) from certain eigenvalue lists.

(6) Does there exist an ITPFI (resp. ATI non-ITPFI) type III_0 factor M such that $M \otimes M \sim M$?

(7) Is there any relation between unitary equivalence of type III_0 ITPFI factors and von Neumann's weak equivalence of product vectors? We formulate a precise version of this question as follows. Let M_j, $j = 1,2$ be type III_0 $ITPFI_2$ factors with $T \in T(M)$, $T > 0$, and let $\lambda = \exp(-2\pi/T)$. Then using weak equivalence one can take the eigenvalue list so that all entries are of the form $((1+\lambda^k)^{-1}, \lambda^k(1+\lambda^k)^{-1})$, $k \in N$ (see [4], lemma 11.2). Let N_{jk} be the number of times such an entry occurs for M_j. The factors M_j are now completely described by the sequences $(N_{jk})_{k\in N}$, $j = 1,2$. The problem is now to determine when M_1 is unitarily equivalent to M_2 in terms of the N_{jk}. If

$$\Sigma_{k\in N} \lambda^k |N_{1k} - N_{2k}| < \infty$$

then it follows easily from the type I condition that $M_1 \sim M_2$. Is it possible that this condition is also necessary?

(8) Give eigenvalue list constructions of type III_0 ITPFI factors M such that T(M) is explicitly determined. Some examples have been given by Hamachi, Oka and Osikawa [20].

(9) Determine the possible values for T(M) with M ITPFI (resp. ATI). The possible values of T(M) for M acting on a separable Hilbert space is also an open question (perhaps more difficult). It is known that all countable subgroups of R occur [9], and that for certain ITPFI factors T(M) can be uncountable [46].

We should remark that we are not aware of any profound consequences that might arise from solving any of the above problems. But it is likely that investigations of these and similar questions should result in improvements in techniques, especially with respect to computing various quantities for <u>given</u> factors. It seems likely that future developments will see applications of operator algebras which ask questions requiring explicit calculations for given factors.

ACKNOWLEDGEMENT. This seems an appropriate time and place to acknowledge the great pleasure of the many months of discussions on this subject with Huzihiro Araki and Alain Connes. Much of this survey has been concerned with the fruits of these discussions.

I would also like to thank the Institut des Hautes Etudes Scientifiques, Bures-sur-Yvette, for its generous hospitality while this survey was being written.

BIBLIOGRAPHY

1. H. Araki, A lattice of von Neumann algebras associated with the quantum theory of a free Bose field, J. Math. Phys. 4 (1963), 1343-1362.

2. H. Araki and E. J. Woods, Representations of the canonical commutation relations describing a nonrelativistic infinite free Bose gas, J. Math. Phys. 4 (1963), 637-662.

3. H. Araki and E. J. Woods, Complete Boolean algebras of type I factors, Publ. Res. Inst. Math. Sci. Ser. A 2 (1966), 157-242.

4. H. Araki and E. J. Woods, A classification of factors, Publ. Res. Inst. Math. Sci. Ser. A 4 (1968), 51-130.

5. H. Araki and W. Wyss, Representations of canonical anticommutation relations, Helv. Phys. Acta 37 (1964), 136-159.

6. D. Bures, Certain factors constructed as infinite tensor products, Comp. Math. 15 (1963), 169-191.

7. D. Bures, Representations of infinite weak product groups, Comp. Math. 22 (1970), 7-18.

8. A. Connes, Calcul des deux invariants d'Araki et Woods par la théorie de Tomita et Takesaki, C. R. Acad. Sci. Paris Sér. A 274 (1972), 175-177.

9. A. Connes, Une classification des facteurs de type III, Ann. Sci. E. N. S. 4ème Série 6 (1973), 133-252.

10. A. Connes, On hyperfinite factors of type III and Krieger's factors, J. Functional Analysis 18 (1975), 318-327.

11. A. Connes, Classification of injective factors, Ann. of Math. 104 (1976), 73-115.

12. A. Connes and M. Takesaki, The flow of weights on factors of type III, Tohoku Math. J. 29 (1977), 473-575.

13. A. Connes and E. J. Woods, Existence de facteurs infinis

asymptotiquement abéliens, C. R. Acad. Sci., Paris, Sér. A 279 (1974), 189-191.

14. A. Connes and E. J. Woods, A construction of approximately finite-dimensional non-ITPFI factors, Can. Math. Bull. 23 (1980), 227-230.

15. A. Connes and E. J. Woods, Approximately transitive flows and ITPFI factors (to appear).

16. J. Dixmier, Les algèbres d'opérateurs dans l'espace hilbertien, Gauthier-Villars, Paris, 1957.

17. E. Effros, The Borel space of von Neumann algebras on a separable Hilbert space, Pacific J. Math. 15 (1965), 1153-1164.

18. G. Elliott and E. J. Woods, The equivalence of various definitions for a properly infinite von Neumann algebra to be approximately finite-dimensional, Proc. Am. Math. Soc. 60 (1976), 175-178.

19. A. Guichardet, Tensor products of C*-algebras, Part II, Mat. Institut, Aarhus Univ., Lecture notes no. 13 (1969).

20. T. Hamachi, Y. Oka and M. Osikawa, A classification of ergodic non-singular transformation groups, Mem. Fac. Sc. Kyushu Univ. Ser. A 28 (1974), 113-133.

21. D. Hill, σ-finite invariant measures on infinite product spaces, Trans. Amer. Math. Soc. 153 (1971), 347-370.

22. R. V. Kadison and J. Ringrose, to appear.

23. Y. Katznelson, The action of diffeomorphisms of the circle on the Lebesgue measure, J. Analyse Math. 36 (1979), 156-166.

24. J. R, Klauder, J. McKenna and E. J. Woods, Direct product representations of the canonical commutation relations, J. Math. Phys. 7 (1966), 822-828.

25. W. Krieger, On non-singular transformations of a measure space I, Z. Wahrscheinlichkeitstheorie verw. Geb. 11 (1969),83-97.

26. W. Krieger, On non-singular transformations of a measure space II, ibid 11 (1969), 98-119.

27. W. Krieger, On constructing non-*isomorphic hyperfinite factors of type III, J. Functional Analysis 6 (1970), 97-109.

28. W. Krieger, On a class of hyperfinite factors that arise from null-recurrent Markov chains, J. Functional Analysis 7 (1971), 27-42.

29. W. Krieger, On the Araki-Woods asymptotic ratio set and nonsingular transformations of a measure space, Lecture notes in

Math. 160, Springer-Verlag, (1970), 158-177.

30. W. Krieger, On the infinite product construction of non-singular transformations of a measure space, Invent. Math. 15 (1972), 144-163.

31. W. Krieger, On ergodic flows and the isomorphism of factors, Math. Ann. 223 (1976), 19-70.

32. O. Maréchal, Une remarque sur un théorème de Glimm, Bull. Soc. Math. 2ème Série 99 (1975), 41-44.

33. C. C. Moore, Invariant measures on product spaces, Proceedings of the Fifth Berkeley Symposium on Mathematical Statistics and Probability, vol. II, part 2, 447-459, University of California, Berkeley, 1967.

34. F. J. Murray and J. von Neumann, On rings of operators, Ann. of Math. 37 (1936), 116-229.

35. F. J. Murray and J. von Neumann, Rings of operators IV, Ann. of Math. 44 (1943), 716-804.

36. J. von Neumann, On infinite direct products, Comp. Math. 6 (1938), 1-77.

37. J. von Neumann, On rings of operators III, Ann. of Math. 41 (1940), 94-161.

38. O. A. Nielsen, The asymptotic ratio set and direct integral decompositions of von Neumann algebras, Can. J. Math. 23 (1971), 598-607.

39. M. Osikawa, Point spectra of non-singular flows, Publ. Res. Inst. Math. Sci. 13 (1977), 167-172.

40. L. Pukanszky, Some examples of factors, Publ. Math. Debrecen, 4 (1955-56), 135-156.

41. R. T. Powers, Representations of uniformly hyperfinite algebras and their associated von Neumann rings, Ann. of Math. 86 (1967), 138-171.

42. E. Størmer, Hyperfinite product factors, Arkiv för matematik 9 (1971), 165-170.

43. E. Størmer, On infinite tensor products of von Neumann algebras, Am. J. Math. 93 (1971), 810-818.

44. O. Takenouchi, On type classification of factors constructed as infinite tensor products, Publ. Res. Inst. Math. Sci. Ser. A 4 (1968), 467-482.

45. L. van Hove, Les difficultés de divergence pour un modelle particulier de champs quantifié, Physica 18 (1952), 145-159.

46. E. J. Woods, The classification of factors is not smooth, Can. J. Math. 25 (1973), 96-102.

47. E. J. Woods, On the construction of infinite tensor products of Hilbert spaces (to appear).

DEPARTMENT OF MATHEMATICS
QUEEN'S UNIVERSITY
KINGSTON, ONTARIO, CANADA

Proceedings of Symposia in Pure Mathematics
Volume 38 (1982), Part 2

CLASSIFICATION DES FACTEURS

par

A. CONNES

Nous traitons dans cet exposé le problème, interne à la théorie, de la classification à isomorphisme près des algèbres de von Neumann et plus spécifiquement des facteurs (i.e. des algèbres à centre trivial).

Le résultat final est l'existence d'une classe remarquable de facteurs : les facteurs moyennables (caractérisés par de nombreuses propriétés équivalentes dont l'analogue de la moyennabilité pour les groupes) et d'une classification complète de ces facteurs, (à l'exception possible de facteurs exotiques de type III_1, problème toujours non résolu) en types :

I_n	$M_n(\mathbb{C})$
I_∞	$\mathcal{L}(H)$
II_1	R facteur hyperfini de Murray et von Neumann
II_∞	$R_{0,1} = R \times \mathcal{L}(H)$
III_λ	$\lambda \in]0,1[$, R_λ facteur de Powers
III_1	R_∞ facteur d'Araki et Woods (+ évent. d'autres)
III_o	R_W facteur de Krieger associé au flot ergodique W .

Le contenu du texte est divisé en paragraphes :

I. Les articles de Murray et von Neumann

II. Représentations des algèbres stellaires.

III. Le cadre algébrique de l'intégration non commutative et la théorie des poids.

IV. Facteurs de Powers, Araki et Woods, et Krieger.

V. Facteurs de type III_λ .

VI. Théorie ergodique non commutative.

VII. Algèbres de von Neumann moyennables.

VIII. Classification des facteurs moyennables et le problème III_1 .

On pourrait facilement illustrer cette classification par les nombreux exemples de facteurs provenant des feuilletages. Notons simplement qu'à tout feuilletage correspond canoniquement une algèbre de von Neumann ([14]), et qu'à des feuilletages aussi naturels que le feuilletage de Kronecker (de pente $\theta \notin Q$) du tore, ou le feuilletage d'Anosov associé à une surface de Riemann (de genre > 1) correspondent (respectivement) les facteurs $R_{0,1}$ et R_∞ de la classification ci-dessus.

La théorie non commutative de l'intégration dans le cadre géométrique devrait donner à la théorie des algèbres de von Neumann un rôle analogue à celui de l'intégration classique en géométrie différentielle.

Parmi les questions ouvertes, internes à la théorie, citons l'étude des facteurs $R(\Gamma)$ associés (par la représentation régulière) aux groupes discrets rigides (disons Γ de covolume fini dans un groupe de Lie simple de rang ≥ 2). Il est facile de voir (en utilisant l'idée de [18]) que pour Γ groupe discret, Γ à la propriété T de Kazhdan si et seulement si $R(\Gamma)$ a la propriété suivante :

Un facteur N de type II_1 a la propriété T si et seulement si il existe $x_1,\dots,x_n \in N$ tels que pour tout hilbert bimodule H(*) sur

(*) (i.e. H est un espace de hilbert et on a deux représentations normales de N, N^o avec $(x\xi)y = x(\xi y)$ $\forall x,y \in N$)

N on ait :

$$\{\exists \xi \in H \ , \quad \|\xi\| = 1 \ , \quad \|x_i\xi - \xi x_i\| \leq 1\} \Rightarrow$$

$$\{\exists \eta \in H \quad \eta \neq 0 \quad , \quad x\eta = \eta x \quad \forall x \in N\}$$

Les facteurs ayant la propriété T jouissent (comme les groupes de Kazhdan) de propriétés de rigidité remarquables, par exemple il n'existe pas de suite non triviale $N_n \subset N_{n+1}$ de sous facteurs avec $\cup N_n$ dense dans N. Le problème principal est d'obtenir :

<u>Problème</u>. Montrer que si Γ_1, Γ_2 sont non isomorphes $R(\Gamma_1)$ et $R(\Gamma_2)$ sont non isomorphes.

Bien sur ici Γ_1 et Γ_2 sont rigides au sens ci-dessus, (pour des groupes moyennables on a une situation radicalement opposée, voir chapitre VII, corollaire 3).

Le seul résultat connu est que si Γ_1 est rigide au sens ci-dessus et Γ_2 est discret dans $SL(2, \mathbb{R})$, il n'existe aucun homomorphisme de $R(\Gamma_1)$ dans $R(\Gamma_2)$.

I. LES ARTICLES DE MURRAY ET VON NEUMANN.

Soient h un espace hilbertien complexe et $\mathcal{L}(h)$ l'algèbre stellaire des opérateurs bornés de h dans h. C'est une algèbre de Banach munie de la norme $\|T\| = \sup_{\|\xi\| \leq 1} \|T\xi\|$ et de l'involution $T \to T^*$ définie par :

$$\langle T^*\xi, \eta\rangle = \langle \xi, T\eta\rangle \ , \ \forall\, \xi, \eta \in h \ , \quad \text{on a} \ : \quad \|T^*T\| = \|T\|^2 \ , \ \forall\, T \in \mathcal{L}(h) \ .$$

Même si h est de dimension dénombrable, l'espace de Banach $\mathcal{L}(h)$ n'est pas de type dénombrable. Il existe un unique sous-espace fermé du dual $\mathcal{L}(h)^*$ dont $\mathcal{L}(h)$ est le dual. Il s'agit de l'espace des formes linéaires sur $\mathcal{L}(h)$ qui s'écrivent :

$$L(T) = \text{Trace} \ \ (\rho T) \quad \forall\, T \in \mathcal{L}(h)$$

où ρ est un opérateur à trace, ce qui signifie que $|\rho| = (\rho^*\rho)^{\frac{1}{2}}$ vérifie $\Sigma < |\rho|\xi_i, \xi_i > = \text{Trace } |\rho| < \infty$ pour toute base orthonormale de $\mathcal{h}$. La norme de la forme linéaire L est égale à Trace $|\rho|$, et muni de cette norme l'espace $\mathfrak{L}(\mathcal{h})_* = \{\rho \in \mathfrak{L}(\mathcal{h}) \text{ , Trace } |\rho| < \infty\}$ est un espace de Banach, le prédual de l'espace de Banach $\mathfrak{L}(\mathcal{h})$.

Quand $\mathcal{h}$ est de dimension dénombrable, l'espace $\mathfrak{L}(\mathcal{h})_*$ est de type dénombrable ; la dualité entre $\mathfrak{L}(\mathcal{h})_*$ et $\mathfrak{L}(\mathcal{h})$ est exactement analogue à la dualité entre $\ell^1(A)$ et $\ell^\infty(A)$ où A est un ensemble. En particulier, dès que $\mathcal{h}$ est de dimension infinie, l'espace $\mathfrak{L}(\mathcal{h})_*$ n'est pas réflexif et la topologie $\sigma(\mathfrak{L}(\mathcal{h}), \mathfrak{L}(\mathcal{h})_*)$ n'est pas compatible avec la topologie normique de $\mathfrak{L}(\mathcal{h})$.

Ainsi, il est beaucoup plus restrictif, pour un sous-espace de $\mathfrak{L}(\mathcal{h})$, d'être fermé pour $\sigma(\mathfrak{L}(\mathcal{h}), \mathfrak{L}(\mathcal{h})_*)$ que pour la topologie normique.

Cette distinction entre ces deux topologies est essentielle pour la suite.

Soit alors M une sous algèbre involutive de $\mathfrak{L}(\mathcal{h})$, contenant l'unité 1, les conditions suivantes sont équivalentes :

Condition topologique M est $\sigma(\mathfrak{L}(\mathcal{h}), \mathfrak{L}(\mathcal{h})_*)$ fermée

Condition algébrique M est égale au commutant $(M')'$ de son commutant M'.

(Le commutant d'une partie $\mathcal{S}$ de $\mathfrak{L}(\mathcal{h})$ est défini par l'égalité

$$\mathcal{S}' = \{T \in \mathfrak{L}(\mathcal{h}) \text{ , } TS = ST \quad \forall\, S \in \mathcal{S}\}) .$$

L'équivalence entre ces deux propriétés est le théorème du bicommutant de von Neumann.

DEFINITION 1. - Une algèbre de von Neumann M dans $\mathcal{h}$ est une sous-algèbre involutive de $\mathfrak{L}(\mathcal{h})$ contenant l'unité et vérifiant les conditions équivalentes ci-dessus.

Citons quelques conséquences immédiates de la définition.

Soit $\mathcal{S} \subset \mathfrak{L}(\mathcal{h})$ une partie telle que $\mathcal{S} = \mathcal{S}^* = \{T^*, T \in \mathcal{S}\}$ alors le commutant $\mathcal{S}'$ de $\mathcal{S}$ est une algèbre de von Neumann.

Soit M une algèbre de von Neumann dans $\mathcal{h}$ et $M_1 = \{T \in M \text{ , } \|T\| \le 1\}$ la boule unité de M, alors comme M_1 est $\sigma(\mathfrak{L}(\mathcal{h}), \mathfrak{L}(\mathcal{h})_*)$ fermé dans la boule unité du

dual de $\mathcal{L}(h)_*$, c'est un compact quand on le munit de la topologie $\sigma(M_1,\mathcal{L}(h)_*)$. En particulier M est le dual d'un espace de Banach M_* , qui est en fait l'unique sous-espace (fermé) de M^* dont M soit le dual - ([53]).
On introduit souvent sous le nom de topologie faible la topologie sur $\mathcal{L}(h)$ provenant de la dualité avec le sous-espace de $\mathcal{L}(h)_*$ formé des opérateurs de rang fini, i.e. la topologie caractérisée par :

$$T_\alpha \to T \Leftrightarrow \forall\ \xi,\eta \in h \quad \text{on a} \quad \langle T_\alpha \xi,\eta\rangle \to \langle T\xi,\eta\rangle .$$

Elle est moins fine que $\sigma(\mathcal{L}(h),\mathcal{L}(h)_*)$ et comme l'espace des opérateurs de rang fini est dense en norme dans $\mathcal{L}(h)_*$ elle coïncide avec $\sigma(\mathcal{L}(h),\mathcal{L}(h)_*)$ sur les parties bornées de $\mathcal{L}(h)$. La terminologie "topologie faible" n'est cependant pas bonne, il s'agit de la topologie de la convergence simple faible.
Les algèbres de von Neumann sont les sous-algèbres involutives de $\mathcal{L}(h)$ contenant 1 qui sont fermées pour la topologie de la convergence simple faible, puisque tout commutant $\mathcal{S}',\mathcal{S}\subset\mathcal{L}(h)$ a cette propriété.

Exemples d'algèbres de von Neumann.

1. Algèbres de von Neumann abéliennes.

La description de cet exemple nous permettra de donner la forme abstraite du théorème spectral et du calcul fonctionnel borélien. Soient (X,B,μ) un espace borélien standard muni d'une mesure de probabilité μ , et $\pi(L^\infty(X,\mu))$ l'algèbre des opérateurs de $L^2(X,\mu)$ dans $L^2(X,\mu)$ définie par :

$$\pi(f)g = fg \qquad f\in L^\infty,\ g\in L^2 .$$

Alors $M = \pi(L^\infty)$ est une algèbre de von Neumann commutative et on a en fait $M = M'$. Le prédual de M est l'espace $L^1(X,\mu)$. Soit alors $x\to n(x)$ une fonction borélienne de X dans $\{1,2,\dots,\infty\}$ et $(\widetilde{X},p)$ le revêtement borélien de X défini par :

$$\widetilde{X} = \{(x,j)\in X\times\mathbb{N},\ 1\le j\le n(x)\},\ p(x,j) = x .$$

A la fonction "multiplicité" n on associe la représentation π_n de $L^\infty(X,\mu)$ dans $L^2(\widetilde{X},\widetilde{\mu})(\int f(x,j)d\widetilde{\mu} = \int \sum_j f(x,j)d\mu)$ donnée par :

$$\pi_n(f)g = (f\circ p).g .$$

L'image $M = \pi_n(L^\infty(X,\mu))$ est une algèbre de von Neumann commutative dans l'espace $\mathcal{h} = L^2(\widetilde{X},\widetilde{\mu})$ et si $n \neq 1$ on a $M' \not\subset M$.

DEFINITION 2. - <u>Soit</u> $M_i, i = 1,2$ <u>une algèbre de von Neumann dans</u> $\mathcal{h}_i, i = 1,2$; <u>on dit que</u> M_1 <u>est spatialement isomorphe à</u> M_2 <u>quand il existe un unitaire</u> $U : \mathcal{h}_1 \to \mathcal{h}_2$ <u>tel que</u> :

$$UM_1U^* = M_2 .$$

Si $\mathcal{h}$ est de type dénombrable, toute algèbre de von Neumann commutative dans $\mathcal{h}$ est spatialement isomorphe à l'algèbre $\pi_n(L^\infty(X,\mu))$ pour un espace X,μ et une fonction n convenables.

Soit alors $T \in \mathcal{L}(\mathcal{h})$ un opérateur normal : $TT^* = T^*T$, soit M l'algèbre de von Neumann engendrée par T, on peut prendre pour (X,B) le spectre $K \subset \mathbb{C}$ de T et la classe de mesure μ donnée par la mesure spectrale de T. L'application qui à la restriction à K de toute fonction polynomiale $\Sigma a_{ij} z^i \bar{z}^j$ associe $\Sigma a_{ij} T^i T^{*j}$ se prolonge en un isomorphisme π de $L^\infty(K,\mu)$ sur M avec $\pi(z) = T$. Ainsi pour toute fonction borélienne bornée f sur $\mathrm{Sp}\, T$ on donne un sens à $f(T)$ et on a :

$$(\lambda_1 f_1 + \lambda_2 f_2)(T) = \lambda_1 f_1(T) + \lambda_2 f_2(T)$$

$$f_1 f_2(T) = f_1(T) f_2(T)$$

$$f \circ g(T) = f(g(T)) .$$

La fonction $x \to n(x)$ de K dans $\{1,\dots,\infty\}$, est unique modulo μ, elle exprime la multiplicité du point x dans le spectre K. Le théorème du bicommutant montre que tout opérateur qui bicommute avec T est une fonction borélienne bornée de T. Enfin, soient N une algèbre de von Neumann non nécessairement commutative, et $T \in N$, alors $T = T_1 + iT_2$, $T_j = T_j^*$ de sorte que T_j étant normal et toute $f(T_j)$, f borélienne étant encore dans N, on voit que N est engendrée par les projecteurs qu'elle contient.

2. Le commutant d'une représentation unitaire.

Soit G un groupe, et π une représentation unitaire de G dans un espace hilbertien $\mathcal{h}_\pi$ (ou, plus généralement, soient $\mathfrak{G}$ une algèbre involutive et π une représentation involutive non dégénérée) Le commutant

$$R(\pi) = \{T \in \mathcal{L}(\mathcal{h}_\pi) \, , \, T\pi(x) = \pi(x)T \quad \forall \, x \in G\}$$

est par construction une algèbre de von Neumann.

L'intérêt de $R(\pi)$ réside dans la proposition suivante :

PROPOSITION 3.- a) Soient $E \subset \mathcal{h}$ un sous-espace fermé et P le projecteur correspondant, alors :

$$E \text{ réduit } \pi \Leftrightarrow P \in R(\pi) \, .$$

b) Soient E_1, E_2 deux sous-espaces fermés réduisant π alors les représentations réduites π^{E_j} sont équivalentes si et seulement si

$$P_1 \sim P_2 \, (R(\pi)) \quad \text{i.e.} \quad \exists \, U \in R(\pi) \, , \, U^*U = P_1 \, , \, UU^* = P_2 \, .$$

c) π^{E_1} est disjointe de π^{E_2} ssi il existe P projecteur, P dans le centre de $R(\pi)$ tel que $PP_1 = P_1, (1-P)P_2 = P_2$.

Il est naturel de dire que la représentation π est isotypique si l'on ne peut trouver deux sous-représentations disjointes. Cela équivaut à dire que $R(\pi)$ est un facteur au sens suivant :

DEFINITION 4. - Un facteur M est une algèbre de von Neumann de centre réduit aux scalaires $\mathbf{C}$.

Un autre corollaire immédiat de la proposition et du calcul fonctionnel borélien est le suivant :

Pour que π soit irréductible il faut et il suffit que $R(\pi) = \mathbf{C}$.

3. Algèbres de von Neumann de dimension finie.

Soit M une algèbre de von Neumann de dimension finie, et considérons-là, en oubliant l'espace de Hilbert dans lequel on la représentait, comme une algèbre semi-simple sur le corps algébriquement clos $\mathbb{C}$. Elle est donc somme directe d'un nombre fini d'algèbres de matrices :

$$M = \bigoplus_{1}^{k'} M_{n_k}(\mathbb{C}) .$$

Ici le signe $=$ correspond à la définition suivante :

DEFINITION 5. - Soit M_i , $i = 1,2$ une algèbre de von Neumann dans l'espace $\mathfrak{h}_i$, $i = 1,2$. On dit que M_1 est algébriquement isomorphe à M_2 , si il existe un isomorphisme algébrique θ de M_1 sur M_2 tel que $\theta(x^*) = \theta(x)^*, x \in M_1$.

Soient G un groupe et π une représentation unitaire de G dans un espace hilbertien de dimension finie : $\mathfrak{h}_\pi$. Alors $R(\pi)$ est de dimension finie et à la décomposition $M = \bigoplus_{1}^{k'} M_{n_k}(\mathbb{C})$ correspond la décomposition de π en composantes isotypiques π_k de multiplicité n_k .

La théorie de réduction (Ecrite en 1939, publiée vers 1949) ([44]).

Soient $\mathfrak{h}$ un espace de Hilbert de type dénombrable et $\mathcal{F}$ l'ensemble des facteurs $M \subset \mathcal{L}(\mathfrak{h})$. Il existe sur $\mathcal{F}$ une structure borélienne qui en fait un espace borélien standard. Soient alors (X,B) un espace borélien standard, μ une mesure de probabilité sur (X,B) et $t \to M(t)$ une application borélienne de X dans $\mathcal{F}$. Alors soit M l'algèbre stellaire dont les éléments $x \in M$ sont les sections boréliennes bornées $t \to x(t) \in M(t)$, identifiées si elles sont égales μ-presque partout, munie des opérations évidentes et de la norme :

$$\|x\| = \operatorname{Sup\,ess} \|x(t)\| .$$

On montre que l'algèbre stellaire M est une algèbre de von Neumann, (i.e. est algébriquement isomorphe à une algèbre de von Neumann représentée dans un espace convenable), par exemple en considérant l'action de M dans l'espace

$$L^2(X,\mu) \otimes \mathfrak{h} = L^2(X,\mu,\mathfrak{h})$$

définie par :

$$(\pi(x)\xi)(t) = x(t)\xi(t) \quad \forall\ \xi \in L^2(X,\mu,h) .$$

On écrit, pour résumer la construction de M , l'égalité :

$$M = \int_X M(t)d\mu(t) . \tag{6}$$

THEOREME 7. - Soit M une algèbre de von Neumann dans un espace de type dénombrable Alors M est algébriquement isomorphe à une intégrale directe de facteurs

$$\int_X M(t)d\mu(t) .$$

Ce théorème de von Neumann montre que les facteurs contiennent déjà l'originalité de toutes les algèbres de von Neumann, ils suffisent à reconstruire toute algèbre de von Neumann comme "somme directe généralisée" de facteurs.
Soient G un groupe et π une représentation unitaire de G dans un espace hilbertien séparable, à la décomposition de $R(\pi)$ comme intégrale directe de facteurs, correspond la décomposition de π comme intégrale directe de représentations isotypiques.

Comparaison des sous-représentations, comparaison des projecteurs et fonction dimension relative.

Soient G un groupe et π une représentation unitaire de G dans l'espace hilbertien h_π . Supposons π isotypique, il est alors naturel d'attendre que, comme en dimension finie, π va être un multiple d'une représentation irréductible π^E , sous représentation de π . Dans la correspondance entre sous-représentations de π et projecteurs $P \in R(\pi)$, on voit facilement que les représentations irréductibles correspondent aux projecteurs minimaux de $R(\pi)$ et on a :
π a une sous-représentation irréductible $\Leftrightarrow$ le facteur $R(\pi)$ a un projecteur minimal $\Leftrightarrow$ Il existe un espace de Hilbert k et un isomorphisme de $\mathcal{L}(k)$ sur $R(\pi)$.

Toute représentation isotypique ayant une sous-représentation irréductible est un multiple de cette représentation. Pour tout facteur M dans h , ayant un projecteur minimal, on peut factoriser $\mathcal{h}$ en produit tensoriel $\mathcal{h} = k \otimes \mathfrak{L}$ de telle sorte que $M = \{T \otimes 1$, T opérant dans $k\}$. Cependant il existe des groupes G ayant une représentation isotypique π sans sous-représentation irréductible. Ce phénomène ne se produit pas si G est un groupe de Lie réel semi-simple et π est continue, ou plus simplement si G est compact et π continue. Cependant il se produit pour la représentation régulière de nombreux groupes discrets : soit G un groupe dénombrable discret et H la réunion des classes de conjuguaison finies de G, alors H est un sous-groupe normal de G, supposons $H = \{1\}$. Soit λ la représentation régulière gauche de G, on a $\mathcal{h}_\lambda = \ell^2(G)$, l'espace de Hilbert ayant pour base $(\varepsilon_g)_{g\in G}$ et on pose :

$$\lambda(g)\varepsilon_k = \varepsilon_{gk} \qquad \lambda'(g)\varepsilon_k = \varepsilon_{kg} .$$

On montre que l'algèbre de von Neumann $R(\lambda)$ est engendrée par les opérateurs $\lambda'(g)$, $g \in G$ et que le vecteur ε_1 est totalisateur et séparateur pour $R(\lambda)$: $R(\lambda)\varepsilon_1$ et $R(\lambda)'\varepsilon_1$ sont denses dans $\mathcal{h}_\lambda$. Les coordonnées de $T\varepsilon_1$, où $T \in R(\lambda) \cap R(\lambda)'$ sont constantes sur les classes de conjuguaison de G, de sorte que comme $H = \{1\}$, on a alors $T\varepsilon_1 \in \mathbb{C}\varepsilon_1$ et comme ε_1 est séparateur, $T \in \mathbb{C}$. Ainsi $R(\lambda)$ est un facteur et λ est isotypique. Pour voir que $R(\lambda)$ n'est pas isomorphe à un facteur $\mathfrak{L}(\mathcal{h})$, on utilise la notion suivante :

DEFINITION 8. - Une trace $\mathfrak{T}$ sur un facteur M est une forme linéaire telle que $\mathfrak{T}(AB) = \mathfrak{T}(BA)$ pour tous $A, B \in M$.

Sur $\mathfrak{L}(\mathcal{h})$ il existe une trace non nulle seulement si $\mathcal{h}$ est de dimension finie, comme on le voit en vérifiant que si $\mathcal{h}$ est de dimension infinie tout élément est combinaison linéaire de commutateurs. Or sur $R(\lambda)$, on peut définir $\mathfrak{T}$ par :

$$\mathfrak{T}(A) = \langle A\varepsilon_1, \varepsilon_1 \rangle \qquad \forall\, A \in R(\lambda) .$$

La propriété $\mathfrak{T}(AB) = \mathfrak{T}(BA)$ se vérifie pour A et B de la forme $\lambda'(g)$ $g \in G$ et se déduit en général par bilinéarité et continuité. Comme $\mathfrak{T}(1) = 1$ on a construit une trace non nulle sur le facteur de dimension infinie $R(\lambda)$.

Ainsi il existe des facteurs de dimension infinie ne possédant aucun projecteur minimal.

Soit M un facteur, la traduction dans le langage des projecteurs des notions de représentations équivalentes et somme directe de représentations, donne comme dans la proposition 3 les notions suivantes :
Pour P projecteur, $P \in M$, on note $[P]$ la classe d'équivalence de P pour la relation $P_1 \sim P_2$ ssi $\exists\, U \in M$, $U^*U = P_1$, $UU^* = P_2$.
Pour P_1 et P_2 tels que $P_1P_2 = 0$, on note $[P_1]+[P_2]$ la classe de $[P_1+P_2]$, elle ne dépend que de celles de P_1 et P_2 .

L'hypothèse : M est un facteur, montre que l'ensemble des classes de projecteurs est totalement ordonné par la relation $[P_1] \leq [P_2]$ quand il existe des représentants $P_1' \leq P_2'$. Cet ensemble totalement ordonné est muni d'une loi de composition partiellement définie qui permet de donner un sens à une égalité comme $[P] = \frac{n}{m}[Q]$, n,m entiers.

DEFINITION 9. - <u>Un projecteur</u> $P \in M$ <u>est dit fini quand</u> $Q \sim P$ <u>et</u> $Q \leq P \Rightarrow Q = P$.

Cette propriété ne dépend que de la classe de P . Dans le langage des représentations, on adopterait la définition suivante : une représentation π est finie quand toute sous-représentation π^E équivalente à π est égale à π . Si π n'est pas finie elle contient une infinité de sous-représentations équivalentes deux à deux, et inversement.

THEOREME 10. - (Murray et von Neumann). <u>Soit</u> M <u>un facteur à prédual séparable.</u> <u>Il existe une application</u> D <u>de l'ensemble des projecteurs de</u> M <u>dans</u> $\bar{R}_+ = [0,+\infty]$, <u>unique à multiplication par</u> $\lambda > 0$ <u>près, telle que</u> :

a) $P_1 \sim P_2 \Leftrightarrow D(P_1) = D(P_2)$

b) $P_1P_2 = 0 \Rightarrow D(P_1)+D(P_2) = D(P_1+P_2)$

c) P fini $\Leftrightarrow D(P) < \infty$.

De plus, à une normalisation près, l'image de D est l'un des sous-ensembles suivants de $\bar{R}_+$:

$\{1,\ldots,n\}$	on dit que M est de type	I_n
$\{1,\ldots,\infty\}$		I_∞
$[0,1]$		II_1
$[0,+\infty]$		II_∞
$\{0,+\infty\}$		III .

On voit que pour que M ait un projecteur minimal, il faut et il suffit qu'il soit de type I , si M est de type I_n , $n<\infty$ c'est $M_n(C)$, et la fonction $D(P)$ est la dimension usuelle du sous-espace de C^n sur lequel le projecteur $P\in M_n(C)$ projette. Si $n=\infty$, on a $M=\mathcal{L}(\mathcal{h})$ avec $\mathcal{h}$ de type dénombrable et $D(P)$ = dimension de l'image de P .

Ce qui est remarquable dans le cas II_1 c'est l'apparition de dimensions à valeurs arbitraires dans $[0,1]$.

On dit qu'un facteur M est fini, s'il ne contient pas de sous-facteur de type I_∞ et cela est équivalent à dire que M est dans l'un des cas I_n , $n<\infty$ ou II_1 . En particulier si M est infinie il n'existe sur M aucune trace $\mathfrak{J}$ telle que $\mathfrak{J}(1)=1$.

Un des résultats remarquables des premiers articles de Murray et von Neumann est la réciproque :

THEOREME 11. - (Murray et von Neumann). <u>Soit</u> M <u>un facteur de type</u> II_1 . <u>Il existe alors une unique trace</u> $\mathfrak{J}$ <u>sur</u> M <u>telle que</u> $\mathfrak{J}(1)=1$.

En outre $\mathfrak{J}\in M_*$. Une démonstration récente, due à F.J. Yeadon [62] ramène ce théorème à un résultat puissant de Ryll Nardzewski : dans tout convexe $\sigma(X,X^*)$ compact d'un Banach X , il y a un point qui est laissé fixe par toute isométrie affine du convexe. On applique ce résultat en montrant que si $\Phi\in M_*$ vérifie $\|\Phi\|=\varphi(1)=1$ l'enveloppe convexe fermée K de l'orbite de Φ sous l'action des automorphismes intérieurs de M (transportés à M_*) est $\sigma(M_*,M)$ compacte.

On obtient alors l'existence de $\mathcal{J} \in M_*, \mathcal{J}(1) = 1$, $\mathcal{J}(uxu^*) = \mathcal{J}(x)$ pour u unitaire de M et $x \in M$.

Isomorphisme algébrique et isomorphisme spatial.

Soient M_1 et M_2 deux algèbres de von Neumann et θ un isomorphisme d'algèbres involutives de M_1 sur M_2. Alors θ est isométrique (car M_1 et M_2 sont des algèbres stellaires, $\|T\|^2 = \|T^*T\|$ = rayon spectral T^*T) et comme le prédual est unique θ est $\sigma(M_i, M_{i*})$ continu. Si M_i, i = 1,2 agit dans l'espace hilbertien h_i, l'isomorphisme θ n'est cependant pas toujours spatial puisque bien qu'isomorphes M_1 et M_2 peuvent avoir des commutants non isomorphes. En fait fixons M, supposons M à prédual de type dénombrable et cherchons à décrire tous les isomorphismes π de M sur une sous-algèbre de von Neumann de $\mathcal{L}(h)$, h espace hilbertien de type dénombrable. Pour simplifier encore ramenons-nous en utilisant la théorie de réduction au cas où M est un facteur.

On est alors ramené à étudier à équivalence près les représentations π de M dans un espace hilbertien h_π qui sont continues quand on munit M de $\sigma(M, M_*)$ et $\mathcal{L}(h_\pi)$ de la topologie de dualité avec $\mathcal{L}(h_\pi)_*$. Comme M est un facteur le commutant $R(\pi) = \pi(M)'$ est aussi un facteur et π est isotypique. Il en résulte en prenant $\pi_1 \oplus \pi_2$ que deux représentations π_1 et π_2 ne sont jamais disjointes et que toute représentation π de M est une sous-représentation d'une représentation infinie fixée une fois pour toutes. Ce résultat se prolonge au cas où M n'est plus un facteur, toute représentation π (continue pour $\sigma(M, M_*)$ et $\sigma(\mathcal{L}(h_\pi), \mathcal{L}(h_\pi)_*)$ de M est une sous-représentation d'une représentation ρ proprement infinie (au sens où le commutant $R(\rho)$ contient un sous-facteur de type I_∞) et fidèle(*), que l'on choisit arbitrairement, par exemple en partant d'un isomorphisme α de M sur une sous-algèbre de von Neumann de $\mathcal{L}(h)$ et en prenant $\rho = \alpha \oplus \alpha \oplus \ldots$

(*) On emploie souvent "représentation fidèle" au lieu de représentation π dont le noyau est réduit à 0 : $\pi(x) = 0 \Rightarrow x = 0$.

Il n'y a donc pas de problème sérieux, une fois M connue algébriquement, pour déterminer tous les isomorphismes de M sur une sous-algèbre de von Neumann de $\mathcal{L}(h)$. Le problème réel étant celui de la classification des algèbres de von Neumann à isomorphisme algébrique près.

Les deux premiers exemples de facteurs de type II_1 , le facteur hyperfini et la propriété Γ .

Rappelons que si G est un groupe discret dénombrable dont toutes les classes de conjuguaison sont infinies et si λ est sa représentation régulière gauche, l'algèbre de von Neumann $R(\lambda)$ est un facteur de type II_1 .

Dans "On rings of operators IV", Murray et von Neumann montrent que tous les facteurs $R(\lambda_G)$ sont isomorphes pour G localement fini (i.e. réunion filtrante croissante de groupes finis) et que si G est le groupe libre à 2 générateurs on obtient un facteur non isomorphe. Soient N un facteur de type II_1 , et $\mathcal{T}$ l'unique trace ($\mathcal{T}(1) = 1$) de N ; on définit sur N la norme de Hilbert Schmidt par :

$$\|x\|_2 = \mathcal{T}(x^*x)^{\frac{1}{2}} .$$

(C'est l'analogue de la norme $(\Sigma|a_{ij}|^2)^{\frac{1}{2}}$ d'une matrice (a_{ij})).

C'est une norme préhilbertienne sur N et on note d_2 la distance correspondante. Le résultat de Murray et von Neumann est alors :

THEOREME 12. (Murray et von Neumann) ([43]). - Il existe à isomorphisme près un facteur de type II_1 et un seul, à prédual de type dénombrable et tel que :

$\forall\, x_1,\dots,x_n \in N$, $\forall\, \varepsilon > 0$, $\exists\, K$ sous-algèbre stellaire de dimension finie telle que $d(x_j, K) \leq \varepsilon \quad \forall\, j$.

Nous désignerons par R cet unique facteur, appelé le facteur hyperfini, pour des raisons évidentes.

Dans leur article, Murray et von Neumann montrent que tout facteur de dimension infinie contient une copie de R .

Pour G discret dénombrable localement fini on voit facilement que $R(\lambda_G)$ vérifie la condition du théorème donc est isomorphe à R . En choisissant convenablement G on voit de plus que R vérifie la condition suivante, appelée propriété Γ :

$$\forall x_1,\dots,x_n \in R \ , \ \forall \varepsilon>0 \ , \ \exists u \text{ unitaire de } R \text{ tel que :}$$

$$\mathcal{J}(u) = 0 \ , \ \|x_j u - u x_j\|_2 \leq \varepsilon \ , \ j = 1,\dots,n \ .$$

Murray et von Neumann démontrent ensuite que si $G = \mathbb{Z}^{*2}$ est le groupe libre à 2 générateurs, le facteur $R(\lambda_G)$ ne vérifie pas Γ. Nous reviendrons plus tard sur cette propriété qui n'était pour Murray et von Neumann qu'un outil technique, ainsi disaient-ils: "Certain algebraic invariants of factors in the case II_1 are formed, (1) and (2) in § 4.6 and the property Γ, of which the first two are probably of greater general significance, but the last one has so far been put to greater practical use". En fait les invariants (1) et (2) qu'ils mentionnent sont :

(1) Savoir si N est antiisomorphe à N (i.e. isomorphe à l'algèbre opposée N°, $x.y = yx$ pour $x,y \in N^{\circ}$).

(2) Le sous-groupe de R_+^* construit de la manière suivante : on prend $\tilde{N} = N \otimes K$ où K est un facteur de type I_∞, alors sur $\tilde{N}$ la fonction dimension relative D a pour image $\bar{R}_+$ et si $\theta \in \mathrm{Aut}\,\tilde{N}$ il existe un unique réel positif $\lambda = \mathrm{mod}\,\theta$ avec : $D(\theta(P)) = \lambda\, D(P)$ $\forall$ P projecteur.

Le groupe $G = \{\mathrm{mod}\,\theta \ , \ \theta \in \mathrm{Aut}\,\tilde{N}\}$ est évidemment un invariant algébrique de N.
En fait un exemple de facteur de type II_1 non antiisomorphe à lui-même n'a été obtenu que récemment (par l'auteur de ces notes) ainsi que l'existence d'un facteur de type II_1 dont le groupe G est distinct de $\mathbb{R}_+^*$ (les seuls exemples calculables donnaient toujours $G = \mathbb{R}_+^*$).

Enfin pour terminer cette revue des résultats de Murray et von Neumann, notons qu'ils avaient réussi à exhiber un facteur dans le cas III mais qu'ils notaient "The purely infinite case i.e. the case III - is the most refractory of all and we have, at least for the time being, scarcely any tools to investigate it".

II. Représentations des algèbres stellaires.

La théorie commence en 1943 avec l'article de Gelfand et Naimark.

DEFINITION 1. - Une algèbre stellaire A est une algèbre de Banach sur $\mathbb{C}$, munie d'une involution antilinéaire $x \to x^*$ telle que :

$$(xy)^* = y^* x^* \quad \text{et} \quad \|x^* x\| = \|x\|^2 \ , \ \text{pour} \quad x, y \in A \ .$$

Soit A une algèbre stellaire commutative, et supposons que A a une unité, alors l'ensemble SpA des homomorphismes de A dans $\mathbb{C}$ tels que $\chi(1) = 1$, muni de la topologie de la convergence simple sur A est compact et on a :

THEOREME 2. - Soit A une algèbre stellaire commutative avec unité, et $X = \mathrm{SpA}$ son spectre, alors la transformation de Gelfand $x \in A \mapsto$ la fonction $\hat{x}(\rho) = \rho(x), \rho \in \mathrm{SpA}$, est un isomorphisme de A sur l'algèbre stellaire $C(X)$ des fonctions complexes continues sur X.

Ainsi, le foncteur contravariant C qui à tout espace compact X associe l'algèbre stellaire $C(X)$ réalise une équivalence entre la catégorie des espaces compacts, applications continues et l'opposée de la catégorie des algèbres stellaires commutatives et homomorphismes préservant l'unité. A l'application continue $f : X \to Y$ correspond l'homomorphisme $C(f) : C(Y) \to C(X)$ qui à $h \in C(Y)$ associe $h \circ f \in C(X)$.

En particulier deux algèbres stellaires commutatives sont isomorphes si et seulement si leurs spectres sont homéomorphes.

Un des théorèmes clef de la théorie de la mesure est le théorème de représentation de Riesz. Enonçons-le seulement pour un compact métrisable X. (On a X métrisable $\Leftrightarrow C(X)$ de type dénombrable).

THEOREME 1. - Soient X un compact métrisable et L une forme linéaire positive sur $C(X)$ i.e. $f \in C(X)$, $f(x) \geq 0 \ \forall x \in X \Rightarrow L(f) \geq 0$. Il existe alors une unique mesure positive μ sur la tribu des boréliens de X (B est la tribu engendrée par les fermés de X), telle que :

$$L(f) = \int f \, d\mu \qquad \forall f \in C(X) \ .$$

En particulier, on peut construire l'espace hilbertien $L^2(X, B, \mu)$ et la représentation π de $C(X)$ dans L^2 par multiplications. On connaît de plus l'algèbre de von Neumann engendrée par $\pi(C(X))$, c'est exactement l'algèbre des multiplications par les éléments de $L^{\infty}(X,B,\mu)$. La propriété de σ - additivité de μ se traduit par l'égalité :

$$\varphi(\sum_{\alpha \in I} e_{\alpha}) = \sum_{\alpha \in I} \varphi(e_{\alpha_i})$$

où φ désigne le prolongement naturel de L à $L^{\infty}(X,B,\mu)$ et où $(e_{\alpha})_{\alpha \in I}$ est une famille dénombrable de projecteurs $e_{\alpha} \in L^{\infty}(X,B,\mu)$. Supposons alors que A soit une algèbre stellaire non commutative avec une unité 1 . Les notions ci-dessus d'élément positif, de forme linéaire positive et de σ - additivité ont un analogue exact, qui est le point de départ de la théorie de l'intégration non commutative.

Eléments positifs dans une algèbre stellaire.

Soient h un espace hilbertien et $T \in \mathcal{L}(h)$, on a équivalence entre les conditions suivantes :

a) $T = T^*$ et Spectre $T \subset [0,+\infty[$

b) $\langle T\xi,\xi \rangle \geq 0$ pour tout $\xi \in h$.

Pour une algèbre stellaire avec unité A et $x \in A$ les propriétés suivantes sont équivalentes :

1) $x = x^*$ et Spectre $x \subset [0,+\infty[$

2) $\exists\, a \in A$, $a^* a = x$

3) $\exists\, a \in A$, $a = a^*$, $a^2 = x$

4) $\exists\, \lambda \geq 0$ tel que $\|x-\lambda 1\| \leq \lambda$.

On dit alors que x est positif et on écrit $x \geq 0$. La condition 4) montre que l'ensemble des éléments positifs est un cône convexe fermé dans A , noté A^+ . Si $A = C(X)$, on a $A^+ = \{f, f(x) \geq 0 , \forall\, x \in X\}$.

Formes linéaires positives sur une algèbre stellaire.

Soit A comme ci-dessus et A^* l'espace de Banach dual. On dit que $L \in A^*$ est positive quand $L(x) \geq 0$, $\forall\, x \geq 0$. On note A^*_+ le cône convexe

$\sigma(A^*,A)$ fermé des formes linéaires positives sur A. L'analogue du théorème de représentation de Riesz consiste en la construction suivante (Gelfand, Naimark, Segal).

Comme L est positive, la condition 2) montre que :

$$L(x^*x) \geq 0 \quad \forall\, x \in A .$$

Il en résulte que la forme sesquilinéaire $\langle x,y \rangle_L = L(y^*x)$ définit sur A une structure préhilbertienne.

Soient $\mathfrak{h}_L$ l'espace hilbertien séparé complété, et pour $x \in A, \pi_L(x)$ l'opérateur de multiplication à gauche défini par :

$$\pi_L(x)y = xy \quad y \in A$$

l'inégalité $L(y^*x^*xy) \leq \|x\|^2 L(y^*y)$ $y \in A$, qui résulte de $\|x\|^2 - x^*x \geq 0$, montre que π_L définit une représentation de A dans l'espace hilbertien $\mathfrak{h}_L$.

De même que, dans le cas commutatif, la forme linéaire L se prolongeait aux fonctions boréliennes $f \in L^\infty(X,B,\mu)$, ici la forme linéaire L se prolonge en une forme linéaire sur l'algèbre de von Neumann $\pi_L(A)''$ engendrée par $\pi_L(A)$ et le prolongement, note $\bar{L}$, a la propriété de σ-additivité suivante :

DEFINITION 2. - Soient M une algèbre de von Neumann dans l'espace hilbertien $\mathfrak{h}$ et ψ une forme linéaire positive sur M. On dit que ψ est normale quand :

$$\psi\Big(\sum_{\alpha \in I} e_\alpha\Big) = \sum_{\alpha \in I} \psi(e_\alpha)$$

pour toute famille $(e_\alpha)_{\alpha \in I}$ de projecteurs 2 à 2 orthogonaux.

Ici Σe_α désigne le plus petit projecteur qui majore toutes les sommes finies $\sum_{i=1}^{n} e_{\alpha_i}$, c'est un élément de M.

On démontre que pour que ψ soit normale il faut et il suffit qu'elle provienne du prédual M_* de M.

Revenons aux algèbres stellaires, soit A une telle algèbre, avec unité. Le théorème de Hahn Banach appliqué au cône convexe d'intérieur non vide A^+, montre que l'ensemble $\mathcal{S} = \{\Phi \in A_+^*, \Phi(1) = 1\}$ des états de A est un convexe non vide qui sépare les points de A . On est donc assuré de l'existence de "mesures positives" et grâce à la construction de Gelfand Naimark Segal, de l'existence d'une représentation isométrique de A comme sous-algèbre stellaire de l'algèbre $\mathcal{L}(h)$, h espace hilbertien.

De plus $\mathcal{S}$ est $\sigma(A^*,A)$ compact, et est donc l'enveloppe convexe fermé de l'ensemble de ses points extrémaux : les états purs, caractérisés par la propriété suivante :

Φ état pur $\Leftrightarrow$ la représentation π_Φ est irréductible. En fait, plus généralement, on a une correspondance bijective entre la face de Φ dans le cône A_+^* et l'ensemble des éléments positifs de l'algèbre de von Neumann $R(\pi_\Phi)$. Elle est donnée par : $\Psi \in A_+^*$ est associé à $y \in R(\pi_\Phi)^+$ quand :

$$\Psi(a) = \langle \pi_\Phi(a)1, y1 \rangle$$

où $1 \in h_\Phi$ est le vecteur associé à l'unité de A .

Ainsi toute algèbre stellaire A admet suffisamment de représentations irréductibles Dans le cas commutatif les représentations irréductibles de A se confondent avec les homomorphismes de A dans $\mathbb{C}$. Dans le cas non commutatif la nature de la relation d'équivalence, entre représentations irréductibles de A dans un hilbert fixe, détermine une classe privilégiée d'algèbre stellaire ; le théorème suivant de J. Glimm est fondamental : (Voir [19]).

THEOREME 3. - Soit A une algèbre stellaire de type dénombrable. Les conditions suivantes sur A sont équivalentes.

1) Toute représentation isotypique π de A dans un espace hilbertien h_π est multiple d'une représentation irréductible.

2) Pour toute représentation irréductible π de A , l'image $\pi(A)$ contient l'idéal $k(h_\pi)$ des opérateurs compacts dans h_π .

3) Soient h un espace hilbertien de type dénombrable et $\mathrm{Rep}(A,h)$ l'espace borélien des représentations irréductibles de A dans h . Alors le quotient par la relation d'équivalence des représentations est dénombrablement séparé.

4) Si π_1 et π_2 sont deux représentations irréductibles de A, ayant même noyau, elles sont équivalentes.

Une algèbre stellaire vérifiant les conditions équivalentes ci-dessus est dite postliminaire.

III. LE CADRE DE L'INTEGRATION NON COMMUTATIVE ET LA THEORIE DES POIDS

Pour pouvoir tenir compte de mesures positives non nécessairement finies, il est nécessaire d'introduire l'analogue non commutatif des mesures positives infinies.

La donnée de départ en intégration non commutative est alors un couple (M,Φ) d'une algèbre de von Neumann M et d'un poids Φ sur M au sens suivant (dû à F. Combes, G. Pedersen et U. Haagerup (Voir [5], [47], [30])).

DEFINITION 1. - On appelle poids sûr une algèbre de von Neumann M, une application additive positivement homogène de M_+ dans $\bar{R}_+ = [0,+\infty]$ telle que :

a) Φ est semi-fini, i.e. $\{x \in M_+, \Phi(x) < \infty\}$ est $\sigma(M,M_*)$ total.

b) Φ est normal, i.e. $\Phi(\operatorname{Sup} x_\alpha) = \operatorname{Sup} \Phi(x_\alpha)$ pour toute famille filtrante croissante majorée d'éléments de M_+.

L'exemple le plus simple de poids infini est celui de la trace usuelle sur les opérateurs bornés dans un espace hilbertien h. On pose $M = \mathcal{L}(h)$ et pour $T \in M_+$ et toute base orthonormale $(\xi_\alpha)_{\alpha \in I}$ de h on a :

$$\Sigma < T\xi_\alpha, \xi_\alpha > = \text{Trace } T = \sup_{0 \leq A \leq T} \text{Trace } (A), \ A \text{ de rang fini.}$$

En fait les premiers poids infinis étudiés étaient des traces au sens suivant :

DEFINITION 2. - Un poids Φ sur l'algèbre de von Neumann M est une trace si il est invariant par les automorphismes intérieurs de M.

Ainsi tout poids sur M qui a une propriété d'unicité est une trace. L'analogue des notions de convergence presque sûre et des espaces L^p $p \in [1,\infty]$, de la théorie classique, a été obtenu, principalement par Dixmier [14] et Segal [43] pour les traces.

Quand Φ est une trace, l'ensemble $C_p = \{x \in M, \Phi(|x|^p) < \infty\}$ est un idéal bilatère de M et $\|x\|_p = (\Phi(|x|^p))^{1/p}$ définit sur C_p une semi-norme. Les espaces complétés $L^p(M,\Phi)$ ont de nombreuses propriétés généralisant le cas commutatif et le cas classique des opérateurs de puissance p - ième-sommable dans un espace hilbertien. En particulier, quand $x \geq 0$, $\Phi(x) = 0 \Rightarrow x = 0$, (On dit alors que Φ est fidèle), le prédual M_* de M s'identifie à l'espace $L^1(M,\Phi)$. De plus l'intersection $L^2(M,\Phi) \cap L^\infty(M,\Phi)$ est une algèbre hilbertienne :

DEFINITION 3. - On appelle algèbre hilbertienne toute algèbre involutive $\mathfrak{a}$ munie d'un produit scalaire préhilbertien séparé tel que :

1) $\langle x,y \rangle = \langle y^*,x^* \rangle \quad \forall\, x,y \in \mathfrak{a}$.

2) La représentation de $\mathfrak{a}$ sur $\mathfrak{a}$ par multiplications à gauche est bornée, involutive et non dégénérée.

La condition 1) définit une isométrie antilinéaire J de l'espace h, complété de $\mathfrak{a}$, sur lui-même. La condition 2) permet de parler de la représentation régulière gauche λ de $\mathfrak{a}$ dans h et donc de lui associer une algèbre de von Neumann : $\lambda(\mathfrak{a})''$ dans h. On a alors :

a) Le commutant de $\lambda(\mathfrak{a})$ est engendré par l'algèbre des multiplications à droite $\lambda'(\mathfrak{a}) = J\lambda(\mathfrak{a})J$.

b) L'algèbre de von Neumann associée à l'algèbre hilbertienne $L^2(M,\Phi) \cap L^\infty$ s'identifie à M.

c) Pour toute algèbre hilbertienne $\mathfrak{a}$, il existe sur l'algèbre de von Neumann $\lambda(\mathfrak{a})''$ une trace fidèle $\mathfrak{J}$ telle que :

$$\mathfrak{J}(\lambda(y^*)\lambda(x)) = \langle x,y \rangle, \quad \forall\, x,y \in \mathfrak{a}$$

et $\mathfrak{a}$ est équivalente à l'algèbre hilbertienne associée à $\mathfrak{J}$.

En général, une algèbre de von Neumann M ne possède pas de trace fidèle, par exemple si M est un facteur, il possède une trace fidèle si et seulement si il n'est pas de type III. On dit que M est semi finie si elle possède une trace fidèle (dans le cas où M agit dans un espace de type dénombrable cela équivaut à dire que M est une intégrale directe de facteurs dont aucun n'est du type III).

Pour M semi-finie, les outils supplémentaires issus de la théorie des algèbres hilbertiennes permettaient de démontrer des résultats inaccessibles dans le cas général comme :

THEOREME 4. - Soit $M_i, i=1,2$ une algèbre de von Neumann dans $\mathcal{H}_i, i=1,2$ alors le commutant $(M_1 \otimes M_2)'$ est engendré par $M_1' \otimes M_2'$.

Pour des algèbres de von Neumann semi-finies M_1 et M_2, ce résultat est conséquence du théorème de commutation a) pour les algèbres hilbertiennes. De même, si G est un groupe localement compact unimodulaire et dg une mesure de Haar sur G, l'algèbre, pour la convolution des fonctions continues à support compact, est une algèbre hilbertienne et le théorème de commutation a) entraîne :

THOREME 5. - Le commutant $R(\lambda)$ de la représentation régulière gauche λ de G dans $L^2(G,dg)$ est engendré par la représentation régulière droite.

Ce théorème a été démontré pour des groupes localement compacts non nécessairement unimodulaires, par J. Dixmier, qui introduisit la notion d'algèbre quasi-hilbertienne. Le théorème 4 a été démontré pour des algèbres de von Neumann non nécessairement semi-finies par M. Tomita en 1967. En fait sa théorie des algèbres hilbertiennes généralisées est le fondement de toute la théorie non commutative de l'intégration pour les poids qui ne sont pas nécessairement des traces.
La théorie de Tomita doit en fait beaucoup à M. Takesaki qui transforma le papier original très difficile à déchiffrer en un texte accessible (Lecture-Notes n° 128).
On peut résumer l'essentiel de la théorie de Tomita-Takesaki, grâce à la définition et au théorème suivant :

DEFINITION 6. - On appelle algèbre hilbertienne à gauche toute algèbre involutive $\mathfrak{A}$, munie d'un produit scalaire préhilbertien séparé tel que :

1) L'opérateur $x \to x^*$ est préfermé.

2) La représentation de $\mathfrak{A}$ dans $\mathfrak{A}$ par multiplications à gauche est bornée, involutive et non dégénérée.

Ainsi la seule différence avec une algèbre hilbertienne et que la fermeture S de l'opérateur $x \to x^*$ peut avoir un module $|S| \neq 1$. Soit alors

$\Delta = (\text{adjoint de } S)\circ S$ le carré du module de S on a : $S = J\Delta^{\frac{1}{2}}$ où J est une involution isométrique et le résultat fondamental :

THEOREME 7. - Soient $\mathcal{G}$ une algèbre hilbertienne à gauche et M l'algèbre de von Neumann engendrée par la représentation régulière gauche de $\mathcal{G}$. Alors $JMJ = M'$ et pour tout $t \in R$ on a :

$$\Delta^{it} M \Delta^{-it} = M \qquad \text{(Voir [58]).}$$

De plus, de même que les algèbres hilbertiennes sont associées aux traces, les algèbres hilbertiennes à gauche sont associées aux poids : ([5]) :
Soit $\mathcal{G}$ une algèbre hilbertienne à gauche et M l'algèbre de von Neumann associée, il existe alors sur M un poids fidèle Φ tel que :

$$\Phi(\lambda(y^*)\lambda(x)) = \langle y^*, x\rangle \qquad x,y \in \mathcal{G} .$$

Inversement, soient M une algèbre de von Neumann et Φ un poids fidèle. Alors $\mathcal{G}_\Phi = \{x \in M, \Phi(x^*x) < \infty, \Phi(xx^*) < \infty\}$ muni du produit de M et du produit scalaire $\langle x,y\rangle = \Phi(y^*x)$ est une algèbre hilbertienne à gauche, l'algèbre de von Neumann associée s'identifie à M et le poids correspondant s'identifie à Φ.
Comme toute algèbre de von Neumann possède un poids fidèle (si M agit dans un espace de type dénombrable elle possède en fait un état fidèle), il en résulte en particulier que toute algèbre de von Neumann est isomorphe à l'algèbre engendrée par la représentation régulière gauche d'une algèbre hilbertienne à gauche.
C'est là que se situe une découverte remarquable de Takesaki et Winnink qui relie la théorie de Tomita et plus exactement le groupe à un paramètre d'automorphismes de l'algèbre de von Neumann M défini par $\sigma_t(x) = \Delta^{it} x \Delta^{-it}$ avec une équation fondamentale en mécanique statistique quantique. Bien sûr le groupe d'automorphisme σ_t n'est pas unique, il dépend de l'algèbre hilbertienne $\mathcal{G}$, c'est-à-dire du poids fidèle Φ sur M :

DEFINITION 8. - Soit φ un poids fidèle sur l'algèbre de von Neumann M on appelle groupe d'automorphismes modulaires sur M le groupe à un paramètre $(\sigma_t^\varphi)_{t\in R}$ d'automorphismes de M associé à l'algèbre hilbertienne à gauche $\mathcal{G}_\varphi$.

En mécanique statistique quantique des systèmes finis, l'état d'équilibre à la température absolue T est l'état de l'algèbre $M = M_n(\mathbb{C})$ donné par l'égalité :

$$\Phi(x) = \frac{\text{Trace } (xe^{-\beta H})}{\text{Trace } (e^{-\beta H})}$$

où Trace est la trace usuelle, $\beta = 1/kT$ où k est la constante de Boltzmann et H est l'hamiltonien, i.e. le générateur du groupe à un paramètre d'évolution du système : $\alpha_t(x) = e^{itH} x e^{-itH}$ $x \in M$, $t \in \mathbb{R}$. Il est facile dans ce cas de montrer que l'état Φ est caractérisé par la condition suivante qui le relie à l'évolution α :

$$\begin{cases} \forall\, x,y \in M, \ \exists\, F_{x,y}(z), \text{ holomorphe pour } 0 < \operatorname{Im} z < 1 \text{ et continue bornée dans} \\ 0 \leq \operatorname{Im} z \leq 1, \text{ telle que :} \end{cases}$$

$$F_{x,y}(t) = \Phi(x\alpha_t(y)), \ F_{x,y}(t+i\beta) = \Phi(\alpha_t(y)x) \quad \forall\, t \in \mathbb{R} .$$

C'est la condition de Kubo-Martin-Schwinger. Pour un système infini on s'attend à ce que la même relation ait lieu entre le groupe d'évolution $(\alpha_t)_{t \in \mathbb{R}}$ de l'algèbre des observables, qui est une algèbre stellaire, et l'état d'équilibre. Ainsi Haag Hugenholtz et Winnink ont été amenés à proposer la condition ci-dessus pour un triplet (A,Φ,α) où A est une algèbre stellaire, Φ un état sur A et α un groupe à un paramètre d'automorphismes de A .

THEOREME 9. (Takesaki-Winnink) [58]. - Soient M une algèbre de von Neumann et Φ un état normal fidèle sur M . Alors le groupe d'automorphismes modulaires $(\sigma_t^{\Phi})_{t \in \mathbb{R}}$ est l'unique groupe à un paramètre d'automorphismes de M qui vérifie les conditions de Kubo-Martin-Schwinger pour $\beta = 1$.

V. FACTEURS DE POWERS, D'ARAKI ET WOODS ET DE KRIEGER.

L'analogue non commutatif d'un espace de probabilité est un couple (M,Φ) où M est une algèbre de von Neumann et Φ un état normal fidèle sur M . L'exemple le plus simple correspond à $M = M_n(\mathbb{C})$, tout état Φ sur M s'écrit

$\Phi = \mathrm{Tr}(\rho.)$ où ρ est une matrice positive dont la somme des valeurs propres vaut 1 :

$$\Phi(x) = \mathrm{Tr}(\rho x) \quad \forall\, x \in M_n(\mathbb{C}) .$$

On peut donc supposer que ρ est diagonale avec la valeur propre $\lambda_i > 0$ à la ligne i et avec $\lambda_1 \geq \lambda_2 \geq \ldots \geq \lambda_n > 0$.

La liste des valeurs propres de ρ est un invariant de Φ . Le groupe d'automorphismes modulaires de Φ est donné par

$$\sigma_t^{\Phi}(x) = e^{itH} x e^{-itH} \quad \text{où } H = \mathrm{Log}\,\rho .$$

En particulier, si e_{ij} désigne l'unité matricielle canonique on a :

$$\sigma_t^{\Phi}(e_{k\ell}) = \left(\frac{\lambda_k}{\lambda_\ell}\right)^{it} e_{k\ell} .$$

Le système (M,Φ) est analogue à un espace probabilisé ayant un nombre fini de points. Pour obtenir des exemples plus intéressants le procédé le plus simple est d'effectuer l'analogue de la construction des produits infinis de mesures.

Soit donc $(M_\nu, \Phi_\nu)_{\nu \in \mathbb{N}}$ une suite de couples (algèbre de matrices, état fidèle), soit A la limite inductive des algèbres stellaires $M_1 \otimes M_2 \otimes \ldots \otimes M_\nu = A_\nu$ où $A_\nu \subset A_{\nu+1}$ grâce à : $x \to x \otimes 1$. Sur A qui est une algèbre stellaire avec unité, on définit un état $\Phi = \overset{\infty}{\underset{1}{\otimes}} \Phi_\nu$ par l'égalité :

$$\Phi(x_1 \otimes x_2 \otimes \ldots \otimes x_\nu \otimes 1 \ldots) = \Phi_1(x_1)\Phi_2(x_2)\ldots\Phi_\nu(x_\nu) .$$

On considère alors le couple (M,Φ) = (algèbre de von Neumann, état normal) associé au couple (A,Φ) comme dans le paragraphe III.

Quand chaque Φ_ν est fidèle il en est de même de Φ et le groupe d'automorphismes modulaires de (M,Φ) est donné par :

$$\sigma_t^{\Phi}(x_1 \otimes \ldots \otimes x_\nu \otimes 1 \ldots) = \sigma_t^{\Phi_1}(x_1) \otimes \ldots \otimes \sigma_t^{\Phi_\nu}(x_\nu) \otimes 1 \ldots$$

En fait, cette construction d'algèbres de von Neumann est due à von Neumann (Compos. Math. 1938), mais il fallut attendre jusqu'en 1967 pour qu'elle se révèle comme fondamentale.

Pendant 30 ans après la naissance des algèbres de von Neumann on ne connaissait que trois facteurs de type III deux à deux non isomorphes. R.T. Powers, qui avait une formation de physicien, réussit à montrer en 1967 que si l'on prend tous les couples (M_ν, Φ_ν) égaux au couple $M_2(C)$, $\Phi((a_{ij})) = \lambda a_{11} + (1-\lambda) a_{22}$ on obtient une famille à un paramètre continu $\lambda \in]0,\frac{1}{2}[$ de facteurs de type III deux à deux non isomorphes : R_α, $\alpha = \lambda/1-\lambda \in]0,1[$ -([35]) . Après la découverte de Powers, H. Araki et E.J. Woods entreprenaient une classification à isomorphisme près des facteurs produits tensoriels infinis d'algèbres de matrices ([2]). Ils démontraient en particulier que l'on peut calculer à partir de la liste des valeurs propres des Φ_ν : $(\lambda_{\nu,j})_{j=1,\ldots,n_\nu}$ les deux invariants suivants :

$$r_\infty(M) = \{\lambda \in]0,1[\text{ , } M \otimes R_\lambda \text{ isomorphe à } M\}$$

$$\rho(M) = \{\lambda \in]0,1[\text{ , } M \otimes R_\lambda \text{ isomorphe à } R_\lambda\} .$$

Ils montraient de plus que $r_\infty(M)$ est un sous-groupe fermé de $\mathbb{R}_+^*$ et que l'égalité $r_\infty(M) = \lambda^{\mathbb{Z}}$ caractérise le facteur de Powers R_λ parmi les produits tensoriels infinis d'algèbres de matrices. En outre grâce à l'étude de $\rho(M)$, E.J. Woods réussit à montrer que la classification des facteurs est impossible par des invariants boréliens à valeurs réelles.

D'un autre côté W. Krieger avait commencé une étude systématique des facteurs associés à la théorie ergodique. Commençons d'abord par expliciter sa construction d'une algèbre de von Neumann à partir d'une relation d'équivalence à orbites dénombrables sur un espace borélien standard (X,B) et d'une mesure quasi-invariante μ . Cette construction généralise la première construction de Murray et von Neumann qui elle-même était inspirée des produits croisés de la théorie des algèbres simples centrales sur un corps. Sous sa forme définitive, elle est due à J. Feldmann et C. Moore (Voir [27]).

Soient donc (X,B,μ) un espace borélien standard probabilisé et $R \subset X \times X$ le graphe, supposé analytique, d'une relation d'équivalence à orbites dénombrables. On suppose μ quasi-invariante au sens où la saturation par R d'un ensemble borélien négligeable est encore négligeable. On considère alors l'algèbre hilbertienne à gauche des fonctions boréliennes bornées de R dans $\mathbb{C}$ telles que pour un $n < \infty$, $\{j, f(i,j) \neq 0\}$ soit de cardinalité $\leq n$ pour tout i .

Le produit scalaire est celui de $L^2(R,\tilde{\mu})$ où $\int f(i,j)d\tilde{\mu} = \int \sum_j f(i,j)d\mu(i)$. Le produit de convolution est donné par

$$(f * g)(\gamma) = \sum_{\gamma_1\gamma_2=\gamma} f(\gamma_1)g(\gamma_2)$$

où pour $\gamma_1,\gamma_2 \in R$ on pose $\gamma_1\gamma_2 = \gamma$ quand $\gamma_1 = (i_1,j_1), \gamma_2 = (i_2,j_2)$, $j_1 = i_2$ et $\gamma = (i_1,j_2)$.

Ainsi l'algèbre hilbertienne à gauche $\mathfrak{a}$ apparaît comme une généralisation de l'algèbre des matrices carrées. Comme elle a une unité : la fonction f : $f(i,j) = 0$ si $i \neq j$, $f(i,i) = 1$ $\forall$ i , il lui correspond un couple (M,Φ) où M est une algèbre de von Neumann et Φ un état normal fidèle sur M . Nous noterons simplement $M = L^\infty(R,\tilde{\mu})$ où R désigne le graphe de la relation d'équivalence, muni de la loi de groupoïde $\gamma_1\gamma_2 = \gamma$. Il est intéressant d'un point de vue heuristique de faire jouer (en intégration non commutative) à R le rôle joué (en intégration classique) par <u>l'espace</u> X , la <u>non commutativité</u> est due à l'existence d'une loi de groupoïde non triviale sur R (la loi triviale sur l'ensemble X est $x.x = x$ pour tout $x \in X$) . A isomorphisme près, l'algèbre de von Neumann $L^\infty(R,\mu)$ ne dépend que de la classe de la mesure μ , on adopte alors la définition :

DEFINITION 1. - Soit (X_i,B_i,μ_i,R_i) , $i = 1,2$. une relation d'équivalence à orbites dénombrables comme ci-dessus ; on dit que R_1 est isomorphe à R_2 si il existe une bijection borélienne θ de X_1 sur X_2 telle que :

$\theta(\mu_1)$ équivalente à μ_2 , et presque sûrement :

θ (classe de x) = classe de $\theta(x)$.

Soit alors T une transformation borélienne de l'espace borélien standard (X,B) et μ une mesure quasi invariante par T . Les orbites de T dans X définissent une relation d'équivalence R_T et on dit que T_1 <u>est faiblement équivalente</u> à T_2 quand R_{T_1} est isomorphe à R_{T_2} .

THEOREME 2. (M. Dye 1958) ([23]). - <u>Deux transformations ergodiques avec mesure invariante sont faiblement équivalentes.</u>

Soit ainsi T une telle transformation, l'état Φ (associé à la mesure invariante) sur l'algèbre de von Neumann $L^\infty(R_T,\mu)$ est une trace, de sorte que $M = L^\infty(R_T,\mu)$ qui est un facteur car R_T est ergodique est de type II_1. H. Dye démontre en outre qu'il s'agit du facteur hyperfini.

Vers 1967, W. Krieger entreprend une étude systématique de l'équivalence faible des transformations (X,B,μ,T) où μ est quasi invariante par T. Il introduit deux invariants : ([38])

$$r(T) = \{\lambda \in [0,+\infty[\ ,\ \forall\ \varepsilon>0\ ,\ \forall\ A\subset X\ ,\ \mu(A)>0\ ,\ \exists\ B\subset A\ ,$$

$$\mu(B)>0 \quad \text{et} \quad n\in\mathbb{Z} \quad \text{tel que} \quad \left|\frac{d\mu(T^n x)}{d\mu(x)} - \lambda\right| \le \varepsilon \ \ \forall\ x\in B \quad \text{et} \quad T^n B\subset A\}$$

$$\rho(T) = \{\lambda\in \mathbb{R}_+^*\ ,\ \exists\ \nu\sim\mu \quad \text{avec} \quad \frac{d\nu(T^k x)}{d\nu(x)} \in \lambda^{\mathbb{Z}}\ \forall\ x\in X\ ,\ \forall\ k\}$$

et il montre que r et ρ sont non seulement des invariants de l'équivalence faible mais en fait que $r(T)$ coïncide avec l'invariant d'Araki-Woods $r_\infty(M) = \{\lambda, M\otimes R_\lambda$ isomorphe à $M\}$ où $M = L^\infty(R_T,\mu)$ et, de même, que $\rho(T) = \rho(M)$.

En fait, il s'agit là d'une généralisation des résultats d'Araki et Woods, en effet, soient $(M_\nu,\Phi_\nu)_{\nu\in\mathbb{N}}$ une suite de couples (algèbre de matrice, état fidèle) et $(\lambda_{\nu,j})_{j=1,\ldots,n_\nu}$ la liste de valeurs propres correspondante. Alors l'algèbre de von Neumann produit tensoriel infini : $\bigotimes_{\nu=1}^{\infty} (M_\nu,\Phi_\nu)$ s'obtient également par la construction de Krieger à partir de l'espace et de la relation d'équivalence suivants :

$X = \prod_1^\infty X_\nu$ où pour chaque ν, $X_\nu = \{1,\ldots,n_\nu\}$

B est la tribu engendrée par la topologie produit.

$\mu = \prod_1^\infty \mu_\nu$ où $\mu_\nu(j) = \lambda_{\nu,j}$, $j\in X_\nu$.

R est la relation $x = y \Leftrightarrow x_i = y_i$ pour tout i assez grand.

La relation R est en fait égale à R_T où T est la transformation qui généralise l'opération d'addition de 1 dans les entiers p-adiques : Pour calculer

Tx on regarde la première coordonnée x_i de x qui n'est pas égale au maximum possible n_i, on la remplace par x_i+1 et on remplace les précédentes $x_j, j<i$ par 1.

Il existe en fait des facteurs de Krieger i.e. de la forme $L^\infty(R_T,\mu)$ qui ne sont pas des produits tensoriels infinis d'algèbres de matrices. Le point culminant de la théorie de Krieger est le théorème suivant, démontré vers 1973, grâce à des invariants des facteurs que nous discuterons plus bas :

THEOREME 3. (Krieger) ([38]). - <u>Soit</u> (X_i,B_i,μ_i,T_i) <u>une transformation ergodique</u> (μ_i <u>quasi invariante et</u> (X_i,B_i) <u>borélien standard), alors</u> T_1 <u>est faiblement équivalente à</u> T_2 (<u>i.e.</u> R_{T_1} <u>isomorphe à</u> R_{T_2}) <u>si et seulement si les facteurs</u> $L^\infty(R_{T_i},\mu_i)$ <u>sont isomorphes.</u>

Ce résultat montrait clairement la nécessité de reconnaître sur une relation d'équivalence R à orbites dénombrable si elle est de la forme R_T pour une certaine transformation borelienne T. En particulier, si R provient de l'action d'un groupe discret Γ, W. Krieger et l'auteur ([16]) ont montré que si Γ est résoluble R est de la forme R_T. Le cas où Γ est un groupe moyennable a été résolu par D. Ornstein et B. Weiss [46] et la réponse définitive au problème a été obtenue dans [15] par J. Feldman, B. Weiss et l'auteur.

<u>Théorème</u>. R <u>est de la forme</u> R_T <u>ssi</u> R <u>est moyennable au sens de Zimmer</u> ([63]).

La définition de Zimmer traduit simplement la moyennabilité de l'algèbre de von Neumann $L^\infty(R,\mu)$ (cf. chapitre VII) de sorte que les deux théorèmes ci-dessus impliquent : ([15]).

<u>Corollaire</u>. <u>Deux sous algèbres de Cartan d'un facteur moyennable</u> $M = L^\infty(R,\mu)$ <u>sont conjuguées par un automorphisme de</u> M.

Ici on appelle sous algèbre de Cartan toute sous algèbre abelienne maximale A de M telle que a) le normalisateur de A engendre M (on dit alors que A

est régulière) b) Il existe une espérance conditionnelle E de M sur A
La condition b) est automatique si M est de type II_1 et l'on voit donc que :

Le facteur hyperfini R ne possède à conjuguaison près qu'une sous algèbre abelienne maximale régulière.

De plus notons, que grâce à un résultat récent de V. Jones et de l'auteur on sait maintenant que pour R_1, R_2 relation d'équivalences non moyennables il est faux que l'isomorphisme $L^\infty(R_1,\mu_1) \approx L^\infty(R_2,\mu_2)$ implique l'isomorphisme des relations R_1 et R_2. On ne peut donc espérer traduire directement les résultats de rigidité de R. Zimmer ([64]) (sur les relations d'équivalence provenant d'actions ergodiques de groupes de Lie semi simples) en un théorème de rigidité sur les facteurs correspondant. Le problème d'adapter sa démonstration ([64]) au cas des facteurs est le même que celui posé dans l'introduction.

VI. FACTEURS DE TYPE III_λ.

Le théorème de Radon Nikodym.

La théorie de Tomita-Takesaki associe à tout poids fidèle Φ sur une algèbre de von Neumann M un groupe à un paramètre σ_t^Φ d'automorphismes de M, le groupe d'automorphismes modulaires, défini par

$$\sigma_t^\Phi(x) = \Delta_\Phi^{it}\, x\, \Delta_\Phi^{-it}$$

où Δ_Φ est l'opérateur modulaire, le carré du module de l'involution $x \to x^*$ considérée comme opérateur non borné dans l'espace $L^2(M,\Phi)$, complété de $\{x \in M, \Phi(x^*x) < \infty\}$ pour le produit scalaire $<x,y> = \Phi(y^*x)$. D'un autre côté, la théorie d'Araki et Woods associe à tout facteur M les deux invariants r_∞ et ρ, $r_\infty(M) = \{\lambda, M \otimes R_\lambda \sim M\}, \rho(M) = \{\lambda, M \otimes R_\lambda \sim R_\lambda\}$.
Or pour les facteurs d'Araki et Woods un calcul direct à partir de leurs travaux montre les égalités suivantes :

1) $r_\infty(M) = \cap$ Spectre Δ_Φ, Φ état normal fidèle sur M

2) $\rho(M) = \{\exp(^{2\pi}/T_o)$, $\exists\, \Phi$ état normal fidèle sur M tel que $\sigma_{T_o}^\Phi = 1\}$.

Ces deux égalités suggèrent bien entendu les définitions suivantes, pour un facteur arbitraire M :

$$S(M) = \bigcap \text{ Spectre } \Delta_\Phi \ , \ \Phi \text{ état normal fidèle sur } M \ .$$

$$T(M) = \{\text{périodes possibles de groupes d'automorphismes modulaires de } M\} \ .$$

La première question étant évidemment de savoir si les égalités $r_\infty = S$ et $\rho = \exp(2\pi/T)$, valables pour les facteurs d'Araki et Woods restent vraies en général. Une question immédiatement reliée est le problème de calcul des invariants S et T . Les définitions ci-dessus de ces invariants montrent que pour calculer S et T on doit passer en revue tous les états normaux fidèles sur M et calculer leurs groupes d'automorphismes modulaires. Or, en général, et ceci est clair pour les facteurs du paragraphe V, un facteur est donné avec un état ou un poids priviligié Φ pour lequel le calcul de Δ_Φ et σ_t^Φ est facile. Le problème était donc posé d'étudier dans quelle mesure exactement le groupe σ^Φ dépend de Φ .

La réponse complète à ce problème constitue en fait exactement la version non commutative du théorème de Radon Nikodym.

THEOREME ([8]). - Soient M une algèbre de von Neumann et Φ un poids fidèle sur M .

a) Pour tout poids fidèle ψ sur M il existe une unique application continue de R dans $\mathcal{U}$ le groupe unitaire de M muni de $\sigma(M,M_*)$ telle que :

$$u_{t+t'} = u_t \ \sigma_t^\Phi(u_{t'}) \quad \forall \ t,t' \in R$$

$$\sigma_t^\psi(x) = u_t \ \sigma_t^\Phi(x) u_t^* \quad \forall \ t \in R \ , \ \forall \ x \in M$$

$$\psi(x) = (\Phi(u_t^* \, x \, u_t))_{t=-i/2} \quad \forall \ x \in M \ .$$

On note $u_t = (D\psi;D\Phi)_t$.

b) Inversement, soit $t \to u_t$ une application continue de R dans $\mathcal{U}$ telle que $u_{t+t'} = u_t \ \sigma_t^\Phi(u_{t'})$, $\forall \ t,t' \in R$, alors il existe un unique poids fidèle ψ sur M tel que $(D\psi : D\Phi) = u$.

Si M est commutative, $M = L^{\infty}(X,B,\mu)$, alors Φ et ψ sont des mesures positives sur X qui sont équivalentes à μ, il existe alors une densité de Radon Nikodym $h : X \to R_+$. Les h^{it}, $t \in R$ sont alors dans $M = L^{\infty}(X,B,\mu)$ et $h^{it} = (D\psi : D\Phi)_t$.

Si M est semi-finie et $\mathfrak{J}$ est une trace fidèle sur M, il existe alors des opérateurs positifs affiliés à M tels que $\Phi = \mathfrak{J}(\rho_\Phi \cdot)$, $\psi = \mathfrak{J}(\rho_\psi \cdot)$ et on a :

$(D\psi : D\Phi)_t = \rho_\psi^{it} \rho_\Phi^{-it}$.

La propriété $\sigma_t^\psi(x) = u_t \sigma_t^\Phi(x) u_t^*$ $\forall x \in M$, montre que bien que le groupe d'automorphismes modulaires change en général avec Φ, sa classe modulo les automorphismes intérieurs ne varie pas. On peut alors se demander si elle n'est pas de toutes façons triviale mais un argument facile à partir du théorème ci-dessus montre que :

$$T(M) = \{T_o, \sigma_{T_o}^\Phi \text{ est un automorphisme intérieur}\}.$$

En outre un théorème de J. Dixmier et M. Takesaki (voir ([58])) montre que, en supposant que M_* est de type dénombrable on a

$$T(M) \neq R \Leftrightarrow M \text{ n'est pas semi-fini.}$$

On introduit alors le groupe $\operatorname{Out} M = \operatorname{Aut} M / \operatorname{Int} M$ des classes d'automorphismes de M modulo les automorphismes intérieurs et on obtient, associée à toute algèbre de von Neumann, un homomorphisme canonique de R dans $\operatorname{Out} M$:

$$\delta(t) = \text{Classe de } \sigma_t^\Phi \text{ (indépendante du choix de } \Phi) .$$

En particulier $T(M) = $ Noyau δ est un sous-groupe de R. En fait l'image de δ est même contenue dans le <u>centre</u> du groupe $\operatorname{Out} M$.
On peut de plus calculer $T(M)$ à partir d'un seul poids Φ fidèle sur M puisque il suffit de déterminer les t pour lesquels σ_t^Φ est un automorphisme intérieur.

Par exemple quand M est un facteur d'Araki Woods : $M = \bigotimes_{\nu=1}^{\infty} (M_\nu, \Phi_\nu)$, on connaît le groupe d'automorphismes modulaires de $\Phi = \bigotimes_{\nu=1}^{\infty} \Phi_\nu$, c'est $\sigma_t^\Phi = \bigotimes_{\nu=1}^{\infty} \sigma_t^{\Phi_\nu}$.

Un calcul simple à partir de la liste des valeurs propres $(\lambda_{\nu,j})_{j=1,\dots,n_\nu}$ de Φ_ν montre alors que :

$$T(M) = T_o , \sum_{\nu=1}^{\infty} (1-|\sum_j \lambda_{\nu,j}^{1+iT_o}|) < \infty\} .$$

On en tire par exemple que $T(R_\alpha) = \{T_o , \alpha^{iT_o} = 1\}$.

Quand M est un facteur de Krieger, ou plus généralement quand $M = L^\infty(R,\mu)$ pour une relation d'équivalence R (Voir paragraphe V on ne suppose pas nécessairement R de la forme R_T) , on vérifie également par un calcul facile que :

$T(M) = {}^{2\pi}/\text{Log}\,\rho(R)$ où $\rho(R)$ est l'ensemble des $\lambda > 0$ pour lesquels il existe $\nu \sim \mu$ telle que les dérivées de Radon Nikodym $\frac{d\nu(Sx)}{d\nu(x)}$ appartiennent à $\lambda^{\mathbb{Z}}$ pour toute transformation borélienne S prescrivant les orbites de R , i.e. l'invariant ρ de Krieger.

Ce deuxième calcul montre facilement que, en général, l'égalité $\rho(M) = \exp({}^{2\pi}/T(M))$ n'a pas lieu.

<u>Les facteurs de type</u> III_λ .

Le théorème de Radon Nikodym permet de calculer l'invariant $S(M)$ à partir d'un unique poids normal fidèle Φ sur M . On définit le centralisateur M_Φ de Φ par l'égalité :

$$M_\Phi = \{x \in M , \sigma_t^\Phi(x) = x \ \forall\ t \in \mathbb{R}\} .$$

Pour tout projecteur $e \neq 0$, $e \in M_\Phi$ on définit un poids fidèle Φ_e sur l'algèbre de von Neumann réduite $eMe = \{x \in M , ex = xe = x\}$ par l'égalité :

$$\Phi_e(x) = \Phi(x) , \ \forall\ x \in eMe , \ x \geq 0 .$$

On a alors la formule :

$$S(M) = \bigcap_{e \neq o} \text{Spectre } \Delta_{\Phi_e}$$

où e varie donc parmi les projecteurs non nuls de M_Φ. De plus comme e commute avec Φ le calcul de $\sigma_t^{\Phi_e}$ (et par conséquent du spectre de Δ_{Φ_e}) est immédiat, on a $\sigma_t^{\Phi_e}(x) = \sigma_t^\Phi(x)$, $\forall\, x \in eMe$. Ainsi la formule ci-dessus permet de calculer, par exemple, $S(M)$ pour $M = L^\infty(R,\mu)$, on obtient l'égalité :

$$S(M) = r(R)$$

où r est l'invariant défini par Krieger comme ensemble des valeurs essentielles des dérivées de Radon Nikodym. En général, donc, $S(M) \neq r_\infty(M)$. En outre on a une interprétation beaucoup plus satisfaisante de $S(M)$ comme spectre de l'homomorphisme modulaire δ.

Supposons que le prédual M_* est de type dénombrable, alors un groupe à un paramètre $(\alpha_t)_{t\in R}$ d'automorphismes de M est de la forme σ_t^Φ, pour un poids fidèle Φ sur M, si et seulement si la classe $\varepsilon(\alpha_t)$ de α_t dans $\mathrm{Out}\,M$ est égale à $\delta(t)$ pour tout t. Dit en d'autres termes cela signifie que l'ensemble de tous les groupes d'automorphismes modulaires de poids fidèles sur M constitue exactement l'ensemble des sections boréliennes multiplicatives de δ.

Pour tout poids fidèle Φ le spectre de Δ_Φ s'identifie au spectre de σ^Φ au sens suivant : Spectre $\sigma^\Phi = \{\lambda \in \text{Groupe dual de } R,\ \hat f(\lambda) = 0$ pour toute $f \in L^1(R)$ telle que $\int f(t)\sigma_t^\Phi\, dt = 0\}$.

(On peut aussi définir le spectre de σ^Φ à partir des supports des distributions transformées de Fourier des $t \to \sigma_t^\Phi(x)$, on obtient ainsi des distributions à valeurs dans M et Spectre σ^Φ est la fermeture de la réunion des supports des $(\sigma^\Phi(x))^\wedge$).

Notons également que dans la formule ci-dessus on identifie R_+^* avec le groupe dual de R par l'égalité $\langle \lambda, t\rangle = \lambda^{it}$, $\lambda \in R_+^*$ et $t \in R$, l'égalité précise est alors $\mathrm{Sp}\Delta\Phi \cap R_+^* = \text{Spectre } \sigma^\Phi$.

On a donc $R_+^* \cap S(M) = \bigcap_{\varepsilon\circ\alpha=\delta} \text{Spectre } \alpha$, et cette formule montre alors que $R_+^* \cap S(M)$ est, quand M est un facteur, un sous-groupe de R_+^*. De plus

$0 \in S(M) \Leftrightarrow M$ est de type III et on a les cas suivants :

III_0 $\quad S(M) = \{0,1\}$

III_λ , $\lambda \in]0,1[: S(M) = \lambda^{\mathbb{Z}} \cup \{0\}$

III_1 $\quad S(M) = [0,+\infty[$.

En un certain sens le λ ci-dessus exprime la distance entre M et les facteurs semi-finis, en fait λ est relié de manière monotone et biunivoque à une grandeur qui mesure l'obstruction à l'existence d'une trace sur M :

$$d(M) = \text{diamètre } {}^{\mathcal{S}}/\text{Int}\, M$$

où $\mathcal{S}$ désigne l'espace métrique (avec la distance $d(\Phi_1,\Phi_2) = \|\Phi_1-\Phi_2\|$) des états normaux sur M et où $\text{Int}\, M$ agit sur $\mathcal{S}$ par $\Phi \to u\Phi u^*$, u unitaire de M . Pour M de type III_1 on a $d(M) = 0$ (voir [10]) de sorte que l'on ne peut distinguer deux états d'un facteur de type III_1 par une propriété fermée et invariante par les automorphismes intérieurs. Citons maintenant une autre interprétation de $S(M)$ d'un point de vue heuristique. Revenons d'abord à l'origine de la terminologie "automorphismes modulaires". Le premier exemple d'algèbre hilbertienne à gauche est l'algèbre de convolution des fonctions continues à support compact sur un groupe localement compact G . Soit dg une mesure de Haar à gauche sur G , le module du groupe est alors l'homomorphisme δ_G de G dans $\mathbb{R}_+^*$ associé à l'action de G à droite sur dg. De plus l'opérateur modulaire de l'algèbre hilbertienne à gauche est l'opérateur de multiplication par la fonction δ_G dans l'espace $L^2(G,dg)$, de sorte que son spectre est la fermeture de l'image de δ_G . On peut alors, toujours d'un point de vue heuristique interpréter pour un facteur M , l'invariant $S(M)$ comme "l'image du module de M" .

Les facteurs de type III_λ , $\lambda \in]0,1[$ sont caractérisés par l'égalité $S(M) = \lambda^{\mathbb{Z}} \cup \{0\}$; or c'est un exercice facile de voir que si G est un groupe localement compact et l'image du module δ_G de G est $\lambda^{\mathbb{Z}}$, G est le produit semi-direct d'un groupe unimodulaire : $H = \text{Noyau } \delta_G$ par un automorphisme

$\alpha \in \mathrm{Aut}\, H$ qui multiplie par λ toute mesure de Haar sur H. Inversement tout couple (H,α), H unimodulaire et α multipliant toute mesure de Haar sur H par λ donne par produit semi-direct un groupe localement compact $G = H \times_{\alpha} \mathbb{Z}$ et $\delta_G(G) = \lambda^{\mathbb{Z}}$.

L'analogie se prolonge par le théorème suivant :

THEOREME 2. [8]. - Soit $\lambda \in]0,1[$.

a) Soit M un facteur de type III_λ, il existe un facteur N de type II_∞ et $\theta \in \mathrm{Aut}\, N$ multipliant toute trace de N par λ (on écrit alors $\mathrm{mod}\,\theta = \lambda$) tels que M soit isomorphe au produit croisé de N par θ.

b) Soient N un facteur de type II_∞, et $\theta \in \mathrm{Aut}\, N$ avec $\mathrm{mod}\,\theta = \lambda$, le produit croisé de N par θ est alors un facteur de type III_λ.

c) Deux couples (N_i,θ_i), $i = 1,2$ donnent des facteurs isomorphes si et seulement si il existe un isomorphisme σ de N_1 sur N_2 tel que les classes de $\sigma\theta_1\sigma^{-1}$ et θ_2, modulo les automorphismes intérieurs de N_2, soient les mêmes.

Avant détudier les implications de ce théorème sur le problème de classification des facteurs, notons quelques précisions importantes quant à la théorie générale des facteurs de type III_λ. La notion de trace est remplacée par la suivante :

DEFINITION 3. - Une trace généralisée Φ sur un facteur M de type III_λ $\lambda \in]0,1[$, est un poids fidèle Φ tel que $\mathrm{Sp}\Delta_\Phi = S(M)$ et $\Phi(1) = +\infty$.

On démontre ([8]) l'existence de traces généralisées sur M en étudiant les relations entre les invariantes $T(M)$ et $S(M)$ et en montrant que sauf quand $S(M) = \{0,1\}$, l'invariant $T(M)$ est déterminé par $S(M)$. (Alors que, dans le cas III_0, $T(M)$ peut être n'importe quel sous-groupe dénombrable, non nécessairement fermé, de $\mathbb{R}$).

De plus on a le résultat d'unicité suivant : si Φ_1 et Φ_2 sont deux traces généralisées sur M il existe un automorphisme intérieur α tel que Φ_2 soit proportionnelle à $\Phi_1 \circ \alpha$.

L'algèbre de von Neumann de type II_∞ , N du théorème n'est autre que le centralisateur M_Φ d'une trace généralisée Φ ; sa position dans M est <u>unique aux automorphismes intérieurs près</u> et on peut la caractériser comme sous-algèbre semi-finie maximale ([8]).

Le théorème 2 montre que le problème de classification des facteurs de type III_λ se ramène à :

1) Classifier les facteurs de type II_∞

2) Etant donné un facteur N de type II_∞ , déterminer dans $\mathrm{Out}\,N = \mathrm{Aut}\,N/\mathrm{Int}\,N$ les classes de conjuguaison de θ tels que $\mathrm{mod}\,\theta = \lambda$.

Ce sont ces deux problèmes qui sont les motivations principales pour les paragraphes VII, VIII ci-dessous, le problème 2) étant englobé dans la théorie ergodique non commutative.

VII. THEORIE ERGODIQUE NON COMMUTATIVE.

Soient (X,B,μ) un espace borélien standard muni d'une mesure de probabilité μ , et T une transformation borélienne de (X,B) laissant μ invariante. Soient $M = L^\infty(X,B,\mu)$ et Φ l'état associé à μ , alors T détermine un automorphisme de M qui préserve Φ , par l'égalité :

$$\theta(f) = f \circ T^{-1} .$$

Inversement tout automorphisme de M préservant Φ est obtenu de cette manière. Ainsi la théorie ergodique classique est-elle, après traduction, la même chose que l'étude à conjuguaison près des automorphismes de M qui fixent Φ . Une des raisons d'être de cette théorie est en fait que tous les triplets (X,B,μ) (et par conséquent tous les couples (M,Φ)) avec $\mu(\{x\}) = 0 \;\; \forall\, x \in X$, sont isomorphes). Ainsi à chaque construction différente d'un tel triplet va correspondre une famille d'automorphismes de (X,B,μ) et il s'agit de les comparer. De même, dans le cadre de l'intégration non commutative. Il existe de nombreuses constructions différentes du facteur hyperfini R , par exemple comme représentation régulière d'un groupe discret localement fini (voir le paragraphe I), comme produit tensoriel infini des couples $(M_n(C), \mathcal{T}_n)$ où $\mathcal{T}_n$ est la trace normalisée par $\mathcal{T}_n(1) = 1$, ou encore à partir du théorème de H. Dye (paragraphe V). A chacune de ces constructions correspondent des automorphismes de R. En fait on peut aussi construire R à partir des relations d'anticommutation canoniques sur un espace hilbertien réel E et obtenir ainsi un homomorphisme injectif du groupe orthogonal de E dans Aut R. Il en résulte que Out R contient en fait tout groupe localement compact de type dénombrable.

Bien entendu on ne veut pas distinguer deux automorphismes de R de la forme θ et $\sigma\theta\sigma^{-1}$ où $\sigma \in \mathrm{Aut}\, R$. Adoptons les définitions générales suivantes :

DEFINITION 1. - Soient M une algèbre de von Neumann, et $\theta_1, \theta_2 \in \mathrm{Aut}\, M$ deux automorphismes de M.

a) On dit que θ_1 et θ_2 sont conjugués quand il existe $\sigma \in \mathrm{Aut}\, M$ tel que $\theta_2 = \sigma\theta_1\sigma^{-1}$.

b) On dit que θ_1 et θ_2 sont extérieurement conjugués quand il existe $\sigma \in \mathrm{Aut}\, M$ tel que $\theta_2 = \sigma\theta_1\sigma^{-1}$ modulo Int M.

Quand M est commutative les deux définitions coïncident car $\mathrm{Int}\, M = \{1\}$. Dans le cas général on a deux problèmes : conjuguaison et conjuguaison extérieure. Commençons par citer deux résultats qui étendent au cas non commutatif des résultats importants de théorie ergodique classique, nous discuterons plus loin les phénomènes spécifiquement non commutatifs, le lecteur intéressé uniquement aux automorphismes de R peut se reporter directement au théorème 8 ci-dessous.

Le théorème de Rokhlin.

Soient (X, B, μ) un espace borélien standard probabilisé et T une transformation borélienne de (X, B) préservant μ. Il existe alors une partition essentiellement unique de X, $X = \bigcup_{i=1}^{\infty} X_i$, où chaque X_i est invariant par T et où :

pour $i > 0$, la restriction de T à X_i est périodique de période i avec :

$$\mathrm{Card}\{T^j x\} = i \quad \forall\, x \in X_i$$

pour $i = 0$, la restriction de T à X_o est <u>apériodique</u>, i.e.

$$\mathrm{Card}\{T^j x\} = \infty \quad \forall\, x \in X_o .$$

Le théorème de Rokhlin s'énonce alors ainsi : soit T une transformation apériodique de (X,B,μ), pour tout $\varepsilon > 0$ et tout $n > 0$ il existe un borélien $E \subset X$, tel que $E, TE, \ldots, T^{n-1}(E)$ soient deux à deux disjoints et

$$\mu(X \setminus \bigcup_{j=o}^{n-1} T^j E) < \varepsilon .$$

Soient alors $(N,\mathcal{T})$ un couple (algèbre de von Neumann, trace fidèle normalisée $\mathcal{T}(1) = 1$) et θ un automorphisme de N préservant $\mathcal{T}$. Il existe alors une partition de l'unité dans N, $\sum_{j=o}^{\infty} e_j = 1$, où chaque e_j est un projecteur du <u>centre</u> de N invariant par θ et où :

- pour $j > 0$, la restriction de θ à l'algèbre de von Neumann réduite N_{e_j} vérifie : θ^j est intérieur et pour $k < j$ et tout projecteur $e \leq e_j, e \neq 0$, il existe un projecteur $f \neq 0$, $f \leq e$ tel que

$$\|f\theta^k(f)\| \leq \varepsilon .$$

- pour $j = 0$, la restriction de θ à l'algèbre de von Neumann N_{e_o} est apériodique : pour tout $k > 0$, tout projecteur non nul $e \leq e_o$ et tout $\varepsilon > 0$, il existe un projecteur $f \neq 0$, $f \leq e$ avec :

$$\|f\theta^k(f)\| \leq \varepsilon .$$

Comme dans le cas commutatif cette décomposition est unique. On dit que θ est apériodique quand $e_o = 1$.

THEOREME 2 [12]. - <u>Soient</u> $(N,\mathcal{T})$ <u>un couple (algèbre de von Neumann, trace fidèle normalisée) et</u> θ <u>un automorphisme apériodique de</u> N <u>préservant</u> $\mathcal{T}$. <u>Pour tout entier</u> $n > 0$ <u>et tout</u> $\varepsilon > 0$, <u>il existe une partition de l'unité</u> $\sum_{j=o}^{n-1} E_j = 1$

dans N , où les E_j sont des projecteurs, telle que :

$$\|\theta(E_j) - E_{j+1}\|_2 \leq \varepsilon \quad j = 0,1,\dots,n-1 \quad (E_n = E_o) .$$

Ici, comme dans le premier paragraphe on pose, pour $x \in N, \|x\|_2 = (\mathcal{J}(|x|^2))^{\frac{1}{2}}$.

L'entropie (voir [11]).

Soient (X,B,μ,T) comme ci-dessus et $\mathcal{P}$ une partition borélienne de X , on appelle entropie de $\mathcal{P}$ relative à T un scalaire $h(T,\mathcal{P})$ qui compte asymptotiquement $\frac{1}{n}$ fois le logarithme du nombre d'éléments dans la partition composée de $\mathcal{P},T\mathcal{P},\dots,T^{n-1}\mathcal{P}$:

$$h(T,\mathcal{P}) = \lim_{n\to\infty} \frac{1}{n} h(\mathcal{P} \vee T\mathcal{P} \vee \dots \vee T^{n-1}\mathcal{P})$$

où $h(Q)$, pour une partition $Q = (q_j)_{j\in\{1,\dots,k\}}$, est $\Sigma\, \eta(N(q_j))$

$\eta(t) = -t \operatorname{Log} t$, $\forall\, t \in [0,1]$.

On définit alors l'entropie de T comme le sup des $h(T,\mathcal{P})$. C'est un invariant calculable grâce au théorème de Kolmogoroff-Sinai : on a $h(T)=h(T,\mathcal{P})$ pour toute partition $\mathcal{P}$ pour laquelle les $T^j\mathcal{P}$ engendrent la tribu B . En particulier l'entropie d'un shift de Bernouilli : la translation de 1 dans $\prod_{\nu\in\mathbb{Z}} (X_\nu,\mu_\nu)$ où tous les X_ν sont égaux à $\{1,\dots,p\}$ et tous les μ_ν à la même mesure $j \in \{1,\dots,p\} \to \lambda_j$, on obtient $h(T) = \sum_1^p \eta(\lambda_j)$.

Or les shifts de Bernouilli ont un analogue dans le cas non commutatif, prenons le plus simple qui est associé à un entier p . On considère le produit tensoriel infini $\bigotimes_{\nu\in\mathbb{Z}} (M_\nu,\Phi_\nu)$ où pour tout ν on a $M_\nu = M_p(\mathbb{C})$, algèbre des matrices $p \times p$, et Φ_ν = Trace normalisée. Le shift S_p donne alors un automorphisme du couple $(R,\mathcal{J})$ où R est le facteur hyperfini et $\mathcal{J}$ sa trace normalisée. On peut alors poser la question : Les S_p sont-ils deux à deux conjugués. Ce problème nous a conduit avec E. Størmer à la généralisation suivante de l'entropie et du théorème de Kolmogoroff-Sinai, qui a permis de distinguer les S_p à conjugaison près :

Le rôle des partitions finies $Q = (q_j)$ de l'expace X est joué par les sous-algèbres de dimension finie $K \subset M$ où M désigne l'algèbre de von Neumann. On définit une fonction $H(K_1,\dots,K_n)$, où $K_1,\dots,K_n$ varient parmi les sous-algèbres de dimension finies de M, qui joue le rôle de $h(P_1 \vee \dots \vee P_n)$. Les propriétés de cette fonction sont établies grâce aux inégalités de E. Lieb concernant l'information non commutative $S(x|y) = \mathcal{T}(x(\mathrm{Log}\, x - \mathrm{Log}\, y))$ où $x,y \in M$ et $\mathcal{T}$ est la trace normalisée. On définit $H(\theta,K) = \lim_{n\to\infty} H(K,\theta(K),\theta^2(K),\dots,\theta^{n-1}(K))$ et

$H(\theta) = \sup_K H(\theta,K)$. L'analogue du théorème de Kolmogoroff-Sinai montre alors

que $H(S_p) = \mathrm{Log}\, p$ ce qui suffit à distinguer les shifts.

Mais passons maintenant aux phénomènes spécifiquement non commutatifs ; nous verrons par exemple que tous les S_p sont extérieurement conjugués.

Automorphismes approximativement intérieurs.

Soient M une algèbre de von Neumann et M_* son prédual. L'action de $\mathrm{Aut}\, M$ sur M_*, muni de la topologie normique, est équicontinue ; il en résulte

que la topologie de la convergence simple normique dans M_* fait de $\mathrm{Aut}\, M$ un groupe topologique.

Dans la suite quand nous parlerons de $\mathrm{Aut}\, M$ comme groupe topologique nous ferons toujours référence à celle-ci. Pour s'assurer que c'est la bonne structure sur $\mathrm{Aut}\, M$ il suffit de noter que si M_* est de type dénombrable le groupe $\mathrm{Aut}\, M$ est alors Polonais.

En général le groupe $\mathrm{Int}\, M \subset \mathrm{Aut}\, M$ n'est pas fermé, par exemple pour le facteur hyperfini R on a $\overline{\mathrm{Int}\, R} = \mathrm{Aut}\, R$. Plus précisément pour que $\mathrm{Int}\, M$ soit fermé dans $\mathrm{Aut}\, M$, en supposant M fini, il faut et il suffit que M ne possède pas la propriété Γ du premier paragraphe (cf [7] et [52]).

Quand M est un facteur de type II_1 les automorphismes approximativement intérieurs de M sont caractérisés par l'équivalence suivante :

THEOREME 3 [9]. – Soit N un facteur de type II_1 , à prédual de type dénombrable, agissant dans l'espace hilbertien $h = L^2(N,\mathfrak{J})$ où $\mathfrak{J}$ désigne la trace normalisée de N . Soit $\theta \in \operatorname{Aut} N$, les conditions suivantes sont équivalentes :

   a) $\theta \in \overline{\operatorname{Int} N}$, i.e. est approximativement intérieur.

   b) $\| \sum_{i=1}^{n} \theta(a_i) b_i \| = \| \sum_{i=1}^{n} a_i b_i \|$ pour $a_i, \ldots, a_n \in N$ et $b_1, \ldots, b_n \in N'$ = commutant de N .

La condition b) montre qu'il existe alors un automorphisme α de l'algèbre stellaire $C^*(N \, N')$ engendrée par N et N' , tel que $\alpha(a) = \theta(a)$ $\forall \, a \in N$ et $\alpha(b) = b$ $\forall \, b \in N'$.

COROLLAIRE 4. – Soient $(N_i)_{i=1,2}$ des facteurs de type II_1 à prédual de type dénombrable et $(\theta_i)_{i=1,2}$ des automorphismes de N_i , alors :

$\theta_1 \otimes \theta_2$ approximativement intérieur $\Leftrightarrow \theta_1$ et θ_2 approximativement intérieurs.

Automorphismes centralement triviaux.

   Soient N un facteur de type II_1 à prédual séparable, $\mathfrak{J}$ sa trace normalisée et θ un automorphisme approximativement intérieur de N : $\theta \in \overline{\operatorname{Int} N}$. Il existe une suite d'unitaires de N : $(u_k)_{k \in \mathbb{N}}$ telle que pour tout $x \in N$ on ait :

$$\theta(x) = \lim_{k \to \infty} u_k \, x \, u_k^*$$

pour la topologie de $L^2(N,\mathfrak{J})$: $\|x-y\|_2 = (\mathfrak{J}|x-y|^2)^{\frac{1}{2}}$.

On traduit cette propriété sous forme d'une égalité :

$$\theta(x) = u \, x \, u^* \quad \forall \, x \in N$$

en introduisant une algèbre de von Neumann contenant N de la manière suivante :

DEFINITION 5. – Pour tout ultrafiltre $\omega \in \beta \, \mathbb{N} \setminus \mathbb{N}$, soit N^ω l'ultraproduit N^ω = algèbre de von Neumann $\ell^\infty(\mathbb{N},N)$ quotientée par l'idéal des suites $(x_n)_{n \in \mathbb{N}}$ telles que $\lim_{n \to \omega} \|x_n\|_2 = 0$.

On démontre que cet ultraproduit est une algèbre de von Neumann finie (cf [22],[61]) bien que, en général l'idéal bilatère mentionné ne soit pas $\sigma(\ell^{\infty}, \ell^{\infty}_{*})$ fermé. De plus N se plonge canoniquement dans l'ultraproduit N^{ω} en associant à $x \in N$ la suite constante $(x)_{n \in \mathbb{N}}$. La suite d'unitaires $(u_k)_{k \in \mathbb{N}}$ définit un unitaire $u \in N^{\omega}$ et on a bien entendu $\theta(x) = uxu^*$, pour tout $x \in N$. Cette égalité détermine u de manière unique modulo le groupe unitaire d'une sous-algèbre de von Neumann de N^{ω} qui joue un rôle crucial dans la suite :

DEFINITION 6. - <u>Soient</u> N <u>et</u> ω <u>comme ci-dessus, on appelle centralisateur asymptotique de</u> N <u>en</u> ω <u>le commutant de</u> N <u>dans</u> N^{ω} :

$$N_{\omega} = \{y \in N^{\omega} , \ yx = xy \quad \forall \ x \in N\} .$$

La construction de N_{ω} est fonctorielle de sorte que chaque automorphisme θ de N définit un automorphisme θ_{ω} de N_{ω}.

Soient alors comme ci-dessus $\theta \in \overline{\operatorname{Int} N}$ et $u \in N^{\omega}$ unitaire tel que :

$$\theta(x) = u x u^* \quad \forall \ x \in N .$$

La question est : peut-on choisir u tel que $\theta^{\omega}(u) = u$.
On peut multiplier u par un unitaire v de N_{ω} sans changer l'égalité $\theta(x) = u x u^*$, $x \in N$, il s'agit donc, en posant $w = u^* \theta^{\omega}(u)$ de trouver $v \in N_{\omega}$, avec $v^* \theta_{\omega}(v) = u$. Par construction w est un unitaire de N_{ω} et il s'agit donc de caractériser les unitaires de N_{ω} de la forme $v^* \theta_{\omega}(v)$, $v \in N_{\omega}$. Le théorème de Rokhlin de théorie ergodique non commutative donne une réponse complète à ce problème sous la forme :

1) La partition de l'unité du centre de N_{ω} associée à l'automorphisme θ_{ω} de N_{ω} est formée d'un unique $e_j = 1$ et θ^k_{ω} est extérieur pour $k < j$ et égal à 1 pour $k = j$.

2) Pour qu'un unitaire $w \in N_{\omega}$ soit de la forme $v^* \theta_{\omega}(v), v \in N_{\omega}$ il faut et il suffit que $w \, \theta_{\omega}(w) \ldots \theta^{j-1}_{\omega}(w) = 1$.

De plus l'entier j ne dépend que de θ et non du choix de $\omega \in \beta\mathbb{N} \setminus \mathbb{N}$ on le note $p_a(\theta)$ = période asymptotique de θ. C'est la période de θ modulo le sous-groupe normal $\mathrm{Ct}\,N$ de $\mathrm{Aut}\,N$ des automorphismes de N qui sont <u>centralement triviaux</u> au sens suivant :

$\theta \in \mathrm{Ct}\,N$ ssi $\theta_\omega = 1$ (pour un $\omega \in \beta\mathbb{N}/\mathbb{N}$ ou de manière équivalente pour tout $\omega \in \beta\mathbb{N}/\mathbb{N}$).

Revenons après cette définition au problème ci-dessus, il s'agit donc de savoir si avec $j = p_a(\theta)$ et $w = u^*\theta^\omega(u)$, on a $w\theta_\omega(w)\ldots\theta_\omega^{j-1}(w) = 1$. Cela revient à savoir si $(\theta^\omega)^j(u) = u$. Or la période de θ^ω est la même que celle de θ et il s'agit de la comparer avec $p_a(\theta)$, on a :

$$\theta^{p_a} = 1 \;,\; \theta \in \overline{\mathrm{Int}\,N} \Rightarrow \exists (u_n)_{n\in\mathbb{N}} \text{ unitaires de } N \text{ tels que}$$

$$\theta(u_n) - u_n \xrightarrow[n\to\infty]{} 0 \,,\; \theta(x) = \lim_{n\to\infty} u_n x u_n^* \quad \forall\, x \in N \,.$$

En particulier si $p_a = 0$ la condition est satisfaite.

C'est là la motivation principale pour chercher à déterminer en général le groupe $\mathrm{Ct}\,N$. Le théorème suivant se déduit de [9].

THEOREME 7. - <u>Soient</u> N <u>un facteur de type</u> II_1, <u>à prédual de type dénombrable, agissant dans</u> $h = L^2(N,\mathfrak{J})$ <u>et</u> $\theta \in \mathrm{Aut}\,N$, U <u>l'unitaire de</u> $L^2(N,\mathfrak{J})$ <u>associé à</u> θ <u>(la construction de</u> L^2 <u>est fonctorielle). Soient</u> $p = p_a(\theta)$ <u>la période asymptotique de</u> θ <u>et</u> $\lambda \in \mathbb{C}, |\lambda| = 1$. <u>Alors, pour que</u> $\lambda^p = 1$, <u>il faut et il suffit qu'il existe un automorphisme</u> α_λ <u>de l'algèbre stellaire engendrée par</u> N, N' <u>et</u> U <u>tel que</u> :

$$\alpha_\lambda(U) = \lambda U \;,\; \alpha_\lambda(A) = A \quad \forall\, A \in C^*(N,N') \,.$$

Pour $N = R$ le facteur hyperfini on a $\mathrm{Ct}\,R = \mathrm{Int}\,R$. Le théorème ci-dessus montre également que $\theta_1 \otimes \theta_2 \in \mathrm{Ct}(N_1 \otimes N_2)$ si et seulement si θ_1 et θ_2 sont centralement triviaux. Une autre caractérisation intéressante de $\mathrm{Ct}\,N$, pour N facteur de type II_1 tel que $\varepsilon(\overline{\mathrm{Int}\,N})$ soit non commutatif, où ε désigne l'application quotient : $\mathrm{Aut}\,N \to \mathrm{Out}\,N$, est la suivante : ([12]).

$\varepsilon(\mathrm{Ct}\,N)$ est le commutant de $\varepsilon(\overline{\mathrm{Int}\,N})$ dans $\mathrm{Out}\,N$.

(Il suffit de connaître $\varepsilon(\mathrm{Ct}\,N)$ pour connaître $\mathrm{Ct}\,N$ car on a toujours $\mathrm{Int}\,N \subset \mathrm{Ct}\,N$).

L'obstruction $\gamma(\theta)$.

Soient M un facteur et $\theta \in \mathrm{Aut}\,M$. Soit $p_o(\theta) \in \mathbb{N}$ la période de θ modulo les automorphismes intérieurs :

$$\theta^j \in \mathrm{Int}\,M \Leftrightarrow j \in p_o\mathbb{Z} .$$

C'est un invariant de conjuguaison extérieure de θ, supposons $p_o \neq 0$, et cherchons θ' extérieurement conjugué à θ tel que $\theta'^{p_o} = 1$. On a un homomorphisme de $\mathbb{Z}/p_o\mathbb{Z}$ dans $\mathrm{Out}\,M$ et il s'agit de le relever dans $\mathrm{Aut}\,M$. Comme le centre du groupe unitaire $\mathcal{U}$ de M est égal au tore $\mathbb{T} = \{z \in \mathbb{C}, |z| = 1\}$, l'obstruction associée à ce problème est un élément de $H^3(\mathbb{Z}/p_o\mathbb{Z}, \mathbb{T})$, où l'action de $\mathbb{Z}/p_o\mathbb{Z}$ sur $\mathbb{T}$ est triviale. Cette obstruction $\gamma(\theta)$ est en fait la racine p_o-ième de 1 dans $\mathbb{C}$ caractérisée par l'égalité :

$$u \in \mathcal{U},\ \theta^{p_o}(x) = u x u^* \quad \forall\, x \in M \Rightarrow \theta(u) = \gamma u .$$

Le point important est alors l'existence d'automorphismes θ, de facteurs comme le facteur hyperfini, dont l'obstruction $\gamma(\theta)$ est $\neq 1$. On peut se convaincre facilement de cette existence par l'exemple suivant : partons de $(X,B,\mu,(F_t)_{t \in R})$ où (X,B,μ) est un espace borélien standard probabilisé et où $(F_t)_{t \in R}$ est un groupe à un paramètre (borélien) de transformations boréliennes préservant la mesure μ. Supposons que chaque $F_t, t \neq 0$ est ergodique (par exemple on peut prendre un flot de Bernouilli ([50])). Alors le produit croisé R de $L^\infty(X,B,\mu)$ par l'automorphisme associé à F_1 est le facteur hyperfini. L'algèbre de von Neumann $L^\infty(X,B,\mu)$ est contenue dans R et l'unitaire $U \in R$ correspondant à F_1 vérifie :

$$U f U^* = f \circ F_1 \quad \forall\, f \in L^\infty(X,B,\mu)$$

$L^\infty(X,B,\mu)$ et U engendrent R.

Comme F_t, $t \in R$, commute avec F_1, il définit un automorphisme θ_t de R tel que $\theta_t(U) = U$ et $\theta_t(f) = f \circ F_t$, $f \in L^\infty(X,B,\mu)$. De plus pour chaque nombre complexe λ de module 1, on définit un automorphisme σ_λ de R tel que $\sigma_\lambda(f) = f \quad \forall f \in L^\infty(X,B,\mu)$ et $\sigma_\lambda(U) = \lambda U$. Par construction les θ et les σ commutent entre eux et $\theta_1(x) = U x U^* \quad \forall x \in R$. Posons $\alpha = \theta_{\frac{1}{p}} \sigma_\gamma$ où $p \in \mathbb{N}$ alors $\alpha^p = \theta_1 \sigma_{\gamma^p}$ de sorte que si $\gamma^p = 1$ on a $\alpha^p(x) = U x U^* \quad \forall x \in R$ et $\alpha(U) = \gamma U$. On en déduit que $\gamma_o(\alpha) = p$ et que $\gamma(\alpha) = \gamma$.

Ce qui est intéressant dans cet invariant $\gamma(\theta)$ c'est que c'est un nombre complexe en général non réel, en particulier si on fait agir θ non sur M mais sur M^c, le facteur obtenu en remplaçant λx, $\lambda \in \mathbb{C}$ $x \in M$ par $\bar{\lambda} x$, on obtient $\gamma(\theta^c) = \bar{\gamma}(\theta)$. En fait c'est le premier invariant qui soit sensible à l'automorphisme $z \to \bar{z}$ de $\mathbb{C}$ sur $\mathbb{R}$ et c'est celui qui permet de construire un facteur (de type III ou de type II_1 ([6])) non antiisomorphe à lui-même (voir paragraphe 1, l'algèbre M^c est isomorphe à M^o pour l'opération $x \to x^*$).

<u>La liste des automorphismes de</u> R <u>à conjuguaison extérieure près.</u>

<u>Pour le facteur hyperfini</u> R, <u>on a</u> $\overline{\mathrm{Int}\, R} = \mathrm{Aut}\, R$ <u>et</u> $\mathrm{Ct}\, R = \mathrm{Int}\, R$.

En particulier $p_a(\theta) = p_o(\theta) \quad \forall \theta \in \mathrm{Aut}\, R$. On dispose donc de deux invariants de conjuguaison extérieure, l'entier $p_o(\theta)$ et la racine p_o-ième de 1, $\gamma(\theta)$, égale à 1 si $p_o(\theta) = 0$.

On a vu d'autre part l'existence d'un automorphisme de R ayant un couple (p_o, γ) d'invariants, donné à priori.

THEOREME 8 ([12]). - <u>Soient</u> θ_1 <u>et</u> θ_2 <u>deux automorphismes de</u> R. <u>Alors</u> θ_1 <u>et</u> θ_2 <u>sont extérieurement conjugués si et seulement si</u> :

$$p_o(\theta_1) = p_o(\theta_2),\ \gamma(\theta_1) = \gamma(\theta_2).$$

Pour $p_o = p \neq 0$ et $\gamma \in \mathbb{C}$, $\gamma^p = 1$, il existe en fait un automorphisme s_p^γ, unique à conjuguaison près, de R, ayant pour invariants $p_o(s_p^\gamma) = p$ et

$\gamma(s_p^\gamma) = \gamma$ et dont la période soit la plus petite compatible avec ces conditions c'est-à-dire égale à p ordre de γ.

En particulier, toutes les symétries extérieures de R, $\theta \in \operatorname{Aut} R$, $\theta^2 = 1$, $\theta \notin \operatorname{Int} R$ sont deux à deux conjuguées. La réalisation la plus simple de la symétrie s_2' consiste à prendre l'automorphisme de $R \otimes R$ qui transforme $x \otimes y$ en $y \otimes x$, pour tous $x, y \in R$.

Pour $p_o = 0$, il existe à conjuguaison extérieure près un seul automorphisme apériodique $\theta \in \operatorname{Aut} R$ (i.e. avec $p_o(\theta) = 0$). En particulier tous les shifts de Bernouilli bien que distingués à conjuguaison près par l'entropie, sont extérieurement conjugués.

COROLLAIRE 9. - Le groupe $\operatorname{Out} R$ est un groupe simple ayant un nombre dénombrable de classes de conjuguaison.

En fait on a un résultat plus général que le théorème ci-dessus, qui montre exactement le rôle joué par les égalités $\overline{\operatorname{Int}} R = \operatorname{Aut} R$ et $\operatorname{Ct} R = \operatorname{Int} R$. D'abord on démontre pour tout facteur M à prédual M_* de type dénombrable l'équivalence entre :

$$\overline{\operatorname{Int} M} / \operatorname{Int} M \text{ est non commutatif} \Leftrightarrow M \text{ isomorphe à } M \otimes R .$$

Soit alors $\theta \in \overline{\operatorname{Int} M}$, pour que θ soit extérieurement conjugué à l'automorphisme $1 \otimes s_p^\gamma$ de $M \otimes R$, pour p et γ convenables, il faut et il suffit que $p_o(\theta) = p_a(\theta)$.

En outre, pour que $\theta \in \operatorname{Aut} M$ soit extérieurement conjugué à $\theta \otimes s_q^1 \in \operatorname{Aut} M \otimes R$ il faut et il suffit que q divise la période asymptotique $p_a(\theta)$. En particulier tout automorphisme θ de M est conjugué extérieurement à $\theta \otimes 1$.

Ces résultats montrent l'intérêt de l'invariant $\chi(M) = \dfrac{\overline{\operatorname{Int} M} \cap \operatorname{Ct} M}{\operatorname{Int} M}$,

qui a permis (cf.[6]) de montrer l'existence d'un facteur de type II_1 non antiisomorphe à lui-même.

Parmi les développements récents sur le sujet citons les suivants : V. Jones [36] a classifié complètement (à conjugaison près) les actions de groupes finis arbitraires sur le facteur R, en introduisant des invariants de nature cohomologique (beaucoup plus élaborés dans le cas général que dans le cas cyclique). Depuis A. Ocneanu [45] a réussi à classifier les actions extérieures de groupes discrets moyennables, à conjugaison extérieure près en utilisant les techniques de pavages (de groupes moyennables) introduites par D. Ornstein et B. Weiss. Un contre exemple de V. Jones montre que l'on ne peut espérer classifier les actions de groupes non moyennables.

Enfin, V. Jones et T. Giordano [28] d'une part et E. Stormer [56] d'autre part ont montré que R possède à conjuguaison près un seul antiautomorphisme involutif.

les appliquer au facteur d'Araki Woods de type II_∞ .

Automorphismes du facteur d'Araki Woods $R_{o,1}$ de type II_∞ .

Le produit tensoriel $R_{o,1}$ du facteur hyperfini R par un facteur de type I_∞ est l'unique facteur d'Araki Woods de type II_∞ (voir [2]). Rappelons que pour tout automorphisme θ d'un facteur de type II_∞ noté N on pose $\operatorname{mod}\theta =$ unique $\lambda \in \mathbb{R}_+^*$ tel que $\mathfrak{J}\circ\theta = \lambda\mathfrak{J}$ pour toute trace sur N . Pour $N = R_{o,1}$ on a :

$$\overline{\operatorname{Int}} R_{o,1} = \text{Noyau de mod} = \{\theta, \operatorname{mod}(\theta) = 1\} .$$

De plus $\operatorname{Ct} R_{o,1} = \operatorname{Int} R_{o,1}$. On en déduit :

THEOREME 10. ([12]). - a) Soient θ_1 et θ_2 deux automorphismes de $R_{o,1}$ tels, pour que θ_1 soit extérieurement conjugué à θ_2 il faut et il suffit que

$$\operatorname{mod}\theta_1 = \operatorname{mod}\theta_2 \quad p_o(\theta_1) = p_o(\theta_2) \quad \gamma(\theta_1) = \gamma(\theta_2) .$$

b) Les relations suivantes sont les seules entre mod, p_o et γ :

$$\operatorname{mod}\theta \neq 1 \Rightarrow p_o = 0 , \gamma = 1 \ ; \ p_o = 0 \Rightarrow \gamma = 1 .$$

Alors que le cas $\operatorname{mod}\theta = 1$ se réduit au cas traité ci-dessus, pour tout $\lambda \neq 1$, il résulte de [13] et du a) du théorème ci-dessus que tous les automorphismes

$\theta \in \mathrm{Aut}\, R_{o,1}$ avec $\mathrm{mod}\,\theta = \lambda$ sont <u>conjugués</u> (et non seulement extérieurement conjugués). C'est là un phénomène remarquable, en effet on peut décrire, pour λ <u>entier</u>, la nature de θ avec précision comme un shift sur un produit tensoriel infini d'algèbres de matrices $\lambda \times \lambda$.

Ainsi, dès que $\mathrm{mod}\,\theta = \lambda$, il existe une algèbre de matrice $\lambda \times \lambda$ dans $R_{o,1}$ notons la $K \subset R_{o,1}$, telle que :

1) Les $\theta^j(K)$ commutent deux à deux.

2) Les $\theta^j(K)$ engendrent l'algèbre de von Neumann $R_{o,1}$.

Et cette propriété reste vraie chaque fois que l'on multiplie θ par un automorphisme de module 1.

En fin de compte, nous avons atteint la réponse au problème 2 du paragraphe sur les facteurs de type III_λ et nous pouvons conclure que pour chaque $\lambda \in]0,1[$ il y a un seul facteur de type III_λ dont le facteur de type II_∞ associé est $R_{o,1}$. On vérifie directement pour le facteur R_λ de Powers que le facteur de type II_∞ associé est $R_{o,1}$. Cela montre l'intérêt du sous-problème suivant du problème 1 du paragraphe des III_λ : caractériser le facteur $R_{o,1}$ parmi les facteurs de type II_∞, on adopte la définition suivante :

DEFINITION 11. - <u>Une algèbre de von Neumann</u> M <u>à prédual de type dénombrable est dite approximativement de dimension finie si elle est engendrée par une suite croissante de sous-algèbres de dimension finie.</u>

(On peut aussi (voir [25]) de manière équivalente demander l'approximation de toute partie finie de M par une sous-algèbre de dimension finie).

Il est immédiat que $R_{o,1}$ et plus généralement tout facteur d'Araki Woods et même de Krieger est approximativement de dimension finie.

On peut alors reformuler le problème ci-dessus comme :

PROBLEME 3. - $R_{o,1}$ <u>est-il le seul facteur de type</u> II_∞ <u>qui soit approximativement de dimension finie.</u>

Ce problème revient à savoir si le commutant d'un facteur ayant cette propriété d'approximation l'a aussi. Vers 1967 V.Ya Golodets proposait une démonstration, malheureusement elle contenait une erreur irréparable. Cependant dans un autre article, le même Golodets utilisait son résultat pour en déduire qu'un produit croisé par un groupe commutatif n'affecte pas la propriété d'approximation ci-dessus. Bien que bâtis sur une hypothèse non démontrée, ses raisonnements montraient cependant (cf [29]) que si un facteur M de type III_{γ} est approximativement de dimension finie, il en est de même du facteur de type II_{∞} associé. Ceci renforçait donc considérablement l'intérêt du problème ci-dessus.

Pour clore ce paragraphe signalons que quand on ne suppose plus que le facteur de type II_{∞}, N, est isomorphe à $R_{0,1}$ il y a en général pour $\lambda \in]0,1[$ une infinité de classes de conjuguaison dans $\mathrm{Out}\,N$, d'automorphismes θ de module λ (cf [7] et [48]), à chacune de ces classes va correspondre un facteur de type III_{λ} et les facteurs correspondant seront deux à deux non isomorphes.

VIII. LES ALGEBRES DE VON NEUMANN MOYENNABLES.

Nous passons en revue dans ce paragraphe les propriétés reliées à l'approximation d'une algèbre de von Neumann M par des algèbres de dimension finie. Nous verrons ci-dessous qu'elles définissent en fait toutes la même classe d'algèbre de von Neumann.

Approximation par des algèbres de dimension finie.

Par définition une algèbre de von Neumann M est approximativement de dimension finie quand elle est engendrée par une suite croissante de sous-algèbres de dimension finie.
Une raison importante de l'intérêt de cette classe est le résultat suivant dû à O. Maréchal, basé sur la démonstration du théorème de Glimm.

THEOREME 1. (O. Maréchal [39]). Soit A une algèbre stellaire de type dénombrable non postliminaire. Alors pour toute algèbre de von Neumann approximativement de dimension finie M, sans trace finie non nulle, il existe un état $\phi \in A^*$, tel que l'algèbre de von Neumann engendrée par $\pi_{\phi}(A)$ (paragraphe III) soit isomorphe à M.

Ainsi une théorie de l'intégration non commutative qui ne se limite pas aux algèbres stellaires postliminaires, c'est-à-dire aux algèbres de von Neumann de type I, voit apparaître nécessairement toutes les algèbres de von Neumann qui sont approximativement de dimension finie.
Inversement, pour l'algèbre stellaire A produit tensoriel infini d'algèbres de matrices 2×2, il est immédiat que toutes les algèbres de von Neumann engendrées par A sont approximativement de dimension finie.

Les propriétés P de Schwartz, E de Hakeda et Tomiyama et l'injectivité.

Avant 1963, on avait seulement deux exemples de facteurs de type II_1, non isomorphes (à prédual de type dénombrable bien entendu). Ainsi la propriété Γ distinguait le facteur hyperfini R du facteur Z engendré par la représentation régulière du groupe libre à 2 générateurs. En 1963, J.T. Schwartz introduisit une propriété permettant de distinguer R de $Z \otimes R$ qui tous deux ont la propriété Γ.
Cette propriété P de R est basée sur la moyennabilité d'un groupe localement fini. En fait J. Schwartz montrait que pour G discret il y a équivalence entre :

1) G est moyennable i.e. il existe un état Φ invariant par translations sur $\ell^\infty(G)$.

2) $M = R(\lambda_G)$ agissant dans $h = \ell^2(G)$ a la propriété P suivante :
Pour tout $T \in \mathcal{L}(h)$ il existe un élément de M dans l'enveloppe convexe $\sigma(\mathcal{L}(h), \mathcal{L}(h)_*)$ fermée des uTu^*, u unitaire de M'. En outre, il construisait pour toute algèbre de von Neumann dans h vérifiant 2) une projection linéaire de norme 1 de $\mathcal{L}(h)$ sur M, vérifiant $P(aTb) = aP(T)b$, $\forall\, a,b \in M, T \in \mathcal{L}(h)$.
On vérifie directement que toute algèbre de von Neumann approximativement de dimension finie vérifie la propriété P de Schwartz.
Un résultat remarquable de J. Tomiyama montre que pour toute projection P de norme 1 d'une algèbre de von Neumann N sur une sous-algèbre de von Neumann M on a automatiquement : (cf [60])

$$P(aTb) = aP(T)b \quad \forall\, a,b \in M, T \in N .$$

Dans [33] Hakeda et Tomiyama définissaient une propriété en apparence plus faible que la propriété P de Schwartz :

DEFINITION 2. - Une algèbre de von Neumann M dans l'espace hilbertien h a la propriété E si il existe une projection de norme 1 de l'espace de Banach $\mathcal{L}(h)$ sur l'espace de Banach $M \subset \mathcal{L}(h)$.

Bien sûr $P \Rightarrow E$, de plus le théorème mentionné ci-dessus de Tomiyama montrait que E joue le même rôle que P pour caractériser la moyennabilité de G discret par une propriété de $R(\lambda_G)$. De plus la propriété E ne dépend pas de l'espace hilbertien h dans lequel M agit et elle caractérise, grâce à un théorème de W. Arveson ([3]) les objets injectifs de la catégorie (algèbres de von Neumann, applications complètement positives). C'est pourquoi les algèbres de von Neumann vérifiant la propriété E sont aussi appelées injectives.

L'intérêt de la propriété E n'est pas descriptif : elle dit très peu en apparence sur l'algèbre de von Neumann M , mais d'un autre côté elle a des propriétés remarquables de stabilité :

1) Soit M une algèbre de von Neumann injective opérant dans l'espace hilbertien h , alors le commutant M' de M est injectif.

2) Soit $(M_\alpha)_{\alpha \in I}$ une famille filtrante décroissante d'algèbres de von Neumann injectives alors $\bigcap_{\alpha \in I} M_\alpha$ est injective.

3) Soit $(M_\alpha)_{\alpha \in I}$ une famille filtrante croissante d'algèbres de von Neumann injectives, alors la fermeture de $\bigcup_{\alpha \in I} M_\alpha$ l'est aussi.

4) Soit M une algèbre de von Neumann à prédual de type dénombrable et $M = \int_X M(t)d\mu(t)$ une désintégration de M en facteurs $M(t)$, alors M est injective $\Leftrightarrow M(t)$ est injectif pour presque tout $t \in X$.

5) Soient M une algèbre de von Neumann, N une sous-algèbre de von Neumann et $\mathcal{G}$ un sous-groupe du normalisateur de N dans M , on suppose que N et $\mathcal{G}$ engendrent M que N est injective et $\mathcal{G}$ moyennable comme groupe discret, alors M est injective.

6) Soient M une algèbre de von Neumann injective et G un groupe discret moyennable agissant par automorphismes sur M, alors

$$N = M^G = \{x \in M,\ gx = x \quad \forall\, g \in G\}$$

est injective.

Les propriétés 2) et 3) montrent que si $\mathcal{h}$ est un espace hilbertien de type dénombrable, la classe monotone engendrée par les algèbres de von Neumann de type I ne contient que des algèbres de von Neumann injectives.
La propriété 4) permet essentiellement de limiter le problème de classification de ces algèbres à celle des facteurs injectifs. La propriété 5) montre que tout groupe moyennable d'unitaires engendre une algèbre de von Neumann injective.
Enfin, la propriété 6) permet de voir facilement que pour qu'un facteur M de type III_λ, $\lambda \in]0,1[$ soit injectif il faut et il suffit que le facteur de type II_∞ associé le soit.
Parmi les exemples les plus importants d'algèbres de von Neumann injectives on a :

a) Le produit croisé d'une algèbre de von Neumann abélienne par un groupe localement compact moyennable.

b) L'algèbre de von Neumann commutant d'une représentation unitaire continue quelconque d'un groupe localement compact connexe.

c) L'algèbre de von Neumann engendrée par une représentation arbitraire d'une algèbre stellaire <u>nucléaire</u> (voir définition ci-dessous).

<u>Algèbres de von Neumann semi-discrètes.</u>

Soit M un facteur de type I dans un espace hilbertien $\mathcal{h}$, il lui correspond alors une décomposition de $\mathcal{h}$ en produit tensoriel $\mathcal{h} = \mathcal{h}_1 \otimes \mathcal{h}_2$ de sorte que $M = \mathcal{L}(\mathcal{h}_1) \otimes 1$, $M' = 1 \otimes \mathcal{L}(\mathcal{h}_2)$. On retrouve alors $\mathcal{L}(\mathcal{h})$ comme produit tensoriel de M par M'. Dans l'un des premiers articles de Murray et von Neumann, ceux-ci montrent que pour tout facteur M dans $\mathcal{h}$, l'homomorphisme η du produit tensoriel <u>algébrique</u> $M \odot M' = \{\sum_{i=1}^{n} a_i \otimes b_i,\ a_i \in M,\ b_i \in M'\}$ dans $\mathcal{L}(\mathcal{h})$ défini par :

$$\eta(\Sigma\, a_i \otimes b_i) = \Sigma\, a_i b_i \in \mathcal{L}(\mathcal{h})$$

est <u>injectif</u> et d'image $\sigma(\mathcal{L}(\mathcal{h}), \mathcal{L}(\mathcal{h})_*)$ dense.

Dans [24], E. Effros et C. Lance ont réussi à pousser l'analyse beaucoup plus loin en étudiant η d'un point de vue <u>métrique</u>. Soit A (resp B) une algèbre stellaire avec unité opérant dans l'espace hilbertien h_A (resp h_B), et considérons sur le produit tensoriel algébrique $A \odot B$ la norme provenant de son action dans $h_A \otimes h_B$. Cette norme sur $A \odot B$ fait du complété une algèbre stellaire, ne dépend pas du choix des représentations (fidèles) de A dans h_A et B dans h_B, et est caractérisée grâce à un très utile théorème de M. Takesaki comme étant <u>la plus petite</u> norme sur $A \odot B$ qui fait du complété une algèbre stellaire. On la note $\| \|_{min}$ et on note $A \otimes_{min} B$ l'algèbre stellaire complétée ([57]). On dit que <u>A est nucléaire</u> si $\| \|_{min}$ est la seule norme préstellaire sur $A \odot B$ pour tout B .

E. Effros et C. Lance ont réussi à caractériser les facteurs M pour lesquels l'application η ci-dessus est <u>isométrique</u> par une propriété qui est un renforcement de la propriété d'approximation métrique pour le prédual M_* de M .

Le prédual M_* est non seulement un espace ordonné (par le cône M_*^+) il est <u>matriciellement ordonné</u>, au sens où l'espace vectoriel produit tensoriel $M_* \otimes M_n(C)$ est ordonné pour tout n , comme prédual de $M \otimes M_n(C)$. Une application complètement positive T de M_* dans M_* est par définition une application linéaire telle que $T \otimes 1_{M_n}$ soit positive pour tout n . Le résultat d'Effros et Lance est alors :

THEOREME 3. - <u>Soit M un facteur opérant dans l'espace hilbertien h, pour que $\eta : M \otimes_{min} M' \to \mathcal{L}(h)$ soit isométrique il faut et il suffit que l'application identique de M_* dans M_* soit une limite simple normique d'applications complètement positives de rang fini.</u>

On définit alors une algèbre de von Neumann <u>semi-discrète</u> par la propriété d'approximation ci-dessus de son prédual M_* .

Notons que quand le facteur M est tel que ni M ni M' ne sont de type I_∞ ou II_∞, et quand h est de type dénombrable, un corollaire du théorème de Takesaki sur la norme min , permet de montrer :

COROLLAIRE 4. - *Un facteur* M *opérant dans* $\mathcal{h}$ *de type dénombrable, tel que ni* M *ni* M' *ne sont de type* I_∞ *ou* II_∞, *est semi-discret si et seulement si l'algèbre* $C^*(M,M')$ *engendrée par* M *et* M' *dans* $\mathcal{h}$ *est simple* (*i.e. sans idéal bilatère non trivial*).

Ce corollaire est très important si on le rapproche de la caractérisation suivante des facteurs de type II_1 ne possédant *pas* la propriété Γ :

THEOREME 5. ([9]). - *Soit* M *un facteur de type* II_1, *à prédual de type dénombrable, opérant dans* $L^2(M,\mathfrak{J}) = \mathcal{h}$ *et soit* $C^*(M,M')$ *l'algèbre stellaire engendrée par* M *et* M' *alors* :

M *n'a pas la propriété* $\Gamma \Leftrightarrow C^*(M,M')$ *contient l'idéal* $k(\mathcal{h})$ *des opérateurs compacts.*

Un exemple de facteur de type II_1 pour lequel $C^*(M,M')$ contient l'idéal $k(\mathcal{h})$ avait été donné auparavant par C. Akemann et P. Ostrand dans [1]. Ainsi, joint au corollaire le théorème montre que tout facteur semi-discret de type II_1, à la propriété Γ.

En fait, dans leur article [24], E. Effros et C. Lance démontraient que tout facteur d'Araki-Woods est semi-discret et l'implication :

Semi-discret $\Rightarrow$ Injectif.

On peut donc résumer les relations entre les diverses propriétés relatives à l'approximation par des algèbres de dimension finie sous forme d'un diagramme :

Approximativement de dimension finie

$\Downarrow$

Propriété P de Schwartz

$\Downarrow$ Semi discret $\swarrow$

Propriété E de Hakeda-Tomiyama = Injectivité

Heureusement la situation est en fait remarquablement simple :

THEOREME 6 ([9]). - Soit h un espace hilbertien de type dénombrable. Pour toute algèbre de von Neumann opérant dans h les quatre propriétés ci-dessus sont équivalentes.

Nous remettons au paragraphe suivant la description des corollaires de ce théorème relatifs au problème de classification. Commençons par un problème de terminologie : on dispose à coup sûr de la bonne classe d'algèbres de von Neumann pour la théorie non commutative de l'intégration vu que d'après le théorème de O. Maréchal, celle-ci doit traiter au moins le cas "approximativement de dimension finie" et d'après les résultats de Effros et Lance le cas "propriété E" suffit à rendre compte

de toutes les algèbres de von Neumann associées aux algèbres stellaires nucléaires. Dans [9] nous avons adopté la terminologie "algèbres de von Neumann injectives" pour désigner la classe ci-dessus, parmi les avantages de ce choix se trouve surtout la simplicité de la définition grâce à la propriété E . Mais cette terminologie a l'inconvénient de ne rendre compte ni qu'il s'agit d'une propriété d'approximation, ni de l'analogie avec la moyennabilité des groupes discrets.
La solution semble donc de choisir la terminologie "algèbres de von Neumann moyennables" qui heureusement est justifiée par l'équivalence entre les quatre propriétés ci-dessus et la cinquième :

DEFINITION 7. - Une algèbre de von Neumann M est moyennable si et seulement si pour tout bimodule de Banach dual normal X sur M toute dérivation de M à coefficients dans X est intérieure.

Nous renvoyons aux articles de Johnson Kadison et Ringrose qui ont établi les bases de la cohomologie des algèbres de von Neumann à coefficients dans des bimodules de Banach ([34] [35]).
Ayant accepté le terme moyennable pour désigner notre classe d'algèbres de von Neumann, on a alors le corollaire suivant, conséquence facile de [24] et [9] (cf. [4]) :

COROLLAIRE 8. - Soit A une algèbre stellaire de type dénombrable, alors A nucléaire $\Leftrightarrow$ l'algèbre de von Neumann $\pi_\Phi(A)''$ engendrée par A est moyennable pour tout état Φ sur A.

Dans [34] B. Johnson a introduit la notion d'algèbre stellaire moyennable, par l'analogue de la définition 7 en enlevant le qualificatif normal. U. Haagerup a réussi grâce à sa démonstration remarquable de l'inégalité de Grothendieck pour les algèbres stellaires arbitraires (qui complète les travaux de Grothendieck et Pisier), à montrer qu'une algèbre stellaire est moyennable si et seulement si elle est nucléaire (cf. [17] et [31], [32])

IX. CLASSIFICATION DES FACTEURS MOYENNABLES ET LE PROBLEME III_1.

Dans tout ce paragraphe toutes les algèbres de von Neumann étudiées sont supposées avoir un prédual de type dénombrable. Le seul obstacle qui reste, à la compréhension idéale des algèbres de von Neumann moyennables est le suivant :

Problème : Existe-t'il à isomorphisme près un seul facteur moyennable de type III_1

Nous allons traiter tous les autres cas un par un et nous reviendrons ensuite à ce problème.

Facteurs de type II_1. Il existe à isomorphisme près un seul facteur moyennable de type II_1, c'est R, le facteur hyperfini. En fait, on en a la caractérisation suivante, qui justifie la terminologie hyperfini et la terminologie moyennable.

THEOREME 1. ([9]). - Soient $\mathcal{h}$ un espace hilbertien de type dénombrable et N un facteur (de dimension infinie) dans $\mathcal{h}$ alors :

N isomorphe à $R \Leftrightarrow \exists\, \Phi$ état sur $\mathcal{L}(\mathcal{h})$ tel que

$$\Phi(xT) = \Phi(Tx) \quad \forall\, x \in N\,,\ T \in \mathcal{L}(\mathcal{h})\ .$$

Cela justifie le terme "hyperfini" car Φ est mieux qu'une trace sur N. Cela justifie le terme "moyennable" et en fait l'analogie avec la "moyennabilité" d'un groupe discret est très utile. Un tel groupe G est moyennable quand il existe un état ψ sur $\ell^\infty(G)$ invariant par translation. Ici le rôle de G est joué par N, celui de $\ell^\infty(G)$ par l'algèbre de von Neumann $\mathcal{L}(\mathcal{h})$ où $\mathcal{h} = L^2(N,\mathfrak{J})$ et

l'existence d'une hypertrace équivaut à l'existence d'une moyenne invariante.
A la condition de Følner caractérisant les groupes discrets moyennables :
$\forall\, g_1,\dots,g_n \in G$, $\forall\, \varepsilon > 0$, $\exists\, F$ partie finie de G avec

$$\|\chi_F - g_i\chi_F\|_2 \leq \varepsilon \|\chi_F\|_2$$

où χ_F désigne la fonction caractéristique de F et $\|\ \|_2$ est la norme de l'espace $\ell^2(G)$, correspond la condition suivante sur N agissant dans $\mathfrak{h} = L^2(N,\mathcal{J})$:

$\forall\, x_1,\dots,x_n \in N$, $\forall\, \varepsilon > 0$, $\exists\, P$ projecteur de dimension finie dans $\mathfrak{h}$ avec :

$$\|x_i P - P x_i\|_{HS} \leq \varepsilon \|P\|_{HS}$$

où $\|T\|_{HS} = (\text{Trace}\ (|T|^2))^{\frac{1}{2}}$ désigne la norme de Hilbert Schmidt. C'est de cette condition que l'on déduit par exemple que N est alors semi-discret, on utilise ensuite les théorèmes de théorie ergodique non commutative pour aboutir au résultat : sur $N \otimes N$ la symétrie de Sakai $\sigma_N \in \text{Aut}(N \otimes N)$ définie par

$$\sigma_N(x \otimes y) = y \otimes x\ ,\ \forall\, x,y \in N \quad \text{vérifie} \quad \sigma_N \in \overline{\text{Int}}(N \otimes N)\ .$$

Nous renvoyons à [9] pour plus de renseignements. Citons maintenant quelques corollaires du théorème :

COROLLAIRE 2. - <u>Tout sous-facteur</u> N <u>de</u> R <u>est soit de dimension finie (i.e. isomorphe à</u> $M_n(\mathbb{C})$), <u>soit isomorphe à</u> R.

Ainsi R est <u>l'unique facteur</u> qui soit contenu dans tous les autres. De plus, et c'est là une conséquence de la <u>théorie de réduction</u>, on connaît à isomorphisme près toutes les sous-algèbres de von Neumann de R, ce sont les algèbres produit d'algèbres de von Neumann de la forme $C \otimes M_n(\mathbb{C})$, $n < \infty$ et $C \otimes R$ où C est abélienne.

Le facteur R restera sans doute le seul pour lequel une telle classification des sous-algèbres de von Neumann est possible.

COROLLAIRE 3. - Soit G un groupe discret dénombrable moyennable et $R(\lambda_G)$ l'algèbre de von Neumann de sa représentation régulière dans $\ell^2(G)$, alors $R(\lambda_G)$ est un produit d'algèbres $C \otimes M_n(\mathbb{C})$, $C \otimes R$, C abélienne.

En particulier tous les groupes discrets dénombrables résolubles n'ayant que des classes de conjuguaison infinies vérifient $R(\lambda_G)$ isomorphe à R.

Facteurs de type II_∞.

Il existe à isomorphisme près un seul facteur moyennable de type II_∞, c'est le facteur d'Araki et Woods $R_{o,1}$. Ceci résout le problème 3 et montre que $R_{o,1}$ est le seul facteur de type II_∞ qui soit approximativement de dimension finie.

La démonstration est très indirecte puisqu'on part de N de type II_∞ on l'écrit $N = M \otimes F$ où M est de type II_1 et F de type I_∞ et on utilise uniquement le fait que M hérite de N la propriété E pour en déduire que M est isomorphe à R, d'où N à $R_{o,1}$.

COROLLAIRE 4. - Soient G un groupe localement compact connexe de type dénombrable, λ la représentation régulière de G dans $L^2(G)$. Alors les seuls facteurs qui apparaissent dans la désintégration de $R(\lambda_G)$ sont soit de type I soit isomorphe à $R_{o,1}$.

Ce résultat découle du théorème ci-dessus et d'un théorème de Dixmier et Pukanszky qui montre qu'aucun facteur de type III n'intervient dans la désintégration de $R(\lambda_G)$.

Facteurs de type III_λ $\lambda \in]0,1[$. Il existe à isomorphisme près un seul facteur moyennable de type III_λ, c'est le facteur de Powers R_λ.

On montre en effet que si M est un facteur de type III_λ ayant la propriété E, le facteur N de type II_∞ qui lui est associé comme au paragraphe VI, vérifie aussi E. Ainsi $N = R_{o,1}$ et les résultats de théorie ergodique non commutative montrent qu'il y a sur $R_{o,1}$. une seule classe d'automorphismes de module λ d'où le résultat.

Facteurs de type III_o .

L'analyse des facteurs moyennables de type III_o résulte de 3 contributions différentes : 1) Les théorèmes d'unicité cités plus haut. 2) La théorie de dualité de Takesaki utilisée par l'invariant des facteurs qu'elle fournit : le flot des poids. 3) La théorie de Krieger sur l'équivalence faible des transformations. La discussion de 2) serait trop longue ici, disons simplement qu'à travers des problèmes précis comme celui de la distinction entre facteurs de Krieger et facteurs d'Araki-Woods, on est arrivé à affiner de mieux en mieux les invariants dans le cas III_o (voir [8] en particulier où on donne une décomposition de tout facteur de type III_o comme produit croisé d'une algèbre de von Neumann semi-finie par un automorphisme. C'est M. Takesaki qui avec sa théorie de la dualité a obtenu l'invariant sous sa forme précise : un groupe à un paramètre d'automorphismes d'une algèbre de von Neumann abélienne. C'est ce qui a conduit W. Krieger à associer à toute transformation T d'un espace (X,B,μ) un flot $W(T)$. Le résultat remarquable de W. Krieger est alors :

THEOREME. (W. Krieger [38]). - Soient (X_i,B_i,μ_i) , $i = 1,2$ un espace borélien standard probabilisé et T_i une transformation ergodique laissant μ_i quasi invariante alors, soit $M_i = L^\infty(R_{T_i},\mu_i)$ le facteur associé, on a :

$$\begin{array}{ccc} T_1 \text{ faiblement équivalente à } T_2 & & \\ \Updownarrow & \Leftrightarrow & W(T_1) \text{ isomorphe à } W(T_2) \\ M_1 \text{ isomorphe à } M_2 & & \end{array}$$

De plus les flots correspondants aux facteurs de type III_o sont exactement les flots ergodiques non transitifs.

Notre contribution se résume alors à :

THEOREME [9]. - Pour qu'un facteur de type III_o soit un facteur de Krieger (i.e. soit de la forme $L^\infty(R_T,\mu)$), il faut et il suffit qu'il soit moyennable.

Facteurs de type III_1 .

Dans [59], M. Takesaki a réussi à montrer, grâce à sa théorie de la dualité pour les produits croisés, que l'on avait pour les facteurs de type III_1 l'analogue exact du théorème décrivant les facteurs de type III_λ, $\lambda \in]0,1[$.

THEOREME [59] (M. Takesaki). - a) Soit M un facteur de type III_1 , il existe un facteur N de type II_∞ et un groupe à un paramètre $(\theta_t)_{t\in R}$ d'automorphismes de N , $\mathrm{mod}\,\theta_t = e^{-t}$, tel que M soit isomorphe au produit croisé de N par les θ_t .

b) Soient N un facteur de type II_∞ et $(\theta_t)_{t\in R}$ un groupe à un paramètre d'automorphismes de N vérifiant $\mathrm{mod}(\theta_t) = e^{-t}$ $\forall\, t$, alors le produit croisé de N par ce groupe est un facteur de type III_1 .

c) Deux couples (N_i, θ_i) comme dans b) donnent des facteurs isomorphes ssi les groupes à un paramètre θ_i sont conjugués par un isomorphisme de N_1 sur N_2 .

En outre, nous avons montré avec M. Takesaki, dans [13], que l'on avait là aussi une classe de poids privilégiée, les poids dominants, unique à automorphisme intérieur près, dont l'algèbre de von Neumann N est le centralisateur. De plus pour que M soit un facteur moyennable, il faut et il suffit que l'algèbre de type II_∞ associée N soit moyennable. On voit donc que le problème de classification des facteurs moyennables de type III_1 revient à la théorie ergodique non commutative des flots d'automorphismes de $R_{0,1}$.

Pour le moment, l'élaboration de cette théorie se heurte à une difficulté sérieuse : soit $(\theta_t)_{t\in R}$ un groupe à un paramètre d'automorphismes disons du facteur fini N , alors la propriété de continuité de l'homomorphisme $t \to \theta_t$ de R dans $\mathrm{Aut}\,N$ qui est toujours requise, n'entraîne pas la continuité de $t \to \theta_t^\omega$ de R dans $\mathrm{Aut}\,N^\omega$, ω ultrafiltre sur N . Cependant on peut, par un chemin détourné, arriver à montrer que tous les facteurs moyennables de type III_1 qui vérifient la condition supplémentaire suivante sont isomorphes au facteur R_∞ d'Araki Woods de type III_1 :

C : Il existe un état Φ normal sur M tel que :

$$(\psi \in M_*, \|[\psi, x_n]\| \to 0 \text{ pour toute suite bornée d'éléments de } M \text{ telle que } \|[\Phi, x_n]\| \to 0) \Rightarrow (\psi = \lambda\Phi \text{ pour un } \lambda \in \mathbb{C}) .$$

Il est facile de montrer que si cette propriété est vraie pour un état normal Φ sur M elle est vraie pour tous, en utilisant [10]. Bien entendu elle est vraie pour $M = R_\infty$ et nous conjecturons qu'elle est vraie pour tout facteur de type III_1.

BIBLIOGRAPHIE.

[1] C. AKEMANN et P. OSTRAND. On a tensor product C^* algebra associated with the free group on two generators (Preprint).

[2] H. ARAKI et E.J. WOODS. A classification of factors. (Publ. Res. Inst. Math. Sci. Kyoto Univ.) t.4. (1968) p. 51-130.

[3] W. ARVESON Subalgebras of C^* algebras Acta Math 123 (1969) 141-224.

[4] M. CHOI et E. EFFROS Separable nuclear C^* algebras and injectivity, (à paraître).

[5] F. COMBES Poids associé à une algèbre hilbertienne à gauche. (Compos. Math. t. 23 (1971) p. 49-77).

[6] A. CONNES Sur la classification des facteurs de type II . (C.R. Acad. Sci. Paris t. 281 (1975) p. 13-15).

[7] A. CONNES Almost periodic states and factors of type III_1 . J. Funct. Analysis 16 (1974) p. 415-445.

[8] A. CONNES Une classification des facteurs de type III . (Ann. Scient. Ecole Norm. Sup. 4ème Serie tome 6 fasc. 2 (1973) p. 133-252).

[9] A. CONNES Classification of injective factors. (Annals of Math. 104 (1976) p. 73-115).

[10] A. CONNES et E. STØRMER Homogeneity of the state space of factors of type III_1 . (Preprint).

[11] A. CONNES et E. STØRMER Entropy for automorphisms of II_1 von Neumann algebras. (Acta Math. t. 134 (1975) p. 289-306).

[12] A. CONNES Outer conjugacy classes of automorphisms of factors. (Annales Scient. Ecole Norm. Sup.).

[13] A. CONNES et M. TAKESAKI The flow of weights on factors of type III . (A paraître).

[14] A. CONNES The von Neumann algebra of a foliation Lecture Notes in Physics 80 (1978) p.145-151

[15] A CONNES, J. FELDMAN, B. WEISS Amenable equivalence relations are generated by a single transformation, Preprint.

[16] A. CONNES, W. KRIEGER Measure space automorphisms, the normalizer of their full groups and approximate finiteness, J. Funct. Analysis 29 (1977) p.336.

[17] A. CONNES On the cohomology of operator algebras, J. Funct. Analysis, Vol.28, N°2 (1978) p.248-253.

[18] A. CONNES A factor of type II_1 with countable fundamental group. J. Operator theory 4 (1980) p.151-153.

[19] J. DIXMIER Formes linéaires sur un anneau d'opérateurs (Bull. Soc. Math. France 81 (1953 p. 9-39).

[20] J. DIXMIER Les algèbres d'opérateurs dans l'espace hilbertien, 2ème édition Paris, Gauthier-Villars 1969.

[21] J. DIXMIER — Les C^* algèbres et leurs représentations Paris Gauthier Villars.

[22] D. Mc. DUFF — Central sequences and the hyperfinite factor Proc. London Math. Soc. XXI (1970) p. 443-461.

[23] H. DYE — On groups of measure preserving transformations I (Am. J. of Math. t. 81 (1959). p. 119-159) et II (t. 85 (1963) p. 551-576.

[24] E. EFFROS et C. LANCE — Tensor products of operator algebras. (A paraître).

[25] G. ELLIOTT et E. J. WOODS — The equivalence of various definitions of hyperfinitences of a property infinite von Neumann algebra (preprint).

[26] G. ELLIOTT — On the classification of inductive limits of sequences of semi-simple finite dim. algebras. J. of Algebra 38 (1976), p. 29-44.

[27] J. FELDMAN et C. MOORE — Ergodic equivalence relations, cohomology and von Neumann algebras I, (à paraître).

[28] T. Giordano et V. Jones — Antiautomorphismes involutifs du facteur hyperfini de type II_1, C.R. Acad. Sci. Série A, 290, (1980).

[29] V. Ya. GOLODETS — Cross products of von Neumann algebras Y. Math. N. 26 N° 5 (1971) p. 3-50, voir (A. Connes, P. Ghez, R. Lima, D. Testard et E.J. Woods : Review of a paper of Golodets).

[30] U. HAAGERUP — Normal weights on W^* algebras. (J. Funct. Analysis t. 19 (1975) p. 302-317).

[31] U. Haagerup — The Grothendieck inequality for bilinear forms on C* algebras. (Preprint)

[32] U. Haagerup — Nuclear C* algebras are amenable. (Preprint).

[33] J. HAKEDA et J. TOMIYAMA — On some extension properties of von Neumann algebras. (Tohoku Math. J. t. 19 (1967) p. 315-323).

[34] B. JOHNSON — Cohomology in Banach Algebras. Memoir A.M.S. 127 (1972).

[35] B. JOHNSON, R. V. KADISON et J. RINGROSE — Cohomology in operator algebras III, Bull. Soc. Math. France 100 (1972) p. 73-96.

[36] V. Jones — Actions of finite groups on the hyperfinite II_1 factor (Thèse à paraître dans les Memoirs A.M.S.).

[37] W. KRIEGER — On the Araki-Woods asymptotic ratio set and non singular transformations of a Measure space. (Dans "contributions to ergodic theory and probability, Lecture notes n° 160 (1970)).

[38] W. KRIEGER — On ergodic flows and the isomorphism of factors. (Math. Ann. 223 (1976) p. 19-70).

[39] O. MARECHAL — Une remarque sur un théorème de Glimm. (Bull. Soc. Math. France 2ème Série 99 (1975) p. 41-44).

[40] F.J. MURRAY et J. von NEUMANN — On rings of operators. (Ann. of Math. t. 37 (1936) p. 116-229).

[41] F.J. MURRAY et J. von NEUMANN — On rings of operators, II (Trans. Amer. Math. Soc. t. 41 (1937) p. 208-248).

[42] J. von NEUMANN — On rings of operators III. (Ann. of Math. t. 41 (1940) p. 94-161).

[43] F.J. MURRAY et J. von NEUMANN — On rings of operators IV. (Ann. of Math. t. 44 (1943) p. 716-808).

[44] J. von NEUMANN — On rings of operators : Reduction theory. (Ann. Math. t. 50 (1949) p. 401-485).

[45] A. OCNEANU — C.R. Acad. Science Paris, t.291 (1980)

[46] D. ORNSTEIN, B. WEISS — Ergodic theory of amenable group actions I. The Rohlin lemma. Bull. A.M.S., Vol.2, N°1 (1980), p.161.

[47] G. PEDERSEN — Measure theory for C* algebras. Math. Scand. 19 (1966) et 22 (1968), 25 (1969)

[48] J. PHILLIPS — Automorphisms of full II_1 factors with applications to type III factors.

[49] R.T. POWERS — Representations of uniformly hyperfinite algebras and their associated von Neumann rings. (Ann. of Math. t. 86 (1967) p. 138-171).

[50] P. SCHIELDS — The theory of Bernouilli shifts (Chicago press).

[51] S. SAKAI — Automorphisms and tensor products of operator algebras (à paraître dans American Journal of Math).

[52] S. SAKAI — On automorphism groups of type II factors Tohôku Math. J. 26 (1974) p. 423-430.

[53] S. SAKAI — C^* and W^* algebras. Ergebnisse der Mathematik und ihrer Grenzgebiete, Band 60.

[54] J. SCHWARTZ — Two finite, non hyperfinite, non isomorphic factors. Comm. Pure Appl. Math. 16 (1963) 19-26.

[55] I. SEGAL — A non commutative extension of abstract integration. Ann. of Math. 57 (1953) p. 401-457.

[56] E. STØRMER — Real structure in the hyperfinite factor (Preprint)

[57] M. TAKESAKI On the cross norm of the direct product of C^* algebras. Tohoku Math. J. 16 (1964) 111-122.

[58] M. TAKESAKI Tomita's theory of modular Hilbert algebras and its applications. (Lecture notes N° 128, Springer Verlag 1970).

[59] M. TAKESAKI Duality in cross products and the structure of von Neumann algebras of type III. Acta Math. 131 (1973) p. 249-310.

[60] J. TOMIYAMA On the projection of norm one in W^* algebras. Proc. Japan Acad. 33 (1957) 608-612.

[61] J. VESTERSTRØM Quotients of finite W^* algebras (J. Funct. Analysis t. 9 (1972) p. 322-335).

[62] F.J. YEADON A new proof of the existence of a trace on a finite von Neumann algebra. (Bull. Am. Math. Soc. t.77 (1971), p.257-260).

[63] R. ZIMMER Hyperfinite factors and amenable ergodic actions. Inv. Math. 41 (1977), p.23-31.

[64] R. ZIMMER Strong rigidity for ergodic actions of semi simple Lie groups. Annals of Math.

Proceedings of Symposia in Pure Mathematics
Volume 38 (1982), Part 2

AUTOMORPHISMS AND VON NEUMANN ALGEBRAS OF TYPE III

Masamichi Takesaki*

ABSTRACT. This article deals with the structure theory of von Neumann algebras and the flow of weights.

0. INTRODUCTION. Since the Baton Rouge Conference in March 1977, remarkable progress has been achieved in operator algebras. The impact of that conference is by now clearly visible. In my opinion, the most important aspect of the meeting was that it provided a sound foundation for an international community of operator algebraists. It is my hope that this Summer Institute will serve a similar purpose, by bringing together the many new researchers in the area, and that once again it will lead to further advances. In my talk I will consider but a few of the great strides that have been made in the last fifteen years.

At the Baton Rouge Conference, R. Powers reported that he had succeeded in distinguishing continuously many non-isomorphic factors of type III. The existence of such a family of factors had been expected, but the proof had eluded mathematicians for over two decades. Powers' work immediately sparked research in the classification theory of factors. On the one hand, Araki and Woods developed a theory for infinite tensor products of type I factors,and on the other, McDuff finally proved that there are continuously many factors of type II_1 and of type II_∞.

Two important developments occurred outside the lectures at the Baton Rouge Conference. In the first of these, R. Haag, N.M. Hughenholtz, and M. Winnink circulated a preprint of their theory of equilibrium states in quantum statistical mechanics. In this paper they introduced the Kubo-Martin-Schwinger boundary condition in an operator algebraic setting. This was used in their study of the representations of the C^*-algebras of observables induced by an equilibrium temperature state. Secondly, M. Tomita distributed his famous, and still unpublished mimeographed notes on modular Hilbert algebras, in which it was asserted that a von Neumann algebra $\{\mathfrak{M},\mathfrak{H}\}$ equipped with a separating cyclic vector ξ, is standard. The physicists gathered at the conference observed the strong similarity between the Haag-Hugenholtz-Winnink theory and that of Tomita. Thus at Baton Rouge an international collaboration of operator algebraists and mathe-

*The author is supported in part by the NSF Grant

matical physicists was begun on what turned out to be the most important results in von Neumann algebra theory since its inception. Within a year it was discovered that these two theories are the same, and that Tomita's theory provided a foundation for operator algebraic quantum statistical mechanics. It was further recognized that Tomita's theory was essential to an understanding of non-commutative integration theory in more general settings than that of J. Dixmier and I.E. Segal. Additional contributions to the subject, such as the theory of weights of G.K. Pedersen, and F. Combes, have led to our seemingly ever deepening knowledge of non-commutative analysis.

The classification theory of factors of type III started by R. Powers was further developed by H. Araki and E.J. Woods by means of two algebraic invariants: the asymptotic ratio set and the ρ-set. This direction eventually led to the theory of W. Krieger. It was A. Connes who recognized in 1971 that the Araki-Woods invariants may be derived from the new noncommutative integration theory. Thus two of the major achievements made after the Baton Rouge Conference were finally united. Subsequently, even faster progress occurred in the subject. It seems, however, that the development after 1975 in von Neumann algebra theory has somewhat levelled. This is now a good time to marshall our recources and to prepare for the next big drive.

1. WEIGHTS ON VON NEUMANN ALGEBRAS. An extended positive real valued additive and homogeneous function φ on the positive cone A_+ of a C^*-algebra A:

$$\varphi(x + y) = \varphi(x) + \varphi(y), \qquad x,y \in A_+ ;$$
$$\varphi(\lambda x) = \lambda\varphi(x), \qquad \lambda \in \mathbb{R}_+ , \quad x \in A_+ ,$$

is called a <u>weight</u> on A. The subset $\{x \in A_+ : \varphi(x) < +\infty\}$ of A_+ where φ takes finite value turns out to be the positive part $\mathfrak{m}_\varphi^+$ of a hereditary self-adjoint subalgebra $\mathfrak{m}_\varphi$ of A where φ is extended to a complex valued linear functional. We call $\mathfrak{m}_\varphi$ the definition domain of φ.

The theory of weights was developed by F. Combes and G. K. Pedersen during the period of 1965 through 1968 mostly for C*-algebras, [6, 36, 55]. The notion of a weight is a simultaneous generalization of those of states and traces. Their work were inspired by a work of J. Dixmier, [20], 1963, on traces on C*-algebras, while one can trace this concept back to an old problem in Dixmier's book, 1957 and to M. Tomita's earlier work, [56], 1959.

In the von Neumann algebra context, one assumes the normality for a weight φ on a von Neumann algebra $\mathfrak{M}$:

$$\varphi(\sup x_i) = \sup \varphi(x_i)$$

for every bounded increasing net $\{x_i\}$ in $\mathfrak{M}_+$; the <u>semi-finiteness</u>: $\mathfrak{m}_\varphi$ is σ-weakly dense in $\mathfrak{M}$. Hereafter, <u>a weight means exclusively a semi-finite</u>

normal weight on a von Neumann algebra.

Given a weight φ on $\mathfrak{M}$, set

$$\mathfrak{n}_\varphi = \{x \in \mathfrak{M} : \varphi(x^*x) < +\infty\}, \qquad N_\varphi = \{x \in \mathfrak{M}_+ : \varphi(x^*x) = 0\} .$$

The inequalities:

$$(x + y)^*(x + y) \leq 2(x^*x + y^*y) , \qquad (ax)^*(ax) \leq \|a\|^2 x^*x$$

guarantee that $\mathfrak{n}_\varphi$ and N_φ are left ideals of $\mathfrak{M}$. It turns out that the definition domain $\mathfrak{m}_\varphi$ of φ is nothing but $\mathfrak{n}_\varphi^*\mathfrak{n}_\varphi$. Let $\eta_\varphi(x) = x + N_\varphi \in \mathfrak{n}_\varphi/N_\varphi$ for each $x \in \mathfrak{n}_\varphi$ and

$$(\eta_\varphi(x) \mid \eta_\varphi(y)) = \varphi(y^*x) .$$

It follows that the left $\mathfrak{M}$-module $\mathfrak{n}_\varphi/N_\varphi$ is a pre-Hilbert space and the completion $\mathfrak{H}_\varphi$ accomodates a representation π_φ of $\mathfrak{M}$ such that

$$\pi_\varphi(a)\eta_\varphi(x) = \eta_\varphi(ax) , \qquad a \in \mathfrak{M}, \quad x \in \mathfrak{n}_\varphi .$$

The representation $\{\pi_\varphi, \mathfrak{H}_\varphi, \eta_\varphi\}$ is called the semi-cyclic representation of $\mathfrak{M}$. This is nothing else but a modified version of the well-known Gelfand-Naimark-Segal construction. The normality of φ guarantees that of π_φ; hence $\pi_\varphi(\mathfrak{M})$ is a von Neumann algebra on $\mathfrak{H}_\varphi$.

Assume now that φ is faithful:

$$\varphi(x^*x) > 0 \quad \text{for every } x \neq 0 .$$

Set $\mathfrak{a}_\varphi = \mathfrak{n}_\varphi \cap \mathfrak{n}_\varphi^*$ and $\mathfrak{A}_\varphi = \eta_\varphi(\mathfrak{a}_\varphi) \subset \mathfrak{H}_\varphi$. The distinction of $\mathfrak{a}_\varphi$ and $\mathfrak{A}_\varphi$ is simply to avoid confusion: the former is a self-adjoint hereditary subalgebra of $\mathfrak{M}$ and the latter is an algebra sitting in the Hilbert space $\mathfrak{H}_\varphi$. It follows that $\mathfrak{A}_\varphi$, equipped with the algebraic structure inherited from $\mathfrak{a}_\varphi$, is a left Hilbert algebra in the sense that

1) $(\xi\eta \mid \zeta) = (\eta \mid \xi^\#\zeta)$;
2) For each $\xi \in \mathfrak{A}_\varphi$, the map $\pi_\ell(\xi) : \eta \in \mathfrak{A}_\varphi \mapsto \xi\eta \in \mathfrak{A}_\varphi$ is bounded;
3) The subalgebra $\mathfrak{A}_\varphi^2$ is dense in $\mathfrak{A}_\varphi$;
4) The involution: $\xi \in \mathfrak{A}_\varphi \mapsto \xi^\# \in \mathfrak{A}_\varphi$ is preclosed.

The closure is denoted by S and its adjoint by F.

The last condition, the preclosedness of the involution, is crucial to the whole theory. This follows from the fact that if φ is a normal weight, then

a) $\Phi = \{\omega \in \mathfrak{M}_*^+ \ (1 + \varepsilon)\omega \leq \varphi$ for some $\varepsilon > 0\}$ is upward directed;
b) $\varphi(x) = \sup\{\omega(x) : \omega \in \Phi\}$, $x \in \mathfrak{M}_+$.

Result (a) was obtained by F. Combes, adapting an argument of G.K. Pedersen. The second result, (b), came a little later. Until U. Haagerup solved the problem in 1973, [23], statement (b) was taken as the definition of a normal weight.

It turns out that the left Hilbert algebra $\mathfrak{A}_\varphi$ is full and the left von Neumann algebra $\mathfrak{R}_{\ell}(\mathfrak{A}_{\varphi}) = \pi_{\ell}(\mathfrak{A}_{\varphi})''$ is exactly $\pi_{\varphi}(\mathfrak{M})$.

Conversely, if $\mathfrak{A}$ is a full left Hilbert algebra, then

$$\varphi(x) = \begin{cases} \|\xi\|^2 & \text{if } x = \pi_{\ell}(\xi)^*\pi_{\ell}(\xi), \quad \xi \in \mathfrak{A}, \\ +\infty & \text{otherwise}, \end{cases}$$

defines a faithful weight on $\mathfrak{R}_{\ell}(\mathfrak{A})$ such that $\mathfrak{A}$ is isomorphic to $\mathfrak{A}_{\varphi}$.

This correspondence of weights and full left Hilbert algebras was established independently by F. Combes [7] and M. Takesaki [49] both in 1970, following Tomita's theory of modular Hilbert algebras.

If $\mathfrak{M}$ is a factor of type I, say $\mathfrak{M} = \mathfrak{L}(\mathfrak{K})$, then every weight φ on $\mathfrak{M}$ is of the form:

$$\varphi(x) = \mathrm{Tr}(x^{1/2} h x^{1/2}), \quad x \in \mathfrak{M}_+,$$

with h a positive self-adjoint (not necessarily bounded) operator on $\mathfrak{K}$. The correspondence of φ and h is bijective. Therefore, the study of weights includes in principle that of positive self-adjoint operators. This fact is also true for a semi-finite von Neumann algebra.

Some of the main difficulties in the study of unbounded operators is lack of algebraic structure in the subject. The same applies for weights. However, we will see that there is a rich structure in weights.

2. MODULAR AUTOMORPHISM GROUPS. The major consequence of Tomita's theory is summarized in the following form:

THEOREM 2.1. Given a left Hilbert algebra $\mathfrak{A}$ with left von Neumann algebra $\mathfrak{M} = \mathfrak{R}_{\ell}(\mathfrak{A})$, let

$$S = J\Delta^{1/2}, \quad \Delta = FS$$

be the polar decomposition of the closure of the involution: $\xi \in \mathfrak{A} \mapsto \xi^{\#} \in \mathfrak{A}$. Then one has

$$J\mathfrak{M}J = \mathfrak{M}'$$

$$\Delta^{it}\mathfrak{M}\Delta^{-it} = \mathfrak{M}, \quad t \in \mathbb{R}.$$

Therefore, if we set

$$\sigma_t(x) = \Delta^{it} x \Delta^{-it}, \quad x \in \mathfrak{M}, \quad t \in \mathbb{R},$$

then $\{\sigma_t\}$ is a one parameter automorphism group of $\mathfrak{M}$, called the <u>modular automorphism group</u>.

To characterize the modular automorphism group, one needs the following condition which was imported from theoretical quantum statistical mechanics by R. Haag, N.M. Hugenholtz and M. Winnink, [22] in 1967 at the same time Tomita's work caught attention from operator algebraists and mathematical physicists:

DEFINITION 2.2. Let $\{\alpha_t\}$ be a one parameter automorphism group of a C^*-algebra A, (where the continuity of $\{\alpha_t\}$ is not assumed). A lower semi-continuous (in norm) weight φ on A is said to satisfy the modular condition (Kubo Martin Schwinger condition) with respect to $\{\alpha_t\}$ if

a) $\varphi \circ \alpha_t = \varphi$, $t \in \mathbb{R}$;

b) For each x, y $\mathfrak{n}_\varphi \cap \mathfrak{n}_\varphi^*$, the functions: $t \in \mathbb{R} \to \varphi(\alpha_t(x)y) = f(t) \in \mathbb{C}$ and $t \in \mathbb{R} \to \varphi(y\alpha_t(x)) = g(t) \in \mathbb{C}$ are continuous and boundary values of an $F \in H^\infty(\mathbb{D})$ in the sense that

$$F(t) = \varphi(\alpha_t(x)y) , \qquad F(t + i) = \varphi(y\alpha_t(x)) ,$$

where

$$\mathbb{D} = \{z \in \mathbb{C} : 0 < \operatorname{Im} z < 1\} .$$

The next result made clear the meaning of Tomita's theory in the context of quantum statistical mechanics:

THEOREM 2.3. A faithful weight φ on $\mathfrak{M}$ satisfies the modular condition with respect to the modular automorphism group $\{\sigma_t\}$ associated with the full left Hilbert algebra $\mathfrak{A}_\varphi$. Such a one parameter automorphism group is unique subject to the modular condition.

Therefore, each faithful weight φ gives rise uniquely to a one parameter automorphism group, which will be denoted by $\{\sigma_t^\varphi\}$, and called the modular automorphism of φ.

To find out the modular automorphism group in visible form, let us consider the case that $\mathfrak{M} = \mathfrak{L}(\mathfrak{H})$. Every faithful weight φ is of the form

$$\varphi(x) = \operatorname{Tr}(x^{1/2} h_\varphi x^{1/2}) , \qquad x \in \mathfrak{M}_+ ,$$

with h_φ a non-singular positive self-adjoint operator on $\mathfrak{H}$. It then follows that

$$\mathfrak{n}_\varphi = \{x \in \mathfrak{L}(\mathfrak{H}) : x^*\mathfrak{H} \subset D(h^{1/2}) \quad \text{and} \quad h^{1/2}x^* \in \mathfrak{H}\mathfrak{S}(\mathfrak{H})\} .$$

The map $U : xh^{1/2} \equiv (h^{1/2}x^*)^* \in \mathfrak{H}\mathfrak{S}(\mathfrak{H}) \to \eta_\varphi(x) \in \mathfrak{H}_\varphi$ provides an isometry implementing the unitary equivalence of the left multiplication representation of $\mathfrak{M}$ on $\mathfrak{H}\mathfrak{S}(\mathfrak{H})$ and π_φ. The modular automorphism group $\{\sigma_t^\varphi\}$ is given by the following:

$$\sigma_t^\varphi(x) = h_\varphi^{it} x h_\varphi^{-it} , \qquad x \in \mathfrak{M}, \quad t \in \mathbb{R} .$$

This phenomena is not special for the type I case as the next result indicates:

THEOREM 2.4. For a von Neumann algebra $\mathfrak{M}$, the following conditions are equivalent:

a) $\mathfrak{M}$ is semi-finite;

b) There exists a faithful weight φ whose modular automorphism group is globally inner in the sense that it is implemented by a continuous one parameter unitary group contained in $\mathfrak{M}$;

c) The modular automorphism group of every faithful weight is globally inner.

In the separable case, Kadison's result [28], modified by R. Kallman [29], shows that the individual innerness of $\{\sigma_t^\varphi\}$ is sufficient to yield the global innerness of $\{\sigma_t^\varphi\}$.

To know the meaning of modular automorphism groups, the following result is somewhat illustrative:

PROPOSITION 2.5. Let $\{\mathfrak{M}_n, \varphi_n\}$ be a sequence of von Neumann algebras equipped with faithful normal states. In the infinite tensor product:

$$\{\mathfrak{M}, \varphi\} = \prod_{n=1}^{\infty}{}^{\otimes}\{\mathfrak{M}_n, \varphi_n\} ,$$

the modular automorphism group $\{\sigma_t^\varphi\}$ is given by the infinite tensor product $\prod_{n=1}^{\infty} \otimes \sigma_t^{\varphi_n}$.

If $\mathfrak{M} = M(k_n : \mathbb{C})$ and φ_n is given by a matrix

$$h_n = \begin{pmatrix} \lambda_1^n & & 0 \\ & \ddots & \\ 0 & & \lambda_{k_n}^n \end{pmatrix} \in M(k_n, \mathbb{C}) , \qquad \sum_{i=1}^{k_n} \lambda_i^n = 1 ,$$

then one has $\sigma_t^{\varphi_n} = \mathrm{Ad}(h_n^{it})$. Although $\prod_{n=1}^{\infty} \otimes h_n^{it}$ does not exist at all in general, $\prod_{n=1}^{\infty}{}^{\otimes}\mathrm{Ad}(h_n^{it})$ makes perfect sense.

Here, one should recall the following historic fact: Heisenberg commutation relation asserts that the fundamental quantities P, Q in quantum physics are self-adjoint operators on a Hilbert space such that

$$PQ - QP = i1 .$$

However, it is not easy to handle the pair P and Q beyond certain formal calculations. By exponentiating P and Q:

$$U(t) = \exp itP , \qquad V(s) = \exp isQ ,$$

the above commutation relation is translated into the new Heisenberg-Weyl commutation relation:

$$U(t)V(s)U(t)^* V(s)^* = e^{ist} 1 .$$

In this form, Stone and von Neumann showed that the solution to the above relation is essentially unique up to multiplicity. If one goes one step further considering

$$\alpha_t = \mathrm{Ad}\, U(t) , \qquad \beta_s = \mathrm{Ad}\, V(s) ,$$

Then the automorphism groups $\{\alpha_t\}$ and $\{\beta_s\}$ of $\mathcal{L}(\mathfrak{H})$ commute now despite the non-commutativity of $\{U(t)\}$ and $\{V(s)\}$. Furthermore, if φ and ψ are the weights given by $\exp P$ and $\exp Q$, then $\alpha_t = \sigma_t^{\varphi}$ and $\beta_s = \sigma_s^{\psi}$. In this setting, the von Neumann algebra generated by $U(t)$ and $V(s)$ is easily identified with $\mathcal{L}(L^2(\mathbb{R}))$ and

$$\begin{cases} U(s)\xi(r) = \xi(r-s) \\ V(t)\xi(r) = e^{itr}\,\xi(r) . \end{cases}$$

For each $f \in L^1(\mathbb{R}^2)$, set

$$W(f) = \iint_{\mathbb{R}^2} f(s,t)\,U(s)\,V(t)\,ds\,dt .$$

Then $W(f)$ is compact. Furthermore, if $f \in \mathcal{S}(\mathbb{R}^2)$, then $W(f)$ is nuclear. Thus, the dualization of the map: $f \in \mathcal{S}(\mathbb{R}^2) \mapsto W(f) \in \mathcal{L}\mathcal{T}(\mathfrak{H})$ gives rise to a description of all operators in $\mathcal{L}(L^2(\mathbb{R}))$ in terms of tempered distributions. In this way, one can relate the theory of weights to the classical Fourier analysis. . More detailed theory was developed by D. Kastler [30] and Loupias-Miracle Sole [33].

3. RADON-NIKODYM THEOREMS. The non-commutative generalization of the Radon-Nikodym theorem in measure theory attracted operator algebraists for quite a long time. The origin of the operator algebraic Radon-Nikodym theorem is the following result of Murray and von Neumann:

THEOREM 3.1. Given a von Neumann algebra $\{\mathfrak{M},\mathfrak{H}\}$, and vectors $\xi,\eta \in \mathfrak{H}$, η belongs to $[\mathfrak{M}\xi]$, the closed subspace spanned by $\{x\xi : x \in \mathfrak{M}\}$, if and only if there exist an operator $b \in \mathfrak{M}$ and a closed densely defined operator t affiliated with $\mathfrak{M}$ such that

$$\eta = bt\,\xi . \tag{3.1}$$

This theorem is called the BT theorem. This theorem yields the following comparability theorem of cyclic projections:

THEOREM 3.2. Given $\{\mathfrak{M},\mathfrak{H}\}$, the following two conditions for a pair ξ,η of vectors in $\mathfrak{H}$ are equivalent:

a) $[\mathfrak{M}\xi] \preccurlyeq [\mathfrak{M}\eta]$; b) $[\mathfrak{M}'\xi] \preccurlyeq [\mathfrak{M}'\eta]$.

COROLLARY 3.3. For a pair φ,ψ of two states on a C^*-algebra A, the following two conditions are equivalent:

a) The cyclic representation π_φ induced by φ is equivalent to a subrepresentation of the cyclic representation π_ψ induced by ψ;

b) The support $s(\varphi)$ in A^{**}, the second dual space which is a von Neumann algebra, is equivalent to a projection majorized by $s(\psi)$; i.e. $s(\varphi) \preccurlyeq s(\psi)$ in A^{**}.

The following result of Dye, [21], 1953, in the improved form by C. Skau [45], 1977, shows the significance of the Radon-Nikodym question:

THEOREM 3.4. Given $\{\mathfrak{M},\mathfrak{H}\}$, the following conditions for a vector $\xi \in \mathfrak{H}$ are equivalent:

a) $[\mathfrak{M}\xi]$ is finite in $\mathfrak{M}'$;

b) Every $\eta \in [\mathfrak{M}\xi]$ is of the form

$$\eta = t\xi \tag{3.2}$$

for some closed densely defined operator t affiliated with $\mathfrak{M}$.

It would be fun for the reader to deduce the classical Radon-Nikodym theorem from the above result. This is called the T-theorem. From this, Dye showed that if $\mathfrak{M}$ is finite and $s(\varphi) \leq s(\psi)$ for $\varphi,\psi \in \mathfrak{M}_*^+$, then there exists a closed densely defined operator t affiliated with $\mathfrak{M}$ such that $\varphi(x) = (xt\xi_\psi \mid t\xi_\psi)$, $x \in \mathfrak{M}$, where ξ_ψ is a vector giving rise to ψ. This can be written in the form:

$$\varphi(x) = \psi(txt) , \quad x \in \mathfrak{M} . \tag{3.3}$$

Immediately after Dye's work, J. Dixmier and I.E. Segal showed that if τ is a faithful semi-finite normal trace on $\mathfrak{M}$, then every $\varphi \in \mathfrak{M}_*$ is uniquely represented by a closed densely defined operator t affiliated with $\mathfrak{M}$ in such a way that

$$\varphi(x) = \tau(tx) , \quad x \in \mathfrak{M} , \quad \|\varphi\| = \tau(|t|) .$$

Therefore, $\mathfrak{M}_*$ for a semi-finite von Neumann algebra $\mathfrak{M}$ is identified with the Banach space $L^1(\mathfrak{M},\tau)$ of certain closed operators affiliated with $\mathfrak{M}$. J. Dixmier constructed $L^1(\mathfrak{M},\tau)$ in an abstract fashion, while I.E. Segal gave a concrete realization of $L^1(\mathfrak{M},\tau)$.

In 1965, S. Sakai showed that if $0 \leq \varphi \leq \psi \in \mathfrak{M}_*^+$, then there exists an operator $t \in \mathfrak{M}$, $0 \leq t \leq 1$, such that (3.3) holds. It turns out, [48], that this t is unique.

After the full recognition of Tomita's theory, it was natural to ask the Radon-Nikodym question for weights on von Neumann algebras. It turns out that Dye-Sakai type generalization of the Radon-Nikodym theorem for weights φ, ψ even with inequality $\varphi \leq \psi$ does not hold. But before going that far, one needed a criteria how to determine the equality of two given weights which behave alike. The following result of Pedersen and Takesaki is useful in many cases:

THEOREM 3.5. Let φ be a faithful weight on $\mathfrak{M}$ and ψ another weight on $\mathfrak{M}$ with $\psi \cdot \sigma_t^\varphi = \psi$. Then $\varphi = \psi$ if and only if there exists a σ-weakly dense $*$-subalgebra of $\mathfrak{m}_\varphi$, which is globally invariant under $\{\sigma_t^\varphi\}$ and on which φ and ψ agree.

One should compare this result with the extension problem of symmetric closed operators; namely with the domain question for unbounded operators.

In the same joint work, the following generalization of Dixmier-Segal's theorem was proved:

THEOREM 3.6. For two faithful weights φ and ψ, the following statements are equivalent:

a) $\psi \cdot \sigma_t^\varphi = \psi$, $t \in \mathbb{R}$;

b) There exists a non-singular positive self-adjoint operator h affiliated with the centralizer $\mathfrak{M}_\varphi$ of φ, the fixed point subalgebra of $\mathfrak{M}$ under $\{\sigma_t^\varphi\}$ such that

$$\psi(x) = \lim_{\varepsilon \to 0} \varphi(x^{1/2} h(1+\varepsilon h)^{-1} x^{1/2}), \qquad x \in \mathfrak{M}_+ ;$$

c) There exists a non-singular positive self-adjoint operator k affiliated with $\mathfrak{M}_\psi$ such that

$$\varphi(x) = \lim_{\varepsilon \to 0} \psi(x^{1/2} k(1+\varepsilon k)^{-1} x^{1/2}), \qquad x \in \mathfrak{M}_+ ;$$

d) $\varphi \cdot \sigma_t^\psi = \psi$, $t \in \mathbb{R}$.

If either φ or ψ is finite, then the commutativity of $\{\sigma_t^\varphi\}$ and $\{\sigma_s^\psi\}$ is also equivalent to the above conditions.

In 1972, there was much remarkable progress in the study of von Neumann algebras of type III. The Radon-Nikodym question was further enriched by H. Araki, [2]. He discovered that if $m\varphi \le \psi \le M\varphi$ for $\varphi, \psi \in \mathfrak{M}_*^+$ then there exists unique $a \in \mathfrak{M}$ such that $\psi(x) = \varphi(a^* x a)$ and $\sigma_{i/4}^\varphi(a) \in \mathfrak{M}_+$: namely a belongs to the domain of $\sigma_{i/4}^\varphi$. Furthermore, writing this $a = (d\psi/d\varphi)$, it satisfies the chain rule: $(d\psi/d\varphi)(d\varphi/d\rho) = d\psi/d\rho$ provided that φ, ψ and ρ are bounded by each other.

The final step in this direction was marked by A. Connes, [10]:

THEOREM 3.7. Let φ be a faithful weight on $\mathfrak{M}$.

a) If ψ is another faithful weight on $\mathfrak{M}$, then there exists a strongly continuous one parameter family $\{u_t\}$ of unitaries in $\mathfrak{M}$ such that

$$u_{s+t} = u_s \sigma_s^\varphi(u_t) ; \tag{3.4}$$

$$\sigma_t^\psi = \mathrm{Ad}(u_t) \cdot \sigma_t^\varphi ; \tag{3.5}$$

for every $x \in \mathfrak{n}_\varphi^* \cap \mathfrak{n}_\psi$ and $y \in \mathfrak{n}_\varphi \cap \mathfrak{n}_\psi^*$, there exists an $F \in H^\infty(\mathbb{D})$ such that

$$\begin{cases} F(t) = \varphi(\sigma_t^\varphi(x) u_t\, y) \\ F(t+i) = \psi(y \sigma_t^\varphi(x) u_t), \qquad t \in \mathbb{R} . \end{cases} \tag{3.6}$$

The family $\{u_t\}$ is uniquely determined by φ and ψ subject to (3.6). Thus we write $u_t = (D\psi : D\varphi)_t$,

b) For another faithful weight ρ on $\mathfrak{M}$, one has

$$(D\psi : D\rho)_t\,(D\rho : D\varphi)_t = (D\psi : D\varphi)_t\,, \qquad t \in \mathbb{R}\,. \tag{3.7}$$

c) If $\{u_t\}$ is a strongly continuous one parameter family of unitaries satisfying (3.4), then there exists uniquely a faithful weight ψ such that

$$u_t = (D\psi : D\varphi)_t\,.$$

In the case that $\mathfrak{M} = \mathcal{L}(\mathfrak{H})$ and

$$\varphi(x) = \mathrm{Tr}(xh)\,, \qquad \psi(x) = \mathrm{Tr}(xk)\,,$$

$(D\psi : D\varphi)_t$ is given by the following:

$$(D\psi : D\varphi)_t = k^{it}\,h^{-it}\,.$$

If one writes

$$h = e^{-H} \quad \text{and} \quad k = e^{-K}\,,$$

then

$$(D\psi : D\varphi)_t = e^{-itH}\,e^{itK}\,,$$

which looks familiar to scattering theorists in mathematical physics.

The last theorem shows the position of weights in the theory of von Neumann algebras. First of all, it says that the Radon-Nikodym theory cannot be self-consistent within the frame of positive linear functionals.

Now, the most important consequence of Theorem 3.7 is that the modular automorphism of a weight on $\mathfrak{M}$ is unique up to perturbations by unitary one cocycles. Namely, the crossed product $\mathfrak{M} \times_{\sigma^\varphi} \mathbb{R}$ of $\mathfrak{M}$ by $\{\sigma_t^\varphi\}$ does not depend on the choice of φ but only on $\mathfrak{M}$ itself. This consideration leads us to the following stage: crossed products!

4. CROSSED PRODUCTS AND DUALITY. We consider for a while a general locally compact abelian group G, and a von Neumann algebra $\mathfrak{M}$. Suppose that a homomorphism $\alpha : s \in G \to \alpha_s \in \mathrm{Aut}(\mathfrak{M})$ is given and continuous in the sense that $\lim_{s\to e} \|\varphi \cdot \alpha_s - \varphi\| = 0$ for every $\varphi \in \mathfrak{M}_*$. We call $\{\mathfrak{M}, G, \alpha\}$ a covariant system. Suppose that $\mathfrak{M}$ acts on $\mathfrak{H}$. The crossed product $\mathfrak{M} \times_\alpha G$ of $\mathfrak{M}$ by α is the von Neumann algebra generated by the following operators on $L^2(G,\mathfrak{H})$:

$$\begin{cases} \pi_\alpha \xi(s) = \alpha_s^{-1}(x)\xi(s), & x \in \mathfrak{M},\ s \in G; \\ u(t)\xi(s) = \xi(s - t), & \xi \in L^2(G,\mathfrak{H})\,. \end{cases} \tag{4.1}$$

On $L^2(G,\mathfrak{H})$, one defines a unitary representation of the dual group $\hat{G}$ by the following:

$$v(p)\xi(s) = \langle s,p\rangle \xi(s), \qquad p \in \hat{G},\ s \in G\,. \tag{4.2}$$

It then follows that

$$(4.3)\qquad \begin{cases} v(p)\pi_\alpha(x)v(p)^* = \pi_\alpha(x), & x \in \mathfrak{M}\,, \\ v(p)u(t)v(p)^* = \langle t,p\rangle u(t), & t \in G, \quad p \in \hat{G} \end{cases}$$

Thus, one has an action $\hat{\alpha}$ of $\hat{G}$ on $\mathfrak{M} \times_\alpha G$:

$$(4.4)\qquad \hat{\alpha}_p(x) = v(p)xv(p)^*, \qquad \alpha \in \mathfrak{M} \times_\alpha G, \quad p \in \hat{G}\,.$$

The new action $\hat{\alpha}$ of $\hat{G}$ is called the dual action.

THEOREM 4.1. (Duality) a) In the above situation, one has

$$\{\mathfrak{M} \times_\alpha G \times_{\hat{\alpha}} \hat{G}, \hat{\hat{\alpha}}\} \cong \{\mathfrak{M} \mathbin{\overline{\otimes}} \mathcal{L}(L^2(G)), \alpha \otimes \rho\},$$

where ρ is the action of G on $\mathcal{L}(L^2(G))$ defined by the unitary representation U of G on $L^2(G)$:

$$(4.5)\qquad U(t)\xi(s) = \xi(s + t), \qquad \xi \in L^2(G), \quad s,t \in G\,.$$

b) The fixed point algebra in $\mathfrak{M} \times_\alpha G$ under the dual action $\hat{\alpha}$ is precisely the image $\pi_\alpha(\mathfrak{M})$ of $\mathfrak{M}$.

There have been considerable works in the theory of crossed products and automorphism groups of operator algebras, throughout the 70's. It is impossible to cite all important results in this area. In any case, there are more lectures on the subject. However, the following result on dual weights, due to T. Digernes in the separable case and U. Haagerup in the general case, [18, 25], is indispensable for the rest of these notes.

Let α be an action of a locally compact abelian group G on $\mathfrak{M}$. If $x : t \in G \to x(t) \in \mathfrak{M}$ is a Bochoner integrable function, then it can be identified with the element of $\mathfrak{M} \times_\alpha G$ given by the formula:

$$(4.6)\qquad x = \int_G \pi_\alpha(x(t))\, u(t)\, dt \in \mathfrak{M} \times_\alpha G\,.$$

Thus, we may consider $L^1(G,\mathfrak{M})$ as a subspace of $\mathfrak{M} \times_\alpha G$. It follows that $L^1(G,\mathfrak{M})$ is a σ-weakly dense $*$-subalgebra of $\mathfrak{M} \times_\alpha G$ and that the arithmetic in $L^1(G,\mathfrak{M})$ is governed by the formula:

$$(4.7)\qquad \begin{cases} (xy)(t) = \displaystyle\int_G x(s)\, \alpha_s(y(t-s))\, ds\,, & x,y \in L^1(G,\mathfrak{M})\,; \\ x^*(t) = \alpha_t(x(-t)^*)\,, & t \in G\,, \end{cases}$$

If $x,y \in \mathcal{K}(G,\mathfrak{M})$, the space of all $\mathfrak{M}$-valued continuous functions with compact support, then $xy \in \mathcal{K}(G,\mathfrak{M})$; in particular, the value $xy(0)$ at the origin of G makes sense. Hence for a given weight φ on $\mathfrak{M}$, the expression

$$(4.8)\qquad \tilde{\varphi}(x^*x) = \varphi(x^*x(0))\,, \qquad x \in \mathcal{K}(G,\mathfrak{M})\,,$$

makes sense.

THEOREM 4.2. (T. Digernes and U. Haagerup). In the above setting, every weight φ on $\mathfrak{M}$ gives rise to a weight $\tilde{\varphi}$ on $\mathfrak{M} \times_\alpha G$. The correspondence:

$\varphi \to \tilde{\varphi}$ maps bijectively weights on $\mathfrak{M}$ onto $\hat{\alpha}$-invariant weights on $\mathfrak{M} \times_\alpha G$. Furthermore, φ is faithful if and only if $\tilde{\varphi}$ is also. The modular automorphism group $\{\sigma_t^{\tilde{\varphi}}\}$ of $\tilde{\varphi}$ is determined by the following formula:

$$(4.9)\quad \begin{aligned} \sigma_t^{\tilde{\varphi}}(\pi_\alpha(x)) &= \pi_\alpha(\sigma_t^{\varphi}(x)), && x \in \mathfrak{M}; \\ \sigma_t^{\tilde{\varphi}}(u(x)) &= u(s)\pi_\alpha((D\varphi \cdot \alpha_s : D\varphi)_t, && s \in G, \quad t \in \mathbb{R}. \end{aligned}$$

If φ and ψ are two faithful weights on $\mathfrak{M}$, then

$$(4.10)\quad (D\tilde{\psi} : D\tilde{\varphi})_t = \pi_\alpha((D\psi : D\varphi)_t), \quad t \in \mathbb{R}.$$

DEFINITION 4.3. The weight $\tilde{\varphi}$ on $\mathfrak{M} \times_\alpha G$ corresponding to a weight φ on $\mathfrak{M}$ is said to be <u>dual</u> to φ.

An important immediate consequence of the above two theorems is the following:

COROLLARY 4.4. For an arbitrary von Neumann algebra $\mathfrak{M}$ and a faithful weight φ, the crossed product $\mathfrak{M} \times_{\sigma^\varphi} \mathbb{R}$ is semi-finite and

$$\mathfrak{M} \bar{\otimes} \mathfrak{L}(L^2(\mathbb{R})) \cong (\mathfrak{M} \times_{\sigma^\varphi} \mathbb{R}) \times_{\widehat{\sigma^\varphi}} \mathbb{R}.$$

Furthermore, $\mathfrak{M} \times_{\sigma^\varphi} \mathbb{R}$ is canonically imbedded into $\mathfrak{M} \bar{\otimes} \mathfrak{L}(L^2(\mathbb{R}))$ as the fixed point subalgebra under the action $\{\sigma_t^\varphi \otimes \rho_t\}$.

Writing $U(t) = h^{it}$, $t \in \mathbb{R}$, in $\mathfrak{L}(L^2(\mathbb{R}))$, and setting

$$\omega(x) = \mathrm{Tr}(x^{1/2} h x^{1/2}), \quad x \in \mathfrak{L}(L^2(\mathbb{R}))_+,$$

one gets

$$\rho_t = \sigma_t^\omega, \quad t \in \mathbb{R},$$

therefore $\mathfrak{M} \times_{\sigma^\varphi} \mathbb{R}$ is canonically identified with the centralizer of $\varphi \otimes \omega$ on $\mathfrak{M} \bar{\otimes} \mathfrak{L}(L^2(\mathbb{R}))$.

Based on Arveson's theory of spectral subspaces for abelian automorphism groups, A. Connes introduced the notion of <u>essential spectrum</u> $\Gamma(\alpha)$ in his thesis, [11]. Given an action α of a locally compact abelian group G on $\mathfrak{M}$, set

$$(4.11)\quad \Gamma(\alpha) = \bigcap \mathrm{Sp}(\alpha^e),$$

where e runs all non-zero projections of the fixed point algebra $\mathfrak{M}^\alpha$ and α^e is the restriction of α to the reduced algebra $\mathfrak{M}_e$. He showed that $\Gamma(\alpha)$ is a closed subgroup of $\hat{G}$ and invariant under the perturbation by a unitary one cocycle. It then turns out that $\Gamma(\alpha)$ is exactly the kernel of the restriction of the dual action $\hat{\alpha}$ to the center of $\mathfrak{M} \times_\alpha G$. Furthermore, if $\mathfrak{M}$ is a factor, then $\Gamma(\alpha)$ is also given by the intersection of $\mathrm{Sp}(\alpha')$ where α' runs over all possible perturbations of α by unitary one cocycles. Here a unitary cocycle means a strongly continuous function: $s \in G \to u_s \in \mathcal{U}(\mathfrak{M})$ into the unitary group of $\mathfrak{M}$ such that

(4.12) $$u_{s+t} = u_s \alpha_s(u_t) , \qquad s,t \in G .$$

In this case, one gets a new action of G by:

(4.13) $${}_u\alpha_s(x) = u_s \alpha_s(x) u_s^* , \qquad x \in \mathfrak{M}, \quad s \in G ,$$

which is called the perturbed action of α by $\{u_s\}$.

5. VON NEUMANN ALGEBRAS OF TYPE III. In his very first paper on this subject, A. Connes introduced an algebraic invariant $S(\mathfrak{M})$ for a factor $\mathfrak{M}$ of type III as follows:

(5.1) $$S(\mathfrak{M}) = \bigcap Sp(\Delta_\varphi) ,$$

where Δ_φ is, of course, the modular operator for φ and φ runs over all possible faithful weights on $\mathfrak{M}$. He and van Daele showed in 1972 that $S(\mathfrak{M})\backslash\{0\}$ is a closed subgroup of the multiplicative group $\mathbb{R}_+^*$ of positive non-zero real numbers.

DEFINITION 5.1. A factor $\mathfrak{M}$ is called of type III_λ, $0 < \lambda < 1$, if $S(\mathfrak{M}) = \{\lambda^n : n \in \mathbb{Z}\} \cup \{0\}$, of type III_0 if $S(\mathfrak{M}) = \{0,1\}$, and of type III_1 if $S(\mathfrak{M}) = \mathbb{R}_+$.

At a glance, it seems not easy to find $S(\mathfrak{M})$ of a given factor. It turns out, however, that $S(\mathfrak{M})$ is relatively easily determined. In fact, if $\mathfrak{M}$ is an infinite tensor product of factors of type I, then $S(\mathfrak{M})$ is precisely the asymptotic ration set $r_\infty(\mathfrak{M})$ of Araki-Woods. This means that the factors of type III which were distinguished by R. Powers in 1967 are indeed factors of type III_λ, $0 < \lambda < 1$. Furthermore, if $\mathfrak{M}$ is the factor generated by a clustering equilibrium state ω, (i.e. a state satisfying the modular condition for a one parameter automorphism group), of an asymptotic abelian C^*-dynamic system, then $S(\mathfrak{M})$ agrees with $Sp(\Delta_\omega)$, a result of E. Stormer [46]. In fact, almost all factors coming from mathematical physics are either known or expected to be of type III_1.

The following result due to A. Connes is useful to determine $S(\mathfrak{M})$:

THEOREM 5.2. If the centralizer $\mathfrak{M}_\varphi$, the fixed point algebra under the modular automorphism group $\{\sigma_t^\varphi\}$ by definition, of a faithful weight φ on $\mathfrak{M}$ is a factor, then one has

$$S(\mathfrak{M}) = Sp(\Delta_\varphi) .$$

Now, another important immediate consequence of the Radon-Nikodym theorem of A. Connes is that the set

(5.2) $$T(\mathfrak{M}) = \{t \in \mathbb{R} : \sigma_t^\varphi \in Int(\mathfrak{M})\}$$

does not depend on the choice of φ and is a subgroup of the additive group $\mathbb{R}$.

Returning to the crossed product, we fix a factor $\mathfrak{M}$ of type III. As we have seen, the crossed product

$$\mathfrak{M}_0 = \mathfrak{M} \times_{\sigma^\varphi} \mathbb{R}$$

does not depend on the choice of φ and is semi-finite. Let $\{\theta_s : s \in \mathbb{R}\}$ be the dual action on $\mathfrak{M}_0$. Since $\theta_s(u(t)) = e^{ist}u(t)$ and $\mathrm{Ad}(u(t)) = \sigma_t^{\tilde{\varphi}}$, $t \in \mathbb{R}$, there exists a faithful trace τ such that

$$\tau \cdot \theta_s = e^{-s}\tau, \quad s \in \mathbb{R}. \tag{5.3}$$

Therefore, to each factor of type III, there corresponds a unique von Neumann algebra equipped with a one parameter automorphism group $\{\theta_s\}$ and a faithful trace τ which is transformed by $\{\theta_s\}$ according to (5.3). Thus, we now start from such a system $\{\mathfrak{M}_0, \theta, \tau\}$.

THEOREM 5.3. Suppose that $\{\mathfrak{M}_0, \theta, \tau\}$ is a system satisfying the above condition, and let $\mathfrak{M} = \mathfrak{M}_0 \times_\theta \mathbb{R}$.

a) If $\mathfrak{M}$ is a factor, then θ is ergodic on the center $\mathcal{C}_0$ of $\mathfrak{M}_0$.

Assume that $\mathfrak{M}$ is a factor.

b) $e^{-t} \in S(\mathfrak{M})$ for $t \in \mathbb{R}$ if and only if $\theta_t = \mathrm{id}$ on $\mathcal{C}_0$. Hence in particular $\mathfrak{M}$ is of type III_1 if and only if $\mathfrak{M}_0$ is a factor.

c) $s \in T(\mathfrak{M})$ for $s \in \mathbb{R}$ if and only if there exists a unitary $v \in \mathcal{C}_0$ such that $\theta_t(v) = e^{ist}v$, $t \in \mathbb{R}$.

d) $\mathfrak{M}$ is of type III if and only if $\{\mathcal{C}_0, \theta\}$ is conservative in the sense that

$$\int_{-\infty}^{\infty} \langle \theta_s(x), \omega \rangle \, ds = +\infty$$

for every non-zero $x \in \mathcal{C}_0^+$ and $\omega \in \mathcal{C}_{0*}^+$. Consequently if $\mathfrak{M}$ is of type III, then $\mathfrak{M}_0$ is of type II_∞.

e) Let $\{\overline{\mathfrak{M}}_0, \overline{\theta}, \tau\}$ be another system satisfying the same condition for $\{\mathfrak{M}_0, \theta, \tau\}$. Set $\overline{\mathfrak{M}} = \overline{\mathfrak{M}}_0 \times_{\overline{\theta}} \mathbb{R}$. Then, $\mathfrak{M} \cong \overline{\mathfrak{M}}$ if and only if there exist an isomorphism π of $\mathfrak{M}_0$ onto $\overline{\mathfrak{M}}_0$ and a $\overline{\theta}$ unitary one cocycle $\{u_s\}$ such that

$$\pi \circ \theta_s \circ \pi^{-1} = \mathrm{Ad}(u_s) \circ \overline{\theta}_s, \quad s \in \mathbb{R}.$$

6. COMPARISON OF WEIGHTS AND FLOW OF WEIGHTS.

We fix a separable <u>infinite</u> factor $\mathfrak{M}$. Let $\mathfrak{W}$ be the set of weights on $\mathfrak{M}$. For each $\varphi \in \mathfrak{W}$, let $s(\varphi)$ denote the support of φ which is characterized by $\mathfrak{M}(1 - s(\varphi)) = \{x \in \mathfrak{M} : \varphi(x^*x) = 0\}$. The modular automorphism group $\{\sigma_t^\varphi\}$ means the modular automorphism group of the restriction of φ to $\mathfrak{M}_e$, and the centralizer $\mathfrak{M}_\varphi$ of φ is also the fixed point subalgebra of $\mathfrak{M}_e$ under $\{\sigma_t^\varphi\}$. For each $\varphi \in \mathfrak{W}$

and a partial isometry u with $e = uu^* \in \mathfrak{M}_\varphi$, we define φ_u by

$$\varphi_u(x) = \varphi(uxu^*), \quad x \in \mathfrak{M}_+ . \tag{6.1}$$

For a projection $e \in \mathfrak{M}_\varphi$, φ_e is called a <u>subweight</u> of φ.

DEFINITION 6.1. For a pair $\varphi_1, \varphi_2 \in \mathfrak{W}$, we say that φ_1 and φ_2 are <u>equivalent</u> and write $\varphi_1 \sim \varphi_2$ if there exists $u \in \mathfrak{M}$ with $uu^* = s(\varphi_1)$ and $u^*u = s(\varphi_2)$ such that $\varphi_2 = \varphi_{1,u}$. We write $\varphi_1 \preccurlyeq \varphi_2$ if φ_1 is equivalent to a subweight of φ_2.

The relation "$\sim$" is an equivalence relation in $\mathfrak{W}$ and is associated with the partial ordering "$\preccurlyeq$".

DEFINITION 6.2. A weight φ is said to be of <u>infinite multiplicity</u> if the centralizer $\mathfrak{M}_\varphi$ is properly infinite. We denote by $\mathfrak{W}_{\text{inf.}}$ the set of all weights of infinite multiplicity.

A weight is of infinite multiplicity if and only if it is the sum of an infinite sequence of mutually equivalent weights with orthogonal support.

THEOREM 6.3. In the above situation, there exist an abelian von Neumann algebra $\mathcal{P}$ (not separable) and a map p from $\mathfrak{W}$ onto the lattice of all σ-finite projections of $\mathcal{P}$ with the following properties:

a) $\varphi \preccurlyeq \psi \Rightarrow p(\varphi) \le p(\psi)$;

b) For $\varphi, \psi \in \mathfrak{W}_{\text{inf.}}$, $\varphi \preccurlyeq \psi \Longleftrightarrow p(\varphi) \le p(\psi)$.

The von Neumann algebra $\mathcal{P}$ and the map p are canonically constructed from $\mathfrak{M}$ and $\mathfrak{W}$. Thus the association of $\{\mathcal{P}, p\}$ to $\mathfrak{M}$ is a functor.

In fact, the coset space $\mathfrak{W}_{\text{inf.}}/\sim$ becomes a σ-complete Boolean lattice which admits sufficiently many completely additive measures, so that it can be realized as the lattice of σ-finite projections of an abelian von Neumann algebra $\mathcal{P}$. The map p is then simply to associate the equivalence class $[\varphi]$ to each $\varphi \in \mathfrak{W}_{\text{inf.}}$. Fixing a sequence $\{u_n\}$ of coisometries in $\mathfrak{M}$ with orthogonal support $\{u_n^* u_n\}$ set

$$\check{\varphi} = \sum_{n=1}^{\infty} \varphi_{u_n} .$$

Since $[\check{\varphi}]$ does not depend on the choice of $\{u_n\}$, one can define

$$p(\varphi) \equiv p(\check{\varphi}) ,$$

and checks that this p does the job.

By definition, the multiplication by a positive member preserves the equivalence and the order relations in $\mathfrak{W}$, so that there exists an automorphism $\mathcal{F}_s$, $s \in \mathbb{R}$, determined by the following:

$$\mathcal{F}_s p(\varphi) = p(e^{-s}\varphi), \quad s \in \mathbb{R} . \tag{6.2}$$

DEFINITION 6.3. We say that $\{\mathcal{P}, \mathfrak{F}\}$ is the global flow of weights on $\mathfrak{M}$.

Now, consider the crossed product decomposition:

(6.3) $$\mathfrak{M} = \mathfrak{M}_0 \times_\theta \mathbb{R}$$

in the previous section. We call this the continuous decomposition of $\mathfrak{M}$. Let $\{u(s)\}$ be the one parameter unitary group in $\mathfrak{M}$ associated with this decomposition (6.3). With τ the trace on $\mathfrak{M}_0$, set $\bar{\omega} = \tilde{\tau}$, the dual weight on $\mathfrak{M}$. Formula (4.9) for $\{\sigma_t^{\bar{\omega}}\}$ shows that

(6.4) $$\begin{aligned} &\sigma_t^{\bar{\omega}}(x) = x\,, \\ &\sigma_t^{\bar{\omega}}(u(s)) = e^{-ist}\,u(s)\,, \qquad s,t \in \mathbb{R}\,, \end{aligned}$$

where $\mathfrak{M}_0$ is identified with the canonical image in $\mathfrak{M}$. This means that the dual action $\{\hat{\theta}_{-t}\}$ of θ on $\mathfrak{M}$ is exactly the modular automorphism group $\{\sigma_t^{\bar{\omega}}\}$. Furthermore, (6.4) yields:

(6.5) $$\begin{aligned} &\bar{\omega}_{u(s)} = e^{-s}\,\bar{\omega}\,; \\ &\mathfrak{M}_{\bar{\omega}} = \mathfrak{M}_0\,. \end{aligned}$$

Hence we conclude that $\bar{\omega} \sim e^{-s}\,\bar{\omega}$ and the centralizer of $\bar{\omega}$ is $\mathfrak{M}_{\bar{\omega}}$. Furthermore, we have the following:

THEOREM 6.4. *In the above situation,* $\bar{\omega}$ *has the following properties:*

a) *If* $\varphi \in \mathfrak{W}_{\mathrm{inf}}$ *and* $\varphi \sim \lambda\varphi$ *for every* $\lambda > 0$, *then* $\varphi \sim \bar{\omega}$.

b) $\bar{\omega} = \varphi \otimes \omega$ *with some decomposition* $\mathfrak{M} = \mathfrak{M}_1 \bar{\otimes} \mathcal{L}(L^2(\mathbb{R}))$, *where* $\varphi \in \mathfrak{W}(\mathfrak{M}_1)$ *and* ω *is the weight on* $\mathcal{L}(L^2(\mathbb{R}))$ *defined after Corollary* 4.4.

DEFINITION 6.5. A weight on $\mathfrak{M}$ equivalent to the $\bar{\omega}$, which is characterized by the above theorem, is called dominant.

Thus, a dominant weight on $\mathfrak{M}$ gives rise to a continuous decomposition of $\mathfrak{M}$.

7. INTEGRABLE WEIGHTS AND SMOOTH FLOW OF WEIGHTS.

The global flow $\{\mathcal{P}, \mathfrak{F}\}$ of weights on a separable properly infinite factor $\mathfrak{M}$ never be continuous in any reasonable topology on $\mathcal{P}$. However, Theorem 6.4 says that $\mathcal{P}$ has a unique σ-finite invariant projection d_0 corresponding to the dominant weight $\bar{\omega}$. Furthermore, we have the following:

THEOREM 7.1. *The following statements for* $\varphi \in \mathfrak{W}$ *are equivalent:*

a) $\varphi \preccurlyeq \bar{\omega}$;

b) *The set* $\mathfrak{q}_\varphi$ *consisting of all those* x *such that*

$$\sup_{K>0} \left\| \int_{-K}^{K} \sigma_t^\varphi(x^*x)\,dt \right\| < +\infty$$

is a σ*-weakly dense hereditary* $*$*-subalgebra of* $\mathfrak{M}$;

c) The map: $s \in \mathbb{R} \mapsto \mathfrak{F}_s p(\varphi) = p(e^{-s}\varphi) \in \mathcal{P}$ is σ-strongly continuous.

DEFINITION 7.2. A weight satisfying any of the above conditions is said to be <u>integrable</u>. The restriction of the flow to $\mathcal{P}_{d_0}$ is called the <u>smooth flow of weights</u> on $\mathfrak{M}$.

THEOREM 7.3. The smooth flow of weights is conjugate to the flow obtained by the restriction of θ to the center C_0 of $\mathfrak{M}_0$, where $\{\mathfrak{M}_0, \theta\}$ is the system appearing in the continuous decomposition of $\mathfrak{M}$.

We denote the smooth flow of weights by $\{C_0, \theta\}$ or $\{\mathbf{C}_0(\mathfrak{M}), \theta^{\mathfrak{M}}\}$ if one needs to indicate the original algebra $\mathfrak{M}$.

The natural question is then how big the set of integrable weights is. This is a sensible question because there is no integrable element in $\mathfrak{M}_*^+$ for example due to condition (b) in Theorem 7.1. To answer this question, one needs regularizations of a given weight.

DEFINITION 7.4. Let $\mathbb{D}_{-\lambda} = \{z \in \mathbb{C} : -\lambda < \operatorname{Im} z < 0\}$ for $\lambda > 0$. For a pair $\varphi_1, \varphi_2 \in \mathfrak{W}$, we write

$$\varphi_1 \leq \varphi_2 \qquad (\lambda)$$

if the map: $t \in \mathbb{R} \to (D\varphi_1 : D\varphi_2)_t = u_t \in \mathfrak{M}$ is extended to an $\mathfrak{M}$-valued continuous function on $\bar{\mathbb{D}}_{-\lambda}$ which is holomorphic in $\mathbb{D}_{-\lambda}$ and $\|u_z\| \leq 1$.

This relation for a fixed $\lambda > 0$ is indeed an ordering on $\mathfrak{W}$. The usual ordering $\varphi_1 \leq \varphi_2$ then means that $\varphi_1 \leq \varphi_2$ (1/2).

DEFINITION 7.5. For a pair φ_1, φ_2 of faithful weights, we set

$$(7.1) \qquad d(\varphi_1, \varphi_2) = \inf\left\{\frac{\alpha}{1+\alpha} : \alpha > 0,\ e^{-\alpha}\varphi_1 \leq \varphi_2 \leq e^{\alpha}\varphi_1 \quad (\infty)\right\}.$$

The function d is a distance function on the set $\mathfrak{W}_0$ of faithful weights on $\mathfrak{M}$. We call the metric topology on $\mathfrak{W}_0$ the <u>uniform topology</u>.

If $d(\varphi_1, \varphi_2) < 1$, then the map: $t \in \mathbb{R} \mapsto (D\varphi_2 : D\varphi_1)_t$ is extended to an entire function on $\mathbb{C}$.

The uniform topology in $\mathfrak{W}_0$ is complete and for each fixed $x \in \mathfrak{M}_+$, the map: $\varphi \in \mathfrak{W}_0 \to \varphi(x) \in [0,\infty]$ is continuous. Furthermore, one has

$$\|\varphi_1 - \varphi_2\| \leq 4\,\frac{d(\varphi_1,\varphi_2)}{1 - d(\varphi_1,\varphi_2)} \qquad \text{for } \varphi_1, \varphi_2 \in \mathfrak{M}_*^+ .$$

Concerning the size of the set $\mathfrak{W}_{\text{int.}}$ of integrable weights, we have the following:

THEOREM 7.5. For every $\varphi \in \mathfrak{W}_{\text{inf.}}$ and $\varepsilon > 0$, there exists $\psi \in \mathfrak{W}_{\text{int.}} \cap \mathfrak{W}_{\text{inf.}}$ commuting with φ such that

$$d(\varphi, \psi) < \varepsilon .$$

COROLLARY 7.6. If $\mathfrak{M}$ is of type III_1, then for any faithful $\varphi_1, \varphi_2 \in \mathfrak{W}_{inf.}$ and $\varepsilon > 0$ there exists a unitary u such that $d(\varphi_{1,u}, \varphi_2) \leq \varepsilon$, i.e. with $\delta = \varepsilon/(1-\varepsilon)$

$$e^{-\delta} \varphi_2 \leq \varphi_{1,u} \leq e^{\delta} \varphi_2 \quad (\infty) .$$

COROLLARY 7.7. If $\mathfrak{M}$ is of type III_λ, $\lambda \neq 0$, then for any faithful $\varphi_1, \varphi_2 \in \mathfrak{W}_{inf.}$ there exists a unitary $u \in \mathfrak{M}$ such that

$$u \mathfrak{m}_{\varphi_1} u^* = \mathfrak{m}_{\varphi_2} .$$

If $\mathfrak{M}$ is of type III_λ, $\lambda \neq 0,1$, then the above conclusion holds for any faithful $\varphi_1, \varphi_2 \in \mathfrak{W}$ with $\varphi_1(1) = \varphi_2(1) = +\infty$.

From this, one conjectures the following:

CONJECTURE 7.8. A factor $\mathfrak{M}$ is of type III_1 if and only if for any faithful normal states φ_1, φ_2 and $\varepsilon > 0$, there exists a unitary $u \in \mathfrak{M}$ such that

$$(1-\varepsilon)\varphi_2 \leq \varphi_{1,u} \leq (1+\varepsilon)\varphi_2 .$$

Based on the stability of the polar decomposition, A. Connes and E. Størmer proved the following, [14]:

THEOREM 7.9. A factor $\mathfrak{M}$ is of type III_1 if and only if for any normal states φ_1, φ_2 and $\varepsilon > 0$ there exists a unitary $u \in \mathfrak{M}$ such that

$$\|\varphi_1 - \varphi_{2,u}\| < \varepsilon .$$

8. EXPLICIT CONSTRUCTION OF THE FLOW OF WEIGHTS. We now consider the case where the factor $\mathfrak{M}$ in question is constructed from an ergodic transformation group.

Let G be a separable locally compact group acting on a standard measure space $\{\Gamma, \mu\}$ as a group of non-singular Borel automorphisms. The action of G on Γ naturally gives rise to an action α of G on $\mathcal{C} = L^\infty(\Gamma, \mu)$ by:

$$(8.1) \qquad \alpha_g(x)(\gamma) = x(g^{-1}\gamma) , \qquad g \in G, \quad \gamma \in \Gamma, \quad x \in \mathcal{C} .$$

A one cocycle on $\{G, \Gamma, \mu\}$ with coefficients in another separable locally compact group H is, by definition, an H-valued Borel function ρ of $G \times \Gamma$ such that

$$(8.2) \qquad \rho(g_1 g_2, \gamma) = \rho(g_1, g_2\gamma)\, \rho(g_2, \gamma) , \qquad g_1, g_2 \in G, \quad \gamma \in \Gamma .$$

Given two such ρ_1 and ρ_2, we say that ρ_1 and ρ_2 are <u>equivalent</u> and write $\rho_1 \sim \rho_2$ if there exists an H-valued Borel function f on Γ such that

$$(8.3) \qquad \rho_2(g, \gamma) = f(g\gamma)\, \rho_1(g, \gamma)\, f(\gamma)^{-1} .$$

Now, fix a one cocycle ρ. Consider the Cartesian product $\{\Gamma\times H, \mu\times\nu\}$, where ν is a left Haar measure on H, and define actions of G and H on $\Gamma\times H$ by

$$(8.4)\qquad \begin{cases} T_g(\gamma,h) = (g\gamma,\rho(g,\gamma)h), & g\in G,\ (\gamma,h)\in\Gamma\times H \\ S_k(\gamma,h) = (\gamma,hk^{-1}), & k\in H\,. \end{cases}$$

The actions of G and H commute. Let $\mathcal{D} = L^\infty(\Gamma\times H, \mu\times\nu)$. The action $\bar{\alpha}$ of G on $\mathcal{D}$ induced by the action T is called the <u>stable kernel</u> of ρ which depends, up to conjugacy, only on the equivalence class $[\rho]$ of ρ. The action β of H on the fixed point algebra $\mathcal{D}^{\bar{\alpha}} = \mathcal{E}$ is called the <u>stable range</u> of ρ.

We apply the above process to the following ρ:

$$(8.5)\qquad \rho(g,\gamma) = \left\{\Delta_G(g)\,\frac{d\mu\cdot g}{d\mu}(\gamma)\right\}^{-1}, \qquad g\in G,\quad \gamma\in\Gamma\,,$$

where Δ_G is the modular function of G. This particular ρ is called the <u>module</u> of $\{G,\Gamma,\mu\}$ and denoted by δ. The values of δ are of course in the multiplicative group $\mathbb{R}_+^*$ of positive real numbers. We consider the Lebesgue measure m on $\mathbb{R}_+^*$ instead of the multiplicative Haar measure on $\mathbb{R}_+^*$.

THEOREM 8.1. If the action of G on Γ is absolutely free in the sense that there is a null set N such that $g\gamma\neq\gamma$ for every $\gamma\in\Gamma - N$ and $g\neq e$, then we have the following conclusions concerning $\mathfrak{M} = \mathcal{C}\times_\alpha G$:

a) $\mathfrak{M}$ is a factor if and only if the action of G is ergodic.

We assume that G is ergodic. Then we have further:

b) $\mathfrak{M}$ is of type I if and only if the action of G is conjugate to the translation of G on G itself.

c) $\mathfrak{M}$ is semi-finite if and only if δ is equivalent to the trivial cocycle constant one.

d) $\mathfrak{M}$ is finite if and only if G is discrete and $\mathcal{C}$ has an invariant normal state.

e) With a continuous decomposition $\mathfrak{M} = \mathfrak{M}_0\times_\theta\mathbb{R}$, $\{\mathfrak{M}_0,\theta\}$ is conjugate to the following system:

$$\mathfrak{M}_0 = \mathcal{D}\times_{\bar{\alpha}} G\,,$$

the action θ is obtained by the lifting action of $\mathbb{R}$ on $\Gamma\times\mathbb{R}_+^*$ defined by:

$$\theta_s^*(\gamma,\lambda) = (\gamma,e^s\lambda), \qquad s\in\mathbb{R}\,,\quad (\gamma,\lambda)\in\Gamma\times\mathbb{R}_+^*\,.$$

f) The smooth flow $\{\mathcal{C}_0,\theta\}$ of weights on $\mathfrak{M}$ is conjugate to the stable range of the module δ under the scale change: $s\in\mathbb{R}\longleftrightarrow e^s\in\mathbb{R}_+^*$.

REMARK 8.2. In the above theorem, the absolute freeness of the action of G is equivalent to the relative commutant statement that $\mathcal{C}$ is maximal abelian in $\mathfrak{M}$. If G is abelian, then this is equivalent to the freeness that the fixed points of each $g \neq 0$ is null.

9. THE AUTOMORPHISM GROUP. Let $\mathfrak{M}$ be a fixed separable factor of type III and $\mathfrak{M} = \mathfrak{M}_0 \times_\theta \mathbb{R}$ a continuous decomposition of $\mathfrak{M}$.

THEOREM 9.1. The one parameter automorphism group $\{\theta_s\}$ on $\mathfrak{M}_0$ is stable in the sense that every one cocycle is a coboundary, i.e. if $\{u_s\}$ is a strongly continuous family of unitaries in $\mathfrak{M}_0$ such that

$$u_{s+t} = u_s\theta_s(u_t) , \tag{9.1}$$

then there exists a unitary v in $\mathfrak{M}_0$ such that

$$u_s = v^*\theta_s(v) . \tag{9.2}$$

COROLLARY 9.2. If $\{\mathfrak{M}_1,\theta^1\}$ and $\{\mathfrak{M}_2,\theta^2\}$ give two continuous decompositions of the same factor, then they must be conjugate. Namely, in Theorem 5.3 (e) the one cocycle $\{u_s\}$ can be eliminated.

THEOREM 9.2. The relative commutant $\mathfrak{M}_0' \cap \mathfrak{M}$ of $\mathfrak{M}_0$ is the center $\mathcal{C}_0$ of $\mathfrak{M}_0$. Hence, the relative commutant $\mathfrak{M}_\varphi' \cap \mathfrak{M}$ of the centralizer $\mathfrak{M}_\varphi$ of an integrable faithful weight φ is the center $\mathcal{C}_\varphi$ of $\mathfrak{M}_\varphi$. If $\mathfrak{M}$ is of type III_0, then the relative commutant $\mathcal{C}_\varphi' \cap \mathfrak{M}$ of the center $\mathcal{C}_\varphi$ of the centralizer $\mathfrak{M}_\varphi$ of any faithful weight is indeed the centralizer $\mathfrak{M}_\varphi$.

The above relative commutant theorem allows us to analyze further the automorphisms leaving $\mathfrak{M}_0$ pointwise fixed. At any rate, the following result is straightforward:

COROLLARY 9.3. If $\mathfrak{M}$ is of type III_1, then every automorphism leaving $\mathfrak{M}_0$ pointwise fixed is on the modular automorphism group $\{\sigma_t^{\bar\omega}\}$ of the dominant weight associated with the continuous decomposition. Hence in this case

$$\mathrm{Aut}(\mathfrak{M}/\mathfrak{M}_0) \cong \mathbb{R} , \tag{9.3}$$

where $\mathrm{Aut}(\mathfrak{M}/\mathfrak{M}_0) = \{\alpha \in \mathrm{Aut}(\mathfrak{M}) : \alpha(x) = x, \ \ x \in \mathfrak{M}_0\}$.

We fix the smooth flow $\{\mathcal{C}_0,\theta\}$ of weights and realize as the restriction of the covariant system $\{\mathfrak{M}_0,\theta\}$ to the center. Let $\{u(s)\}$ be the one parameter unitary group in $\mathfrak{M}$ associated with the decomposition $\mathfrak{M} = \mathfrak{M} \times_\theta \mathbb{R}$. Let $Z_\theta^1(\mathbb{R}, \mathcal{C}_0)$ be the set of all strongly continuous unitary one cocycles in $\mathcal{C}_0$, i.e. the set of all strongly continuous functions on $\mathbb{R}$ with values in the unitary group $\mathcal{U}(\mathcal{C}_0)$ satisfying (9.1). It follows that $Z_\theta^1(\mathbb{R}, \mathcal{C})$ is an abelian group by pointwise multiplication. If $a \in Z_\theta^1(\mathbb{R}, \mathcal{C})$ admits $b \in \mathcal{U}(\mathcal{C}_0)$ such that $a_s = b^*\theta_s(b)$, then a is called a <u>coboundary</u>. Let $B_\theta^1(\mathbb{R}, \mathcal{C}_0)$ be the set of all coboundaries. Then $B^1_\theta(\mathbb{R}, \mathcal{C}_0)$ is a subgroup of $Z^1_\theta(\mathbb{R}, \mathcal{C}_0)$. Set

$$H^1_\theta(\mathbb{R}, \mathcal{C}_0) = Z^1_\theta(\mathbb{R}, \mathcal{C}_0)/B^1_\theta(\mathbb{R}, \mathcal{C}_0) . \tag{9.4}$$

For short, we simply denote them by Z^1, B^1 and H^1.

THEOREM 9.4. a) To each $a \in Z^1$, there corresponds uniquely an automorphism $\bar{\sigma}_a \in \mathrm{Aut}(\mathfrak{M}/\mathfrak{M}_0)$ such that

$$\begin{cases} \bar{\sigma}_a(x) = x , & x \in \mathfrak{M}_0; \\ \bar{\sigma}_a(u(s)) = a_s u(s) , & s \in \mathbb{R} . \end{cases} \tag{9.5}$$

b) The map: $a \in Z^1 \mapsto \bar{\sigma}_a \in \mathrm{Aut}(\mathfrak{M}/\mathfrak{M}_0)$ is an automorphism.

c) $\bar{\sigma}_a \in \mathrm{Int}(\mathfrak{M})$ if and only if $a \in B^1$, where $\mathrm{Int}(\mathfrak{M})$ means the group of all inner automorphism groups. Therefore, we have

$$H^1 \cong \mathrm{Aut}(\mathfrak{M}/\mathfrak{M}_0) . \tag{9.6}$$

Let $\bar{\omega}$ be the dominant weight associated with the decomposition $\mathfrak{M} = \mathfrak{M}_0 \times_\theta \mathbb{R}$. Making use of the fact that $\varphi \precsim \bar{\omega}$ for every $\varphi \in \mathfrak{W}_{\mathrm{int.}}$, one has the following:

THEOREM 9.5. a) For each faithful $\varphi \in \mathfrak{W}_{\mathrm{int.}}$, there exists an isomorphism: $a \in Z^1 \mapsto \bar{\sigma}^\varphi_a \in \mathrm{Aut}(\mathfrak{M}/\mathfrak{M}_\varphi)$ from Z^1 onto the group $\mathrm{Aut}(\mathfrak{M}/\mathfrak{M}_\varphi)$ of all automorphisms leaving the centralizer $\mathfrak{M}_\varphi$ of φ pointwise fixed such that

i) $\varphi \cdot \bar{\sigma}^\varphi_a = \varphi$;

ii) $\bar{\sigma}^\varphi_a(x) = p_\varphi^{-1}(a_s p(\varphi))x$ for every $x \in \mathfrak{M}$ such that $\sigma^\varphi_t(x) = e^{ist}x$, $t \in \mathbb{R}$, where p_φ is an isomorphism of $\mathfrak{M}_\varphi$ onto the reduced algebra $\mathfrak{M}_{0,p(\varphi)}$ determined by the subequivalence: $\varphi \precsim \bar{\omega}$ and $p(\varphi)$ is the projection in $\mathcal{C}_0$ corresponding to φ;

iii) If $\bar{t} : s \in \mathbb{R} \mapsto e^{ist} \in \mathcal{U}(\mathcal{C}_0)$ is considered as an element of Z^1, then

$$\bar{\sigma}^\varphi_{\bar{t}} = \sigma^\varphi_t , \quad t \in \mathbb{R} .$$

Although we assumed $\mathfrak{M}$ to be of type III, let us consider for the moment the case that $\mathfrak{M}$ is semi-finite. In this case, one has

$$\begin{cases} \mathcal{C}_0 = L^\infty(\mathbb{R}^*_+) , \\ \theta_s(f)(\lambda) = f(e^{-s}\lambda), \quad s \in \mathbb{R} , \ \lambda \in \mathbb{R}^*_+ , \ f \in \mathcal{C}_0 . \end{cases} \tag{9.6}$$

Since $Z^1 = B^1$ in this case, every $a \in Z^1$ is of the form:

$$a_s = f\theta_s(f^*) , \quad f \in \mathcal{U}(\mathcal{C}_0). \tag{9.7}$$

If τ is a trace, then every $\varphi \in \mathfrak{W}_{\mathrm{int.}}(\mathfrak{M})$ is of the form

$$\varphi = \tau(h_\varphi \cdot) , \tag{9.8}$$

where h_φ is a positive self-adjoint operator affiliated with $\mathfrak{M}$ and has an absolutely continuous spectrum. In this situation, we get

(9.2) $$\bar{\sigma}_a^{\varphi} = \mathrm{Ad}(f(h_\varphi)) \in \mathrm{Int}(\mathfrak{M}) \cap \mathrm{Aut}(\mathfrak{M}/\mathfrak{M}_0) .$$

Returning to the original case of type III, we can prove the following:

THEOREM 9.6. Let φ_1 and φ_2 be faithful integrable weights on $\mathfrak{M}$. Set $\mathfrak{M}_2 = \mathfrak{M} \otimes M(2;\mathbb{C})$, the algebra of 2×2 matrices over $\mathfrak{M}$, and

(9.10) $$\varphi\Big(\sum_{i,j=1}^{2} x_{ij} \otimes e_{ij}\Big) = \varphi_1(x_{11}) + \varphi_2(x_{22}), \quad x = \sum_{i,j=1}^{2} x_{ij} \otimes e_{ij} \in \mathfrak{M}_2 .$$

a) To each $a \in Z^1$, there corresponds a unique unitary $u_a = (D\varphi_2 : D\varphi_1)_a$ in $\mathfrak{M}$ such that

(9.11) $$\bar{\sigma}_a^{\varphi}(1 \otimes e_{21}) = u_a \otimes e_{21} ;$$

b) We have

(9.12) $$\begin{aligned} &\bar{\sigma}_a^{\varphi_2} = \mathrm{Ad}(u_a)\, \bar{\sigma}_a^{\varphi_1}, \\ &u_{a_1 a_2} = u_{a_1} \bar{\sigma}^{\varphi_1}(u_{a_2}) , \qquad a_1, a_2 \in Z^1 . \end{aligned}$$

c) If $a \in Z^1$ is continuously differentiable in norm, then $\bar{\sigma}_a^{\varphi}$ and $(D\varphi : D\psi)_a$ are defined for any $\varphi, \psi \in \mathfrak{W}$ and behave naturally.

We now analyze the group $\mathrm{Aut}(\mathfrak{M}/\mathfrak{M}_0)$. By Corollary 9.2, the subalgebra $\mathfrak{M}_0$ is a characteristic subalgebra of $\mathfrak{M}$ in the strong sense that if $\alpha \in \mathrm{Aut}(\mathfrak{M})$, then there exists $u \in \mathfrak{U}(\mathfrak{M})$ such that

i) $\mathrm{Ad}(u) \cdot \alpha$ leaves $\mathfrak{M}_0$ globally invariant,

ii) The restriction of $\mathrm{Ad}(u) \cdot \alpha$ on $\mathfrak{M}_0$ leaves τ invariant and commutes with θ, and furthermore

iii) The restriction of $\mathrm{Ad}(u) \cdot \alpha$ does not depend on the choice of u, up to the perturbation by θ-one cocycles. Hence, setting

(9.13) $$\begin{cases} \mathrm{Aut}_{\tau,\theta}(\mathfrak{M}_0) = \{\alpha \in \mathrm{Aut}(\mathfrak{M}_0) : \tau \cdot \alpha = \tau, \ \alpha\theta_s = \theta_s\alpha, \ s \in \mathbb{R}\}, \\ \mathrm{Out}_{\tau,\theta}(\mathfrak{M}_0) = \mathrm{Aut}_{\tau,\theta}(\mathfrak{M}_0)/\mathrm{Int}(\mathfrak{M}_0) \cap \mathrm{Aut}_{\tau,\theta}(\mathfrak{M}_0) , \end{cases}$$

we get the following:

THEOREM 9.7. For each $\dot{\alpha} \in \mathrm{Out}(\mathfrak{M}) = \mathrm{Aut}(\mathfrak{M})/\mathrm{Int}(\mathfrak{M})$, let $\bar{\gamma}(\dot{\alpha})$ be the image of $\mathrm{Ad}(u) \cdot \alpha|_{\mathfrak{M}_0}$ in $\mathrm{Out}_{,\theta}(\mathfrak{M}_0)$, where α is a representative of $\dot{\alpha}$ and u is a unitary in $\mathfrak{M}$ such that $u\alpha(\mathfrak{M}_0)u^* = \mathfrak{M}_0$. Then one has the following exact sequence:

$$\{1\} \to H_\theta^1(\mathbb{R}, C_0) \xrightarrow{\gamma} \mathrm{Out}(\mathfrak{M}) \xrightarrow{\bar{\delta}} \mathrm{Out}_{\tau,\theta}(\mathfrak{M}_0) \to \{1\} ,$$

where $\delta([a]) = [\bar{\sigma}_a^{\omega}]$, $a \in Z^1$. This exact sequence does not split.

10. CONCLUDING WORDS. In my opinion, the theory of operator algebras should provide an algebraic frame for analysis, at least for L^2-analysis if not for all. The algebraic content of non-commutative integration theory became much clearer during the last decade. The lack of traces on von Neumann algebras no longer creates any problem. Not only this, one can describe the algebra completely in terms of a semi-finite von Neumann algebra and a one parameter automorphism group. On the analytic side, there are still many problems left. For example, U. Haagerup defined L^p-spaces for an arbitrary von Neumann algebra. However, there is no counter part of measurable operators of I.E. Segal in the context of type III. Is there any associated "regular" ring of "unbounded" operators? The 2×2 matrix technique of A. Connes provides the relative modular operator $\Delta_{\psi,\varphi}$ for any weight ψ and faithful weight φ. It is shown recently that fixing φ, $\Delta_{\psi_1,\varphi} + \Delta_{\psi_2,\varphi}$ is essentially self-adjoint for any $\psi_1, \psi_2 \in \mathfrak{M}_*^+$, but this does not hold for weights, a result of Hilsum [27] and H. Kosaki [31]. Their proof however involves the structure theorem for von Neumann algebras of type III. I would like to see a direct proof of this fact without making use of crossed products. Furthermore, the spectral analysis of the modular operator Δ_φ of any fixed weight should yield the structure theorem for a von Neumann algebra of type III without appealing to the duality for crossed products.

To strengthem the analytic side of the theory, it seems that we need to develop hard analysis techniques adapted in our situation.

REFERENCES

No effort was made to list all important literatures. The total volume of literatures on related topics is prohibitive. The references cited here are only those quoted in these notes. The C^*-News and the recently published books are good sources for literature hunting.

1. H. Araki, "Structure of some von Neumann algebras with isolated modular spectrum", Publ. R.I.M.S., Kyoto Univ., 9 (1973), 1-44.

2. ________, "Some properties of the modular conjugation operator of von Neumann algebras and a non-commutative Radon-Nikodym theorem with a chain rule", Pacific J. Math., 50 (1974), 309-354.

3. ________, Introduction to relative Hamiltonian and relative entropy, Seminar Notes, Marseille (1975).

4. H. Araki and E.J. Woods, "A classification of factors", Publ. R.I.M.S., Kyoto Univ., 3 (1968), 51-130.

5. W.B. Arveson, "On groups of automorphisms of operator algebras", J. Functional Analysis, 14 (1974), 217-243.

6. F. Combes, "Poids sur C^*-algèbres", J. Math. Pures et Appl., 47 (1968), 57-100.

7. ________, "Poids associés à une algèbre hilbertienne à gauche," Composition Math., 23 (1971), 49-77.

8. A. Connes, "Un nouvel invariant pour les algèbres de von Neumann", C.R. Acad. Sci. Paris, 273 (1971), 900-903.

9. A. Connes, "Calcul des deux invariants d'Araki et Woods par la théorie de Tomita et Takesaki", C.R. Acad. Sci. Paris, 274 (1972), 175-178.

10. ________, "Groupe modulaire d'une algèbre de von Neumann", C.R. Acad. Sci. Paris, 274 (1972), 1923-1926.

11. ________, "Une classification des facteurs de type III", Ann. Ec. Norm. Sup., 6 (1973), 133-252.

12. ________, "On a spatial theory of von Neumann algebras", preprint, 1977.

13. A. Connes and A. Van Daele, "The group property of the invariant S", Math. Scand., 32 (1973), 187-192.

14. A. Connes and E. Størmer, "Homogeneity of the state space of factors of type III_1", J. Functional Analysis, 28 (1978), 187-196.

15. A. Connes and M. Takesaki, "Flot des poids sur les facteurs de type III", C.R. Acad. Sci. Paris, 278 (1974), 945-948.

16. ________, "The flow of weights on factors of type III", Tôhoku Math. J., 29 (1977), 473-477.

17. A. Van Daele, "A new approach to the Tomita-Takesaki theory of generalized Hilbert algebras", J. Functional Analysis, 15 (1974), 378-393.

18. T. Digernes, Duality for weights on covariant systems and its applications, Thesis (1975), U.C.L.A.

19. J. Dixmier, "Algèbres quasi-unitaires", Comment. Math. Helv., 26 (1952), 275-322.

20. ________, "Traces sur C^*-algebras, I, II", Ann. Inst. Fourier, 13 (1963), 219-262; Bull. Sci. Math., 88 (1964), 39-57.

21. H.A. Dye, "The Radon-Nikodym theorem for finite rings of operators", Trans. Amer. Math. Soc., 72 (1952), 243-380.

22. R. Haag, N.M. Hugenholtz and M. Winnink, "On the equilibrium states in quantum statistical mechanics", Comm. Math. Phys., 5 (1967), 215-236.

23. U. Haagerup, "Normal weights on W^*-algebras", J. Functional Analysis, 19 (1975), 302-318.

24. ________, L_p-spaces associated with an arbitrary von Neumann algebra, Marceille (1977).

25. ________, "On the dual weight for crossed products of von Neumann algebras", I, Math. Scand., 43 (1978), 99-118; II, Math. Scand., 43 (1978), 119-140.

26. ________, "An example of a weight with type III centralizer", Proc. Amer. Math. Soc., 62 (1977), 278-280.

27. M. Hilsum, "Les espaces L_p d'une algèbre de von Neumann, preprint.

28. R.V. Kadison, "Transformations of states in operator theory and dynamics", Topology, 3 (1965), 177-198.

29. R.R. Kallman, "Groups of inner automorphisms of von Neumann algebras", J. Functional Analysis, 7 (1971), 43-60.

30. D. Kastler, "C^*-algebras of a free Boson field", Comm. Math. Phys., 1 (1965), 14-48.

31. H. Kosaki, "Application of the complex interpolation method to a von Neumann algebra", preprint (1980).

32. W. Krieger, "On ergodic flows and the isomorphism of factors", Math. Ann., 223 (1976), 19-70.

33. G. Loupias and S. Miracle-Sole, "C^*-algèbres de systems canoniques", I, Comm. Math. Phys., 2 (1966), 31-48; II, Ann. Inst. H. Poincaré, 6 (1966), 39-54.

34. D. McDuff, "Uncountable many II_1 factors", Ann. Math., 90 (1969), 372-377.

35. E. Nelson, "Notes on non-commutative integration", J. Functional Analysis, 15 (1974), 103-116.

36. G.K. Pedersen, "Measure theory for C^*-algebras, I, II, III, IV", Math. Scand., 19 (1966), 131-145; 22 (1968), 63-44; 25 (1969), 71-93; 121-127.

37. ____________, "Weights on operator algebras", Int. School Phys. "E. Fermi", Varenna (1973).

38. ____________, C^*-algebras and their automorphism groups, Academic Press, 1979.

39. G.K. Pedersen and M. Takesaki, "The Radon-Nikodym theorem for von Neumann algebras", Acta Math., 130 (1973), 53-88.

40. R.T. Powers, "Representations of uniformly hyperfinite algebras and their associated von Neumann algebras", Ann. Math., 86 (1967), 138-171.

41. L. Pukanszky, "On the theory of quasi-unitary algebras", Acta Sci. Math., 16 (1955), 103-121.

42. M.A. Rieffel and A. Van Daele, "A bounded operator approach to Tomita-Takesaki theory", Pacific J. Math., 69 (1977), 187-221.

43. S. Sakai, "A Radon-Nikodym theorem in W^*-algebras", Bull. Amer. Math. Soc., 71 (1965), 149-151.

44. I.E. Segal, "A non-commutative extension of abstract integration", Ann. Math., 57 (1953), 401-457.

45. C. Skau, "Finite subalgebras of a von Neumann algebra", J. Functional Analysis, 25 (1977), 211-235.

46. E. Størmer, "Spectra of states and asymptotically abelian C^*-algebras", Comm. Math. Phys., 28 (1972), 279-294; correction, 33 (1974), 341-343.

47. C.E. Sutherland, "Crossed products, direct integrals and Connes classification of type III factors", Math. Scand., 40 (1977), 209-214.

48. M. Takesaki, "Tomita's theory of modular Hilbert algebras and its applications", Lecture Notes in Math., No. 128, Springer-Verlag, 1970.

49. __________, "The theory of operator algebras", Lecture Notes, U.C.L.A., 1969/70.

50. __________, "Duality for crossed products and the structure of von Neumann algebras of type III", Acta Math., 131 (1973), 249-310.

51. __________, Theory of operator algebras, I, Springer-Verlag, 1979.

52. M. Tomita, "Quasi standard von Neumann algebras", Mimeographed Notes, 1967.

53. ________, "Standard forms of von Neumann algebras", the Vth Functional Analysis Symposium of the Math. Soc. of Japan, Sendai, 1967.

54. S.L. Woronowicz, "Operator systems and their applications to the Tomita-Takesaki theory", J. Operator Theory, 2 (1979), 167-209.

55. F. Combes, Étude des représentations tracées d'une C*-algebre, C. R. Acad. Sc. Paris 262(1966), 114-117; Étude des poids définis sur une C*-algèbre, C. R. Acad. Sci. Paris, 265 (1967), 340-343.

56. M. Tomita, Spectral Theory of operator algebras, I, Math. J. Okayama Univ. 9 (1959), 63-98.

University of California
Los Angeles, California 90024

Proceedings of Symposia in Pure Mathematics
Volume 38 (1982), Part 2

A SURVEY OF W^*-CATEGORIES

-o-o-

P. GHEZ

ABSTRACT : A W^*-category is the categorical counterpart of a von Neumann algebra with an abstract definition equivalent to a concrete definition in terms of operators between Hilbert spaces. We give here a brief description of the matter involved with emphasis on definitions and general results needed by R. Lima in the next talk. A complete exposition of the subject will be founded in [1].

A C^*-category is a category such that each set of arrows (A,B) is a Banach space, the composition of arrows is bilinear and there is an involution $(A,B) \ni x \to x^* \in (B,A)$ such that x^*x is a positive element in the $*$-algebra (A,A). We assume also $\|xy\| \leqslant \|x\| \|y\|$ whenever xy is defined and $\|x^*x\| = \|x\|^2$. Note that it follows that the mapping $(x,y) \to x^*y$ is an (A,A)-valued inner product on the right module (A,B) over (A,A) in the sense of Rieffel [2] and Paschke [3].

FIRST EXAMPLES :

1) The category $\mathcal{H}$ of Hilbert spaces.
2) The category Rep(A) of non degenerate representations of a C^*-algebra A.
3) The category Rep(G) of continuous unitary representations of a locally group G .
4) The category $\mathcal{H}$ (M) of Hilbert spaces inside a von Neumann algebra M [4].
5) The category End(M) of endomorphisms of a von Neumann algebra M [4].
6) The category of cocycles in the context of non Abelian 1-cohomology or in the context of local cohomology (see for instance [5]).

If $\mathcal{A}$ and $\mathcal{B}$ are C^*-categories, one can construct a new C^*-category $(\mathcal{A}, \mathcal{B})$ whose objects are linear $*$-functors and whose arrows are all bounded natural transformations t between them, with $\|t\| = \sup_A \|t_A\|$.

A representation of a C^*-category $\mathcal{A}$ is a linear $*$-functor $F:\mathcal{A}\to\mathcal{H}$ and the commutant F' of F is the von Neumann algebra (F,F) of loops at the point F in $(\mathcal{A}, \mathcal{H})$.

0082-0717/80/0000-0527/$01.75

PROPOSITION. Any C^*-category $\mathcal{A}$ may be realized as a concrete one, i.e. there is a faithful embedding $*$-functor $F: \mathcal{A} \to \mathcal{H}$.

A W^*-category is a C^*-category where each (A,B) has a predual. It turns out that this predual is unique.

EXAMPLES. $\mathcal{H}$, Rep(G), $\mathcal{H}(M)$, End(M).

The appropriate notion of normality for $*$-functors F between W^*-categories is to say that F is normal if it induces a normal morphism of (A,A) into $(F(A),F(A))$ for each object A. Now we have the following:

PROPOSITION. Any W^*-category $\mathcal{A}$ may be realized as a concrete one, i.e. there is a faithful embedding normal $*$-functor $F: \mathcal{A} \to \mathcal{H}$.

REMARK. In a W^*-category, each (A,B) is self-dual as a right-Hilbert module over (A,A) in the sense of Rieffel [2] and Paschke [3].

A weight on a W^*-category can be viewed as a field of weights on the W^*-algebras (A,A). Adapting K.M.S. condition to our context, we have the following result :

PROPOSITION. Any faithful normal semi-finite weight Φ on a W^*-category satisfies K.M.S.-condition with respect to a unique action σ^{Φ} of $\mathbb{R}$ on $\mathcal{A}$, called modular action.

The center $\mathcal{Z}(\mathcal{A})$ is the set of bounded natural transformations from the identity functor of $\mathcal{A}$ to itself. It turns out that $\mathcal{Z}(\mathcal{A})$ is an Abelian von Neumann algebra. Now we have the following important notion of central support.

DEFINITION. Let $x \in (A,B)$ in a W^*-category $\mathcal{A}$. The central support of x is the smallest projection e in $\mathcal{Z}(\mathcal{A})$ such that $xe_A = x = e_Bx$. We denote it by $c(x)$ and put $c(A)$ for $c(1_A)$.

COMPARISON THEORY. Let A,B be two objects in a W^*-category. A is equivalent to B if (A,B) contains a unitary u $(u^*u = 1_A,\ uu^* = 1_B)$. A is a sub-object of B if (A,B) contains an isometry v $(v^*v = 1_A,\ vv^* =$ some projection in $(B,B))$.

EXAMPLES.

1) Let M be a W^*-algebra and let $\mathcal{P}(M)$ denote the W^*-category whose objects are all projections in M and where arrows are given by : $(e,e') = \{t \in M / e't = te = t\}$.
The comparison theory in $\mathcal{P}(M)$ is just the usual Murray and von Neumann comparison theory of projections.
2) Let M be a W^*-algebra and consider the W^*-category $\mathcal{V}(M)$ whose objects are all normal semi-finite weights on M and whose arrows are given by: $(\varphi,\varphi') = \{t \in M / s(\varphi')t = ts(\varphi) = t\}$.

This category is equivalent to $\mathcal{P}(M)$ but it also carries a canonical weight Φ : $\Phi_\varphi = \varphi \restriction M_{(\varphi)}$. The W^*-category $\mathcal{W}(M)$ of weights on M is then defined to be the fixed points of $\mathcal{V}(M)$ under the modular action of Φ. Now the comparison of objects in $\mathcal{W}(M)$ is just the comparison theory of weights of Connes and Takesaki [6].

QUASI-EQUIVALENCE. A is <u>quasi-equivalent</u> to B if $c(A) = c(B)$. Quasi-maximal objects, that is such that $c(A) = 1$, are called <u>generators.</u>

PROPOSITION. A is a generator iff the map $z \longrightarrow z_A$ is an isomorphism of $\mathcal{Z}(\mathcal{O})$ with $\mathcal{Z}(A,A)$.

THEOREM. The representation theory of a W^*-category $\mathcal{O}$ is equivalent to the representation theory of the W^*-algebra (A,A) provided A is a generator in $\mathcal{O}$.

REMARK. If A and B are quasi-equivalent objects, then the W^*-algebras (A,A) and (B,B) are Morita equivalent.

BIBLIOGRAPHY

1 P. GHEZ, R. LIMA, and J.E. ROBERTS : W^*-categories (Preprint Osnabrück).

2 M. RIEFFEL : Morita Equivalence for C^*- and W^*-Algebras. Journal of Pure and Appl. Alg. 5 (1974) 51-96.

3 W.L. PASCHKE : Inner Product Modules arising from Compact Automorphism Groups of von Neumann Algebras. Trans. Amer. Math. Soc., Vol. 224, 1 (1976).

4 J.E. ROBERTS : Cross Product of von Neumann Algebras by Group Duals. Symposia Matematica 22 (1976) 335-363.

5 J.E. ROBERTS : Local Cohomology and its Structural Implications for Field Theory, R.C.P. n° 25, Strasbourg, Novembre 1976.

6 A. CONNES, M. TAKESAKI : The Flow of Weights on Factors of Type III, Tohoku Math. Journal, Vol. 29, n° 4 (1977).

DEPARTEMENT DE MATHEMATIQUES
UNIVERSITE DE TOULON
LA GARDE, FRANCE

and

CENTRE DE PHYSIQUE THEORIQUE
CNRS - LUMINY - CASE 907
13288 MARSEILLE CEDEX 2 (FRANCE)

Proceedings of Symposia in Pure Mathematics
Volume 38 (1982), Part 2

ACTIONS OF NON-ABELIAN GROUPS AND INVARIANT Γ (*)

R. LIMA

This seminar is a report on a joint work with P. Ghez and J.E. Roberts. We use standard notions of W*-categories described in [1].(1)

The problem of defining an invariant Γ goes back to [3] where such an invariant was defined in the case of the action of an Abelian group. The connection with the study of crossed products was present in [4]. Finally, in [5], the case of an action of a compact non abelian group was studied in connection with Quantum Field Theory and some of the ideas and concepts developed in the present seminar were already implicit in [5].

Let G be a separable locally compact group, M a σ-finite von Neumann algebra and α a σ-continuous integrable action of G on M, i.e. a σ-continuous action such that the set :

(1) $$\{x \in M \text{ s.t. } \int_G (x^*x)\, dg \text{ exists in } M\}$$

is σ-weakly dense in M. Here dg stays for the left Haar measure on G and the integral is defined as the limit of the increasing net $\int_K \alpha_g(x^*x)\, dg$ indexed by the compacts K of G. [3].

We consider the W*-category $\mathcal{H}(G) \otimes M$ with arrows $t \in \mathcal{B}(H_\sigma, H_\tau) \otimes M$, h_σ, h_τ, ... , being separable Hilbert spaces carrying continuous unitary representations $\sigma, \tau, \ldots$ of G.

$\mathcal{H}(G) \otimes M$ carries a natural action of G namely :

(2) $$t \in \mathcal{B}(h_\sigma, h_\tau) \otimes M \longrightarrow (\tau_g \otimes 1)\,(i \otimes \alpha_g(t))\,(\sigma_g{}^* \otimes 1).$$

The fixed points under this action will be a W*-category denoted

---

(1) A survey of the basic definitions and results in the theory of W*-categories may be founded in these proceedings, [2].

(*) Talk given at the AMS Conference on Operator Algebras and Applications, Kingston 1980.

0082-0717/80/0000-0545/$02.00

$Sp(M,\alpha)$ and called Spectral W*-category of the action α.

The objects of this W*-category will be denoted $\sigma\otimes\alpha$, $\tau\otimes\alpha,\dots$ so that $t\in(\sigma\otimes\alpha,\tau\otimes\alpha)$ if and only if

$t \in \mathcal{B}(h_\sigma,h_\tau)\otimes M$ and

(3) $$(\tau_g\otimes 1)\,(i\otimes\alpha_g(t))\,(\sigma_g^*\otimes 1) = t \qquad \forall g\in G$$

Remark that this W*-category contains the "point spectral-subspaces" $M_\sigma = (\sigma\otimes\alpha,\ i\otimes\alpha)$ where i stands for the trivial representation of G on $\mathbb{C}$. It contains also the fixed-point subalgebra of M, i.e. $M^\alpha = (i\otimes\alpha, i\otimes\alpha)$ and the crossed product $W^*(M,\alpha,G) = (\rho\otimes\alpha\ ,\ \rho\otimes\alpha)$ where ρ stands for the right regular representation of G.

The object $\rho\otimes\alpha$ is a generator of $Sp(M,\alpha)$, see [2], and we say that α is quasi-dominant, see also [4], if $i\otimes\alpha$ is a generator.

If α is quasi-dominant, by [1], (see also [2]) we have that ZM^α (the center of M^α) is isomorphic to $ZSp(M,\alpha)$ and $ZW^*(M,\alpha,G)$. Furthermore M^α and $W^*(M,\alpha,G)$ are Morita equivalent.

The idea is now to study $Sp(M,\alpha)$ to produce some useful invariants of the action.

If we want to treat problems of equivalence up to multiplicity or, say, isomorphism up to Morita equivalence, then certainly the first invariant we can think about is the monoïdal quasi-spectrum of α :

(4) $$\underset{\sim}{Q}Sp(\alpha) = \{\sigma\ :\ c(\sigma\otimes\alpha) = c(i\otimes\alpha)\}$$

Therefore, a unitary representation σ of G is in $\underset{\sim}{Q}Sp(\alpha)$ if and only if the object $\sigma\otimes\alpha$ is quasi-equivalent to $i\otimes\alpha$ in $Sp(M,\alpha)$. Clearly

(5) $$\sigma,\sigma'\in \underset{\sim}{Q}Sp(\alpha) \Rightarrow \sigma\otimes\sigma'\in\underset{\sim}{Q}Sp(\alpha)$$ and

(6) $$\sigma,\sigma'\in \underset{\sim}{Q}Sp(\alpha) \Rightarrow \sigma\oplus\sigma'\in\underset{\sim}{Q}Sp(\alpha)$$

Note, that in general

(7) $$\underset{\sim}{Q}Sp(\alpha) \neq \{\sigma\ :\ M_\sigma\neq\{0\}\} = Sp(\alpha)$$

We now come to the problem of defining the analog of the invariant $\Gamma(\alpha) = \bigcap_{0\neq e\in M^\alpha} Sp(\alpha_e)$ defined by A. Connes for the action of an Abelian group. Here α_e denotes the reduced action by a projection e of M^α.

We define

(8) $$\Gamma_1(\alpha) = \bigcap_{0\neq e\in M^\alpha} \underset{\sim}{Q}Sp(\alpha)$$

and we get the following characterization.

PROPOSITION 1 : Let M , G , α be as above, then the following are equivalent :

(1) $\sigma \in \Gamma_1(\alpha)$

(2) $\sigma \in \bigcap_{0 \neq e \in ZM^{\alpha}} \mathbb{Q}\,Sp(\alpha_e)$, e projection

(3) $\forall$ e projection of ZM^{α} $c(e)_{\sigma\otimes\alpha} = 1_{\sigma} \otimes e$

(4) $\forall$ e projection of ZM^{α} $0 \neq f \leq 1_{\sigma} \otimes e$ and $f \in (\sigma\otimes\alpha, \sigma\otimes\alpha)$ then $\exists$ $t \in M_{\sigma}$ s.t. $etf \neq 0$.

The last property shows the difference between $\Gamma_1(\alpha)$ and $\bigcap Sp(\alpha_e)$. It was implicit in [5].

$\Gamma_1(\alpha)$ is closed under direct sums, sub-representations and tensor products.

If G is abelian then $\Gamma_1(\alpha)$ coincides with the Connes invariant $\Gamma(\alpha)$. But the last one is also stable under perturbation of the action by unitary cocycles $g \longrightarrow a(g) \in M$, i.e.

(9) $$\Gamma(\alpha) = \Gamma({}_a\alpha)$$

where

(10) $${}_a\alpha_g(x) = a(g)\,\alpha_g(x)a(g)^*$$

It is also invariant by trivial extension of the action, i.e.

(11) $$\Gamma(\alpha) = \Gamma(\bar{i}\otimes\alpha)$$

where $\bar{i}$ is the trivial action on $\mathbb{B}(H)$, H separable.

It is not very difficult to see that this is not the case for $\Gamma_1(\alpha)$ if G is not abelian.

So we define a stabilized invariant :

(12) $$\Gamma_0(\alpha) = \bigcap_{\sim} \Gamma_1(\tilde{\alpha})$$

where $\sim$ stands for all perturbations or extensions as above. Nevertheless we have the following :

PROPOSITION 2 : If α is quasi-dominant then

(13) $$\Gamma_1(\alpha) = \Gamma_0(\alpha)$$

Therefore the computation of $\Gamma_0(\alpha)$ reduces to the computation of $\Gamma_1(Ad\,\lambda\otimes\alpha)$ where λ is the left-regular representation of G on $L^2(G)$.

The reason why it is so is clear after the following :

PROPOSITION 3 : $\sigma \in \Gamma_0(\alpha)$ if and only if

$$z_{\sigma\otimes\tau\otimes\alpha} = 1_{\sigma}\otimes z_{\tau\otimes\alpha} \quad \forall z \in ZSp(M,\alpha) \text{ and } \forall \tau$$

Actually comparing (3) of Proposition 1 with Proposition 3 we can see that $\Gamma_1(\alpha)$ and $\Gamma_0(\alpha)$ differ only because we replace ZM^{α} by $Z\,Sp(M,\alpha)$, but these two von Neumann algebras are isomorphic if α is quasi-dominant.

Proposition 3 can be understood as the generalization of the property stated in [4] that $\Gamma(\alpha)$ is the kernel of the restriction of the dual action of $\hat{G}$ on $W^*(M,\alpha,G)$ to the center of $W^*(M,\alpha,G)$.

In particular if G is abelian we check $\Gamma_1(\alpha) = \Gamma_0(\alpha) = \Gamma(\alpha)$

Now we use these invariants to state some results relating the algebras M^{α}, M and $W^*(M,\alpha,G)$.

PROPOSITION 4 : Let M,G,α be as above, then the following conditions are equivalent :

(1) Any σ is in $\Gamma_0(\alpha)$

(2) Any σ is in $\Gamma_1(\alpha)$

(3) $ZM^{\alpha} = ZM \cap M^{\alpha}$ and α is quasi-dominant.

(4) $ZW^*(M,\alpha,G) \subseteq \pi_{\alpha}(ZM)$.

(5) $ZW^*(M,\alpha,G) = \pi_{\alpha}(ZM^{\alpha})$.

Here π_{α} is the representation of M in $W^*(M,\alpha,G)$.

From that we easily can prove the following :

COROLLARY 5 : $W^*(M,\alpha,G)$ is a factor if and only if all σ are in $\Gamma_0(\alpha)$ and α is ergodic on ZM.

Finally we consider some properties of relative commutant each of them implies the conditions of Proposition 4 :

(1) $\pi_{\alpha}(ZM)' \cap W^*(M,\alpha,G) = \pi_{\alpha}(M)$.

(2) $\pi_{\alpha}(M) \cap W^*(M,\alpha,G) = \pi_{\alpha}(ZM)$.

(3) $(M^{\alpha})' \cap M = ZM$ and α is quasi-dominant.

REFERENCES

1 P. Ghez, R. Lima, J.E. Roberts, W*-Categories.

2 P. Ghez, A Survey of W*-Categories, these proceedings.

3 A. Connes, Une classification des facteurs de type III, Ann. Sc. de l'Ecole Normale Supérieure, 4è Série, t.6, fasc. 2, (1973).

4 A. Connes, M. Takesaki, The Flow of Weights on Factors of Type III, Tohoku Math. Journal, Vol.29, n°4 (1977).

5 J.E. Roberts, Some Applications of Dilation Invariance to Structural Questions in the Theory of Local Observables, Comm. math. Phys., 37, 273-286 (1974).

C.N.R.S. - LUMINY - CASE 907
Centre de Physique Théorique
F-13288 MARSEILLE CEDEX 2 (France)

and

Fakultät für Physik
Universität Bielefeld
(West-Germany)

Proceedings of Symposia in Pure Mathematics
Volume **38** (1982), Part 2

COHOMOLOGICAL INVARIANTS FOR GROUPS OF OUTER AUTOMORPHISMS OF VON NEUMANN ALGEBRAS

Colin E. Sutherland

Throughout, Q denotes a separable locally compact group, and M a von Neumann algebra (on a separable Hilbert space $\mathcal{H}$) with unitary group $U(M)$ and centre $Z(M)$. The group $\mathrm{Aut}\, M$ of $*$-automorphisms of M is given the standard Borel structure inherited from the topology of pointwise norm convergence against M_*, and $\mathrm{Out}\, M$ denotes the group of outer automorphisms.

A homomorphism $\theta : Q \to \mathrm{Out}\, M$ is a Q-kernel if there is a Borel map $\alpha : Q \to \mathrm{Aut}\, M$ with $\varepsilon \circ \alpha = \theta$, where $\varepsilon : \mathrm{Aut}\, M \to \mathrm{Out}\, M$ is the quotient map. If θ is a Q-kernel, the cohomology group $H^n_\theta(Q;UZ(M))$ are as defined in [M]

H^3-obstructions Let (M,θ) be a Q-kernel and choose $\alpha : Q \to \mathrm{Aut}\, M$, Borel, with $\varepsilon \circ \alpha = \theta$. Then there are Borel maps $u : Q \times Q \to U(M)$ and $\gamma : Q \times Q \times Q \to UZ(M)$ with

$$\alpha_g \circ \alpha_h = \mathrm{Ad}u(g,h)\alpha_{gh}, \quad \text{and}$$

$$\alpha_g(u(h,k))\, u(g,hk) = \gamma(g,h,k)\, u(g,h)\, u(gh,k)$$

Note $\gamma \in Z^3_\theta(Q;UZ(M))$ and its class γ_θ in H^3_θ depends only on the conjugacy class of θ.

A realization of the Q-kernel θ is a system $\{I,\lambda\}$ where I is a representation of M on $\mathcal{K}$, $\lambda : Q \to U(\mathcal{K})$ with

a) $\mathrm{Ad}\lambda(g)(I(M)) = I(M)$ and
$\theta_g = \varepsilon(I^{-1} \circ \mathrm{Ad}\lambda(g) \circ I)$

b) $\lambda(g)\, \lambda(h) \in I(M)\, \lambda(gh)$

THEOREM 1 [S,II] A Q-kernel (M,θ) admits a realization if and only if γ_θ vanishes. Further, every element of $H^3_\theta(Q;UZ(M))$ is of the form γ_θ for some kernel (M,θ).

If Q is discrete, each element of $H^3(Q,\mathbb{T})$ may be realized in the hyperfinite II_1 factor R (see [J,I]); for $Q = \mathbb{Z}_n$, γ_θ is just Connes' invariant γ for periodic automorphisms ([C,I]). In many cases (i.e. $M = R$ and $Q = \mathbb{Z}_n$, finite or amenable), γ_θ is a complete outer conjugacy invariant (see [C,I], [J,I] and [O]).

0082-0717/80/0000-0578/$01.75

Splitting results If (M,θ) is a Q-kernel for which there is a Borel homomorphism $\alpha : Q \to \text{Aut } M$ with $\varepsilon \circ \alpha = \theta$, we say (M,θ) splits. Clearly if (M,θ) splits, γ_θ vanishes.

THEOREM 2 ([S,II]) If (M,θ) is a Q-kernel with γ_θ vanishing, then (M,θ) splits if either

a) M is properly infinite (and Q arbitrary); or

b) M is finite and Q is finite

The proof depends on the observation that in either case, $M = N \otimes B(L^2(Q))$ for some N, and a version of Shapiro's Lemma appropriate to non-abelian cohomology. Ocneanu, [O], has extended the conclusion to the case $M = R$, Q discrete amenable.

Regular extensions If (M,θ) is a Q-kernel with γ_θ vanishing, one may construct a realization via

$$(I(x)\xi)(g) = \alpha_{g^{-1}}(x)\ \xi(g)$$

$$(\lambda(h)\xi)(g) = u(g^{-1},h)\ \xi(h^{-1}g)$$

for $\xi \in L^2(Q;\mathcal{H})$, $x \in M$, $h \in Q$, where we choose α, u with $\varepsilon \circ \alpha = \theta$ and $\alpha_g(u(h,k))\, u(g,hk) = u(g,h)\, u(gh,k)$. The von Neumann algebra generated by $I(x)$ and $\lambda(h)$, for $x \in M$ and $h \in G$, is called the regular extension of M by α and u, denoted $W^*(M,\alpha,u)$.

THEOREM 3 ([S,II]) If (M,θ) is a Q-kernel with γ_θ vanishing, $W^*(M,\alpha,u)$ is isomorphic with $W^*(M,\beta,1)$ for some action β of Q on M (i.e. an ordinary crossed product) if either

a) M is properly infinite (and Q arbitrary); or

b) M is finite and Q is finite.

Note that the examples of [C,II] are regular extensions of abelian algebras by discrete groups; also if $1 \to N \to G \to Q \to 1$ is an exact sequence of topological groups, the group von Neumann algebra of G is a regular extension of that of N by Q.

The notion of extension of a von Neumann algebra by a group of automorphisms appears in [N,T] for finite groups and factors, and includes Zeller-Meier's twisted crossed products ([Z-M], [S,I]). In [J,I], [J,II], Jones has refined the invariant γ_θ to produce an invariant of cocycle conjugacy (for finite groups), and Ocneanu, [O], has extended this to the case $M = R$, Q discrete amenable, (see also [J,II]). Finally, we note that in the case $M = R$, $Q = \mathbb{R}$, there are no substantive results at present.

REFERENCES

[C,I] Connes, A; Outer conjugacy classes of automorphisms of factors, Ann. Sci. Ecole Norm. Sup. (4), 8, (1975), 383-419.

[C,II] Connes, A; A factor not antiisomorphic to itself, Ann. Math., 101, (1975) 536-553.

[J,I] Jones, V.F.R; An invariant for group actions, in "Algebras d'operateurs", Springer Lecture notes in Math, No. 725, (1980).

[J,II] Jones, V.F.R; Talk given at 1980 A.M.S. Summer congress, this volume.

[M] Moore, C.C; Group extensions and cohomology for locally compace groups III T.A.M.S. 221, (1968), 1-34.

[N,J] Nakamura, M and Takeda, Z; On the extensions of finite factors; Proc. Japan Academy, 35 (1959), 149-154.

[O] Ocneanu, A; Actions des groupes moyennables sur les algebres de von Neumann; Increst preprint No. 35/1980.

[S,I] Sutherland, C.E; Cohomology and extensions of von Neumann algebras I. Publ. R.I.M.S. Vol. 16, No. 1 (1980), 105-133.

[S,II] Sutherland, C.E; Cohomology and extensions of von Neumann algebras II. Publ. R.I.M.S. Vol. 16, No. 1 (1980), 135-174.

[Z-M] Zeller-Meier, G; Produits croises d'une C*-algebra par un groupe d'automorphisms; J. Math. Pures et Appliques, 47 (1968), 101-239.

Colin E. Sutherland,
School of Mathematics,
The University of New South Wales,
KENSINGTON, N.S.W. 2033.
AUSTRALIA.

Proceedings of Symposia in Pure Mathematics
Volume 38 (1982), Part 2

AUTOMORPHISM GROUPS AND INVARIANT STATES

Erling Størmer

1. INTRODUCTION. The study of automorphism groups of operator algebras may be viewed as nonabelian ergodic theory, and in fact many parts of classical (abelian) ergodic theory generalize to the nonabelian setting. A major difference between the abelian and the nonabelian theories arise when we want to describe the invariant states when they exist, and the corresponding operator algebras. While in the abelian case these two problems are in a sense trivial; an invariant state is an integral defined by an invariant probability measure, and the operator algebra is some $C(X)$ or $L^\infty(X,\mu)$, this is far from the case in the nonabelian situation. In the present survey I shall discuss the main aspects of this latter part of the theory.

2. GENERALITIES. Let A be a C^*-algebra, G a group and α a representation of G as $*$-automorphisms of A. In most of our discussion topologies on G are irrelevant, so we may presently assume G is discrete. Suppose ρ is a G-invariant state, i.e. $\rho(\alpha_g(x)) = \rho(x)$ for all $g \in G$, $x \in A$. Then by the GNS-representation there are a Hilbert space H_ρ, a $*$-representation π_ρ of A on H_ρ, and a unit vector $\xi_\rho \in H_\rho$ cyclic for $\pi_\rho(A)$ such that $\rho(x) = (\pi_\rho(x)\xi_\rho,\xi_\rho)$ for $x \in A$. Moreover, there is a unitary representation u of G on H_ρ such that $u_g\xi_\rho = \xi_\rho$ and $\pi_\rho(\alpha_g(x)) = u_g\pi_\rho(x)u_g^{-1}$ for all $g \in G$, $x \in A$. In particular α induces a representation of G as $*$-automorphisms of the weak closure $\pi_\rho(A)^-$ of $\pi_\rho(A)$ by $y \to u_g y u_g^{-1}$. Thus in order to study properties of ρ we may study the von Neumann algebra $\pi_\rho(A)^-$, the vector state ω_{ξ_ρ}, and the unitary representation u. We shall therefore mostly restrict attention to von Neumann algebras and normal invariant states, and whenever desired assume that the states and the automorphisms are as above.

0082-0717/80/0000-0575/$04.25

Let now M be a von Neumann algebra acting on a Hilbert space H and α a representation of the group G into the group $\operatorname{Aut} M$ of *-automorphisms of M. We say G is <u>ergodic</u>, or that α is an <u>ergodic representation</u>, if $\alpha_g(x) = x$ for all $g \in G$ implies $x = \lambda 1$ for some complex number λ. In this case if ω is a normal G-invariant state the support projection of ω; which is clearly a G-invariant operator in M, is a scalar, so is the identity. Thus ω is faithful. Suppose furthermore ω is a vector state defined by a cyclic vector ξ_0 and that $\alpha_g(x) = u_g x u_g^{-1}$ with u a unitary representation of G on H with $u_g \xi_0 = \xi_0$. Since ω is faithful ξ_0 is separating for M, hence by the Tomita theorem [23] there exist a positive self-adjoint operator Δ on H and a conjugation J such that $\Delta\xi_0 = \xi_0$, $J\Delta^{\frac{1}{2}} x\xi_0 = x^*\xi_0$ for all $x \in M$, $JMJ = M'$, and $\Delta^{it} M \Delta^{-it} = M$ for all real t. By uniqueness of polar decomposition of the operator $x\xi_0 \to x^*\xi_0$, $u_g J u_g^{-1} = J$ and $u_g \Delta u_g^{-1} = \Delta$ for all $g \in G$ [19,24]. Notice that the last identity immediately implies that $\alpha_g \circ \sigma_t = \sigma_t \circ \alpha_g$ for all $g \in G$ and real t, where $\sigma_t(x) = \Delta^{it} x \Delta^{-it}$.

Even though we shall usually assume the existence of normal G-invariant states I would like to say a few words on this and its consequences. Following [10] we say M is <u>G-finite</u> if for all nonzero positive $x \in M$ there is a normal G-invariant state ω with $\omega(x) \neq 0$. Let $M^G = \{x \in M : \alpha_g(x) = x, \text{ for all } g \in G\}$ be the fixed point algebra of G in M, and let e_0 be the projection onto the subspace $\{\xi \in H : u_g\xi = \xi \text{ for all } g \in G\}$, where we assume α is unitarily implemented. If M is G-finite and we take GNS-representations due to all normal G-invariant states we may assume e_0 is separating for M, or rather its central support in M is 1, i.e. 1 is the only central operator in M majorizing e_0. A consequence of this and the Alaoglu-Birkhoff fixed point theorem [16] is that e_0 belongs to the strong closure of the convex hull $\operatorname{conv}(u_g : g \in G)$ of the u_g. If we choose a net $\{\sum_i \lambda_i^\alpha u_{g_i^\alpha}\}_{\alpha\in I}$ in $\operatorname{conv}(u_g)$ converging strongly to e_0 then for all $x \in M$

$$e_0 x e_0 = \text{strong} \lim_\alpha \sum_i \lambda_i^\alpha u_{g_i^\alpha} x u_{g_i^\alpha}^{-1} e_0 .$$

Since the central support of e_0 is 1 the map Φ defined by

$$\Phi(x) = \text{strong} \lim_{\alpha} \sum_i \lambda_i^{\alpha} u_{g_i^{\alpha}} x \, u_{g_i^{\alpha}}^{-1}$$

is well defined [5]. From this it is straightforward to obtain the Kovàcs-Szücs theorem.

THEOREM 2.1, [10]. Let M be a von Neumann algebra, G a group and α a representation of G in $\mathrm{Aut}\, M$ so that M is G-finite. Then there exists a faithful normal positive G-invariant map Φ of M onto the fixed point algebra M^G which is the identity on M^G. For $x \in M$, $\Phi(x)$ in the unique element in $M^G \cap \overline{\mathrm{conv}(\alpha_g(x) : g \in G)}$. The map $\rho \to \rho \circ \Phi$ is an affine isomorphism between the normal states on M^G and the normal G-invariant states on M. Conversely, if such a map Φ exists then M is G-finite.

Note that in particular, if G is ergodic then there exists a unique normal G-invariant state ω on M, and $\omega(x)1 = \Phi(x)$. Note also that if G is the group of inner automorphisms of M then G-finiteness is the same as finiteness of M in the usual sense of von Neumann algebras, and Φ is the center valued trace.

It should be remarked that G-finiteness is not very far from compactness of α_G in $\mathrm{Aut}\, M$. Indeed if α_G is compact the map Φ is just the integral over α_g with respect to the normalized Haar measure on α_G. More generally, let $L(M)$ (resp. $L_*(M)$) denote the space of bounded (resp. ultraweakly continuous) linear maps of M into itself, and let $L(M)$ have the point-open topology where M is given the weak topology. If $L_*(M)$ is given the relative topology from $L(M)$ it can be shown [20] that M is G-finite if and only if α_G is relatively compact in $L_*(M)$.

3. SEMIFINITE VON NEUMANN ALGEBRAS.

If the von Neumann algebra M is semifinite and G is ergodic on M we can often give quite explicit description of the invariant states. The first such result is due to Hugenholtz and was directly motivated by quantum statistical mechanics. With some extra physically natural assumptions he proved the following result.

THEOREM 3.1, [9,17]. Suppose M is a factor, G a group and α an ergodic representation of G in $\mathrm{Aut}\, M$. Suppose ω is a normal G-invariant state. Then either M is of type III or M is finite with ω a trace.

It is possible to give a quite simple proof of this theorem. Suppose M is semifinite and let τ be a faithful normal semifinite trace on M. Then τ is G-invariant. Indeed, if this is not the case then by uniqueness of the trace, for all $g \in G$ there is $\lambda(g) > 0$ such that $\tau(\alpha_g(x)) = \lambda(g)\tau(x)$, for $x \in M$, and for some $h \in G$, $\lambda(h) \neq 1$. It is immediate that λ is a multiplicative homomorphism of G into $\mathbb{R}^+$, so there is $h \in G$ with $\lambda(h) < 1$. Choose $x \in M^+$ with $0 < \tau(x) < \infty$. Then $\tau(\alpha_{h^n}(x)) = \lambda(h)^n \tau(x) \to 0$, and it is easily concluded as for normal states that $\alpha_{h^n}(x) \to 0$ strongly. But then $\omega(x) = \omega(\alpha_{h^n}(x)) \to 0$, hence $\omega(x) = 0$, a contradiction. Thus τ is constant on $\operatorname{conv}(\alpha_g(x) : g \in G)$. By Theorem 2.1 $\omega(x)1 \in \operatorname{conv}(\alpha_g(x) : g \in G)^-$, so by lower semicontinuity of τ, $\tau(x) = \omega(x)$, proving the theorem.

If we use the center valued trace instead of τ we may extend the above arguments to prove the following global result.

THEOREM 3.2, [18]. Let M be a semifinite von Neumann algebra with center Z. Let G be a group and α a representation of G in $\operatorname{Aut} M$ such that $M^G \supset Z$. If ω is a faithful normal G-invariant state on M we have:

(1) M^G is generated by projections which are finite in M.

(2) ω is β-invariant for every $\beta \in \operatorname{Aut} M$ such that $\beta | M^G$ is the identity.

(3) $(M^G)' \cap M$ is finite.

If M is of type I then in (3) $(M^G)' \cap M$ is of type I and its center C is totally atomic over Z, i.e. C is generated by projections p such that $pZ = pC$.

In a sense the opposite case to the one above is when Z contains M^G. The extreme such situation is taken care of in the following theorem.

THEOREM 3.3, [19]. Let M be a semifinite von Neumann algebra with center Z. Let G be a group and α a representation of G in $\operatorname{Aut} M$ such that the restriction of α to Z is ergodic. Suppose ω is a faithful normal G-invariant state. Then there exist a unique faithful normal G-invariant trace τ_0 and a positive self-adjoint operator $b \in L^1(M, \tau_0)$ affiliated with M^G such that $\omega(x) = \tau_0(bx)$ for all $x \in M$.

The main ideas in the proof are contained in the following formal proof, which ignores that most operators involved are unbounded.

Let τ be a faithful normal semifinite trace on M and let $h \in L^1(M, \tau)$ be such that $\omega(x) = \tau(hx)$ for $x \in M$. We may assume we are in the GNS-representation due to ω, so α is implemented by u and ω is a vector state. Let Δ and J be the corresponding operators from Tomita theory, cf. section 2. Then

$\Delta = hJh^{-1}J$ [23]. Thus we have by the invariance of Δ and J by u

$$hJh^{-1}J = \Delta = u_g \Delta u_g^{-1} = u_g h u_g^{-1} u_g J h^{-1} J u_g^{-1}$$
$$= u_g h u_g^{-1} J u_g h^{-1} u_g^{-1} J.$$

Thus $J h^{-1} u_g h u_g^{-1} J = h^{-1} u_g h u_g^{-1}$ for all $g \in G$. Therefore by the construction of the map Φ in Theorem 2.1 we have

$$J h^{-1} \Phi(h) J = h^{-1} \Phi(h) \in M \cap M' = Z.$$

Thus $\Phi(h) = ch$ with $c \in Z$, so if we let $\tau_0(x) = \tau(c^{-1}x)$ then τ_0 is easily seen to be a faithful normal G-invariant trace, and $\omega(x) = \tau(hx) = \tau(c^{-1}chx) = \tau_0(\Phi(h)x)$, completing the proof.

In the rigorous proof we must consider $\Delta^{it} = h^{it} J h^{it} J$ for all real t, and as above conclude $\Phi(h^{it}) = c_t h^{it}$. Then c_t depends on t and more work is needed to complete the proof. While in the formal proof we don't need the assumption that G is ergodic on Z this is needed for technical reasons in the rigorous proof. If separability assumptions are added the ergodicity hypothesis is redundant [6, 13], but in the general case it is an open problem whether the theorem is true without this extra assumption.

Note that if G is ergodic on M then $M^G = \mathbb{C}1$ and a normal G-invariant state ω is automatically faithful. Thus the operator b in Theorem 3.3 is a scalar, and we have the following extension of Theorem 3.1.

COROLLARY 3.4. Let M be a semifinite von Neumann algebra and G an ergodic group of $*$-automorphisms. If ω is a normal G-invariant state then M is finite and ω is a trace.

In applications to C*-algebras it is often advantageous to formulate the last corollary in the setting of the GNS-representation due to an invariant state. Since a separating and cyclic vector for a von Neumann algebra M is a trace vector if and only if it is a trace vector for M' the following result is quite easy to prove from Corollary 3.4.

COROLLARY 3.5. Let M be a von Neumann algebra acting on a Hilbert space H. Let G be a group and u a unitary representation of G on H such that $u_g M u_g^{-1} = M$ for all $g \in G$. Suppose there exists a unit vector ξ_0 in H which is cyclic for M and such that $\mathbb{C}\xi_0 = \{\xi \in H : u_g \xi = \xi \text{ for all } g \in G\}$. Then M is semifinite if and only if ξ_0 is a trace vector for M'.

4. ASYMPTOTICALLY ABELIAN C*-ALGEBRAS. The results in the previous section were inspired by the theory of asymptotically abelian C*-algebras, which again were directly motivated by quantum statistical mechanics. Let A be a C*-algebra, G a group and α a representation of G in $\operatorname{Aut} A$. Let ρ be a G-invariant state on A and use the notation in section 2. Let $M = \pi_\rho(A)^-$ and e_0 be the projection onto the G-invariant vectors in H_ρ. If $e_0 M e_0$ is an abelian set of operators for all G-invariant ρ then A is called G-abelian [11]. This is the most general of the many definitions of asymptotically abelian C*-algebras, see [5]. One of the more special ones, which I find more intuitive is the following. We say A is (norm) asymptotically abelian with respect to G if there exists a sequence (g_n) in G such that for all $x,y \in A$

$$\lim_{n\to\infty} \|\alpha_{g_n}(x)y - y\alpha_{g_n}(x)\| = 0.$$

Many theorems in the theory can be proved in the general situation of G-abelianness. The basic one, [5,11], is that the G-invariant states on A form a Choquet simplex, hence it is natural to study the extreme points of this convex set called the ergodic states. With ρ and e_0 as above it can be shown, see e.g. [5], that ρ is an ergodic state if and only if e_0 is 1-dimensional. We are thus in the situation of Corollary 3.5, hence we have

THEOREM 4.1, [19]. Let A be a C*-algebra which is G-abelian with respect to a group G of $*$-automorphisms. Suppose ρ is a G-invariant ergodic state with GNS-representation $(\pi_\rho, H_\rho, \xi_\rho)$. Then $\pi_\rho(A)^-$ is a semifinite von Neumann algebra if and only if ξ_ρ is a trace vector for $\pi_\rho(A)'$.

While it is easy from Theorem 4.1 to deduce most of the information we want on ρ and $M = \pi_\rho(A)^-$ in the semifinite case, it tells us nothing about ρ if M is of type III. Let $e_\rho = [M'\xi_\rho]$ be the support of the vector state ω_{ξ_ρ} on M. Then ξ_ρ is separating and cyclic for M_{e_ρ} on $e_\rho H_\rho$, and thus has a modular operator Δ_ρ relative to M_{e_ρ}. Define the modular operator of ρ on all of H_ρ by letting it be 0 on $(1-e_\rho)H_\rho$, and continue to denote it by Δ_ρ. We define the spectrum of ρ to be $\operatorname{spec}(\rho) = \operatorname{spec} \Delta_\rho$, see [21], where a more intrinsic definition is also given. As a slight variation of the invariant $S(M)$ of Connes [3] we define the invariant $S'(M)$ as

$$S'(M) = \cap \{\mathrm{spec}(\omega) : \omega \text{ normal state on } M\}.$$

THEOREM 4.2, [2,21]. Let A be a C*-algebra which is asymptotically abelian with respect to a group G. Let ρ be a G-invariant ergodic state on A with GNS-representation $(\pi_\rho, H_\rho, \xi_\rho)$. Then $\mathrm{spec}(\rho) = S'(\pi_\rho(A)'')$, and the nonzero elements in $\mathrm{spec}(\rho)$ form a closed multiplicative subgroup of $\mathbb{R}^+$.

Generalizations of the above theorem can also be found in [7,12]. Let me indicate the first proof of the theorem, which was given in the case when ρ is strongly clustering [21], since the main points in that proof are quite transparent. First we recall that ρ is said to be <u>strongly clustering</u> if for all $x, y \in A$

$$\lim_n \rho(\alpha_{g_n}(x)y) = \rho(x)\rho(y),$$

where (g_n) is the sequence in the definition of asymptotic abelianness. Then ρ is in particular an ergodic state, as is rather easily proved. For simplicity we also assume ξ_ρ is separating and cyclic for $M = \pi_\rho(A)''$. Let $\Delta = \Delta_\rho$ and J be the corresponding operators from Tomita theory. Note that if $x\xi_\rho$ is an eigenvector for Δ with eigenvalue λ, $x \in M$, we have

$$\lambda^{\frac{1}{2}} x\xi_\rho = \Delta^{\frac{1}{2}} x\xi_\rho = J x^*\xi_\rho = J x^* J\xi_\rho$$

and similarly for $x^*\xi_\rho$. More generally it can be shown [3] that $\lambda \in \mathrm{spec}\,\Delta$ if and only if for each $\varepsilon > 0$ there is $x \in M$ such that $\|x\xi_\rho\| = 1$ and

$$(4.1) \qquad \|\lambda^{\frac{1}{2}} x\xi_\rho - J x^* J\xi_\rho\| < \varepsilon, \quad \|x^*\xi_\rho - \lambda^{\frac{1}{2}} J x J\xi_\rho\| < \varepsilon.$$

Let now $\lambda \in \mathrm{spec}\,\Delta (=\mathrm{spec}(\rho))$ and let φ be a normal state on M. By density arguments we may choose x as above in $\pi_\rho(A)$, and we may assume φ a faithful vector state ω_ψ with $\psi = y\xi_\rho$, $y \in \pi_\rho(A)$, and $J\psi = \psi$. We shall show $\lambda \in \mathrm{spec}\,\varphi$. If u is the unitary representation on H_ρ implementing α we let $x_n = u_{g_n} x u_{g_n}^{-1}$ and choose n_0 so large that $\|[x_n, y]\| < \varepsilon$ for $n \geq n_0$. Then we have for $n \geq n_0$

$$\begin{aligned}
\|\lambda^{\frac{1}{2}} x_n\psi - J x_n^* J\psi\| &= \\
&= \|\lambda^{\frac{1}{2}} x_n y\xi_\rho - J x_n^* J y\xi_\rho\| \\
&\leq \lambda^{\frac{1}{2}} \|[x_n, y]\xi_\rho\| + \|\lambda^{\frac{1}{2}} y x_n \xi_\rho - y J x_n^* J\xi_\rho\| \\
&< \lambda^{\frac{1}{2}}\varepsilon + \|y(\lambda^{\frac{1}{2}} u_{g_n} x\xi_\rho - J u_{g_n} x^* J\xi_\rho)\|
\end{aligned}$$

$$= \lambda^{\frac{1}{2}}\varepsilon + \|y\, u_{g_n}\, (\lambda^{\frac{1}{2}}\, x\, \xi_\rho - Jx^*J\,\xi_\rho)\|$$
$$\leq \lambda^{\frac{1}{2}}\varepsilon + \|y\|\,\varepsilon\,,$$

and similarly for $\|x_n^*\psi - \lambda^{\frac{1}{2}} J x_n J \psi\|$. By strong clustering we have

$$\begin{aligned}\lim_n \|x_n\psi\| &= \lim_n\, (u_{g_n} x^*x\, u_{g_n}^{-1} y\,\xi_\rho, y\xi_\rho)\\ &= (x^*x\,\xi_\rho, \xi_\rho)\,(y\,\xi_\rho, y\,\xi_\rho)\\ &= \|x\,\xi_\rho\|^2\,\|y\,\xi_\rho\|^2 = 1\,.\end{aligned}$$

Thus by (4.1) $\lambda \in \operatorname{spec}\varphi$ as asserted. It is possible to use an extension of the above argument to conclude $\operatorname{spec}\rho\setminus\{0\}$ is a multiplicative group, but this can also be deduced from the fact that $S(M)$ is a group.

From Theorems 4.1 and 4.2 it is now an easy matter to write down the following description of ergodic states.

COROLLARY 4.3. Let A be a C*-algebra which is asymptotically abelian with respect to a group G. Let ρ be a G-invariant ergodic state with GNS-representation $(\pi_\rho, H_\rho, \xi_\rho)$. Let $M = \pi_\rho(A)^-$ and $e = [M'\xi_\rho]$. Let $\omega = \omega_{\xi_\rho}$ on M. Then the following possibilities may occur.

1) M is of type I_∞ or II_∞ and $\omega|M_e$ is a trace.

2) M is finite and ω is a trace.

3) M is of type III_λ, $0 < \lambda < 1$, and $\operatorname{spec}\rho = \{\lambda^n : n \in \mathbb{Z}\}^-$.

4) M is of type III_1.

Note that M cannot be of type III_0 and we obtain no information on ρ in the type III_1 case. If we further assume G is abelian, if M is of type I, then we have M is either 1-dimensional and ρ is a homomorphism onto the complex numbers, or M is of type I_∞, in which case ρ is a pure state.

At this point it is natural to apply the preceding to the following important example [3,15].

For each $\mu \in [0,\frac{1}{2}]$ let $\lambda = \mu/(1-\mu)$, and let ω_λ be the state on the complex 2×2 matrices M_2 defined by

$$\omega_\lambda\left(\begin{pmatrix} a & b\\ c & d\end{pmatrix}\right) = \mu a + (1-\mu)d.$$

Let A be the infinite tensor product $A = \bigotimes_{i=1}^{\infty} N_i$, where each $N_i = M_2$ and let ρ_λ be the product state $\rho_\lambda = \bigotimes_1^\infty \rho_i$ on A, where $\rho_i = \omega_\lambda$ for each i. If G is the group of finite permutations on the natural numbers then G is represented in $\mathrm{Aut}\,A$ by permuting the factors in the tensor product. It is easy to see that A is asymptotically abelian with respect to G and that ρ_λ is a strongly clustering state. Let $M_\lambda = \pi_{\rho_\lambda}(A)^-$. Then M_λ is a factor - a so-called Powers factor. We shall now show the famous result of Powers [15] that the factors M_λ are mutually nonisomorphic by modifying an argument of Connes [3]. If $\lambda = 0$ then ρ_λ is a pure state, so we assume $0 < \lambda \leq 1$. Then for each n $\bigotimes_1^n \rho_i$ is a faithful state on $\bigotimes_1^n N_i$ with spectrum $\{\lambda^j : j=0,\pm 1,\ldots,\pm n\}$, as follows from the fact that $\Delta_{\bigotimes_1^n \rho_i} = \bigotimes_1^n \Delta_{\rho_i}$. A little argument now shows that $\mathrm{spec}\,\rho_\lambda = \{\lambda^j : j \in \mathbb{Z}\}^-$, which by Corollary 4.3 shows that M_λ is of type III_λ if $0 < \lambda < 1$, and of type II_1 if $\lambda = 1$.

5. COMPACT GROUPS. While up to this point the group G in question has been treated as a discrete group we shall now study a situation in which the topology on G is of main importance. We shall assume G is a compact group and that the representation is continuous. If $\alpha : G \to \mathrm{Aut}\,A$ with A a C*-algebra, we assume α is point-norm continuous, i.e. $\lim_{g\to e}\|\alpha_g(x)-x\| = 0$ for all $x \in A$, where e is the identity in G. If $\alpha : G \to \mathrm{Aut}\,M$ and M is a von Neumann algebra we assume the functions $g \to \varphi(\alpha_g(x))$ are continuous for all $x \in M$ and all ultraweakly continuous linear functionals φ on M. The case of abelian groups were studied first and is much easier than the nonabelian case. Then by equation (5.2) below ergodic actions are closely related to the Weyl commutation relations and are in a weak sense classified, even though it is not yet known which compact abelian groups have faithful ergodic representation in $\mathrm{Aut}\,M$ for M not of type I. Most of the known general results for abelian groups are summarized in the following theorem.

THEOREM 5.1, [1,14,22]. Let M be a von Neumann algebra with center Z, G a compact abelian group, and α a continuous ergodic representation of G in $\mathrm{Aut}\,M$. Then we have:

(1) The unique normal G-invariant state ω is a trace and is given by the formula

$$\omega(x)1 = \int_G \alpha_g(x)\,dg, \quad x \in M,$$

where dg is the normalized Haar measure on G.

(2) If G is second countable and $G_Z = \{g \in G : \alpha_g(x) = x \text{ for } x \in Z\}$ then $M \simeq L^\infty(G/G_Z) \otimes N$, where either $N = M_n$ or N is the hyperfinite II_1-factor.
Furthermore, M is a factor if and only if the subgroup of $g \in G$ for which α_g is inner on M is dense in G.

The proof of (1) is quite instructive and shows how the theory of spectral subspaces enters and becomes easy because $\hat{G}$ is a discrete group. The first observation is that ω defined by the integral formula in (1) is the unique normal invariant state. Let $x \in M$, $p \in \hat{G}$, and define

$$\hat{x}(p) = \int_G \overline{\langle p,g\rangle}\alpha_g(x)\, dg.$$

Then $\hat{x}(p) \in M$ and by left invariance of Haar measure, if $h \in G$ then

$$(5.1) \qquad \alpha_h(\hat{x}(p)) = \langle p,h\rangle \hat{x}(p).$$

Thus $\hat{x}(p)$ belongs to the spectral subspace $M(p)$ of elements satisfying (5.1). Since G is ergodic $M(p)$ is either 0 or 1-dimensional. In the latter case it is easily seen that there exists a unitary operator $u(p) \in M(p)$ and that we have

$$\alpha_g(u(p)\, u(q)) = \alpha_g(u(p))\alpha_g(u(q)) = \langle p,g\rangle\, u(p)\langle q,g\rangle\, u(q)$$
$$= \langle p+q,g\rangle\, u(p)\, u(q),$$

hence

$$(5.2) \qquad u(p)\, u(q) = \lambda(p,q)\, u(p+q)$$

for some complex number $\lambda(p,q)$ of modulus 1. In particular the set of p with $M(p) \neq 0$ form a subgroup Λ of $\hat{G}$. From the theory of spectral subspaces, or by using the GNS-representation of ω and Stone's theorem, we conclude

$$(5.3) \qquad M = \Big(\sum_{p\in\Lambda} M(p)\Big)^- = \Big(\sum_{p\in\Lambda} \mathbb{C}\, u(p)\Big)^- .$$

We have for $0 \neq p \in \hat{G}$, $x \in M$,

$$\omega(\hat{x}(p)) = \int \overline{\langle p,g\rangle}\omega(\alpha_g(x))dg = \omega(x)\int \overline{\langle p,g\rangle} dg = 0$$

Thus $\omega(u(p)) = 0$ if $p \neq 0$ and 1 if $p = 0$. Therefore by (5.2) $\omega(u(p)\, u(q)) = \omega(u(q)u(p))$ for all $p,q \in \hat{G}$, hence ω is a trace by (5.3).
Concerning (2) we only show M is injective, hence if M is a factor then

M is either finite dimensional or the hyperfinite II_1-factor [4]. Let $K = \{Ad\, u(p) : p \in \Lambda\}$. By (5.2) K is an abelian subgroup of $Aut\, B(H)$, where $B(H)$ denotes the bounded operators on the underlying Hilbert space H. By (5.3) the fixed points of K in $B(H)$ is $B(H)^K = M'$. Since K is abelian it is amenable so has an invariant mean. If we average $u(p)x\,u(p)^*$ over this mean for $x \in B(H)$, we obtain an idempotent positive linear map of $B(H)$ onto M'. Thus M' is injective and therefore M.

If the compact group is nonabelian the situation is much more complicated. Even to construct nontrivial examples seems to be a major obstacle. The only ones I know at present are finite extensions of abelian examples and infinite products $\prod_{i=1}^{\infty} G_i$ of compact groups acting on an infinite tensor product $\bigotimes_{i=1}^{\infty} A_i$, where for each i G_i acts ergodically on A_i. In the general theory we can still get the analogues of the spectral subspaces $M(p)$, now defined by irreducible representations of the group G, but the subspaces are no longer 1-dimensional. Furthermore, we have only a very weak form of (5.2), which now corresponds to taking tensor products of irreducible representations. Since the analogue of (5.3) is still true, it is possible to derive a large part of Theorem 5.1. This can be done both for C*- and von Neumann algebras. I state it first for C*-algebras.

THEOREM 5.2, [8]. Let A be a C*-algebra, G a compact group and α a continuous ergodic representation of G in $Aut\, A$. Then the formula

$$\omega(x)1 = \int_G \alpha_g(x)dg$$

defines the unique G-invariant state on A. ω is a trace, and A is a nuclear C*-algebra.

COROLLARY 5.3. Let M be a von Neumann algebra, G a compact group and α a continuous ergodic representation of G in $Aut\, M$. Then the unique normal G-invariant state is a trace, so M is finite, Furthermore, M is injective.

Note that by Corollary 3.4 the proof of the last result is really an argument on type III von Neumann algebras. The corollary follows from the observation that the set A of $x \in M$ for which the function $g \to \alpha_g(x)$ is norm continuous, is a C*-algebra which is dense in M and globally invariant under G. Since the unique normal G-invariant state is a trace on A by Theorem 5.2 it is by continuity a trace on M.

Finally I'll state a variation of Corollary 5.3 which is more in the spirit of the earlier sections in that it avoids topological considerations. Its proof can be read out of the proof of Theorem 5.2. Recall that if ω is a normal state its <u>centralizer</u> is the set $M_\omega = \{x \in M : \omega(xy-yx) = 0$ for all $y \in M\}$. If ω is faithful $M_\omega = \{x \in M : \Delta_\omega^{it} x \Delta_\omega^{-it} = x$ for all real $t\}$, [23].

COROLLARY 5.4. Let M be a von Neumann algebra, G a group and α an ergodic representation of G in Aut M. Suppose ω is a normal G-invariant state and that V is a finite dimensional subspace of M globally invariant under G. Then V is contained in the centralizer M_ω of ω.

REFERENCES

1. S. ALBEVERIO and R. HØEGH-KROHN, Ergodic actions by compact groups on C*-algebras, Math. Zeitschrift, 174 (1980), 1-18.

2. H. ARAKI, Remarks on spectra of modular operators of von Neumann algebras, Comm. Math. Phys., 28 (1972), 267-278.

3. A, CONNES, Un nouvel invariant pour les algèbres de von Neumann, Compt. Rend. Ser. A, 273 (1971), 900-903.

4. A. CONNES, Classification of injective factors, Ann. of Math., 104 (1976), 73-115.

5. S. DOPLICHER, D. KASTLER and E. STØRMER, Invariant states and asymptotic abeliannes, J. Funct. Anal., 3 (1969), 419-434.

6. A. GUICHARDET, Système dynamiques non commutatifs, Asterisque 13-14 (1974).

7. R. HERMAN, Spectra of automorphism groups of operator algebras, Duke Math. J., 41 (1974), 667-674.

8. R. HØEGH-KROHN, M. LANDSTAD and E. STØRMER, Compact ergodic groups of automorphisms, Ann. of Math., to appear.

9. N. HUGENHOLTZ, On the factor type of equilibrium states in quantum statistical mechanics, Comm. Math. Phys., 6 (1967), 189-193.

10. I. KOVÁCS and J. SZÜCS, Ergodic type theorems in von Neumann algebras, Acta Sci. Math., 27 (1966), 233-246.

11. O. LANFORD and D. RUELLE,Integral representations of invariant states on B*-algebras, J. Math. Phys., 8 (1967), 1460-1463.

12. R. LONGO, Notes on algebraic invariants for non-commutative dynamical systems, Comm. Math. Phys., 69 (1979), 195-207.

13. R. NEST, A non-commutative version of the maximal ergodic theorem for invariant traces, preprint, Copenhagen 1976.

14. D. OLESEN, G.K. PEDERSEN and M. TAKESAKI, Ergodic actions of compact abelian groups, J. Operator Theory, to appear.

15. R. POWERS, Representations of uniformly hyperfinite algebras and their associated von Neumann rings, Ann. of Math., 86 (1967), 138-171.

16. F. RIESZ and B. SZ. NAGY, Leçons d'analyse fonctionelle, Akadémiai Kiado, Budapest 1955.

17. E. STØRMER, Types of von Neumann algebras associated with extremal invariant states, Comm. Math. Phys., 6 (1967), 194-204.

18. E. STØRMER, States and invariant maps of operator algebras, J. Funct. Anal., 5 (1970), 44-65.

19. E. STØRMER, Automorphisms and invariant states of operator algebras, Acta math., 127 (1971), 1-9.

20. E. STØRMER, Invariant states of von Neumann algebras, Math. Scand., 30 (1972), 253-256.

21. E. STØRMER, Spectra of states, and asymptotically abelian C*-algebras, Comm. Math. Phys., 28 (1972), 279-294, and 38 (1974), 341-343.

22. E. STØRMER, Spectra of ergodic transformations, J. Funct. Anal., 15 (1974), 202-215.

23. M. TAKESAKI, Tomita's theory of modular Hilbert algebras and its applications, Springer-Verlag, Lecture notes in mathematics, 128 (1970).

24. M. WINNINK, An application of C*-algebras to quantum statistical mechanics of systems in equilibrium, Thesis, Groningen 1968.

DEPARTMENT OF MATHEMATICS,
UNIVERSITETET I OSLO
BLINDERN, OSLO 3,
NORWAY

Proceedings of Symposia in Pure Mathematics
Volume 38 (1982), Part 2

COMPACT ERGODIC GROUPS OF AUTOMORPHISMS

Magnus B. Landstad

A C*-dynamical system (G,α,A) can be considered as a generalization of a projective representation of G in the following way: If U is a projective representation of the group G on a Hilbert space H, consider the corresponding action α on B(H) by

$$\alpha_x(a) = U_x a U_x^*.$$

Irreducible projective representations is seen to correspond to ergodic actions (i.e. $\alpha_x(a) = a$ for all $x \in G$ only if a is a scalar) and are therefore of particular interest. If the group G is compact and abelian a complete description of all its ergodic actions has been made by S. Albeverio, R. Høegh-Krohn and D. Olesen, G. K. Pedersen, M. Takesaki, see [3] for details. The following result which is joint work with R. Høegh-Krohn and E. Størmer would be the first step in a similar classification for compact non-abelian groups:

THEOREM. Let (A,α,G) be a C*-dynamical system with A unital and G compact. If the action is ergodic A has a unique G-invariant state ω which is a trace. Furthermore A is nuclear.

COROLLARY. If (M,α,G) is an ergodic W*-dynamical system with G compact, then M is finite and injective.

If the group is abelian the existence of a trace follows from the existence of unitary eigenoperators and was first proved by E. Størmer in [4]. The nuclearity (and injectivity) was proved for abelian groups by D. Olesen, G. K. Pedersen, M. Takesaki in [3]. I shall now give a sketch of the proof in the general case, for details see [1].

First define a state ω on A by

$$(*) \qquad \omega(a) = \int \alpha_g(a)dg.$$

Then ω is faithful and is the only G-invariant state on A. By the GNS-construction we have

0082-0717/80/0000-0544/$01.75

$$\omega(a) = \langle a\,\xi_0, \xi_0\rangle$$

over a Hilbert space H. Let Δ be the modular operator over H associated with ω.

LEMMA 1. For each $D \in \hat{G}$, D occurs at most dim D times in α. Furthermore, there are finite dimensional subspaces $V_j(D)$ of A such that

(1) $H = \sum \oplus V_j(D)\xi_0$

(2) $\alpha_g(V_j(D)) = V_j(D)$ for $g \in G$

(3) $\alpha|V_j(D) \cong D$

(4) $\Delta\xi = \lambda_j(D)\xi$ for $\xi \in V_j(D)\xi_0$ for some $\lambda_j(D) > 0$, i.e. Δ is diagonalizable

(5) $\lambda_j(\bar{D}) = \lambda_j(D)^{-1}$

(6) $\lambda_j(D) \leq \dim D$.

Proof. Careful choice of a maximal orthonormal set in the D-spectral subspace A(D) of A.

Now fix one of these subspaces $V = V_j(D)$, and let $\lambda = \lambda_j(D)$, $n = \dim V = \dim D$. Take

$$V^m = \text{lin span}\,\{a_1 \cdots a_m | a_i \in V\},$$

this will be a G-invariant subspace of A.

LEMMA 2. If U is a G-invariant irreducible subspace of V^m then

$$\dim U \leq (1 + m)^{n(n-1)/2}.$$

Proof. The multiplication map $V \otimes \cdots \otimes V \to V^m$ transfers this statement to one about irreducible subspaces of a tensor-product representation. The dimension estimate then follows from H. Weyl's classification of the finite dimensional irreducible representations of $GL(n, \mathbb{C})$ occurring in a tensor-product representation.

By (5) we may suppose $\lambda \geq 1$, and we take $U \neq 0$ as in Lemma 2 (this is possible). Now

$$\Delta a\,\xi_0 = \lambda a\,\xi_0 \quad \text{for} \quad a \in V, \quad \text{so}$$

$$\Delta b\,\xi_0 = \lambda^m b\xi_0 \quad \text{for} \quad b \in V^m.$$

If $b \in U$ we therefore have by (6) and Lemma 2

$$\lambda^m \leq \dim U \leq (1 + m)^{n(n-1)/2}$$

$$\lambda \leq (1+m)^{n(n-1)/2m} \to 1 \quad \text{as} \quad m \to \infty.$$

So $\lambda = 1$, $\Delta = I$ and ω is a trace.

The nuclearity of A is easiest seen by the following argument communicated to me by U. Haagerup and which also has been independently made by Y. Katayama, G. Song in [2]: Let $\{f_i\}$ be an approximate identity of $L^1(G)$ consisting of trigonometric polynomials f_i with $f_i(x) \geq 0$ for all $x \in G$. The maps $p_i : A \to A$ defined by

$$p_i(a) = \int f_i(x)\alpha_x(a)dx$$

are then completely positive, by Lemma 1 p_i is of finite rank and $p_i(a) \to a$ in norm. So A is nuclear.

Finally the Corollary follows from the observations that

$$A = \{a \in M \mid x \to \alpha_x(a) \text{ is norm continuous}\}$$

is a σ-weakly dense C*-subalgebra of M and that the state ω defined on M by (*) will be normal.

BIBLIOGRAPHY

1. R. Høegh-Krohn, M. B. Landstad, E. Størmer, "Compact ergodic groups of automorphisms", Ann. of Math. (to appear).

2. Y. Katayama, G. Song, "Ergodic co-actions of discrete groups", preprint 1980.

3. D. Olesen, G. K. Pedersen, M. Takesaki, "Ergodic actions of compact abelian groups", J. of Operator Theory 3 (1980), 237-269.

4. E. Størmer, "Spectra of ergodic transformations", J. Functional Analysis 15 (1974), 202-215.

DEPARTMENT OF MATHEMATICS
UNIVERSITY OF TRONDHEIM, NLHT
7055 DRAGVOLL, NORWAY

Department of Mathematics
University of California, Los Angeles
Los Angeles, CA 90024

Proceedings of Symposia in Pure Mathematics
Volume **38** (1982), Part 2

ACTIONS OF DISCRETE GROUPS ON FACTORS

V. F. R. JONES

ABSTRACT. Beginning with Connes' work on single automorphisms, we survey recent progress in the study of actions of countable discrete groups on factors. Further topics for research and some open problems are indicated.

1. SINGLE AUTOMORPHISMS. The original motivation for the detailed study of actions of discrete groups came from knowledge of the structure of type III factors. In his thesis [CT], Connes defined the invariant $\lambda \in [0,1]$ for a type III factor (thereafter called type III_λ) and, among other things, showed that for $\lambda \in (0,1)$, a type III_λ factor M decomposes as the crossed product of a type II_∞ factor N by an action of the group $\mathbb{Z}$ coming from an automorphism α of N which scales the trace on N by the module λ. The uniqueness properties of this decomposition are rather strong: the factor N is determined up to isomorphism (indeed up to inner conjugacy in M), and the automorphism α is determined up to outer conjugacy (two automorphisms α and β are outer conjugate when there is a third automorphism θ and an inner automorphism Ad u with $\theta\alpha\theta^{-1} = \mathrm{Ad}\, u\beta$).

To be sure that this discrete crossed product decomposition was of more than aesthetic value, it was necessary to show that, at least for some II_∞ factor N, something can be said about outer conjugacy classes of trace-scaling automorphisms. The best possible would be to completely determine the set of all trace-scaling automorphisms up to outer conjugacy. The obvious candidate for such an N was the hyperfinite II_∞ factor $R_{0,1}$, which is the tensor product of the hyperfinite II_1 factor R by a type I_∞ factor.

Connes' first important observation must have been that the problem of outer conjugacy for $R_{0,1}$ can be reduced to the trace-preserving case by a trick which I shall now describe. Since $R_{0,1}$ is isomorphic to $R_{0,1} \otimes R_{0,1}$, one can define a binary operation on outer conjugacy classes $[\alpha]$ of automorphisms α with α^n outer for $n \neq 0$: $[\alpha]\cdot[\beta] = [\alpha \otimes \beta]$. Moreover the module λ defines a homomorphism from the resulting semigroup to the real line, known to be surjective by results of Murray and von Neumann [MvN]. It is also a (non-trivial) fact that this semigroup has an identity

---

1980 Mathematics Subject Classification. 46L40, 46L55, 20C25, 20F29

There is a trace preserving automorphism α of $R_{0,1}$ with α^n outer for $n \neq 0$ such that if β is any automorphism all of whose powers are outer, then $[\alpha \otimes \beta] = [\beta]$. So if we could show that there is only one outer conjugacy class of trace-preserving automorphisms then the semigroup described above would be a group isomorphic, via the modulus λ, to the real line.

What has been achieved by reducing the problem to trace preserving automorphisms? The important fact about such an automorphism α of $R_{0,1}$ is that it is approximately inner, i.e. there is a sequence u_n of unitaries with $\lim_{n\to\infty} \mathrm{Ad}\, u_n = \alpha$ in the appropriate topology. Thus one might hope to classify trace-preserving automorphisms of $R_{0,1}$ by playing off an increasing sequence of matrix algebras generating $R_{0,1}$ with the u_n's satisfying $\mathrm{Ad}\, u_n \to \alpha$. This is what Connes does in [C 0]. In fact he proves that if M is any factor with separable predual isomorphic to $M \otimes R$ and α and β are approximately inner automorphisms such that α^n and β^n act non-trivially on central sequences(*) for $n \neq 0$, then α and β are outer conjugate. The proof is delicate to the end and uses MacDuff's algebra of central sequences ([McD]) suitably generalized for infinite factors.

With characteristic thoroughness, Connes also looked at periodic automorphisms. In this case one can actually go from outer conjugacy to conjugacy, but the classification is more difficult to describe as some invariants appear. For simplicity let us restrict ourselves to periodic automorphisms of the hyperfinite II_1 factor R. If α is such an automorphism, let p be the first power of α such that α^p is an inner automorphism, say $\alpha^p = \mathrm{Ad}\, v$ for a unitary $v \in R$. Then since α commutes with $\mathrm{Ad}\, v$ and R is a factor, there is a scalar γ with $|\gamma| = 1$ such that $\alpha(v) = \gamma v$. This γ does not depend on the choice of v with $\alpha^p = \mathrm{Ad}\, v$. Also $\gamma^p = 1$. In fact one obtains the same γ if α is changed by an inner automorphism so that γ is an outer conjugacy invariant. The main result of [CP] is that the pair (p,γ) is a complete outer conjugacy invariant for periodic automorphisms of R. Thus if $\mathrm{Out}\, R$ is the quotient of $\mathrm{Aut}\, R$ by the subgroup of inner automorphisms, one has a complete list of its conjugacy classes. There is one aperiodic class and a countable family of classes indexed by (p,γ) where $p \in \mathbb{N}$ and γ runs through the pth roots of unity.

Note in passing that Connes seized the complex number γ as an invariant sensitive to changing from an algebra to its opposite and gave in [CA] and

---

*To say that an automorphism α acts non-trivially on central sequences means there is a sequence x_n in the unit ball of M with $\lim_{n\to\infty} \sup_{\|y\|\leq 1} |\varphi(x_n y - y x_n)| = 0$ for any $\varphi \in M_*$ (i.e. a central sequence), and such that $\alpha(x_n) - x_n$ does not tend *-strongly to zero.

[CC] the first examples of factors not antiisomorphic to themselves, solving a problem posed by Murray and von Neumann.

Coming back to an automorphism α of R with period n and invariants (p,γ), one sees that another invariant is needed to go from outer conjugacy to conjugacy. For if α is an inner automorphism then $p = 1$ and all periodic automorphisms are certainly not conjugate. The required invariant is obtained as follows. Let r be the order of γ as a pth root of unity. Then v^r satisfies $\alpha^{pr} = \mathrm{Ad}\, v^r$ and $\alpha(v^r) = 1$. By changing v by a scalar one may also suppose that $v^{rq} = 1$ where $prq = n$. Thus the spectral projections of v^r are indexed by the qth roots of unity. Evaluating the trace on these spectral projections gives a probability measure on the qth roots of unity but note that this measure is not perfectly defined by $\mathrm{Ad}\, v^r$ as we may change v^r by any qth root of unity. So the <u>inner invariant</u> $\varepsilon(\alpha)$ of α is a probability measure on the qth roots of unity, defined up to a rotation. In [CP] it is shown that a periodic automorphism of R is determined up to conjugacy by p, n, γ and $\varepsilon(\alpha)$.

This highly satisfactory classification of automorphisms does not persist for nonhyperfinite factors. In [P], Phillips gives examples of II_∞ factors with many (in fact a non-smooth family of) outer conjugacy classes of automorphisms scaling the trace by λ for each $\lambda \in (0,1)$. The invariant he uses to distinguish these automorphisms is basically the topology on $\mathbb{Z}$ obtained when $\mathbb{Z}$ is embedded in Out M as powers of α, where M is the factor under consideration. Thus there is a non-smooth family of type III_λ factors for every $\lambda \in (0,1)$.

Before going on to discuss actions of more general discrete groups, let me remind the reader that this detailed study of automorphisms was imposed by the structure theory of type III factors. Connes tied things up by finding the intrinsic condition (injectivity) which guarantees that the II_∞ factor in the crossed product decomposition be hyperfinite. He also determined the outer conjugacy classes for injective type III_λ factors ($\lambda \in (0,1)$) and the Araki-Woods type III_1 factor.

2. COCYCLE CONJUGACY. While casting around for a serious thesis topic, the author asked Connes if it would be interesting to look at actions of other groups than cyclic ones on R. His answer was unequivocal: yes and do it. Several things were not clear. The first was to find the appropriate generalization of outer conjugacy. Rather than retrace the painful steps that led to the answer, let me just say that it is cocycle conjugacy which I shall now define. If $\alpha : G \to \mathrm{Aut}\, M$ is an action of the group G on the von Neumann algebra M, an α-cocycle consists of a unitary w_g for each $g \in G$ such

that $w_g\alpha_g(w_h) = w_{gh}$. If α and β are two actions, they are said to be <u>cocycle conjugate</u> if there is an α-cycle $\{w_g\}$ such that β is conjugate to the action $g \to \operatorname{Ad} w_g \alpha_g$. This notion coincides with outer conjugacy for cyclic group actions.

The second problem was to find an invariant of cocycle conjugacy to generalize the invariant γ described above. This was easy when G was abelian. Let M be a factor and α be an action of the <u>abelian</u> group G on M. Let $N(\alpha)$, be the subgroup of G acting by inner automorphisms and for each $h \in N(\alpha)$, choose v_h with $\operatorname{Ad} v_h = \alpha_h$. Then for each $g \in G$, α_g commutes with α_h so that $\alpha_g(v_h) = \lambda(g,h)v_h$ for some scalar λ, $|\lambda| = 1$. The function thus defined $\lambda : G \times N(\alpha) \to \mathbb{T}$ ($\mathbb{T}$ = circle group) is $\mathbb{Z}$-bilinear and antisymmetric when restricted to $N(\alpha) \times N(\alpha)$. It is independent of the choice of v_h's and an invariant of cocycle conjugacy. In the case of a cyclic group it may be identified with the γ defined above. Moreover using a crossed product construction it is easy to produce actions α of any countable discrete abelian group on R realizing an arbitrary $N(\alpha)$ and λ satisfying the above conditions.

Unfortunately when G is not abelian, life is complicated by the fact that if $N(\alpha)$ and v_h are defined in the same way, α_g does not always commute with $\operatorname{Ad} v_h$. In fact it turns out that $\alpha_g(v_{g^{-1}hg}) = \lambda(g,h)v_h$ for a function $\lambda : G \times N(\alpha) \to \mathbb{T}$. Now λ <u>does</u> depend on the choice of the v_h's. Indeed if we change v_h to $\eta(h)v_h$ where $\eta : N(\alpha)\ \eta\ \mathbb{T}$, then λ is changed by the function $\partial\eta : G \times N(\alpha) \to \mathbb{T}$ where

(a) $$\partial\eta(g,h) = \eta(h)\overline{\eta(g^{-1}hg)}.$$

Another difference from the abelian case is that one may no longer suppress the 2 cocycle $\mu : N(\alpha) \times N(\alpha) \to \mathbb{T}$ defined by $v_h v_k = \mu(h,k)v_{hk}$. In the abelian case it was determined by the function λ. Also the algebraic relations corresponding to bilinearity involve both λ and μ. Thus one is led to consider the pair (λ,μ) as the correct generalization of Connes' γ invariant.

The cocycle μ is also affected by changing the v_h's with $\alpha_h = \operatorname{Ad} v_h$. Indeed if we choose $\eta(h)v_h$ instead, μ is changed by $\delta\eta : N(\alpha) \times N(\alpha) \to \mathbb{T}$ where

(b) $$\delta\eta(h,k) = \eta(hk)\overline{\eta(h)}\,\overline{\eta(k)}.$$

The relations satisfied by λ and μ are (for $f,h,k \in N(\alpha), g, g' \in G$).

(1) $\mu(f,h)\mu(fh,k) = \mu(h,k)\mu(f,hk)$

(2) $\lambda(g,hk)\mu(g^{-1}hg,g^{-1}kg) = \mu(h,k)\lambda(g,h)\lambda(g,k)$

(3) $\lambda(gg',h) = \lambda(g,h)\lambda(g',g^{-1}hg)$

(4) $\lambda(h,k)\mu(k,h) = \mu(h,h^{-1}kh)$

So, to set up a space for the pair (λ,μ) to live in one defines the cocycle group $Z(G,N)$ for a normal subgroup N of G as the set of all pairs $\lambda : G \times N \to \mathbb{T}$, $\mu : N \times N \to \mathbb{T}$ satisfying (1) to (4). It is a group under pointwise multiplication. To take care of the dependence on the choice of the v_h's, one defines the coboundaries $B(G,N)$ as the subgroup of $Z(G,N)$ consisting of those pairs of the form $(\partial\eta,\delta\eta)$ for some function $\eta : N \to \mathbb{T}$, where ∂ and δ are defined by (a) and (b). The cohomology group $\Lambda(G,N)$ is defined as the quotient $Z(G,N)/B(G,N)$.

With $\Lambda(G,N)$ thus defined, it is clear that, given an action α of G on a factor with $N(\alpha) = N$, the image $[\lambda,\mu]$ of the corresponding cocycle (λ,μ) in $\Lambda(G,N)$ is independent of the choice of the v_h's. By a simple calculation, $[\lambda,\mu]$ is an invariant of cocycle conjugacy. It is called the <u>characteristic invariant</u>.

The most general result on discrete group actions is Ocneanu's Theorem. The hypotheses are that M is a factor with M_* separable, isomorphic to $M \otimes R$ and G is a countable amenable discrete group. The theorem says that the characteristic invariant is a complete cocycle conjugacy invariant for actions α of G on M such that $\alpha(g)$ is always approximately inner and only centrally trivial when $g \in N(\alpha)$ (see [O]). To say that the characteristic invariant is complete also means that every element in $\Lambda(G,N)$ occurs as the characteristic invariant for some action α. This is not too hard and had already been shown in [JI]. Ocneanu's theorem is proved by the same means as Connes' theorem on outer conjugacy classes. It involves the detailed study of the action of the group on central sequences and a result of Ornstein and Weiss on amenable groups [OW].

All automorphisms of R are approximately inner and Connes showed in [CP] that no outer automorphism of R is centrally trivial. So, when applied to R, Ocneanu's theorem says that the characteristic invariant is a complete cocycle conjugacy invariant. For $R_{0,1}$ one may play the same game as Connes played with trace-scaling automorphisms. Given an action α of G on $R_{0,1}$ one obtains a homomorphism $\mathrm{mod} : G \to \mathbb{R}$ given by how α_g scales the trace. The characteristic invariant and mod form a complete set of cocycle conjugacy invariants for actions of an amenable G on $R_{0,1}$.

Thus we have a satisfactory generalization of Connes' work on outer conjugacy classes of automorphisms. But remember that for periodic automorphisms of R, Connes gave a complete classification up to conjugacy. So we might expect to be able to extend this classification to all finite groups. This is indeed the case.

3. FINITE GROUPS. Finite groups are of course amenable so Ocneanu's theorem solves the problem of cocycle conjugacy for actions on the hyperfinite II_1 factor R. To get from cocycle conjugacy to conjugacy one needs another invariant. The natural generalization of Connes' inner invariant is a little complicated but it works so let me explain it. Suppose we are given a finite group G, an action α of G on R with $N(\alpha) = N \triangleleft G$ and unitaries v_h for $h \in N$ satisfying $\alpha_h = \mathrm{Ad}\, v_h$, $v_h v_k = \mu(h,k)v_{hk}$ and $\alpha_g(v_{g^{-1}hg}) = \lambda(g,h)v_h$ for $(\lambda,\mu) \in Z(G,N)$. Then μ is a 2-cocycle so one may form the twisted group algebra $\mathbb{C}_\mu N$. It is a finite dimensional von Neumann algebra. Let $\{z_h\}$ be the unitary basis of $\mathbb{C}_\mu N$ satisfying $z_h z_k = \mu(h,k)z_{hk}$. This algebra is a free model for the subalgebra of R generated by the v_h's so we can "pull back" the action of G on R and let it act on $\mathbb{C}_\mu N$ by defining $g(z_{g^{-1}hg}) = \lambda(g,h)z_h$. We also get an equivariant *-homomorphism $\Phi : \mathbb{C}_\mu N \to R$ by setting $\Phi(z_h) = v_h$. The fixed point algebra $(\mathbb{C}_\mu N)^G$ is actually abelian so one may consider the finite set P of minimal projections in $(\mathbb{C}_\mu N)^G$. The inner invariant $\iota(\alpha)$ is the probability measure on P defined by $\iota(\alpha)(p) = \tau(\Phi(p))$ where τ is the normalized trace on R.

This cannot be the whole story though as if we think back to the cyclic case, the inner invariant was only defined up to a rotation on some roots of unity. The ambiguity was due to the possibility of choosing several different v's with $\alpha^p = \mathrm{Ad}\, v$. Here also there will be several choices of v_h with $\alpha_h = \mathrm{Ad}\, v_h$ for $h \in N$, even if we demand that $v_h v_k = \mu(h,k)v_{hk}$ and $\alpha_g(v_{g^{-1}hg}) = \lambda(g,h)v_h$ for fixed (λ,μ). In fact all other choices are given by changing v_h to $\eta(h)v_h$ where $\eta : N \to \mathbb{T}$ is a G-invariant homomorphism, i.e., and element of $H^1(N,\mathbb{T})^G$. Pulled back to the algebra $\mathbb{C}_\mu N$, these choices may be accounted for by saying that $H^1(N,\mathbb{T})^G$ acts on $\mathbb{C}_\mu N$ so as to leave $(\mathbb{C}_\mu N)^G$ globally invariant, and the inner invariant is only defined modulo the induced action on P. Thus if we think of probability measures on P as a simplex, the inner invariant takes its values in the quotient of that simplex by a simplicial action of $H^1(N)^G$ coming from permutations of the vertices.

It is proved in [JF] that this inner invariant and the characteristic invariant are a complete set of conjugacy invariants for the actions of a finite group on R. So if G was given and one wanted to look at all of its actions on R, one would look at all normal subgroups $N \triangleleft G$, all elements of the group $\Lambda(G,N)$ and for a representative (λ,μ) of each element, one would form the algebra $\mathbb{C}_\mu N$ and calculate the set P of minimal projections in $(\mathbb{C}_\mu N)^G$. One would then claculate the action of $H^1(N,\mathbb{T})^G$ on P and the quotient of the simplex with vertices indexed by P would then parametrize conjugacy classes of actions with characteristic invariant $[\lambda,\mu]$.

The position of the inner invariant in the simplex is interesting. If it is on a q-dimensional face then there are exactly $q + 1$ minimal projections in the centre of the fixed point algebra R^G.

The reader will probably have noticed that the whole discussion of the characteristic and inner invariants had precious little to do with the hyperfinite II_1 factor R and would have worked for any II_1 factor and, with suitable modifications, any factor at all. Of course it is no longer true that the invariants are a complete set of conjugacy invariants. But all the purely algebraic properties of the action are determined by these invariants. Let me illustrate by calculating the spectrum of an action, i.e., those irreducible representations which occur in the decomposition of the factor as a direct sum of irreducible vector subspaces.

If $\alpha : G \to \mathrm{Aut}\, M$ is an action of the finite group G on the factor M, one can arrange to have a G-invariant faithful normal semifinite weight φ on M and the decomposition of the ensuing unitary representation of G on the Hilbert space $\mathcal{H}_\varphi$ of the G. N. S. construction is the same as the decomposition of M itself. So if $g \mapsto w_g$ is this unitary representation, we just want to calculate the kernel of the extension w of $g \to w_g$ to $\mathbb{C}G$. To do this let me first construct some other maps. If $\{u_g\}$ denotes the usual basis of the crossed product $M \times_\alpha G$ over M, one may construct a W*-homomorphism of $M \times_\alpha G$ to the algebra generated by M and the w_g's on $\mathcal{H}_\varphi$ by sending M identically onto M and u_g onto w_g. Call this homomorphism I (see [A]). One may also send $\mathbb{C}G$ isomorphically into $M \times_\alpha G$ in the obvious way and the following diagram commutes

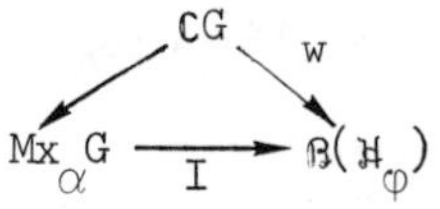

So to calculate the kernel of w it suffices to calculate the kernel of I. Since I is a W*-homomorphism, its kernel will be given by a projection in the centre. Now the centre of $M \times_\alpha G$ is isomorphic to $(\mathbb{C}_\mu N)^G$, the isomorphism being obtained by sending z_h to $v_h^* u_h$ (see [JF]). Call this isomorphism Ψ. Then if q is the projection giving the kernel of Φ (the W*-homomorphism defining the inner invariant, $\Phi(z_h) = v_h$), it turns out that the projection giving the kernel of I is just $\Psi(q)$. In terms of the inner invariant, q is just the sum of all minimal projections p in $(\mathbb{C}_\mu N)^G$ with $\iota(\alpha)(p) = 0$. We see that a character $\chi : G \to \mathbb{C}$ is in the spectrum of the action iff $(1 - \Psi(q))(\sum_{g \in G} \chi(g) u_g) \neq 0$. If we write $1 - q = \sum_{h \in N} c_h z_h$, this condition becomes

$$\chi \in \text{Spectrum}(\alpha) \Longleftrightarrow \sum_{h \in N} \chi(h) |c_h|^2 \neq 0.$$

Nothing could be more explicit.

4. FURTHER TOPICS. What are the directions for further research in discrete group actions?

First one might try to extend the class of factors on which amenable group actions can be satisfactorily classified up to cocycle conjugacy. This is feasible for the known injective type III_λ factors. But the results of Phillips mentioned already ([P]) suggest that even single automorphisms are not easy to classify up to outer conjugacy on a general factor. On the other hand, Connes' recent result [CF] shows that the outer automorphism group of a II_1 factor can be discrete, which suggests that discrete group actions might be reasonable on such "rigid" factors.

One might also try to get from order conjugacy to actual conjugacy for single automorphisms of R. Connes and Størmer ([CS])defined the entropy of an automorphism and used it to give a large family of non-conjugate aperiodic automorphisms of R. In fact, using a crossed product construction it is easy to construct a family of single automorphisms parametrized up to conjugacy by conjugacy classes of ergodic measure preserving transformations of a probability space. Thus the best possible result in this direction would be to reduce the classification of single automorphisms to ergodic theory.

One might also try to extend Ocneanu's theorem to more general groups. The following example indicates that this is unlikely to be fruitful. I shall exhibit two outer actions of $SL(3,\mathbb{Z})$ (or any infinite discrete group with Kazhdan's property T) on R which are not cocycle conjugate. Call an action α of G on R "ergodic on central sequences" if any central sequence $\{x_n\}$ with $\tau(x_n) = 0$ and $\alpha_g(x_n) - x_n \to 0$ strongly for all g also satisfies $x_n \to 0$ strongly. This property is an invariant of cocycle conjugacy. Moreover property T implies that an action is ergodic on central seqeunces if it is ergodic, i.e., if its fixed point algebra is the scalars. To obtain an ergodic action, just let R_g be copies of R indexed by G and let G act on $\bigotimes_{g \in G} R_g$ by permutation of the tensor product components. To obtain an action that is not ergodic on central sequences, simply form the tensor product of one that is by a trivial action on R. One may then exhibit central sequences x_n with $\tau(x_n) = 0$, $\alpha_g(x_n) = x_n$ and $x_n \not\to 0$ strongly.

Another direction of research in discrete group actions is to use them to help in the classification of continuous group actions. For instance compact abelian group actions should be classified by looking at the action of the dual group (discrete) on the crossed product. A step in this direction is outlined in [JM].

One might also consider actions of discrete groups by both automorphisms and antiautomorphisms. For R, this was done when $G = \mathbb{Z}_2$, independently by [GJ] and [S]. The two proofs are different, the one in [GJ] is based on [CO] and the one in [S] on [CI]. The answer is simple-there is only one period 2 antiautomorphism of R up to conjugacy. Giordano has extended this work to cover the known injective type III factors. No doubt the work can be extended to actions of amenable groups up to cocycle conjugacy and finite groups up to conjugacy.

No survey of discrete group actions would be complete without a discussion of Q-kernels, which will lead to the thorniest open problem in the field. One is led to Q-kernels in a different attempt to extend Connes' γ invariant. If α is an automorphism of the factor M and p is the first power such that α^p is inner, say $\mathrm{Ad}\, u$, then one obtains a homomorphism θ from $\mathbb{Z}_p$ to $\mathrm{Out}\, M (= \mathrm{Aut}\, M/\mathrm{Int}\, M)$. Since γ is an invariant of outer conjugacy, it is not possible to change α by an inner automorphism so that $(\mathrm{Ad}\, x\, \alpha)^p = \mathrm{id}$, if $\gamma \neq 1$. Another way of saying this is to say that γ is an obstruction to lifting θ to $\mathrm{Aut}\, M$. Given a group Q more general than $\mathbb{Z}_p$, we are led to consider homomorphisms $\theta = Q \to \mathrm{Out}\, M$. An almost identical situation was encountered by Eilenberg and MacLane [EM] in their study of group extensions. Following their recipe one associates with such a θ, called a Q-kernel, an element $\mathrm{Ob}(\theta) \in H^3(Q, \mathbb{T})$ (trivial action). We know that $H^3(\mathbb{Z}_p, \mathbb{T}) \cong \mathbb{Z}_p$ so it is not surprising that, if $Q = \mathbb{Z}_p$, $\mathrm{Ob}(\theta)$ may be identified with Connes' invariant γ.

The obstruction $\mathrm{Ob}(\theta)$ may be adequately interpreted as the obstruction to forming an "extension of M by Q". This is thoroughly treated in [NT] and [S]. But $\mathrm{Ob}(\theta)$ is also an obstruction to lifting θ to $\mathrm{Aut}\, M$, i.e., finding an action $\alpha : Q \to \mathrm{Aut}\, M$ with $\varepsilon \circ \alpha = \theta$ where $\varepsilon : \mathrm{Aut}\, M \to \mathrm{Out}\, M$ is the canonical projection. In [S], Sutherland proved that if M is properly infinite or if Q is finite, then $\mathrm{Ob}(\theta)$ is the <u>only</u> obstruction to lifting θ to $\mathrm{Aut}\, M$. This result is surprising from the point of view of abstract group extensions where it is almost never true. The question remains as to whether $\mathrm{Ob}(\theta)$ is the only obstruction when Q is infinite and M is finite. A partial answer was given by Ocneanu in the course of the proof of his theorem. He shows that if Q is amenable and $M = R$, then $\mathrm{Ob}(\theta)$ is the only obstruction to lifting. His proof invokes all the machinery of central sequences and combinatorics of amenable groups. At this stage there are no known examples of non-amenable Q-kernels $\theta : Q \to \mathrm{Out}\, R$ with $\mathrm{Ob}(\theta) = 0$ which do not lift to $\mathrm{Aut}\, R$. Nor are there any known II_1 factors M which admit a Q-kernel $\theta : Q \to \mathrm{Out}\, M$, $\mathrm{Ob}(\theta) = 0$ which does not lift, for Q amenable or otherwise. The author of this article will not be able to sleep soundly as long as this problem remains unsolved.

The relationship between the characteristic invariant and $Ob(\theta)$ is covered in [JI] and [JF].

BIBLIOGRAPHY

[A] P. L. Aubert: "Théorie de Galois pour une W*-algèbre", Comment. Math. Helv. 39 (51) (1976) 411-413.

[CA] A. Connes: "A factor not antiisomorphic to itself", Ann. of Math. 101 (1975) 536-554.

[CC] A. Connes: "Sur la classification des facteurs de type II", C. R. Acad. Sci. Paris t. 281 (1975) 13-15.

[CF] A. Connes: "A type II_1 factor with countable fundamental group", Journ. Op. Theory 4(1980) 151-153.

[CI] A. Connes: "Classification of injective factors", Ann. of Math. 104 (1976) 73-115.

[CO] A. Connes: "Outer conjugacy classes of automorphisms of factors", Ann. Sci. Ec. Norm. Sup. 4 ème série, t. 8 (1975) 383-420.

[CP] A. Connes: "Periodic automorphisms of the hyperfinite factor of type II_1" Acta. Sci. Math. 39 (1977) 39-66.

[CS] A. Connes and E. Størmer: Entropy for automorphisms of II_1 von Neumann algebras, 134 (1975) 239-306.

[CT] A. Connes: "Une classification des facteurs de type III", Ann. Sci. Ec. Norm. Sup., 4 ème série t. 6. (1973) 133-252.

[EM] S. Eilenberg and S. MacLane: "Cohomology Theory in Abstract Groups II", Ann. of Math. 48 (1947) 326-341.

[GJ] T. Giordano and V. Jones, "Antiautomorphismes involutifs du facteur hyperfini de type II_1", C. R. Acad. Sc. Paris t. 290 (1980) 29-31.

[JI] V. Jones: "An invariant for gorup actions", in Springer lecture notes in mathematics. Vol. 725, 237-253.

[JF] V. Jones: "Actions of finite gorups on the hyperfinite II_1 factor", Memoirs A. M. S. no. 237 (1980).

[JM] V. Jones: "Minimal actions of compact abelian groups on the hyperfinite II_1 factor". To appear: J. Operator Theory.

[McD] D. McDuff: "Central sequences and the hyperfinite factor", Proc. London Math. Soc. XXI (1970) 443-461.

[MuN] Murray and J. von Neumann: "Rings of operators IV", Ann. Math. 44 (1963) 716-809.

[NT] N. Nakamura and Z. Takeda: "On the extensions of finite factors II", Proc. Jap. Acad. 35 (1959) 215-220.

[O] A. Ocneanu: "Actions of amenable groups on factors". Preprint.

[OW] D. Ornstein and B. Weiss: "Ergodic theory of amenable group actions I", Bull. A. M. S. (1979) 161-164.

[P] J. Phillips: "Automorphisms of full II_1 factors, with applications to factors of type III", 43 (1976) 375-385.

[St] E. Størmer: "Real Structure in the hyperfinite II_1 factor". Preprint.

[S] C. Sutherland: "Cohomology and extensions of operator algebras, II". Preprint.

DEPARTMENT OF MATHEMATICS
UNIVERSITY OF CALIFORNIA, LOS ANGELES
LOS ANGELES, CALIFORNIA 90024

Proceedings of Symposia in Pure Mathematics
Volume **38** (1982), Part 2

ERGODIC THEORY AND VON NEUMANN ALGEBRAS

Calvin C. Moore[1]

Table of Contents

1980 Mathematics Subject Classification 28D99, 46L05, 22D40

[1]Supported in part by NSF Grant MCS 77-13070.

1. INTRODUCTION

Our objective here is to explain the fascinating and productive interplay between ergodic theory and von Neumann algebras that has developed during the last twenty years. The ergodic theory that enters into this interplay is not the classical ergodic theory of single transformations but rather it is ergodic theory based on orbit equivalence (rather than conjugacy) -- a notion that we explain in Section 2.1. It is also the ergodic theory that is implicit or explicit in George Mackey's notion of virtual groups [59]. Indeed another objective here is to explain (and popularize) Mackey's point of view embodied in his work on virtual groups; we do this in a slightly different language and also we concentrate on what is really a special case of virtual groups (the measured equivalence relations), which we feel, however, already captures the essence of the matter and which is much more accessible. This interplay we describe between ergodic theory and von Neumann algebras goes back to the very origins of the study of these algebras, and is seen in the group-measure space construction used by Murray and von Neumann to construct examples of factors [66]. In a certain sense what we are going to describe is an elaboration on this theme.

Historically perhaps the first result in the spirit of the orbit equivalence point of view in ergodic theory was the Ambrose-Kakutani theorem [1] that any flow is a flow built under a function; however, some of this point of view was implicit for many years in the stability theory of ordinary differential equations cf. [87]. But it was not until Henry Dye's papers [23], [24] in 1959 and 1963 and that orbit equivalence and ergodic theory and its connection with von Neumann algebras really became established. Dye proved two remarkable theorems; one classifying integer measure preserving actions up to orbit equivalence, and the other showing integer actions were approximately finite -- see below for definitions. Mackey published a short announcement in 1963 [58] of his virtual group program and in 1966 a longer account appeared [59]. These results were not, however, followed up upon for several years until W. Krieger who, in part guided by the work of Araki and Woods [3] and Powers [69], began a systematic study of orbit equivalence and the classification of non-measure preserving transformations up to orbit equivalence and their connections with von Neumann algebras ([45], [46], [47], [48], [49], [50]) which culminated in a far-reaching extension of Dye's original results. Since 1970 there has been intense activity in the study of measured equivalence relations and orbit theory in ergodic theory by a number of authors -- see the references at the end. There have also been some very interesting developments in a different direction where one places some extra requirements on the orbit equivalences, such as some kind of differentiability. We shall not be able to go into this at all for lack of space.

At about the same time in the early part of the decade Connes and Takesaki were making enormous progress in understanding the structure and classification of von Neumann algebras; Connes' numerical invariants introduced in [8] are motivated by the interplay of ergodic theory and von Neumann algebras and also the use of cross products in [8] and [92] to unravel the structure of type III algebras is yet another example of this interplay. Moreover, thanks to the work of Connes [12], one now knows the equivalence of the properties of injectivity, approximate finiteness, amenability, and several others for von Neumann algebras, thus establishing this class of von Neumann algebras as one of major importance and interest. The interplay of ergodic theory and von Neumann algebras in this context leads to the Krieger-Connes [12], [50] classification of the factors in this class (with one difficult but small point left unresolved). This result can be viewed as the culmination of Dye's original classification theorem: A key role in the interplay of ergodic theory and von Neumann algebras is played by the Cartan subalgebras; the recent results of Connes, Feldman and Weiss [18] and Ornstein-Weiss [68] provide beautifully elegant answers to the natural conjugacy question for Cartan subalgebras of approximately finite factors and provide the best possible extension of Dye's original result on approximate finiteness of integer actions. Hence this seems a natural point to present an exposition of this material.

The outline of the paper is as follows: in section two we discuss and axiomatize the notion of measured equivalence relations as the principal ingredient from ergodic theory; as noted these are a special case of Mackey's virtual groups. We discuss how they arise from group actions and how they capture the notion of orbit equivalence for such actions. We then go on to discuss a variety of structural questions, constructions, and examples for these objects, including the important notion of stable isomprhism, a notion that we propose to use in place of similarity as discussed in [59], [71], and [28]. A very elegant and useful result of Ramsay [74], which is also announced in these proceedings, makes the treatment here considerably simpler than it was previously. We remark parenthetically that our point of view in this section is analogous to the point of view about measure and integration where one has an abstract space with a σ-field of sets and a fixed measure or measure class and where one works modulo null sets of this measure class. The reader should bear in mind that there is a complementary approach to equivalence relations analogous to point of view about measure and integration where one fixes a space with a topology and allows consideration of many different measures, and where one cannot delete null sets. This point of view is developed in [13], and we see these two points of view as complementing each other.

Section three discusses the analogue of Haar measure for measured equivalence relations and discusses homomorphisms of these objects into groups, and also sketches the outlines of a cohomology theory. In the last part of this section we discuss the very important notions of the Poincaré flow and the asymptotic range of a homorphism (or cocycle). In the context of this section and the next it is helpful to think of measured equivalence relations as groupoids since many of the constructions are completely analogous to what one would do for a group.

In section four we specialize from general homomorphisms to the particular case of unitary representations of a measured equivalence relation (homomorphisms into the unitary group on a Hilbert space). Subsequent circumstances force us to consider also cocycle or projective representations, just as one is forced to do likewise for groups as in [55]. We then discuss what can profitably be thought of as the regular (or cocycle-regular) representation of a measured equivalence relation. As with groups there is an associated enveloping von Neumann algebra which is none other than the old Murray-von Neumann group measure space algebra. This algebra can also profitably be thought of as the algebra of matrices over the measured equivalence relation as in [27]. In the final portion of this section we introduce the notion of Cartan subalgebras and discuss the converse theorem of [27] to the effect that a von Neumann algebra with a Cartan subalgebra necessarily comes from a (unique) measured equivalence relation with a possible twist as the enveloping von Neumann algebra of the regular representation. All von Neumann algebras will be assumed to have separable predual.

In section five we discuss the notions of amenability and approximate finiteness for measured equivalence relations and show how these are related to corresponding properties of the von Neumann algebra of matrices over the equivalence relation. We then discuss the Connes-Krieger classification of approximately finite factors and conclude with a discussion of the Connes-Feldman-Weiss and Ornstein-Weiss results which imply deep conjugacy theorems for Cartan subalgebras of approximately finite factors and show, as one hoped all along that actions of amenable groups always give rise to approximately finite measured equivalence relations.

In section six we discuss some recent developments concerning an analogue of Kazhdan's property T for groups [44] in the context of measured equivalence relations. Specifically, we discuss the results of Zimmer [110] and Connes [14] together with some complements which also show how the formulations of Zimmer and Connes are related. The proofs of some of the results in section six, especially those involving some technical calculations with cohomology groups, are contained in a separate appendix.

One should remember that the entire discussion here concerns von Neumann algebras and group actions and equivalence relations viewed *measure theoretically*. There is an extraordinarily rich parallel theory for C^* algebras and group actions, and equivalence relations from the topological point of view that is discussed in many other articles in these proceedings. For the most part this theory has concentrated so far mostly on group actions; the topological groupoid or topological equivalence relation point of view, which would be the strict analogue in the C^* context of the approach here, is beginning to emerge and will play an increasingly important role, cf. [75], [51], [52], [20], and also [13] for C^* algebras associated with foliations.

2. MEASURED EQUIVALENCE RELATIONS

2.1 DEFINITIONS. Let G be a locally compact second countable topological group, X a standard Borel space [54], and suppose that G acts on X as a Borel transformation group; that is we have given a map $G \times X \to X$, written $(g,x) \to g \cdot x$ which is jointly Borel and which defines a homomorphism of G into the automorphisms of X. Finally as part of the structure we shall assume that X has a σ-finite measure μ which is quasi-invariant under the action of G in the sense that for each $g \in G$; $\mu(g \cdot E) = 0$ iff $\mu(E) = 0$. Such actions are sometimes termed 'non singular.' Actually, of course, any measure μ' equivalent to μ in the sense of absolute continuity has the same property and we should really be talking about a quasi-invariant equivalence class $C(= C_\mu)$ of measures rather than an individual measure. From the point of view of group actions the natural notion of equivalence is that of conjugacy; that is if (G,X,C) and (G,Y,B) are two such actions of G on spaces X and Y with quasi-invariant measure classes C and B respectively, we say that they are conjugate if there is a Borel isomorphism ϕ of a G invariant conull set X' of X onto a G-invariant conull set Y' of Y which sends measures in C to measures in B and which intertwines the actions in that $\phi(g \cdot x) = g \cdot \phi(x)$ for $g \in G$, $x \in X'$. One may relax these conditions a bit without changing the end result; for instance X' and Y' need not be G-invariant and the intertwining identity need only be assumed to hold for almost all pairs (g,x) in $G \times X'$ (Haar measure understood on G) (cf. [71] Theorem 3.5, and [56] for details).

An often useful equivalent point of view is to note that if (G,X,C) is given, then G operates naturally on the measure algebra $M = M(C)$ of (X,C) as a transformation group of automorphisms of this σ-Boolean algebra.

Indeed M comes equipped with the topology of convergence in measure and under Boolean addition this is a polonais abelian topological group ([71], [63], [56]), and the map $G \times M \to M$ defining the action is easily seen to be Borel. It follows then that it is continuous [63], so that G acts on M as a topological transformation group (of σ-Boolean ring automorphisms). Conversely, if we are provided with an action of a second countable locally compact group G on a measure algebra M of some standard measure space (X,C) as a topological (or Borel) transformation group of ring automorphisms of M, then one may construct a (point) action of G on (X,C) which realizes the action on M, [71], [56]. Moreover, conjugacy of point actions is equivalent to the existence of a continuous ring isomorphism of the measure algebras which intertwines the actions of G.

However, in addition to the notion of conjugacy of group actions, there is a complementary notion of *orbit equivalence* of group actions that has become of ever increasing importance during the previous twenty years. Moreover it is this notion of orbit equivalence that emerges as by far the more significant one for the interaction of ergodic theory and operator algebras. Specifically if G acts on (X,C) it defines an equivalence relation $R = R_G$ on X where $x \in X$ is defined to be equivalent to x' if and only if there is a $g \in G$ with $g \cdot x = x'$. Not only do we have an equivalence relation, but in addition each equivalence class or leaf ℓ of the relation comes equipped with a natural measure class B^{ℓ}; specifically, the leaf $\ell(x)$ of x can be identified with G/G_x where G_x is the isotropy group at x and we project any finite measure μ_G on G equivalent to Haar measure onto the space G/G_x. The equivalence class of this measure is independent of the choice of x and μ_G, and is the desired class. In the sequel we will often refer to such measures on the leaves as fiber measures or orbit measures.

Furthermore the entire equivalence relation B_G carries a distinguished class B_G of measures; specifically if μ is a finite measure on X in the class C, and if as before μ_G is a finite measure on G equivalent to Haar measure, we form the product $\mu_G \times \mu$ on $G \times X$. Then noting that $R = R_G$ is the image of the map $(g,x) \to (g \cdot x, x)$ of $G \times X$ into $X \times X$, we let ν be the direct image of the measure $\mu_G \times \mu$ and let B_G be its equivalence class, as measure on R_G. It is not difficult to see that R_G is a Borel subset of $X \times X$, and is hence a standard Borel space. This pair $(R_G, B_{G'})$ as an example of what we call a measured equivalence relation -- a notion that we will axiomatize in a moment. But first let us define *orbit equivalence* of group actions -- it will be, in a word, just isomorphism of the associated measured equivalence relation. But more concretely, an action of G on (X,C) is orbit equivalent to an action of G_1 on (X_1,C_1) if there is a Borel isomorphism ϕ of a conull subset X' of X onto a conull subset X_1' of X_1 which maps C to C_1 and such that for almost all $x \in X'$, ϕ maps the measure class $B^{\ell(x)}$ on the leaf $\ell(x)$ of X onto the measure class B^k, $k = \ell_1(\phi(x))$ on the leaf $\ell_1(\phi(x))$ of $\phi(x)$ in X_1. Clearly for almost all x, ϕ maps almost all of the leaf $\ell(x)$ into the leaf $\ell_1(\phi(x))$ pointwise. Of course, an equivalent and perhaps more elegant formulation is that the Cartesian square of ϕ should map the measure class B_G on R_G onto the measure class B_{G_1} on R_{G_1}. It is useful in this context to know from 5.1 and 5.2 of [71] that one can then pinch the conull sets X' and X_1' a bit if necessary so that the Cartesian square of ϕ actually maps $R_G \cap (X' \times X')$ onto $R_{G_1} \cap (X_1' \times X_1')$ pointwise.

We note the very significant point that one can have actions of G and G_1 orbit equivalent where G and G_1 can be quite different as groups. We should also note that passage to R_G washes out any information about the isotropy groups of the action; alternatively one could opt to carry this information along in which case the object under consideration would be a *measure(d) groupoid* which would have the structure of a fiber space whose base is R_G and which has as fiber over a pair $(x,y) \in R$, the set of group elements of G which map x to y -- that is a coset of the isotropy group of x. This is eactly what Mackey has done in his theory of virtual groups [59]; what we are doing here is in effect to examine a special case -- a measured equivalence relation -- of a virtual group. In our view this is by far the most important special case of virtual groups, and we restrict our attention to it because it is much easier to understand geometrically and also because one can readily build up the most general case from this special case. We will generally not use the language of virtual groups.

Let us now give the formal definition of a measured equivalence relation; if (X,C) is a standard Borel space with measure class C, a pair (R,B) consisting of an equivalence relation R on X and a measure class B on R, the object (X,C,R,B) is called a measured equivalence relation (with base (X,C)) if

(1) R is a Borel subset of $X \times X$

(2) B is invariant under the flip θ, $\theta(x,y) = (y,x)$ and the projection of R onto one (and hence both) factor(s) maps the class B onto the class C.

(c) If $\nu \in B$ and $\mu \in C$ is its image by projection to the first factor and if

$$\nu = \int \nu^x \, d\mu(x)$$

is the disintegration [54] of ν under this projection, then the fiber measures ν^x (each of which, of course, lives on the equivalence class or leaf of x) satisfy $\nu^x \sim \nu^y$ for B-almost pairs $(x,y) \in R$. A similar condition for disintegration under projection to the second factors is a consequence of the above together with (2).

We often use a symbol such as R to denote the object (X,C,B,R). The notion of isomorphism of two such objects R and R_1 is by now clear -- one demands a Borel isomorphism of a conull set of the base X of one onto a conull set of the base X_1 of the other which sends B to B_1. As above, from section 5 of [71], we can pass to smaller conull sets so that the equivalence relations restricted to these conull sets are preserved set wise by ϕ. It is clear that the object (R_G,B) defined by a group action as described previously satisfies these conditions; as we pointed out, the measure ν^x on the R_G equivalence class of x -- that is, the G-orbit of x -- is in the unique quasi-invariant measure class on this orbit. The condition in (2) of invariance under the flip θ is the manifestation in this picture that measures in the class C are quasi-invariant under the action of G. We shall denote the measured equivalence relation coming from an action by a symbol such as R_G.

2.2 OPERATIONS AND EXAMPLES. There are several elementary operations and constructions on measured equivalence relations which we now discuss, as well as various examples. First if $R_i = (X_i,C_i,R_i,B_i)$ are two of them, we can form the product $R_1 \times R_2$ which has base $(X_1 \times X_2, C_1 \times C_2)$ and the relation R is just the product $R_1 \times R_2$ so $(x_1,x_2) \sim (x_1',x_2')$ if $x_1 \sim x_1'$ and $x_2 \sim x_2'$. The measure class B is just the product $B_1 \times B_2$. This is an obvious analogue of products for group actions. Secondly if $R = (X,C,R,B)$ is given, let Y be a subset of the base X of R of positive C-measure. Then we can "relativize" R to Y, which we write as $R|_Y$ where the base is $(Y,Y|_C)$ and the equivalence relation is $R|_Y = R \cap (Y \times Y)$ and $B|_Y$ is the restriction of B to $R|_Y$. It is easy to see that the axioms for a measured equivalence relation are satisfied in this case. In general there is no analogue of this for group actions, but for $G = \mathbb{Z}$, the integers, it is simply Kakutani's notion of the induced transformation on a subset [43].

Suppose that $R = (X,C,R,B)$ is a measured equivalence relation, and let $Y \subset X$ be a Borel subset of the base space X. We say that Y is *invariant* if the measure class B lives on $R \cap ((Y \times Y) \cup (Y^c \times Y^c))$ where Y^c is the complement of Y. It is not difficult to see that this is equivalent to the assertion that for almost all $x \in Y$, the disintegration measure ν^x in the definition lives on Y; this property of Y is invariant under perturbations by null sets. In case R comes from a group action this is saying that for almost all $x \in Y$, $g \cdot x \in Y$ for almost all $g \in G$ -- which is a standard notion of invariant set for group actions. There is the stronger notion of strict invariance which means that Y is saturated with respect to the equivalence relation R. The relation between these is somewhat subtle -- it is known for instance that one can find a Borel conull $X_0 \subset X$ such that if Y is invariant, there is a Borel $Y_0 \subset X_0$, strictly invariant for $R|_{X_0}$ and which differs from Y by a null set. In particular, the saturation $\tilde{Y}_0^{X_0}$ of this set Y_0 is analytic, strictly invariant, and differs from Y by a null set (cf. Theorem 4.2 of [71]). Hence the existence of invariant sets yields the existence of closely related strictly invariant sets.

In terms of this notion of invariant set, one may define the notion of ergodicity of a measured equivalence relation $\mathcal{R} = (X,C,R,B)$; namely $\mathcal{R}$ is *ergodic* if any invariant set is null or conull, in exact analogy with group actions. We point out that the collection of invariant sets for $\mathcal{R}$ when projected into the measure algebra of the base space (X,C) forms a sub σ-algebra of this measure algebra. Using these facts one likewise shows that any measured equivalence relation may be decomposed as a continuous sum or integral of ergodic ones [37]. As both the exact formulation and proof of this result parallel the group action case, we shall not say any more about this. In any case our study is effectively reduced to the ergodic ones; these are so to speak, the building blocks for the general case.

Let us describe some ways that measured equivalence relations can arise -- other than directly from group actions. Suppose that X is a C^∞ manifold and let $\mathcal{F}$ be a folition of X, cf. [53]. Thus we are given for each $x \in X$ a subspace L_x of the tangent space T_x to X at x such that the $\{L_x\}$ fit together to form an involutive distribution [53]. The leaves ℓ of the foliation are maximal submanifolds (not necessarily closed) with tangent space equal to L_x at each point x of ℓ. It is a standard fact of elementary differential geometry that there is a unique leaf ℓ through each $x \in X$, and one defines an equivalence relation R by $(x,y) \in R$ if x and y belong to the same leaf. As X is a manifold, it has a distinguished equivalence class C of smooth measures which we use for the base. It remains only to define a measure class B on R, and one could do this essentially by reversing (3) of the definition of measured equivalence relations and specify a $\nu \in B$ by giving μ (which we have) and the disintegration products ν^x on each leaf. Since each leaf is a manifold in its own right, we could just specify ν^x to be in the class of smooth measures on the leaf through x, and it is not hard to see that this constructs B satisfying the necessary conditions. There is, however, a somewhat more interesting and intrinsic construction which we describe, cf. [100]. One would like the set R to have itself the structure of a manifold,and then one could take B to be the class of smooth measures. Unfortunately R is not in general a manifold; there is an obstruction owing to the presence of non-trivial holonomy in the foliation. We shall not discuss this in detail, but we observe that what happens is that R has a "covering" $\tilde{R}$ that is a manifold [100], cf. also [13]. Specifically, one lets $\tilde{R}$ be the set of triples $(x,y,[\gamma])$ with $(x,y) \in R$ and with $[\gamma]$ a homotopy class of paths from x to y in the common leaf of x and y. Then it is possible to give $\tilde{R}$ the structure of a manifold in a nice way such that the projection of $\tilde{R}$ onto $R \subset X \times X$ is an immersion. If the leaves are simply connected, then of course $R = \tilde{R}$. The measure class on R one wants is simply the image of the smooth measure class on $\tilde{R}$.

In fact it should be noticed that $\tilde{R}$ is itself a measure groupoid in the sense of [59]. The base equivalence relation is R and the Borel group "sitting over" a point (x,x) is the homotopy group of the leaf based at x, viewed as a discrete group. This situation is completely analogous to adding the extra data of isotropy groups in the case of measure groupoids coming from group actions.

A derivative type of example is discussed in the recent work of Zimmer [108] where he defines the notion of a measurable foliation. In essence this is a measured equivalence relation R on a base X with the additional structure that each equivalence class is to have the structure of a connected C^∞ manifold. Ordinary foliations are a special case as are equivalence relations R_G arising from Borel actions of any connected *Lie* group G on measure spaces (X,C).

At the other extreme from foliations (in one sense, but not in another), is the class of *discrete measured equivalence* relations. These, as it turns out, play a very special and significant role in the theory and have the additional feature that many measure theoretic difficulties that obscure the picture in general evaporate in this special case. To be more exact, we say that a measured equivalence $R = (X,C,R,B)$ relation with base (X,C), is *discrete* if each equivalence class of R is countable and if almost all distintegration products ν^x in (2) of the definition are equivalent to the counting measure on the equivalence class of x. For instance, if G is a discrete group acting on (X,C), then the associated measured equivalence relation (X,C,R_G,B_G) is rather clearly discrete. It is useful and significant that the converse of this is true; namely if (R,B) is discrete with base (X,C) then there is a discrete (countable) group G acting on (X,C) with $(R,B) = (R_G,B_G)$ (Theorem 1 of [26]). The G is highly non-unique and there are in general many quite different choices. It is not known whether or not one can always find a G that acts freely.

Another, somewhat trivial but nevertheless important example of measured equivalence relations is the following; let (X,C) be a measure space and let $R = X \times X$, and let B be the product measure class $C \times C$. The axioms are trivially verified and these examples are called *transitive* as they correspond to transitive group actions. Up to isomorphism the above depends only on the isomorphism type of (X,C) and these are known. In the special case when C has no atoms, in which case $(X,C) \simeq [0,1]$ with Lebesgue measure, we denote the measured equivalence relation by I -- this will play an important role. In the special case when C is atomic, in which case we can take (X,C) to be, up to isomorphism, a discrete set of cardinality n with counting measure $(n = 1,2,\ldots,\infty)$. The corresponding measure equivalence relation is denoted by I_n.

2.3. STABLE ISOMORPHISM. In addition to the notion of isomorphism of measured equivalence relations, Mackey and others have introduced and studied a companion and somewhat weaker notion of similarity of measured equivalence relations [59], [71]. If $\mathcal{R} = (X,C,R,B)$ is given,one can form the orbit space X/R with the corresponding equivalence class of measure C/R obtained by projecting C onto X/R. Of course X/R is a terrible space in general -- it is usually not countably separated,and if $\mathcal{R}$ is ergodic, C/R has only null sets or conull sets. (In the ergodic case Connes has coined the term 'tiny space' to describe this kind of measure space $(X/R,C/R)$.) In essence, similarity of two measured equivalence relations $\mathcal{R}_1$ and $\mathcal{R}_2$ means simply that these measure spaces $(X_1/R_1, C_1/R_1)$ and $(X_2/R_2,C_2/R_2)$ are isomorphic. More specifically one demands that there be Borel maps $\theta_1: X_1 \to X_2$, and $\theta_2: X_2 \to X_1$ defined on conull subsets with θ_1 carring R_1 equivalence classes to R_2 equivalences (and similarly for θ_2) such that the induced map $\tilde{\theta}_1$ from X_1/R_1 to X_2/R_2 is bijective from a conull set of one to a conull subset of the other, and carries C_1/R_1 to C_2/R_2; finally one asks that $\tilde{\theta}_2$ be the set theoretic inverse of $\tilde{\theta}_1$. Actually it is not too hard to see that once one has θ_1, then one produces a companion map θ_2, cf. [71], [28]. Two similarities θ and ψ are said to be equivalent if $\tilde{\theta} = \tilde{\psi}$ a.e. It would be interesting to see if these notions could be formulated somewhat more intrinsically.

The prime example of a similarity that one should keep in mind is the following: for any $\mathcal{R}$, $\mathcal{R}$ is similar to $\mathcal{R} \times I$ where I is the transitive measured equivalence relation on the unit interval I defined above in 2.2. Indeed if X is the base space of $\mathcal{R}$, the maps $\theta_1(x) = (x,0)$ from X to $X \times I$ and $\theta_2((x,t)) = x$ from $X \times I$ to X satisfy the above conditions. In [28] this observation was used as the starting point for introducing a special kind of similarity of R_1 and R_2 which were called concretely similarities. The definition is that $R_1 \times I$ should be isomorphic to $R_2 \times I$. If R_1 and R_2 satisfy this they are clearly similar in the above sense by transitivity of the relation of similarity. One can profitably think of this notion as *stable isomorphism* of R_1 and R_2 for this is precisely what it means for the von Neumann algebras associated to R_1 and R_2, and we shall use this terminology instead of that in [78], i.e. R_1 and R_2 are *stably isomorphic* if $R_1 \times I$ is isomorphic to $R_2 \times I$.

It turns out in fact, thanks to a very recent result of Ramsay [74] that similarity is in fact the same as stable isomorphism -- a fact that makes for considerable simplification in the exposition. We explain briefly how this comes about; in [28] we studied those measured equivalence relations R such that R is stably isomorphic to a discrete measured equivalence -- i.e. $R \times I \simeq D \times I$ for some discrete D. (Such R's could be termed stably discrete.) It was proved (Theorem 2.8 of [28]) that any R_G coming from a group action has this property. This was proved in a special case by Forrest [29] and for the case $G = \mathbb{R}$, the real line the result is equivalent to the classic Ambrose-Kakutani Theorem [1] that any flow is isomorphic to a flow built under a function. It is clear also that measured equivalence relations coming from foliations have this property -- just look at transverse submanifolds. Ramsay's result is that all measured equivalence relations are stably discrete, so in fact one doesn't need the distinction any more.

One combines this result with the result from [28] that if R_1 and R_2 are similar and one of them is stably discrete, then both are and there is a stable isomorphism equivalent to given similarity (Theorems 4.6 and 5.5 of [28]). Consequently, similarity is completely equivalent in general to stable isomorphism, and to be exact we have the following.

THEOREM 2.3.1. Given a similarity θ, of R_1 and R_2 there is isomorphism ϕ of $R_1 \times I$ with $R_2 \times I$ such that the induced map $\tilde{\phi}$ on the space of orbits coincides with $\tilde{\theta}$.

In fact in many cases stable isomorphism already implies isomorphism. Let us agree to say that a measured equivalence relation $R = (X,C,R,B)$ has *continuous orbits* if the disintegration products ν^x (the fiber orbit measures) which appear in the definition of a measured equivalence relation are almost all free of atoms. If R_1 and R_2 have continuous orbits and are stably isomorphic, it can be shown that they are actually isomorphic, Theorem 5.6 of [28]. Most interesting examples of measured equivalence relations which occur in nature are either discrete or have continuous orbits; the general case can be built up as a mixture of the two, but situations where the ν^x have mixed type are really artifacts.

As an added corollary of Ramsay's result and [28] one now knows that any R with continuous orbits is isomorphic to an R_G for some group action so we have in fact not extended the original class of examples. To see this, we note that R and $R \times I$ being stably isomorphic, both with continuous orbits, are isomorphic. Hence $R \simeq R \times I \simeq D \times I$, with D discrete, and as noted in 2.2 from [26] it follows that, $D = R_D$ for some discrete group, and as $I = R_T$ for a transitive action of the circle T, it follows that $R \simeq R_D \times R_T \simeq R_{D\times T}$. As in the discrete case the choice of G is highly non-unique, but is useful in some context to know simply that it exists.

There is one final point that needs to be mentioned. If R is given then $R \times I \simeq D \times I$ for some discrete D. But also we might have $R \times D \simeq D' \times I$ for some other discrete D'. Of course D and D' are stably isomorphic, but to be useful we often want a stronger relation between D and D' that involves only discrete relations. In fact it is true that $D \times I_\infty \simeq D' \times I_\infty$ where I_∞ was defined above to be the discrete transitive measured equivalence relation on a countably infinite set, cf. Theorem 3 of [26]. Just as in the continuous case there are a number of instances in which one can strengthen this conclusion to the assertion that D and D' are already isomorphic (cf. [26] pp. 299 for details).

2.4 THE FULL GROUP. Associated with a measured equivalence relation R, there are two groups of automorphisms of interest. The first of these is the automorphism group Aut(R) of the structure, as implicitly described when we talked about isomorphisms. Of perhaps somewhat more interest is the subgroup of these which map almost every orbit to itself. This is called the *full group* of R and written [R], and can be thought of productively as the *inner automorphisms* of R. If $R = R_G$ comes from a group action, then G sits naturally inside [R] at least if G acts effectively. The theorem that any discrete R is R_D for some discrete group can be viewed as saying that [R] is "sufficiently large" in some sense. Moreover, if R has continuous orbits so that $R \simeq D \times I$, then [D] is "large" and as [I], which is the entire automorphism group of measure algebra of the unit interval is also quite "large," it follows that [R] is also "large." This observation will be of importance to us later.

3. HAAR MEASURE AND HOMOMORPHISMS

3.1 HAAR MEASURE. Having treated a variety of internal structural questions about measured equivalence relations in the previous section, we shall now look at how these objects interact with their surroundings. An important theme here is that measured equivalence relations, at least the ergodic ones, have a somewhat tenuous, but nevertheless real resemblance to locally compact groups. In fact they are groupoids -- but we promised not to dwell on that. In particular, here we want to discuss Haar measure for R, homomorphisms of an R into groups and representations; we also briefly discuss cohomology groups analogous to those one defines for a group and a module.

We first recall from the definition of a measured equivalence relation $R = (X,C,R,B)$ that we disintegrated a measure $\nu \in B$ with respect to projection to the first coordinate to obtain fiber measures ν^x, $x \in X$. It was demanded that $\nu^x \sim \nu^y$ for almost all pairs (x,y), and with respect to the analogy with locally compact groups such as ν is analogous to a quasi-invariant measure [54]. However, locally compact groups possess not just quasi-invariant measures but invariant (Haar) measures. This leads one to seek a definition of Haar measure on R and it is to be found in [37], but cf. also [13], [16], [82], [95]. For this we fix a measure μ in the class C on the base space X and can then talk about a (right or left) Haar measure ν *on* R *relative to* μ. The condition on ν is simply that if we write the disintegration

$$\nu = \int_X \nu^x \, d\mu(x)$$

of ν over μ (projection to first coordinate), then $\nu^x = \nu^y$ for B-almost all pairs $(x,y) \in R$. In [37] Hahn establishes the existence and unicity of such a ν corresponding to any given μ; it is unique up to multiplication by a function $b(x,y) = b(y)$ on R which depends only on the second variables. This defines right or left Haar measure (depending on one's parity), and the same condition except with projection to the second coordinate instead of the first gives the Haar measure of the other parity. The flip θ $\theta(x,y) = (y,x)$ obviously carries right Haar measure to left Haar measure and as the class B is θ invariant, the Radon-Nikodym derivative $d\theta(\nu)/d\nu = (\Delta(x,y))^{-1}$ defines a function on R into the positive reals, called, naturally, the modular function. The invariance condition on ν forces Δ to be a homomorphism of R into $\mathbb{R}^+$ in the sense that

$$(*) \qquad \Delta(x,y)\Delta(y,z) = \Delta(x,z)$$

for almost all triples (x,y,z) with $x \sim y \sim z$, - cf. [37], [26]. This almost everywhere condition requires a slight explanation; if we form the set $R^{(n)} = \{(x_1,x_2,\ldots,x_{n+1})\ x_1 \sim x_2 \sim \ldots \sim x_{n+1}\} \subset X^{n+1}$, one easily sees that there are unique measure classes $B^{(n)}$ on $R^{(n)}$ which upon disintegration over projection to any coordinate yield disintegration products equivalent to the n-fold Cartesian product of ν^x above (see [26], p. 295). Then (*) is to hold almost everywhere in this sense on $R^{(2)}$. Of course $(R^{(1)},B^{(1)}) = (R,B)$. The reader will also recognize (*) as the manifestation in this picture of the chain rule for differentiation.

In the case of a discrete measured equivalence relation, one may describe Haar measure rather more simply and directly. For any choice of base measure μ, Haar measure ν is specified by giving the fiber measures ν^x on the leaf of x. In this case the leaf of x is a countable set and ν^x is simply the counting measure. We clearly have $\nu^x = \nu^y$ if $x \sim y$. This is discussed in more detail in [26].

Of course ν and μ above depend on the choice of the measure μ, but any other μ' is of the form $d\mu'(x) = b(x)d\mu(x)$. Then Haar measure ν' for μ' is of the form $d\nu'(x,y) = b(x)d\nu(x,y)$ as one sees immediately. Then it is easy to verify that the new modular function Δ' differs from Δ by the factor $b(x)b(y)^{-1}$ and so is equivalent (or cohomologous in terminology to be introduced shortly) to Δ. One says that R has an *invariant* measure if Δ for one (equivalently for all) choice(s) of μ has the form $\Delta(x,y) = c(x)c(y)^{-1}$ for some c. This means that one can choose μ so that ν satisfies $\theta_*(\nu) = \nu$. One says that a measure μ is invariant if $\Delta(x,y) \equiv 1$ for this μ. That this parallels the group action situation is fairly clear, for if $R = R_G$ for a *free* action of a *unimodular* group G and if $\mu \in C$, then one easily calculates that $\Delta(x,y) = dg_*(u)/d\mu(y)$ for that unique group element $g \in G$ such that $g(x) = y$. It is interesting to note then that existence of an invariant measure for unimodular groups is an orbit equivalence invariant, not just a conjugacy invariant. Also, as defined above, invariant measures are the usual things. For free actions of *non unimodular* groups, the situation is more involved and we have $\Delta(x,y) = (dg_*(u)/du(y))\Delta_G(g)^{-1}$ where $g \cdot x = y$ and Δ_G is the usual modular function of G [16], [85]. Hence invariant measures in the sense of the equivalence relation are not quite invariant measures for the group action in the usual sense. It will become apparent, however, that this is still the proper definition. For non-free actions there are correspondingly more complicated formulas.

3.2 HOMOMORPHISMS AND COHOMOLOGY. The discussion above suggests that a homomorphism of R into a topological group A (not necessarily abelian) should be simply a Borel map ϕ of R into A (given the Borel structure corresponding to its topology) such that $\phi(x,y)\phi(y,z) = \phi(x,z)$ for almost all triples $(x,y,z) \in R^{(2)}$ (definition of $R^{(2)}$ as above). We agree to identify two such if they agree a.e. on R. We find this definition a bit more tractable than the parallel one that demands that the formula above hold for *all* $(x,y,z) \in R^2$ but with x,y,z all restricte to be in some conull subset of X, but see [71]. Similarly one defines ϕ_1 to be equivalent to ϕ_2 if $\phi_2(x,y) = b(x)\phi_1(x,y)b(y)^{-1}$ a.e. for some Borel function

$b: X \to A$. The set of equivalence classes of homomorphisms is usually written $H^1(\mathcal{R},A)$ in analogy with group theory and group cohomology.

In fact it is straightforward now to introduce cohomology groups in all dimensions, and we indicate this briefly. Consider an abelian polonais group A and suppose it is an $\mathcal{R}$-module in the sense that we have a homomorphism of $\mathcal{R}$ into Aut(A) -- the proper continuity assumption is that $\phi(x,y)a$ is jointly Borel on $R \times A$. Then the group of n-cochains $C^n(\mathcal{R},A)$ is the set of equivalence classes of functions modulo null functions from $(\mathcal{R}^{(n)},\mathcal{B}^{(n)})$ into A, denoted $U(R^{(n)},A))$; a typical element is a function of $n+1$ variables in X, all equivalent under R. The coboundary operation δ_n is defined in an obvious way $(\delta_n f)(x_0,x_1,\ldots,x_{n+1}) =$

$$= \phi(x_0,x_1)\cdot f(x_1,\ldots,x_{n+1}) + \sum_{i=1}^{n+1} (-1)^i f(x_0,\ldots,\hat{x}_i,\ldots,x_{n+1})$$

See [26], [97], [84] for more details -- actually the only groups other than H^1's that will really be used in the sequel now are $H^2(\mathcal{R},T)$ where T is the circle group.

The alternate approach to the 'almost everywhere' conditions to be imposed on homomorphisms or cocycles hinted at above can be shown to lead to the same results. More specifica ly, one could define strict homomorphisms or cocycles to be those functions $\phi(x_0,\ldots,x_n)$ which satisfy the appropriate cocycle identity for *all* $(x_0,\ldots,x_{n+1}) \in R^{(n+1)} \cap (X_0 \times \ldots \times X_0)$ with X_0 a fixed conull set; one then agrees to identify two such functions ϕ_1 and ϕ_2 if they agree on $R^{(n)} \cap (X_0 \times \ldots \times X_0)$ again for some conull X_0. Finally one defines two cocycles to be equivalent (or cohomologous) if their difference is equal to the boundary of an n-1 cochain everywhere on $R^n \cap (X_0 \times \ldots \times X_0)$ again for some conull set X_0. There is clearly a natural mapping of equivalence classes of such strict homomorphisms or cocycles into the equivalence classes of 'almost everywhere' homomorphisms or cocycles defined previously. It is of considerable use to know that these maps are bijective and this was shown in [81]. For the case $n = 1$, and non abelian A the result is in [71], Theorem 5.1.

We add a remark that if G acts freely on X and A is an ordinary G module, then U(X,A) (the equivalence classes need null functions of Borel functions from X to A) becomes in a natural way a G module in the usual sense and A becomes an $\mathcal{R}_G$ module. Then it may be shown that the group cohomology $H^n(G,U(X,A))$ is isomorphic to the equivalence relation cohomology $H^n(\mathcal{R}_G,A)$ (see [26], [97]). This shows roughly speaking that cohomology is an orbit equivalence invariant and this can be used to draw some stricking consequences (cf. Theorem 7 of [26]).

It is important to note that if A is an R module, and if I is, as above, the transtive equivalence relation on the unit interval, then A becomes an $R \times I$ module in a natural way with I acting trivially and there a natural map $H^n(R,A)$ to $H^n(R \times I,A)$. In fact, it can be shown that this map is an isomorphism, and moreover that every $R \times I$ module is equivalent to a module arising in this way. One proves this by combining Proposition 7.5 of [26], which is in essence the desired result for discrete relations $\mathcal{D}$, with the techniques of [28], all making use of the axiomatic treatment of cohomology in [84]. The final assertion about $R \times I$ modules is the same result in dimension one for non-abelian coefficient groups. This result can in effect be viewed as the proper version of Shapiro's Lemma (cf. [63]) in this context. Therefore we see that all of these cohomological objects introduced previously are stable isomorphism invariants, and hence that any cohomological question is immediately reduced to the discrete case.

It is evident from this discussion that the modular functions Δ associated with various Haar measures constitute a single cohomology class in $H^1(R, \mathbb{R}^+)$ called the modular class or the Radon-Nikodym class, $[\Delta]$. We will often want to use its logarithm instead, $\log[\Delta]$ in $H^1(R,\mathbb{R})$ where $\mathbb{R}$ is viewed as additive group.

3.3. POINCARÉ FLOW AND ASYMPTOTIC RANGE. We now describe two extremely important constructions associated with an $R = (X,C,R,B)$ and a homomorphism ϕ of R into a *locally compact* group A. The first of these constructions associates to the pair (R,ϕ) an action or flow of the group A on a measure space (Z,D), D a measure class on Z. This non-singular action, called the Poincaré flow of (R,ϕ), is defined up to conjugacy -- not just orbit equivalence -- and will depend only on the equivalence class (cohomology class) of the homomorphism ϕ. This flow was introduced in this context by Mackey [59] although some vestiges can be traced back quite far, and subsequent treatments, for instance in [45] and [26], derived from this paper. For the construction we first form the product space $X \times A$ and equip it with the product measure class -- Haar measure on A understood on the second factor. Then one defines a (measured) equivalence relation R_ϕ on $X \times A$ which is just like the skew product construction in ergodic theory [2]. The pair (x,a) is defined to be equivalent to (x',a') iff $x \sim x'$ and $a = \phi(x,x')a'$. It is easy to see how to make this into a measured equivalence relation, and we note that for each $b \in A$ $\theta(b)(x,a) = (x,ab)$ is an automorphism of R_ϕ. The measured equivalence relation R_ϕ may not be ergodic even if R is, and we consider its ergodic decomposition. The parameter space for this decomposition will be denoted by (Z,D) where D is a measure class on Z. By definition, the measure algebra of (Z,D) is the measure algebra of all R_ϕ invariant sets, and as A acts on R_ϕ by automorphisms, it acts on this measure algebra of invariants and hence by the point realization theorem, [56], we obtain an action of A on (Z,D) which is called the *Poincaré flow* associated to R and ϕ. This is evidently an ergodic action if R is ergodic, and an easy calculation shows that up to conjugacy, it depends only on the equivalence class of the cocycle ϕ. A case of particular interest is that when $A = \mathbb{R}$ and $\phi = \log \Delta$, the logarithm of the Radon-Nikodym cocycle. The action of $\mathbb{R}$ obtained in this case is called simply the Poincaré flow of R.

It is an easy exercise to see that if R_1 and R_2 are stably isomorphic and ϕ_1 and ϕ_2 are cocycles such that their classes are mapped to each other by the induced isomorphism on cohomology, then their Poincaré flows are the same.

The second and closely related construction associated to the pair (R,ϕ) is what is variously called its asymptotic range or asymptotic ratio set, denoted $r(\phi)$. This will be a closed subset of $\bar{A}$, the one point compactification of A (where A is again assumed to be locally compact); $r(\phi) \cap A$ is to be a closed subgroup of A. Although one can say certain things if A is non-abelian, this concept is really only used (so far) when A abelian and R is ergodic, so we stick to that case. We view ϕ as a map from R into $\bar{A}$ and then for each Borel $Y \subset X$ of positive measure we let $r(\phi,Y)$ be the essential range in $\bar{A}$ of the restriction of ϕ to $R \cap (Y \times Y)$. (The essential range of a map ψ is the smallest closed subset F such that $\psi^{-1}(F^c)$ is a null set.) Then the asymptotic range of ϕ is by definition the intersection $\cap_Y r(\phi,Y)$ over all subsets Y. The original idea here comes from Araki-Woods [3] and was developed and expanded by Krieger [45], [46], [49]. The form of the definition above comes from the treatments by Schmidt [79] and in [26]. Again it is easy to see that $r(\phi)$ depends only on the equivalence class of ϕ and that it is the same for cocycles ϕ_1 of R_1 and ϕ_2 of R_2 whose classes correspond to each other under a stable isomorphism cf. [79], [26] and [28]. If $\phi = \Delta$, the Radon-Nikodym or modular class, $r(\Delta) \subset \mathbb{R}^+ \cup \{\infty\}$ is usually called the asymptotic ratio set or just the ratio set of R [3], [49].

There is a simple relation between $r(\phi)$ and the Poincaré flow of ϕ; namely the Poincaré flow, being an ergodic action of an abelian group has isotropy groups that are almost everywhere constant ([4], p. 70) say equal to A_0. Then $r(\phi) \cap A = A_0$ (see [79] and Theorem 8 of [26] for the discrete case; then *mutatis mutandi* by stable isomorphism arguments, one can obtain the result in the general case.

The **ratio set of** R (Radon-Nikodym class understood) was used by Araki-Woods and Krieger to introduce a type classification, initially for single transformations, but then by obvious extension the same classification applies to equivalence relations. First of all it is a fact cf. [79], [26], [49] that $r(\phi) \subset \mathbb{R}^+$, i.e. $\infty \notin r(\phi)$ is equivalent to R being measure preserving and then in that case $r(\phi) = \{1\}$. This one calls the type III case. If $\infty \in r(\phi)$ we call R type III; in this case $r(\phi)$ is determined by $R(\phi) \cap \mathbb{R}^+$ which is a closed subgroup of $\mathbb{R}^+$ and hence is of one of three types as follows; if $r(\phi) \cap \mathbb{R} = \{1\}$, one says that R is type III_0; if $r(\phi) \cap \mathbb{R} = \{\lambda^n, n \in \mathbb{Z}\}$ with $\lambda < 1$, one says that R is type III . If $r(\phi) \cap \mathbb{R}^+ = \mathbb{R}^+$, one says that R is type III_1, cf. [3], [49], [50].

4. REPRESENTATIONS

4.1 GENERALITIES. Perhaps the most important special case of a homomorphism of a measured equivalence relation R into a group is that of a unitary representation; specifically if H is a fixed Hilbert space, a representation of R on H is simply a homomorphism ϕ of R into the unitary group $U(H)$; unitary equivalence is simply equivalence of homomorphisms as defined previously. Often a minor refinement of this notion is useful: assume one has a 'field' of Hilbert spaces H_x, one for each x where x runs over the base space of R, and suppose we are given $\phi(x,y)$, a unitary map from H_y to H_x satisfying the obvious composition property that $\phi(x,z) = \phi(x,y)\phi(y,z)$. If all the H_x's have the same dimension, which is usually the case, we can replace them by a fixed Hilbert space H of the common dimension and then we have an ordinary representation. In general, the dimension of H_x is an invariant function and hence is constant in the ergodic case. As one doesn't gain much, except a lot of notational difficulty in treating the variable dimension case, we shall stick to the constant dimension case.

A more subtle variation on this theme introduced in [13] is as follows: suppose that for each leaf ℓ (or equivalence class) of R we have given a Hilbert space H_ℓ (which for simplicity we again assume to be of constant dimension). We assume that this "field" $\{H_\ell\}$ is measureable in the sense that for almost all $x \in X$, we can find a unitary operator $V(x)$ mapping the Hilbert space $H_{\ell(x)}$ of the leaf of x onto a fixed standard Hilbert space H such that $V(x)V(y)^{-1}$, which is defined when $\ell(x) = \ell(y)$ i.e. when $(x,y) \in R$ is Borel on R. Then quite evidently $\phi(x,y) = V(x)V(y)^{-1}$ is a unitary representation of R, on H and if we arbitrarily modify the choice of the $V(x)$'s by $V'(x) = K(x)V(x)$ where K is any Borel map of X into $U(H)$, then the resulting ϕ' is clearly equivalent to ϕ. Thus the class of the representation depends (somewhat mysteriously) only on the field $\{H_\ell\}$ and how the H_ℓ vary with ℓ. This is essential to the point of view in [13].

A special case of this construction is the *regular representation* of R which can be defined by fixing a measure μ of class C on the base space of R and constructing the corresponding Haar measure ν on R. The disintegration of ν with respect to projection to the first coordinate gives measures ν^x on the leaf $\ell(x)$ of x which by definition of Haar measure satisfy $\nu^x = \nu^y$ if $\ell(x) = \ell(y)$; hence we can write ν^ℓ for this measure. Then we define H_ℓ to be $L^1(\ell,\nu^\ell)$; it is easy to see that if these L^2 spaces have constant dimension (which is the case in all but artificial and pathological examples) then the unitary identifications $V(x)$ as required above can be easily produced. This defines what is called the regular representation of R [38], [13]; it is easily seen to be independent of the choice of Haar measure. A related kind of example, also of considerable importance is obtained by specifying a subspace M_ℓ of $H_\ell = L^2(\ell,\nu^\ell)$ depending only on ℓ and varying measurably from leaf to leaf. For instance, if R comes from a C^∞ foliation of the base manifold X and if D is some differential operaton on X acting only in the leaf directions, D defines an operator D_ℓ on $L^2(\ell,\nu^\ell)$ for each ℓ, and one can take M_ℓ to be kernel of D_ℓ; this is the key to the approach in [13] to the index theorem for foliations. One can also define a "calculus" of unitary representations as in [13] but we shall not elaborate this point here.

There is one further twist on these definitions that is useful and in fact necessary, and this is the notion of a projective representation. Specifically, let us fix a 2-cocycle σ on R with coefficients in the circle group T. Thus σ is function $\sigma(x,y,z)$ of 3 variables with $(x,y,z) \in R^{(2)}$ which satisfies a four term cocycle identity obtained by specializing the definition of δ_2 in Section 3.2. A σ-representation ϕ of R on H is a map of R into $U(H)$ such that $\phi(x,y)\phi(y,z) = \sigma(x,y,z)\phi(x,z)$ almost everywhere. Equivalently, one could talk about homomorphisms of into the projective unitary group $U(H)/T$ on H and come to the same thing -- see [38], [26] for further details. As an example, given any $\sigma \in Z^2(R,T)$, one can form the σ-regular representation. Specifically for each x one chooses as before an isomorphism $V(x)$ of $L^2(\ell(x),\nu^x)$ with a fixed H. Then one defines $\phi(x,y) \in U(H)$ by $\phi(x,y) = V(x)S(x,y)V(y)^{-1}$ where $S(x,y)$ is the unitary operator on $L^2(\ell,\nu^\ell)$, $\ell = \ell(x) = \ell(y)$, defined by multiplication by the function $z \to \sigma(x,y,z)$.

For our purposes it is important to be able, just as for ordinary groups, to be able to "integrate" a representation or a σ-representation to get a representation of an associative algebra, and by taking the weak closure of the image to obtain an enveloping von Neumann algebra. There will be a difference in that this algebra representation and associated von Neumann algebra will live on a *different* Hilbert space, and not on the Hilbert space H of the original representation of $\mathcal{R}$. Specifically, let ϕ be a representation (or a σ-representation) on H and let $\bar{H} = L^2(X,H)$ be the space of L^2 H-valued functions on X_ℓ or equivalently if we have instead a field H_ℓ of Hilbert spaces, one for each leaf, we can think of $\bar{H}$ as the direct integral $\bar{H} = \int H_{\ell(x)} du(x)$ of the spaces $H_{\ell(x)}$ using the unitary identifications $V(x)$ implicit in the field $\{H_\ell\}$.) It is on the Hilbert space $\bar{H}$ that one constructs algebra representations and a von Neumann algebra. More precisely let k be a complex valued function on R integrable with respect to the Haar measure ν associated to μ on the base X; then one wants to define an operator $(M_k f)(x) = \int k(x,y)\phi(x,y)f(y)d\nu^x(y)$ on $\bar{H} = L^2(X,H)$. This is treated in [38]. Also there is a suitable convolution product on the functions k on R,

$$(k_1 * k_2)(x,y) = \int k_1(x,z)k_2(z,y)d\nu^{\ell(x)}(z)$$

(or with a factor $\sigma(x,z,y)$ added if we have a σ-representation) and one obtains a suitable (σ) convolution algebra on $\mathcal{R}$ and then the map $k \to M_k$ becomes a homomorphism. The weak closure of the image of M in $\mathcal{B}(\bar{H})$ is the enveloping von Neumann algebra $R(\phi)$ of the representation ϕ. For the details of this, see [38]. The algebra $R(\phi)$ can be shown to contain a distinguished abelian subalgebra $A(\phi)$ isomorphic to $L^\infty(X)$ of operators of the form $(A_h f)(x) = h(x)f(x)$ where $h \in L^\infty(X)$. This would corresponding to an operator M_k but with k a "Dirac function" on the diagonal, $k(x,y) = \delta(x,y)h(x)$ (Theorem 3.8 of [38]). Also, if α is the full group $[\mathcal{R}]$ of $\mathcal{R}$, one defines a natural unitary operator U_α by

$$(U_\alpha f)(x) = \phi(x,\alpha^{-1}(x))f(\alpha^{-1}(x))\Delta(x,\alpha^{-1}(x))^{1/2}$$

for $f \in \bar{H}$. (Actually one has to be slightly careful in this definition and first change ϕ on a null set so that its multiplicative property holds for all triples (x,y,z) in $R^{(2)}$ with x, y, and z in a fixed conull set.) Note that U_α yields a representation (or a σ-representation suitably defined of $[\mathcal{R}]$ and that each U_α normalizes the abelian algebra $A(\phi)$; hence one has a system of imprimitivity. Finally it can easily be shown that U_α belongs to $R(\phi)$ and of course resembles an operator M_k where k is a "Dirac function" supported on the graph of α. Of course, one would like to know that the von Neumann algebra $R(\phi)$ is generated by the subalgebra $A(\phi)$ and the U_α's for α $[\mathcal{R}]$. This is easily seen to be the case when $\mathcal{R}$ is discrete, or when $\mathcal{R}$ has continuous orbits; the discrete case follows from the discussion in [26], and the second case follows by noting from Section 2.3 that $\mathcal{R}$ is of the form $\mathcal{D} \times I$ with $\mathcal{D}$ discrete and then using the discrete case together with obvious facts about $\mathcal{I}$. Moreover, any $\mathcal{R}$ of the two types above comes from a group action; so if $\mathcal{R} = \mathcal{R}_G$ with G acting effectively then G may be viewed as a subgroup of the full group $[\mathcal{R}_G]$ and one easily shows by the same ideas that $R(\phi)$ is generated by $A(\phi)$ and the U_α with $\alpha \in G$.

One can easily identify the commutant of $R(\phi)$; by elementary methods an operator T in $R(\phi)'$ can be diagonalized with respect to $A(\phi)$ and hence can be written as a direct integral $(Tf)(x) = T(x)f(x)$ with $T(x)$ a bounded operator on H. By Theorem 3.8 of [38] the necessary and sufficient condition for T to $R(\phi)'$ is that

$$\phi(x,y)T(y) = T(x)\phi(x,y)$$

for almost all pairs $(x,y) \in R$.

Also if $\mathcal{R}_1$ and $\mathcal{R}_2$ are measured equivalence relations and ϕ_2 and ϕ_2 are representations on H_1 and H_2, one may form in a natural way the product representation $\phi = \phi_1 \otimes \phi_2$ of $\mathcal{R} = \mathcal{R}_1 \times \mathcal{R}_2$ on $H_1 \otimes H_2$. The Hilbert space $\bar{H}$ for the integrated version of ϕ is immediately seen to be isomorphic to $\bar{H}_1 \otimes \bar{H}_2$ (where $\bar{H}_i$ is the Hilbert space for the integrated version of ϕ_i); moreover $R(\phi)$ is simply the von Nuemann tensor product $R(\phi_1) \otimes R(\phi_2)$. If $\mathcal{R} = \mathcal{I}$ is the transitive relation on the unit interval I, then by the version of Shapiro's lemma in Section 3.2 and our comments about the cohomology of $\mathcal{R} \times \mathcal{I}$, all cohomology groups and sets of $\mathcal{I}$ are trivial and in particular any unitary representation is equivalent to the trivial one $\phi(x,y) = 1 \; \forall \; x,y$, on an n-dimensional space H_n for some n. It is quite easy to see then that $\bar{H} = L^2(I,H_n) = L^2(I) \otimes H_n$ and that the operators M_k defined above are ordinary integral kernal operators operating on the first factor; hence it follows that $R(\phi)$ is the algebra of all bounded operators on $L^2(I)$ (tensored with H_n) and that $A(\phi)$ is the algebra of multiplication operators. Similarly if $\mathcal{R} = \mathcal{I}_\infty$ the transitive relation on an infinite discrete set, say $\mathbb{Z}^+$, any representation is trivial and $R(\phi)$ is modulo multiplicities all bounded operators on $\ell^2(\mathbb{Z}^+)$ realized as infinite matrices. The subalgebra $A(\phi)$ consists of diagonal matrices -- see [27].

From Section 2.3 we know that for any $\mathcal{R}$ there is a discrete $\mathcal{D}$ with $\mathcal{R} \times \mathcal{I} \simeq \mathcal{D} \times \mathcal{I}$, and since we have noted that there is a bijection between equivalence classes of representations of $\mathcal{R}$ and $\mathcal{R} \times \mathcal{I}$, there is a bijection $\phi_{\mathcal{R}} \leftrightarrow \phi_{\mathcal{D}}$ between classes of representations of $\mathcal{R}$ and those of $\mathcal{D}$; the corresponding von Neumann algebras satisfy

$$R(\phi_{\mathcal{R}}) \otimes \mathcal{B}(L^2(I)) \simeq R(\phi_{\mathcal{D}}) \otimes \mathcal{B}(L^2(I));$$

i.e. they are stably isomorphic. Moreover this stable isomorphism preserves the distinguished abelian subalgebras $A(\phi)$; specifically

$$A(\phi_R) \otimes L^\infty(I) \simeq A(\phi_D) \otimes L^\infty(I).$$

The point of this is that now just about any question concerning the algebras $R(\phi)$ can be reduced to the discrete case.

4.2 THE REGULAR REPRESENTATION AND ITS VON NEUMANN ALGEBRA. By far the most important representation of a measured equivalence relation R is its regular representation (or its regular σ-representation if one wants to mix in a cocycle). Recall that this representation is defined by a field of Hilbert spaces $H_\ell = L^2(\ell, \nu^\ell)$ where $\nu = \int \nu^x d\mu(x)$ is a Haar measure on R, $\nu^\ell = \nu^{\ell(x)}$. The integrated version of the representation takes place on the Hilbert space $\bar{H} = \int H_{\ell(x)} d\mu(x)$ and it is clear that $\bar{H}$ can be directly identified as $L^2(R,\nu)$. The operators M_k introduced in the general case for k a function on R take the form

$$(M_k f)(x,z) = \int k(x,y) f(y,z) d\nu^x(y)$$

of convolution operators on R. In the case of the regular σ representation there is an added factor of $\sigma(x,y,z)$ in the integrand, and of course, k has to satisfy appropriate conditions so that the integral formula above defines a bounded operator. The enveloping von Neumann algebra will be denoted $M(R)$ or $M(R,\sigma)$ if there is a two cocycle present, and the abelian algebra $A(R)$.

The commutant of $M(R,\sigma)$ is easily determined in this case and it turns out to be the algebra generated by right convolutions

$$(N_k f)(x,z) = \int f(x,y) k(y,z) \sigma(x,y,z) d\nu^x(y)$$

for appropriate functions k cf. [26], [38]. In fact in complete analogy with the group situation, one can define an appropriate Hilbert algebra in this context and invoke the Tomita-Takesaki theory. One obtains a normal weight w [34] on $M(R,\sigma)$ which has the form

$$w(M_k) = \int k(x,x) d\mu(x)$$

for suitable k and in particualr for k of the $k(x,z) = \int h(x,y)\overline{h(z,y)}\Delta(z,y) d\nu^x(y)$ which correspond to operators of the form $M_h M_h^*$. This weight has the property that the abelian algebra $A(R)$ is contained in its centralizer [91]; i.e. $\omega(aTa^*) = \omega(T)$ for all unitary elements a of $A(R)$. The corresponding modular automorphism group supplied by the Tomita-Takesaki apparatus can be spatially implemented on $\bar{H} = L^2(R,\nu)$ by the one parameter group of unitary operators u_t where $(u_t f)(x,y) = \Delta^{it}(x,y) f(x,y)$ is a multiplication operator, [16], [38], [27].

What makes u_t an automorphism of $M(\mathcal{R},\sigma)$ is the fact that the function Δ^{it} is a one-cocycle of $\mathcal{R}$ i.e. $\Delta^{it} \in Z^1(\mathcal{R},T)$ where T is the circle group. More generally if $\theta \in Z^1(\mathcal{R},T)$, then one sees immediately that if M_θ is multiplication by θ on $\bar{H} = L^2(R,\nu)$, then $C_\theta: x \to M_\theta \to M_\theta^{-1}$ defines an automorphism of $M(\mathcal{R},\sigma)$. This automorphism also has the property that it is the identity on the distinguished abelian subalgebra $A(\mathcal{R},\sigma)$. It is not difficult to show cf. [27], [86], [16] that any automorphism of $M(\mathcal{R},\sigma)$ which is the identity on $A(\mathcal{R},\sigma)$ is uniquely of the form C_θ, for some $\theta \in Z^1(\mathcal{R},T)$. Thus $\theta \to C_\theta$ is an algebraic isomorphism of $Z^1(\mathcal{R},T)$ onto a subgroup of the automorphism group of $M(\mathcal{R},\sigma)$, and it is immediate that this is a topological isomorphism as well from the topology of convergence in measure on $Z^1(\mathcal{R},T)$ (cf. [26], [16]) to the usual topology of predual convergence on Aut $(M(\mathcal{R},\sigma))$. Finally it is evident cf. [27], [16] that C_θ is inner if and only if $\theta \in B^1(\mathcal{R},T)$. Hence the first cohomology group $H^1(\mathcal{R},T)$ can be interpreted topologically and algebraically as a subgroup of the outer automorphism group of $M(\mathcal{R},\sigma)$.

The Poincaré flow of $\mathcal{R}$ and its asymptotic ratio set coincide, as one might suspect, with familiar invariants of $M(\mathcal{R},\sigma)$ defined in terms of von Neumann algebras. In [16] Connes and Takesaki introduce the 'flow of weights' of a von Neumann algebra M; this is an intrinsically defined action of the line $\mathbb{R}$ on a measure space, ergodic if M is a factor. As one expects, for $M(\mathcal{R},\sigma)$ this coincides with the Poincaré flow of $\mathcal{R}$, [16], [38], [78]. Moreover, for $\mathcal{R}$ ergodic, the asymptotic ratio set of the measured equivalence relation $\mathcal{R}$ defined above coincides (as it "must!) with A. Connes' ratio set [8] of the von Neumann algebra $M(\mathcal{R},\sigma)$; this was the motivating idea for introducing the ratio set of a von Neumann algebra from [3] and [70].

If the relation $\mathcal{R} = (X,C,R,B)$ is discrete in the sense of Section 2, then the structure of $M(\mathcal{R},\sigma)$ becomes particularly transparent. In this case let's take, as we may, the measure μ on the base X to be finite; then as noted in Section 3.1, the Haar measure for μ is given by $\nu = \int \nu^x du(x)$ where ν^x is counting measure on the leaf or orbit $\ell(x)$ of x. We follow the discussion in [27]; the "convolution" operators M_k take the form

$$(M_k f)(x,z) = \sum k(x,y)f(y,z)\sigma(x,y,z)$$

where the sum is extended over the common class of x and z, and $f \in L^2(R,\nu)$. The product of two such operators $M_k M_h$ is given as another one of the same kind M_{k*h} with

$$(k*h)(x,z) = \sum k(x,y)h(y,z)\sigma(x,y,z).$$

Moreover every operator in the weak closure $M(\mathcal{R},\sigma)$ is easily seen to be representable as such a kernel operator M_k for some scalar function k on $\mathcal{R}$ ([27], p. 330). Thus $M(\mathcal{R},\sigma)$ literally is the "algebra of matrices" over the measured equivalence relation $\mathcal{R}$, making a complete analogy with the situation when $\mathcal{R} = I_\infty$, the transitive equivalence on a countably infinite set say $\mathbb{Z}^+$ where $R(\mathcal{R},\sigma) = \mathcal{B}(H_\infty)$. In this case the presentation of an operator as an M_k, k a function on $R = \mathbb{Z}^+ \times \mathbb{Z}^+$ is its traditional presentation as an infinite matrix. For general $\mathcal{R}$, as in the special case of I_∞, there is no simple necessary and sufficient condition on the function k which ensures that the formula above for M_k defines a bounded operator.

The distinguished abelian subalgebra $A(\mathcal{R},\sigma)$ consists of the M_k with $k(x,y) = 0$ if $x \neq y$ (recall that the diagonal now is a subset of positive ν-measure on R). The weight w defined in general in this case becomes a state and is given by the formula

$$\omega(M_k) = \int k(x,x)d\mu(x)$$

which is always finite as k is necessarily bounded and μ is finite. If μ where σ-finite instead of finite, then w would be a proper weight, but it would have the key additional property that it is semi-finite on $A(\mathcal{R},\sigma)$. We have noted already that $A(\mathcal{R},\sigma)$ is in the centralizer of this state; but according to [91] this holds if and only if there is a conditional expectation E of $M(\mathcal{R},\sigma)$ onto $A(\mathcal{R},\sigma)$. In the present context this conditional expectation is given explicitly by $F(M_k) = M_{I(D)k}$ where $I(D)$ is the characteristic function of the diagonal in R. Note that $I(D)k$ lives on the diagonal and is necessarily bounded (this is an easy consequence of the fact that M_k is a bounded operator) and hence defines an element of $A(\mathcal{R},\sigma)$.

Moreover each unitary U_α $\alpha \in [\mathcal{R}]$, the full group, normalizes $A(\mathcal{R},\sigma)$ and as we have noted (cf. [27] for more details) the U's together with $A(\mathcal{R},\sigma)$ generate $M(\mathcal{R},\sigma)$.

4.3 CARTAN SUBALGEBRAS. The next step is to try to obtain in some sense a converse of the construction of $M(\mathcal{R},\sigma)$ above. Toward this goal the discussion above strongly suggests the following definition, cf. [27], [94].

DEFINITION. If M is a von Neumann algebra, an abelian subalgebra A of M is called a *Cartan subalgebra* of M if

(1) A is maximal abelian in M

(2) A is contained in the centralizer of a state ω of M-equivalently [91]) there is a conditional expectation E of M onto A

(3) M is generated by the normalizer $N(A)$ of A $(= \{u \in M,\ u \text{ unitary } uAu^{-1} = A\})$.

(Of course, if M is finite, (2) is superfluous.) It turns out that these conditions characterize pairs $(M(\mathcal{R},\sigma),A(\mathcal{R},\sigma))$ and the desired converse theorem is as follows, [27] Theorem 1.

THEOREM. If A is a Cartan subalgebra of M, then there exists a unique discrete equivalence relation $\mathcal{R}$ and a $[\sigma] \in H^2(\mathcal{R},T)$ such that the pair (M,A) is isomorphic to the pair $(M(\mathcal{R},\sigma),A(\mathcal{R},\sigma))$. Specifically if $(\mathcal{R}_1,[\sigma_1])$ and $(\mathcal{R}_2,[\sigma_2])$ are two such pairs doing this, then there is an isomorphism of $\mathcal{R}_1$ onto $\mathcal{R}_2$ carrying $[\sigma_1]$ onto $[\sigma_2]$.

If on the other hand we consider a pair $(M(\mathcal{R},\sigma),A(\mathcal{R},\sigma))$ where $\mathcal{R}$ has continuous orbits, then (1) and (3) still hold, but in place of (2) one has something rather weaker -- namely the existence of an unbounded conditional expectation, or an operator valued weight in the sense of [35], and then one should be able to prove a theorem analogous to the one above in the continuous case. A version of such a theorem appears in [28] but this is not as satisfactory as it should be. Since $\mathcal{R}$ is of the form $\mathcal{D} \times \mathcal{I}$, $\mathcal{D}$ discrete we can decompose $(M(\mathcal{R},\sigma),A(\mathcal{R},\sigma))$ as a product $(M(\mathcal{D},\sigma') \otimes \mathcal{B}(L^2(I)),A(\mathcal{R},\sigma') \otimes L^\infty(0,1))$ where σ' is a suitable cocycle on $\mathcal{D}$. Then if we replace $L^\infty(0,1)$ by a Cartan subalgebra A of $\mathcal{B}(L^2(I))$. Then $A(\mathcal{R},\sigma')$ A will be a Cartan subalgebra of $M(\mathcal{R},\sigma)$ and in particular $M(\mathcal{R},\sigma)$ will always have a Cartan subalgebra for any $\mathcal{R}$.

A helpful way to rephrase this is as follows: Consider two pairs (M_1,A_1) (M_2,A_2) with A_i satisfying (1) and (3); say that they are stably isomorphic if after tensoring both sides with $(\mathcal{B}(L^2(I)), L^\infty(I))$, the pairs become isomorphic. Then the map $(\mathcal{R},\sigma)$ going to $(M(\mathcal{R},\sigma),A(\mathcal{R},\sigma))$ is a bijection between pairs $(\mathcal{R},\sigma)$ taken up to stable isomorphism of measured equivalence relations (carrying over cohomology classes), with stable isomorphism classes of pairs (M,A) which contain a pair (M_1,A_1) with A_1 a Cartan subalgebra of M_1.

At this point one may ask whether any von Neumann algebra has a Cartan subalgebra -- we guess that the answer is most probably negative in general. However, once one has Cartan subalgebras, one can ask about conjugacy. Conjugacy within the group of inner automorphisms is clearly out of the question as it fails for the hyperfinite II_1 factor M given the structure of M from [11]. One should rather ask for conjugacy in the full automorphism group. We hazard no guesses on this in general. If, however, M is approximately finite or equivalently injective, [12], the situation is quite good and is discussed below in Section 5.3.

5. AMENABILITY

5.1 APPROXIMATE FINITENESS AND AMENABILITY. With the correspondence of the second paragraph above established, we now explore the connection between properties of the von Neumann algebra M and properties of the equivalence relation $\mathcal{R}$. We have seen examples of this already in the correspondence between the flow of weights of M and the Poincaré flow of $\mathcal{R}$, and the matching up of the ratio set of M and the asymptotic range of the modular homomorphism of $\mathcal{R}$.

First of all, in view of the central role that amenable groups are known or expected to play in ergodic theory, it is natural and important to seek an analogue of amenability for measured equivalence relations which parallels the notion for groups. Zimmer [103] has formulated just such a notion.

DEFINITION. A measured equivalence relation $\mathcal{R}$ is *amenable* if for every separable Banach space E, and for every homomorphism ϕ of $\mathcal{R}$ into the group of isometries Iso(E) of E, given the strong operator topology, and for every measurable field $x \to K_x$ of ω^*-compact convex sets in E^* which is invariant under ϕ in the sense that

$$\phi^*(x,y)^{-1}K_y = K_x \quad \text{a.e. } (x,y) \in R$$

there exists a fixed point in K -- that is -- a Borel function e of X in E^* such that $e(x) \in K_x$ a.e. x and

$$\phi(x,y)^{*-1}\, e(y) = e(x) \quad \text{a.e.}$$

As for what one means by K_x being measurable, one puts a suitable Borel structure on the set of ω^*-compact convex sets in E and demands that $x \to K_x$ be measurable in the usual sense.

Since it is known that amenability of a second countable G can be characterized by the fixed point property in E^* for actions of G on E by isometries for all separable Banach spaces, the definition really does parallel quite closely the notion for groups. Actually this is a special case of Zimmer's definition which he gives for measured groupoids; but following our introductory comments we stick with the special case of measured equivalence relations.

Zimmer obtains a number of results which reinforce the parallel with groups.

PROPOSITION (1) If G is amenable, and acts on (X,C), then $\mathcal{R}_G$ is amenable in the sense above ([103]).

(2) If G is discrete and acts freely on (X,C) preserving a finite measure and if $\mathcal{R}_G$ is amenable, then G is amenable. ([103]).

It is also easy to see that this notion is invariant under stable isomorphism. But most importantly it turns out that amenability of $\mathcal{R}$ reflects itself in important properties of $M(\mathcal{R})$.

Recall that a von Neumann algebra is called approximately finite if there is an increasing sequence M_n of sub von Neumann algebras which are finite dimensional and whose union is weakly dense in M; it comes to the same thing to require that the M_n be type I algebras [24'], [19] The notion is venerable and goes back to the very beginnings of the subject [67], and has continued to play a key role in subsequent developments [12]. From [12] one knows the equivalence of this condition with a number of others including injectivity. One could also view the condition as an "amenability" condition also. Hence the following is not totally unexpected.

THEOREM. (Zimmer [105]). The measured equivalence relation $\mathcal{R}$ is amenable if and only if $M(\mathcal{R})$ is approximately finite.

It is not difficult to see that the same is true for $M(\mathcal{R},\sigma)$ [40] but as it subsequently turns out, all higher cohomology vanishes (see below).

At this point what has to be added to the picture (at least in the case of a discrete measured equivalence relation) is an analogue of another equivalent formulation of amenability for groups, namely the Følner condition [32] or its equivalent formulation in the limit as the existence of an invariant mean. This will come in a moment; first we outline the setting for this which is the notion of approximate finiteness for $\mathcal{R}$.

The notion of approximate finiteness for $\mathcal{R}$ goes back to Dye [23], and can be formulated for discrete $\mathcal{R}$ as follow; there should exist an increasing sequence of subrelations $\mathcal{R}_n$ of $\mathcal{R}$ such that $\mathcal{R}$ is the union of the $\mathcal{R}_n$ and such that all equivalence classes of $\mathcal{R}_n$ are finite. It is quite easy to see that if all classes are finite, then the quotient space X/R is countably separated, and this observation leads naturally to the formulation in the general (non-discrete) case. Following standard terminology we call a measured equivalence relation $\mathcal{R}$ *smooth* or type I if the quotient measure space $(X/R, C/R)$ is a standard measure space; this means equivalently that X/R contains a standard Borel subset which is conull, or that the equivalence R is smooth mod null sets.

DEFINITION (cf. [28]). A measured equivalence relation $\mathcal{R} = (X,C,R,B)$ is approximately finite if there is an ascending sequence of measured equivalence subrelations $\mathcal{R}_n = (X,C,R_n,B_n)$ such that $\mathcal{R} = \cup\mathcal{R}_n$; more exactly the R_n are increasing as subsets to R (mod null sets) with B_n absolutely continuous relative to B and with $B = \lim B_n$ and with each $\mathcal{R}_n$ *smooth*.

We make a number of comments about the definition; first the notion of approximate finiteness is readily seen to be invariant under stable isomorphism cf. [28]. In the context of the original formulation in the finite measure preserving discrete case, Dye proved the very important theorem [23] that such an $\mathcal{R}$ was approximately in his sense if and only if $\mathcal{R} = \mathcal{R}_{\mathbb{Z}}$ for some finite measure preserving action of the integers. It is a simple remark that the same argument works whether or not it is measure preserving.

For general R it is natural to restrict to the case of continuous orbits, and then a combination of the Dye theorem and the structural results in Section 2 leads one to the conclusion that if R has continuous orbits, then it is approximately finite if and only if $R = R_{\mathbb{R}}$ for some measure class preserving action of the real line ([28]). Hence one has an intrinsic characterization of the measured equivalence relations arising from $\mathbb{R}$ and $\mathbb{Z}$ actions; we shall shortly discuss the situation for other kinds of groups and clarify the meaning of this observation.

The 'only if' part of the statement above has an important cohomological consequence; if R is approximately finite and A is a Polonais group, then R is stably isomorphic to a discrete approximately finite $\mathcal{D}$, $R \times I \simeq \mathcal{D} \times I$; and as noted in Section 3, $H^n(R,A) \simeq H^n(\mathcal{D},A)$ (trivial action on A); but now $\mathcal{D} = R_{\mathbb{Z}}$ with $\mathbb{Z}$ acting freely and so $H^n(R,A) \simeq H^n(\mathcal{D},A) \simeq H^n(\mathbb{Z},U(X,A)) = 0$ for $n \geqslant 1$ because $\mathbb{Z}$ is free; hence higher cohomology groups $H^n(R,A)$ vanish for $n \geqslant 1$, for R approximately at least if the action of R on A is trivial [97], [26]. In particular $H^2(R,T) = 0$ and so there can be no non-trivial projective representations and there is no way to "twist" the regular representation with a cocycle. In this case we write simply $M(R)$ for the von Neumann algebra generated by the regular representation. Since $\mathbb{Z}$ and $\mathbb{R}$ are (trivially) amenable groups it also follows immediately from Zimmer's results quoted earlier in this section that if R is approximately finite then it is amenable in Zimmer's sense.

Finally, we observe that if R is a smooth equivalence relation, then one sees easily, [38], that $M(R)$ is a type I algebra and conversely. Now recallying the definition of approximate finiteness of R, we have $R = \cup R_n$ (ascending); then one easily deduces that $M(R)$ is the weak closure of the $M(R_n)$ which are all of type I. As this is one of the several equivalent definitions of approximately finiteness (or injectivity) one finds that if R is approximately finite, then $M(R)$ is approximately finite.

5.2 THE CLASSIFICATION THEOREM. One of the highlights of Dye's work [23], [24] on orbit equivalence was his remarkable theorem that if one looks at all discrete approximately finite ergodic measured equivalence relations with finite invariant measure (i.e. ergodic finite measure preserving transformations of 10.1] taken up to orbit equivalence) that then there is exactly *one* isomorphism class. After a sequence of developments Krieger [50], found a remarkable and beautiful generalization of this result to the non pressure-preserving case, using the Poincaré flow (cf. 3.3. above); part of Krieger's result is the following statement; the rest we will discuss presently.

KRIEGER-DYE CLASSIFICATION THEOREM. The map which associates to any approximately finite ergodic (non transitive) measure equivalence relation its Poincaré flow defines a bijection from all stable isomorphism classes of such measured equivalence relations to all conjugacy classes of measure class preserving flows of the real line.

It is easy to see that any measure preserving R has as Poincaré flow, a transitive and free action of the real line. Since up to conjugacy there is exactly one such transitive and free action, the Dye-Krieger theorem says in this case that there is one stable isomorphism class of measure preserving approximately finite R's; and this is just a restatement of Dye's original result. If R is of type III_1 so that $r(R) = (0,\infty) \cup \{\infty\}$, it follows that the Poincaré flow is the identity and then by ergodicity, it must be on a one point space. Since there is trivially only one such flow, the theorem says there is just one approximately finite III_1 R. If R is type III_λ, $\lambda > 0$, the Poincaré flow is periodic with fixed period, and since there is just one such ergodic flow, the theorem is saying that there is a unique approximately finite R of type III_λ for each $\lambda > 0$. If R is of type III_0, the corresponding Poincare flow is easily seen to be non-transitive and the theorem provides a bijection of all III_0 approximately finite R's with all ergodic non-transitive flows.

Krieger's results also provide the classification of the algebras $M(R)$ for R approximately finite and his argument makes use at a crucial point of Takesaki's results on duality [92]; when all of this is combined with Connes' results on injective (= approximately finite) factors and his work on the III_λ and III_0 cases, there results an essentially complete classification.

THEOREM (Connes-Krieger). The map $R \to M(R)$ provides a bijective map from all approximately finite ergodic non transitive measured equivalence relations, mod stable isomorphism to classes of approximately finite (= injective) factors mod stable isomorphism (excluding the type I case and any "strange" (see below) approximately finite III_1 factors.

The factors in the range of this map are called *Krieger factors*. For simplicity we have excluded the single class of type I factors although they, of course, correspond to the class of transitive R's. In the above, we mean by a "strange" approximately finite III_1 factor any such factor which is not the Krieger factor $M(R)$ where R is the the unique III_1 approximately finite equivalence relation. Such factors conjecturally do not exist [12], but this is a most difficult question. Of course, it is clear that an approximately finite III_1 will be a Krieger factor if and only if it has a Cartan subalgebra and hence this is a question of whether approximately finite factors always have Cartan subalgebras. In conclusion, modulo the non-existence of such strange III_1's the above is a complete classification of injective = approximately finite factors, and in fact provides bijective correspondences between three classes of objects

(1) approximately finite ergodic non-transitive measured equivalence relations/stable isomorphism

(2) ergodic flows of $\mathbb{R}$ mod conjugacy

(3) approximately finite factors/stable isomorphism.

We have described the maps from (1) to (2) and from (1) to (3). The direct map from (3) to (2) is the map which assigns to M its flow of weights [16].

The use of stable isomorphism in (3) is something of an illusion for if we exclude the II_1 algebras, all the algebras are infinite and so stable isomorphism is the same as isomorphism. There is an element of this in (1) also for if we assume, as is natural, that the orbits of R are either all discrete or are all continuous and if we again exclude the II_1 case, then in the stable isomorphism class we are left with just two isomorphism classes, the continuous ones, and the discrete ones. We have already discussed this point in the continuous case (cf. [28]); see [26] for the discrete case.

5.3 RESULTS OF CONNES-FELDMAN-WEISS AND ORNSTEIN-WEISS. At this point there is one very important element missing in the picture, and one is led to this question from two rather different starting points. On the one hand one can raise the conjugacy question for Cartan subalgebras; given M approximately finite and two Cartans A_1 and A_2, are they conjugate in Aut(M)? If one assumes that both A_1 and A_2 are approximately finite in the sense that the corresponding measured equivalence relations are approximately finite, then Krieger's results [50] quoted above already provide an affirmative answer, and we conclude automatically that R_1 and R_2 are isomorphic. But what if say A_2 is not approximately finite? In view of the above, the conjugacy problem is equivalent to asking whether an approximately finite M can ever have a Cartan A with corresponding *not* approximately finite. However, according to the result of Zimmer [105] quoted above on amenability, M approximately finite will imply that if A is a Cartan of M, then the corresponding equivalence relation R is amenable in his sense. Therefore, the question becomes whether R amenable will imply that R is approximately finite (and hence that the conditions are equivalently what we said already).

On the other hand, starting from the different point of view of group actions, we have already noted that any measured equivalence relation $R_{\mathbb{Z}}$ or $R_{\mathbb{R}}$ coming from an action of the integers or the real line is approximately finite, and one is led to ask for what other kinds of groups G can one conclude that any R_G is automatically approximately finite. It is not difficult to see that this holds for any abelian group G [25]. Subsequently, Krieger and Connes showed by a very ingenious argument [15] that R_G is approximately finite for any countable solvable G. (See also [69] for a simplified version of this.) Also Series [83] showed that for any connected amenable group G acting in a finite measure preserving way, R_G is approximately finite. In view of all of this, the obvious conjecture for group actions is (or was) that R_G should be approximately finite for any amenable G. Since we know that G amenable implies R_G amenable, we are led to exactly the same question discussed above as to whether R amenable implies R approximately finite. The answer is now known to be affirmative:

THEOREM. (Connes-Feldman, Weiss [18]). If R is amenable then it is approximately finite (and hence the two properties are equivalent).

This theorem builds on immediately preceding work of Ornstein and Weiss [68] in which they prove that if G is discrete amenable and acts as a group of finite measure preserving transformations, then R_G is approximately finite. What is essential in general is a kind of Følner condition for amenable measured equivalence relations analogous to the Følner condition for amenable group [32] and then an analogue of the Rohlin lemma for amenable measured equivalence relations. The philosophy is that essentially the only way one ever proves approximate finiteness is via a Rohlin lemma.

6. PROPERTY T

6.1 PROPERTY T FOR EQUIVALENCE RELATIONS. The previous section discussed a class of measured equivalence relations (and their associated von Neumann algebras) that are clearly analogues of amenable groups. In 1966 D. Kazhdan [44] formulated a property for groups which he termed property (T) that can profitably be thought of an characterizing the opposite extreme from amenability. Recently R. Zimmer [110] saw how to formulate a corresponding property for measure groupoids that again plays the role of the opposite extreme of amenability for these objects. Klaus Schmidt had pursued some similar ideas in [80] and then in [81] he arrived at some of the same results obtained by Zimmer. Also independently B. Weiss and A. Connes [17] and the present author had arrived at about the same point in terms of formulating a property T. In this discussion we will follow Zimmer's approach but in accord with out philosophy we will stick to the case of measured equivalence relations rather than general measure groupoids. Zimmer considers only the case of discrete orbits, but as one easily sees, the same arguments work for general orbit types as well.

Hence let $R = (X,C,R,B)$ be a measured equivalence relation with base space X and let U be a unitary representation of R on a Hilbert space H, i.e. a homomorphism of R into the unitary group of H. One says that U *weakly contains the identity representation* if there is a sequence $v_n(x)$ of functions from X to unit vectors in H such that

$$(*) \qquad |U(y,x)v_n(x) - v_n(y)| \to 0 \text{ on } R$$

as $n \to \infty$, where convergence means convergence in measure on R with respect to any finite measure m the class B. Then following Zimmer, R is said to have *property* T if whenever a representation U weakly contains the identity, then it strongly contains the identity; that is there is a measurable function v from X to unit vectors in H such that

(**) $$U(y,x)\, v(x) = v(y) \quad \text{a.e. on } R.$$

Of course (**) simply means that after a unitary equivalence U can be written as a direct sum $U = U_1 \oplus \tilde{U}$ where U_1 is the one dimensional trivial representation -- $U_0(x,y) = id_1$ and $\tilde{U}$ is some other representation. The analogy with Kazhdan's property T for groups (that identity representation be an isolated point in $\hat{G}$ for the Fell topology [44]) is evident.

Let R be a measured equivalence relation and let I be the transitive equivalence on [0,1]; we mentioned in Section 2 that there is a natural bijection between (equivalence classes of) unitary representations of R and those of $R \times I$. If we view this correspondence in the light of the above definition, it becomes clear that property T holds for R iff it holds for $R \times I$, and hence property T is preserved under stable isomorphism.

Further we note that Property T implies that R has an invariant measure; to see this, observe that if Δ is the modular function of R relative to some measure μ on X, then for every real λ, $\Delta^{i\lambda t}$ defines a (one dimensional) unitary representation. One forms the direct integral of these representations $U = \int \Delta^{i\lambda t} d\lambda$ using Lebesgue measure in λ. Then U pretty clearly weakly contains the identity representation. By Property T, U strongly contains the identity and this in turn implies that $\Delta^{i\lambda t}$ is equivalent to the identity for a set of positive Lebesgue measure. By the results of [65], we conclude that Δ is the trivial cohomology class, and that is just the definition of R being measure preserving. In fact this argument shows that $H^1(R, \mathbb{R}) = (0)$ if R has property T, which is one of Zimmer's results [110].

In order to justify this definition of property T as an analogue of the Kazhdan property for groups, one should show an explicit connection as was done with amenability. Zimmer [110] obtains the following results.

PROPOSITION 6.1.1 Let a locally compact group G act as a group of finite measure preserving transformations of (X,μ) and let R_G be the corresponding measured equivalence relation if

(1) If G has property T, then R_G has property T

(2) If in addition G acts freely and is weak mixing in that $L^2(X,\mu)$ contains no finite dimensional G invariant subspaces (cf. [57]), then if R_G has property T, G has property T.

We turn now to consideration of some consequences of property T; one of the most significant is that property T limits rather severely homomorphism of the measured equivalence R into certain types of groups; that is, we obtain information on $H^1(R,A)$. Recall that one of Kazhdan's observations is that property T for groups implies that the commutator factor group is compact, or equivalently that Hom(G,T), T the circle group, is discrete in its natural topology. Zimmer [110] and independently Schmidt [81] have shown that this carries over exactly.

THEOREM 6.1.2 Let R have property (T) and be ergodic; then

(1) $B^1(R, T)$ is open in $Z^1(R,T)$ in the topology of convergence in measure, and hence $H^1(R,T)$ is discrete and therefore countable (T is the circle group).

(2) If A is locally compact and amenable, any $\sigma \in Z^1(R,A)$ is equivalent to a σ' which lies in $Z^1(R,K)$ where K is some compact subgroup.

(3) Hence if A is amenable and has no compact subgroups, $H^1(R,A) = 0$.

As a sidelight to item (1) in the theorem above, we shall show how (in the appendix) to construct examples of R's with property T so that $H^1(R,T)$ is any preassigned finite abelian group, including the one element group. This provides an example of an R where $H^1(R,T) = (0)$, answering a question raised in [80]. These calculations of some simple concrete examples we feel add insight into the abstract statements of the propositions.

The result in (1) above can also be interpreted in terms of von Neumann algebras for suppose we have a von Neumann algebra M with a subalgebra A. Let us recall that the group Out(M,A) is the group of all automorphisms of M which are the identity on A, modulo the inner ones with this property. This is clearly isomorphic to a subgroup of Out(M) (all automorphisms mod the inner ones). Now suppose that A is a Cartan subalgebra of M; then by the main theorem in Section 4.3, there is a discrete equivalence relation R, a $\sigma \in H^2(R,T))$ such that $(M,A) \simeq (M(R,),A(R,\sigma))$. Moreover, as we also saw in Section 4.3, Out(M,A) in this case is canonically isomorphic, topologically and algebraically, to $H^1(R,T)$. Hence making the natural definition that a Cartan subalgebra A of M has property T (*or better that the pair* (M,A) *has property* T *if the corresponding discrete measured equivalence relation has property* T, we obtain

PROPOSITION 6.1.3 If A is a Cartan subalgebra of a factor M, then Out(M,A) is a discrete countable group in its natural topology.

The examples in the appendix show that Out(M,A) = (e) is possible.

Another remarkable consequence of property T for groups is that a discrete group with property T is automatically finitely generated [44]. We indicate now how this also carries over to measured equivalence relations.

Recall from [28] p. 205 the notion of an open subrelation $\mathcal{R}' = (X,C,R',B')$ of a measured equivalence relation $\mathcal{R} = (X,C,R,B)$. We require that the base measure (X,C) spaces be the same, and that $R' \subset R$ as sets with B' absolutely continuous with respect to B. This means that up to sets of measure zero, each equivalence class or leaf ℓ of R is a union of a countable number of equivalence classes or leaves of R', and that the measure class on this leaf is the "sum" of the measure classes on the leaves of R' which constitute it. If $\mathcal{R}(n) = (X,C,R(n),B(n))$ is an increasing sequence of open subrelations of $\mathcal{R}$, we say that $\mathcal{R}$ is the limit of the $\mathcal{R}(n)$ if $R = \cup R(n)$ up to null sets, and $B = \lim B(n)$. Recall from Section 5.1 that in this context $\mathcal{R}$ is approximately finite iff $\mathcal{R} = \lim \mathcal{R}(n)$ of an increasing sequence of smooth (type I) open subrelations $\mathcal{R}(n)$. The following reinforces the intuition that property T and amenability (= approximate finiteness) are exact opposite and provides the analogue of Kazhdan's result on finite generation.

PROPOSITION 6.1.4. If $\mathcal{R}$ is ergodic and has property T, and if $\mathcal{R} = \lim \mathcal{R}(n)$ for an increasing sequence of open subrelations, then $\exists N$ such that $\mathcal{R} = \mathcal{R}(N)$.

For a proof, see the appendix.

This result can be easily and rather interestingly recast in terms of von Neumann algebras. Suppose that A is a Cartan subalgebra of M so that we have $(M,A) = (M(\mathcal{R},\sigma), A(\mathcal{R},\sigma))$ for some discrete $\mathcal{R}$. Then if $\tilde{R}$ is an (open) subrelation of $\mathcal{R}$, the restriction $\tilde{\sigma}$ of σ to $\tilde{R}$ defines a class there, and one may construct $(M(\tilde{R},\tilde{\sigma}),A(\mathcal{R},\tilde{\sigma}))$. It is clear that $A(\mathcal{R},\sigma) \simeq A(\mathcal{R},\sigma)$ naturally and moreover that $M(\tilde{R},\tilde{\sigma})$ naturally embeds in $M(\mathcal{R},\sigma)$ as a von Neumann subalgebra which contains $A(\mathcal{R},\sigma)$. In fact C. Sutherland [90] shows that $\tilde{R} \to M(\tilde{R},\tilde{\sigma})$ is a bijection from all (open) subrelations $\tilde{R}$ of $\mathcal{R}$ to all von Neumann subalgebras of $M(\mathcal{R},\sigma) = M$ which contain $A(\mathcal{R},\sigma) = A$ -- see also [5], [6], [7]. Thus we obtain the following statement.

PROPOSITION 6.1.5. If M is a factor, A a Cartan in M such that (M,A) has property T, then if M_n is an increasing sequence of subalgebras of M all containing A, then $\exists N$ with $M_N = M$.

We conclude this section with a discussion of a remarkable result of Zimmer in a closely related direction -- namely his strong rigidity theorem for ergodic actions of semi-simple groups [109]. Suppose that G is a centerless semi-simple Lie group without compact factors [36], and suppose that we have a free finite measure preserving action of G on a space X with associated measured equivalence relation $\mathcal{R}_G$. We also suppose, as is natural, that the action is irreducible in the sense that the action restricted to any non-trivial connected normal subgroup of G remains ergodic.

(This is not an important restriction as one can build up a general action from irreducible ones.) Observe that R_G is a virtual subgroup of G in Makey's language [59] and the fact that R_G has a finite invariant measure can be interpreted as saying that R_G has finite covolume in G. For an honest subgroup H of G (instead of a virtual one) the corresponding condition is that the coset space G/H should have a finite G-invariant measure. Such subgroups are called lattices -- see [70] for an exposition of this subject -- and hence one can then profitably think of R_G as a virtual lattice in G. Moreover the notion of irreducibility for R_G described above corresponds exactly to the usual notion of irreducibility for lattices [60].

Margulis in [60] established a very deep result to the effect that under a mild restriction on the rank of G (see below) lattices in semi-simple groups are strongly rigid in the sense that if Γ is an irreducible lattice in G and Γ' is an irreducible lattice in G' and if ϕ is an isomorphism of Γ onto Γ', then ϕ extends to an isomorphism of G onto G'. The condition on rank is that the real rank of G_1 should be at least two [60]. (To define the real rank of G, let G = KAN be an Isawasa decomposition for G, then the real rank is the dimension of A, cf. [36].) Zimmer's result is in a word that the same thing is true for virtual lattices under the same condition on rank. First, what does it mean for two virtual lattices R_G and $R_{G'}$ coming from actions of G and G' on X and X', to be isomorphic? From the point of view of virtual groups [59], the answer is clear -- it should be that the actions are orbit equivalent, or what is the same thing, that R_G and $R_{G'}$ are isomorphic measured equivalence relations. Hence one hopes for the following.

THEOREM 6.1.6 (Zimmer [109]). Let G, G' be semi-simple, centerless with no compact factors and let the real rank of G be at least two. Suppose that R_G and $R_{G'}$ are irriducible virtual lattices in G and G' defined by actions on X and X', and suppose that $R_G \simeq R_{G'}$. Then there is an isomorphism ϕ of G onto G' and a map θ of X to X' with $\theta(g\cdot x) = \phi(g)\theta(x)$ a.e. (the actions on X and X' are assumed free).

In other words, orbit equivalence of actions implies first that G and G' are necessarily isomorphic and then that the actions are in fact conjugate (up to the choice of an isomorphism of G with G'). Clearly nothing could be more diffierent from the approximately finite case described in Section 5.2.

The rigidity property in the Zimmer theorem closely resembles the rigidity that is characteristic of property T that we have been discussing. Indeed, the measured equivalence relations R_G coming from semi-simple groups do have property T if G satisfies a condition closely related to but not identical to the condition on rank necessary for the Margulis-Zimmer result. Specifically if R_G is irreducible in the above sense and if G has at least one simple factor which is not locally isomorphic to SO(n,1) or SU(n,1) for any n, then one easily proves that R_G has property T.

6.2 PROPERTY T FOR VON NEUMANN ALGEBRAS. Connes in an article in these proceedings, cf. also [14], introduces the notion of a Property T for a von Neumann algebra M, and we shall now address the question of how his property T is related to the discussion above. His property T is an intrinsic property of a von Neumann algebra; the property T discussed above is really a property of a measured equivalence relation R, which was then transported over to the situation of a Cartan subalgebra A of a von Neumann algebra, or rather, as we put it , of a pair (M,A). What we want is a notion embracing both, and more precisely what we want is to formulate a property T for a pair (M,B) consisting of a von Neumann algebra M and any sub- von Neumann algebra B of M. This is to be thought of intuitively as M having property T relative to B and also to the rather great extent that Connes' property T can be viewed as a rigidity property of M, one should think of property T for the pair (M,B) as M being rigid relative to B. This definition should, of course, coincide with the Zimmer notion in case B is a Cartan subalgebra of M, and also property T for the pair $(M,\mathbb{C})$ ($\mathbb{C}$ = complex numbers viewed as a one dimensional subalgebra of M) should coincide with Connes' property T for M. Also, for any M, the pair (M,M) should always have property T. For simplicity let us stick to the case when M is a factor for this discussion. We clearly owe a debt to Connes for this entire discussion.

Recall that in defining property T for a algebra M, Connes considers M, M^* bimodules where M^* is the same real algebra as M but has a different multiplication by complex scalars $\lambda \circ x = \bar{\lambda} x$. (Of course M^* is also isomorphic to the opposed algebra.) By a bimodule (H,ρ,ρ^*) for (M,M^*), Connes means a pair of normal representations ρ of M and ρ^* of M^* on a Hilbert space H with the images commuting. Clearly there are notions of unitary equivalence of representations, subrepresentations, and also the notion of weak containment of one (M,M^*) bimodule in another, cf. Connes' article. Note that the standard realization of a von Neumann algebra M and of $M' \simeq M^*$ on $L^2(M,\phi)$, ϕ any faithful normal semi-finite weight (cf. [33]), provides a unique and intrinsically defined equivalence class of bimodules called the trivial bimodule. Connes then says that M *has property* T *if whenever a bimodule weakly contains the trivial bimodule then it contains the trivial bimodule as a summand.* The modification necessary for pairs is now clear; if $B \subset M$, instead of all (M,M^*) bimodules let $J(M,M^*;B)$ denote the set of those (M,M^*) bimodules (H,ρ,ρ^*) such that the restriction of the bimodule to (B,M^*) i.e. $(H, \rho|_B,\rho^*)$ is unitarily equivalent to multiple of the restriction of the trivial bimodule to (B,M^*). Then we say that *the pair* (M,B) *has property* T *if whenever a bimodule in* $J(M,M^*;B)$ *weakly contains the trivial bimodule, then it contains the trivial bimodule as a summand.* This is clearly a stronger property as B gets smaller and if $B = \mathbb{C}$, it is clearly the same as Connes' property; if $B = M$, the condition becomes vacuous and is always satisfied, as expected.

It remains to unravel the situation when $B = A$ is a Cartan subalgebra of a factor M. We then write $M = M(R,\sigma)$ with $A = A(R,\sigma)$ for a (unique) discrete measured ergodic equivalence relation $R = (X,C,R,B)$ and some $\sigma \in H^2(R,T)$. The trivial bimodule of (M,M^*) is realized on the Hilbert space $H = L^2(R,\nu)$, $(\nu \in B)$ and $M = M(R,\sigma)$ acting by left σ-twisted convolution as in Section 4.2 and with $M^* \simeq M'$ acting by right σ-twisted convolution; a typical operator in M has the form

$$(L_k\phi)(x,z) = \sum_y k(x,y)\phi(y,z)\sigma(x,y,z)$$

where the sum is over y in the common equivalence class of x and z and where $\phi \in L^2(R,\nu)$; k here ranges over suitable complex valued functions on R.

Now suppose that U is a unitary representation of R on a Hilbert space H_0; that is U is a map of R into the unitary group on H_0 satisfying $U(x,y)U(y,z) = U(x,z)$ when $x \sim y \sim z$. Now form the Hilbert space of H_0-valued functions on R, $H_1 = L^2(R,\nu,H_0) \simeq L^2(R,\nu) \otimes H_0$ and write down formally the operator $N(U)_k$ on H_1

$$(L(U)_k\psi)(x,z) = \sum_y k(x,y)U(x,y)\psi(y,z)\sigma(x,y,z) \tag{*}$$

for $\psi \in L^2(R,\nu,H_0)$ and $L_k \in M$. Moreover, if $R_h \in M^* \simeq H'$ we define $R(U)_h = R_h \otimes \mathrm{id}(H_0)$. We claim that this defines an (M,M^*) bimodule.

PROPOSITION 6.2.1. For $L_k \in M$, formula (*) gives a well defined bounded operator on H_1, and the map $\rho\colon L_k \to L(U)_k$ is in fact a normal * isomorphism of M onto a von Neumann algebra on H_1. The map $\rho^*\colon L_h \to L(U)_h$ is also a normal * isomorphism; the ranges of ρ and ρ^* commute and hence (H_1,ρ,ρ^*) is an (M,M^*) bimodule, which we denote $b(U)$. On $A \times M^*$, $b(U)$ is a multiple of the restriction of the trivial (M,M^*) bimodule to $A \times M^*$ and so $b(U) \in J(M,M^*;A)$. Conversely, every element of $J(M,M^*;A)$ arises in this way and the map $U \to b(U)$ is bijective from unitary equivalence of representations of R to $J(M,M^*;A)$.

The proofs of all of these assertions are quite straightforward and we omit the details. It is clear that the trivial representation U of R

$(U(x,y) = \mathrm{id})$ goes over to the trivial (M,M^*) bimodule; it is also clear upon inspection that a representation U of $\mathcal{R}$ will weakly contain the trivial representation in Zimmer's sense iff $b(U)$ weakly contains the trivial (M,M^*) bimodule. Hence we obtain what we obtain what we desired.

PROPOSITION 6.2.2. Let A be a Cartan subalgebra of the factor M. Then (M,A) has property T in the sense that the corresponding measured equivalence relation has property T iff the pair (M,A) has property T in the bimodule sense above.

Several of the previously examined consequences of property T have analogues in this setting. For instance, the discreteness of outer automorphism groups carries over exactly.

PROPOSITION 6.2.3. If the pair (M,B) has property T, the group $\mathrm{Out}(M,B)$ of all automorphisms centralizing B mod the inner ones of M doing this is discrete.

PROOF. As in Connes, any automorphism θ of M can be used to twist an (M,M^*) bimodule (H,ρ,ρ^*) by setting $H^\theta = H$, $(\rho^*)^\theta = \rho^*$, $\rho^\theta = \rho\circ\theta$. One applies this to the trivial bimodule and if $\theta \in \mathrm{Out}(M,B)$, the resulting bimodule lies in $\mathcal{J}(M,M^*;B)$ and the result follows readily as in Connes' situation.

The result that groups with property T are unimodular, or that equivalence relations with Property T are measure preserving and Connes' result on semifiniteness become the following.

PROPOSITION 6.2.4. Let B be a subalgebra of M with a normal conditional expectation of M to B. Then if the pair (M,B) has property T and if B is semi-finite, then M is semi-finite.

PROOF. Composing the conditional expectation with a faithful trace on B, we obtain a faithful normal semi-finite weight on M which by [91] has modular automorphism group σ^t centralizing B. Then by the discreteness of $\mathrm{Out}(M,B)$ and continuity of σ^t in t, σ^t is inner for all t, and hence M is semi-finite.

Finally we remark that it should be possible to obtain an ascending chain condition concerning an increasing sequence $M(n)$ of sub von Neumann algebras, $B \subset M(n) \subset M$ with dense union when the pair (M,B) satisfies property T, hence unifying results of Connes and the result in Section 6.2 for the case when B is a Cartan subalgebra.

APPENDIX

The first item for consideration here is the proof of Proposition 6.1.4 in Section 6.1. For this let $\mathcal{R}$ be ergodic and have property T, and let $\mathcal{R}(n)$ be an ascending sequence of open subrelations whose limit is $\mathcal{R}$. We wish to show that $\mathcal{R}(N) = \mathcal{R}$ for some N. For each leaf ℓ of $\mathcal{R}$, and

and each n, ℓ is a countable union of a set of leaves of $\mathcal{R}(n)$. We define a Hilbert space $H_\ell(n)$ which is to consist of measurable function f on ℓ, constant on the leaves of $\mathcal{R}(n)$ contained in ℓ such that $\Sigma|f(y)|^2 = |f|^2 < \infty$ where the sum is extended over all representatives y of the countable number of leaves of $\mathcal{R}(n)$ in ℓ. The space $H_\ell(n)$ is just a ℓ_2 sequence space, and by ergodicity, has for fixed n, almost everywhere constant dimension. According to our discussion in Section 4.1, such an assignment of a Hilbert space $H_\ell(n)$ to each leaf defines a representation $U(n)$ of $\mathcal{R}$ for each fixed n. (The technical conditions on $H_\ell(n)$ written down in Section 4.2 which are necessary to make this work are easily verified.)

We form the direct sum $U = \Sigma U(n)$ of these representations for all n and observe that the fact that $\mathcal{R} = \lim \mathcal{R}(n)$ yields at once the conclusion that U weakly contains the trivial representation. By property T, U must contain the trivial representation as a summand, and then by projection, some $U(n)$ must contain the trivial representation as a summand. Rephrasing, we see that for this value of n, we can find for almost all leaves ℓ of $\mathcal{R}$ a unit vector $v_\ell(n)$ in $H_\ell(n)$ which is invariant in the sense of being invariant under permutations of the distinguished orthonormal basis in $H_\ell(n)$ of functions supported on one $\mathcal{R}(n)$ leaf. But now such vectors exist only if almost all leaves ℓ of $\mathcal{R}$ are unions of a *finite* number of leaves of $\mathcal{R}(n)$. By ergodicity, this finite number must be constant a.e. and so we obtain the result that for some n_0, almost all leaves of $\mathcal{R}$ are the union of precisely $k(n_0)$ leaves of $\mathcal{R}(n_0)$ for some integer $k(n_0)$. Moreover if $n \geqslant n_0$, then since $\mathcal{R}(n) \supset \mathcal{R}(n_0)$, the same argument shows that almost all leaves of $\mathcal{R}$ are unions of $k(n)$ leaves of $\mathcal{R}(n)$ where evidently k is *decreasing* function of n. The condition that $\mathcal{R} = \lim \mathcal{R}(n)$ clearly tells us that $\lim k(n) = 1$ as $n \to \infty$, but then $k(N) = 1$ for some finite N, and this means that $\mathcal{R} = \mathcal{R}(N)$ as desired. This completes the proof.

As promised in 6.1, we now turn to the explicit calculation of $H^1(\mathcal{R},T)$ for some specific cases of $\mathcal{R}$'s with property T. For background here, see [36], [70], and [63]. Suppose that G_1 and G_2 are semi-simple Lie groups with G_1 having property T and G_2 simply connected; let Γ be a discrete finite volume subgroup of $G_1 \times G_2$ with the properties that $\Gamma \cap G_1 = (e)$, and $\overline{\Gamma \cdot G_1} = G_1 \times G_2$. Let $X = (G_1 \times G_2)/\Gamma$ as $G_1 \times G_2$ space. We consider X as G_1 space by forgetting about the action of G_2 for the moment; as such it has a finite invariant measure, is ergodic as $\overline{G_1 \cdot \Gamma} = G_1 \times G_2$ (cf. [54]), and is free since $\Gamma \cap G_1 = (e)$. Let $\mathcal{R}$ denote the corresponding measured equivalence relation; then as G_1 has property T, $\mathcal{R}$ has property T. We remark that the fact that Γ is a lattice in

$G_1 \times G_2$, $\overline{\Gamma \cdot G_1} = G_1 \times G_2$ and the fact that G_1 has property T implies by a slight variant of Kazhdan's original argument [44] that Γ also has property T. Moreover, as the action of G_1 is free, the analogue of Theorem 5 of [26] for continuous groups tells us that $H^1(R,T) \simeq H^1(G_1,U(X,T))$ where $U(X,T)$ is the polonais G_1 module of equivalence classes of measurable functions from X to T.

We can calculate this cohomology group by making use of the restriction-inflation sequence of group cohomology [61], [63] for the normal subgroup G_1 of $G = G_1 \times G_2$ with quotient $G_2 \simeq (G_1 \times G_2)/G_1$. We also remember that X and $U(X,T)$ are G-modules and write

$$0 \to H^1(G_2,U(X,T))^{G_1} \to H^1(G_1,U(X,T))$$
$$\to H^1(G_1,U(X,T))^{G_2} \to H^2(G_2,U(X,T)^{G_1}$$
$$\to \cdots$$

where as usual A^G denotes the G fixed elements in A. Since G_1 is ergodic on X, $U(X,T)^{G_1} = T$ consists only of constant functions. Then as G_2 is semi simple, $H^1(G_2,T) = (0)$ and additionally as it is also simply connected, $H^2(G_2,T) = (0)$ (cf. [61]). Finally as X is a transitive G space with isotropy group Γ, Shapiro's lemma (cf. [63]), says that $H^1(G,U(X,T)) \simeq H^1(\Gamma,T) = \mathrm{Hom}(\Gamma,T)$, which as Γ has property T, is a finite group.

The restriction-inflation sequence becomes

$$0 \to \mathrm{Hom}(\Gamma,T) \to H^1(G_1,U(X,T))^{G_2} \to 0$$

But since R has property T, the group $H^1(G_1,U(X,T))$ is discrete in its natural topology and since G_2 is connected and operates continuously (cf. [64]) this action must be trivial and everything is fixed. Thus we have shown the following fact.

PROPOSITION A.1. If G_1, G_2 are semi simple, G_1 with property T, G_2 simply connected, Γ a lattice in $G = G_1 \times G_2$ with $\Gamma \cap G_1 = (e)$ and $\Gamma \cdot G_1$ dense in G, then the equivalence relation defined by the G_1 space $X = (G_1 \times G_2)/\Gamma$ has property T and $H^1(R,T)$ is a finite group isomorphic to the dual group of the finite group $\Gamma/[\Gamma,\Gamma]$.

Since it is an easy matter to concoct examples of triples (Γ,G_1,G_2) satisfying the conditions of the proposition with $\Gamma/[\Gamma,\Gamma]$ being any prescribed finite abelian group, we obtain the class of examples described in the text.

There is another somewhat different looking example of ergodic non transitive relations with property T which is actually quite similar and can be analyzed similarly. Consider a semi simple group G and a lattice Γ in G and suppose that there is a descending sequence Γ_n of subgroups of Γ all lattices (i.e. each is of finite index in Γ). Then for $n > m$, the natural maps $G/\Gamma_n \to G/\Gamma_m$ and these spaces make up a projective system of G-spaces, and one can form the projective limit $X = \varprojlim G/\Gamma_n$ as G space. This is ergodic, non transitive, and free except in trivial cases. Let R denote the corresponding measured equivalence relation. It is not difficult to see, although we shall not go into the details, that if we replace Γ_n be the largest normal subgroup of Γ contained in Γ_n (which is also of finite index) that we obtain a measured equivalence relation stably isomorphic, hence isomorphic to R. Therefore we shall henceforth assume that each Γ_n is normal in Γ.

Then we can view X from a slightly different point of view; namely we form the projective limit $\bar{\Gamma}$ of the inverse system of finite groups Γ/Γ_n. Then $\bar{\Gamma}$ is a compact totally disconnected group with $\bar{\Gamma}$ embedded as a dense subgroup of $\bar{\Gamma}$. The natural action of Γ on $\bar{\Gamma}$ by left translations is ergodic and one sees quickly that the G-space X is simply the induced action of this Γ action (a special case of the Poincaré flow construction in Section 3.3). More concretely we form the product $G \times \bar{\Gamma}$ and embed Γ as a (closed) subgroup diagonally. Then the G-space X can be canonically identified with $(G \times \bar{\Gamma})/\Gamma$ with G acting by left translation. Hence X now looks almost exactly like the examples considered in Proposition A.1. Let us impose the further condition

$$(*) \qquad \forall n,\ \exists m,\ \Gamma_m \subset [\Gamma_n, \Gamma_n]$$

- a condition that is usually satisfied in typical examples of such sequences Γ_n as for instance with congruence subgroups of arithmetic groups, cf. [70].

PROPOSITION A.2. If G has property T, Γ a lattice with a sequence Γ_n of normal subgroups of Γ satisfying (*) and R the measured equivalene relation defined by the G-space $X = \lim (G/\Gamma_n)$, then $H^1(R,T) = H^1(G,U(X,T)) = (0)$.

PROOF. As we have noted, the action of G on X is free so that the first isomorphism in the statement above holds. Further if L is an open subgroup of $\bar{\Gamma}$, the space X can also be represented as $X = (G \times \bar{\Gamma})/\Gamma \simeq (G \times L)/\Gamma_L$ where $\Gamma_L = (L \cap \Gamma)$. We write out the restriction inflation sequence just as in Proposition A.1 for L as normal subgroup of $G \times L$

$$0 \to H^1(L,T) \to H^1(G \times L, U(X,T)) \to H^1(G,U(X,T))^L$$

$$\to H^2(L,T) \quad \dots$$

As before, the second term in the sequence is $H^1(\Gamma,T) = \mathrm{Hom}(\Gamma,T)$ and condition (*) is exactly what one needs in order to conclude that the map from the first term to the second is surjective. Therefore we can abbreviate matters and write

$$0 \to H^1(G,U(X,T))^L \to H^2(L,T) \to$$

We now take $K < L$ and obtain a similar sequence with K in place of L together with natural "vertical" maps from the sequence for L to the sequence for K where the first "vertical" map is the natural inclusion of the L-fixed points in the K fixpoints, while the second one is restriction of cohomology to a subgroup. However, by the results of [61], for any given class in $H^2(L,T)$, there is some K small enough so that the restriction to K vanishes. This says that $H^1(G,U(X,T))^L = (0)$ for any L. But on the other hand, G has property T and so by Zimmer's results $H^1(G,U(X,T))$ is discrete in its natural topology. But L is compact and acts continuously and so every L orbit is finite. This says exactly that any element of $H^1(G,U(X,T))$ is fixed by a sufficiently small open subgroup K. Combining these two results,we conclude that $H^1(G,U(X,T)) = (0)$ as desired.

REFERENCES

1. W. Ambrose, Representation of ergodic flows, Ann. of Math. 42 (1941), 723-729.

2. H. Anzai, Ergodic skew product transformations on the torus, Osaka Math. J. 3 (1951), 83-89.

3. H. Araki and E. J. Woods, A classification of factors, Publ. RIMS, Kyoto University Ser. A (4), 51-130.

4. L. Auslander and C. C. Moore, Unitary representations of solvable Lie groups, Mem. Math. Soc. No. 62, Amer. Math. Soc., Providence, R. I. 1966.

5. H. Choda, A Galois correspondence in a von Neumann algebra, Tohoku Math. J. 30 (1978), 491-504.

6. M. Choda, Normal expectations and crossed products of von Neumann algebras, Proc. Japan Acad. 50 (1974), 738-742.

7. ______, Some relations on II_1 factors on free groups, Math. Japon. 22 (1977), 383-394.

8. A. Connes, Une classification des facteurs de type III, Ann. Scient Ec. Norm. Sup. 4ème Série, 6, (1973), 133-252.

9. ______, Almost periodic states and factors of type III_1, J. Funct. Anal. 16 (1974), 415-445.

10. ______, On hyperfinite factors of type III_0 and Krieger's factors, J. Funct. Anal. 18 (1975), 318-327.

11. ______, Outer conjugacy classes of automorphisms of factors, Ann. Scient. Ec. Norm. Sup., 4ème Série, 8 (1975) 383-419.

12. ______, Classification of injective factors, Cases II_1, II_∞, III_λ, $\lambda \neq 1$, Ann. of Math. 104 (1976), 73-115.

13. A. Connes, Sur la theorie non commutative de l'integration, Springer Lecture Notes in Math. 725 (1979), 19-143.

14. ________, A factor of type II_1 with countable fundamental group, preprint 1980.

15. ________ and W. Krieger, Measure space automorphisms, the normalizers of their full groups, and approximate finiteness. J. Funct. Anal. 24 (1974), 336-352.

16. ________ and M. Takesaki, The flow of weights on factors of type III, Tohoku Math. J. 29 (1977), 473-575.

17. ________ and B. Weiss, Property T and almost invariant sequences, to appear in Israel J. Math.

18. ________, J. Feldman, B. Weiss, An amenable equivalence relation is generated by a single transformation, preprint.

19. ________, P. Ghez, R. Lima, D. Testard, E. J. Woods, Review of crossed products of von Neumann algebras, Preprint 1973.

20. J. Cuntz and W. Krieger, A class of C^* algebras and topological Markov chains, Invent. Math. 56 (1980), 251-268.

21. Dang Ngoc Nghiem, On the classification of dynamical systems, Ann. Inst. H. Poincare, 9 (1973).

22. ______________, Decomposition et classification des systems dynamiques, Bull. Soc. Math. France 103 (1975), 149-175.

23. H. Dye, On groups of measure preserving transformations I, Amer. J. Math. 81 (1959), 119-159.

24. _____, On groups of measure preserving transformations II, Amer. J. Math. 85 (1963) 551-576.

24'. G. Elliot and E. J. Woods, The equivalence of various definitions for a property infinite von Neumann algebra to be approximately finite, Proc. Amer. Math. Soc. 60 (1976), 175-178.

25. J. Feldman and D. Lind, Hyperfiniteness and the Halmos-Rohlin theorem for non-singular abelian actions, Proc. Amer. Math. Soc. 55(1976), 339-344.

26. J. Feldman and C. C. Moore, Ergodic equivalence relations, cohomology, and von Neumann algebras I, Trans. Amer. Math. Soc. 234 (1977), 289-324.

27. ________________________, Ergodic equivalence relations, cohomology, and von Neumann algebras II, Trans. Amer. Math. Soc. 234 (1977), 325-359.

28. J. Feldman, P. Hahn, and C. C. Moore, Orbit structure and countable sections for actions of continuous groups, Adv. in Math. 28 (1978), 186-230.

29. P. Forrest, Virtual subgroups of R^n and Z^n, Advances in Math. 14 (1974) 187-207.

30. P. Ghez, R. Lima, D. Testard, Une extension d'une théorème de A. Connes sur la facteurs constructibles, Comm. Math. Phys. 32 (1973), 305-311.

31. V. Ja. Golodets, Crossed products of von Neumann algebras, Uspekhi Mat. Nauk 26 (1971), 3-50.

32. F. Greenleaf, Invariant Means on Topological Groups; Van Nostrand Mathematical Studies 16 (Van Nostrand, Princeton, N.J. 1969).

33. U. Haagerup, The standard form of von Neumann algebras, Math. Scand. 37 (1975), 271-283.

34. U. Haagerup, Normal weights on W^* algebras, J. Funct. Anal. 19 (1975), 302-317.

35. ________, Operator valued weights I, J. Funct. Anal. 32 (1979), 176-206 II, ibid. 33(1979), 339-361.

36. S. Helgason, <u>Differential Geometry, Lie Groups and Symmetric Spaces</u>, Academic Press (1978), New York.

37. P. Hahn, Haar measure for measure groupoids, Trans. Amer. Math. Soc. 242 (1978), 1-33.

38. ______, The regular representations of measure groupoids, Trans. Amer. Math. Soc. 242 (1978), 35-72.

39. ______, Reconstruction of a factor from measures on Takesaki's unitary equivalence relation, J. Funct. Anal. 31 (1979), 263-271.

40. ______, The σ-representation of amenable groupoids, Rocky Mtn. J. Math. 9 (1979), 631-639.

41. T. Hamachi, Y. Oka, and M. Osikawa, Flows associated with ergodic non-singular transformation groups, Publ. RIMS (Kyoto), 11 (1975), 31-50.

42. ____________________, A classification of ergodic non-singular transformation groups, Mem. Fac. Sci. Kyushu Univ. Ser. A 28 (1974), 113-133.

43. S. Kakutani, Induced measure preserving transformations, Proc. Imp. Acad. Tokyo 19 (1943), 635-641.

44. D. Kazhdan, Connection of the dual space of a group with the structure of its closed subgroups, Funct. Anal. Appl. 1 (1967), 63-65.

45. W. Krieger, On non-singular transformations of a measure space I, Z. Wahr. und Ver. W. Gebiete 11 (1969), 83-97.

46. ________, On singular transformations of a measure space II, Z. Wahr. und. Verw. Gebiete 11 (1969) 98-119.

47. ________, On constructing non * isomorphic hyperfinite factors of type III, J. Funct. Anal. 6 (1970), 97-109.

48. ________, On a class of hyperfinite factors that arise from null-recurrent Markov chains, J. Funct. Anal. 7 (1971), 27-42.

49. ________, On the Araki-Woods ratio set and non-singular transformations, Springer Lecture Notes in Math. 160 (1970), 158-177.

50. ________, On ergodic flows and isomorphism of factors, Math. Ann. 223 (1976), 19-70.

51. ________, On dimension functions and topological Markov chains, Invent. Math. 56 (1980), 239-250.

52. ________, On a dimension function for a class of homeomorphism groups, to appear.

53. B. Lawson, The quantitative theory of foliations, CBMS regional conference monograph no. 27 (1977).

54. G. W. Mackey, Borel structures in groups and their duals, Trans. Amer. Math. Soc. 85 (1957), 134-165.

55. _________, Unitary representations of group extensions I, Acta Math. 99 (1958), 265-311.

56. _________, Point realizations of transformation groups, Ill. J. Math. 6 (1962), 327-335.

57. _________, Ergodic transformations with a pure point spectrum, Ill. J. Math. 8 (1964), 593-600.

58. G. W. Mackey, Ergodic theory, group theory, and differential geometry, Proc. Nat. Acad. Sci. U.S.A. 50 (1963), 1184-1191.

59. __________, Ergodic theory and virtual groups, Math. Ann. 166 (1966), 187-207.

60. G. A. Margulis, Discrete groups of motions of manifolds of non-positive curvature, Amer. Math. Soc. Translations 109 (1977), 33-45.

61. C. C. Moore, Extensions and cohomology theory of locally compact groups I, Trans. Amer. Math. Soc. 113 (1964), 40-63.

62. __________, II, ibid., 113 (1964), 64-86.

63. __________, III, ibid., 221 (1976), 1-33.

64. __________, IV, ibid., 221 (1976), 35-58.

65. __________, and K. Schmidt, Coboundaries and homomorphisms for non-singular actions and a problem of H. Helson, Proc. London Math. Soc. 40 (1980), 443-475.

66. F. J. Murray and J. v. Neumann, On rings of operators, Ann. of Math. 37 (1936), 116-229.

67. ______________________________, Rings of Operators IV, Ann. of Math. 44 (1943), 716-804.

68. D. Ornstein and B. Weiss, Ergodic theory of amenable group actions I: the Rohlin Theorem, Bull. Amer. Math. Soc. (New Series) 2 (1980), 161-164.

69. R. Powers, Representations of uniformly hyperfinite algebras and the associated von Neumann rings, Ann. of Math. 86 (1967), 138-171.

70. M. S. Raghunathan, Discrete Subgroups of Lie Groups, Ergebrisse der. Math. 68, Springer Verlag 1972, New York.

71. A. Ramsay, Virtual groups and group actions, Adv. in Math. 6 (1971), 253-322.

72. ________, Boolean duals of virtual groups, J. Funct. Anal. 15 (1974), 56-101.

73. ________, Subobjects of virtual groups, Pac. J. Math. 87 (1980).

74. ________, Constructing topologies for measured groupoids, preprint.

75. J. Renault, A groupoid approach to C^*-algebras, Springer Lecture Notes in Mathematics, 793 (1980).

76. M. Samuelides, Mesures de Haar et W^* couple d'un groupoïde mesuré (preprint).

77. __________ and J.-L. Sauvageot, Algèbre de Krieger d'un système dynamique, C.R. Acad. Sci. Paris, Ser. A-B 280 (1975), A709-A712.

78. J.-L. Sauvageot, Image d'un homomorphisme et flot des poids d'une relation d'equivalence mesuré, C.R. Acad. Sci. Paris Ser. A-B, 282.

79. K. Schmidt, Cocycles on Ergodic Transformation Groups, Macmillan Lectures in Mathematics, No. 1, Macmillan of India, New Delhi, 1977.

80. _________, Asymptotically invariant sequences and an action of SL(2,Z) on the 2-sphere, to appear Israel J. Math.

81. _________, Amenability, Kazhdan's property T, strong ergodicity and invariant means for ergodic group actions, preprint.

82. A. K. Seda, Un concept de mesures invariant pour les groupoïdes topologiques, C. R. Acad. Sci. Paris, Ser. A-B, 280 (1975) A1603-A1605.

83. C. Series, The Rohlin tower theorem and hyperfiniteness for actions of continuous groups, Israel J. Math. 30(1978), 99-122.

84. ________, An application of groupoid cohomology, to appear Pac. J. Math. 89 (1980).

85. ________, The Poincaré flow of a foliation, Amer. J. Math. 102 (1980), 93-128.

86. I. M. Singer, Automorphisms of finite factors, Amer. J. Math. 77 (1955), 117-133.

87. S. Smale, Differentiable dynamical systems, Bull. Amer. Math. Soc. 73 (1973), 747-817.

88. C. Sutherland, Cohomology and extensions of von Neumann algebras I, to appear Publ. RIMS Kyoto, Japan.

89. ________, II, to appear Publ. RIMS, Kyoto, Japan.

90. ________, Maximal subalgebras of von Neumann algebras and representation of equivalence relations, preprint.

91. M. Takesaki, Conditional expectations in von Neumann algebras, J. Funct. Anal. 9 (1972), 306-321.

92. ________, Duality in cross products of and the structure of von Neumann algebras of type III, Acta Math. 131 (1973), 249-310.

93. J. Tomiyama, On some types of maximal abelian subalgebras, J. Funct. Anal. 10 (1972), 373-386.

94. A. M. Versik, Non measurable decompositions, orbit theory, algebras of operators, Dokl. Akad. Nauk SSR 199 (1971), 1004-1007.

95. J. Westman, Harmonic analysis on groupoids, Pac. J. Math. 27 (1968), 621-632.

96. ________, Cohomology for ergodic groupoids, Trans. Amer. Math. Soc. 146 (1969), 465-471.

97. ________, Cohomology for the ergodic actions of countable groups, Proc. Amer. Math. Soc. 30 (1971), 318-320.

98. ________, Virtual group homomorphisms with dense range, Ill. J. Math. 20 (1976), 41-47.

99. ________, Skew products and the similarity of virtual groups, Ill. J. Math. 21 (1977), 332-346.

100. H. Winkelnkemper, The graph of a foliation, preprint.

101. G. Zeller-Meier, Products croisés d'un C^* algèbre par un groupe d'automorphismes, J. Math. Pure Appl., 47 (1968), 101-239.

102. R. J. Zimmer, Amenable ergodic actions, hyperfinite factors, and Poincare flows, Bull. Amer. Math. Soc. 83(1977), 1078-1080.

103. ________, Amenable ergodic group actions and an application to Poisson boundaries of random walks, J. Funct. Anal. 27 (1978), 350-372.

104. ________, On the von Neumann algebra of an ergodic group action, Proc. Amer. Math. Soc. 66 (1977), 289-293.

105. ________, Hyperfinite factors and amenable ergodic actions, Invent. Math. 41 (1977), 23-31.

106. ________, Induced and amenable ergodic actions of Lie groups, Ann. Sci. Ecole Norm. Sup. 11 (1978), 407-428.

107. R. J. Zimmer, Amenable pairs of groups and ergodic actions and the associated von Neumann algebras, Trans. Amer. Math. Soc. 243 (1978), 271-286.

108. ___________, Global structure and measure theory, a new cohomology theory for foliations and ergodic Lie group actions, to appear Bull. Amer. Math. Soc.

109. ___________, Strong rigidity for ergodic actions of semi-simple Lie groups, to appear Ann. Math.

110. ___________, On the cohomology of ergodic actions of semi-simple Lie groups and discrete subgroups, preprint.

DEPARTMENT OF MATHEMATICS
UNIVERSITY OF CALIFORNIA
BERKELEY, CALIFORNIA 94720

Proceedings of Symposia in Pure Mathematics
Volume 38 (1982), Part 2

Topologies on Measured Groupoids

Arlan Ramsay[1]

According to a theorem of G.W. Mackey [3], if an analytic Borel group G has a non-zero quasiinvariant measure μ, then G has a locally compact group topology which generates the given Borel sets and for which the Haar measure is equivalent to μ. The topology can be described as the topology G inherits via its imbedding into the unitary group of $L^2(\mu)$ under the right regular representation. We have a way to prove this map is Borel without first getting an invariant measure, and also prove that G is locally compact by using the existence of a σ-compact conull subset of the image of G. This method extends to show that every analytic Borel measured groupoid has a σ-compact metric groupoid as an inessential reduction. From this we take a further inessential reduction to get local compactness. It is proved that on a σ-compact measured groupoid every "almost" homomorphism agrees a.e. with a strict homomorphism. This can be used to simplify parts of the theory of measured groupoids [4]. It is also proved that every locally compact equivalence relation has a complete transversal [1] and hence that every measured groupoid has a complete countable section [2]. These facts are used to show that some results of J. Feldman, P. Hahn and C. Moore on measured groupoids and operator algebras hold in general [2], and that a locally compact groupoid with a continuous Haar system has sufficiently many non-singular Borel G-sets provided that the orbit measures are atom-free [5]. This latter result is of use in studying C^*-algebras associated with locally compact groupoids [5].

The details and related results are expected to appear elsewhere.

BIBLIOGRAPHY

1. A. Connes, Sur la théorie non commutative de l'intégration, Springer Lecture Notes in Math. 725 (1979), 19-143.

1980 Mathematics Subject Classification 22.

[1]Supported by the N.S.F.

2. J. Feldman, P. Hahn and C.C. Moore, Orbit structure and countable sections for actions of continuous groups, Adv. Math. 28 (1978), 186-230.

3. G.W. Mackey, Borel structures in groups and their duals, Trans. Amer. Math. Soc. 85 (1957), 265-311.

4. A. Ramsay, Virtual groups and group actions, Adv. Math. 6 (1971), 253-322.

5. J. Renault, "A Groupoid Approach to C*-Algebras, Springer Lecture Notes in Math. 793 (1980), New York, N.Y., 1980.

DEPARTMENT OF MATHEMATICS
UNIVERSITY OF COLORADO
BOULDER, COLORADO 80309

Proceedings of Symposia in Pure Mathematics
Volume 38 (1982), Part 2

COHOMOLOGY THEORY FOR OPERATOR ALGEBRAS

J.R. Ringrose

1. INTRODUCTION. Cohomology theory for operator algebras was initiated in a series of three papers [20, 21, 16]. A detailed expository account of the material in those papers, together with some additional results and a discussion of various problems which motivate the theory, is given in [23]. The purpose of the present article is, in some respects, complementary to that of [23]. It is intended to provide a brief introduction to the main ideas of the subject (and to its literature), and to display those ideas by giving outline proofs of simple special cases when these seem more illuminating than detailed proofs of general results.

We begin by introducing some terminology. When we refer to an (associative linear) algebra $\mathfrak{A}$ and a two-sided $\mathfrak{A}$-module $\mathfrak{M}$, it will always be assumed (for simplicity) that $\mathfrak{A}$ has a unit element I and that $\mathfrak{M}$ is unital (in the sense that Im = m = mI, for each m in $\mathfrak{M}$).

1.1 DEFINITION. Suppose that $\mathfrak{A}$ is a complex Banach algebra and $\mathfrak{M}$ is a two-sided $\mathfrak{A}$-module. We say that $\mathfrak{M}$ is a <u>Banach $\mathfrak{A}$-module</u> if $\mathfrak{M}$ is a Banach space and the bilinear mappings

$$(A, m) \to Am, \qquad (A, m) \to mA : \mathfrak{A}\times\mathfrak{M} \to \mathfrak{M}$$

are bounded.

To provide some examples, we note that $\mathfrak{M}$ is a Banach $\mathfrak{A}$-module in each of the following cases:

(a) $\mathfrak{M} = \mathfrak{A}$,

(b) $\mathfrak{M}$ is a closed two-sided ideal in $\mathfrak{A}$,

(c) $\mathfrak{M}$ is a Banach algebra (with unit I) and $\mathfrak{A}$ is a closed subalgebra containing I.

In each of these examples, the left and right actions of $\mathfrak{A}$ on $\mathfrak{M}$ are simply left and right multiplication in the appropriate algebra.

Given a complex Banach algebra $\mathfrak{A}$ and a Banach $\mathfrak{A}$-module $\mathfrak{M}$, for each positive integer n we denote by $C_c^n(\mathfrak{A},\mathfrak{M})$ the (normed) linear space which consists of

---

1980 Mathematics Subject Classification. 46-02, 46L05

all bounded n-linear mappings from $\mathfrak{A}^n (= \mathfrak{A} \times \mathfrak{A} \times \dots \times \mathfrak{A})$ into $\mathcal{M}$. We also define $C_c^0(\mathfrak{A},\mathcal{M})$ to be $\mathcal{M}$. When $n \geq 0$, we refer to elements of $C_c^n(\mathfrak{A},\mathcal{M})$ as <u>n-cochains</u>. For each positive integer n, we define the <u>coboundary operator</u> $\Delta: C_c^{n-1}(\mathfrak{A},\mathcal{M}) \to C_c^n(\mathfrak{A},\mathcal{M})$ as follows (allowing the slight imprecision of using the same symbol Δ for each of these mappings):-

When $m \in C_c^0(\mathfrak{A},\mathcal{M})(=\mathcal{M})$, Δm in $C_c^1(\mathfrak{A},\mathcal{M})$ is defined by

$$(\Delta m)(A) = Am-mA \quad (A \in \mathfrak{A}).$$

When $\xi \in C_c^1(\mathfrak{A},\mathcal{M})$, $\Delta\xi$ in $C_c^2(\mathfrak{A},\mathcal{M})$ is defined by

$$(\Delta\xi)(A,B) = A\xi(B)-\xi(AB) + \xi(A)B \quad (A, B \in \mathfrak{A}).$$

When $\eta \in C_c^2(\mathfrak{A},\mathcal{M})$, $\Delta\eta$ in $C_c^3(\mathfrak{A},\mathcal{M})$ is defined by

$$(\Delta\eta)(A,B,C) = A\eta(B,C) - \eta(AB,C) + \eta(A,BC) - \eta(A,B)C \quad (A,B,C \in \mathcal{M}).$$

Following the above pattern, the general definition is as follows:-

When $\rho \in C_c^n(\mathfrak{A},\mathcal{M})$, $\Delta\rho$ in $C_c^{n+1}(\mathfrak{A},\mathcal{M})$ is defined by

$$\begin{aligned}(\Delta\rho)(A_0,A_1,\dots,A_n) &= A_0\rho(A_1,\dots,A_n)\\ &+ \sum_{j=1}^{n}(-1)^j\rho(A_0,\dots,A_{j-2},A_{j-1}A_j,A_{j+1},\dots,A_n)\\ &+ (-1)^{n+1}\rho(A_0,\dots,A_{n-1})A_n \quad (A_0,\dots,A_n \in \mathfrak{A}).\end{aligned}$$

In this way, we obtain a sequence of linear mappings

$$\mathcal{M} = C_c^0(\mathfrak{A},\mathcal{M}) \xrightarrow{\Delta} C_c^1(\mathfrak{A},\mathcal{M}) \xrightarrow{\Delta} C_c^2(\mathfrak{A},\mathcal{M}) \xrightarrow{\Delta} C_c^3(\mathfrak{A},\mathcal{M}) \xrightarrow{\Delta} \dots;$$

and it can be shown that, at each stage, $\Delta^2 = 0$ in the sequence

$$\dots\ C_c^{n-1}(\mathfrak{A},\mathcal{M}) \xrightarrow{\Delta} C_c^n(\mathfrak{A},\mathcal{M}) \xrightarrow{\Delta} C_c^{n+1}(\mathfrak{A},\mathcal{M})\ \dots$$

For $n = 1, 2, \dots$, we introduce two linear subspaces of $C_c^n(\mathfrak{A},\mathcal{M})$. The range of $\Delta : C_c^{n-1}(\mathfrak{A},\mathcal{M}) \to C_c^n(\mathfrak{A},\mathcal{M})$ is denoted by $B_c^n(\mathfrak{A},\mathcal{M})$, and its elements are called <u>n-coboundaries</u>; the kernel of $\Delta : C_c^n(\mathfrak{A},\mathcal{M}) \to C_c^{n+1}(\mathfrak{A},\mathcal{M})$ is denoted by $Z_c^n(\mathfrak{A},\mathcal{M})$, and its elements are called <u>n-cocycles</u>. Since $\Delta^2 = 0$, it follows that $B_c^n(\mathfrak{A},\mathcal{M}) \subseteq Z_c^n(\mathfrak{A},\mathcal{M})$; and (ignoring topologies on these spaces) we can form the quotient linear space

$$H_c^n(\mathfrak{A},\mathcal{M}) = Z_c^n(\mathfrak{A},\mathcal{M})/B_c^n(\mathfrak{A},\mathcal{M}).$$

We call $H_c^n(\mathfrak{A},\mathcal{M})$ the <u>n-dimensional (continuous) cohomology group</u> of $\mathfrak{A}$, with coefficients in $\mathcal{M}$.

In the above definitions of $C_c^n(\mathfrak{A},\mathcal{M})$, $B_c^n(\mathfrak{A},\mathcal{M})$, $Z_c^n(\mathfrak{A},\mathcal{M})$ and $H_c^n(\mathfrak{A},\mathcal{M})$, the suffix "c" is included to indicate the norm continuity (that is, boundedness) conditions imposed both on the module action of $\mathfrak{A}$ on $\mathcal{M}$ and also on the n-linear mappings which are used as n-cochains. As regards algebraic structure, the definitions are analogous to those used by Hochschild in his series of papers [9, 10, 11] which founded the cohomology theory of associative linear algebras. In that theory there is, of course, no requirement of any topological structure, either on the algebra $\mathfrak{A}$ or on the two-sided $\mathfrak{A}$-module $\mathfrak{M}$. The cochain spaces $C^n(\mathfrak{A},\mathfrak{M})$, the coboundary operator Δ, the coboundary and cocycle spaces $B^n(\mathfrak{A},\mathfrak{M})$ and $Z^n(\mathfrak{A},\mathfrak{M})$, and the cohomology groups

$$H^n(\mathfrak{A},\mathfrak{M}) = Z^n(\mathfrak{A},\mathfrak{M})/B^n(\mathfrak{A},\mathfrak{M}),$$

are defined as above, except that references to boundedness (of the module action, and of the cochains) are deleted.

When $\mathfrak{A}$ is a Banach algebra and $\mathfrak{M}$ is a Banach $\mathfrak{A}$-module, both the (purely algebraic) Hochschild theory, and also the norm continuous theory, are available. Note that, in this case,

$$C_c^o(\mathfrak{A},\mathfrak{M}) = C^o(\mathfrak{A},\mathfrak{M}) = \mathfrak{M};$$

from the definition of the coboundary operator from 0-cochains to 1-cochains,

$$B_c^1(\mathfrak{A},\mathfrak{M}) = B^1(\mathfrak{A},\mathfrak{M}) = \text{the set of all inner derivations from } \mathfrak{A} \text{ into } \mathfrak{M};$$

and, from the definition of the coboundary operator from 1-cochains to 2-cochains,

$$Z_c^1(\mathfrak{A},\mathfrak{M}) = \text{the set of all continuous derivations from } \mathfrak{A} \text{ into } \mathfrak{M},$$

$$Z^1(\mathfrak{A},\mathfrak{M}) = \text{the set of all derivations from } \mathfrak{A} \text{ into } \mathfrak{M}.$$

Hence

$$H_c^1(\mathfrak{A},\mathfrak{M}) = \{\text{continuous derivations}\}/\{\text{inner derivations}\},$$

$$H^1(\mathfrak{A},\mathfrak{M}) = \{\text{derivations}\}/\{\text{inner derivations}\}.$$

Thus $H^1(\mathfrak{A},\mathfrak{M}) = \{0\}$ if and only if all derivations from $\mathfrak{A}$ into $\mathfrak{M}$ are inner; $H_c^1(\mathfrak{A},\mathfrak{M}) = \{0\}$ if and only if all continuous derivations from $\mathfrak{A}$ into $\mathfrak{M}$ are inner and $H^1(\mathfrak{A},\mathfrak{M}) = H_c^1(\mathfrak{A},\mathfrak{M})$ if and only if all derivations from $\mathfrak{A}$ into $\mathfrak{M}$ are continuous.

In defining the (continuous) cohomology group $H_c^n(\mathfrak{A},\mathfrak{M})$ as $Z_c^n(\mathfrak{A},\mathfrak{M})/B_c^n(\mathfrak{A},\mathfrak{M})$, we indicated that this quotient is to be regarded as one of linear spaces, <u>ignoring their topologies</u>. It is, of course, possible to view $H_c^n(\mathfrak{A},\mathfrak{M})$ as a quotient of normed linear spaces, and as such it will itself be a normed space (in fact, a Banach space) if $B_c^n(\mathfrak{A},\mathfrak{M})$ is a closed subspace of $Z_c^n(\mathfrak{A},\mathfrak{M})$. Unfortunately, this is not always the case; while $Z_c^n(\mathfrak{A},\mathfrak{M})$ is always a closed subspace of $C_c^n(\mathfrak{A},\mathfrak{M})$, $B_c^n(\mathfrak{A},\mathfrak{M})$ is not always closed. Indeed, there are norm separable C*-algebras $\mathfrak{A}$ for which $B_c^1(\mathfrak{A},\mathfrak{A})$ is not closed in $C_c^1(\mathfrak{A},\mathfrak{A})$ [19 : Example 6.3].

So far, we have introduced the basic concepts needed to define the (norm continuous) cohomology of a Banach algebra $\mathfrak{A}$, with coefficients in a Banach $\mathfrak{A}$-module $\mathfrak{M}$. For technical reasons, it is often desirable to impose further restrictions on $\mathfrak{M}$. For example, by requiring $\mathfrak{M}$ to be a dual $\mathfrak{A}$-module, in the sense defined below, it is possible to make use of the weak* compactness of bounded weak* closed subsets of $\mathfrak{M}$, and this permits the development of a fuller theory than would be possible in the context of general Banach $\mathfrak{A}$-modules [13, 14].

1.2 DEFINITION. Suppose that $\mathfrak{A}$ is a complex Banach algebra and $\mathfrak{M}$ is a Banach $\mathfrak{A}$-module. We say that $\mathfrak{M}$ is a <u>dual $\mathfrak{A}$-module</u> if $\mathfrak{M}$ is the Banach dual space of some Banach space and, for each A_o in $\mathfrak{A}$, the linear mappings

$$m \to A_o m, \quad m \to mA_o : \mathfrak{M} \to \mathfrak{M}$$

are weak* continuous.

When $\mathfrak{A}$ is a C*-algebra represented on a specific Hilbert space, (in particular, when $\mathfrak{A}$ is a von Neumann algebra), it is useful to impose conditions of ultraweak continuity (as well as norm continuity) both on the module action and also on the n-linear mappings that are used as n-cochains. In this way we obtain a "normal" (ultraweakly continuous) cohomology theory for represented C*-algebras, in addition to the norm continuous theory and the purely algebraic Hochschild theory. While it turns out that (for the appropriate class of modules) the norm continuous and normal theories produce isomorphic cohomology groups, the interplay between these two theories is fruitful for both.

1.3 DEFINITION. Suppose that $\mathfrak{A}$ is a C*-algebra acting on a Hilbert space $\mathcal{H}$, and $\mathfrak{M}$ is a dual $\mathfrak{A}$-module. We say that $\mathfrak{M}$ is a <u>dual normal $\mathfrak{A}$-module</u> if, for each m_o in $\mathfrak{M}$, the linear mappings

$$A \to Am_o, \quad A \to m_o A : \mathfrak{A} \to \mathfrak{M}$$

are continuous, from $\mathfrak{A}$ (with its ultraweak topology as a subset of the algebra $\mathcal{B}(\mathcal{H})$ of all bounded linear operators on $\mathcal{H}$) into $\mathfrak{M}$ (with its weak* topology).

To provide some simple examples we note that, if $\mathfrak{A}$ is a C*-subalgebra of a von Neumann algebra $\mathfrak{M}$ acting on the Hilbert space $\mathcal{H}$, then $\mathfrak{M}$ is a dual normal $\mathfrak{A}$-module (with multiplication in $\mathfrak{M}$ providing the left and right actions of $\mathfrak{A}$ on $\mathfrak{M}$). Indeed, $\mathfrak{M}$ is the dual space of a (unique) Banach space $\mathfrak{M}_*$, its weak* topology is the ultraweak topology, and multiplication in $\mathfrak{M}$ is (separately) ultraweakly continuous in both variables.

When $\mathfrak{M}$ is a dual normal module for a represented C*-algebra $\mathfrak{A}$, the normal cochain spaces $C_w^n(\mathfrak{A},\mathfrak{M})$ are defined as follows: $C_w^o(\mathfrak{A},\mathfrak{M})$ is $\mathfrak{M}$ and, when $n \geq 1$, $C_w^n(\mathfrak{A},\mathfrak{M})$ consists of those elements ρ of $C_c^n(\mathfrak{A},\mathfrak{M})$ which are (separately) continuous in each of their n arguments, from $\mathfrak{A}$ (with its ultraweak topology) into $\mathfrak{M}$ (with its weak* topology). The coboundary operator, defined as before, provides a sequence of linear mappings,

$$\mathfrak{M} = C_w^o(\mathfrak{A},\mathfrak{M}) \overset{\Delta}{\to} C_w^1(\mathfrak{A},\mathfrak{M}) \overset{\Delta}{\to} C_w^2(\mathfrak{A},\mathfrak{M}) \overset{\Delta}{\to} C_w^3(\mathfrak{A},\mathfrak{M}) \ldots$$

From this sequence, the coboundary and cocycle spaces $B_w^n(\mathfrak{A},\mathfrak{M})$ and $Z_w^n(\mathfrak{A},\mathfrak{M})$, and the (normal) cohomology group

$$H^n_w(\mathfrak{A}, \mathfrak{M}) = Z^n_w(\mathfrak{A}, \mathfrak{M})/B^n_w(\mathfrak{A}, \mathfrak{M}),$$

are defined just as in the case of norm continuous cohomology.

In section 2, we describe a number of problems which can naturally be presented in cohomological terms. In section 3, we state the main results obtained so far, in the cohomology of operator algebras, and discuss some open questions. The remaining four sections are devoted to a description of a number of techniques that have proved useful, with an indication as to how they have contributed to the proofs of the main results.

2. SOME COHOMOLOGICAL PROBLEMS.

In this section we give a brief outline of certain questions that can be formulated naturally in cohomological terms.

2.1 THE LIFTING PROBLEM FOR DERIVATIONS. Suppose that $\mathfrak{A}$ is a C*-algebra, $\delta : \mathfrak{A} \to \mathfrak{A}$ is a derivation, and $\mathcal{J}$ is a closed two-sided ideal in $\mathfrak{A}$. Then $\mathcal{J} = \mathcal{J}^2$, in the sense that each A in $\mathcal{J}$ can be expressed as a product BC of two elements of $\mathcal{J}$. Thus

$$\delta(A) = \delta(BC) = B\delta(C) + \delta(B)C \in \mathcal{J},$$

and $\delta(\mathcal{J}) \subseteq \mathcal{J}$. From this, δ induces a derivation δ_o of the quotient C*-algebra $\mathfrak{A}/\mathcal{J}$; δ_o is defined by the condition $\delta_o q = q\delta$, where $q : \mathfrak{A} \to \mathfrak{A}/\mathcal{J}$ is the quotient mapping.

The "lifting problem" for derivations arises naturally from the preceding discussion. If $\mathcal{J}$ is a closed two-sided ideal in a C*-algebra $\mathfrak{A}$, $q : \mathfrak{A} \to \mathfrak{A}/\mathcal{J}$ is the quotient mapping, and $\delta_o : \mathfrak{A}/\mathcal{J} \to \mathfrak{A}/\mathcal{J}$ is a derivation, is there a derivation $\delta : \mathfrak{A} \to \mathfrak{A}$ such that $q\delta = \delta_o q$? When such a derivation δ exists, we say that δ_o lefts (to δ).

We shall see that there is an "obstruction" to the lifting process, an element of the cohomology group $Z^2(\mathfrak{A}, \mathcal{J})$; moreover, the obstruction lies in $Z^2_c(\mathfrak{A}, \mathcal{J})$ if (as occurs in some cases, [3 : Theorem 7], [28 : Theorem 12]) $\mathcal{J}$ is complemented as a closed subspace of the Banach space $\mathfrak{A}$. To this end, note that $\mathcal{J}$ <u>is</u> complemented in the <u>purely algebraic</u> sense, as a linear subspace of $\mathfrak{A}$. Hence, there is a linear mapping $\xi : \mathfrak{A} \to \mathfrak{A}$ which is a lifting of δ_o (that is, $q\xi = \delta_o q$); and when $\mathcal{J}$ has a closed complementary subspace, we may assume that ξ is continuous. Since $\xi \in C^1(\mathfrak{A}, \mathfrak{A})$, $\Delta\xi$ is an element ρ of $C^2(\mathfrak{A}, \mathfrak{A})$. When $A, B \in \mathfrak{A}$, since $q\xi = \delta_o q$, we have

$$\begin{aligned} q\rho(A, B) &= q(A\xi(B) - \xi(AB) + \xi(A)b) \\ &= q(A)q\xi(B) - q\xi(AB) + q\xi(A)q(B) \\ &= q(A)\delta_o(q(B)) - \delta_o(q(A)q(B)) + \delta_o(q(A))q(B) = 0. \end{aligned}$$

Thus $\rho(A,B) \in \mathcal{J}$, and $\rho \in C^2(\mathfrak{A},\mathcal{J})$; since $\Delta\rho = \Delta^2\xi = 0$, we have $\rho \in Z^2(\mathfrak{A},\mathcal{J})$. Note that the corresponding element $[\rho] = \rho + B^2(\mathfrak{A},\mathcal{J})$ of $H^2(\mathfrak{A},\mathcal{J})$ is independent of the choice of ξ. Indeed, if ξ_1 (in $C^1(\mathfrak{A},\mathfrak{A})$) is another lifting of δ_o, and $\rho_1 = \Delta\xi_1$, then $q\xi_1 = \delta_o q = q\xi$, whence $q(\xi-\xi_1) = 0$ and $\xi - \xi_1$ takes values in $\mathcal{J}$. Hence $\xi - \xi_1 \in C^1(\mathfrak{A},\mathcal{J})$, and $\rho - \rho_1 = \Delta(\xi-\xi_1) \in B^2(\mathfrak{A},\mathcal{J})$.

We assert that δ_o lifts to a derivation δ of $\mathfrak{A}$ if and only if $[\rho] = 0$. (Note that the equation $\rho = \Delta\xi$ implies that $\rho \in B^2(\mathfrak{A},\mathfrak{A})$, since $\xi \in C^1(\mathfrak{A},\mathfrak{A})$, but does <u>not</u> imply that $\rho \in B^2(\mathfrak{A},\mathcal{J})$.) In fact, if $\rho = \Delta\eta$, where $\eta \in C^1(\mathfrak{A},\mathcal{J})$ $(\subseteq C^1(\mathfrak{A},\mathfrak{A}))$, then $\xi - \eta$ is a derivation δ of $\mathfrak{A}$ (that is, $\xi-\eta \in Z^1(\mathfrak{A},\mathfrak{A})$), since $\Delta(\xi-\eta) = \rho-\Delta\eta = 0$; moreover, since η takes values in $\mathcal{J}$, we have $q\eta = 0$, and

$$q\delta = q(\xi-\eta) = q\xi = \delta_o q,$$

whence δ is a lifting of δ_o. Conversely, if δ_o lifts to a derivation δ of $\mathfrak{A}$, it is not difficult to verify that $\xi-\delta$ is an element η of $C^1(\mathfrak{A},\mathcal{J})$, and that $\rho = \Delta\eta \in B^2(\mathfrak{A},\mathcal{J})$.

The preceeding discussion shows that δ_o gives rise to an element $[\rho]$ of $H^2(\mathfrak{A},\mathcal{J})$, and that δ_o lifts to a derivation of $\mathfrak{A}$ if and only if $[\rho] = 0$. When $\mathcal{J}$ has a closed complementary subspace in $\mathfrak{A}$, the above analysis can be reformulated in terms of norm continuous cohomology, since ξ can be assumed continuous, while derivations of $\mathfrak{A}$ are automatically continuous.

There are examples of derivations which do not lift. The first (due to Kadison and Ringrose, and announced in [23: p.369]) was the result of a rather tortuous ad hoc construction, and was never published. A more natural example was given in [2: Theorem 6.2]. A beautiful theorem of G.K. Pederson [22] asserts that every derivation of $\mathfrak{A}/\mathcal{J}$ lifts to a derivation of $\mathfrak{A}$, when $\mathfrak{A}$ is a norm separable C*-algebra.

Suppose next that $\mathcal{J}$ is a norm closed two-sided ideal in a von Neumann algebra $\mathfrak{R}$, and δ_o is a derivation of $\mathfrak{R}/\mathcal{J}$. Since $\mathfrak{R}$ has only inner derivations, it is apparent that δ_o lifts to a derivation of $\mathfrak{R}$ if and only if δ_o is inner. Accordingly, the lifting problem can in this case be reformulated as follows: does $\mathfrak{R}/\mathcal{J}$ have outer derivations, when $\mathcal{J}$ is a norm closed two-sided ideal in a von Neumann algebra $\mathfrak{R}$? When the ideal $\mathcal{J}$ satisfies certain additional conditions, it is known that $\mathfrak{R}/\mathcal{J}$ has only inner derivations ([25]; see also the note at the end of [8]). The general problem remains open.

2.2 SOME EXTENSIONS OF BANACH ALGEBRAS. Let $\mathfrak{B}$ be a Banach algebra with a closed two-sided ideal $\mathcal{M}$ such that $\mathcal{M}^2 = \{0\}$ (that is, $m_1m_2 = 0$ whenenever $m_1, m_2 \in \mathcal{M}$). Suppose also that, as a closed subspace of the Banach space $\mathfrak{B}$, $\mathcal{M}$ has a complementary closed subspace. What additional conditions are needed to ensure that $\mathcal{M}$ has a complementary subspace which is, at the same time, a closed <u>subalgebra</u> of $\mathfrak{B}$? Of course, such a subalgebra would necessarily be isomorphic to the Banach algebra $\mathfrak{A} = \mathfrak{B}/\mathcal{M}$.

This question arises naturally in the investigation of the Wedderburn property, in the context of a Banach algebra $\mathfrak{B}$ with a finite-dimensional radical $\mathcal{M}$ [12]. The problem is to determine whether $\mathcal{M}$ has a "closed complementary subalgebra". The first Wedderburn theorem asserts the existence of such a subalgebra when $\mathfrak{B}$ is finite-dimensional. In general, the problem can be reduced, by algebraic devices, to the case in which $\mathcal{M}^2 = \{0\}$; it then falls within the more general question set out in the preceding paragraph, since the finite-dimensionality of $\mathcal{M}$ ensures the existence of a closed complementary subspace.

In the context of an associative linear algebra $\mathfrak{B}$ over a general field, the Wedderburn property was investigated by Hochschild [9]. His analysis, in terms of two-dimensional cohomology, made use of the fact that $\mathcal{M}$ is complemented, as a linear subspace of $\mathfrak{B}$. In the setting of Banach algebras, the existence of a closed complementary subspace must be postulated, or ensured by other assumptions (such as the finite-dimensionality of $\mathcal{M}$).

Returning to the question set out in the first paragraph of the present subsection, we suppose that $\mathfrak{B}$ is a Banach algebra containing a closed two-sided ideal $\mathcal{M}$, such that $\mathcal{M}^2 = \{0\}$ and $\mathcal{M}$ has a complementary closed subspace $\mathfrak{X}$ in $\mathfrak{B}$. Let $\mathfrak{A}$ be the Banach algebra $\mathfrak{B}/\mathcal{M}$, and note that $\mathcal{M}$ becomes a Banach $\mathfrak{A}$-module if we define the left and right actions of $\mathfrak{A}$ on $\mathcal{M}$ by

(1) $$(b+\mathcal{M}).m = bm, \quad m.(b+\mathcal{M}) = mb \quad (b \in \mathfrak{B},\ m \in \mathcal{M});$$

the definitions are unambiguous, since $\mathcal{M}^2 = \{0\}$. We shall show that a natural construction now gives rise to an element $[\rho]$ of $H_c^2(\mathfrak{A},\mathcal{M})$, and that $\mathcal{M}$ has a closed complementary subalgebra if and only if $[\rho] = 0$.

If $q : \mathfrak{B} \to \mathfrak{B}/\mathcal{M}$ $(=\mathfrak{A})$ is the quotient mapping, the restriction $q|\mathfrak{X}$ is a one-to-one continuous linear operator from $\mathfrak{X}$ onto $\mathfrak{A}$, and so has a continuous inverse. In this way, we obtain a continuous linear operator $\lambda : \mathfrak{A} \to \mathfrak{B}$, such that $q\lambda$ is the identity mapping on $\mathfrak{A}$. Since

$$A = q(\lambda(A)) = \lambda(A) + \mathcal{M},$$

for each A in $\mathfrak{A}$, it follows from (1) that

(2) $$A.m = \lambda(A)m, \quad m.A = m\lambda(A) \quad (A \in \mathfrak{A},\ m \in \mathcal{M}).$$

When $A_1, A_2 \in \mathfrak{A}$,

$$q(\lambda(A_1A_2) - \lambda(A_1)\lambda(A_2)) = (q\lambda)(A_1A_2) - (q\lambda)(A_1)(q\lambda)(A_2)$$
$$= A_1A_2 - A_1A_2 = 0.$$

Accordingly, $\lambda(A_1A_2) - \lambda(A_1)\lambda(A_2) \in \mathcal{M}$, and the equation

(3) $$\rho(A_1, A_2) = \lambda(A_1)\lambda(A_2) - \lambda(A_1A_2)$$

defines an element ρ of $C_c^2(\mathfrak{A},\mathcal{M})$. When $A_1, A_2, A_3 \in \mathfrak{A}$, it follows from (2) and (3) that

$$(\Delta\rho)(A_1, A_2, A_3) = A_1.\rho(A_2, A_3) - \rho(A_1A_2, A_3) + \rho(A_1, A_2A_3) - \rho(A_1, A_2).A_3$$
$$= \lambda(A_1)[\lambda(A_2)\lambda(A_3)-\lambda(A_2A_3)] - [\lambda(A_1A_2)\lambda(A_3)-\lambda(A_1A_2A_3)]$$
$$+ [\lambda(A_1)\lambda(A_2A_3)-\lambda(A_1A_2A_3)] - [\lambda(A_1)\lambda(A_2)-\lambda(A_1A_2)]\lambda(A_3)$$
$$= 0;$$

so $\rho \in Z_c^2(\mathfrak{A}, \mathfrak{M})$.

The above reasoning shows that (2) is satisfied, and that (3) defines an element ρ of $Z_c^2(\mathfrak{A}, \mathfrak{M})$, whenever $\lambda : \mathfrak{A} \to \mathfrak{B}$ is a continuous linear operator such that $q\lambda$ is the identity mapping on $\mathfrak{A}$. We assert that the corresponding element $[\rho] = \rho + B_c^2(\mathfrak{A}, \mathfrak{M})$ of $H_c^2(\mathfrak{A}, \mathfrak{M})$ is independent of the choice of λ. Indeed, suppose that $\lambda_1 : \mathfrak{A} \to \mathfrak{B}$ is another such operator, and that ρ_1 (in $Z_c^2(\mathfrak{A}, \mathfrak{M})$) is derived from λ_1 as in (3). Since $q\lambda = q\lambda_1$, it follows that $\lambda-\lambda_1$ takes values in $\mathfrak{M}$, and thus $\lambda-\lambda_1 \in C_c^1(\mathfrak{A}, \mathfrak{M})$. When $A, B \in \mathfrak{A}$, by use of (2) for both λ and λ_1, we obtain

$$(\Delta(\lambda-\lambda_1))(A, B)$$
$$= A.[\lambda(B)-\lambda_1(B)] - [\lambda(AB)-\lambda_1(AB)] + [\lambda(A)-\lambda_1(A)].B$$
$$= \lambda(A)[\lambda(B)-\lambda_1(B)] - [\lambda(AB)-\lambda_1(AB)] + [\lambda(A)-\lambda_1(A)]\lambda_1(B)$$
$$= [\lambda(A)\lambda(B)-\lambda(AB)] - [\lambda_1(A)\lambda_1(B)-\lambda_1(AB)]$$
$$= \rho(A, B) - \rho_1(A, B).$$

Thus $\rho - \rho_1 = \Delta(\lambda-\lambda_1) \in B_c^2(\mathfrak{A}, \mathfrak{M})$, and $[\rho] = [\rho_1]$.

If $\mathfrak{M}$ has a complementary closed subspace $\mathfrak{X}_1$ in $\mathfrak{B}$, which is a subalgebra of $\mathfrak{B}$, the restriction $q|\mathfrak{X}_1$ is a one-to-one continuous multiplicative linear mapping from $\mathfrak{X}_1$ onto $\mathfrak{A}$. The inverse of $q|\mathfrak{X}_1$ can be viewed as a continuous multiplicative linear mapping $\lambda_1 : \mathfrak{A} \to \mathfrak{B}$, and $q\lambda_1$ is the identity mapping on $\mathfrak{A}$. From the preceding paragraph, $[\rho] = [\rho_1]$, where ρ_1 is derived from λ_1 as in (3). However, $\rho_1 = 0$ (and hence $[\rho] = 0$), since λ_1 is multiplicative.

Conversely, suppose that $[\rho] = 0$, so that $\rho = \Delta\xi$ for some ξ in $C_c^1(\mathfrak{A}, \mathfrak{M})$. Then $\lambda - \xi$ is a continuous linear operator $\lambda_1 : \mathfrak{A} \to \mathfrak{B}$; and, since ξ takes values in $\mathfrak{M}$, $q\lambda_1$ coincides with $q\lambda$, the identity mapping on $\mathfrak{A}$. With ρ_1 derived from λ_1 as in (3), we have (as before)

$$\rho-\rho_1 = \Delta(\lambda-\lambda_1) = \Delta\xi = \rho;$$

so $\rho_1 = 0$, and (from the analogue for λ_1 of (3)) λ_1 is multiplicative. Since λ_1 is a continuous multiplicative linear operator from $\mathfrak{A}$ into $\mathfrak{B}$, and $q\lambda_1$ is the identity mapping on $\mathfrak{A}$, it follows that the range of λ_1 is a closed subalgebra of $\mathfrak{B}$ and a complementary subspace for $\mathfrak{M}$. We have now shown that $\mathfrak{M}$ has a closed complementary subalgebra in $\mathfrak{B}$ if and only if $[\rho] = 0$.

Finally, we look at the problem from a different point of view, in which attention is focussed on the Banach algebra $\mathfrak{A}$. Given a Banach $\mathfrak{A}$-module $\mathfrak{M}$ and an element ρ of $Z_c^2(\mathfrak{A},\mathcal{M})$, the Cartesian product set $\mathfrak{A} \times \mathcal{M}$ becomes an associative linear algebra when the algebraic operations are defined by

$$(A_1, m_1) + (A_2, m_2) = (A_1+A_2,\ m_1+m_2),$$

$$a(A, m) = (aA, am),$$

$$(A_1, m_1)(A_2, m_2) = (A_1A_2,\ A_1m_2+m_1A_2+\rho(A_1, A_2)).$$

Associativity of the multiplication just defined is a consequence of the cocycle condition $\Delta\rho = 0$. From the relations

$$(\Delta\rho)(A, I, I) = 0 = (\Delta\rho)(I, I, A),$$

it follows that the element $(I, -\rho(I, I))$ of $\mathfrak{A} \times \mathcal{M}$ acts as a unit, for the given multiplication on $\mathfrak{A} \times \mathcal{M}$. With any of the usual norms, $\mathfrak{A} \times \mathcal{M}$ is a Banach space, and its multiplication is jointly continuous; so, with a suitable (equivalent) norm, it becomes a Banach algebra $\mathfrak{B}$. The set

$$\mathcal{M}_o = \{(0, m) : m \in \mathcal{M}\}$$

is a closed two-sided ideal in $\mathfrak{B}$, is complemented by the closed subspace

$$\{(A, 0) : A \in \mathfrak{A}\}$$

of $\mathfrak{B}$, and satisfies $\mathcal{M}_o^2 = \{0\}$. Moreover $\mathcal{M}_o$ and $\mathfrak{B}/\mathcal{M}_o$ $(= \mathfrak{A}_o)$ are (algebraically and topologically) isomorphic to $\mathcal{M}$ and $\mathfrak{A}$, respectively.

The process described earlier in this section can now be applied to the Banach algebra $\mathfrak{B}$ and its ideal $\mathcal{M}_o$. It gives rise to a Banach $\mathfrak{A}_o$-module structure on $\mathcal{M}_o$, and an element of $H_c^2(\mathfrak{A}_o, \mathcal{M}_o)$; and these turn out to be the given Banach $\mathfrak{A}$-module structure on $\mathcal{M}$ and the element $[\rho]$ of $H_c^2(\mathfrak{A},\mathcal{M})$, if we identify $\mathfrak{A}_o$ with $\mathfrak{A}$ and $\mathcal{M}_o$ with $\mathcal{M}$. This shows that, up to (algebraic and topological) isomorphism, any Banach algebra $\mathfrak{A}$, any Banach $\mathfrak{A}$-module $\mathcal{M}$, and any element $[\rho]$ of $H_c^2(\mathfrak{A},\mathcal{M})$, can arise in the manner described earlier in this subsection.

2.3 SOME EARLIER RESULTS ON DERIVATIONS. The norm continuity of all derivations $\delta : \mathfrak{A} \to \mathfrak{A}$, when $\mathfrak{A}$ is a C*-algebra, was proved by Sakai [26]. The result was later generalised to the context of derivations from $\mathfrak{A}$ into a Banach $\mathfrak{A}$-module [24]. In the broader form, it can be restated as follows.

2.3.1 THEOREM. If $\mathfrak{A}$ is a C*-algebra and $\mathcal{M}$ is a Banach $\mathfrak{A}$-module, then $H^1(\mathfrak{A},\mathcal{M}) = H_c^1(\mathfrak{A},\mathcal{M})$.

The "derivation theorem" [18, 27] asserts that a von Neumann algebra has only inner derivations; it can be expressed in the following form

2.3.2 THEOREM. If $\mathfrak{R}$ is a von Neumann algebra, then $H^1(\mathfrak{R}, \mathfrak{R}) = \{0\}$.

Of course this implies (and, given Theorem 2.3.1, is equivalent to) the assertion that $H_c^1(\mathfrak{R}, \mathfrak{R}) = \{0\}$. One of the most interesting unsolved problems, in the cohomology theory of von Neumann algebras, is to determine whether $H_c^n(\mathfrak{R},\mathfrak{R}) = \{0\}$, for all von Neumann algebras and all $n = 1, 2, 3, \ldots$. In the case of hyperfinite von Neumann algebras, Corollary 3.3 (or Theorem 3.5) below gives an affirmative answer. An example is known [15] of a factor $\mathfrak{R}$ of type II_1, which is not hyperfinite, such that $H_c^2(\mathfrak{R},\mathfrak{R}) = \{0\}$.

3. THE MAIN RESULTS IN N-DIMENSIONAL COHOMOLOGY. Suppose that $\mathfrak{R}$ is a von Neumann algebra acting on a Hilbert space $\mathcal{H}$, and $\mathcal{A}$ is a maximal abelian self-adjoint subalgebra of the commutant $\mathfrak{R}'$. Then $\mathfrak{R} \cup \mathcal{A}$ generates the type I von Neumann algebra $\mathcal{A}'$, and $\mathfrak{R} \subseteq \mathcal{A}' \subseteq \mathcal{B}(\mathcal{H})$. Both $\mathcal{A}'$ and $\mathcal{B}(\mathcal{H})$ are dual normal $\mathfrak{R}$-modules. The following result appears in [23 : Theorem 8.1]; the proof given in [23] is essentially the same as the argument used in proving [21 : Theorem 2.4], and is sketched in section 7 of the present article.

3.1 THEOREM. If $\mathfrak{R}$ is a von Neumann algebra acting on a Hilbert space $\mathcal{H}$, and $\mathcal{A}$ is a maximal abelian self-adjoint subalgebra of $\mathfrak{R}'$, then

$$H_c^n(\mathfrak{R}, \mathcal{A}') = \{0\} \qquad (n = 1, 2, \ldots).$$

We note two consequences of Theorem 3.1.

3.2 COROLLARY. If $\mathfrak{R}$ is a von Neumann algebra acting on a Hilbert space $\mathcal{H}$, n is a positive integer, and $\rho \in Z_c^n(\mathfrak{R},\mathfrak{R})$, then $\rho = \Delta\xi$ for some ξ in $C_c^{n-1}(\mathfrak{R},\mathcal{B}(\mathcal{H}))$.

PROOF. If $\mathcal{A}$ is a maximal abelian self-adjoint subalgebra of $\mathfrak{R}'$, then $C_c^n(\mathfrak{R},\mathfrak{R}) \subseteq C_c^n(\mathfrak{R},\mathcal{A}')$ and $C_c^{n-1}(\mathfrak{R},\mathcal{A}') \subseteq C_c^{n-1}(\mathfrak{R},\mathcal{B}(\mathcal{H}))$, since $\mathfrak{R} \subseteq \mathcal{A}' \subseteq \mathcal{B}(\mathcal{H})$. Thus

$$Z_c^n(\mathfrak{R},\mathfrak{R}) \subseteq Z_c^n(\mathfrak{R},\mathcal{A}'), \quad B_c^n(\mathfrak{R},\mathcal{A}') \subseteq B_c^n(\mathfrak{R},\mathcal{B}(\mathcal{H}));$$

and, since $Z_c^n(\mathfrak{R},\mathcal{A}') = B_c^n(\mathfrak{R},\mathcal{A}')$ by Theorem 3.1, we have

$$\rho \in Z_c^n(\mathfrak{R},\mathfrak{R}) \subseteq B_c^n(\mathfrak{R},\mathcal{B}(\mathcal{H})).$$

We shall say that a von Neumann algebra $\mathfrak{R}$ is <u>hyperfinite</u> if it is the ultraweak closure of the union of an increasing net of finite-dimensional self-adjoint subalgebras. Note that type I von Neumann algebras are hyperfinite, in this sense.

3.3 COROLLARY. If $\mathfrak{R}$ is a hyperfinite von Neumann algebra, then

$$H_c^n(\mathfrak{R},\mathfrak{R}) = \{0\} \qquad (n = 1, 2, \ldots).$$

PROOF. If $\rho \in Z_c^n(\mathfrak{R},\mathfrak{R})$ then, by Corollary 3.2, $\rho = \Delta\xi_o$ for some ξ_o in $C_c^{n-1}(\mathfrak{R},\mathcal{B}(\mathcal{H}))$. Since $\mathfrak{R}$ is hyperfinite, there is a projection p of norm one, from $\mathcal{B}(\mathcal{H})$ onto $\mathfrak{R}$ (and, for any such projection, $p(A_1TA_2) = A_1p(T)A_2$, whenever $T \in \mathcal{B}(\mathcal{H})$ and $A_1, A_2 \in \mathfrak{R}$). We have $p\rho = \rho$ (since ρ takes values in $\mathfrak{R}$), and $p\xi_o$ is an element ξ of $C_c^{n-1}(\mathfrak{R},\mathfrak{R})$. When $A_1, \ldots, A_n \in \mathfrak{R}$,

$$\rho(A_1,\dots,A_n) = p\rho(A_1,\dots,A_n) = p(\Delta\xi_o)(A_1,\dots,A_n)$$

$$= p(A_1\xi_o(A_2,\dots,A_n)) - p(\xi_o(A_1A_2,A_3,\dots,A_n)) + \dots$$

$$+(-1)^{n-1}p(\xi_o(A_1,\dots,A_{n-2},A_{n-1}A_n)) + (-1)^n p(\xi_o(A_1,\dots,A_{n-1})A_n)$$

$$= A_1p\xi_o(A_2,\dots,A_n) - p\xi_o(A_1A_2,A_3,\dots,A_n) + \dots$$

$$+ (-1)^{n-1}p\xi_o(A_1,\dots,A_{n-2},A_{n-1}A_n) + (-1)^n p\xi_o(A_1,\dots,A_{n-1})A_n$$

$$= (\Delta\xi)(A_1,\dots,A_n),$$

since $\xi = p\xi_o$. Thus $\rho = \Delta\xi \in B^n_c(\mathfrak{R},\mathfrak{R})$.

The following theorem describes the connection between the normal and norm continuous cohomology theories, for a represented C*-algebra $\mathfrak{A}$ and its ultraweak closure $\mathfrak{A}^-$. The assumption that $\mathfrak{M}$ is a dual normal module for $\mathfrak{A}^-$ (not just for $\mathfrak{A}$) is to be expected in part (i) of the theorem; for technical reasons it is needed also in part (ii), even though the cohomology groups appearing in (ii) are for the algebra $\mathfrak{A}$ only, since the proof of (ii) involves interplay between $\mathfrak{A}$ and $\mathfrak{A}^-$. Detailed proofs of the theorem are given in [16], and in [23 : section 6]; the main ideas involved are sketched in sections 4, 5 and 6 of the present article.

3.4 THEOREM. If $\mathfrak{A}$ is a C*-algebra acting on a Hilbert space $\mathcal{H}$, and $\mathfrak{M}$ is a dual normal $\mathfrak{A}^-$-module, then

(i) $$H^n_w(\mathfrak{A},\mathfrak{M}) \simeq H^n_w(\mathfrak{A}^-,\mathfrak{M}),$$

(ii) $$H^n_w(\mathfrak{A},\mathfrak{M}) \simeq H^n_c(\mathfrak{A},\mathfrak{M}).$$

3.5 COROLLARY. Under the conditions of Theorem 3.4,

$$H^n_c(\mathfrak{A},\mathfrak{M}) \simeq H^n_c(\mathfrak{A}^-,\mathfrak{M}).$$

The following result generalises Corollary 3.3; as noted in Remark 4.5 below, it is a fairly straightforward consequence of Corollary 3.5 (see also [16], [23 : section 7]).

3.6 THEOREM. If $\mathfrak{R}$ is a hyperfinite von Neumann algebra, and $\mathfrak{M}$ is a dual normal $\mathfrak{R}$-module, then

$$H^n_c(\mathfrak{R},\mathfrak{M}) = \{0\} \qquad (n = 1, 2, \dots).$$

A result of Connes [6] asserts that the property established in Theorem 3.6 characterises hyperfinite algebras, if attention is confined to von Neumann algebras with separable preduals. A shorter proof of this result has subsequently been given by Bunce and Paschke [4]. In both proofs it is shown that, for a suitably chosen dual normal $\mathfrak{R}$-module, $\mathfrak{M}$, the condition $H^1_c(\mathfrak{R},\mathfrak{M})=\{0\}$

implies that there is a projection of norm one from $\mathcal{B}(\mathcal{H})$ onto $\mathcal{R}$ (that is, $\mathcal{R}$ is injective). This in turn implies that $\mathcal{R}$ is hyperfinite (when $\mathcal{R}$ has separable predual), by a deep result of Connes [5].

3.7 THEOREM. If a von Neumann algebra $\mathcal{R}$ has separable predual, and $H_c^1(\mathcal{R},\mathcal{M}) = \{0\}$ for each dual normal $\mathcal{R}$-module $\mathcal{M}$, then $\mathcal{R}$ is hyperfinite.

Some of the arguments used in [20, 21, 16, 23] require quite a lot of algebraic manipulation. An axiomatic approach to the cohomology theory of operator algebras, introduced by Craw [7], reduces the necessary calculations substantially.

4. ADJUSTMENT OF COCYCLES BY "AVERAGING".

The following simple result turns out to be useful at various points in the cohomology theory of operator algebras. It is proved in detail in [23 : Theorem 4.3] and also, in more general form, in [16 : Theorem 4.1]. Here, we illustrate the ideas involved by giving the proof for the case in which n = 2.

4.1 THEOREM. Suppose that $\mathfrak{A}$ is a C*-algebra with unitary group $\mathcal{U}$, $\mathcal{V}$ is a subgroup of $\mathcal{U}$ and is amenable (as a topological group with the norm topology), $\mathcal{B}$ is the C*-subalgebra generated (as a closed linear subspace of $\mathfrak{A}$) by $\mathcal{V}$, and $\mathcal{M}$ is a dual $\mathfrak{A}$-module. If $n \geq 1$ and $\rho \in Z_c^n(\mathfrak{A},\mathcal{M})$, there exists ξ in $C_c^{n-1}(\mathfrak{A},\mathcal{M})$, such that $\rho - \Delta\xi$ vanishes whenever any of its arguments lies in $\mathcal{B}$.

PROOF IN THE CASE n=2. Let $\mathcal{M}_*$ be a Banach space whose dual space is $\mathcal{M}$, and write $\langle m, x\rangle$ for the value of $m(\in\mathcal{M})$ at $x(\in \mathcal{M}_*)$. Let μ be a (normalised, two-sided) invariant mean on $\mathcal{V}$. When f is a norm continuous function, from $\mathcal{V}$ into a bounded subset of $\mathcal{M}$, we denote by $\int_{\mathcal{V}} f(V)d\mu(V)$ the element of $\mathcal{M}$ whose value, at a vector x in $\mathcal{M}_*$, is obtained by applying μ to the continuous complex -valued function $\langle f(.), x\rangle$ on $\mathcal{V}$.

The equation

$$\xi_1(A) = \int_{\mathcal{V}} V^*\rho(V, A)d\mu(V) \qquad (A \in \mathfrak{A})$$

defines a linear mapping $\xi_1 : \mathfrak{A} \to \mathcal{M}$, and $\|\xi_1(A)\| \leq \|\rho\|\|A\|$; so $\xi_1 \in C_c^1(\mathfrak{A},\mathcal{M})$. When $R, S \in \mathfrak{A}$,

$$(\Delta\xi_1)(R, S) = R\xi_1(S) - \xi_1(RS) + \xi_1(R)S$$
$$= \int_{\mathcal{V}} [RV^*\rho(V, S) - V^*\rho(V, RS) + V^*\rho(V, R)S]d\mu(V).$$

By using the fact that

$$0 = (\Delta\rho)(V, R, S)$$
$$= V\rho(R, S) - \rho(VR, S) + \rho(V, RS) - \rho(V, R)S,$$

it now follows that

$$(\Delta\xi_1)(R,S) = \int_{\mathcal{V}} [RV^*\rho(V,S) + \rho(R,S) - V^*\rho(VR,S)]d\mu(V)$$
$$= \rho(R,S) + \int_{\mathcal{V}} [RV^*\rho(V,S) - V^*\rho(VR,S)]d\mu(V).$$

When $R \in \mathcal{V}$, it results from the invariance of μ that

$$\int_{\mathcal{V}} RV^*\rho(V,S)d\mu(V) = \int_{\mathcal{V}} R(VR)^*\rho(VR,S)d\mu(V)$$
$$= \int_{\mathcal{V}} V^*\rho(VR,S)d\mu(V);$$

so $(\rho - \Delta\xi_1)(R,S) = 0$, when $R \in \mathcal{V}$ and $S \in \mathfrak{A}$. From the linearity and continuity of $\rho - \Delta\xi_1$ in its first variable, it now follows that $(\rho - \Delta\xi_1)(R,S) = 0$ when $R \in B$ and $S \in \mathfrak{A}$.

Upon replacing ρ by $\rho - \Delta\xi_1$, we may now suppose that

(1) $$\rho(B,S) = 0 \qquad (B \in \mathfrak{B},\ S \in \mathfrak{A}).$$

When $B \in \mathfrak{B}$ and $R, S \in \mathfrak{A}$, we have

$$0 = (\Delta\rho)(B,R,S)$$
$$= B\rho(R,S) - \rho(BR,S) + \rho(B,RS) - \rho(B,R)S.$$

From (1), the last two terms on the right-hand side vanish, so

(2) $$\rho(BR,S) = B\rho(R,S) \qquad (B \in \mathfrak{B};\ R,S \in \mathfrak{A}).$$

We now define an element ξ of $C_c^1(\mathfrak{A},\mathcal{M})$ by

$$\xi(A) = -\int_{\mathcal{V}} \rho(AV^*,V)d\mu(V) \qquad (A \in \mathfrak{A}).$$

From (1) and (2), and since $\mathcal{V} \subseteq \mathfrak{B}$, we have

(3) $$\xi(B) = 0, \qquad \xi(BA) = B\xi(A) \qquad (A \in \mathfrak{A},\ B \in \mathfrak{B}).$$

From (1) and (3),

$$(\Delta\xi)(B,A) = B\xi(A) - \xi(BA) + \xi(B)A = 0,$$

and hence $(\rho - \Delta\xi)(B,A) = 0$, when $B \in \mathfrak{B}$ and $A \in \mathfrak{A}$.

It remains to prove that $(\rho - \Delta\xi)(A,B) = 0$, when $A \in \mathfrak{A}$ and $B \in \mathfrak{B}$; and, from the linearity and continuity of $\rho - \Delta\xi$ in the second variable, it suffices to consider the case in which $B \in \mathcal{V}$. Now

$$(\Delta\xi)(A,B) = A\xi(B) - \xi(AB) + \xi(A)B$$
$$= -\xi(AB) + \xi(A)B$$
$$= \int_{\mathcal{V}} [\rho(ABV^*,V) - \rho(AV^*,V)B]d\mu(V)$$

Since

$$0 = (\Delta\rho)(AV^*,V,B)$$
$$= AV^*\rho(V,B) - \rho(A,B) + \rho(AV^*,VB) - \rho(AV^*,V)B$$
$$= -\rho(A,B) + \rho(AV^*,VB) - \rho(AV^*,V)B,$$

we have

$$(\Delta\xi)(A,B) = \int_{\mathcal{V}} [\rho(ABV^*,V) + \rho(A,B) - \rho(AV^*,VB)]d\mu(V)$$

$$= \rho(A,B) + \int_{\mathcal{V}} [\rho(ABV^*,V) - \rho(AB(VB)^*,VB)]d\mu(V).$$

From the invariance of μ, and since $B \in \mathcal{V}$, we now have

$$(\Delta\xi)(A,B) = \rho(A,B).$$

Thus, $\rho - \Delta\xi$ vanishes whenever its second (or first) argument lies in $\mathfrak{B}$.

4.2 REMARK. Suppose that $\mathfrak{A}$ is a C*-algebra, $\mathfrak{B}$ is a C*-subalgebra of the centre of $\mathfrak{A}$, and $\mathcal{V}$ is the unitary group of $\mathfrak{B}$. Since $\mathcal{V}$ is an abelian group, it is amenable. Accordingly, if $\mathcal{M}$ is a dual $\mathfrak{A}$-module, $n \geq 1$ and $\rho \in Z_c^n(\mathfrak{A},\mathcal{M})$, we can choose ξ in $C_c^{n-1}(\mathfrak{A},\mathcal{M})$ to satisfy the conclusion of Theorem 4.1. With σ the element $\rho - \Delta\xi$ of $Z_c^n(\mathfrak{A},\mathcal{M})$, σ vanishes whenever any of its arguments lies in $\mathfrak{B}$. From this, together with the cocycle condition $\Delta\sigma = 0$, it follows easily that

$$\sigma(A_1,\ldots,A_{j-1},A_jB,A_{j+1},\ldots,A_n) = \sigma(A_1,\ldots,A_n)B$$

whenever $1 \leq j \leq n$, $B \in \mathfrak{B}$, and $A_1,\ldots,A_n \in \mathfrak{A}$. For example, in the case $n = 2$, we have

$$0 = (\Delta\sigma)(A_1,A_2,B)$$

$$= A_1\sigma(A_2,B) - \sigma(A_1A_2,B) + \sigma(A_1,A_2B) - \sigma(A_1,A_2)B$$

$$= \sigma(A_1,A_2B) - \sigma(A_1,A_2)B;$$

and $\sigma(A_1,A_2B) = \sigma(A_1,A_2)B$. Also,

$$0 = (\Delta\sigma)(A_1,B,A_2)$$

$$= A_1\sigma(B,A_2) - \sigma(A_1B,A_2) + \sigma(A_1,BA_2) - \sigma(A_1,B)A_2$$

$$= -\sigma(A_1B,A_2) + \sigma(A_1,BA_2);$$

and $\sigma(A_1B,A_2) = \sigma(A_1,BA_2) = \sigma(A_1,A_2B) = \sigma(A_1,A_2)B$.

4.3 REMARK. If, in Theorem 4.1, $\mathfrak{A}$ is a represented C*-algebra, $\mathcal{M}$ is a dual normal $\mathfrak{A}$-module, $\rho \in Z_w^n(\mathfrak{A},\mathcal{M})$, and $\mathcal{V}$ is a finite subgroup of the unitary group of $\mathfrak{A}$, then $\xi \in C_w^{n-1}(\mathfrak{A},\mathcal{M})$. This follows from the proof of Theorem 4.1, bearing in mind that the invariant mean μ is simply a finite sum, since $\mathcal{V}$ is a finite group.

4.4 COROLLARY. If a C*-algebra $\mathfrak{A}$ is the norm closure of the union $\mathfrak{A}_o$ of an increasing net of finite-dimensional self-adjoint subalgebras, then

$$H_c^n(\mathfrak{A},\mathcal{M}) = \{0\} \quad (n = 1,2,\ldots)$$

for every dual $\mathfrak{A}$-module $\mathcal{M}$.

PROOF. The unitary group $\mathcal{V}$ of $\mathfrak{A}_o$ is amenable, since it is the union of an increasing net of finite-dimensional unitary groups, each of which is compact (and hence amenable) in the norm topology. Since the linear span $\mathfrak{A}_o$ of $\mathcal{V}$ is norm dense in $\mathfrak{A}$, the conditions of Theorem 4.1 are satisfied, with $\mathfrak{B}$ the whole of $\mathfrak{A}$. From that theorem, each ρ in $Z_c^n(\mathfrak{A},\mathcal{M})$ has the form $\Delta\xi$, where $\xi \in C_c^{n-1}(\mathfrak{A},\mathcal{M})$; so $H_c^n(\mathfrak{A},\mathcal{M}) = \{0\}$.

4.5 REMARK. Suppose that $\mathcal{R}$ is a hyperfinite von Neumann algebra, and $\mathcal{M}$ is a dual normal $\mathcal{R}$-module. There is an increasing net of finite-dimensional self-adjoint subalgebras of $\mathcal{R}$, whose union $\mathfrak{A}_o$ is ultraweakly dense in $\mathcal{R}$. The norm closure $\mathfrak{A}$ of $\mathfrak{A}_o$ is a C*-algebra, and $\mathfrak{A}^- = \mathcal{R}$. From Corollary 4.4,

$$H_c^n(\mathfrak{A},\mathcal{M}) = \{0\} \qquad (n = 1, 2, \dots);$$

and Corollary 3.5 implies that

$$H_c^n(\mathcal{R},\mathcal{M}) = \{0\} \qquad (n = 1, 2, \dots).$$

Thus Theorem 3.6 is an immediate consequence of Corollaries 4.4 and 3.5.

5. EXTENSION OF MULTILINEAR MAPPINGS. The following result ([16 : Theorem 2.3], [23 : Theorem 5.3]) is needed at several points, in developing the cohomology theory of operator algebras.

5.1 THEOREM. Suppose that $\mathfrak{A}_1, \dots, \mathfrak{A}_n$ are C*-algebras acting on Hilbert spaces $\mathcal{H}_1, \dots, \mathcal{H}_n$, respectively, $\mathcal{M}$ is the dual space of a Banach space $\mathcal{M}_*$, and $\rho : \mathfrak{A}_1 \times \dots \times \mathfrak{A}_n \to \mathcal{M}$ is a bounded multilinear mapping which is separately continuous relative to the ultraweak topologies on $\mathfrak{A}_1, \dots, \mathfrak{A}_n$ and the weak* topology on $\mathcal{M}$. Then ρ extends uniquely, without change of norm, to a bounded multilinear mapping $\bar{\rho} : \mathfrak{A}_1^- \times \dots \times \mathfrak{A}_n^- \to \mathcal{M}$, which is separately ultraweak-weak* continuous

OUTLINE OF PROOF. Note first that it suffices to prove the corresponding result for multilinear <u>functionals</u> on $\mathfrak{A}_1 \times \dots \times \mathfrak{A}_n$ (that is, with $\mathcal{M}$ replaced by the complex field $\mathbb{C}$). Indeed, suppose that $x \in \mathcal{M}_*$. We can define a multilinear functional $l_x : \mathfrak{A}_1 \times \dots \times \mathfrak{A}_n \to \mathbb{C}$ by

$$l_x(A_1, \dots, A_n) = \langle \rho(A_1, \dots, A_n), x \rangle,$$

where $\langle m, x \rangle$ denotes the value of m at x, when $m \in \mathcal{M}$. Then l_x is bounded, with $\|l_x\| \le \|\rho\|\|x\|$, and is separately ultraweakly continuous. If the theorem is true, for multilinear functionals, then l_x extends (uniquely, without change of norm, and retaining separate ultraweak continuity) to a bounded multilinear functional L_x on $\mathfrak{A}_1^- \times \dots \times \mathfrak{A}_n^-$. For each $(A_1, \dots, A_n)$ in $\mathfrak{A}_1^- \times \dots \times \mathfrak{A}_n^-$,

$$|L_x(A_1,\dots,A_n)| \le \|1_x\|\|A_1\|\dots\|A_n\|$$
$$\le \|x\|\|\rho\|\|A_1\|\dots\|A_n\| \qquad (x \in \mathfrak{M}_*);$$

so the mapping $x \to L_x(A_1,\dots,A_n)$ is a bounded linear functional on $\mathfrak{M}_*$ (that is, an element, which we may denote by $\bar{\rho}(A_1,\dots,A_n)$, of $\mathfrak{M}$). Moreover,

$$\|\bar{\rho}(A_1,\dots,A_n)\| \le \|\rho\|\|A_1\|\|\dots\|A_n\|.$$

It is not difficult to verfy that $\bar{\rho} : \mathfrak{A}_1^- \times \dots \times \mathfrak{A}_n^- \to \mathfrak{M}$ is a multilinear mapping with the properites stated in the theorem. We may assume, henceforth, that $\mathfrak{M} = \mathbb{C}$.

An ultraweakly continuous linear functional ω on a represented C*-algebra $\mathfrak{A}$ extends (uniquely, by continuity) to such a functional $\bar{\omega}$ on $\mathfrak{A}^-$; and, by the Kaplansky density theorem, $\|\bar{\omega}\| = \|\omega\|$. For each fixed A_2 in $\mathfrak{A}_2$, A_3 in $\mathfrak{A}_3$, ..., A_n in $\mathfrak{A}_n$, the mapping

$$A_1 \to \rho(A_1, A_2,\dots,A_n) : \mathfrak{A}_1 \to \mathbb{C}$$

is an ultraweakly continuous linear functional on $\mathfrak{A}_1$, and so extends (uniquely, and without change of norm) to an ultraweakly continuous linear functional on $\mathfrak{A}_1^-$. If we write $\rho_1(A_1, A_2,\dots,A_n)$ for the value taken by this extended linear functional, at an element A_1 of $\mathfrak{A}_1^-$, then

$$\rho_1 : \mathfrak{A}_1^- \times \mathfrak{A}_2 \times \dots \times \mathfrak{A}_n \to \mathbb{C}$$

is a bounded multilinear functional extending ρ, and $\|\rho_1\| = \|\rho\|$. By construction, ρ_1 is ultraweakly continuous in A_1, when the other variables are fixed. When $k > 1$, it is ultraweakly continuous in A_k when the other variables are fixed, and $A_1 \in \mathfrak{A}_1$ (because $\rho_1|\mathfrak{A}_1 \times \dots \times \mathfrak{A}_n = \rho$). It is not immediately obvious, however, that ultraweak continuity of ρ_1 in A_k is retained, when $A_1 \in \mathfrak{A}_1^- \setminus \mathfrak{A}_1$. In order to overcome this difficulty, we require the following result, which is equivalent to the particular case of Theorem 5.1 in which $\mathfrak{M} = \mathbb{C}$ and $n = 2$.

5.2 LEMMA. Suppose that $\mathfrak{A}$ and $\mathfrak{B}$ are C*-algebras acting on Hilbert spaces $\mathcal{H}$ and $\mathcal{K}$, respectively, and $\tau : \mathfrak{A} \times \mathfrak{B} \to \mathbb{C}$ is a bounded bilinear functional and is separately ultraweakly continuous. Then τ extends to a separately ultraweakly continuous bilinear functional $\bar{\tau} : \mathfrak{A}^- \times \mathfrak{B} \to \mathbb{C}$.

Let us assume, for the moment, that Lemma 5.2 has been proved. Suppose that $1 < k \le n$ and that, for each j other than 1 and k, a fixed element A_j of $\mathfrak{A}_j$ has been selected. The mapping

$$(A_1, A_k) \to \rho(A_1,\dots,A_n) : \mathfrak{A}_1 \times \mathfrak{A}_k \to \mathbb{C}$$

is a bounded bilinear functional τ, and is separately ultraweakly continuous; so it extends (retaining separate ultraweak continuity) to a bilinear functional $\bar{\tau} : \mathfrak{A}_1^- \times \mathfrak{A}_2 \to \mathbb{C}$. since

$$\bar{\tau}(A_1, A_k) = \tau(A_1, A_k)$$
$$= \rho(A_1, \ldots, A_n) = \rho_1(A_1, \ldots, A_n)$$

for all A_1 in $\mathfrak{A}_1$ and A_k in $\mathfrak{A}_k$ (the other variables taking the fixed values already chosen), it follows from the ultraweak continuity of $\bar{\tau}$ and ρ_1 in A_1 that

$$\bar{\tau}(A_1, A_k) = \rho_1(A_1, \ldots, A_n)$$

when $A_1 \in \mathfrak{A}_1^-$ and $A_k \in \mathfrak{A}_k$. Since $\bar{\tau}$ is ultraweakly continuous in A_k, for all A_1 in $\mathfrak{A}_1^-$, it now follows that ρ_1 is ultraweakly continuous as a function of A_k ($\in \mathfrak{A}_k$), when A_1 is fixed in $\mathfrak{A}_1^-$ and (for $j \neq 1, k$) A_j is fixed in $\mathfrak{A}_j$.

So far, we have shown (subject to proving Lemma 5.2) that

$$\rho : \mathfrak{A}_1 \times \mathfrak{A}_2 \times \mathfrak{A}_3 \times \ldots \times \mathfrak{A}_n \to \mathbb{C}$$

extends (without change of norm and retaining separate ultraweak continuity) to a bounded multilinear functional

$$\rho_1 : \mathfrak{A}_1^- \times \mathfrak{A}_2 \times \mathfrak{A}_3 \times \ldots \times \mathfrak{A}_n \to \mathbb{C}.$$

Repetition of the same argument now permits successive extension, one variable at a time, to bounded multilinear functionals

$$\rho_2 : \mathfrak{A}_1^- \times \mathfrak{A}_2^- \times \mathfrak{A}_3 \times \ldots \times \mathfrak{A}_n \to \mathbb{C}.$$

$$\cdot \quad \cdot \quad \cdot \quad \cdot \quad \cdot \quad \cdot \quad \cdot \quad \cdot \quad \cdot \quad \cdot \quad \cdot \quad \cdot$$

$$\rho_n : \mathfrak{A}_1^- \times \mathfrak{A}_2^- \times \mathfrak{A}_3^- \times \ldots \times \mathfrak{A}_n^- \to \mathbb{C}.$$

Since the norm is unchanged, and separate ultraweak continuity is retained, at each stage of the extension process, it follows that the conclusions of the theorem are satisfied, with $\bar{\rho} = \rho_n$.

It now remains to prove Lemma 5.2. Given $\mathfrak{A}$, $\mathfrak{B}$ and τ as in that lemma, for each fixed B in $\mathfrak{B}$, the mapping $A \to \tau(A, B)$ is an ultraweakly continuous linear functional $S(B)$ on $\mathfrak{A}$; and $S(B)$ extends, uniquely and without change of norm, to an ultraweakly continuous linear functional $T(B)$ on $\mathfrak{A}^-$. The mapping $B \to T(B)$ is a bounded linear operator T, from $\mathfrak{B}$ into the predual $(\mathfrak{A}^-)_*$ of $\mathfrak{A}^-$. Moreover

$$\tau(A, B) = \langle T(B), A \rangle \qquad (A \in \mathfrak{A},\ B \in \mathfrak{B}),$$

where $\langle \omega, S \rangle = \omega(S)$, for ω in $(\mathfrak{A}^-)_*$ and S in $\mathfrak{A}^-$. Since τ is ultraweakly continuous in its second argument, T is continuous as a mapping from $\mathfrak{B}$ (with the ultraweak topology) into $(\mathfrak{A}^-)_*$ (with the weak topology $\sigma((\mathfrak{A}^-)_*, \mathfrak{A})$ induced by elements of $\mathfrak{A}$).

By [1 : Corollary II.9], T maps the unit ball $(\mathfrak{B})_1$ of $\mathfrak{B}$ onto a subset Ω of $(\mathfrak{A}^-)_*$, which is relatively compact in the weak topology $\sigma((\mathfrak{A}^-)_*, \mathfrak{A}^-)$. On Ω, therefore, $\sigma((\mathfrak{A}^-)_*, \mathfrak{A}^-)$ coincides with the coarser Hausdorff topology $\sigma((\mathfrak{A}^-)_*, \mathfrak{A})$. This, together with the final assertion of the preceding paragraph, shows that T is continuous as a mapping from $(\mathfrak{B})_1$ (with the ultraweak topology) into $(\mathfrak{A}^-)_*$ (with the topology $\sigma((\mathfrak{A}^-)_*, \mathfrak{A}^-)$. Accordingly, for each fixed A in $\mathfrak{A}^-$, the linear functional $B \to \langle T(B), A\rangle$ on $\mathfrak{B}$ is ultraweakly continuous, on $(\mathfrak{B})_1$, and hence on $\mathfrak{B}$. The conclusions of Lemma 5.2 are therefore satisfied if we define

$$\overline{\tau}(A,B) = \langle T(B), A\rangle \quad (A \in \mathfrak{A}^-, \ B \in \mathfrak{B}).$$

5.3 REMARK. We are now in a position to prove Theorem 3.4(i). If $\mathfrak{A}$ is a C*-algebra acting on a Hilbert space $\mathcal{H}$, and $\mathcal{M}$ is a dual normal $\mathfrak{A}^-$-module, it follows from Theorem 5.1, that there is a bijection

$$e_n : \rho \to \overline{\rho} : C_w^n(\mathfrak{A}, \mathcal{M}) \to C_w^n(\mathfrak{A}^-, \mathcal{M}),$$

in which $\overline{\rho}$ denotes the unique separately ultraweakly continuous extension of ρ. With e_o the identity mapping on $\mathcal{M}$, the diagram

$$\begin{array}{ccccccc}
\mathcal{M} = C_w^o(\mathfrak{A}, \mathcal{M}) & \xrightarrow{\Delta} & C_w^1(\mathfrak{A}, \mathcal{M}) & \xrightarrow{\Delta} & C_w^2(\mathfrak{A}, \mathcal{M}) & \longrightarrow & \cdots \\
\downarrow e_o & & \downarrow e_1 & & \downarrow e_2 & & \\
\mathcal{M} = C_w^o(\mathfrak{A}^-, \mathcal{M}) & \xrightarrow{\Delta} & C_w^1(\mathfrak{A}^-, \mathcal{M}) & \xrightarrow{\Delta} & C_w^2(\mathfrak{A}^-, \mathcal{M}) & \longrightarrow & \cdots
\end{array}$$

commutes, and induces an isomorphism between $H_w^n(\mathfrak{A}, \mathcal{M})$ and $H_w^n(\mathfrak{A}^-, \mathcal{M})$.

6. NORM CONTINUOUS AND NORMAL COHOMOLOGY.

6.1 PROPOSITION. If $\mathfrak{A}$ is a C*-algebra acting on a Hilbert space $\mathcal{H}$, $\mathcal{M}$ is a dual normal $\mathfrak{A}^-$-module, and $\rho \in Z_c^n(\mathfrak{A}, \mathcal{M})$, there exists ξ in $C_c^{n-1}(\mathfrak{A}, \mathcal{M})$ such that $\rho - \Delta\xi \in Z_w^n(\mathfrak{A}, \mathcal{M})$.

PROOF. The given representation of $\mathfrak{A}$, acting on $\mathcal{H}$, is quasi-equivalent to a subrepresentation of the universal representation π of $\mathfrak{A}$. Accordingly, there is a projection P in the centre of $\pi(\mathfrak{A})^-$, and a * isomorphism

$$\alpha : \pi(\mathfrak{A})^- P \text{ onto } \mathfrak{A}^-,$$

such that

$$\alpha(\pi(A)P) = A \quad (A \in \mathfrak{A}), \tag{1}$$

equivalently,

$$\alpha(BP) = \pi^{-1}(B) \quad (B \in \pi(\mathfrak{A})). \tag{2}$$

Since α is a * isomorphism between von Neumann algebras, it is a homeomorphism for the ultraweak topologies.

Since $\mathcal{M}$ is a dual normal $\mathfrak{A}^-$-module, it becomes a dual normal $\pi(\mathfrak{A})^-$-module if we define

$$B.m = \alpha(BP)m, \quad m.B = m\alpha(BP) \quad (B \in \pi(\mathfrak{A})^-, \ m \in \mathcal{M}). \tag{3}$$

Note that

$$P.m = m.P = m \quad (m \in \mathcal{M}). \tag{4}$$

The equation

$$\rho_1(B_1,\dots,B_n) = \rho(\alpha(B_1P),\dots,\alpha(B_nP)) \quad (B_1,\dots,B_n \in \pi(\mathfrak{A})) \tag{5}$$

defines an element ρ_1 of $C_c^n(\pi(\mathfrak{A}),\mathcal{M})$. Since π is the universal representation of $\mathfrak{A}$, ρ_1 is ultraweak-weak* continuous in each of its arguments. Thus $\rho_1 \in C_w^n(\pi(\mathfrak{A}),\mathcal{M})$; and, by Theorem 5.1, ρ_1 extends to an element $\bar{\rho}_1$ of $C_w^n(\pi(\mathfrak{A})^-,\mathcal{M})$. An easy calculation, based on (3) and (5), shows that

$$(\Delta\rho_1)(B_o,\dots,B_n) = (\Delta\rho)(\alpha(B_oP),\dots,\alpha(B_nP)) = 0,$$

when $B_o,\dots,B_n \in \pi(\mathfrak{A})$; so $\rho_1 \in Z_w^n(\pi(\mathfrak{A}),\mathcal{M})$, and $\bar{\rho}_1 \in Z_w^n(\pi(\mathfrak{A})^-,\mathcal{M})$.

The unitary group $\{I, 2P-I\}$ generates a two-dimensional subalgebra C of the centre of $\pi(\mathfrak{A})^-$. By Theorem 4.1 and Remark 4.3, there is an element ξ_1 of $C_w^{n-1}(\pi(\mathfrak{A})^-,\mathcal{M})$, such that $\bar{\rho}_1 - \Delta\xi_1$ vanishes whenever any of its arguments lies in C. Moreover, by (4) and Remark 4.2,

$$(\bar{\rho}_1 - \Delta\xi_1)(B_1,\dots,B_n) = (\bar{\rho}_1 - \Delta\xi_1)(B_1,\dots,B_n).P,$$

and thus,

$$(\bar{\rho}_1 - \Delta\xi_1)(B_1,\dots,B_n) = (\bar{\rho}_1 - \Delta\xi_1)(B_1P,\dots,B_nP), \tag{6}$$

whenever $B_1,\dots,B_n \in \pi(\mathfrak{A})^-$.

Define ξ in $C_c^{n-1}(\mathfrak{A},\mathcal{M})$ by

$$\xi(A_1,\dots,A_{n-1}) = \xi_1(\pi(A_1),\dots,\pi(A_{n-1})).$$

When the B_j's lie in $\pi(\mathfrak{A})$, it follows from (2) that

$$\xi_1(B_1,\dots,B_{n-1}) = \xi(\alpha(B_1P),\dots,\alpha(B_{n-1}P)); \tag{7}$$

and a straightforward calculation, based on (5), (7), (3) and (6), shows that

$$\begin{aligned}(\rho - \Delta\xi)(\alpha(B_1P),\dots,\alpha(B_nP)) &= (\bar{\rho}_1 - \Delta\xi_1)(B_1,\dots,B_n)\\ &= (\bar{\rho}_1 - \Delta\xi_1)(B_1P,\dots,B_nP).\end{aligned}$$

Thus, when $A_1,\dots,A_n \in \mathfrak{A}$,

$$(\rho - \Delta\xi)(A_1,\dots,A_n) = (\bar{\rho}_1 - \Delta\xi_1)(\alpha^{-1}(A_1),\dots,\alpha^{-1}(A_n)).$$

Since $\bar{\rho}_1$, ξ_1 and α^{-1} are ultraweakly continuous, $\rho - \Delta\xi \in Z_w^n(\mathfrak{A},\mathcal{M})$.

6.2 PROPOSITION. If $\mathfrak{A}$ is a C*-algebra acting on a Hilbert space $\mathfrak{H}$ and $\mathcal{M}$ is a dual normal $\mathfrak{A}^-$-module, then

$$Z_w^n(\mathfrak{A},\mathcal{M}) \cap B_c^n(\mathfrak{A},\mathcal{M}) = B_w^n(\mathfrak{A},\mathcal{M}).$$

PROOF. We have to show that, if $\rho \in Z_w^n(\mathfrak{A},\mathfrak{M})$, and $\rho = \Delta\xi$ for some ξ in $C_c^{n-1}(\mathfrak{A},\mathfrak{M})$, then $\rho = \Delta\eta$ for some η in $C_w^{n-1}(\mathfrak{A},\mathfrak{M})$.

Just as in the proof of Proposition 6.1, we consider $\mathfrak{M}$ as a dual mormal $\pi(\mathfrak{A})^-$-module, with the structure given by (3), where π is the universal representation of $\mathfrak{A}$. The equations

$$\rho_1(B_1,\ldots,B_n) = \rho(\alpha(B_1P)\ \ldots,\alpha(B_nP)), \tag{8}$$

$$\xi_1(B_1,\ldots,B_{n-1}) = \xi(\alpha(B_1P),\ldots,\alpha(B_{n-1}P)) \qquad (B_1,\ldots,B_n \in \pi(\mathfrak{A})) \tag{9}$$

define elements ρ_1 of $C_c^n(\pi(\mathfrak{A}),\mathfrak{M})$ and ξ_1 of $C_c^{n-1}(\pi(\mathfrak{A}),\mathfrak{M})$.
A routine calculation, based on (3), (8), (9) (and the fact that $\rho = \Delta\xi$), shows that

$$\rho_1 = \Delta\xi_1. \tag{10}$$

Since π is the universal representation of $\mathfrak{A}$, both ρ_1 and ξ_1 are separately ultraweak-weak* continuous; and by hypothesis, the same is true of ρ. By Theorem 5.1, ρ_1, ξ_1 and ρ extend to elements $\bar{\rho}_1$ of $C_w^n(\pi(\mathfrak{A})^-,\mathfrak{M})$, $\bar{\xi}_1$ of $C_w^{n-1}(\pi(\mathfrak{A})^-,\mathfrak{M})$ and $\bar{\rho}$ of $C_w^n(\mathfrak{A}^-,\mathfrak{M})$, respectively; and, by (10),

$$\bar{\rho}_1 = \Delta\bar{\xi}_1. \tag{11}$$

From (8), together with the ultraweak continuity of $\bar{\rho}_1$, $\bar{\rho}$ and α, we have

$$\bar{\rho}_1(B_1,\ldots,B_n) = \bar{\rho}(\alpha(B_1P),\ldots,\alpha(B_nP)) \qquad (B_1,\ldots,B_n \in \pi(\mathfrak{A})^-). \tag{12}$$

It follows from (12) that

$$\bar{\rho}_1(B_1,\ldots,B_n) = \bar{\rho}_1(B_1P,\ldots,B_nP) \qquad (B_1,\ldots,B_n \in \pi(\mathfrak{A})^-). \tag{13}$$

Since α^{-1} is ultraweakly continuous, and carries $\mathfrak{A}$ onto $\pi(\mathfrak{A})P$ ($\subseteq \pi(\mathfrak{A})^-$), the equation

$$\eta(A_1,\ldots,A_{n-1}) = \bar{\xi}_1(\alpha^{-1}(A_1),\ldots,\alpha^{-1}(A_{n-1})) \qquad (A_1,\ldots,A_{n-1} \in \mathfrak{A})$$

defines an element η of $C_w^{n-1}(\mathfrak{A},\mathfrak{M})$. When the B_j's lie in $\pi(\mathfrak{A})$, it follows from (2) that $\alpha(B_jP) \in \mathfrak{A}$, and

$$\eta(\alpha(B_1P),\ldots,\alpha(B_{n-1}P)) = \bar{\xi}_1(B_1P,\ldots,B_{n-1}P); \tag{14}$$

and this, together with (3), (11), (13) and (8), gives

$$\begin{aligned}(\Delta\eta)(\alpha(B_1P),\ldots,\alpha(B_nP)) &= (\Delta\bar{\xi}_1)(B_1P,\ldots,B_nP)\\ &= \bar{\rho}_1(B_1P,\ldots,B_nP)\\ &= \bar{\rho}_1(B_1,\ldots,B_n)\\ &= \rho_1(B_1,\ldots,B_n)\\ &= \rho(\alpha(B_1P),\ldots,\alpha(B_nP)).\end{aligned}$$

Thus $\rho = \Delta\eta$.

6.3 REMARK. The two preceding propositions, together, give a proof of Theorem 3.4(ii). Since $B_w^n(\mathfrak{A},\mathfrak{M}) \subseteq B_c^n(\mathfrak{A},\mathfrak{M})$, the mapping

$$\rho + B_w^n(\mathfrak{A},\mathfrak{M}) \to \rho + B_c^n(\mathfrak{A},\mathfrak{M})$$

(where ρ runs through $Z_w^n(\mathfrak{A},\mathfrak{M})$) is well-defined, and is a homomorphism from $H_w^n(\mathfrak{A},\mathfrak{M})$ into $H_c^n(\mathfrak{A},\mathfrak{M})$. It is one-to-one, by Proposition 6.2; and its range is the whole of $H_c^n(\mathfrak{A},\mathfrak{M})$, by Proposition 6.1.

7. COHOMOLOGY WITH COEFFICIENTS IN $\mathcal{A}'$. In this section, we sketch a proof of Theorem 3.1; a more detailed account is given in [23 : section 8].

Suppose that $\mathcal{R}$ is a von Neumann algebra with centre $\mathcal{Z}$, acting on a Hilbert space $\mathcal{H}$, and $\mathcal{A}$ is a maximal abelian self-adjoint subalgebra of the commutant $\mathcal{R}'$. Let $\mathfrak{A}_o$ be the linear span of the set of all operators of the form AE, where $A \in \mathcal{R}$ and E is a projection in $\mathcal{A}$; and let $\mathfrak{A}$ be the norm closure of $\mathfrak{A}_o$. Then $\mathfrak{A}_o$ and $\mathfrak{A}$ are self-adjoint subalgebras of $\mathcal{A}'$, $\mathfrak{A}$ being a C*-algebra. Since $\mathcal{A}'$ is generated, as a von Neumann algebra, by $\mathcal{R} \cup \mathcal{A}$, it follows easily that $\mathfrak{A}^- = \mathcal{A}'$.

Suppose that $\rho \in Z_c^n(\mathcal{R}, \mathcal{A}')$. From Reamrk 4.2, with $\mathcal{B}$ equal to the centre $\mathcal{Z}$ of $\mathcal{R}$. there is an element ξ_1 of $C_c^{n-1}(\mathcal{R}, \mathcal{A}')$ such that $\rho - \Delta\xi_1$ $(=\rho_1)$ vanishes whenever any of its arguments lies in $\mathcal{Z}$. Moreover

(1) $$\rho_1(A_1, \ldots, A_n)Z = \rho_1(A_1, \ldots, A_{j-1}, A_jZ, A_{j+1}, \ldots, A_n)$$

whenever $A_1, \ldots, A_n \in \mathcal{R}$, $Z \in \mathcal{Z}$ and $1 \leq j \leq n$.

For any ρ_1 in $C_c^n(\mathcal{R}, \mathcal{A}')$, the condition (1) is sufficient to ensure that ρ_1 extends, without increase of norm, to an element ρ_2 of $C_c^n(\mathfrak{A}, \mathcal{A}')$. Moreover, $\rho_2 \in Z_c^n(\mathfrak{A}, \mathcal{A}')$ if (as obtains in the present case) $\rho_1 \in Z_c^n(\mathcal{R}, \mathcal{A}')$. We sketch the proof of these facts, essentially in the form given in [18], for n = 1 only; the argument for general n involves greater notational complexity but no additional ideas. When n = 1, we have

(2) $$\rho_1 \in C_c^1(\mathcal{R}, \mathcal{A}'), \qquad \rho_1(A)Z = \rho_1(AZ) \quad (A \in \mathcal{R},\ Z \in \mathcal{Z}).$$

Each element S of $\mathfrak{A}_o$ can be expressed in the form

(3) $$S = A_1E_1 + \cdots + A_kE_k,$$

where $A_1, \ldots, A_k \in \mathcal{R}$ and $E_1, \ldots, E_k$ are projections in $\mathcal{A}$. In general, there are many such expressions for S, in some of which the projections $E_1, \ldots, E_k$ are mutually orthogonal. We assert that the operator

(4) $$\rho_o(S) = \rho_1(A_1)E_1 + \cdots + \rho_1(A_k)E_k$$

depends only on S (not on the choice of its representation in the form (3)), and satisfies

(5) $$\|\rho_o(S)\| \le \|\rho_1\|\|S\|.$$

Once this has been proved, it follows easily that $\rho_o : \mathfrak{A}_o \to \mathcal{A}'$ is a bounded linear mapping which extends ρ_1; and it is not difficult to check that ρ_o is a derivation (that is, 1-cocycle) when ρ_1 is a derivation. By norm continuity, ρ_o extends to a bounded linear mapping $\rho_2 : \mathfrak{A} \to \mathcal{A}'$, and ρ_2 is a derivation if ρ_1 is a derivation.

To check that the operator on the right-hand side of (4) does not depend on the particular representation of S in the form (3), it suffices to show that it vanishes when $A_1E_1 + \dots + A_kE_k = 0$. By [17 : Lemma 3.1.1], the condition $\Sigma A_jE_j = 0$ entails the existence of a $k \times k$ matrix $[Z_{rs}]$ of elements of $\mathcal{Z}$, such that

$$\sum_{r=1}^{k} A_rZ_{rs} = 0 \quad (s = 1,\dots,k),$$

$$\sum_{s=1}^{k} Z_{rs}E_s = E_r \quad (r = 1,\dots,k).$$

From this, together with (2),

$$\begin{aligned}\sum_{r=1}^{k} \rho_1(A_r)E_r &= \sum_{r=1}^{k}\sum_{s=1}^{k} \rho(A_r)Z_{rs}E_s \\ &= \sum_{s=1}^{k}\sum_{r=1}^{k} \rho(A_rZ_{rs})E_s \\ &= \sum_{s=1}^{k} \rho(\sum_{s=1}^{k} A_rZ_{rs})E_s = 0.\end{aligned}$$

This proves that the definition (4) of $\rho_o(S)$ is unambiguous.

To prove (5), we may assume that S is expressed in the form (3), with $E_1,\dots,E_k$ mutually orthogonal projections in $\mathcal{A}$. Since $\rho_1(A_r)$ commutes with E_r,

$$\|\rho_o(S)\| = \|\sum_{r=1}^{k} \rho_1(A_r)E_r\| = \max_{1\le r\le k} \|\rho_1(A_r)E_r\|.$$

With P_r the central carrier of E_r as an element of $\mathcal{R}'$, $P_r \in \mathcal{Z}$, and the mapping $AP_r \to AE_r : \mathcal{R}P_r \to \mathcal{R}E_r$ is isometric, since it is a * isomorphism. Since $E_r \le P_r$, it now follows from (2) that

$$\begin{aligned}\|\rho_o(S)\| &= \max_{1\le r\le k} \|\rho_1(A_r)E_r\| \le \max_{1\le r\le k} \|\rho_1(A_r)P_r\| \\ &= \max_{1\le r\le k} \|\rho_1(A_rP_r)\| \le \|\rho_1\| \max_{1\le r\le k} \|A_rP_r\| \\ &= \|\rho_1\| \max_{1\le r\le k} \|A_rE_r\| = \|\rho_1\|\|S\|.\end{aligned}$$

We now return to the n-dimensional situation, and assume that ρ_1 has been extended to an element ρ_2 of $Z_c^n(\mathfrak{A}, \mathcal{A}')$. From Proposition 6.1, there is an element ξ_2 of $C_c^{n-1}(\mathfrak{A}, \mathcal{A}')$, such that $\rho_2 - \Delta\xi_2 \in Z_w^n(\mathfrak{A}, \mathcal{A}')$; By Theorem 5.1, and since $\mathfrak{A}^- = \mathcal{A}'$, $\rho_2 - \Delta\xi_2$ extends to an element ρ_3 of $Z_w^n(\mathcal{A}', \mathcal{A}')$. Finally, since $\mathcal{A}'$ is a type I von Neumann algebra, $H_c^n(\mathcal{A}', \mathcal{A}') = \{0\}$ by Theorem 3.6, and thus $\rho_3 = \Delta\xi_3$ for some ξ_3 in $C_c^{n-1}(\mathcal{A}', \mathcal{A}')$.

Since $\xi_1 \in C_c^{n-1}(\mathfrak{R}, \mathcal{A}')$, $\xi_2 \in C_c^{n-1}(\mathfrak{A}, \mathcal{A}')$ and $\xi_3 \in C_c^{n-1}(\mathcal{A}', \mathcal{A}')$, the equation

$$\xi(A_1, \ldots, A_{n-1}) = \xi_1(A_1, \ldots, A_{n-1}) + \xi_2(A_1, \ldots, A_{n-1}) + \xi_3(A_1, \ldots, A_{n-1})$$

$(A_1, \ldots, A_{n-1} \in \mathfrak{R})$ defines an element ξ of $C_c^{n-1}(\mathfrak{R}, \mathcal{A}')$. Since $\Delta\xi_3$ $(=\rho_3)$ extends $\rho_2 - \Delta\xi_2$, while ρ_2 extends ρ_1 $(= \rho - \Delta\xi_1)$, it follows that $\rho = \Delta\xi$. Thus $H_c^n(\mathfrak{R}, \mathcal{A}') = \{0\}$.

BIBLIOGRAPHY

1. C.A. Akemann, "The dual-space of an operator algebra", Trans. Amer. Math. Soc., 126 (1967), 286-302

2. C.A. Akemann, G.A. Elliott, G.K. Pederson and J. Tomiyama, "Derivations and multipliers of C*-algebras", Amer. J. Math., 98 (1976), 679-708.

3. T.B. Anderson, "Linear extensions, projections, and split faces", J. Functional Analysis, 17 (1974), 161-173.

4. J.W. Bunce and W.L. Paschke, "Quasi-expectations and amenable von Neumann algebras", Proc. Amer. Math. Soc., 71 (1978), 232-236.

5. A. Connes, "Classification of injective factors", Ann. of Math., 104 (1976), 73-115.

6. A. Connes, "On the cohomology of operator algebras", J. Functional Analysis, 28 (1978), 248-253.

7. I.G. Craw, "Axiomatic cohomology of operator algebras", Bull. Soc. Math. France, 100 (1972), 449-460.

6. G.A. Elliott, "On derivations of AW*-algebras", Tôhoku Math. J., 30 (1978), 263-276.

9. G. Hochschild, "On the cohomology groups of an associative algebra", Ann. of Math., 46 (1945), 58-67.

10. G. Hochschild, "On the cohomology theory for associative algebras", Ann. of Math., 47 (1946), 568-579.

11. G. Hochschild, "Cohomology and representations of associative algebras", Duke Math. J., 14 (1947), 921-948.

12. B.E. Johnson, "The Wedderburn decomposition of Banach algebras with finite dimensional radical", Amer. J. Math., 90 (1968), 866-876.

13. B.E. Johnson, "Cohomology in Banach algebras", Memoirs Amer. Math. Soc., no. 127 (1972).

14. B.E. Johnson, "Approximate diagonals and cohomology of certain annihilation Banach algebras", Amer. J. Math., 94 (1972), 685-698.

15. B.E. Johnson, "A class of II_1 factors without property P but with zero second cohomology", Arkiv för Matematik, 12 (1974), 153-159.

16. B.E. Johnson, R.V. Kadison and J.R. Ringrose, "Cohomology of operator algebras, III. Reduction to normal cohomology", Bull. Soc. Math. France, 100 (1972), 73-96.

17. R.V. Kadison, "Unitary invariants for representations of operator algebras", Ann. of Math., 66 (1957), 304-379.

18. R.V. Kadison, "Derivations of operator algebras", Ann. of Math., 83 (1966), 280-293.

19. R.V. Kadison, E.C. Lance and J.R. Ringrose, "Derivations and automorphisms of operator algebras II", J. Functional Analysis, 1(1967), 204-221.

20. R.V. Kadison and J.R. Ringrose, "Cohomology of operator algebras I. Type I von Neumann algebras". Acta Math., 126 (1971), 227-243.

21. R.V. Kadison and J.R. Ringrose, "Cohomology of operator algebras II. Extended cobounding and the hyperfinite case", Arkiv för Matematik, 9 (1971), 53-63.

22. G.K. Pedersen, "Lifting derivations from quotients of separable C*-algebras", Proc. Nat. Acad. Sci. U.S.A., 73 (1976), 1414-1415.

23. J.R. Ringrose, "Cohomology of operator algebras", in Lectures on Operator Algebras 355-434, (Springer Lecture Notes in Mathematics, no. 247).

24. J.R. Ringrose, "Automatic continuity of derivations of operator algebras", J. London Math. Soc., 5 (1972), 432-438.

25. J.R. Ringrose, "Derivations of quotients of von Neumann algebras", Proc. London. Math. Soc., 36 (1978), 1-26.

26. S. Sakai, "On a conjecture of Kaplansky", Tôhoku Math. J., 12 (1960), 31-33.

27. S. Sakai, "Derivations of W*-algebras", Ann. of Math., 83 (1966), 273-279.

28. J. Vesterström, "Positive linear extension operators for spaces of affine functions", Israel J. Math., 16 (1973), 203-211.

School of Mathematics,
The University of Newcastle upon Tyne,
NEWCASTLE UPON TYNE, NE1 7RU, England

Proceedings of Symposia in Pure Mathematics
Volume 38 (1982), Part 2

LOW DIMENSIONAL COHOMOLOGY OF BANACH ALGEBRAS

B.E. Johnson

1. COHOMOLOGICAL NOTATION. The main reference for this lecture is [8].

Throughout this lecture $\mathfrak{A}$ is a Banach algebra over $\mathbb{C}$ and $\mathfrak{X}$ a Banach $\mathfrak{A}$-bimodule, that is a Banach space which is an $\mathfrak{A}$-bimodule with $\|ax\| \leq \|a\|\|x\|$ and $\|xa\| \leq \|x\|\|a\|$ $(x \in \mathfrak{X},\ a \in \mathfrak{A})$. $\mathfrak{L}^n(\mathfrak{A},\mathfrak{X})$ is the space of continuous n-linear maps from $\mathfrak{A}$ into $\mathfrak{X}$ and for n = 0, 1, 2, ... we define $\delta^{n+1}\colon\ \mathfrak{L}^n(\mathfrak{A},\mathfrak{X}) \to \mathfrak{L}^{n+1}(\mathfrak{A},\mathfrak{X})$ by

$$(\delta^{n+1}T)(a_0,\dots,a_n) = a_0T(a_1,\dots,a_n) + \sum_{i=1}^{n} (-1)^i T(a_0,\dots,a_{i-1}a_i,\dots,a_n)$$
$$+ (-1)^{n+1}T(a_0,\dots,a_{n-1})a_n .$$

Straightforward calculation shows $\delta^{n+1}\delta^n = 0$, that is $\operatorname{Im}\delta^n \leq \operatorname{Ker}\delta^{n+1}$ and we put

$$H^n(\mathfrak{A},\mathfrak{X}) = \operatorname{Ker}\delta^{n+1}/\operatorname{Im}\delta^n$$

where the quotient is a quotient of abelian groups or complex vector spaces. The norm on $\mathfrak{L}^n$ gives a seminorm on H^n which is not necessarily a norm as $\operatorname{Im}\delta^n$ need not be closed. We shall not calculate H^n but introduce it as $H^n = 0$ is a convenient way of saying $\operatorname{Im}\delta^n = \operatorname{Ker}\delta^{n+1}$.

For n = 1 we have

$$\operatorname{Ker}\delta^2 = \text{set of derivations } \mathfrak{A} \to \mathfrak{X}$$
$$\operatorname{Im}\delta^1 = \text{set of inner derivations}$$

so $H^1(\mathfrak{A},\mathfrak{X}) = 0$ means that every derivation $\mathfrak{A} \to \mathfrak{X}$ is inner.

For n = 2 we see $S \in \operatorname{Ker}\delta^3$ if

$$\text{(i)}\quad a\,S(b,c) - S(ab,c) + S(a,bc) - S(a,b)c = 0 \qquad a,b,c \in \mathfrak{A}$$

whereas $S \in \operatorname{Im}\delta^2$ if there is $T \in \mathfrak{L}(\mathfrak{A},\mathfrak{X})$ with

$$\text{(ii)}\quad S(a,b) = aT(b) - T(ab) + T(a)b \qquad a,b \in \mathfrak{A} .$$

Thus $H^2(\mathfrak{A},\mathfrak{X}) = 0$ means that every solution of the functional equation (i) is of the form (ii).

Helemskiĭ [3] has shown that, at least if $\mathfrak{A}$ has a unit 1, the groups H^n are examples of the groups in a relative cohomology theory. Starting with a Banach

---

1980 Mathematics Subject Classification. 46 H25

algebra $\mathfrak{B}$ with unit, a continuous $\mathfrak{B}$-linear map f from a left unital $\mathfrak{B}$-module X into another Y is called <u>admissible</u> if Im f is closed and Ker f is a complemented subspace of X. An exact sequence of left Banach $\mathfrak{B}$-modules is relatively exact if the maps are all admissible. A left Banach $\mathfrak{B}$-module P is relatively projective if in any diagram

$$\begin{array}{ccccc} & f & P & & \\ & \swarrow & \downarrow g & & \\ Y & \underset{j}{\rightarrow} & Z & \rightarrow & 0 \end{array} \qquad \text{(iii)}$$

where the bottom line is relatively exact and g is a continuous $\mathfrak{B}$-linear map, the dashed arrow can be filled in with a continuous $\mathfrak{B}$-linear map f making the diagram commute. A relative projective resolution of a left unital $\mathfrak{B}$ module X is a relative exact sequence of relatively projective modules

$$\cdots \; P_2 \xrightarrow{\varepsilon_2} P_1 \xrightarrow{\varepsilon_1} P_0 \xrightarrow{\varepsilon_0} X \rightarrow 0$$

Every module has such a resolution because we can take $P_0 = \mathfrak{B} \hat{\otimes} X$, where $\hat{\otimes}$ is the completion in the greatest cross norm, $b(c \otimes x) = bc \otimes x$ and $\varepsilon_0(b \otimes x) = bx$. ε_0 is admissible because $x \mapsto 1 \otimes x$ is a right inverse for ε_0 and it is relatively projective because in the diagram (iii) we can define f by $f(b \otimes x) = bk\,g(1 \otimes x)$ where $k \in \mathfrak{L}^1(Z,Y)$ is a right inverse for j. ε_1 is defined in the same way replacing X by ker ε_0 and so on.

If X,Y are two left Banach $\mathfrak{B}$-modules we take a relative projective resolution of X

$$\rightarrow P_1 \xrightarrow{\varepsilon_1} P_0 \xrightarrow{\varepsilon_0} X \rightarrow 0$$

which gives a sequence

$$\cdots \xleftarrow{\varepsilon_1'} \mathrm{Hom}_{\mathfrak{B}}(P_0,Y) \xleftarrow{\varepsilon_0'} \mathrm{Hom}_{\mathfrak{B}}(X,Y) \leftarrow 0$$

where $\mathrm{Hom}_{\mathfrak{B}}$ is the space of continuous $\mathfrak{B}$-linear maps between the modules specified. The cohomology groups are given by

$$\mathrm{Ext}^n_{\mathfrak{B}}(X,Y) = \mathrm{Ker}\ \varepsilon_{n+1}/\mathrm{Im}\ \varepsilon_n \ .$$

These groups depend only on $\mathfrak{B}$, X, Y and not on the projective resolution used.

Returning to the case of a Banach $\mathfrak{A}$-bimodule $\mathfrak{X}$ where we assume $\mathfrak{A}$ has a unit, with $1x = x = x1$ we form the algebra $\mathfrak{A} \hat{\otimes} \mathfrak{A}^o$ where $\mathfrak{A}^o$ is the Banach algebra obtained by taking the Banach space $\mathfrak{A}$ with the multiplication $\sigma(a,b) = ba$. Defining $(a \otimes b)x = axb$, $\mathfrak{X}$ is a left Banach $\mathfrak{A} \hat{\otimes} \mathfrak{A}^o$-module and we get $H^n(\mathfrak{A},\mathfrak{X}) = \mathrm{Ext}^n_{\mathfrak{A} \hat{\otimes} \mathfrak{A}^o}(\mathfrak{A},\mathfrak{X})$.

Given a short relative exact sequence of Banach $\mathfrak{A}$-bimodules

$$0 \to \mathfrak{Y} \to \mathfrak{X} \to \mathfrak{Z} \to 0$$

we get a long exact sequence

$$\cdots \to H^{n-1}(\mathfrak{A},\mathfrak{Z}) \to H^n(\mathfrak{A},\mathfrak{Y}) \to H^n(\mathfrak{A},\mathfrak{X}) \to H^n(\mathfrak{A},\mathfrak{Z}) \to H^{n+1}(\mathfrak{A},\mathfrak{Y}) \to \cdots$$

2. COHOMOLOGICALLY TRIVIAL ALGEBRAS. How can one show $H^1(\mathfrak{A},\mathfrak{X}) = 0$? In the case of $\mathfrak{A} = \mathbb{C}^n$ (with pointwise multiplication) we can use a device due to Hochschild [5, p.61] and form the element $d = \sum_j e_j \otimes e_j \in \mathfrak{A} \otimes \mathfrak{A}$ where the e_j are the minimal idempotents in $\mathfrak{A}$. This element satisfies

$$\text{(iv)} \quad ad = da \qquad a \in \mathfrak{A}$$
$$\text{(v)} \quad \pi d = 1$$

where $\mathfrak{A} \otimes \mathfrak{A}$ is an $\mathfrak{A}$-bimodule under the actions $a(b \otimes c) = ab \otimes c$ and $(b \otimes c)a = b \otimes ca$ and π is the map $\pi(a \otimes b) = ab$. If now D is a derivation $\mathfrak{A} \to \mathfrak{X}$ then put $x = \sum_j e_j D(e_j)$. Then $ax = \sum ae_j D(e_j) = \sum e_j D(e_j a) = \sum e_j D(a) + \sum e_j D(e_j)a = D(a) + xa$ (if $1y = y$ for all $y \in \mathfrak{X}$ - if not replace x by $x - D(1)1$). So $D = \delta^1 x$. For infinite dimensional algebras we could allow d to lie in $\mathfrak{A} \hat{\otimes} \mathfrak{A}$, the above argument works and has a converse - $H^1(\mathfrak{A},\mathfrak{X}) = 0$ for all Banach $\mathfrak{A}$-modules $\mathfrak{X}$ if and only if $\mathfrak{A}$ has a unit and an element d of $\mathfrak{A} \hat{\otimes} \mathfrak{A}$ satisfying (iv) and (v). However the class of algebras satisfying these conditions is not very interesting - if $\mathfrak{A}$ is commutative or satisfies the metric approximation property then these conditions imply that $\mathfrak{A}$ is finite dimensional and semi-simple [3]

When $\mathfrak{A}$ is the algebra c_0 of sequences which are zero at infinity we need formally to take $d = \sum_{i=1}^{\infty} e_i \otimes e_i$. This does not exist in $\mathfrak{A} \hat{\otimes} \mathfrak{A}$ but is an element of $(\mathfrak{A} \hat{\otimes} \mathfrak{A})^{**}$. For any Banach $\mathfrak{A}$-bimodule $\mathfrak{X}$, $\mathfrak{X}^*$ is an $\mathfrak{A}$-bimodule if we define

$$(af,x) = (f,xa)$$
$$(fa,x) = (f,ax) \qquad a \in \mathfrak{A} \quad x \in \mathfrak{X} \quad f \in \mathfrak{X}^* .$$

$\mathfrak{A} \hat{\otimes} \mathfrak{A}$ is an $\mathfrak{A}$-bimodule with the same action as described for $\mathfrak{A} \otimes \mathfrak{A}$ above and π extends to a map $\mathfrak{A} \hat{\otimes} \mathfrak{A} \to \mathfrak{A}$. An element d of $(\mathfrak{A} \hat{\otimes} \mathfrak{A})^{**}$ is a <u>virtual diagonal</u> if $ad = da$ and $\pi^{**}(d)a = a = a\pi^{**}(d)$ $(a \in \mathfrak{A})$. Then THEOREM [9] $\mathfrak{A}$ has a virtual diagonal if and only if $H^1(\mathfrak{A},\mathfrak{X}^*) = 0$ for all $\mathfrak{A}$-bimodules $\mathfrak{X}$. When $\mathfrak{A}$ satisfies the hypothesis of this theorem we say $\mathfrak{A}$ is <u>amenable</u>. The following algebras are amenable

Type I	C* algebras
AF C* algebras	
$L^1(G)$	where G is a locally compact amenable group
$\mathcal{LC}(\mathfrak{X})$	the compact operators on $\mathfrak{X}$ when $\mathfrak{X} = L^P$, ℓ^P or $C(0,1)$.

The following are not amenable

$L^1(G)$ where G is locally compact and not amenable

Non nuclear C* algebras

Any algebra with no bounded approximate unit.

3. UNSOLVED PROBLEMS ABOUT H^1. If $\mathfrak{A}$ is not amenable or $\mathfrak{X}$ is not the dual of an $\mathfrak{A}$-bimodule then it can be very difficult to decide whether $H^1(\mathfrak{A},\mathfrak{X}) = 0$. Interesting outstanding cases are

(i) $\mathfrak{A} \subseteq \mathfrak{L}(H)$, a C* algebra. Is $H^1(\mathfrak{A},\mathfrak{L}(H)) = 0$? Christensen [1 and 2] has shown that in most cases the answer is yes but it is not known for example if $\mathfrak{A}$ is the factor M generated by the left regular representation of the free group on two generators on $\ell^2(G)$ and $H = \ell^2(G) \otimes K$ for an infinite dimensional Hilbert space K.

(ii) If $\mathfrak{A}$ is a von Neumann algebra acting on H, is $H^1(\mathfrak{A},\mathfrak{LC}(H)) = 0$? [10]. Again the answer is yes in most cases but the exceptional cases are not quite the same as (i) as the answer is not known when $\mathfrak{A} = M$, $H = \ell^2(G)$.

(iii) If G is a locally compact group is $H^1(L^1(G), M(G)) = 0$? Here $M(G)$, the set of bounded Radon measures on X, is an algebra under convolution and the module product is convolution. Again the answer is yes in many cases [8; §4] but not all. More generally if G is a group of homeomorphisms of a locally compact space X then G acts on $C_0(X)$ by

$$(fg)(x) = f(gx) \qquad f \in C_0(X), \quad g \in G, \quad x \in X$$

and on $M(X)$ by the adjoint action

$$\int f \, d(g\mu) = \int fg \, d\mu \qquad f \in C_0(X), \quad g \in G, \quad \mu \in M(X) .$$

$M(X)$ is then a bimodule over $\ell^1(G)$ under the action

$$(\Sigma\, a_g g)\mu = \Sigma\, a_g (g\mu)$$

$$\mu(\Sigma\, a_g g) = (\Sigma\, a_g)\mu .$$

Again we ask whether $H^1(\ell^1(G), M(X)) = 0$. A special case of interest is $G = S\ell(2,\mathbb{Z})$ which acts as usual on $\mathbb{R}^2$ leaving $\mathbb{Z}^2$ invariant and so acts on $X = \mathbb{T}^2 = \mathbb{R}^2/\mathbb{Z}^2$.

4. HIGHER COHOMOLOGY. Questions concerning H^n $n = 2, 3, \ldots$ can be replaced by questions about H^1 by the technique of reduction of dimension. $\mathfrak{L}^1(\mathfrak{A},\mathfrak{X})$ is an $\mathfrak{A}$-bimodule under the actions

$$(aT)(b) = aT(b)$$

$$(Ta)(b) = T(ab) - T(a)b \qquad a,b \in \mathfrak{A} ,$$

and there is a natural identity between $\mathcal{L}^{n+1}(\mathfrak{A},\mathfrak{X})$ and $\mathcal{L}^n(\mathfrak{A}, \mathcal{L}^1(\mathfrak{A},\mathfrak{X}))$ which gives $H^{n+1}(\mathfrak{A},\mathfrak{X}) \approx H^n(\mathfrak{A}, \mathcal{L}^1(\mathfrak{A},\mathfrak{X}))$. If $\mathfrak{X}$ is the dual of an $\mathfrak{A}$-bimodule $\mathfrak{Y}$ then $\mathcal{L}(\mathfrak{A},\mathfrak{X})$ is the dual of $\mathfrak{A} \hat{\otimes} \mathfrak{Y}$ which can be made an $\mathfrak{A}$-bimodule giving the correct products on $\mathcal{L}(\mathfrak{A},\mathfrak{X})$ so if $\mathfrak{A}$ is amenable then $H^n(\mathfrak{A},\mathfrak{Y}^*) = 0$ for all $\mathfrak{Y}$. Reduction of dimension is not always as effective as this however as it can be shown that if $\mathfrak{A}$ is a C* algebra and $\mathfrak{X}$ a reflexive $\mathfrak{A}$-bimodule then $H^1(\mathfrak{A},\mathfrak{X}) = 0$ but this does not imply $H^n(\mathfrak{A},\mathfrak{X}) = 0$, $n = 2, 3, \ldots$ for reflexive $\mathfrak{X}$.

We have seen that the only commutative algebras with $H^1(\mathfrak{A},\mathfrak{X}) = 0$ for all $\mathfrak{X}$ are $\mathbb{C}^n$. A similar result holds for H^2. If $\mathfrak{A}$ is commutative, semi-simple, amenable and $H^2(\mathfrak{A},\mathfrak{X}) = 0$ for all $\mathfrak{X}$ then $\mathfrak{A} = \mathbb{C}^n$. It is significant that for $n = 3$ non trivial examples exist.

THEOREM [9]. If $\mathfrak{A} = L^1(G)$ (G compact), c_0 or a restricted direct sum of matrix algebras then $H^3(\mathfrak{A},\mathfrak{X}) = 0$ for all $\mathfrak{A}$-bimodules $\mathfrak{X}$.

I do not know whether this also holds for $\mathfrak{A} = C[0,1]$, $\mathcal{LC}(H)$ or $\ell^1(\mathbb{Z})$. Other problems involving higher cohomology are suggested by results or conjectures about H^1 - for example if $\mathfrak{A}$ is a von Neumann algebra is $H^2(\mathfrak{A},\mathfrak{A}) = 0$? The automatic continuity of derivations on a commutative semi-simple Banach algebra [7] can be expressed as $H^1(\mathfrak{A},\mathfrak{A}) = H^1_{alg}(\mathfrak{A},\mathfrak{A})$ where the latter group is constructed by replacing $\mathcal{L}^n(\mathfrak{A},\mathfrak{X})$ by the space $L^n(\mathfrak{A},\mathfrak{X})$ of n linear functions $\mathfrak{A} \to \mathfrak{X}$ as would be done in algebraic cohomology theory. It would be interesting to know whether $H^2(\mathfrak{A},\mathfrak{A}) = H^2_{alg}(\mathfrak{A},\mathfrak{A})$ but the situation is obscure even for the case $\mathfrak{A} = c_0$.

5. USES OF COHOMOLOGY.

(i) Lifting derivations. If $\mathfrak{X} \supseteq \mathfrak{Y}$ are Banach $\mathfrak{A}$-bimodules we have in some cases (for example if $\mathfrak{Y}$ is complemented in $\mathfrak{X}$ as a Banach space) a long exact sequence

$$\ldots \to H^1(\mathfrak{A},\mathfrak{X}) \to H^1(\mathfrak{A}, \mathfrak{X}/\mathfrak{Y}) \to H^2(\mathfrak{A},\mathfrak{Y}) \to H^2(\mathfrak{A},\mathfrak{X}) \to \ldots$$

If $H^1(\mathfrak{A},\mathfrak{X}) = 0 = H^2(\mathfrak{A},\mathfrak{X})$ then $H^1(\mathfrak{A}, \mathfrak{X}/\mathfrak{Y}) = H^2(\mathfrak{A}, \mathfrak{Y})$ so that derivations into $\mathfrak{X}/\mathfrak{Y}$ will all lift to derivations into $\mathfrak{X}$ (which in this case happens if and only if they are all inner) if and only if $H^2(\mathfrak{A},\mathfrak{Y}) = 0$. When the long exact sequence does not apply, $H^2(\mathfrak{A},\mathfrak{Y})$ is related to the subset of $H^1(\mathfrak{A}, \mathfrak{X}/\mathfrak{Y})$ consisting of derivations which lift as linear operators.

(ii) Radical extensions (See [6]). The general theory of Banach algebras separates into the theory of semi-simple algebras, the theory of radical algebras and a description of how the general algebra can be constructed from these. Under the third heading we study the algebras $\mathfrak{A}$ which can arise in

$$0 \to \mathfrak{R} \to \mathfrak{A} \to \mathfrak{B} \to 0$$

for given $\mathfrak{R}$ (radical) and $\mathfrak{B}$ (semi-simple). $\mathfrak{R}$ is an $\mathfrak{A}$-module and, if $\mathfrak{R}^2 = 0$ then $\mathfrak{R}$ is a $\mathfrak{B}$-module as $(b+r)s = bs$ $(b \in \mathfrak{A}, r,s \in \mathfrak{R})$. Suppose now we are given a Banach algebra $\mathfrak{B}$ and a Banach $\mathfrak{B}$-bimodule $\mathfrak{R}$ which we make a radical algebra by defining all products to be zero and look for all algebras $\mathfrak{A}$ with

$$0 \to \mathfrak{R} \to \mathfrak{A} \to \mathfrak{B} \to 0$$

where the induced $\mathfrak{B}$-module structure on $\mathfrak{R}$ is the given one. One way of constructing such algebras is to take $\mathfrak{A} = \mathfrak{B} \oplus \mathfrak{R}$ and define a product by

$$(b,r)(c,s) = (bc, rs + bs + rc + T(b,c))$$

where $T \in \mathfrak{L}^2(\mathfrak{B},\mathfrak{R})$. For this to be associative we need $\delta^3 T = 0$. Two extensions of this kind are "the same" if there is an isomorphism α

$$\begin{array}{ccccccc} & & & \mathfrak{A}_1 & & & \\ & & \nearrow & & \searrow & & \\ 0 \to & \mathfrak{R} & & \alpha\downarrow & & \mathfrak{B} & \to 0 \\ & & \searrow & & \nearrow & & \\ & & & \mathfrak{A}_2 & & & \end{array} .$$

If $\mathfrak{A}_1$ and $\mathfrak{A}_2$ are given by T_1 and T_2 then α is of the form $\alpha(br) = (b,r+S(b))$ where $S \in \mathfrak{L}(\mathfrak{B}, \mathfrak{R})$ satisfies $T_1 - T_2 = \delta^2 S$. Thus these extensions correspond to elements of $H^2(\mathfrak{B},\mathfrak{R})$.

(iii) Perturbations (see [11] and [12]). Let $\mathfrak{A}$ be a Banach algebra with multiplication π and suppose ρ is another associative multiplication on $\mathfrak{A}$ near π in $\mathfrak{L}^2(\mathfrak{A},\mathfrak{A})$ norm - is $(\mathfrak{A},\pi)$ isomorphic with $(\mathfrak{A},\rho)$? More generally which properties of $\mathfrak{A}$ are inherited by $(\mathfrak{A},\rho)$? If $T : \mathfrak{A} \to \mathfrak{A}$ is the isomorphism then

$$T\rho(a,b) = (Ta)(Tb) .$$

Conversely if T is an invertable element of $\mathfrak{L}(\mathfrak{A})$ then

$$\rho_T(a,b) = T^{-1}(Ta)(Tb)$$

is a multiplication on $\mathfrak{A}$ near π if T is near I.

The associative law determines an "algebraic" subset of $\mathfrak{L}^2(\mathfrak{A},\mathfrak{A})$ and the ρ_T form a subset of these. The tangent space to the manifold at π consists of $\operatorname{Ker} \delta^2$ and the tangent space to the subset consisting of the ρ_T is $\operatorname{Im} \delta^1$. Thus for the ρ_T to cover the whole of a neighbourhood of π we expect $\operatorname{Im} \delta^1 = \operatorname{Ker} \delta^2$.

DEFINITION. $\mathfrak{A}$ is (strongly) stable if (for each $\delta > 0$) there is $\varepsilon > 0$ such that for all multiplications with $\|\rho-\pi\| < \varepsilon$ there is an invertible $T \in \mathfrak{L}(\mathfrak{A})$ with $\rho = \rho_T$ (and $\|I-T\| < \delta$).

THEOREM. If $H^2(\mathfrak{A},\mathfrak{A}) = 0$ and $\operatorname{Im} \delta^3$ is closed in $\mathfrak{L}^3(\mathfrak{A},\mathfrak{A})$ (in particular if $H^3(\mathfrak{A},\mathfrak{A}) = 0$) then $\mathfrak{A}$ is strongly stable.

Strongly stable Banach algebras

$\mathfrak{L}(\mathfrak{X})$ any $\mathfrak{X}$

$\mathfrak{LC}(\mathfrak{X})$ if it has a bounded approximate unit

$C(X)$ for all compact spaces X

C^1, the algebra of trace class operators on Hilbert space.

The ℓ^P and C^P algebras for $1 < p < \infty$ are not stable. The disc algebra is stable

(Rochberg [13]) but not strongly stable. I do not know which other C* algebras are strongly stable.

BIBLIOGRAPHY

1. E. Christensen, "Extensions of derivations", J. Functional Analysis, 27 (1978), 234-247.

2. E. Christensen, "Extensions of derivations II", to appear.

3. A. Ja. Helemskii, "On the homological dimension of normal modules over Banach algebras",Mat. Sb., 81 (123) (1970), 430-444 = Math. USSR Sb. 10 (1970), 399-412.

4. A. Ja. Helemskii, "On a method for calculating and estimating the global homological dimension of Banach algebras", Mat. Sb. 87 (1972) = Math. USSR Sb. 16 (1972), 125-138.

5. G. Hochschild, "On the cohomology groups of an associative algebra", Ann. of Math., 46 (1945), 58-67.

6. B.E. Johnson, "The Wedderburn decomposition of Banach algebras with finite dimensional radical", Amer. J. Math., 90 (1968), 866-876.

7. B.E. Johnson, "Continuity of derivations on commutative algebras", Amer. J. Math., 91 (1969), 1-10.

8. B.E. Johnson, "Cohomology of operator algebras", Mem. Amer. Math. Soc., 127 (1972).

9. B.E. Johnson, "Approximate diagonals and cohomology of certain annihilator Banach algebras", Amer. J. Math., 94 (1972), 685-698.

10. B.E. Johnson and S.K. Parrott, "Operators commuting with a von Neumann algebra modulo the set of compact operators", J. Functional Analysis, 11 (1972), 39-61.

11. B.E. Johnson, "Perturbations of Banach algebras", Proc. London Math. Soc. (3), 34 (1977), 439-458.

12. I. Raeburn and J.L. Taylor, "Hochschild cohomology and perturbations of Banach algebras", J. Functional Analysis, 25 (1977), 258-267.

13. R. Rochberg, "The disc algebra is rigid", Proc. London Math. Soc. (3), 39 (1979), 119-129.

DEPARTMENT OF PURE MATHEMATICS
THE UNIVERSITY
NEWCASTLE UPON TYNE NE1 7RU
ENGLAND

Proceedings of Symposia in Pure Mathematics
Volume **38** (1982), Part 2

DERIVATIONS AND THEIR RELATION TO PERTURBATIONS OF OPERATOR ALGEBRAS

Erik Christensen

1. INTRODUCTION. Suppose A and B are C*-algebras on a Hilbert space H; is it then possible that A and B will have a lot of common algebraic properties, if they are sufficiently close? Or alternatively do the algebraic structures change smoothly when one passes through the set of subalgebras of B(H)?

In order to answer such questions we have taken up the study intiated by Kadison and Kastler in [20] and we have studied norm perturbations of operator algebras.

To be more specific we say that an algebra A on H is γ contained $(\gamma \in \mathbb{R}_+)$ in an algebra B on H, if for each a in A there is a b in B such that $\|a-b\| \leq \gamma\|a\|$. This is written $A \overset{\gamma}{\subset} B$. If $A \overset{\gamma}{\subset} B$ and $B \overset{\gamma}{\subset} A$ we say that the distance between A and B is less than or equal to γ, and we write $\|A-B\| \leq \gamma$.

Suppose now that $A \overset{\gamma}{\subset} B$ and that γ is fairly small, then we would like to be able to show, that B has a subalgebra, which is isomorphic to A. Moreover we would like to show that there exists a unitary u close to the identity I on H such that uAu^* is contained in B. It turns out - as we shall see in section 5 - that we can get such results in a number of cases.

In 1974, in [5], we developed a strategy for proving such results, and we did also show that it worked in the case where $\|A-B\| < 169^{-1}$ and both A and B are injective von Neumann algebras.

The strategy is divided into 3 steps.

0082-0717/80/0000-0512/$04.25

1. Find a linear embedding of A into B which is close to the identity mapping on A.

Such a linear map is nearly multiplicative so the next step is quite obvious.

2. Show that if a hermitian linear map of a C*-algebra A into a C*-algebra B is nearly multiplicative then it is close to a *-homomorphism of A into B.

The third problem now presents itself

3. Show that if a *-homomorphism α of a C*-algebra $A \subseteq B(H)$ into $B(H)$ is close to the identity on A, then α is implemented by a unitary close to the identity.

Below in the sections 2,3,4 we have tried to describe how far we have pushed each of the problems for then in section 5 to combine some of these to present results on perturbations.

Section 2 is the major part of the paper because it deals with derivations and they do interfer a lot with the problems mentioned above.

2. DERIVATIONS INTO $B(H)$. From the description of the strategy above we ought to treat the linear embedding problem first, but it turns out - as we shall see later on - that the partial solution we can get to that problem is very much dependent upon results answering questions on derivations of operatoralgebras. The second problem from the introduction raises the question whether all derivations of a subalgebra into an algebra are all inner.

Suppose now a C*-algebra A on a Hilbert space H is given, then it turns out that isomorphisms of A into $B(H)$ which are close to the identity mapping are implemented by unitaries close to the identity, if and only if every derivation δ of A into $B(H)$ is implemented by an operator x in $B(H)$. We have therefore studied derivations of A into $B(H)$ intensively, and we have got the general result that all derivations of A into $B(H)$ are inner if A has a cyclic vector.

In order to connect the original problem No. 3 with the derivation problem just described we set up the following Theorem 2.1, which also yields some other equivalent conditions. Among these we would like to mention number iv) which gives some insight on the structure of the space of ultraweakly continuous functionals on $B(H)$ which vanish on a C*-subalgebra of $B(H)$.

2.1. THEOREM. Let A be a C*-algebra on a Hilbert space H. The following conditions are equivalent.

i) $H^1(A,B(H)) = 0$

ii) $\exists k > 0 \ \ \forall x \in B(H) \ \ d(x,A') \leq k\|\operatorname{ad}(x)|A\|$.

iii) There exists a $k > 0$ such that to each *-isomorphism α of A into $B(H)$ with $d = \|\alpha - \mathrm{id}|A\| < k^{-1}$ there exists a unitary v in $B(H)$ such that $\mathrm{Ad}(v)|A = \alpha$ and $\|I-v\| \leq \sqrt{2}kd$.

iv) There exists an $h > 0$ such that any ultraweakly continuous functional φ which vanishes on A' can be expressed as a sum $\varphi = \Sigma\omega_{\xi_n\eta_n}$ of vectorfunctionals each vanishing on A' and such that $\Sigma\|\omega_{\xi_n\eta_n}\| \leq h\|\varphi\|$.

COMMENTS. The equivalence between i) and ii) is the main content of the paper [7]. You will get an idea about this equivalence along with the comments to Theorem 2.2.

The equivalence between ii) and iii) is not written down in detail in other papers so we would like to take this opportunity to do so. Suppose ii) is fulfilled for some positive k, and α is a *-isomorphism of A into $B(H)$ as described in iii). Let B denote the algebra on $H \oplus H$ given by

$$B = \{\begin{pmatrix} a & 0 \\ 0 & \alpha(a) \end{pmatrix} | a \in A\}$$

and let e and f be the projections $\begin{pmatrix} 1 & 0 \\ 0 & 0 \end{pmatrix}$ and $\begin{pmatrix} 0 & 0 \\ 0 & 1 \end{pmatrix}$ respectively. We will first show that e and f are equivalent projections in B', to this end we use the comparison lemma which tells that e and f have decompositions $e = r+p$, $f = s+q$ where (in B'); $p \sim q$ and r and s are centrally orthogonal. Suppose $r \neq 0$, then $I-c(s) \neq 0$ and we get if we restrict ourselves to $(I-c(s))B$ a situation where we can redefine A,e,f and B such that we have a situation similar to the original, but with the extra assumption that $f \prec e$ in B'. Choose v, a partial isometry in B', such that $v^*v \leq e$ and $vv^* = f$ then there exists a coisometry w in $B(H)$ such that $v = \begin{pmatrix} 0 & 0 \\ w & 0 \end{pmatrix}$ and $\alpha(a) = waw^*$. This implies that $\|\operatorname{ad} w|A\| \leq d$ so there exists an operator x in A' with $\|w-x\| \leq kd$, and $\|I-xw^*\| \leq kd < 1$. This implies, since w^*w belongs to A', that xw^*w when polar-decomposed will show that w^*w and I are equivalent in A', and we get that $e \sim f$. We can then return to the original situation and look at this when cutted down by $c(s)$. A usual argument then shows that $e \sim f$ in the original situation as well. We do now know that there exists a unitary u in $B(H)$ implementing α. As above we get $\|ad(u)|A\| \leq d$ and by ii) there exists an x in A' such that $\|u^*-x\| \leq kd$, hence $\|I-ux\| \leq kd$ and we get when polar decomposing $x = wh$ that uwh is the polardecomposition of ux so by [4, Lemma 2.7] we have $\|I-uw\| \leq \sqrt{2}kd$ and obviously $Ad(uw)|A = \alpha$, so we have proved that ii) implies iii).

Suppose now that h is a selfadjoint operator in $B(H)$ then e^{ith} is a unitary for each real t and we have that for each a in A and all t's in a neighbourhood of zero;

$$e^{ith}ae^{-ith} - a = t[h,a] + 0(t^2).$$

Under the assumption iii) we can then - for t's sufficiently close to zero - find unitaries u_t in A' such that $\|e^{ith}u_t^*-I\| \leq \sqrt{2}k(t\|adh|A\| + 0(t^2))$. From this we get, if we let t decrease to zero, that the net $(e^{ith}-u_t)t^{-1}$ has an ultraweak contact point of norm less than or equal to $\sqrt{2}k\|adh|A\|$. On the other hand $(e^{ith}-u_t)t^{-1} = (e^{ith}-I)t^{-1} - (I-u_t)t^{-1}$ so we find that there must exist an operator x in A' such that $\|h-x\| \leq \sqrt{2}k\|adh|A\|$.

Since we can prove the statement for hermitian operators we get the general result by simple manipulations.

The equivalence between ii) and iv) is proved in [11, Theorem 3.1], the idea in the proof being that when assuming ii) an operator x is close to A' if and only if it nearly commutes with all projections in A, but this happens if and only if $(1-p)xp$ is small for all projections p in A. On the other hand a vector-functional $\omega_{\xi\eta}$ vanishes on A' if and only if there exists a projection p in A such that $p\xi = \xi$ and $(1-p)\eta = \eta$. Hence it follows that you can estimate the distance to A' from x only by looking at $\omega_{\xi\eta}(x)$ where $\|\omega_{\xi\eta}\| = 1$ and $\omega_{\xi\eta}$ vanishes on A', and one finds by a convexity argument that ii) implies iv). The other implication iv) $\Rightarrow$ ii) follows easily from the Hahn-Banach Theorem.

The problem, whether the statement ii) in Theorem 2.1 holds, presents itself when one wants to ask the question whether the commutants of two close algebras are also close. This question can - if the algebra is injective - be solved very easily by the averaging technique, which was used intensively in the study of cohomology of operator algebras. Suppose namely that A is a von Neumann algebra with property P on a Hilbert space H and x belongs to B(H), then an operator z in $\overline{\mathrm{conv}}\{uxu^* | u \text{ unitary in } A\} \cap A'$ will satisfy $z \in A'$ and $\|x-z\| \leq \|\mathrm{ad}x|A\|$.

In [6] we did extend a method Arveon had used for nest algebras to cover properly infinite von Neumann algebras, and we found that for a properly infinite von Neumann algebra A on a Hilbert space H and any x in B(H), we have $d(x,A') \leq 3/2\|\mathrm{ad}x|A\|$.

Having these results and Theorem 2.1 we know that all derivations of a C*-algebra A into B(H) are inner in a lot of cases. The next theorem states our knowledge so far, but it may be easier to state which cases we do not know so far. We do not know the result if the ultraweak closure of A is a non injective type II_1 von Neumann algebra with properly infinite commutant.

2.2. THEOREM. Let A be a C*-algebra on a Hilbert space H. If one of the following conditions i),...,iv) holds then $H^1(A,B(H)) = 0$.

i) $\bar{A}$ is an injective von Neumann algebra.

ii) $\bar{A}$ is a properly infinite von Neumann algebra.

iii) $\bar{A}$ is of type II_1 and isomorphic to $\bar{A} \bar{\otimes} R$ where R is the hyperfinite II_1 factor.

iv) A has a cyclic vector.

COMMENTS. The first statement is a consequence of Connes characterization of injective algebras which show that these do have amenability properties such that the averaging technique can be applied.

The second and third are proved by first showing, that these algebras satisfy ii) of Theorem 2.1. Next one shows that this condition is also fulfilled for unbounded operators implementing bounded derivations and the result follows as soon as one can get an unbounded, closed and densely defined operator which implements a given derivation of A into B(H), [7].

The fourth result was very interesting to the author because at that time we had problems with derivations of von Neumann algebras of type II_1, just as Johnson & Parrott and Johnson & Akemann had when working with derivations into the compacts and sequences of derivations converging simply in norm, respectively.

The statement iv) namely shows that the solution of the problem is not linked to algebraic properties of A, but rather to properties of the commutant.

The proof of iv) is contained in [11] and it is based upon the non commutative Grothendieck inequality due to Pisier-Ringrose, [29] [26] [15]. The idea is that this inequality together with an algebraic trick makes it possible to prove that a derivation δ of A into $B(H)$ is completely bounded. In fact we prove that $\delta \otimes \mathrm{id}$ is defined on $A \otimes C(L^2(\mathbb{N})$ and has norm dominated by $8\|\delta\|$, and one can then use ii) to obtain the general result.

We would like to close this part of the section with mentioning the connection between this problem on derivations and the so-called similarity problem for C*-algebras.

Suppose Φ is a bounded algebra homomorphism of a C*-algebra A into $B(H)$ for some Hilbert space H. The similarity question then asks whether there exists an invertible operator x on H such that $\mathrm{Ad}(x)\circ\Phi$ is a star representation.

In order to see the connection with the derivation problem we suppose that A is a C*-algebra on H and δ a derivation of A into $B(H)$. Then $\Phi(a) = \begin{pmatrix} a & \delta(a) \\ 0 & a \end{pmatrix}$ is a bounded homomorphism of A into $B(H \oplus H)$ and we see immediately that if Φ is similar to a star representation then δ is implemented by a bounded operator.

For this reason and because the similarity problem is very interesting in itself we have studied the problem. Both we and Haagerup [12], [15] have proved that if $\Phi(A)$ has a cyclic vector then Φ is similar to a star-homomorphism. Unfortunately this gives no new insigth to the derivation problem except for the new proof of the known results. The powerful tool which is used here is again the non-commutative Grothendieck inequality.

We stated in the introduction that the cohomology question $H^1(A,B) = 0$? for $A \subseteq B$ is related very much to questions on whether mappings which are nearly homomorphisms are in fact close to homomorphisms.

In order to be able to obtain some results we have to assume that B is a von Neumann algebra on a Hilbert space H.

In this set up we then have:

2.3. THEOREM.

i) If $\bar{A}$ is injective then $H^1(A,B) = 0$.

ii) If B is injective and $H^1(A,B(H)) = 0$ then $H^1(A,B) = 0$.

iii) B is a finite von Neumann algebra then $H^1(A,B) = 0$.

COMMENTS. These results do not give very much insight in the general problem on the eventual vanishing of $H^1(A,B)$ for arbitrary von Neumann algebras A,B with $A \subseteq B$. If one looks back at Theorem 2.2 and tries to do similar things here one finds that one reason why this is not possible, is that the correspondance between subalgebras of B and their relative commutants in B is not one to one.

Before we close this section we would like to mention a cohomology problem which we could solve. In the study of generators of norm continuous one parameter semigroups of completely positive mappings on C*-algebras, Lindblad [21], and Evans & Lewis [14] had proved-up to a question on derivations of operator algebras - that such a generator L has the form $L(a) = \psi(a) + x^*a + ax$ where ψ is a completely positive map of A into $\bar{A}$ (we do assume A concretely represented) and $x \in \bar{A}$. The question which was left over was solved in [9] where we proved that a derivation δ of A into B - where $A \subseteq$ B, B is a von Neumann algebra and for each a in A $\delta(a)^*\delta(a) \in \bar{A}$ is always implemented by an operator in B. The assumptions on δ made it possible to solve this problem by methods similar to the ones used when proving $H^1(A,A) = 0$ in [19].

3. THE LINEAR EMBEDDING PROBLEM. Let us suppose that γ is a small positive number and that two C*-subalgebras A and B of B(H) satisfy $A \overset{\gamma}{\subset} B$.

It is obvious that a linear embedding of A in B is possible if there exists a projection π of norm one from B(H) onto B - B is an injective von Neumann algebra - and A is arbitrary [5], [6], [8], [10].

In a lot of important cases we thus find that the problem has a very simple solution. On the other hand if B is not injective, the problem is in general very difficult, but it turns out that one can get linear embeddings if A is a finite dimensional C*-algebra or if B is a von Neumann algebra and $\bar{A}$ is an injective von Neumann algebra. The basic lemma upon which these results are based is presented below, it says simply that in a lot of cases the relation $A \overset{\gamma}{\subset} B$ can be tensored by any nuclear C*-algebra E such that one get $A \otimes E \overset{6k\gamma}{\subset} B \otimes E$ for some constant k depending upon A. When examining this result one finds that at least in some cases this implies that it is possible simultaneously to approximate several elements in A with elements in B such that certain linear conditions are fulfilled.

3.1. LEMMA. Let $A \overset{\gamma}{\subset} B$ be C*-subalgebras of a C*-algebra C. Suppose that there exists a $k > 0$ such that for each representation π of A on a Hilbert space K and each operator x in B(K); $d(x,\pi(A)') \leq k \| \mathrm{ad}x | \pi(a) \|$, then for any nuclear C*-algebra E

$$A \otimes E \overset{6k\gamma}{\subset} B \otimes E.$$

COMMENTS. Suppose A and B are subalgebras of B(H) and K is an infinite dimensional Hilbert space then $A \otimes \mathcal{C}_K \overset{\gamma}{\subset} B \otimes \mathcal{C}_K$. The assumption on A easily implies that $B' \bar{\otimes} B(K) \otimes \mathcal{C}_K \overset{2k\gamma}{\subseteq} A' \bar{\otimes} B(K) \otimes \mathcal{C}_K$ (see 5.1), but now the left hand algebra is properly infinite and hence by the theorems 2.1 and 2.2 satisfies a condition, similar to the one A satisfies, with $k = 3/2$. Therefore $A'' \bar{\otimes} B(K) \overset{6k\gamma}{\subset} B'' \bar{\otimes} B(K)$. When playing around with results due to Choi & Effros [3] on nuclear C*-algebras it is possible to deduce the stated result.

When dealing with abelian algebras E we can get an even better result namely that $A \otimes E \overset{\gamma}{\subset} B \otimes E$. Let us now suppose that A is a finite-dimensional abelian C*-algebra with minimal projections $p_1,\dots,p_k$, and choose integers $n_1,\dots,n_k$ such that the functions $z^{n_1},\dots,z^{n_k}$ on the torus $\mathbb{T}$ satisfy any wanted degree of independence. If one then considers the element

$$f(z) = p_1 z^{n_1} + p_2 z^{n_2} + \dots + p_k z^{n_k}$$

in $A \otimes C(\mathbb{T})$ one has got a function, the image of which is practically all unitaries in A. When one then finds a function $g(z)$ in $B \otimes C(\mathbb{T})$ with $\| f-g \| \leq \gamma$ it turns out that the linear mapping of A into B defined by $\Phi(p_i) = a_{n_i}{}^* a_{n_i}$ where a_{n_i} is the Fourier coefficient for g corresponding to the number n_i, then Φ becomes a completely positive linear embedding of A into B. If one continues by standard techniques from here one easily gets

3.2. THEOREM. If $A \overset{\gamma}{\subset} B$ and A is finite dimensional, then there exists a completely positive linear mapping $\Phi: A \to B$ such that $\|\Phi - \mathrm{id}|A\| \leq (2\gamma+\gamma^2)$.

COMMENTS. The proof is found in [10, Proof of Proposition 5.2].

This result can be extended to an approximately finite-dimensional C*-algebra A in the way that it tells that to any given finite-dimensional subspace F of A, we can get a completely positive linear mapping Φ of A into B which satisfies $\|\Phi - \mathrm{id}|F\| \leq (2\gamma+\gamma^2)$.

When trying to extend this result to "bigger" C*-algebras we have to assume that B is ultraweakly closed.

3.3. THEOREM. Let $A \overset{\gamma}{\subset} B$ be C*-subalgebras of B(H) where H is a separable Hilbert space.

If B is a von Neumann algebra and $\bar{A}$ is an injective von Neumann algebra then there exists a completely positive linear mapping Φ of A into B such that $\|\Phi - \mathrm{id}|A\| \leq (2\gamma+\gamma^2)$.

COMMENTS. It is clear that $\bar{A} \overset{\gamma}{\subset} B$. By Connes' result [13] $\bar{A}$ is approximately finite-dimensional, Theorem 3.2 and a compactness argument yeild the result.

4. NEARLY MULTIPLICATIVE MAPPINGS. Suppose Φ is a completely positive linear mapping of a unital C*-algebra A into a C*-algebra B acting on a Hilbert space H and that $\Phi(I) = I_H$. Then we know from Stinesprings work [31] that there exists a Hilbert space K containing H and a representation π of A on K such that if p is the orthogonal projection from K onto H then

$$\Phi(a) = p\,\pi(a)|H. \tag{1}$$

If Φ is nearly multiplicative then there exists a small positive number γ such that for each unitary u in A $\|\Phi(u^*)\Phi(u) - I\| \leq \gamma$. When looking at (1) this inequality turns into

$$\forall u \in A:\ \|p\pi(u^*)(p-I)\pi(u)p\| \leq \gamma \tag{2}$$

therefore p nearly commutes with all operators in $\pi(A)$.

We have now reached the point where the problem under consideration turns into a cohomology problem. To this end we do suppose that B is a von Neumann algebra and define C as the von Neumann algebra generated by $\pi(A)$ and p. If now $H^1(\pi(A), C) = 0$, then one finds by an easy closed graph argument - as in Theorem 2.1 (i ⇒ ii) - that there exists a constant $k > 0$ such that for each x in C $d(x,\pi(A)'\cap C) \leq k\| \mathrm{ad}\, x | \pi(A) \|$, [7, Theorem 2.4]. Let us then assume this, hence there exists an operator z in $\pi(A)' \cap C$ such that $\| z-p\| \leq k\gamma^{\frac{1}{2}}$, and if $k\gamma^{\frac{1}{2}} < 1$, we find that there is a projection q in $\pi(A)'\cap C$ which is close to p. Since these two projections are close in C they are equivalent in C and there exists a partial isometry v in C with $\| p-v\|$ small, $v^*v = q$ and $vv^* = p$. It is now fairly easy to see that the mapping Ψ, described below, is a homomorphism of A into B and close to Φ;

$$\Psi(a) = v\pi(a)v^* | H.$$

In Theorem 2.3 we proved that certain cohomology groups $H^1(A,B)$ do vanish, so we get when combining this with the arguments just above the following:

4.1. THEOREM. Let Φ be a completely positive mapping of a von Neumann algebra A into a von Neumann algebra B such that $\Phi(I) = I$. Let $d = \sup\{\| I-\Phi(u^*)\Phi(u)\| \mid u$ unitary in $A\}$.

If A is injective and $d < \frac{1}{4}$ or B is finite and $d < \frac{1}{4}$, then there exists a star homomorphism Ψ of A into B such that $\| \Phi-\Psi\| \leq 6d^{\frac{1}{2}}$.

COMMENTS. For proofs see [5, Lemma 3.3] and [6, Proposition 1.1].

It is also possible to prove similar results if one knows that $\Phi(A)''$ is an injective algebra and $H^1(\pi(A),B(K)) = 0$. With respect to the condition on B being finite, we would like to say, that this condition makes a fixed point argument - with Ryll-Nardzewskis fixed point theorem - possible. Furthermore one can extend this result to semifinite algebras in the following sense

4.2. COROLLARY. Let Φ be a completely positive linear, injective mapping of a semifinite von Neumann algebra A onto a von Neumann algebra B. If $\Phi(I) = I$, $\| \Phi^{-1}\| \leq 1+t$ and $t \leq 3000^{-1}$, then there exists an isomorphism Ψ of A onto B such that $\| \Phi\circ\Psi^{-1}-\mathrm{id}\| \leq 49t^{\frac{1}{2}}$.

5. PERTURBATIONS OF ALGEBRAS. In this context we will start with the problem concerning the distance between the commutants of close algebras. The following simple lemma transfers this problem to the cohomology discussion from section 2.

5.1. LEMMA. Let $A \overset{\gamma}{\subset} B$ be C*-algebras on a Hilbert space H.

Suppose there exists a constant $k > 0$ such that for each x in $B(H)$ $d(x,A') \leq k\| \mathrm{ad}x|A\|$, then $B' \overset{2k\gamma}{\subset} A'$.

PROOF. Take z in B' an a in A and a b in B such that $\|b-a\| \leq \gamma$ then

$$\|[z,a]\| = \|[z,b-a]\| \leq 2\gamma\|z\| .$$

We thus find that the commutant operation behaves nicely with respect to near inclusions - $A \overset{\gamma}{\subset} B$ - in a lot of cases, in particular if A has a cyclic vector we get $B' \overset{24\gamma}{\subset} A'$ [11, Corollary 5.4], if A is properly infinite $B' \overset{3\gamma}{\subset} A'$ and if $\bar{A}$ is injective $B' \overset{2\gamma}{\subset} A'$.

When combining results from the previous sections we can get

5.2 THEOREM. Suppose $A \overset{\gamma}{\subset} B$ are von Neumann algebras on a Hilbert space H.

If $\gamma < 10^{-2}$ and A is injective then there exists a unitary u in $(A \cup B)''$ such that $uAu^* \subseteq B$ and $\|I-u\| \leq 150\gamma$.

If B is injective and for some k and any x in $B(H)$; A satisfies $d(x,A') \leq k\|\mathrm{ad}(x)|A\|$, then, provided $\gamma(6k^2+2k) < 1$ there exists a unitary u in $B(H)$ such that $uAu^* \subseteq B$ and $\|I-u\| \leq (9k^2+3k)\gamma$.

For proofs see [10, Theorem 4.3 and Theorem 4.1].

Moreover we would like to mention that the first part of the above theorem is valid at the "C*-level" if A is a finite-dimensional C*-algebra. To be more precise we do not have to assume that B is a von Neumann algebra in this case.

This "C*-result" makes it possible to prove the following:

5.3 THEOREM. Let A and B be approximately finite-dimensional C*-algebras on a Hilbert space H.

If $\|A-B\| < 1/16$ then A and B are unitarily equivalent.

PROOF. [10, Corollary 6.3], [24].

In [5] we proved similar results for commutative and ideal algebras. It has however, not been possible to get general results for nuclear C*-algebras, and it follows from [17] that one cannot expect similar nice result as those valid for injective von Neumann algebras. On the other hand we do get very easily the following:

5.4. THEOREM. Let A be a nuclear and B an arbitrary C*-algebra on a Hilbert space H. If $\|A-B\| < 10^{-2}$ then B is nuclear and the biduals A^{**} and B^{**} are isomorphic von Neumann algebras.

PROOF. [10, Theorem 6.5].

REFERENCES

1. W.B. Arveson, Interpolation problems in nest algebras. J. Funct. Anal. 20 (1975), 208-233.

2. C. Cecchini, On neighbouring type I von Neumann factors. Bollettino U.M.I. (4) 11 (1975), 103-106.

3. M.D. Choi & E. Effros, Nuclear C*-algebras and the approximation property. Amer. J. Math. 100 (1978), 61-79.

4. E. Chistensen, Perturbations of type I von Neumann algebras. J. London Math. Soc. 9 (1975), 395-405.

5. E. Christensen, Perturbations of operator algebras, Invent. Math. 43 (1977), 1-13.

6. E. Christensen, Perturbations of operator algebras II. Indiana Univ. Math. J. 26 (1977), 891-904.

7. E. Christensen, Extensions of derivations. J. Funct. Anal. 27 (1978), 234-247.

8. E. Christensen, Subalgebras of a finite algebra. Math. Ann. 243 (1979), 17-29.

9. E. Christensen & D.E. Evans, Cohomology of operator algebras and quantum dynamical semigroups. J. London Math. Soc. 20 (1979), 358-368.

10. E. Christensen, Near inclusion of C*-algebras. Acta Math. 1980

11. E. Christensen, Extensions of derivations II. Math. Scand.

12. E. Christensen, On non self-adjoint representations of C*-algebras. Amer. J. Math.

13. A. Connes, Classification of injective factors. Ann. Math. 104 (1976), 73-115.

14. D.E. Evans & J.T. Lewis, Dilations of irreversible evolutions in algebraic quantum theory. Commun. Dubl. Inst. Adv. Stud. Ser. A 24 (1977).

15. U. Haagerup, Solution of the similarity problem for cyclic representations of C*-algebras. To appear.

16. B.E. Johnson, Perturbations of Banach algebras. J. London Math. Soc. 34 (1977), 439-458.

17. B.E. Johnson, A counterexample in the perturbation theory of C*-algebras. Invent. Math.

18. R.V. Kadison, Derivations of operator algebras. Ann. Math. 83 (1966), 280-293.

19. R.V. Kadison, A note on derivations of operator algebras. Bull. London Math. Soc. 7 (1975), 41-44.

20. R.V. Kadison & D. Kastler, Perturbations of von Neumann algebras, I stability of type. Amer. J. Math. 94 (1972), 38-54.

21. G. Lindblad, Dissipative operators and cohomology of operator algebras. Letters in Math. Phys. 1 (1976), 219-224.

22. J. Phillips, Perturbations of type I von Neumann algebras. Pacific J. Math. 52 (1974), 505-511.

23. J. Phillips, Perturbations of C*-algebras. Indiana Univ. Math. J. 23 (1974), 1167-1176.

24. J. Phillips & I. Raeburn, Perturbations of A. F. Algebras. Canad. J. Math. 31 (1979), 1012-1016.

25. J. Phillips, & I. Raeburn, Perturbations of operator algebras II. Proc. London Math. Soc.

26. G. Pisier, Grothendieck's theorem for non-commutative C*-algebras with an appendix on Grothendieck's constants. J. Funct. Anal. 29 (1978), 397-415.

27. I. Raeburn & J. Taylor, Hochschild cohomology and perturbations of Banach algebras. J. Funct. Anal. 25 (1977), 258-266.

28. J.R. Ringrose, Cohomology of operator algebras. Lecture Notes in Math. Vol. 247, Springer, Berlin 1972.

29. J.R. Ringrose, Linear mappings between operator algebras. Symposia Mathematica XX (1976), 297-316.

30. S. Sakai, Derivations of W*-algebras. Ann. Math. 83 (1966), 273-279.

31. W.F. Stinespring, Positive functions on C*-algebras. Proc. Amer. Math. Soc. 6 (1955), 211-216.

32. J. Tomiyama, Tensorproducts and projections of norm one in von Neumann algebras. Seminar University of Copenhagen 1970.

Matematisk Institut

Universitetsparken 5

DK - 2100 Copenhagen Ø

Proceedings of Symposia in Pure Mathematics
Volume 38 (1982), Part 2

INVARIANTS OF C*-ALGEBRAS STABLE UNDER PERTURBATIONS

John Phillips

In this section we consider those isomorphism invariants of C*-algebras which remain stable under small perturbations. The reason for this approach is obvious; if enough invariants can be shown to be stable then it may be possible to show that certain neighbouring C*-algebras are isomorphic. Of course, the isomorphism class of a C*-algebra is an isomorphism invariant, but it is seldom amenable to a direct approach.

In the paper [20] by R.V. Kadison and D. Kastler which initiated the study of perturbations, they showed that the type of a von Neumann algebra is stable under small perturbations. In particular, they proved the following theorem.

THEOREM. ([20], theorem A): If R and S are von Neumann algebras acting on a Hilbert space H, $||R - S|| < a(\leq \frac{1}{26,000})$, P_d, P_c, P_{c_∞}, P_∞, P_n, $n = 1, 2, \ldots, \dim(H)$ are the maximal central projections in R of types I, II_1, III_∞, III and I_n respectively, and Q_d, Q_{c_1}, Q_{c_∞}, Q_∞, and Q_n, $n = 1, 2, \ldots, \dim(H)$ are the corresponding central projections for S; then $||P_d - Q_d|| < 2\alpha(a)$, $||P_{c_1} - Q_{c_1}|| < \alpha(5a + 3\alpha(a))$, $||P_{c_\infty} - Q_{c_\infty}|| < 4a + 3\alpha(a) + \alpha(5a + 3\alpha(a))$, $||P_\infty - Q_\infty|| < \alpha(a)$, and $||P_n - Q_n|| < \alpha(a + 2\alpha(a))$. In particular the same types occur in the type decomposition of R and S.

The function α occuring in this theorem is given by $\alpha(a) = a + \frac{1}{2} - (\frac{1}{4} - 2a)^{1/2}$. A more refined argument by E. Christensen [4] showed that $\alpha(a)$ could be replaced by $2a$.

Subsequently, it was shown simultaneously and independently by E Christensen [4] and J. Phillips [22] that if R is type I and S is sufficiently close to R then R' and S' are also close. Combining this with the theorem of R.V. Kadison and D. Kastler, and with the observation that $Z(R)$ and $Z(S)$ are isomorphic via an isomorphism close to the identity, J. Phillips [22] showed that if R is type I then S is unitarily equivalent to R when $||R - S||$ is sufficiently small. By a much deeper and more delicate analysis, E. Christensen [4] showed that in fact, the above mentioned unitary could be chosen close to 1 in the algebra $(R \cup S)''$.

At approximately the same time, J. Phillips showed in another paper, [23],

0082-0717/80/0000-0554/$02.25

that if A and B are close C*-subalgebras of a C*-algebra C and if π is a representation of A on H then there are representations π' and ρ of A,B respectively, on some Hilbert space K so that π' is quasi-equivalent to π and $\pi'(A)$ is close to $\rho(B)$. By looking at the kernels of these representations one can show if I is an ideal in A then there is a unique ideal J in B so that I is close to J. A slightly more detailed analysis then yielded the following result.

THEOREM. (1.5 and 1.6 of [23]): Let A and B be C*-algebras on H such that $||A - B|| < a(\leq \frac{1}{100})$. The map from ideals of A to ideals of B defined above is a lattice isomorphism. Moreover, the restriction of this map to Prim A yields a homeomorphism onto Prim B in the Jacobson topology.

Since R.V. Kadison and D. Kastler had already shown in [20] corollary C, that a C*-algebra near a commutative C*-algebra was also commutative, an immediate corollary of this theorem is that a C*-algebra near a commutative C*-algebra is *-isomorphic to it. This result was later subsumed by E. Christensen's a more delicate analysis [5] of the "close" isomorphism $A^{-} \to B^{-}$, involving the Stone-Weirstrass theorem. This argument shows that if A is commutative and B is close to A then A and B are unitarily equivalent via a unitary close to 1 in $(A \cup B)''$.

Also in the paper [23], J. Phillips showed by an easy argument that the compact operators are "imperturbable". This, together with a careful look at the representation ideas introduced earlier, showed that if A and B are close C*-algebras then B is liminal, postliminal or antiliminal respectively, precisely when A is. Finally, a slightly more careful analysis of the representation trick in the irreducible case yielded the following theorem.

THEOREM. (3.1 of [23]): Let A and B be C*-algebras on H such that $||A - B|| < a(\leq \frac{1}{1932})$. Then there is a bijection $\psi:\hat{A} \to \hat{B}$ such that $\psi(\hat{A}_n) = \hat{B}_n$ for each cardinal n. Moreover, $\ker(\psi(\pi)) = \phi(\ker \pi)$ where $\phi:\text{Prim } A \to \text{Prim } B$ is the previously defined homeomorphism. Hence, ψ is a homeomorphism.

Then in a paper [5] (the deeper results of which are discussed in another section of this survey), E. Christensen used this technique concerning ideals, and his result on type I von Neumann algebras [4], to show that a C*-algebra B near an ideal C*-algebra A is unitarily equivalent to it via a unitary close to 1 in $(A \cup B)''$.

In a slightly different vein, B.E. Johnson [16] showed that certain properties of a Banach algebra remain stable when the multiplication is perturbed slightly. In particular, he proved the following theorem.

THEOREM. (6.2 of [16]): Let A be an amenable Banach algebra with multiplication π. Then there exists $\epsilon > 0$ such that, for any multiplication ρ on A with $||\rho - \pi|| < \epsilon$, (A,ρ) is amenable

The proof of this result uses a modification of the algebra cohomology ideas outlined elsewhere in this survey together with B.E. Johnson's characterization; (A,ρ) amenable $\iff$ (A,ρ) has a virtual diagonal.

Another stability result in this paper is the following.

THEOREM. (6.3 of [16]): Let (A,π) be a commutative Banach algebra with $H^2(A,A) = 0$. Then there exists $\epsilon > 0$ such that any multiplication ρ on A with $||\pi - \rho|| < \epsilon$ is commutative. Both of these theorems lean heavily on the fact that for Banach spaces, the perturbation of an exact sequence is exact: this is the content of lemma 6.1 of [16].

If one defines the <u>local semigroup</u> of a C*-algebra A to be the set $S(A)$ of equivalence classes of projections of A together with the partially defined addition:

$[e] + [f] = [e' + f']$ if there are $e' \sim e$ and $f' \sim f$ with $e' \perp f'$, then this is clearly an invariant of A. By refining the ideas of R.V. Kadison and Kastler, J. Phillips and I. Raeburn proved the following theorem.

THEOREM. (2.6 of [24]): Let A and B be unital C*-algebras on H with $||A - B|| < a(\leq \frac{1}{8})$. Then, $\rho:S(A) \to S(B)$ defined by $\rho: S(A) \to S(B)$ defined by $\rho[e] = [f]$ where f is a projection in B with $||e - f|| < a$ is an isomorphism of local semigroups.

By a result of G. Elliott, who formalized the notion of local semigroup in order to classify A.F. algebras, one gets an immediate corollary.

COROLLARY (2.7 of [24]): Let A and B be unital A.F. algebras on H such that $||A - B|| < a(\leq \frac{1}{8})$. Then there is an isomorphism ϕ of A onto B such that $[\phi(e)] = \rho([e])$ for all projections e in A.

By modifying techniques of R. Powers and O. Bratteli, J. Phillips and I. Raeburn proved the following theorem.

THEOREM. (1.3 of [24]): Let A and B be A.F. algebras acting on a separable Hilbert space H, and suppose that $A^- = B^- = M$. If there is an isomorphism $\phi: A \to B$ such that $\phi(e) \sim e$ in M for each projection e in A, then there is a unitary operator u in M such that $uAu^* = B$.

Combining this theorem together with the previous corollary and a result of E. Christensen on neighbouring von Neumann algebras (4.1 of [5]) J. Phillips and I. Raeburn showed:

THEOREM (2.8 of [24]): Let A and B be (unital) A.F. algebras on H such that $||A - B|| < a(\leq \frac{1}{(305)^2})$. Then, there is a unitary u in $(A \cup B)''$ such that $uAu^* = B$.

At about the same time, E. Christensen proved a number of stability results in his paper [10]. To state these results, we first need a definition. Let A be a C*-algebra and $k > 0$. A is said to have property D_k if for any representation π of A on a Hilbert space H and any operator x in $\mathcal{B}(H)$

$$\inf\{||x - m|| \mid m \text{ in } \pi(A)'\} \leq k||ad(x)|\pi(A)||.$$

If A is nuclear, so that $\pi(A)'$ is injective, E. Christensen projects $\mathcal{B}(H)$ onto $\pi(A)'$ and observes that A has property D_1(2.6 of [10]). If a C*-algebra A has a co-infinite isometry, then earlier work of the same author (2.4 of [6]) using the distance function of W. Arveson shows that A has property $D_{3/2}$. Finally, putting these ideas together with a clever use of the Hahn-Banach theorem, E. Christensen proves the following theorem.

THEOREM. (4.1 of [10]): Let C be a C*-algebra with C*-subalgebras $A \overset{\gamma}{\subset} B$ and let D be a nuclear C*-algebra. If A has property D_k then

$$A \otimes D \overset{6k\gamma}{\subset} B \otimes D.$$

Using this theorem together with the fact that the flip automorphism on $M_n \otimes M_n$ is inner, he proves the following result.

THEOREM. (6.5 of [10]): Let $A \overset{\gamma}{\subset} B$ be C*-subalgebras of a C*-algebra C. Suppose A is finite-dimensional of type I_n. If $\gamma < 2.10^{-4}$ then there exists a partial isometry v in C such that $vAv^* \subseteq B$ and $||1_A - v|| \leq 57\gamma^{1/2}$. If A,B and C have a common unit 1 and $\gamma < 10^{-3}$, then there is a unitary u in C such that $uAu^* \subseteq B$ and $||1 - u|| \leq 28\gamma^{1/2}$.

This result is perhaps surprising in that the constants obtained are independent of the dimension n which is not the case with J. Glimm's results in the paper on the classification of U.H.F. algebras. E. Christensen also proves a similar result for an arbitrary finite dimensional C*-algebra A (6.4 of [10]), where the hardest step (and it is hard) is proving it for the commutative case! Using this latter result, he is able to show that a C*-algebra near an A.F. algebra is also A.F. and by using Bratteli diagrams is able to show they are isomorphic. This, together with the observation of J. Phillips and I. Raeburn which yields unitary equivalence, is the strongest result known for A.F. algebras.

Turning to the case of continuous trace C*-algebras, J. Phillips and I. Raeburn investigated the stability of the Dixmier-Douady invariant in [25]. In particular, they proved the following theorem.

THEOREM. (2.13 of [25]): Suppose A and B are C*-subalgebras of a C*-algebra C and $||A - B|| < \frac{1}{64}$. If A is separable with continuous trace then so is B and there is a homeomorphism $\hat{A} \leftarrow \hat{B}$ such that the induced isomorphism $H^3(\hat{A},\mathbb{Z}) \to H^3(\hat{B},\mathbb{Z})$ takes $\delta(A)$ to $\delta(B)$.

The proof of this result involves a careful construction of 2-cocycles representing $\delta(A)$ and $\delta(B)$ respectively, and showing that, after identifying $\hat{A}$ and $\hat{B}$, these cocycles are close to each other. An elementary sheaf-theoretic argument completes the proof. An immediate corollary to this result and the Dixmier-Douady classification, is that if A and B satisfy the hypotheses of this theorem, then A is stably isomorphic to B $(A \otimes K(H) \cong B \otimes K(H))$.

Now, Dixmier and Douady showed that the property of being stable $(A \cong A \otimes K(H))$ was a "local" result for continuous trace C*-algebras and so that if one wishes to see that being stable is stable under small perturbations one need only work locally (we're not sure if the pun is intended, or not). This amounts to showing that a C*-algebra near $C(X) \otimes K(H)$ is isomorphic to it. This latter result requires constructing an approximate identity for B consisting of an increasing sequence of projections, near a similar approximate identity for A(actually, a little more is required). This is done and the following result is deduced.

THEOREM. (3.10 of [25]): Let A and B be C*-subalgebras of a C*-algebra C with $||A - B|| < \frac{1}{309}$. If A is separable, stable and has continuous trace, then $A \cong B$.

As shown by B. E. Johnson [17], one cannot show, even in the case $A \cong C[0,1] \otimes K(H)$, that the above isomorphism is close to the identity.

More recently, in some (as yet unpublished) work, Mahmood Khoskam, a Ph.D. student at Dalhousie University, has shown a number of stability results. In particular, if A and B are close C*-algebras and A is nuclear then the K-theories of A and B are naturally isomorphic. Also, if A and B are close, then their unitary groups have the same homotopy type. Finally, in a number of cases, he has been able to show that if A and B are close then their centres are isomorphic and has used this to deduce that if A and B are close AW* algebras and A is of type I then $A \cong B$.

DEPARTMENT OF MATHEMATICS
DALHOUSIE UNIVERSITY
HALIFAX, NOVA SCOTIA B3H 4H8

Proceedings of Symposia in Pure Mathematics
Volume **38** (1982), Part 2

THE MAP Ad: U(B) → InnB

John Phillips

The following theorem was proven independently by E.C. Lance [7] and M.S.B. Smith [12].

THEOREM. Let X be a compact metrizable space and let $A = C(X,\mathcal{B}(H))$. Then there is an exact sequence:

$$0 \to \text{Inn}A \to \{{\pi(A) \atop \text{PInn}A}\} \xrightarrow{\zeta} H^2(X,\mathbb{Z}) \to 0 .$$

Where $\pi(A)$ is the group of π-inner automorphisms of A and PInnA is the group of pointwise inner automorphisms of A and $H^2(X,\mathbb{Z})$ is the second Cech cohomology group of X with coefficients in $\mathbb{Z}$.

However, to our knowledge, the first use of algebraic topology in analyzing automorphisms of C*-algebras came in the 1967 paper of R.V. Kadison and J.R. Ringrose [6] where they used homotopy theory to distinguish various subgroups of AutA where $A = C(X,M_n)$ for certain specific spaces X.

Two years ago, J. Phillips and I. Raeburn extended the result of Lance and Smith in a couple of directions, [10]. The following theorems appear in that paper.

1. THEOREM. Let X be a compact metrizable space and let B be a von Neumann algebra. Let $A = \Gamma(E)$ be the C*-algebra of sections of some locally trivial bundle E over X with fibre B and structure group InnB (norm topology). Then there is an exact sequence:

$$0 \to \text{Inn}A \to \{{\pi(A) \atop \text{PInn}A}\} \xrightarrow{\zeta} H^2(X,G)$$

where $G = H^0(Z(B)\hat{}\,,\mathbb{Z})$. Moreover, if B is purely infinite, the ζ is onto.

2. THEOREM. Let A be a separable continuous trace C*-algebra and let $\hat{A} = X$. Then there is an exact sequence:

$$0 \to \text{Inn}A \to \left\{\begin{matrix} \pi(A) \\ \text{PInn}A \\ \text{Aut}_{C_b(X)}A \end{matrix}\right\} \xrightarrow{\zeta} H^2(X,\mathbb{Z}) .$$

If A is stable, the ζ is onto.

Besides the usual sort of sheaf-theoretic arguments, there were two crucial analytic elements that entered the proofs of these results. In theorem 1. the result of G. Elliott [3] on pointwise convergence of automorphism was needed while in both theorems certain continuous selections had to be made.

0082-0717/80/0000-0555/$01.75

I. If B is a C*-algebra, when does $Ad:U(B) \to InnB$ have local sections (in the norm topology)?

II. If B is a von Neumann factor, when does $Ad:U(B) \to InnB$ have local sections (*-strong topology $\to$ u-topology)?

Question I arises in theorem 1. and question II arises in theorem 2. so that theorem 1. is the C*-theorem and theorem 2. is the von Neumann theorem!

The answer to question I is the following theorem whose proof will appear in [9].

I. THEOREM. Let B be a unital C*-algebra. Then $Ad:U(B) \to InnB$ has local sections in the norm topology if and only if the space of inner derivations on B is norm closed. If B is separable, this is equivalent to each of :

(a) InnB is norm closed,

(b) Every uniformly central sequence in B is trivial.

Remarks:

1. That (a) is equivalent to the space of inner derivations being norm closed when B is separable, is a result of R.V. Kadison, E.C. Lance and J.R. Ringrose [5].

2. The equivalence of (a) and (b) is the C*-algebra analogue of a result of A. Connes [1], and S. Sakai [11] for von Neumann algebras.

3. In order to obtain the local sections, one takes logs to change the problem to one about derivations and then uses the Open Mapping Theorem to set up an application of the Bartle-Graves selection theorem [8]. Exponentiating completes that part of the proof.

4. Techniques for the rest of the proof are mainly adaptations of those of A. Connes [1] and S. Sakai [11] to the C*-algebra setting. This result together with some other techniques yields the following improvement on theorem 1.:

1.' THEOREM: Let B be the quotient of an AW*-algebra and X a compact metrizable space. Let E be a locally trivial bundle over X with fibre B and structure group InnB. Then we have an exact sequence:

$$0 \to Inn\Gamma(E) \to PInn\Gamma(E) \xrightarrow{\zeta} H^2(X,G)$$

where $G = H^0(Z(B)^\wedge, \mathbb{Z})$. If E is trivial and B is the quotient of a purely infinite AW*-algebra then ζ is onto.

Now, in answer to question II we have the following result.

II. THEOREM. Let B be a factor on a separable Hilbert space. Then, $Ad:U(B) \to InnB$ has local sections (*-strong topology $\to$ u-topology) if and only if B is full (i.e., InnB is closed in AutB in the u-topology).

Remarks

1. In order to get local sections, one uses elementary polish theory to see that $U(B)/S^1 \to InnB$ is a homeomorphism and then [4] to deduce that $U(B) \to U(B)/S^1$ has local sections.

2. The other half of the argument is a straightforward application of a result of A. Connes [1].

This result together with some other techniques allows us to improve theorem 2. to obtain:

2.' THEOREM. Let X be a compact space and let R be a full factor with R separable. Let $A = \Gamma(E)$ be the C*-algebra of sections of a locally trivial bundle E over X with fibre R and structure group $\mathrm{Inn}R$ (in the u-topology). Then there is an exact sequence:

$$0 \to \mathrm{Inn}A \to \mathrm{LInn}A \xrightarrow{\zeta} H^2(X,\mathbb{Z}).$$

If R is purely infinite, then ζ is onto. If R is not of type III then $\pi(A) = \mathrm{LInn}A$. Where $\mathrm{LInn}A$ is the group of locally inner automorphisms of A.

Remark: It is also shown in this paper [9] that when $X = S^2$ so that $H^2(X,\mathbb{Z}) = \mathbb{Z}$ and B is a II_1 factor, then the image of ζ in theorem 1. (or 1'.) is $\{0\}$ so that ζ is not onto. In a forthcoming joint paper with I. Craw, I. Raeburn and J.L. Taylor [2] this result will be greatly improved to show that for B a II_1 factor we have the following exact sequence:

$$0 \to \mathrm{Inn}C(X,B) \to \pi(C(X,B)) \xrightarrow{\zeta} \mathrm{tor}H^2(X,\mathbb{Z}) \to 0 ,$$

where tor stands for "torsion subgroup of".

BIBLIOGRAPHY

1. A. Connes, Almost periodic states and factors of type III_1, Journal of Functional Analysis, 16(1974) 415-445.

2. I. G. Craw, J. Phillips, I. Raeburn, J.L. Taylor, Automorphisms of certain C*-algebras and torsion in second Cech cohomology, preprint.

3. G.A. Elliott, Convergence of automorphisms in certain C*-algebras, Journal of Functional Analysis, 11 (1972) 204-206.

4. A.M. Gleason, Spaces with a compact Lie group of transformations, Proc. Amer. Math. Soc., 1(1950)35-43.

5. R.V. Kadison, E.C. Lance, J.R. Ringrose, Derivations and automorphisms of operator algebras, II, Journal of Functional Analysis, 1(1967) 204-221.

6. R.V. Kadison, J.R. Ringrose, Derivations and automorphisms of operator algebras, Commun. Math. Phys. 4(1967) 32-63.

7. E.C. Lance, Automorphisms of certain operator algebras, Amer. J. Math., 91 (1969) 160-174.

8. E. Michael, Continuous selections I, Annals of Math., 63(1956) 361-382.

9. J. Phillips, Automorphisms of C*-algebra bundles, preprint.

10. J. Phillips and I. Raeburn, Automorphisms of C*-algebras and second Cech cohomology, Indiana Univ. Math. J., to appear.

11. S. Sakai, On automorphism groups of II_1 factors, Tôhoku Math.J., 26(1974) 423-430.

12. M.S.B. Smith, On automorphism groups of C*-algebras, Trans. Amer. Math. Soc., 152 (1970)623-648.

DEPARTMENT OF MATHEMATICS, DALHOUSIE UNIVERSITY, HALIFAX, N.S. B3H 4H8

Proceedings of Symposia in Pure Mathematics
Volume 38 (1982), Part 2

AUTOMORPHISMS OF CONTINUOUS TRACE C*-ALGEBRAS

Iain Raeburn

Let A be a separable continuous trace C*-algebra with spectrum X, and let $C(X)$ denote the algebra of bounded continuous functions on X, so that $C(X)$ acts naturally on A. We denote by $\mathrm{Aut}_{C(X)}A$ the group of $C(X)$-module *-automorphisms of A, by $M(A)$ the multiplier algebra of A and by $\mathrm{Inn}\, A$ the group of those $\alpha \in \mathrm{Aut}_{C(X)}A$ implemented by unitary elements of $M(A)$. There are two recent descriptions of the group $\mathrm{Aut}_{C(X)}A/\mathrm{Inn}\, A$: the first, a special case of a very general result due to Brown, Green and Rieffel, represents automorphisms algebraically as elements of a Picard group, and the second, due to Phillips and Raeburn, represents them topologically in the second Cech cohomology group $H^2(X,Z)$. Recall that A is stable if $A \otimes K(H) \cong A$.

THEOREM 1. [2, corollary 3.5]. There is an injection $\zeta\colon \mathrm{Aut}_{C(X)}A/\mathrm{Inn}\, A \to \mathrm{Pic}_{C(X)}A$, which is surjective if A is stable.

THEOREM 2. [3, theorem 2.1]. There is an injection $\eta\colon \mathrm{Aut}_{C(X)}A/\mathrm{Inn}\, A \to H^2(X,Z)$, which is surjective if A is stable.

Even though we have not yet defined the Picard group of a C*-algebra, it is clear that these two theorems should be equivalent. Here we shall merely indicate how to go about proving this: the details can be found in [4].

We begin by considering the case when A is n-homogeneous and unital, so that A is finitely generated over its center $C(X)$. An A-B bimodule M is called invertible if there is a B-A bimodule N such that $M \otimes_B N \cong A$ and $N \otimes_A M \cong B$ as bimodules. The Picard group $\mathrm{Pic}\, A$ consists of the isomorphism classes of invertible A-A bimodules under $\otimes_A$, and $\mathrm{Pic}_{C(X)}A$ consists of those for which the actions of $C(X)$ on the left and right are the same--we call these $A\text{-}_{C(X)}A$ bimodules. In the trivial case where $A = C(X)$ a theorem of Swan identifies the invertible $C(X)\text{-}_{C(X)}C(X)$ bimodules as the modules $\Gamma(L)$ of sections of line bundles L over X; it is a standard fact in topology that $H^2(X,Z)$ is isomorphic to the group of line bundles under $\otimes$, and so we have $H^2(X,Z) \cong \mathrm{Pic}_{C(X)}C(X)$. Thus we want to show that $\mathrm{Pic}_{C(X)}A \cong \mathrm{Pic}_{C(X)}C(X)$. Since we are in a purely algebraic situation, we might expect to find this in the literature: on page

0082-0717/80/0000-0558/$01.75

108 of his Tata notes [1], Bass shows that $M \to \{m \in M: am = ma \text{ for all } a \in A\}$ is an isomorphism with inverse $P \to A \otimes_{C(X)} P$.

In general, A will not have an identity and purely algebraic arguments do not work. The analogue of invertible A-B bimodules for C*-algebras are the A-B imprimitivity bimodules of Rieffel [5]; these are A-B bimodules equipped with A- and B-valued inner products $\langle\cdot,\cdot\rangle_A$ and $\langle\cdot,\cdot\rangle_B$ which are required to satisfy a whole string of compatibility conditions (see either Rieffel's article in these proceedings, or section 1 of [4]). The inverse of an A-B imprimitivity bimodule is the dual B-A bimodule $\tilde{X}$, which setwise is X itself with the module actions $b\tilde{x} = (xb^*)^\sim$, $\tilde{x}a = (a^*x)^\sim$; the map $\tilde{x} \otimes y \to \langle x,y\rangle_B$ defines an "isomorphism" of $\tilde{X} \otimes_A X$ onto B (we are glossing over some technicalities here: we have to complete the tensor product in an appropriate sense). A good example to illustrate the idea is a Hilbert space H : H is a $K(H) - \mathbb{C}$ imprimitivity bimodule with the obvious module actions and the inner products

$$\langle h|k\rangle_{\mathbb{C}} = (h|k) \ , \langle h,k\rangle_{K(H)} = h \otimes \bar{k} \ . \qquad \ldots\ldots(*)$$

In the same vein, if $\mathcal{H}$ is a continuous field of Hilbert spaces over X and A is the C*-algebra $\Gamma_0(K(\mathcal{H}))$ defined by $\mathcal{H}$, then the space $\Gamma_0(\mathcal{H})$ is an $A - C_0(X)$ imprimitivity bimodule with the inner products defined pointwise by (*). The Picard group $\mathrm{Pic}_{C(X)}A$ consists of the $A-_{C(X)}A$ bimodules which are also imprimitivity bimodules; much as before, the $C_0(X)-_{C(X)}C_0(X)$ imprimitivity bimodules all have the form $\Gamma_0(L)$ for some (Hermitian) line bundle L, and again we have $\mathrm{Pic}_{C(X)}C_0(X) \cong H^2(X,Z)$. So the result we want is the following:

THEOREM 3. [4, theorem 2.1].

$$\mathrm{Pic}_{C(X)}A \cong \mathrm{Pic}_{C(X)}C_0(X) \ .$$

The proof is easy if $A = \Gamma(K(\mathcal{H}))$ for some field $\mathcal{H}$: then $Y \to \Gamma(\mathcal{H})^\sim \otimes_A Y \otimes_A \Gamma(\mathcal{H})$ defines an isomorphism of $\mathrm{Pic}_{C(X)}A$ onto $\mathrm{Pic}_{C(X)}C_0(X)$. For general A, each $x \in X$ has a closed neighborhood N such that the quotient A^N of A by the ideal corresponding to N is isomorphic to $\Gamma(K(\mathcal{H}_N))$ for some field $\mathcal{H}_N$ over N. It is not possible to piece together the local imprimitivity bimodules $\Gamma(\mathcal{H}_N)$ to obtain an $A-_{C(X)}C_0(X)$ imprimitivity bimodule, just as it is not always possible to build a field $\mathcal{H}$ from the $\mathcal{H}_N$'s with the property that $A = \Gamma(K(\mathcal{H}))$ --the obstruction in both cases is the Dixmier-Douady class $\delta(A) \in H^3(X,Z)$. However, given $Y \in \mathrm{Pic}_{C(X)}A$ and $N \subset Y$, we can form $A^N-_{C(N)}A^N$ imprimitivity bimodules Y^N, and then use the $C(N)-_{C(N)}C(N)$ bimodules $\Gamma(\mathcal{H}_N)^\sim \otimes Y^N \otimes \Gamma(\mathcal{H}_N)$ to construct a $C_0(X)-_{C(X)}C_0(X)$ imprimitivity bimodule $\gamma(Y)$. Then $\gamma\colon \mathrm{Pic}_{C(X)}A \to \mathrm{Pic}_{C(X)}C_0(X)$ turns out to be the required isomorphism.

BIBLIOGRAPHY

1. H. Bass, Topics in algebraic K-theory, Tata Institute of Fundamental Research, Bombay, 1967.

2. L. G. Brown, P. Green and M. A. Rieffel, "Stable isomorphism and strong Morita equivalence of C*-algebras", Pacific J. Math, 71(1977), 349-363.

3. J. Phillips and I. Raeburn, "Automorphisms of C*-algebras and second Cech cohomology", Indiana Univ. Math. J., to appear.

4. I. Raeburn, "On the Picard group of a continuous trace C*-algebra", Trans. Amer. Math. Soc., to appear.

5. M. A. Rieffel, "Induced representations of C*-algebras", 13(1974), 176-257.

DEPARTMENT OF MATHEMATICS
UNIVERSITY OF UTAH
SALT LAKE CITY, UTAH 84112

Usual address:
School of Mathematics,
University of New South Wales,
P.O. Box 1, Kensington, NSW 2033,
Australia

Proceedings of Symposia in Pure Mathematics
Volume 38 (1982), Part 2

ON LIFTING AUTOMORPHISMS

Dennis Sullivan and J.D. Maitland Wright

This is a summary of results which will appear shortly, with complete proofs, in [7].

Let S be the σ-algebra of all Borel subsets of the unit interval. When N is the σ-ideal of Borel sets of Lebesgue measure zero, it follows from the work of von Neumann [2], that each automorphism of S/N is induced by a Borel bijection of $[0, 1]$. Answering a question posed by Kakutani, it was shown, recently, by Maharam and Stone [1], that when M is the σ-ideal of meagre Borel subsets of $[0, 1]$ then, given an automorphism β of S/M, there can be found a dense G_δ-set $X \subset [0, 1]$ such that β is induced by a homeomorphism of X. Their result is more general than this since they replace the unit interval by an arbitrary complete metric space. When specialized to commutative algebras our results give an alternative proof of the Maharam-Stone Theorem, but only for separable complete metric spaces.

Let A be a separable C*-algebra with Borel* envelope A^∞, (as defined by Pedersen [5]. Let J be a (two-sided) σ-ideal of A^∞ and, for each $x \in A^\infty$, let $[x]$ denote the image of x under the quotient map from A^∞ onto A^∞/J.

THEOREM. *Let* β *be an automorphism of* A/J. *Then there exists an automorphism* α *of* A^∞ *such that*

$$\beta([x]) = [\alpha(x)]$$

for all $x \in A^\infty$.

COROLLARY. *Let* A_1, A_2 *be separable* C*-*algebras with Borel* envelopes* A_1^∞ *and* A_2^∞, *respectively. Let* J_1 *and* J_2 *be* σ-*ideals in* A_1^∞ *and* A_2^∞ *respectively. Let*

$$A_1^\infty/J_1 \text{ and } A_2^\infty/J_2 \text{ be isomorphic.}$$

Then there exist central projections p_1, p_2 *in* A_1^∞ *and* A_2^∞, *respectively, such that* $1 - p_j \in J_j$ $(j = 1, 2)$ *and*

$$p_1A_1^\infty \text{ is isomorphic to } p_2A_2^\infty.$$

0082-0717/80/0000-0577/$01.50

When our main theorem is specialized to commutative algebras we recover a theorem of Sikorski [6] for Boolean algebras. We lean heavily on the results and methods of Pedersen on Borel* algebras [3, 4]. In the special case where A^{∞}/J is a von Neumann algebra the above theorem already follows from his results [3]. We remark that wherever A is simple, unital, separable and infinite-dimensional then there exists a σ-ideal J in A^{∞} such that A^{∞}/J is a monotone complete AW*-factor which is never a von Neumann algebra [8].

BIBLIOGRAPHY

1. Dorothy Maharam and A.H. Stone, "Realizing isomorphisms of category algebras", Bull. Australian Math. Soc. 19 (1978), 5-10.

2. J. von Neumann, "Einige Sätze über messbare Abbildungen", Ann. of Math. (2) 33 (1932), 574-586.

3. G.K. Pedersen, "Borel structure in operator algebras", Danske Vid. Selsk. Mat. Fys. Medd. 39, 5 (1974).

4. G.K. Pedersen, "A non-commutative version of Souslin's theorem", Bull. London Math. Soc. 8 (1976), 87-90.

5. G.K. Pedersen, C*-algebras and their Automorphism Groups, London Math. Soc. Monographs 14, Academic Press, (1979).

6. R. Sikorski, Boolean Algebras, second edition, Springer, (1964).

7. D. Sullivan and J.D.M. Wright, "On lifting automorphisms of monotone σ-complete C*-algebras", Q.J. Math. (Oxford), to appear.

8. J.D.M. Wright, "Wild AW*-factors and Kaplansky-Rickart algebras", J. London Math. Soc. (2) 13, (1976), 83-89.

I.H.E.S.
91440 BURES SUR YVETTE
FRANCE

DEPARTMENT OF MATHEMATICS
UNIVERSITY OF READING
READING, BERKS, RG6 2AX

Proceedings of Symposia in Pure Mathematics
Volume 38 (1982), Part 2

DERIVATIONS AND AUTOMORPHISMS OF JORDAN C*-ALGEBRAS

Harald Upmeier[1]

ABSTRACT. Some results are given about derivations and automorphisms of JB-algebras introduced by Alfsen, Shultz and Størmer.

1. JB-ALGEBRAS. In connection with operator algebras and quantum physics, Jordan algebras have been studied by several authors:

Jordan, v. Neumann and Wigner [4] classified all finite dimensional "formally real" Jordan algebras,

Segal [9] proposed to use non-associative algebras for the quantum mechanical formalism,

Kadison [5] clarified the Jordan algebraic structure of C*-algebras,

Alfsen, Shultz and Størmer [1] proved a Gelfand-Neumark type embedding theorem for a certain class of Banach Jordan algebras (JB-algebras).

1.1. DEFINITION. A JB-algebra is a real unital Jordan algebra X with product $x \circ y$ (satisfying the Jordan identities $x \circ y = y \circ x$ and $x^2 \circ (x \circ y) = x \circ (x^2 \circ y)$) which is a Banach space such that $\|x \circ y\| \leqslant \|x\| \circ \|y\|$ and $\|x^2 + y^2\| \geqslant \|x\|^2$.

1.2. DEFINITION. Let $\mathrm{Aut}(X)$ be the group of all Jordan algebra automorphisms of a JB-algebra X . Each automorphism of X is an isometry with respect to the unique JB-norm on X . Denote by $\mathrm{aut}(X)$ the real Lie algebra of all derivations of X (i.e. linear mappings $D : X \longrightarrow X$ satisfying $D(x \circ y) = Dx \circ y + x \circ Dy$). As for C*-algebras it can be shown that (everywhere defined) derivations of JB-algebras are bounded.

1980 Mathematics Subject Classification. 46 L 40.

[1] Supported by Deutsche Forschungsgemeinschaft.

0082-0717/80/0000-0583/$02.50

2. EXAMPLES. The self-adjoint part A_{sa} of a unital C*-algebra A is a JB-algebra with operator norm and anticommutator product

(2.1) $$x \circ y := (xy+yx)/2 .$$

2.2. The JB-factors of type I can be classified as follows:

Let E be a Hilbert space over the field $\mathbb{K}$ of real numbers, complex numbers or quaternions, respectively. Then the set $H(E)$ of all bounded $\mathbb{K}$-linear self-adjoint operators on E is a JB-algebra with product (2.1). Let Y be a real Hilbert space. The orthogonal sum $V := \mathbb{R} \oplus Y$ becomes a JB-factor (called spin factor) with product $(a+y) \circ (b+z) := (ab+(y,z))+(az+by)$. The set $H_3(\mathbb{O})$ of all self-adjoint 3x3 matrices over the octonions is a JB-algebra with product (2.1) which cannot be realized as a Jordan algebra of Hilbert space operators (exceptional Jordan algebra).

2.3. The main result of [1] shows that the study of JB-algebras can be reduced to the class of JC-algebras (i.e. unital Jordan subalgebras of $H(E)$ for some complex Hilbert space E) and $H_3(\mathbb{O})$. For JB-algebras with predual (JBW-algebras) there is a decomposition [10] $X = C(S,H_3(\mathbb{O})) \oplus Y$, where S is a hyperstonian space, C denotes continuous mappings and Y is a JW-algebra, i.e. a weakly closed JC-algebra.

3. RELATIONS TO COMPLEX ANALYSIS. JB-algebras play an important role in the algebraic description of bounded symmetric domains in complex Banach spaces. In finite dimensions these domains have been classified by E. Cartan using Lie theoretic methods. An example is the open unit disk $D = \{ z \in \mathbb{C} : |z| < 1 \}$. It turns out that in infinite dimensions Jordan theoretic methods are more efficient.

3.1. THEOREM [2]. There exists a natural correspondence between bounded symmetric domains (of tube type) D and JB-algebras X given by $D = \{ z \in X \oplus iX : \|z\| < 1 \}$. Moreover by this correspondence holomorphic properties of D can be described algebraically by the JB-algebra X.

3.2. EXAMPLE. The group $\mathrm{Aut}(D)$ of all biholomorphic automorphisms of a bounded symmetric domain D is a Banach Lie group with Lie algebra $\mathrm{aut}(D)$ consisting of all complete holomorphic vector fields on D [12]. Denoting by X the JB-algebra associated with D, there is a unique decomposition

$$\mathrm{aut}(D) = \{\text{infinitesimal transvections}\} \oplus \{iL_x : x \in X\} \oplus \mathrm{aut}(X) .$$

Here $L_x y := x \circ y$ denotes Jordan multiplication.

4. EXTENSION OF DERIVATIONS. The Lie algebra $\mathrm{aut}(H_3(\mathbb{O}))$ is the exceptional Lie algebra F_4 of dimension 52. By 2.3, we may therefore concentrate on the case of JC-algebras X realized on a complex Hilbert space E. Let A be the C*-algebra generated by X on E. Problem: Which derivations of X have an extension to a derivation of A?

4.1. EXAMPLE. For a spin factor $V = \mathbb{R} \oplus Y$, $\mathrm{aut}(V)$ can be identified with the Lie algebra of all skew-symmetric bounded operators on Y. If $\dim(V)$ is infinite, then V generates the CAR-algebra A in each representation. It can be shown that $D \in \mathrm{aut}(V)$ has an extension to a derivation of A if and only if D is of trace class.

4.2. DEFINITION. A JC-algebra X is called reversible if $x_1 \cdot\cdot x_n + x_n \cdot\cdot x_1 \in X$ whenever $x_1, \cdot\cdot, x_n \in X$.

Størmer [11] has shown that reversible JC-algebras are in some sense "complementary" to spin factors. A JW-algebra X has a decomposition

$$X = X_{\mathrm{rev}} \oplus X_{\mathrm{spin}} , \tag{4.3}$$

where X_{rev} is a reversible JW-algebra and X_{spin} is (roughly) a direct integral of spin factors.

In view of 4.1 and (4.3) we can now formulate our

4.4. EXTENSION THEOREM [13]. Let X be a reversible JC-algebra. Then each derivation of X has an extension to a *-derivation of the C*-algebra A generated by X.

4.5. REMARK [15]. Similar extension theorems hold for certain unbounded derivations, namely for JC-dynamical systems on simple JC-algebras and for JW-dynamical systems on JW-factors.

5. INNER DERIVATIONS. Let X be a JB-algebra and let $x_1, \cdot\cdot, x_n$, $y_1, \cdot\cdot, y_n$ be elements of X. Then the commutator sum

$$(5.1) \qquad D = \sum_{k=1}^{n} [L_{x_k}, L_{y_k}]$$

is a derivation of X, called inner derivation. Let $\operatorname{int}(X)$ be the ideal of all inner derivations.

If X is the self-adjoint part of a C*-algebra A, then D given as in (5.1) has the form $Dx = [a,x]$, where

$$4a = \sum_{k=1}^{n} [x_k, y_k] \in A .$$

Hence the Jordan theoretic notion of inner derivation is more restrictive than the C*-algebraic notion.

For a spin factor V we have

$$\operatorname{int}(V) = \{ D \in \operatorname{aut}(V) : \operatorname{rank} D < \infty \} .$$

In particular, there exist JBW-factors having outer derivations.

5.2. DEFINITION. A JBW-algebra X is said to have bounded spin part if

$$\sup \{ \dim \beta(X) : \beta \text{ spin factor representation of } X \} < \infty .$$

For the canonical decomposition (4.3) of X this means that the dimensions of the spin factors occuring in the direct integral representation of X_{spin} remain bounded.

5.3. THEOREM [13]. Let X be a JBW-algebra. Then $\operatorname{aut}(X) = \operatorname{int}(X)$ if and only if X has bounded spin part. In particular, purely exceptional JBW-algebras and reversible JW-algebras have only inner derivations.

The proof uses a result of T. Fack and P. de la Harpe [3] on commutators in finite von Neumann algebras. For JB-algebras in general we obtain the following approximation theorem:

5.4. THEOREM [13]. Let X be a JB-algebra. Then $\operatorname{aut}(X)$ is the closure of $\operatorname{int}(X)$ in the topology of pointwise convergence.

6. INVOLUTIONS IN JBW-ALGEBRAS. Let X be a JB-algebra and $a \in X$. The operator

$$P_a := 2 L_a^2 - L_{(a^2)}$$

is called the quadratic representation of a. An element $s \in X$ is called a symmetry if $s^2 = 1 \in X$. In this case P_s is an involutive automorphism of X.

If X is the self-adjoint part of a C*-algebra A, then

$P_a x = axa$. For a spin factor $V = \mathbb{R} \oplus Y$ and a unit vector $s \in Y$ P_s is the hyperplane reflection associated with s . The following theorem can be viewed as a generalization of the classical Cartan-Dieudonné Theorem (applying to spin factors of finite dimension).

6.1. THEOREM [15]. Let X be a JBW-algebra with bounded spin part. Then the identity component $\mathrm{Aut}^o(X)$ is algebraically generated by the involutions P_s , where s is a symmetry in X .

Our proof uses the following result of independent interest:

6.2. THEOREM [15]. Let A be a real W*-algebra without I_{fin}-part. Then each unitary in A is a finite product of symmetries in A .

As a final remark, the above results have geometric and holomorphic consequences to the description of the automorphism group of bounded symmetric domains.

BIBLIOGRAPHY

1. E. Alfsen, F. Shultz, E. Størmer, "A Gelfand-Neumark Theorem for Jordan algebras", Adv. Math. 28 (1978), 11-56.

2. R. Braun, W. Kaup, H. Upmeier, "A holomorphic characterization of Jordan C*-algebras", Math. Z. 161 (1978), 277-290.

3. T. Fack, P. de la Harpe, "Sommes de commutateurs dans les algèbres de von Neumann finies continues", Ann. Inst. Four. (to appear).

4. P. Jordan, J. von Neumann, E. Wigner, "On an algebraic generalization of the quantum mechanical formalism", Ann. Math. 35 (1934), 29-64.

5. R. Kadison, "Isometries of operator algebras", Ann. Math. 54 (1951), 325-338.

6. W. Kaup, "Algebraic characterization of symmetric complex Banach manifolds", Math. Ann. 228 (1977), 39-64.

7. W. Kaup, H. Upmeier, "An infinitesimal version of Cartan's uniqueness theorem", Manuscripta math. 22 (1977), 381-401.

8. W. Kaup, H. Upmeier, "Jordan algebras and symmetric Siegel domains in complex Banach spaces", Math. Z. 157 (1977), 179-200.

9. I. Segal, "Postulates for general quantum mechanics", Ann. Math. 48 (1947), 930-948.

10. F. Shultz, "On normed Jordan algebras which are Banach dual spaces", J. Func. Anal. 31 (1979), 360-376.

11. E. Størmer, "Jordan algebras of type I", Acta Math. 115 (1966), 165-184.

12. H. Upmeier, "Ueber die Automorphismen-Gruppen von Banach-Mannigfaltigkeiten mit invarianter Metrik", Math. Ann. 223 (1976), 279-288.

13. H. Upmeier, "Derivations of Jordan C*-algebras", Math. Scand. (to appear).

14. H. Upmeier, "Derivation algebras of JB-algebras", Manuscripta math. 30 (1979), 199-214.

15. H. Upmeier, "Automorphism groups of Jordan C*-algebras", Preprint Tuebingen 1980.

DEPARTMENT OF MATHEMATICS
UNIVERSITY OF TUEBINGEN
AUF DER MORGENSTELLE 10
D-7400 TUEBINGEN
FED. REP. GERMANY

Proceedings of Symposia in Pure Mathematics
Volume 38 (1982), Part 2

ORDER STRUCTURE AND JORDAN BANACH ALGEBRAS*

J. BELLISSARD - B. IOCHUM

Some years ago, A. Connes has proved a theorem connecting the algebraic structure and the order structure in a W*-algebra [9]. Here we generalize this result by a characterization of the ordered space given by Jordan algebras, namely (cf [2] and [14] for definitions) :

THEOREM. The category of facially homogeneous self-dual cones is isomorphic to the category of Jordan-Banach algebras with predual (J.B.W. alg.).

Recall some definitions.

A convex cone H^+ in a real Hilbert space H is self-dual if $H^+ = \{\xi \in H / \langle \xi, \eta \rangle \geq 0, \forall \eta \in H^+\}$. The set of derivations of the cone H^+ is $D(H^+) = \{\delta \in B(H) / e^{t\delta} H^+ \subset H^+, \forall t \in \mathbb{R}\}$. $D(H^+)$ is a real Lie algebra. If P_F is the orthogonal projection on the closed subspace generated by a face F of H^+, then H^+ is facially homogeneous if $(P_F - P_{F^\perp}) \in D(H^+)$ where $F^\perp = \{\xi \in H^+ / \langle \xi, F \rangle = 0\}$ is the orthogonal face of F in H^+. (cf.[9]).

The following theorem was first proved in a particular case by M. Ajlani [1].

THEOREM 1. (In collaboration with R. Lima [4]). If H^+ is finite dimensional and self-dual, the facial homogeneity is equivalent to the transitive homogeneity in the sense of Vindberg [17] (i.e., the group of transformations preserving H^+ acts transitively on the interior of H^+).

COROLLARY 2. All facially homogeneous self-dual cones in a finite dimensional Hilbert space are decomposable. The indecomposable cones (i.e., $H^+ \neq F \oplus F^\perp, \forall F$) are of the form : $M_{n\times n}(\mathbb{R})_s$, $M_{n\times n}(\mathbb{C})_{s.a}$, $M_{n\times n}(\mathbb{K})_{s.a.}$, $M_{3\times 3}(\mathbb{O})_{s.a.}$ (w.r. to the Hilbert structure defined by the trace) or the ice cream cone $V_n^+ = \{(x_1,..,x_n) \in R^n / x_1 \geq (x_2^2 + .. + x_n^2)^{\frac{1}{2}}\}$.

* Lecture given by B. Iochum at the 28th Summer Research Institute on Operator algebras and Applications, Kingston (Canada), July 14 - August 2, 1980.

0082-0717/80/0000-0502/$01.75

Here $\mathbb{K}$ is the quaternionic field and $\mathbb{O}$ the Cayley algebra. The following theorem is proved by using a spectral decomposition on the derivations. Every $\delta = \delta^* \in D(H^+)$ as a spectral representation as $\int \lambda \, d\, \delta_{F(\lambda)}$ (weak integral) where $F(\lambda)$ is a spectral family of complete faces (i.e. $F(\lambda) = F(\lambda)^{\perp\perp}$) and $\left\{ \delta_F = \frac{1}{2}(1 + P_F - P_{F'})/F \text{ face of } H^+ \right\}$ is exactly the set of extreme points of the set of positive, less than one derivations (cf. [7]).

THEOREM 3. [6] If H^+ is a facially homogeneous self-dual cone in H, the hermitian part of $D(H^+)$ has a natural structure of J.B.W. Algebra.

Conversely, recall the Størmer and Shultz's central decomposition for a J.B.W. algebra A : ([13] and [15], [16])

$A =$	A_1	$\oplus$	A_2	$\oplus$	A_3
	Non reversible part		Self-adjoint part of a W*-algebra		Fix point set of the self-adjoint part of a W* by an antiautomorphism of order 2.

Using the fact that A_1 has a trace ([5], [8] [13], and that A_2 and A_3 can be treated by the usual Tomita-Takesaki theory and Connes's result, we have

THEOREM 4. Let A be a J.B.W. algebra. Then there is a real Hilbert space H and a facially homogeneous self-dual cone H^+ in H such that A is isomorphic to $D(H^+)_{s.a.}$. Moreover if ω is a normal state on A, then there is a vector ξ_ω in H^+ such that ω coincides with the state : $\delta \in D(H^+)_{s.a.} \longrightarrow \langle \xi_\omega, \delta \xi_\omega \rangle$ by this isomorphism.

REMARK 5. U. Haagerup and H. Hanche-Olsen have developed a Tomita-Takesaki theory by defining self-polar forms on any J.B.W. algebras from which this self-dual cone can be reconstructed and it should be possible to prove directly that it is facially homogeneous ([11], [18]). See also [3] [9] and [10] for the W* case.

A link between this framework and the work of Alfsen-Shultz is given by the following

THEOREM 6. ([12]) Let K be a compact convex set. The following are equivalent :

i) K is affinely isomorphic and homeomorphic to the state space of a J.B. algebra (with the W* topology).

ii) K is strongly spectral and $e^{t(P-P')}$ preserves the order of $A^b(K)$ for all real t and all P-projection P.

BIBLIOGRAPHY

[1] M. AJLANI : Séminaire Choquet 1974/75, n° 18.

[2] E.M. ALFSEN, F.W. SHULTZ, and E. STØRMER : Adv. in Math. 28, (1978) 11-56.

[3] H. ARAKI : Pacific Journ. Math. 50, 2 (1974), 309-354.

[4] J. BELLISSARD, B. IOCHUM, and R. LIMA : Linear Alg. and Appl., 19 (1978) 1-16.

[5] J. BELLISSARD and B. IOCHUM : Annales Inst. Fourier, Grenoble, 28 (1978) 27-67.

[6] J. BELLISSARD and B. IOCHUM : C.R.Acad.Sc. Paris, 288 (1979) 229-232.

[7] J. BELLISSARD and B. IOCHUM : Math. Scand., 45 (1979) 118-126.

[8] J. BOS : The Structure of Finite Homogeneous Cones and Jordan Algebras, Osnabrück Preprint (1976).

[9] A. CONNES : Annales Inst. Fourier, Grenoble, 24 (1974) 121-155.

[10] U. HAAGERUP : Math. Scand., 37 (1975) 271-283.

[11] U. HAAGERUP and H. HANCHE-OLSEN : In this issue.

[12] B. IOCHUM and F.W. SHULTZ : To appear.

[13] G. JANSSEN : Journ. reine u angew. Math., 249 (1971) 143-200.

[14] F.W. SHULTZ : J. Funct. Anal., 31 (1979) 360-376.

[15] E. STØRMER : Acta Math., 115 (1966) 165-184.

[16] E. STØRMER : Trans. Amer. Math. Soc., 130 (1968) 153-166.

[17] E.B. WINBERG : Trans. Moscow Math. Soc., 12 (1963) 340-403

[18] S.L. WORONOWICZ : Reports on Math. Phys. 6 (1974) 487-495.

UNIVERSITE DE PROVENCE, and
CENTRE DE PHYSIQUE THEORIQUE, CNRS MARSEILLE

C.N.R.S. - Luminy - Case 907
Centre de Physique Théorique
F-13288 MARSEILLE CEDEX 2
France

Proceedings of Symposia in Pure Mathematics
Volume 38 (1982), Part 2

A TOMITA-TAKESAKI THEORY FOR JBW-ALGEBRAS

Harald Hanche-Olsen

ABSTRACT. It is explored which notions of Tomita-Takesaki theory can be carried over to the domain of JBW-algebras. This happens for the "symmetrization" of the modular automorphism group, and the self-polar form associated with a normal state ϕ.

1. INTRODUCTION. This is a report on joint work with U. Haagerup (not yet published in any form).

When trying to extend the notions of Tomita-Takesaki theory to Jordan algebras, one immediately runs into a major obstacle: One can not, in general, associate a "GNS-representation" with a state. So there is no Hilbert space on which to define the operators of Tomita-Takesaki theory. Moreover, if the product in a von Neumann algebra is reversed then the modular automorphism group is also reversed; hence it cannot be determined in terms of Jordan structure alone.

The "symmetrization" $\theta_t = (\sigma_t + \sigma_{-t})/2$, however, is left unharmed by this reversal of product. We will see that the analogue of θ_t can be defined on any JBW-algebra with a faithful normal state. Our definition of θ_t uses the structure theory of JBW-algebras; however, the characterization of θ_t does not depend on this structure. In particular, this yields a new characterization of $\sigma_t + \sigma_{-t}$ in the von Neumann algebra case. For the basic theory of JBW-algebras, see [1],[2],[4].

2. COSINE FAMILIES. A (one-parameter) <u>cosine family</u> on a linear space M is a family $(v_t)_{t \in \mathbb{R}}$ of linear operators on M satisfying $v_o = I$ and the cosine identity,

---

1980 Mathematics Subject Classification. 17C65, 46L40.

$$2v_s v_t = v_{s+t} + v_{s-t} .$$

Notice that if (u_t) is a one-parameter group, then $((u_t+u_{-t})/2)$ is a cosine family. A partial converse is contained in [3]: If (v_t) is a norm bounded weakly continuous cosine family of self-adjoint operators on a Hilbert space H, there is a (unique) positive operator d on H such that

$$v_t = \cos(td) \qquad (t \in \mathbb{R})$$

3. SELFPOLAR FORMS. Let M be a JBW-algebra, ϕ a normal state on M. A selfpolar form associated with ϕ is a bilinear, symmetric, positive semidefinite form s on M satisfying

(i) $s(a,b) \geq 0 \quad (a \geq 0,\ b \geq 0)$

(ii) $s(1,a) = \phi(a) \quad (a \in M)$

(iii) If $0 \leq \psi \leq \phi$, there is $0 \leq b \leq 1$ so that

$$\psi(a) = s(a,b) \qquad (a \in M).$$

There is at most one self-polar form associated with ϕ.

4. THE MAIN THEOREM: Let M be a JBW-algebra and ϕ a faithful normal state on M. Then there is a unique cosine family (θ_t) of positive, unital linear mappings of M into itself, satisfying:

(i) for each $a \in M$, $t \to \theta_t(a)$ is weakly continuous.

(ii) $\phi(\theta_t(a) \circ b) = \phi(a \circ \theta_t(b))$

(iii) $s(a,b) = \int \phi(a \circ \theta_t(b)) \cosh(\pi t)^{-1}\, dt$ defines a self-polar form associated with ϕ.

We outline the proof, starting with the uniqueness part. Let H be the Hilbert space obtained by completing M in the norm $\|a\|_\# = \phi(a^2)^{\frac{1}{2}}$. We extend θ_t to a self-adjoint operator v_t on H, with $\|v_t\| \leq 1$. Applying the result of §2, we find $v_t = \cos(td)$ for some positive operator d. Using (iii) and Fourier analysis, we get:

$$s(a,b) = (a | \textstyle\int v_t(b) \cosh(\pi t)^{-1} dt)_\# = (a | \cosh(d/2)^{-1} b)_\#$$

The uniqueness of s (§3) now implies the uniqueness of d

and hence that of (θ_t).

The existence proof is more ugly. We are allowed to split M along its center and prove existence of θ_t for each part; thus, we may assume either that (1) M has only a type I_2 part and an exceptional part, or (2) M can be represented as a JW-algebra on some Hilbert space, and has no I_2 part. (See [4],[6]).

In case (1) there are lots of tracial states. We may chop a little more along the center if necessary, and assume there is a positive, invertible $h \in M$ and a tracial state T so that $\phi(a) = T(a \circ h)$. Put

$$\theta_t(a) = \{h^{it} a h^{-it}\} = \{\cos(ht) a \cos(ht)\} + \{\sin(ht) a \sin(ht)\}$$

(Here we have employed the triple product $\{b a c\} = (b \circ a) \circ c + (c \circ a) \circ b - (b \circ c) \circ a$, which generalizes $\frac{1}{2}(bac + cab)$).

In case (2) there is a von Neumann algebra $\mathcal{U}$, and an antiautomorphism α of $\mathcal{U}$ of order 2 so that M is (isomorphic to)

$$\{x \in \mathcal{U} \mid x = x^* = \alpha(x)\}.$$

(See [5]). Extend ϕ to an α-invariant state $\tilde{\phi}$ on $\mathcal{U}$, and suddenly we have all the tools of Tomita-Takesaki theory available. We find $\alpha\sigma_t = \sigma_{-t}\alpha$, so we may safely define $2\theta_t = (\sigma_t + \sigma_{-t})|_M$.

BIBLIOGRAPHY

1. E.M. Alfsen, F.W. Shultz, E. Størmer, "A. Gelfand-Neumark theorem for Jordan algebras", Adv. Math. 28(1978), 11-56.

2. H. Hanche-Olsen, "A note on the bidual of a JB-algebra", Math. Zeitschr. (to appear).

3. S. Kurepa, "A cosine functional equation in Hilbert space", Can. J. Math. 12(1960), 45-50.

4. F.W. Shultz, "On normed Jordan algebras which are Banach dual spaces", J. Functional Anal. 31(1979), 360-376.

5. E. Størmer, "Jordan algebras of type I", Acta Math. 115(1966), 165-184.

6. D. Topping, "Jordan algebras of self-adjoint operators", Mem. Amer. Math. Soc. 53 (1965).

Matematisk Institutt
Boks 1053 Blindern
Oslo 3, Norway.

Proceedings of Symposia in Pure Mathematics
Volume 38 (1982), Part 2

FAITHFUL NORMAL STATES ON JBW-ALGEBRAS

G.G. Emch and W.P.C. King

§1. MOTIVATION. The purpose of this note is to draw attention to an interpretation, involving only physical concepts, for that part of the Tomita-Takesaki theorem [6] which establishes an (anti-)isomorphism between the von Neumann algebras N and N' when N admits a cyclic and separating vector. We further indicate how this interpretation leads to a genuine mathematical extension of the theory into the realm of JBW-algebras.

§2. NOTATION. Throughout this note A will stand for a JBW-algebra, i.e. A will be a real dual Banach space, equipped with a bilinear multiplication $a \circ b$ satisfying: $a^2 \circ (b \circ a) = a \circ (b \circ a^2)$; $\|a \circ b\| \leq \|a\|\|b\|$; $\|a^2\| = \|a\|^2$; $\|a^2\| \leq \|a^2 + b^2\|$. A will have a unit, I; and we will write E for the predual of A. With f denoting any bounded linear functional on A, f will be said to be normal if $f \in E$; to be positive if $\langle f:a^2\rangle \geq 0$ for all a in A; to be a state if it is positive and if $\langle f:I\rangle = 1$; and to be faithful if $\langle f;a^2\rangle = 0$ implies $a = 0$. With f a state; we write

$$V_f = \{g \in A^* | \exists\, x \in \mathbb{R} \text{ with } -xf \leq g \leq xf\}.$$

Note that V_f is an order ideal in A^*, and [4] that when f is normal, f is faithful iff V_f is dense in E iff the support of f in A is I.

§3. EXAMPLE. Let N be a W*-algebra. Then

$$a \circ b = \tfrac{1}{2}\{(a+b)^2 - a^2 - b^2\}$$

implements a structure of JBW-algebra on

$$A = N_{sa} = \{a \in N | a = a^*\}$$

equipped with the norm inherited from N. Let f be a state on N; $(\mathcal{H},\pi,\Omega)$ be the GNS representation associated to f; $W = \pi(N)'_{sa}$ be the self-adjoint part of the commutant of $\pi(N)$ in $\mathcal{H}$. There exists then a bipositive, bijective map μ from V_f onto W such that

Mathematics Subject Classification: 46-L10, 46-L99

$$\langle g;a\rangle = (\mu(g)\pi(a)\Omega,\Omega) \quad \forall\, g \in V_f, \quad a \in A.$$

When f is a faithful normal state, $\pi(N)$ is a von Neumann algebra isomorphic to N [5], and the Tomita-Takesaki theory [6] establishes in particular the existence of an anti-isomorphism ν from $\pi(N)$ onto $\pi(N)'$. Upon restricting π to A we conclude that there exists an order isomorphism $\lambda = \mu^{-1} \circ \nu \circ \pi$ from A onto V_f such that $\lambda(I) = f$.

From a physical point of view N (with its ordinary product) lacks the direct interpretation that one can give to A (with its Jordan product defined above), the elements of which are the observables of the theory. A fortiori $\pi(N)'$ only gains a physical interpretation through the primitive physical object V_f via the map μ. The "upstairs" of the diagram

$$\begin{array}{ccc} \pi(N)_{sa} & \xrightarrow{\ \nu\ } & \pi(N)'_{sa} \\ \uparrow{\scriptstyle \pi} & & \uparrow{\scriptstyle \mu} \\ A & \xrightarrow{\ \lambda\ } & V_f \end{array}$$

appears therefore as a mathematical construct above the "downstairs", where the physical meaning primarily lies [1,3]. This suggests trying for a direct proof of the bottom line of the diagram. The remainder of this note shows some cases where this is indeed possible without appeal to the physically less primitive assumption that A could eventually be realized as the self-adjoint part of a W*-algebra.

§4. THE ATOMIC CASE. An idempotent $p(=p^2)$ in A is said to be an atom if $0 \leq q \leq p$ with $q = q^2$ implies that q can only be 0 or p. Two idempotents p and q are said to be orthogonal if $p \circ q = 0$. A JBW-algebra A is said to be atomic when every idempotent $p \in A$ is the least upper bound of orthogonal atoms. Upon exploiting then the fact that the atoms of A are in one-to-one correspondence with the extreme points of the set of all normal states on A, one obtains [4]:

THEOREM A: Let f be a faithful normal state on an atomic JBW-algebra A. Then there exists an order isomorphism λ from A onto V_f with $\lambda(I) = f$.

We remark that since the algebra M_3^8 of all hermitian matrices of order 3 over the Cayley numbers (or quasi-quaternions) is an atomic JBW-algebra which is not covered by the example of section 3, this theorem shows (in the atomic case) that the bottom of the diagram discussed in that section can be obtained independently of the tools provided by the putative existence of an "upstairs". The mathematical interest of this theorem is thus to indicate that a genuine extension into the realm of JBW-algebras is possible for the isomorphism theorem appearing in the Tomita-Takesaki theory.

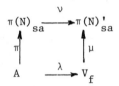

§5. THE FINITE CASE. With $a,b,c \in A$, let

$$\{abc\} = (a \circ b) \circ c + (b \circ c) \circ a - (a \circ c) \circ b.$$

A state t on A is said to be a trace if for every pair (p,q) of idempotents in A we have $\langle t;\{pqp\} - \{qpq\}\rangle = 0$. A JBW-algebra which admits a faithful normal trace is said to be finite. The following result has then been obtained [4] upon making use of the spectral theory developed by Alfsen and Schultz [2].

THEOREM B: Let t be a faithful normal trace on a finite JBW-algebra. Then there exists an order isomorphism λ from A onto V_t such that $\lambda(I) = t$. Moreover for every positive $f \in V_t$ there exists a positive element $b \in A$ such that $\langle f;a\rangle = \langle t;\{bab\}\rangle$.

The first part of this theorem shows that the isomorphism theorem can be extended further into the realm of JBW-algebras. One verifies that the Radon-Nikodym property of the second part of the theorem implies that if f is furthermore faithful then V_f and V_t are order isomorphic. We say that a JBW-algebra is of class N if given any pair (f,g) of normal positive linear functionals on A with g dominated by f, there exists a positive element $b \in A$ such that $\langle g;a\rangle = \langle f;\{bab\}\rangle$. One obtains thus (again without requiring existence of the "upstairs" of section 3):

COROLLARY C: Let f be a faithful normal state on a finite JBW-algebra A of class N. Then A is order isomorphic to V_f.

REFERENCES

1. Araki, H.: On the State Space of Quantum Mechanics, preprint, 1980.

2. Alfsen, E.M. and F. Shultz: Proc. London Math. Soc. 38 (1979) 497.

3. Emch, G.G.: Algebraic Methods in Statistical Mechanics and Quantum Field Theory, Interscience, New York, 1972.

4. King, W.P.C.: Ph.D. thesis, University of Rochester, 1980.

5. Sakai, S: C*- and W*-algebras, Springer, 1971.

6. Takesaki, M.: Tomita's Theory of Modular Operator Algebras and its Applications, Springer, 1970.

DEPARTMENT OF MATHEMATICS
UNIVERSITY OF ROCHESTER
RIVER CAMPUS STATION
ROCHESTER, NY 14627

Proceedings of Symposia in Pure Mathematics
Volume 38 (1982), Part 2

DEVELOPMENTS IN THE THEORY OF UNBOUNDED DERIVATIONS IN C*-ALGEBRAS

Shôichirô Sakai*)

§1. INTRODUCTION. The study of derivations in C*-algebras began at the fairly early stage in the young field of C*-algebras, and has always been one of the central branches in the field. It may be devided into two parts. One is for bounded derivations and the other is for unbounded derivations with dense domain. About twenty years ago, Kaplansky [94] wrote an excellent survey on derivations in which he brought together two apparently unrelated results which stimulated research on continuous derivations. One is related to quantum mechanics; in 1949, Wielandt [95] proved that the commutation relation AB-BA=1 can not be realized by bounded operators. The other is related to differentiation ; in 1947, Šilov [96] proved that if a Banach algebra A of continuous functions on the unit interval contains all infinitely differentiable functions, then A contains all n-fold differentiable functions for some n.

It is noteworthy that the observation of Kaplansky two decades ago is still applicable to the main part of recent developments in the theory of unbounded derivations, though one has to replace quantum mechanics by all of quantum physics. From Batty's and Goodman's talks, the audiences will find that the work of Šilov has strong influence on the study of unbounded derivations in commutative C*-algebras. At the early stage, mathematicians devoted their effort to the study of bounded derivations, though the works of Šilov and Wielandt had already suggested the importance of unbounded derivations. This is understandable, because bounded derivations can be more easily handled than unbounded ones so that one can expect a beautiful theory as mathematics and also knowledge on bounded derivations may contribute to the study of unbounded derivations. In fact, the study of bounded derivations is now approaching to completion. On the other hand, the study of unbounded derivations occurred much later and was initially motivated by the problem of the construction of dynamics in statistical mechanics.

Soon it became apparent that the work of Šilov [96] also has strong influence on the study of unbounded derivations in commutative C*-algebras.

1980 Mathematics Subject Classification.
*) This research is supported by N.S.F.

For bounded derivations, the main theme is when they are inner. On the other hand, for unbounded ones, it is rich in variety, because they include dynamical systems in quantum physics and differentiations in manifolds. Since Drs. Batty, Bratteli, Goodman, Herman, Jørgensen and Takai will talk about each of specialized subjects, I am going to restrict my talk to some of topics which are not covered by them.

§2. Statistical mechanics and derivations. There is a good possibility that the theory of quantum lattice systems in statistical mechanics may be well-developed within the theory of unbounded derivations in C*-algebras, although its phase transition has not been established even for the three-dimensional Heisenberg ferromagnet (for the anti-ferromagnet, it has been proved by Dyson, Lieb and Simon). In fact, many theorems in the theory of quantum lattice systems have been formulated for normal *-derivations in UHF algebras. One of the most ambitious schemes in the theory of unbounded derivations is to develop statistical mechanics within the C*-theory. Especially the generalization of phase transition theory in lattice systems is one of the most important subjects. The prospect is not necessarily gloomy, because it has succeeded in formulating the absence of phase transition in one-dimensional lattice systems with bounded surface energy in a quite general setting of UHF algebras ([1], [43], [67], [68]). In this section, I will give a brief survey on developments in the theory of derivations in UHF algebras from the point of view of quantum statistical mechanics and will state some related problems.

Let $\mathfrak{A}$ be a C*-algebra. A linear mapping δ in $\mathfrak{A}$ is said to be a derivation if it satisfies the following conditions : (1) the domain $\mathcal{D}(\delta)$ of δ is a dense subalgebra of $\mathfrak{A}$; (2) $\delta(ab) = \delta(a)b + a\delta(b)$ $(a,b \in \mathcal{D}(\delta))$. A derivation δ is said to be a *-derivation if it satisfies : (3) $a \in \mathcal{D}(\delta)$ implies $a^* \in \mathcal{D}(\delta)$ and $\delta(a^*) = \delta(a)^*$. δ is said to be closed if $x_n \in \mathcal{D}(\delta)$, $x_n \longrightarrow x$ and $\delta(x_n) \longrightarrow y$ implies $x \in \mathcal{D}(\delta)$ and $\delta(x) = y$. δ is said to be closable if $x_n \in \mathcal{D}(\delta)$, $x_n \longrightarrow 0$ and $\delta(x_n) \longrightarrow y$ implies $y = 0$. If δ is closable, then the closure $\bar{\delta}$ of δ is a closed derivation. If $\mathcal{D}(\delta) = \mathfrak{A}$, then δ is closed, so that by the closed graph theorem, it is bounded ([64]). Conversly if δ is bounded, then $\mathcal{D}(\bar{\delta}) = \mathfrak{A}$. Therefore everywhere defined derivations are equivalent to bounded derivations. There are extensive literatures on bounded derivations. We will mention some of them for future use.

Let $\mathfrak{A}$ be a C*-algebra on a Hilbert space $\mathcal{H}$ and let δ be a bounded derivation on $\mathfrak{A}$; then there is an element d in the weak closure $\bar{\mathfrak{A}}$ of $\mathfrak{A}$ such that $\delta(a) = [d,a]$ $(a \in \mathfrak{A})$. In particular, any bounded derivation of a W*-algebra is inner ([97], [98]). Let $\mathfrak{A}$ be a simple C*-algebra with identity ; then any bounded derivation is inner ([99]). In quantum physics, simple C*-algebras with identity are often appearing.

Let $\mathfrak{A}$ be a simple C*-algebra without identity and let δ be a bounded derivation on $\mathfrak{A}$; then there is an element d in the multiplier C*-algebra $M(\mathfrak{A})$ of $\mathfrak{A}$ such that $\delta(a) = [d,a]$ $(a \in \mathfrak{A})$. ([100]).

There is a complete characterization of separable C*-algebras whose all bounded derivations are defined by multipliers ([101], [102], [103])

From now on we shall discuss unbounded derivations. If $\mathcal{D}(\delta) \subsetneqq \mathfrak{A}$, then δ is not necessarily closable ([9]). Therefore it is an interesting problem to study the closability conditions of derivations. In fact, there are many literatures on the problem (cf. [92]). However, we shall mention here only one of them for future use. Let δ be a *-derivation in a C*-algebra $\mathfrak{A}$. An element x of the self-adjoint portion $\mathcal{D}(\delta)^s$ of $\mathcal{D}(\delta)$ is said to be well-behaved if there is a state ψ_x on $\mathfrak{A}$ such that

$$| \psi_x(x) | = \| x \| \quad \text{and} \quad \psi_x(\delta(x)) = 0.$$

Let $W(\delta)$ be the set of all well-behaved elements ; then it is dense in $\mathfrak{A}^s$ (the self-adjoint portion of $\mathfrak{A}$). δ is said to be well-behaved (resp. quasi well-behaved) if $W(\delta) = \mathcal{D}(\delta)^s$ (resp. the interior $W(\delta)^0$ of $W(\delta)$ in $\mathcal{D}(\delta)^s$ is dense in $\mathcal{D}(\delta)^s$).

Any (quasi) well-behaved *-derivation is closable and its closure is again (quasi) well-behaved ([50], [84]). If δ is well-behaved, then

$$\| (1 \pm \delta)(a) \| \geqq \| a \| \quad (a \in \mathcal{D}(\delta)) \quad ([60]).$$

In mathematical physics, unbounded derivations are often defined by Hamiltonians. In those cases, it is not so difficult to see that the derivations are closable. If $\mathfrak{A}$ is commutative, δ is closed and $f \in C^1(R)$ (continuously differentiable), then we can easily see that $f(a) \in \mathcal{D}(\delta)$ and $\delta(f(a)) = f'(a)\,\delta(a)$ for $a \in \mathcal{D}(\delta)$. On the other hand, for a non-commutative $\mathfrak{A}$, one has to replace $C^1(R)$ by $C^2(R)$ to get $f(a) \in \mathcal{D}(\delta)$ ([9],[52],[59]).

Here we shall mention several important examples of closable unbounded *-derivations in their simplest forms.

EXAMPLES: (a) DIFFERENTIATION. Let $C([0,1])$ be the C*-algebra of all continuous functions on the unit interval [0,1] and let $C^1([0,1])$ be the algebra of all continuously differentiable functions on [0,1]. Define $\delta(f)(t) = \frac{d}{dt} f(t)$ for $f \in C^1([0,1])$; then δ is a quasi well-behaved closed *-derivation in $C([0,1])$.

(b) ISING MODELS. Let Z be the group of all integers and $Z^n = \overbrace{Z\times..\times Z}^{n}$.

For $p \in Z^n$, let B_p be the full matrix algebra of 2×2 and let $\mathfrak{A} = \bigotimes_{p \in Z^n} B_p$ be the infinite C*-tensor product. Let $\sigma_p = \begin{pmatrix} 1 & 0 \\ 0 & -1 \end{pmatrix} \in B_p$ and let H be the total potential defined by $H = \sum_{p,q \in Z^n} -J(|p-q|)\sigma_p\sigma_q$, where $J(1) = 1$, $J(n) = 0$ $(n=0,2,3,\dots)$ and $|p-q| = \sum_{i=1}^{n} |p_i - q_i|$ with $p = (p_1,p_2,\dots,p_n)$ and $q = (q_1,q_2,\dots,q_n)$.

Let
$$\mathfrak{A}_m = \bigotimes_{p \in \Lambda_m} B_p \text{ , where } \Lambda_m = \{p \in Z^n \ ; \ |p| \leq m\} \ ;$$
then $\mathfrak{A}_m$ can be canonically embedded into $\mathfrak{A}$ so that $1 \in \mathfrak{A}_1 \subset \mathfrak{A}_2 \subset \mathfrak{A}_3 \subset \cdots \subset \mathfrak{A}_m \cdots \subset \mathfrak{A}$ and $\bigcup_{m=1}^{\infty} \mathfrak{A}_m$ is dense in $\mathfrak{A}$. Let $\mathcal{D}(\delta) = \bigcup_{m=1}^{\infty} \mathfrak{A}_m$ and define $\delta(a) = i[H,a]$ $(a \in \mathcal{D}(\delta))$.

If $a \in \mathfrak{A}_m$, then $i[H,a] = i[\sum_{p,q \in \Lambda_{m+1}} - J(|p-q|)\sigma_p\sigma_q , a]$. Hence δ is well-defined. It is easily seen that δ is a well-bahaved *-derivation.

(c) HEISENBERG MODELS. In the same notations with the (b), let
$$\sigma_{p,1} = \begin{pmatrix} 1 & 0 \\ 0 & -1 \end{pmatrix}, \ \sigma_{p,2} = \begin{pmatrix} 0 & 1 \\ 1 & 0 \end{pmatrix}, \ \sigma_{p,3} = \begin{pmatrix} 0 & i \\ -i & 0 \end{pmatrix} \in B_p \text{ and let}$$
$H = \sum_{b,q \in Z^n} -J(|p-q|) \sum_{j=1}^{3} \sigma_{p,j}\sigma_{q,j}$ and define $\delta(a) = i[H,a]$ $(a \in \mathcal{D}(\delta))$, where $\mathcal{D}(\delta) = \bigcup_{m=1}^{\infty} \mathfrak{A}_m$. Then δ is a well-behaved *-derivation.

(d) QUASI-FREE DERIVATIONS. Let $\mathfrak{A}(\mathcal{H})$ be the canonical anticommutation relation algebra over a separable Hilbert space $\mathcal{H}$ ([58]). Let S be a symmetric operator in $\mathcal{H}$ and put $\delta_{iS}\, a(f) = ia(Sf)$ $(f \in \mathcal{D}(S))$, where $f \longrightarrow a(f)$ is the basic isometric linear mapping of $\mathcal{H}$ into $\mathfrak{A}(\mathcal{H})$ which defines $\mathfrak{A}(\mathcal{H})$ canonically. Then δ_{iS} extends (uniquely) to a *-derivation, defined on a *-subalgebra $\mathfrak{A}_0(\mathcal{D}(S))$ of $\mathfrak{A}(\mathcal{H})$ generated by $\{a(f) \mid f \in \mathcal{D}(S)\}$. If $\mathcal{K}(\subset \mathcal{D}(S))$ is a finite-dimensional subspace of $\mathcal{H}$, then $\mathfrak{A}(\mathcal{K})$ is a finite type I subfactor of $\mathfrak{A}(\mathcal{H})$ and there is a self-adjoint element h in $\mathfrak{A}(\mathcal{H})$ such that $\delta_{iS}(x) = i[h,x]$ $(x \in \mathfrak{A}(\mathcal{K}))$. From this, we can easily conclude that δ_{iS} is well-behaved.

We shall introduce a class of C*-algebras (UHF algebras) which is important for quantum lattice systems and Fermion field theory. A C*-algebra $\mathfrak{A}$ is said to be a uniformly hyperfinite C*-algebra (UHF) if there is an increasing sequence $\{\mathfrak{A}_n\}$ of full matrix subalgebras (finite type I subfactors) in $\mathfrak{A}$ such that $1 \in \mathfrak{A}_1 \subset \mathfrak{A}_2 \subset \cdots \subset \mathfrak{A}_n \subset \cdots$ and $\bigcup_{n=1}^{\infty} \mathfrak{A}_n$ is dense in $\mathfrak{A}$. The C*-algebras $\mathfrak{A}$'s in the above examples (b) ~ (d) are UHF algebras. Let $\{\rho(t)\ ;\ t \in R\}$ be a strongly continuous one-parameter group of *-automorphisms on $\mathfrak{A}$. $\{\mathfrak{A}, \rho(t)\ (t \in R)\}$ is called a C*-dynamics. Let $\rho(t) = \exp t\delta$; then δ is a well-behaved closed *-derivation in $\mathfrak{A}$ and $\mathcal{D}(\delta)$ consists of all elements a for which $\lim_{t\to 0} \frac{\rho(t)a-a}{t}$ exist ; moreover $\delta(a) = \lim_{t\to 0} \frac{\rho(t)(a)-a}{t}$ $(a \in \mathcal{D}(\delta))$ and $(1 \pm \delta)\mathcal{D}(\delta) = \mathfrak{A}$. Let $\{\rho(t)\ ;\ t \in R\}$, $\{\rho_n(t)\ ;\ t \in R\}$ be a family of one-parameter groups of *-automorphisms on $\mathfrak{A}$. Let $\rho(t) = \exp t\delta$, $\rho_n(t) = \exp t\delta_n$. $\{\rho(t)\ ;\ t \in R\}$ is said to be a strong limit of $\{\rho_n(t)\ ;\ t \in R\}$ if $\| \rho(t)(a) - \rho_n(t)(a) \| \to 0$ uniformly on every compact subset of R for each $a \in \mathfrak{A}$. By Baire category theorem, "strong lim $\rho_n(t) = \rho(t)$" is equivalent to "$\| \rho(t)(a) - \rho_n(t)(a) \| \longrightarrow 0$ (simple convergence) for $t \in R$ and $a \in \mathfrak{A}$". By Kato-Trotter theorem, strong lim $\rho_n(t) = \rho(t) \Leftrightarrow (1 \pm \delta_n)^{-1} \longrightarrow (1 \pm \delta)^{-1}$ (strongly) $\Leftrightarrow (1 - \delta_n)^{-1} \longrightarrow (1 - \delta)^{-1}$ (strongly), because of the isometry of $\rho(t)$ $(t \in R)$.

$\{\rho(t)\ ;\ t \in R\}$ is said to be approximately inner if there is a sequence $\{\rho_n(t)\ ;\ t \in R\}$ of uniformly continuous one-parameter groups of *-automorphisms on $\mathfrak{A}$ such that strong lim $\rho_n(t) = \rho(t)$. Since a UHF algebra is a simple C*-algebra with identity and δ_n is a bounded derivation, there is a self-adjoint element h_n in $\mathfrak{A}$ such that $\delta_n = \delta_{ih_n}$, where $\delta_{ih_n}(a) = i[h_n, a]$ $(a \in \mathfrak{A})$.

In the following, we shall show that all C*-dynamics appearing in quantum lattice systems and Fermion field theory are approximately inner. Let δ be a *-derivation in a UHF algebra $\mathfrak{A}$. δ is said to be normal if there is an increasing sequence $\{\mathfrak{A}_n\}$ of finite type I subfactors in $\mathfrak{A}$ such that $1 \in \mathfrak{A}_1 \subset \mathfrak{A}_2 \subset \cdots \subset \mathfrak{A}_n \subset \cdots$, $\bigcup_{n=1}^{\infty} \mathfrak{A}_n$ is dense in $\mathfrak{A}$ and the domain $\mathcal{D}(\delta)$ of δ is $\bigcup_{n=1}^{\infty} \mathfrak{A}_n$. Let δ be a normal *-derivation in $\mathfrak{A}$ and let $\{e_{p,q}^n\ ;\ p,q = 1,2,\ldots, m_n\}$ be a matrix unit of $\mathfrak{A}_n$. Set $ih_n = \sum_{p=1}^{m_n} \delta(e_{p,1}^n)e_{1,p}^n$; then we have $\delta(a) = i[h_n, a]$ $(a \in \mathfrak{A}_n)$ and $h_n^* = h_n$ $(n=1,2,\ldots,)$. Therefore a normal *-derivation is well-behaved and so

$$\| (1 \pm \delta)(a) \| \geqq \| a \| \quad (a \in \mathcal{D}(\delta)).$$

Let $\bar{\delta}$ be the closure of δ. If $\bar{\delta}$ is the generator of a strongly continuous one-parameter group of *-automorphisms on $\mathfrak{A}$, δ is said to be a pregenerator. From the theory of semi-groups, δ is a pregenerator if and only if $(1 \pm \delta)\mathcal{D}(\delta)$ are dense in $\mathfrak{A}$ (note $\|(1 \pm \delta)(a)\| \geqq \|a\|$ $(a \in \mathcal{D}(\delta))$).

In quantum lattice systems, it is not so easy to check the density of $(1 \pm \delta)\mathcal{D}(\delta)$ directly ; instead, the analytic method is often used. Suppose that $\delta(\mathcal{D}(\delta)) \subset \mathcal{D}(\delta)$ (finite range interactions) ; then one can consider the iterations δ^n (n=1,2,3,...) on $\mathcal{D}(\delta)$. An element a in $\mathcal{D}(\delta)$ is said to be analytic if there exists a positive number r such that

$\sum_{n=0}^{\infty} \frac{\|\delta^n(a)\|}{n!} r^n < +\infty$. Let $A(\delta)$ be the set of all analytic elements in $\mathcal{D}(\delta)$. If $A(\delta)$ is dense in $\mathfrak{A}$, then we can easily see that $(1 \pm \delta)\mathcal{D}(\delta)$ are dense in $\mathfrak{A}$, so that δ is a pregenerator. ([10], [60])

If a quantum lattice system has a translation-invariant, finite range interaction, then $A(\delta) = \mathcal{D}(\delta)$ and so δ is a pregenerator ([63]). In particular, the examples (b) and (c) are pregenerators. For general normal *-derivations, we will use bounded perturbations.

PROPOSITION 2.1 ([24], [84]). Let δ be a normal *-derivation in a UHF algebra $\mathfrak{A}$ with $\mathcal{D}(\delta) = \bigcup_{n=1}^{\infty} \mathfrak{A}_n$. Then for $\varepsilon > 0$, there is a normal *-derivation δ_ε in $\mathfrak{A}$ such that $\mathcal{D}(\delta_\varepsilon) = \mathcal{D}(\delta)$, $\delta_\varepsilon(\mathcal{D}(\delta)) \subset \mathcal{D}(\delta)$ and $\delta - \delta_\varepsilon$ is a bounded *-derivation with $\|\delta - \delta_\varepsilon\| < \varepsilon$

Proof. Let τ be the unique tracial state on $\mathfrak{A}$ and let p_n be the canonical conditional expectation of $\mathfrak{A}$ onto $\mathfrak{A}_n$ defined by $\tau(xy) = \tau(p_n(x)y)$ for $x \in \mathfrak{A}$ and $y \in \mathfrak{A}_n$. Let $\delta(a) = i[h_n, a]$ $(a \in \mathfrak{A}_n)$. Take P_{n_1} with $n_1 \geqslant 1$ such that $\| h_1 - P_{n_1}(h_1) \| < \varepsilon/2^2$ and P_{n_2} with $n_2 > n_1$ such that $\| (h_2 - h_1) - P_{n_2}(h_2 - h_1)\| < \varepsilon/2^3$. Continuing this process, take P_{n_j} with $n_j > n_{j-1}$ such that $\| (h_j - h_{j-1}) - P_{n_j}(h_j - h_{j-1}) \| < \varepsilon/2^{j+1}$, (j=3,4,...). Put $\ell_j = (h_j - h_{j-1}) - P_{n_j}(h_j - h_{j-1})$; then $\sum_{j=1}^{\infty} \| \ell_j \| < \varepsilon/2$, where $\ell_1 = h_1 - P_{n_1}(h_1)$, and so $\sum_{j=1}^{\infty} \ell_j = d$ is a self-adjoint element of $\mathfrak{A}$. For $a \in \mathfrak{A}_{j_0}$, $[i \sum_{j=1}^{\infty} \ell_j, a] = \sum_{j=1}^{\infty} i[(h_j - h_{j-1}) - P_{n_j}(h_j - h_{j-1}), a] = i \sum_{j=1}^{j_0} [(h_j - h_{j-1}), a] - \sum_{j=1}^{\infty} [P_{n_j}(h_j - h_{j-1}), a]$.

For $j > j_0$, $[P_{n_j}(h_j - h_{j-1}), a] = (P_{n_j}(h_j - h_{j-1}))a - a(P_{n_j}(h_j - h_{j-1}))$

$= P_{n_j}((h_j - h_{j-1})a - a(h_j - h_{j-1})) = P_{n_j}[h_j - h_{j-1}, a] = 0.$

Hence $i[\sum_{j=1}^{\infty} \ell_j, a] = i[h_{j_0}, a] - i[P_{n_j}(h_{j_0}), a] = i[h_j - P_{n_{j_0}}(h_{j_0}), a].$

Now put $\delta_\varepsilon = \delta - \delta_{id}$; then $\mathcal{D}(\delta_\varepsilon) = \bigcup_{n=1}^{\infty} \mathfrak{A}_n$ and $\delta_\varepsilon(a) = i[P_{n_{j_0}}(h_{j_0}), a]$ $(a \in \mathfrak{A}_{j_0})$, and so $\delta_\varepsilon(\mathcal{D}(\delta)) \subset \mathcal{D}(\delta)$ and $\|\delta - \delta_\varepsilon\| < \varepsilon$. This completes the proof.

Since the property of "pregenerator" is invariant under bounded perturbations, it is enough to study the generation problem for normal *-derivations under the assumption of $\delta(\mathcal{D}(\delta)) \subset \mathcal{D}(\delta)$. The audiences will find many interesting results on the generation problem in Bratteli-Jørgensen's talk. The speaker feels, the generation problem for normal *-derivations has not thoroughly studied yet. This is one important problem in the theory. Next suppose that a normal *-derivation δ in $\mathfrak{A}$ is a pregenerator ; then

$$\|(1 \pm \delta_{ih_n})^{-1}(1 \pm \delta)a - (1 \pm \delta)^{-1}(1 \pm \delta)a\|$$

$$\leq \|(1 \pm \delta_{ih_n})^{-1}\{(1 \pm \delta)a - (1 \pm \delta_{ih_n})\}a\|$$

$$\leq \|(\delta - \delta_{ih_n})(a)\| = 0 \qquad (a \in \mathfrak{A}_n).$$

Hence $(1 \pm \delta_{ih_n})^{-1} \longrightarrow (1 \pm \delta)^{-1}$ (strongly) so that $\exp t\bar{\delta} = \text{strong lim} \exp t\delta_{ih_n}$ - namely $\{\exp t\bar{\delta}\}$ is approximately inner. Therefore it is approximately inner whenever a quantum lattice system defines a C*-dynamics.

Next, we shall consider quasi free derivations in the example (d). Let H be a self-adjoint operator in $\mathcal{H}$ and let $H = H_1 + T$ be the Weyl's decomposition, where H_1 is a diagonalizable self-adjoint operator and T is of Hilbert-Schmidt class ; then $H_1 = \sum_{i=-\infty}^{\infty} \lambda_i E_i$, $\lambda_i \in R$, $E_i E_j = 0 (i \neq j)$. $E_i^2 = E_i$, $E_i^* = E_i$ and $\dim(E_i) = 1$ $(i.j=1,2,\ldots)$. Let $P_n = \sum_{i=-n}^{n} E_i$ and $\mathcal{H}_0 = \bigcup_{n=1}^{\infty} P_n\mathcal{H}$. Let δ be the restriction of δ_{iH} to $\mathfrak{A}_0(\mathcal{H}_0)$; then δ is a normal *-derivation. Since $\overline{H|_{\mathcal{H}}} = H$, $\bar{\delta} = \bar{\delta}_{iH}$, and moreover $\overline{(1 \pm \delta)\mathcal{D}(\delta)} \supset (1 \pm iH)a(\mathcal{D}(H)) = a(\mathcal{H})$; hence $\overline{(1 \pm \delta)\mathcal{D}(\delta)} = \mathfrak{A}(\mathcal{H})$, and so δ_{iH} is a pregenerator and $\{\exp t\bar{\delta}_{iH}\}$ is approximately inner. More generally, if S is a symmetric operator, we can easily show that δ_{iS}

has a normal *-derivation δ as a core (i.e. $\bar{\delta} = \bar{\delta}_{iS}$), by using the polar decomposition of S.

If S has no self-adjoint extension, then δ_{iS} does not necessarily have a generator extension. The audiences will find some results on the extension problem in Bratteli-Jøgensen's talk.

Let us introduce an important class of normal *-derivations, which have always generator extensions. A normal *-derivation δ is said to be commutative if one can choose a sequence (k_n) in $\mathfrak{A}$ such that $k_n{}^* = k_n$, $\delta(a) = i[k_n, a]$ $(a \in \mathfrak{A}_n)$, $k_n k_m = k_m k_n$ (m,n=1,2,3,...). All derivations defined by classical lattice systems and Ising models are commutative (see the example (b) and [5], [63]).

PROPOSITION 2.2 ([66]). Suppose that δ is a commutative normal *-derivation and let $\mathcal{D}_n$ be the C*-subalgebra of $\mathfrak{A}$ generated by $\mathfrak{A}_n$ and $k_1, k_2, \ldots, k_{n-1}, k_n$; then δ can be extended to a pregenerator δ_1 with

$$\mathcal{D}(\delta) = \bigcup_{n=1}^{\infty} \mathcal{D}_n \quad \text{and} \quad (\exp t\bar{\delta}_1)(a) = \exp t\delta_{ik_n}(a) \quad (a \in \mathcal{D}_n \ ; \ n=1,2,\ldots).$$

In particular, $\{\exp t\bar{\delta}_1\}$ is approximately inner.

Proof For $a \in \mathfrak{A}_n$, $i[k_m, k_n a] = i[k_m, k_n] a + ik_n [k_m, a]$

$= k_n\delta(a)$ $(m \geq n)$. Analogously $i[k_m, ak_n] = \delta(a)k_n$ $(m \geq n)$.

From this we can easily conclude that

$$i[k_m, b] = i[k_n, b] \quad (m \geq n) \quad (b \in \mathcal{D}_n).$$

Since $[ik_n, \mathcal{D}_n] \subset \mathcal{D}_n$, $\delta_{ik_m}^p(b) = \delta_{ik_n}^p(b)$ (p=1,2,3,...) for $b \in \mathcal{D}_n$; hence $\exp t\delta_{ik_m}(b) = \exp t\delta_{ik_n}(b)$ $(b \in \mathcal{D}_n)$.

Define $\delta_1(x) = \lim_{n\to\infty} i[k_n, x]$ for $x \in \bigcup_{n=1}^{\infty} \mathcal{D}_n$.

Since $(1 \pm \delta_1) \bigcup_{n=1}^{\infty} \mathcal{D}_n = \bigcup_{n=1}^{\infty} \mathcal{D}_n$, δ_1 is a pregenerator, and so $(\exp t\bar{\delta}_1)(b) = \exp t\delta_{ik_n}(b)$ $(b \in \mathcal{D}_n)$ and $\exp t\bar{\delta}_1 = \text{strong} \lim \exp i\delta_{ik_n}$. This completes the proof.

This proposition is so powerful that various results for commutative normal *-derivations are much sharper than general ones. (see later considerations).

There is a relation between commutative normal *-derivations and general ones as follows.

PROPOSITION 2.3 ([24], [84]). Let δ be a normal *-derivation with $\delta(\mathcal{D}(\delta)) \subset \mathcal{D}(\delta)$; then there are an increasing sequence $\{\mathfrak{A}_n\}$ of type I subfactors in $\mathfrak{A}$ and two normal *-derivations δ_1, δ_2 such that $\bigcup_{n=1}^{\infty} \mathfrak{A}_n = \mathcal{D}(\delta)$, $\mathcal{D}(\delta) = \mathcal{D}(\delta_1) = \mathcal{D}(\delta_2)$, $\delta_1(\mathfrak{A}_{2n}) \subset \mathfrak{A}_{2n}$, $\delta_2(\mathfrak{A}_{2n+1}) \subset \mathfrak{A}_{2n+1}$ and $\delta = \delta_1 + \delta_2$. In particular, δ_1 , δ_2 are commutative.

Commutative normal *-derivations are generalizations of classical lattice systems, and non-commutative normal *-derivations are generalizations of quantum lattice systems. Therefore it is a quite interesting problem how much the study of normal *-derivations can be reduced to the one of commutative normal *-derivation.

We have seen that all C*-dynamics appearing in quantum lattice systems and Fermion field theory are approximately inner. On the other hand, if a C*-dynamics $\{ \mathfrak{A}, \rho(t) \ (t \in R) \}$ ($\mathfrak{A}$, UHF algebra) is approximately inner, then it has many physical properties (see later considerations), so that a physical theory can be developed for approximately inner dynamics. Therefore the following conjecture is very important.

The Powers-Sakai conjecture ([60], [65]). Any C*-dynamics $\{\mathfrak{A}, \rho(t) \ (t \in R)\}$ with a UHF algebra $\mathfrak{A}$ is approximately inner.

Let $\rho(t) = \exp t\delta$; then the following problem is important.

The core problem ([65]). Does every generator δ have a normal *-derivation as a core ?

The affirmative solution to the core problem implies the affirmative solution to the conjecture. The affirmative solution to the core problem would be more desirable than the affirmative solution to the conjecture from the point of view of quantum lattice systems.

In the following, we shall give some considerations for the conjecture and the core problem. Let $\rho(t) = \exp t\delta$. An element a in $\mathfrak{A}$ is said to be analytic if there is a positive number r such that

$$\sum_{n=0}^{\infty} \frac{\|\delta^n(a)\|}{n!} r^n < +\infty .$$

Let $A(\delta)$ be the set of all analytic elements in $\mathfrak{A}$.

THEOREM 2.4 ([65]). There is an increasing sequence $\{\mathfrak{A}_n\}$ of finite type I subfactors in $\mathfrak{A}$ such that $1 \in \mathfrak{A}_1 \subset \mathfrak{A}_2 \subset \cdots \subset \mathfrak{A}_n \subset \cdots$, $\bigcup_{n=1}^{\infty} \mathfrak{A}_n \subset A(\delta)$ and $\bigcup_{n=1}^{\infty} \mathfrak{A}_n$ is dense in $\mathfrak{A}$.

By the previous considerations, there is a sequence (h_n) in $\mathfrak{A}$ such that

$$h_n^* = h_n, \ \delta(a) = i[h_n, a] \quad (a \in \mathfrak{A}_n; n=1,2,3,\ldots).$$

If $(1 - \delta) \bigcup_{n=1}^{\infty} \mathfrak{A}_n$ is dense in $\mathfrak{A}$, then δ is the closure of restriction of δ to $\bigcup_{n=1}^{\infty} \mathfrak{A}_n$ and so has a normal *-derivation as a core ; hence $\exp t\delta = \text{strong} \lim \exp t\delta_{ih_n}$ - namely $\{\exp t\delta\}$ is approximately inner. However, from the considerations on quasi-free derivations, one can easily see that the density of $(1 - \delta)\mathcal{D}(\delta)$ can not be expected under a general selection of $\bigcup_{n=1}^{\infty} \mathfrak{A}_n$. An important, difficult problem is whether or not one can choose $\{\mathfrak{A}_n\}$ such that $(1 - \delta) \bigcup_{n=1}^{\infty} \mathfrak{A}_n$ is dense in $\mathfrak{A}$.

On the other hand, if the conjecture is true, then a general $\{\mathfrak{A}_n\}$ may have all necessary informations to prove it in a sense. In fact, if $(1 - \delta_{i\ell_n})^{-1} \longrightarrow (1 - \delta)^{-1}$ (strongly) for some (ℓ_n), then by the density of $\bigcup_{n-1}^{\infty} \mathfrak{A}_n$, there is a sequence (s_n) of self-adjoint elements in $\bigcup_{n=1}^{\infty} \mathfrak{A}_n$ such that $\| \ell_n - s_n \| < 1/n$. Then,

$$\begin{aligned}
\| \{(1-\delta_{is_n})^{-1} - (1-\delta)^{-1}\}(a) \| &\leqq \|\{(1-\delta_{is_n})^{-1}-(1-\delta_{i\ell_n})^{-1}\} (a)\| \\
&\quad + \| \{(1- \delta_{i\ell_n})^{-1} - (1-\delta)^{-1}\} (a) \| \\
&\leqq \| \{(\delta_{is_n}-\delta_{i\ell_n})\}(1-\delta_{i\ell_n})^{-1}a \| + \| \{(1-\delta_{i\ell_n})^{-1}-(1-\delta)^{-1}\}(a) \| \\
&\leqq \frac{2\|a\|}{n} + \| \{(1-\delta_{i\ell_n})^{-1}-(1-\delta)^{-1}\}(a) \| \longrightarrow 0 .
\end{aligned}$$

Hence there is a sequence (s_n) in $\bigcup_{n=1}^{\infty} \mathfrak{A}_n$ such that

$$(1 - \delta_{is_n})^{-1} \longrightarrow (1 - \delta)^{-1} \quad \text{(strongly)}.$$

Moreover

PROPOSITION 2.5. $(1 - \delta_{i\ell_n})^{-1} \longrightarrow (1 - \delta)^{-1}$ (strongly) if and only if for each $a \in \mathcal{D}(\delta)$, there is a sequence (a_n) in $\mathfrak{A}$ such that $a_n \longrightarrow a$ and $\delta_{i\ell_n}(a_n) \longrightarrow \delta(a)$

Proof. Suppose that $(1-\delta_{i\ell_n})^{-1} \longrightarrow (1-\delta)^{-1}$; then $\delta_{i\ell_n}(1-\delta_{i\ell_n})^{-1} = (1-\delta_{i\ell_n})^{-1} - 1 \longrightarrow \delta(1-\delta)^{-1}$ (strongly).
For $a \in \mathcal{D}(\delta)$, take $b \in \mathfrak{A}$ such that $a = (1-\delta)^{-1} b$ (note $(1-\delta)^{-1}\mathfrak{A} = \mathcal{D}(\delta)$), and put $a_n = (1-\delta_{i\ell_n})^{-1} b$; then $a_n \longrightarrow a$ and $\delta_{i\ell_n}(a_n) \longrightarrow \delta(a)$.

Conversely, for $a \in \mathcal{D}(\delta)$,

$$\| (1-\delta_{i\ell_n})^{-1}(1-\delta)\, a - (1-\delta)^{-1}(1-\delta)\, a \|$$
$$= \| (1-\delta_{i\ell_n})^{-1}(1-\delta)\, a - a \|$$
$$\leqq \| (1-\delta_{i\ell_n})^{-1}(1-\delta_{i\ell_n})a_n - a \|$$
$$+ \| (1-\delta_{i\ell_n})^{-1} (\delta_{i\ell_n}(a_n) - \delta(a) + a - a_n) \|$$
$$\leqq 2 \| a - a_n \| + \| \delta_{i\ell_n}(a_n) - \delta(a) \| \longrightarrow 0.$$

Since $(1-\delta)\mathcal{D}(\delta) = \mathfrak{A}$, $(1-\delta_{i\ell_n})^{-1} \longrightarrow (1-\delta)^{-1}$ (strongly)

This completes the proof.

Now take $x_n \in \bigcup_{r=1}^{\infty} \mathfrak{A}_r$ such that $\| x_n - a_n \| < \dfrac{1}{n\{\|\ell_n\| + 1\}}$; then

$$\| a - x_n \| \longrightarrow 0 \text{ and } \| \delta(a) - \delta_{i\ell_n}(x_n)\| \leqq \| \delta(a) - \delta_{i\ell_n}(a_n) \|$$
$$+ \| \delta_{i\ell_n}(x_n) - \delta_{i\ell_n}(a_n) \| \longrightarrow 0.$$ Therefore we have

PROPOSITION 2.6. $\{ \exp t\delta \}$ is approximately inner if and only if for a general increasing sequence $\{ \mathfrak{A}_n \}$ of finite type I subfactors in $\mathfrak{A}$ with $\overline{\bigcup_{n=1}^{\infty} \mathfrak{A}_n} = \mathfrak{A}$, there is a sequence (s_n) of self-adjoint elements in $\bigcup_{n=1}^{\infty} \mathfrak{A}_n$ such that for each $a \in \mathcal{D}(\delta)$, there exists a sequence (a_n) in $\bigcup_{n=1}^{\infty} \mathfrak{A}_n$ with $a_n \longrightarrow a$ and $\delta_{is_n}(a_n) \longrightarrow \delta(a)$.

However this proposition is not so useful, because it does not suggest how to construct (s_n) and (a_n). In the following, we shall formulate a problem to attak the conjecture.

Let $\mathcal{D} = \bigcup_{n=1}^{\infty} \mathfrak{A}_n$, where $\{ \mathfrak{A}_n \}$ is the one given in Theorem 2.4 ; since $\mathcal{D} \subset A(\delta)$, $\delta^n(\mathcal{D}) \subset A(\delta)$ $(n=1,2,3,\ldots)$. Let $\mathcal{C}$ be the linear subspace of $\mathfrak{A}$ spanned by $\bigcup_{n=1}^{\infty} \delta^n(\mathcal{D})$ $(\delta^0(\mathcal{D}) = \mathcal{D})$.

PROPOSITION 2.7. $(1 - \delta)\mathcal{C}$ is dense in $\mathfrak{A}$.

Proof. Suppose that $(1 - \delta)\mathcal{C}$ is not dense in $\mathfrak{A}$; then there is an element f in the dual $\mathfrak{A}^*$ of $\mathfrak{A}$ such that

$f((1 - \delta)\mathcal{C}) = 0$; $f(x) = f(\delta(x))$ $(x \in \mathcal{C})$ and so $f(a) = f(\delta^m(a))$

$(a \in \bigcup_{n=1}^{\infty} \mathfrak{A}_m$; $m=1,2,3,\ldots)$. Since $a \in A(\delta)$, there is an $r > 0$ such that

$$\sum_{m=0}^{\infty} \frac{\|\delta^m(a)\|}{m!} r^m < +\infty .$$

$$f(\rho(t)(a)) = \sum_{m=0}^{\infty} \frac{f(\delta^m(a))}{m!} t^m = (\exp t)f(a) \quad (|t| < r).$$

$t \longrightarrow f(\rho(t)(a))$; $t \longrightarrow (\exp t)f(a)$ are real analytic on R ; hence

$$f(\rho(t)(a)) = (\exp t)f(a) \ (t \in R \ ; \ a \in \bigcup_{n=1}^{\infty} \mathfrak{A}_n).$$

$|f(\rho(t)(a))| \leqq \|f\| \ \|a\|$; hence $f(a) = 0$ and so $f = 0$.

This completes the proof.

PROPOSITION 2.8. Let $\{k_n\}$ be a sequence of self-adjoint elements in $\mathfrak{A}$ and suppose that for each $a \in \mathcal{C}$, there is a sequence $\{a_n\}$ of elements in $\mathfrak{A}$ such that

$a_n \longrightarrow a$ and $\delta_{ik_n}(a_n) \longrightarrow \delta(a)$; then $\exp t\delta = \text{strong lim} \exp t\delta_{ik_n}$.

Proof. $\| (1 - \delta_{ik_n})^{-1} (1 - \delta)(a) - (1 - \delta)^{-1}(1 - \delta)\, a \|$

$$= \| (1 - \delta_{ik_n})^{-1} (1 - \delta)(a) - a \|$$

$$= \| (1 - \delta_{ik_n})^{-1} \{(1 - \delta_{ik_n})(a_n) + (1 - \delta)(a) - (1 - \delta_{ik_n})(a_n)\} - a \|$$

$$= \| a_n - a + (1 - \delta_{ik_n})^{-1} \{ a - a_n + \delta_{ik_n}(a_n) - \delta(a) \} \|$$

$$\leqq 2 \| a_n - a \| + \| \delta_{ikn}(a_n) - \delta(a) \| \longrightarrow 0.$$

Hence by Proposition 2.7, $\exp t\delta = \text{strong lim} \exp t\delta_{ik_n}$ $(t \geqslant 0)$.

$$\| (\exp - t\delta)(x) - (\exp - t\delta_{ik_n})(x) \| = \| \exp t\delta_{ik_n} (\exp(-t\delta))(x) - x \|$$

$$\longrightarrow \| (\exp t\delta)(\exp - t\delta)(x) - x \| = 0 \quad (t \geqslant 0,\ a \in \mathfrak{A}).$$

Hence, strongly lim $\exp t\delta_{ik_n} = \exp t\delta$ $(t \in R)$.

This completes the proof.

COROLLARY 2.9. If there is a sequence $\{k_n\}$ of self-adjoint elements in $\mathfrak{A}$ such that for $a \in \bigcup_{n=1}^{\infty} \mathfrak{A}_n$, $\delta^m(a) = \lim_{n\to\infty} \delta^m_{ik_n}(a)$ $(m=1,2,\ldots,)$, then $\exp t\delta = \text{strong} \lim \exp t\delta_{ik_n}$.

Proof. $\delta^m_{ik_n}(a) \longrightarrow \delta^m(a)$ and $\delta_{ik_n}(\delta^m_{ik_n}(a)) \longrightarrow \delta(\delta^m(a))$ $(m=0,1,2,\ldots)$. Hence for $a \in \mathscr{C}$ there is a sequence of elements $\{a_n\}$ in $\mathfrak{A}$ such that $a_n \longrightarrow a$ and $\delta_{ik_n}(a_n) \longrightarrow \delta(a)$. By Proposition 2.8, strong $\lim \exp t\delta_{ik_n} = \exp t\delta$. This completes the proof.

An important fact is that all generators considered in the examples (b) $\sim$ (d) satisfy the assumption of the Corollary. Therefore it would be an interesting problem whether or not the assumption of the corollary is always true. We have seen that the assumption of the corollary is true for $m = 1$. To attack the problem, the following remark might be useful (cf. [65]): A mapping of $\mathfrak{D}(\delta)$ into $\mathfrak{D}(\delta) \otimes B_{m+1}$ (B_{m+1} is the full matrix algebra of $(m+1) \times (m+1)$) :

$$a \longrightarrow \begin{pmatrix} a & \delta(a) & \cdots & \frac{\delta^m(a)}{m!} \\ 0 & a & \delta(a) \cdots & \frac{\delta^{m-1}(a)}{(m-1)!} \\ 0 & 0 & a \cdots & \delta(a) \\ \vdots & \vdots & \vdots & \vdots \\ 0 & 0 & 0 \cdots & a \end{pmatrix}$$

is *-isomorphic.

Now we shall discuss physical states on the C*-dynamics $\{\mathfrak{A}, \rho(t)(t \in R)\}$ ($\mathfrak{A}$, UHF algebra). A state ψ on $\mathfrak{A}$ is said to be a ground state if

$$-i\psi(a^*\delta(a)) \geqslant 0 \quad \text{for} \quad a \in \mathfrak{D}(\delta), \text{ where } \rho(t) = \exp t\delta .$$

A ground state is an abstraction of the state of the lowest energy. A state $\psi_\beta(-\infty < \beta < \infty)$ on $\mathfrak{A}$ is said to be a KMS state at inverse temperature β, if for $a,b \in \mathfrak{A}$, there is a continuous bounded function $F_{a,b}$ on the strip $0 \leq I_m(z) \leq \beta$ (or $\beta \leq I_m(z) \leq 0$) in the complex plane such that $F_{a,b}$ is analytic on $0 < I_m(z) < \beta$ (or $\beta < I_m(z) < 0$) and $F_{a,b}(t) = \psi_\beta(a\rho(t)(b))$ and $F_{a,b}(t + i\beta) = \psi_\beta(\rho(t)(b)a)$.

A KMS state is an abstraction of the state of equilibrium ([28]). If $\rho(t) = \exp t\delta_{ih}$ $(h = h^*, h \in \mathfrak{A})$, then we may assume that $h \geqq 0$ and h is not invertible ; then there is a pure state ψ on $\mathfrak{A}$ such that $\psi(h) = 0$.

One can easily see that ψ is a ground state for $\{ \rho(t) \}$.

Let $\psi_\beta(x) = \dfrac{\tau(xe^{-\beta h})}{\tau(e^{-\beta h})}$ $(x \in \mathfrak{A})$, where τ is the unique tracial state on $\mathfrak{A}$; then ψ_β is a KMS state at inverse temperature β for $\{ \rho(t) \}$. Furthermore, in this case, a KMS state at β is unique. Next suppose that $\{ \rho(t) \}$ is approximately inner and $(1-\delta_{ih_n})^{-1} \longrightarrow (1 - \delta)^{-1}$ (strongly). Let ψ_n be the ground state for $\{ \exp t\delta_{ih_n} \}$ constructed above and let ψ be an accumulation point of $\{ \psi_n \}$ in the state space of $\mathfrak{A}$; then ψ is a ground state for $\{ \exp t\delta \}$ ([60]). Let $\psi_{\beta,n}$ be the KMS state for $\{ \exp t\delta_{ih_n} \}$ at β constructed above and let ψ_β be an accumulation of $\{ \psi_{\beta,n} \}$ in the state space of $\mathfrak{A}$; then ψ_β is a KMS state at β for $\{ \exp t\delta \}$ ([60]). Namely we have the following theorem

THEOREM 2.10. ([60]). Let $\{ \mathfrak{A}, \rho(t)\ (t \in R)\}$ be an approximately inner C*-dynamics with a UHF algebra $\mathfrak{A}$; then $\{ \rho(t) \}$ has at least one ground state and at least one KMS state at each inverse temperature $\beta(-\infty < \beta < +\infty)$.

PHASE TRANSITION. Let $\{ \mathfrak{A}, \rho(t)\ (t \in R) \}$ be an approximately inner C*-dynamics with a UHF algebra $\mathfrak{A}$. For $\beta(-\infty < \beta < +\infty)$, $\{\rho(t)\}$ is said to have no phase transition at inverse temperature β if it has a unique KMS state at β; $\{\rho(t)\}$ is said to have phase transition at inverse temperature β, if it has at least two KMS states at β.

It is known that Ising model $(n \geq 2)$ has phase transition at some $\beta(> 0)$ ([63]) and Heisenberg antiferromagnet $(n \geq 3)$ has phase transition at some β (> 0) - i.e. Heisenberg model $(m \geq 3)$ in the example (c) has phase transition at some $\beta(< 0)$. (Dyson-Lieb-Simon).

It is a big open question whether or not Heisenberg ferromagnet $(n \geq 3)$ has phase transition at some $\beta(> 0)$ - i.e., does Heisenberg models $(m \geq 3)$ in the example (c) have phase transition at some $\beta(> 0)$?

Now let δ be a normal *-derivation in $\mathfrak{A}$ with $\mathfrak{D}(\delta) = \bigcup_{n=1}^{\infty} \mathfrak{A}_n$. If $\delta(a) = i[h_n, a]$ $(a \in \mathfrak{A}_n ; n=1,2,\ldots)$ and $\| h_n - P_n(h_n) \| = 0(1)$, then δ is a pregenerator and $\exp t\bar{\delta} = \text{strong} \lim \exp t\delta_{ih_n}$ (Kishimoto [42], cf. Bratteli-Jørgensen's talk).

THEOREM 2.11. ([1], [67], [68]). If $\| h_n - P_n(h_n) \| = 0(1)$, then $\{ \exp t\bar{\delta} \}$ has no phase transition at every $\beta(-\infty < \beta < +\infty)$.

COROLLARY 2.12. Ising model (n=1) and Heisenberg model (n=1) has no phase transition at every $\beta(-\infty < \beta < +\infty)$.

THEOREM 2.11 implies that if a normal *-derivation has a bounded surface energy, then it has no phase transition at every $\beta(-\infty < \beta < +\infty)$.

For the 2-dimensional Heisenberg model, Mermin and Wagner had proved the model to have no spontaneous magnitization at every $\beta(> 0)$.
This is almost equivalent to "no phase transition at every $\beta(> 0)$ for the model". The following problem would be interesting.

PROBLEM 1. Let $\{ \mathfrak{A}, \rho(t)\ (t \in R)\}$ be an approximately inner C*-dynamics with a UHF algebra $\mathfrak{A}$ and let ψ_β be a KMS state at β for $\{ \rho(t) \}$. Then does there exists a sequence $\{ \ell_n \}$ of self-adjoint elements in $\mathfrak{A}$ such that $\psi_\beta(a) = \lim_n \dfrac{\tau(ae^{-\beta\ell_n})}{\tau(e^{-\beta\ell_n})}$ $(a \in \mathfrak{A})$ and $\rho(t) = \text{strong} \lim \exp t\delta_{i\ell_n}$?

For commutative normal *-derivations, we have much more exact results.

THEOREM 2.13 ([67]). Let δ be a commutative normal *-derivation in a UHF algebra $\mathfrak{A}$ such that $\mathcal{D}(\delta) = \bigcup_{n=1}^{\infty} \mathfrak{A}_n$. $\delta(a) = i[h_n, a]$ $(a \in \mathfrak{A}_n;$ $n=1,2,3,\ldots)$, $h_n h_m = h_m h_n$ $(m,n,=1,2,3,\ldots)$ and $h_m \in \bigcup_{n=1}^{\infty} \mathfrak{A}_n$ $(m=1,2,3,\ldots)$. Let $\mathcal{L}_n$ be the C*-subalgebra of $\mathfrak{A}$ generated by $\mathfrak{A}_n$ and $h_1, h_2, \ldots, h_{n-1}$, h_n, and put $\mathcal{L}_n = \sum_{j=1}^{m_n} \oplus \mathcal{L}_n z_{n,j}$, where $\{ z_{n,j}\ ;\ j=1,2,\ldots,m_n \}$ is the family of mutually orthogonal minimal central projections of $\mathcal{L}_n$ with $\sum_{j=1}^{m_n} z_{n,j} = 1$. Then for any KMS state ψ_β at β for $\{ \exp t\bar{\delta} \}$, there is a unique family $\{ \lambda_{n,j}\ ;\ j=1,2,\ldots,m_n \}$ of positive numbers for each n such that $\sum_{j=1}^{m_n} \lambda_{n,j} = 1$, and $\psi_\beta(b) = \sum_{j=1}^{m_n} \lambda_{n,j} \dfrac{\tau(be^{-\beta h_n} z_{n,j})}{\tau(e^{-\beta h_n} z_{n,j})}$ $(b \in \mathcal{L}_n\ ;\ n=1,2,\ldots)$.

Theorem 2.13 is applicable to all classical lattice systems with finite range interactions - in particular, Ising models in the example (b). For classical lattice systems with infinite range interactions, we need minor changes ([68]).

In Theorem 2.13, put $\dfrac{\lambda_{n,j}}{\tau(e^{-\beta h_n} z_{n,j})} = e^{-\beta\alpha_{n,j}}$ and $h_{n,\beta} = h_n + \sum_{j=1}^{m_n} \alpha_{n,j} z_{n,j}$; then $\psi_\beta(b) = \tau(b\, e^{-\beta h_{n,\beta}})$ $(b \in \mathcal{L}_n\ ;\ n=1,2,\ldots)$ and $\delta(a) = i[h_n, b] = i[h_{n,\beta}, b]$ $(b \in \mathcal{L}_n)$. Now let Q_n be the canonical conditional expectation of

$\mathfrak{A}$ onto $\mathcal{L}_n$ - i.e. $\tau(Q_n(x)b) = \tau(xb)$ $(b \in \mathcal{L}_n, x \in \mathfrak{A})$. Then $Q(e^{-\beta h_{m,\beta}})$ $= e^{-\beta h_{n,\beta}}$ $(m \geqq n)$, $\tau(e^{-\beta h_{n,\beta}}) = 1$, $i[-\frac{1}{\beta}\log e^{-\beta h_{n,\beta}}, a]$ $= i[h_{n,\beta}, a] = i[h_n, a] = \delta(a)$ $(a \in \mathfrak{A}_n)$, and $\{h_{n,\beta}; n=1,2,\ldots\}$ is a mutually commuting family. Conversely suppose that there is a sequence $(A_{n,\beta})$ of mutually commuting, positive invertible elements in $\mathfrak{A}$ such that $Q_n(A_{m,\beta}) = A_{n,\beta}(m \geqq n)$, $\tau(A_{n,\beta}) = 1$, $i[-\frac{1}{\beta}\log A_{n,\beta}, a] = \delta(a)$ $(a \in \mathfrak{A}_n)$. Then for $a \in \bigcup_{n=1}^{\infty} \mathfrak{A}_n$, take an n_0 with $a \in \mathfrak{A}_{n_0}$ and define $\Psi_\beta(a) = \tau(a A_{n_0,\beta})$. Since $Q_n(A_{m,\beta}) = A_{n,\beta}(m \geqq n)$, $\psi_\beta(a)$ does not depend on a special choice of n_0; hence Ψ_β is well-defined on $\bigcup_{n=1}^{\infty} \mathfrak{A}_n$.

Since Ψ_β is a state on each $\mathfrak{A}_n$, it can be uniquely extended to a state on $\mathfrak{A}$ (denoted by Ψ_β again)

Since $\delta^m(a) = \delta^m_{i(-\frac{1}{\beta}\log A_{n,\beta})}(a)$ $(a \in \mathfrak{A}_n, m=1,2,3,\ldots)$, $\exp t\bar{\delta}(a) = \exp t\delta_{i(-\frac{1}{\beta}\log A_{n,\beta})}(a)$ $(a \in \mathfrak{A}_n)$.

Hence one can easily see that a state Ψ_β is a KMS state at β for $\{\exp t\bar{\delta}\}$. Therefore, there is a one-to-one correspondence between KMS-states Ψ_β for $\{\exp t\bar{\delta}\}$ and families $\{A_{n,\beta}\}$. It would be an interesting problem to study the constructions of $\{A_{n,\beta}\}$ and to analyse phase transition, using $\{A_{n,\beta}\}$.

Now suppose that $\{\rho(t)\ (t \in R)\}$ is a general approximately inner dynamics. Since a KMS state ψ_0 at 0 for $\{\rho(t)\}$ is tracial, $\psi_0 = \tau$; hence it has no phase transition at $\beta = 0$.

Let Ψ_β be a KMS state at β_n for $\{\rho(t)\}$, and $\beta_n \longrightarrow \beta$ and $\Psi_{\beta_n} \longrightarrow \Psi$ in the state space of $\mathfrak{A}$; then Ψ is a KMS state at β for $\{\rho(t)\}$ ([90], [92]). If $\beta \longrightarrow \infty$ and $\Psi_\beta \longrightarrow \Psi$, then Ψ is a ground state for $\{\rho(t)\}$([90], [92]). Now let $\mathfrak{S}_g$ be the set of all ground states on $\mathfrak{A}$ for $\{\rho(t)\}$ and let $\widetilde{\mathfrak{S}}_g$ be the set of all ground states ψ on $\mathfrak{A}$ for $\{\rho(t)\}$ such that there is a sequence $\{\psi_{\beta_n}\}$ of KMS states at β_n for $\{\rho(t)\}$ with $\beta_n \longrightarrow \infty$ and $\psi_{\beta_n} \longrightarrow \psi$. In most cases, $\widetilde{\mathfrak{S}}_g \subsetneqq \mathfrak{S}_g$ (even for Ising models). In many cases (for example, Ising models), the dynamics has phase transition at some $\beta\ (> 0)$ if and only if $\widetilde{\mathfrak{S}}_g$ consists of at least two points. Therefore the following problem would be interesting.

PROBLEM 2. Characterize ground states belonging to $\tilde{\mathfrak{S}}_g$.

Let ψ be a ground state for $\{ \rho(t) \}$; then ψ is invariant under $\rho(t)$ - i.e., $\psi(\rho(t)(a)) = \psi(a)$ ([60]). Let $\{ \pi_\psi, U_\psi, \mathcal{H}_\psi \}$ be the covariant representation of $\{ \mathfrak{A}, \rho(t) \ (t \in R) \}$ on a Hilbert space $\mathcal{H}_\psi$ constructed via ψ .

Then $t \longrightarrow U_\psi(t)$ is a strongly continuous one-parameter group of unitary operators on $\mathcal{H}_\psi$ and $\pi_\psi(\rho(t)(a)) = U_\psi(t)\, \pi_\psi(a)\, U_\psi(-t)$ $(a \in \mathfrak{A})$. Let $U_\psi(t) = \exp{itH_\psi}$ be the Stone representation of $\{ U_\psi(t) \}$; then H_ψ becomes a positive self-adjoint operator in $\mathcal{H}_\psi$. Let $\mathcal{K}_\psi = \{ \xi \ ; H_\psi \xi = 0, \xi \in \mathcal{H}_\psi \}$; then $1_\psi \in \mathcal{K}_\psi$ so that lim $(\mathcal{K}_\psi) \geqq 1$. If dim $(\mathcal{K}_\psi) = 1$, then the ground state ψ is said to be physical.

Generally $U_\psi(t) \in$ the weak closure of $\pi_\psi(\mathfrak{A})$ (Araki) and so dim $(\mathcal{K}_\psi) = 1$ implies that $\pi_\psi(\mathfrak{A})'$ is one-dimensional ; hence ψ is pure.

PROBLEM 3. Suppose $\psi \in \tilde{\mathfrak{S}}_g$ is a pure state ; then can we conclude that ψ is physical ?

In general, there is an exact correspondence between KMS states at β for $\{ \exp t\delta \}$ and KMS states at β for $\{ \exp t(\delta+\delta_{ih}) \}$, where δ is a generator and δ_{ih} is a bounded *-derivation ([91]) ; in particular the existence of phase transition at β is invariant under bounded perturbations. About many nice properties of KMS states and ground states, the reader should refer the references ([90], [91], [84]).

§3. Differentiations. Concerning differentiation, the audiences will learn details from Batty's and Goodman's talks. Therefore I am going to state just a scheme for future study. Batty [75] has proved that any quasi well behaved closed *-derivation δ in $C([0,1])$ is equivalent to an extension of $\lambda \frac{d}{dt}$ $(\lambda \in C([0,1]))$. Let $\{ \theta_t \ ; t \in R \}$ be a one-parameter group of homeomorphisms on $[0,1]$ such that $t \longrightarrow \theta_t(s)$ is continuous on R for each $s \in [0,1]$; then $\{ \theta_t \ ; t \in R \}$ defines canonically a strongly continuous one-parameter group of *-automorphisms on $C([0,1])$ (denoted by $\{ \theta_t \ ; t \in R \}$ again). Let $\theta_t = \exp t\delta$; then δ is a generator so that it is well-behaved. Therefore $\alpha^{-1}\delta\alpha$ is an extension of $\lambda \frac{d}{dt}$, where α is a *-automorphism on $C([0,1])$. This may be interpreted that continuous dynamical systems are equivalent to differential dynamical systems. Therefore it would be an interesting problem to extend this result to higher dimensional differential dynamical systems. For this problem, we have to deal with a family of derivations. This problem may also be closely related to the generalized Hilbert's fifth problem on transformation groups.

§4. Quantization of space-time (non-commutative differential geometry).

As a future important subject, we may mention the quantization of space-time within the C*-framework. This subject may become a central subject in future, because it is closely related to quantum gravity.

Theoretical physicists suggest that the theory of quantum gravity will require eventually the quantization of space-time [93]. If this is the case, the quantization of space-time within the C*-framework (alias non-commutative differential geometry in C*-algebras) would be one of the most hopeful methods for quantization. To quantize spaces within the C*-framework, it would be desirable to construct a class of simple C*-algebras (with or without identity) with non-commutative differential structure which include classical spaces as special cases.

Totally disconnected spaces have no non-trivial closed derivations. Therefore, a natural quantization of totally disconnected spaces may be a class of simple C*-algebras which have only closed *-derivations approximated by bounded *-derivations. It is known that UHF algebras have such property in some sense so that they may be considered quantization of totally disconnected spaces. Now we shall call a closed *-derivation δ an approximately bounded *-derivation, if there is a sequence (δ_n) of bounded *-derivations such that

$$\| \delta_n(a) - \delta(a) \| \longrightarrow 0 \quad (n \longrightarrow \infty) \quad (a \in \mathcal{D}(\delta)).$$

As a quantization of the real line, we may ask the following question : Is there a simple C*-algebra $\mathfrak{A}$ which has a not approximately bounded generator δ_0 such that for any closed *-derivation δ in $\mathfrak{A}$ with $(\exp t\delta_0)\mathcal{D}(\delta) \subset \mathcal{D}(\delta)$ and $(\exp t\delta_0)\delta = \delta(\exp t\delta_0)$ $(t \in R)$, we have an expression $\delta = k\delta_0 + \delta_1$, where k is a constant and δ_1 is approximately bounded ?

Also the following problem whould be interesting. For an arbitrary $n \geq 1$, is there a simple C*-algebra $\mathfrak{A}$ satisfying the following conditions : (1) there are generators $\delta_1, \delta_2, \ldots, \delta_n$ which are not approximately bounded ; (2) there is a dense *-subalgebra $\mathcal{D}$ of $\mathfrak{A}$ such that $\mathcal{D} \subset \bigcap_{i=1}^{\infty} \mathcal{D}(\delta_i)$ and $\delta_i(\mathcal{D}) \subset \mathcal{D}$ $(i=1,2,\ldots n)$ and $\mathcal{D}$ is the core of δ_i $(i=1,2,\ldots,n)$; (3) for any closed *-derivation δ with $\mathcal{D}(\delta) = \mathcal{D}$, there are constants $k_1, k_2, \ldots, k_n$ such that $\delta = \sum_{i=1}^{n} k_i\delta_i + \tilde{\delta}$, where $\tilde{\delta}$ is approximately bounded ; (4) $\delta_i - \sum_{j \neq i} \alpha_j \delta_j$ is not approximately bounded for any real numbers $\alpha_1, \alpha_2, \ldots, \alpha_n$

(i=1,2,...,n)?

For the convenience of the audiences, many articles on unbounded derivations, which are not refered in my talk, are included in the references.

REFERENCES

1. H. Araki, On the uniqueness of KMS states of one-dimensional quantum lattice system, Comm. Math. Phys., 44 (1975), 1-7.

2. H. Araki and G.L. Swell, KMS conditions and local Thermo-dynamical stability of Quantum lattice system, Comm. Math. Phys., 52 (1977), 103-110.

3. W. Arveson, On groups of automorphisms of operator algebras, J. Functional Analysis, 15 (1974), 217-243.

4. H.J. Borchers, Energy and momentum as observables in quantum field theory, Comm. Math. Phys., 2 (1966), 49-54.

5. H.J. Brascamp, Equilibrium states for a classical lattice gas, Comm. Math. Phys., 18 (1970), 82-96.

6. O. Bratteli, Self-adjointness of unbounded derivations of C*-algebras. Proceedings of Meeting on C*-algebras and their application to Theoretical Physics, Rome, 1974.

7. O. Bratteli and U. Haagerup, Unbounded derivations and invariant states, preprint.

8. O. Bratteli, R. Herman and D.W. Robinson, Perturbations of Flows on Banach Spaces and Operator Algebras, Math. Ann.

9. O. Bratteli and D.W. Robinson, Unbounded derivations of C*-algebras, Comm. Math. Phys. 42 (1975) 253-268.

10. ——, Unbounded derivations of C*-algebras II, Comm. Math. Phys., 46 (1976), 11-30.

11. ——, Unbounded derivations of von Neumann algebras, to appear in Ann. Inst. H. Poincare.

12. ——, Unbounded derivations and invariant trace state, Comm. Math. Phys., 46 (1976), 31-35.

13. G. Brink, On a class of approximately inner automorphisms of the C.A.R.-algebra, a preprint.

14. G. Brink and M. Winnink, Spectra of Liouville operators, Comm. Math. Phys., 51 (1976) 135-150.

15. O. Buchholtz and J.E. Roberts, Bounded perturbations of dynamics, Comm. Mathe. Phys., 49 (1976), 161-177.

16. D.P. Chi, Derivations in C*-algebras, Dissertation, University of Pennsylvania.

17. I. Colojoara and C. Foias, Theory of generalized spectral theory, Gordon and Breach, 1968.

18. J. Cuntz, Locally C*-equivalent algebras, J. Functional Analysis, 23 (1976).

19. G.F. Dell'Antonio, On some groups of automorphisms of physical observables, Comm. Math. Phys., 2 (1966), 384-397.

20. S. Doplicher, An algebraic spectrum condition, Comm. Math. Phys., 1 (1965), 1-5.

21. ——, A remark on a theorem of Powers and Sakai, Comm. Math. Phys., 45 (1975), 59.

22. S. Doplicher, R.V. Kadison, D. Kastler and D.W. Robinson, Asymptotically abelian systems, Comm. Math. Phys., 6 (1957), 101-120.

23. S. Doplicher, D. Kastler, and E. Størmer, Invariant states and asymptotic abelianness, J. Functional Analysis, 3 (1969), 21-26.

24. G. Elliott, Derivations of matroid C*-algebras, Inventions Math., 9 (1970), 253-269.

25. ——, Some C*-algebras with outer derivations III, Ann. of Math.

26. G. Gallavotti and M. Pulvirenti, Classical KMS condition and Tomita-Takesaki theory, Comm. Math. Phys., 46 (1976), 1-9.

27. L. Gross, Existence and Uniqueness of physical ground states, J. Functional Analysis, 10 (1972), 52-109.

28. R. Haag, N. Hugenholtz and M. Winnink, On the equilibrium states in quantum statistical mechanics, Comm. Math. Phys., 5 (1967), 215-236.

29. R. Haag, D. Kastler and E.B. Trych-Pohlmeyer, Stability and equilibrium states, Comm. Math. Phys., 38 (1974), 173-193.

30. A. Helenski and Ya. Sinai, A description of differentiations in algebras of the type of local observables of Spin systems, Functional Anal. Appl., 6 (1973), 343-344.

31. R. Herman, Unbounded derivations, J. Functional Analysis, 20 (1975), 234-239.

32. ——, Unbounded derivations, Amer. J. Math.

33. E. Hile and R. Philips, Functional analysis and semi-groups, Amer. Math. Soc. Colloquium publication, Vol.31, Rhode Island, 1957.

34. B. Johnson and A. Sinclair, Continuity of derivations and a problem of Kaplansky, Amer. J. Math., 90 (1968), 1067-1073.

35. P. Jøgensen, Approximately reducing subspaces for unbounded linear operators, J. Functional Analysis, 23 (1976), 392-414.

36. ——, Trace states and KMS states for approximately inner dynamical one-parameter group of *-automorphisms, Commun. Math. Phys., 53 (1977), 135-142.

37. ——, Approximately invariant subspaces for unbounded linear operators, II, Math. Ann., 227 (1977), 177-182.

38. ——, Unbounded derivations in operator algebras and extensions of states, to appear in Tohoku Math. J.

39. P. Jørgensen and C. Radin, Approximately inner dynamics, preprint.

40. S. Kantorovitz, Classification of operators by means of their operator calculus, Trans. Amer. Math. Soc., 115 (1965), 192-214.

41. T. Kato, Perturbation theory for linear operators, Berlin-Heiderberg-New York, Springer-Verlag, 1966.

42. A. Kishimoto, Dissipation and derivations, Comm. Math. Phys., 47 (1976), 25-32.

43. ——, On Uniqueness of KMS states of one-dimensional quantum lattice systems, Comm. Math. Phys., 47 (1976), 167-170.

44. ——, Equilibrium states of a semi-quantum lattice system, preprint.

45. P. Kruszynski, On existence of KMS states for invariantly approximately inner dynamics, Bull. Acad. Pol. Sci., vol. xxiv, No.4 (1976)

46. C. Lance and A. Niknam, Unbounded derivations of group C*-algebras, Proc. Amer. Math. Soc., 6 (1976), 310-314.

47. O. Lanford and D. Ruelle, Integral representations of invariant states on a B*-algebra, J. Math. Phys., 8 (1967), 1460-1463.

48. G. Lindblad, On the generators of quantum dynamical semi-groups, Comm. Math. Phys., 48 (1976), 147.

49. R. Longo, On perturbed derivations of C*-algebras, Reports on Mathematical Physics, 12 (1877), 119-124.

50. G. Lumer and R.S. Philips, Dissipative operators in a Banach space, Pacific J. Math., 11 (1961), 679-698.

51. R. McGovern, Quasi-free derivations on the canonical anti-commutation relation algebra, J. Functional Analysis, 11 (1961), 679-698.

52. A. McIntosh, Functions and derivations of C*-algebras, J. Functional Analysis.

53. B. Sz-Nagy, On uniformly bounded linear transformations in Hilbert space, Acta. Sci. Math. (Szaged) 11 (1947), 152-157.

54. E. Nelson, Analytic vectors, Ann. of Math., 70 (1958), 572-615.

55. D. Olesen and G. Pedersen, Groups of automorphisms with spectrum condition and the lifting problem, Comm. Math. Phys., 51 (1976), 85-95.

56. S. Ota, Certain operator algebras induced by *-derivations in C*-algebras on an indefinite inner product space, J. Functional Analysis.

57. G. Pedersen, Lifting derivations from quotients of separable C*-algebras, Proc. Nat. Acad. Sci., USA, 73 (1976), 1414-1415.

58. R.T. Powers, Representations of the canonical anti-commutations relations, Thesis, Princeton (1967).

59. R.T. Powers, A remark on the domain of an unbounded derivation of a C*-algebra, J. Functional Analysis, 18 (1975), 85-95.

60. R.T. Powers and S. Sakai, Existence of ground states and KMS states for approximately inner dynamics, Comm. Math. Phys., 39 (1975), 273-288.

61. ——, Unbounded derivations in operator algebras, J. Functional Analysis, 19 (1975), 81-95.

62. D.W. Robinson, The approximation of flow, J. Functional Analysis, 24 (1977), 280-290.

63. D. Ruelle, Statistical Mechanics: Rigorous results, W.A. Benjamin, New York (1969).

64. S. Sakai, C*-algebras and W*-algebras, Berlin-Heiderberg-New York, Springer-Verlag, 1971.

65. ——, On one-parameter subgroups of *-automorphisms on operator algebras and the corresponding unbounded derivations, Amer. J. Math., 98 (1976), 427-440.

66. ——, On commutative normal *-derivations, Comm. Math. Phys., 43 (1975), 39-40.

67. ——, On commutative normal *-derivations II, J. Functional Analysis, 21 (1976), 203-208.

68. ——, On commutative normal *-derivations III, Tohoku Math. J., 28 (1976), 583-590.

69. J.G. Stamfli, Derivations on B($\mathcal{H}$): the range, Illinois J. Math., 17 (1973), 518-524.

70. H.F. Trotter, Pacific J. Math., 8 (1958), 887-919.

71. K. Yoshida, Functional Analysis, Berlin-Heiderberg-New York, Springer-Verlag, 1974 (Fourth edition).

72, C.J.K. Batty, Dissipative mappings with approximately invariant subspaces, J. Functional Analysis.

73. ——, Dissipative mappings and well-behaved derivations, J. London Math. Soc. (2) 18 (1978), 527-533.

74. ——, Unbounded derivations of commutative C*-algebras, Comm. Math. Phys., 61 (1978), 261-266.

75. ——, Derivations on compact spaces, to appear.

76. ——, Ground states of uniformly continuous dynamical system, to appear.

77. ——, Simplexes of invariant states and G-ablian C*-algebras, to appear.

78. ——, Small perturbations of C*-dynamical system, to appear.

79. O. Bratteli, G. Elliott and R. Herman, On the possible temperatures of a dynamical system, to appear.

80. O. Bratteli and A. Kishimoto, Generations of semi-groups, and two-dimensional quantum lattice system, J. Functional Analysis, 35 (1980), 344-368.

81. F. Goodman, Closed derivations in commutative C*-algebras.

82. R. Herman and O. Takenouchi, Extensions of d/dt, to appear.

83. P.E.T. Jørgensen, Approximately invariant subspaces for unbounded linear operators III, semibounded operators.

84. S. Sakai, The theory of unbounded derivation in C*-algebras, Lecture Notes, Univ. of Copenhagen and Newcastle upon Tyne, 1977.

85. ——, Developments in the theory of derivations in C*-algebras, Proceedings of International conference on operator algebras, ideals and their applications in theoretical physics, Leibzig, 1977.

86, ——, Recent developments in the theory of unbounded derivations in C*-algebras, Proceedings of International congress of mathematicians, Helsinki, 1978.

87. ——, Derivations in operator algebras, to appear in the proceeding on functional analysis and its applications on the occasion of professor Segal's sixtieths birthday.

88. R.D. Mosak, Banach algebras, Univ. of Chicago Press, 1975.

89. H. Araki, J. Math. Phys. 5, 1, 1964.

90. ——, On KMS states of a C*-dynamical system, in Springer Lecture Notes in mathematics, V. 650, Springer-Verlag, 1978.

91. ——, Publ. RIMS, Kyoto Univ. A4 (1968), 361-371.

92. S. Sakai, Recent developments in the theory of unbounded derivations in C*-algebras, in Springer Lecture Notes in Mathematics, V. 650, Springer-Verlag, 1978.

93. Quantum gravity, An Oxford conference, 1975.

94. I. Kaplansky, John Wiley & Sons, New York, 1958.

95. H. Wielandt, Math. Ann. 121 (1949), 21.

96. G. Šilov, Doklady Akad. Nauk SSSR 58 (1947), 985-988.

97. R. Kadison, Ann. of Math. (2) 83 (1966), 280-293.

98. S. Sakai, Ann. of Math., (2) 83 (1966) 273-279.

99. ——, J. Functional Anal. (2) (1968), 202-206.

100. —, Bull. Soc. Math. France 99 (1971), 258-263.

101. C. Akemann G., Elliott, G, Pederson and J. Tomiyama, Amer. J. Math. 98 (1976), 679-708.

102. C. Akemann, G. Pedersen, to appear in Amer. J. Math.

103. G. Elliott, Ann. of Math. 106 (1977), 121-143.

DEPARTMENT OF MATHEMATICS
UNIVERSITY OF PENNSYLVANIA
PHILADELPHIA,PA. 19104
U.S.A.

Department of Mathematics
Faculty of Humanities and Sciences
Nihon University
Sakurajosui, Setagaya-ku
Tokyo, Japan

Proceedings of Symposia in Pure Mathematics
Volume 38 (1982), Part 2

DERIVATIONS OF ABELIAN C*-ALGEBRAS

C. J. K. Batty

INTRODUCTION. Unlike many topics in operator algebras (including bounded derivations), the theory of unbounded derivations reduces neither to a triviality nor to classical results for abelian C*-algebras. Sakai [12,13,14] obtained several results in this context and he posed a great number of problems, some of which were taken up by various workers, specially Batty [2,3] and Goodman [8.9,10]. In this lecture the answers to these and some further extensions will be discussed. The techniques of proof usually involve inspection not so much of the ablian C*-algebra as its spectrum, and it therefore seems unlikely that they will be of great use in non abelian cases. However it is to be hoped that the results themselves will suggest the type of theorems that may be valid for general C*-algebras, and to some extent this expectation has already been fulfilled.

§1. Closed derivations

Let Ω be a compact Hausdorff space, and $C(\Omega)$ be the abelian C*-algebra of continuous complex-valued functions on Ω (extensions to locally compact spaces are usually valid and trivial to prove). There are two basic examples of closed *-derivations on $C(\Omega)$.

EXAMPLE 1. Let $\{\theta_t : t \in R\}$ be a (continuous) flow on Ω, so that θ is a continuous homomorphism of R into the group of homeomorphisms of Ω (equipped with the topology of pointwise convergence). Let $\{\alpha_t\}$ be the corresponding strongly continuous one-parameter group of *-automorphisms of $C(\Omega)$ given by $\alpha_t(f) = f\circ\theta_t$, and let δ_θ be the generator of $\{\alpha_t\}$, so that

$$\delta_\theta(f)\ (\omega) = \frac{d}{dt}(f(\theta_t(\omega)))\Big|_{t=0} .$$

Then δ_θ is a closed *-derivation on $C(\Omega)$.

EXAMPLE 2. Let I be the unit interval $[0,1]$. There is a very familiar closed *-derivation D on $C(I)$ with domain $C^1(I)$, given by

1980 Mathematics Subject Classification.

0082-0717/80/0000-0499/$02.50

$$Df = f' .$$

There is a more detailed discussion of derivations on $C(I)$ in §3.

Now suppose that δ is a closed *-derivation on $C(\Omega)$ with domain $\mathcal{D} = \mathcal{D}(\delta)$, and let $\| \ \|_\delta$ be the graph norm on $\mathcal{D}$, so

$$\| f \|_\delta = \| f \| + \| \delta(f) \| .$$

Then $(\mathcal{D}, \| \ \|_\delta)$ is a semi-simple commutative Banach *-algebra with structure space Ω, and has the following additional properties:

(i) $\mathcal{D}$ has a C^1-functional calculus, for it can easily be shown that for real f in $\mathcal{D}$ and g in $C^1(\mathbb{R})$, $g \circ f \in \mathcal{D}$ and $\delta(g \circ f) = (g' \circ f)\,\delta(f)$ [6].

(ii) $\mathcal{D}$ is "regular", i.e. given a closed subset Ω_0 of Ω and a point ω in $\Omega \backslash \Omega_0$, there is a function f in $\mathcal{D}$ with $f = 0$ on Ω_0 and $f(\omega) = 1$. Such algebras have been the subject of some study especially by Shilov (See [7]).

(iii) $\mathcal{D}$ is "local" in that if, for some f in $C(\Omega)$ and each ω in Ω, there is a function f_ω in $\mathcal{D}$ such that $f_\omega = f$ in a neighborhood of ω, then $f \in \mathcal{D}$ (and $\delta(f)(\omega) = \delta(f_\omega)(\omega)$).

These and similar facts are the basis for the proofs of most of the following results.

The first simplification which can be made is to eliminate those spaces which have no closed *-derivations at all. Sakai [12] showed that there are no non-zero closed *-derivations on $C(\Omega)$ if Ω is totally disconnected. This was improved in the following way.

THEOREM 1 (Johnson (unpublished), [2]). Suppose Ω has a dense open totally disconnected subset. Then $C(\Omega)$ has no non-zero closed *-derivations.

This raises the following problem.

PROBLEM 1. Find necessary and sufficient topological (or differential) conditions on Ω that $C(\Omega)$ has non-zero closed *-derivations.

§2. Quasi well-behaved derivations

Generators are well-behaved; the derivation D of Example 2 is quasi well-behaved, but not well-behaved, since any real function in $C^1(I)$ which attains its maximum in $(0,1)$ is well-behaved, but functions attaining their maxima at 0 or 1 may not be.

For a general *-derivation δ on $C(\Omega)$ with domain $\mathcal{D}$, it is also advantageous to study the behaviour of points of Ω rather than functions in $\mathcal{D}$. Let

$$W_p(\delta) = \{\omega \in \Omega : \delta(f)(\omega) = 0 \text{ whenever } f = \bar{f} \in \mathcal{D} \text{ and } f(\omega) = \|f\|\}.$$

THEOREM 2 [2]. Let δ be a *-derivation on $C(\Omega)$. Then
(a) δ is well-behaved if and only if $W_p(\delta) = \Omega$.
(b) δ is quasi well-behaved if and only if $W_p(\delta)$ has dense interior in Ω.

In view of THEOREM 2 and the fact that any *-derivation has a dense set of well-behaved elements [1], one might expect that $W_p(\delta)$ is always dense in Ω. However there is an extension of D on $C(I)$ for which $W_p(\delta)$ is empty. Such an extension of D is necessarily not closable (See PROBLEM 2 and THEOREM 4 below).

THEOREM 2 simplifies the task of finding a closed *-derivation which is not quasi well-behaved, and this was originally achieved in [2] by constructing a compact subset of $I \times I$ on which the *-derivation given by $\frac{\partial}{\partial x}$ is not quasi well-behaved. Goodman obtained the following refinement of this example.

EXAMPLE 3 [8,10]. There is a compact subset Ω of $I \times I$ such that $C(\Omega)$ has non-zero closed *-derivations, but none of these are quasi well-behaved.

Goodman also showed that a closed *-derivation may also fail to be quasi well-behaved because of the pathological nature of $\mathcal{D}$ rather than of Ω.

EXAMPLE 4 [8.10]. There is a closed *-derivation on $C(I \times I)$ extending $\frac{\partial}{\partial x}$ which is not quasi well-behaved.

For general Ω, there is one problem, which, although probably not particularly profound, has proved to be diffucult to resolve.

PROBLEM 2. Let δ be a closed *-derivation on $C(\Omega)$. Is $W_p(\delta)$ dense in Ω?

§3. Derivations on C(I)

Let δ be a derivation on $C(\Omega)$. For a function λ in $C(\Omega)$, there is a derivation $\lambda \cdot \delta$ with domain $\mathcal{D}(\delta)$, given by

$$(\lambda \cdot \delta)f = \lambda\, \delta(f).$$

For a homeomorphism θ of Ω, there is a derivation δ^{θ} with domain $\{f\circ\theta : f \in \mathcal{D}(\delta)\}$, given by

$$\delta^{\theta}(f\circ\theta) = \delta(f)\circ\theta .$$

The following theorem shows that all quasi well-behaved derivations on $C(I)$ can be constructed by these methods from the standard derivation D of EXAMPLE 2.

THEOREM 3 [3,8,12] Let δ be a closed *-derivation on $C(I)$. The following are equivalent:

(i) δ is quasi well-behaved

(ii) $\mathcal{D}(\delta)$ contains a strictly monotone function

(iii) There is a function λ in $C(I)$ and a homeomorphism θ of I such that δ extends $(\lambda.D)^{\theta}$.

One can specify some closed extensions of $\lambda.D$, where λ is a real function in $C(I)$.

EXAMPLE 5. Let U be a dense open subset of the cozero set $U_{\lambda} = \{t : \lambda(t) \neq 0\}$, and define

$$D_{\lambda U}(f) = \begin{cases} \lambda f' & \text{on } U \\ 0 & \text{on } I\backslash\bar{U} \end{cases}$$

(where $\mathcal{D}(D_{\lambda U})$ is the set of all f for which a (unique) function $D_{\lambda U}(f)$ in $C(\Omega)$ with these properties exists). Then $D_{\lambda U}$ is a closed quasi well-behaved *-derivation extending $\lambda.D$.

If $\lambda = 1$ and U is the complement of the Cantor set, then $\mathcal{D}(D_{1U}) = C^{1}(I) + B$, where B is the C^*-algebra generated by the Cantor function, and

$$D_{1U}(f + g) = f' \qquad (f \in C^{1}(I), \ g \in B).$$

This example of a proper closed extension of D and an example of an extension of D which is not closable were originally given in [11].

EXAMPLE 5 is almost universal.

THEOREM 4 [3,9]. Let δ be a closed *-derivation on $C(I)$ extending $\lambda.D$. There is a dense open subset U of the cozero set U_{λ} such that $D_{\lambda U}$ extends δ. If $U_{\lambda} = I$, then U can be chosen so that $\delta = D_{\lambda U}$.

The last statement of THEOREM 4 may fail if λ has zeros. However if $U_{\lambda} = I$, it is not even necessary to assume that δ is self-adjoint.

It is also possible to give a complete description of all flows on I whose generators extend $\lambda.D$, and hence (by THEOREM 3) of all flows on I [4].

The mathematical details of this characterization are rather technical, but the principle is simple. The flow has a basic speed λ, but may be subject to some exceptional delays at zero of λ. These delays are measured by a non-atomic measure on the zero set whose local finiteness is related to the local integrability of $1/\lambda$.

In view of THEOREMs 3 and 4, the following question becomes very important.

PROBLEM 3. Is every closed *-derivation on $C(I)$ quasi well-behaved ?

§4. Derivations with equal domains

In perturbation theory, one is concerned with two C*-dynamical systems $(A.\mathbb{R}\ \alpha)$ and $(A, \mathbb{R}, \beta)$ whose generators have equal domains. For commutative C*-algebras, the generators have a very close relationship.

THEOREM 5 [3]. Let δ_0 be a closed *-derivation on $C(\Omega)$, and δ be any derivation on $C(\Omega)$ with $\mathcal{D}(\delta) = \mathcal{D}(\delta_0)$. Then there is a bounded function $\lambda:\Omega \longrightarrow \mathbb{R}$ such that $\delta(f) = \lambda\delta_0(f)$ $(f \in \mathcal{D}(\delta))$. In particular, δ is closable. Moreover if δ_0 is (quasi) well-behaved, so is δ.

The function λ of THEOREM 5 is uniquely determined and continuous at any point of Ω where $\delta_0(f)$ does not vanish for some f in $\mathcal{D}(\delta_0)$. However it may not be possible to arrange even locally that λ is continuous.

EXAMPLE 6. Let Ω be the unit circle in $\mathbb{C}$, and

$$\lambda(\omega) = 2 + \sin\frac{1}{|\omega|} \qquad (\omega \neq 0)$$

where $\lambda(0)$ is arbitrary. Let θ and θ' be the flows given by:

$$\theta_t(\omega) = \omega\exp(it) \qquad \theta'_t(\omega) = \omega\exp(it\lambda(\omega)).$$

Then $\mathcal{D}(\delta_{\theta'}) = \mathcal{D}(\delta_{\theta})$, and $\delta_{\theta'}(f) = \lambda\delta_\theta(f)$ $(f \in \mathcal{D}(\delta_\theta))$.

THEOREM 5 is proved by means of Johnson's THEOREM that a derivation of the commutative semi-simple Banach algebra $(\mathcal{D}(\delta_0), \|\ \|_{\delta_0})$ into $C(\Omega)$ is automatically continuous (See [5]). Minor modifications of this argument show that if δ has domain $\bigcap_{n=1}^{\infty}\mathcal{D}(\delta_n)$ (assuming this is dense in $C(\Omega)$), where (δ_n) is a sequence of closed derivations on $C(\Omega)$, then there is an integer N and functions $\lambda_n:\Omega \longrightarrow \mathbb{R}$ $(1 \le n \le N)$ such that $\delta(f) = \sum_{n=1}^{N}\lambda_n\delta_n(f)$ $(f \in \mathcal{D}(\delta))$. More major modifications of the argument show that THEOREM 5 remains valid if δ has domain $\bigcap_{n=1}^{\infty}\mathcal{D}(\delta_0^n)$ (assuming this is dense). It seems likely that the following question has an affirmative answer .

PROBLEM 6. Is the conclusion of Theorem 5 valid if $\mathcal{D}(\delta) = \mathcal{D}(\delta_0^n)$ $(1 < n < \infty)$?

Some analogues of various aspects of THEOREM 5 for non-abelian C*-algebras will be found in Section 13.

REFERENCES

1. C.J.K. Batty, Dissipative mappings and well-behaved derivations. J. London Math. Soc. 18 (1978), 527-533.

2. C.J.K. Batty, Unbounded derivations of commutative C*-algebras. Commun. Math. Phys. 61 (1978), 419-425.

3. C.J.K. Batty, Derivations on compact spaces, Proc. London Math. Soc., to appear.

4. C.J.K. Batty, Delays to flows on the real line, preprint.

5. F.F. Bonsall and J. Duncan, Complete normed algebras, Springer, Berlin-Heidelberg-New York, 1973.

6. O. Bratteli and D.W. Robinson, Unbounded derivations of C*-algebras, Commun. Math. Phys. 42 (1975), 253-268.

7. I.M. Gelfand, D.A. Raikov and G.E. Shilov, Commutative normed rings, Chebea, New York, 1964.

8. F. Goodman, Closed derivations in commutative C*-algebras. Ph. D. thesis, University of California. Berkeley, 1979.

9. F. Goodman, Closed derivations in commutative C*-algebras. J. Functional Analysis, to appear

10. F. Goodman, Trans, Aner. Math. Soc., to appear

11. R.H. Herman and O. Takenouchi, Extensions of $\frac{d}{dx}$, preprint.

12. S. Sakai, The theory of unbounded derivations in C*-algebras, Copenhagen and Newcastle-upon-Tyne University Lecture Notes, 1977.

13. S. Sakai, Recent developments in the theory of unbounded derivations in C*-algebras, Proceedings of International Congress of mathematicians Helsinki, 1978, pp. 709-714.

14. S. Sakai, Derivation in operator algebras, preprint, 1979.

Department of Mathematics
University of Edinburgh
Edinburgh, Scotland

Proceedings of Symposia in Pure Mathematics
Volume 38 (1982), Part 2

CLOSED *-DERIVATIONS COMMUTING WITH AN AUTOMORPHISM GROUP

F. Goodman

Let (A,G,α) be a C*-dynamical system and let δ be a closed *-derivation in A commuting with α (i.e. $\alpha_g^{-1}\,\delta\,\alpha_g = \delta$ for all $g \in G$). Under what conditions does δ generate a C*-dynamics and how is this dynamics related to α ?

Investigation of this question was begun by Sakai, who proved:

THEOREM 1 [6]. A non-zero closed *-derivation in $C(\mathbb{T})$ commuting with translations by elements of $\mathbb{T}$ has domain $C^1(\mathbb{T})$ and is a constant multiple of differentiation.

Batty [1] and Goodman [2] obtained tha same result for $C_0(\mathbb{R})$. Subsequent results have also been for abelian A, but it is likely that theorems for non-abelian A will be forthcoming.

A C*-dynamical system $(C_0(X),G,\alpha)$ is equivalent to a continuous action of G on X. Goodman [3] and Nakazato [5] have considered the situation where G acts transitively. Let $L(G)$ denote the set of closed *-derivations in $C_0(G)$ which commute with left translations by elements of G.

THEOREM 2 [3,5]. Let G be a locally compact group and let $\delta \in L(G)$. Then δ is a generator, and there is a continuous one-parameter subgroup γ_t of G such that $\exp(t\delta)(f)(x) = f(x\gamma_t)$ $(f \in C_0(G),\ x \in G)$.
If G is a Lie group, then $C_c^\infty(G)$ is a core for δ. Then $L(G)$ is identified with the Lie algebra of G.

In general, all elements of $L(G)$ have a common core $\mathcal{D}$, and each left-invariant *-derivation with domain $\mathcal{D}$ is closable. It follows that $L(G)$ has a natural Lie algebra structure and can be identified with the Lie algebra of G as defined in [4].

1980 Mathematics Subject Classification.

THEOREM 3 [3]. Let G be a locally compact group which is either

(a) a projective limit of lie groups, or

(b) separable.

Let H be a closed subgroup and let δ be a closed *-derivation in $C_0(G/H)$ which commutes with left translations by elements of G. Then δ generates a C*-dynamics of $C_0(G/H)$.

Again, if G is a Lie group, then $C_c^\infty(G/H)$ is a core for δ.

Batty considered actions of $\mathbb{R}$ on X with general orbit structure. For a derivation δ in $C_0(X)$ let $U(\delta)$ denote $\{x \in X : \exists\, f \in \mathcal{D}(\delta),$ $\delta(f)(x) \neq 0\}$.

THEOREM 4 [1]. Let $\{\alpha_t\}$ be a C*-dynamics of $C_0(X)$ with generator δ_0 and let $\{\theta_t\}$ be the corresponding continuous flow on X. Let δ be a closed *-derivation in $C_0(X)$ commuting with $\{\alpha_t\}$, such that

(i) δ is well behaved, and

(ii) if $U \subseteqq X$ is open and $f \in \mathcal{D}(\delta)$ is constant on $M \cap U$ for each orbit M of $\{\theta_t\}$, then $\delta(f)\big|_U = 0$.

Then there is a $\lambda : X \longrightarrow \mathbb{R}$, continuous on $U(\delta_0)$ and constant on each orbit of $\{\theta_t\}$, such that

$$\delta(f)(x) = \lambda(x)\,\delta_0(f)(x) \quad (f \in \mathcal{D}(\delta) \cap \mathcal{D}(\delta_0),\ x \in U(\delta_0)).$$

If X is compact and $U(\delta_0) = U(\delta) = X$, then $\mathcal{D}(\delta) = \mathcal{D}(\delta_0)$ and δ is a generator.

COROLLARY. Assume $\{\theta_t\}$ has an orbit which is dense in X. Then δ is a constant multiple of δ_0.

From THEOREM 3 and the proof of THEOREM 4 one can establish.

THEOREM 5. Let $(C\ (X),G,\alpha)$ be a C*-dynamical system with G compact. Let δ be a closed *-derivation in $C_0(X)$ commuting with α, such that

(i) δ is well behaved, and

(ii) if $f \in \mathcal{D}(\delta)$ and $\alpha_s(f) = f$ for all $s \in G$, then $\delta(f) = 0$. Then there is a generator δ_1 such that δ_1 extends δ and δ_1 commutes with α.

REFERENCES

[1] C.J.K. Batty, Derivations in compact spaces, Proc. Lond. Math. Soc., to appear.

[2] F. Goodman, Closed derivations in commutative C*-algebras, J. Functional

Anal., to appear.

[3] F. Goodman, Translation invariant closed *-derivations, Pacific J. Math., to appear.

[4] R.K. Lashof, Lie algebras of locally compact groups, Pacific J. Math., 7 (1957), 1145-1162.

[5] H. Nakazato, Closed *-derivations on compact groups, preprint, 1980.

[6] S. Sakai, The theory of unbounded derivations in operator algebras, Lecture notes, University of Copenhagen and University of Newcastle upon Tyne, 1977.

Department of Mathematics
University of Pennsylvania
Philadelphia, Pa. 19104
U.S.A.

Proceedings of Symposia in Pure Mathematics
Volume 38 (1982), Part 2

DIFFERENTIABLE STRUCTURE ON NON-ABELIAN C*-ALGEBRAS

H. Takai

One of the most important problems of unbounded derivations of C*-algebras is to determine their structure in terms of more explicit manners. However once we start with, we definitely encounter with various pathological behaviors of their domains. Nevertheless some remarkable results have been obtained in non-abelian cases.

In this chapter, we describe some properties of unbounded derivations of C*-algebras with the same domains as a given closed *-derivation. First we show that any *-derivation as above is relatively bounded, which is due to Longo [4]. We then treat a result of Batty [2] using the above result of Longo. Namely any *-derivation with the same domain as a well-behaved (resp. quasi well-behaved) closed *-derivation is well-behaved (resp. quasi well-behaved). Finally combining with these results, we prove that in the irrational rotation algebra there are two non-approximately bounded pregenerators δ_1, δ_2 with the property that any *-derivation δ with $\mathcal{D}(\delta) = \mathcal{D}(\delta_1) \cap \mathcal{D}(\delta_2)$ can be expressed as $\delta = k_1\delta_1 + k_2\delta_1 + \delta_0$ where k_1, k_2 are real numbers and δ_0 is approximately bounded, which is a solution of a problem of Sakai for the 2-dimensional case. (cf: [6], [8])

13.1 RELATIVE BOUNDEDNESS.

Let $\mathfrak{A}$ be a C*-algebra and δ_0 be a closed *-derivation of $\mathfrak{A}$. Then the domain $\mathcal{D}(\delta_0)$ of δ_0 becomes a Banach *-algebra $\mathfrak{A}_0$ with respect to the graph norm $\| \ \|_0$ of δ_0. Let δ be a derivation of $\mathfrak{A}$ with $\mathcal{D}(\delta) = \mathcal{D}(\delta_0)$. We shall show that it is continuous from $\mathfrak{A}_0$ into $\mathfrak{A}$, in other words δ is relatively bounded with respect to δ_0. Let (a_n) be a sequence of $\mathfrak{A}_0$ and $a \in \mathfrak{A}$ such that $\| a_n \|_0 \longrightarrow 0$, $\delta(a_n) \longrightarrow a$. Put

$$\mathcal{S} = \{ x \in \mathfrak{A}_0 \mid y \in \mathfrak{A}_0 \longrightarrow \delta(yx) \text{ is bounded } \}.$$

1980 Mathematics Subject Classification.

0082-0717/80/0000-0580/$03.50

Then $\delta(a_n x) \longrightarrow 0$ for each $x \in \mathscr{J}$. On the other hand,

$$\delta(a_n x) = a_n \delta(x) + \delta(a_n)x \longrightarrow ax$$

for each $x \in \mathscr{J}$. Therefore $ax = 0$ for each $x \in \mathscr{J}$. So a is in the left annihilator $L(\mathscr{J})$ of $\mathscr{J}$ in $\mathfrak{A}$. Since it is a closed two-sided ideal of $\mathfrak{A}$ and $L(\mathscr{J}) \cap \bar{\mathscr{J}} = (0)$, then $L(\mathscr{J})$ is finite dimensional where $\bar{\mathscr{J}}$ is the closure of $\mathscr{J}$ in $\mathfrak{A}$. In fact suppose the dimension of $\mathfrak{A}/\bar{\mathscr{J}}$ is infinite, so is $\mathfrak{A}_0/\bar{\mathscr{J}} \cap \mathfrak{A}_0$. Since it is a reduced Banach *-algebra, there exists $h = h^* \in \mathfrak{A}_0$ such that the spectrum $sp(\pi(h))$ of $\pi(h)$ is infinite where π is the quotient mapping of $\mathfrak{A}$ onto $\mathfrak{A}/\bar{\mathscr{J}}$. Then there are $h_n \in \mathfrak{A}_0$ such that

$$h_n h_m = 0 \quad (n \neq m) \quad \text{and} \quad h_n^2 \notin \bar{\mathscr{J}} \cap \mathfrak{A}_0.$$ Moerover one can assume $\| h_n \| \leq 1$. Since $h_n^2 \notin \mathscr{J}$, there exist a sequence (k_n) of $\mathfrak{A}_0$ such that

$$\| k_n \|_0 \leq 2^{-n} \quad \text{and} \quad \| \delta(k_n h_m{}^2) \|_0 \geq \| \delta(h_n) \|_0 + n .$$

Let $\ell = \sum_n k_n h_n \in \mathfrak{A}_0$. Then $\| \ell \| \leq 1$ and $\ell h_n = k_n h_n^2$.

Therefore $\| \delta(\ell)h_n \|_0 = \| \delta(\ell h_n) - \ell\delta(h_n) \|_0 \geq \| \delta(k_n h_n^2) \| - \| \ell \|_0 \| \delta(h_n) \| \geq n$,

which is a contradiction. Therefore $L(\mathscr{J})$ is finite dimensional. So there exists a projection p in the center of $\mathfrak{A}$ such that $L(\mathscr{J}) = \mathfrak{A}p$. Since $\mathfrak{A}_0$ is dense in $\mathfrak{A}$, there is a projection $p_0 \in \mathfrak{A}_0$ such that $\| p_0 - p \| < 1$. Since p_0 commutes with p, we have $p_0 = p \in \mathfrak{A}_0$. Since $L(\mathscr{J})$ is finite dimensional,

$$x \in \mathfrak{A}_0 \longrightarrow xp \in L(\mathscr{J}) \longrightarrow \delta(xp) \in \mathfrak{A}$$

is bounded. So $p \in \mathscr{J} \cap L(\mathscr{J}) = (0)$ and $L(\mathscr{J}) = (0)$. Therefore $a = 0$, which implies the following theorem:

THEOREM 1 (Longo) Let δ_0 be a closed *-derivation of a C*-algebra $\mathfrak{A}$. Suppose δ is a derivation of $\mathfrak{A}$ with $\mathcal{D}(\delta) = \mathcal{D}(\delta_0)$, then it is relatively bounded with respect to δ_0.

13.2 SMALL PERTURBATION

Let δ, δ_0 be as in 13.1. we assume that δ is a *-derivation. Let $\mathscr{J}$ be a closed ideal of $\mathfrak{A}$ and $a = a^* \in \mathcal{D}(\delta)$. Let $\lambda \in \mathbb{R}$ be the smallest point of $sp(\pi(a))$ where π is the quotient mapping of $\mathfrak{A}$ onto $\mathfrak{A}/\mathscr{J}$. Adding a unit to $\mathfrak{A}$, we may assume that $\mathcal{D}(\delta) \ni 1$ and $\delta(1) = \delta_0(1) = 0$. Replacing a by $a - \lambda 1$, we also assume $\pi(a) \geq 0$. Let p be the spectral projection of a in $\mathfrak{A}^{**}$ corresponding to the interval $(-\infty, 0]$ and let q be the central projection of $\mathfrak{A}^{**}$ such that $\overline{\mathscr{J}}^{w^*} = \mathfrak{A}^{**}(1-q)$ in $\mathfrak{A}^{**}$ where $\overline{\mathscr{J}}^{w^*}$ is the weak* closure of $\mathscr{J}$ in $\mathfrak{A}^{**}$. Then $p\,a\,p\,q = 0$.

For any state ψ of $\mathfrak{A}$ with $\psi(pq) > 0$, let

$$\tilde{\psi}(b) = \psi(pq)^{-1}\,\psi(pbpq) \quad (b \in \mathfrak{A}) .$$

Then $\tilde{\psi}$ annihilates $\mathcal{J}$ and $\tilde{\psi}(a) = 0$. Now suppose $\phi \cdot \delta_0(a) = 0$ whenever ϕ is a state of $\mathfrak{A}$ annihilating $\mathcal{J}$ with $\phi(a) = 0$. Then $\tilde{\psi} \cdot \delta_0(a) = 0$. So $p\delta_0(a)\,p\,q = 0$. Let g_1, g_2 and h be real valued C^2-functions on $\mathbb{R}$ such that

$$g_1(0) = g_2(0) = 0 , \quad g_1(t)g_2(t) = t \ (|t| \geq 1) , \quad h(t) = 0 \quad (t \leq 1)$$

and $h(t) = 1$ $(t \geq 2)$. Put $f(t) = t - g_1(t)g_2(t)$. Then f is a C^2-function of compact support, and $f(nt)h(rt) = 0$ for $t \in \mathbb{R}$ and integers n and r with $n \geq r \geq 0$. Let $a_n = n^{-1} f(na)$, so that $\| a_n \| \longrightarrow 0$ as $n \to \infty$. Then a_n, $g_1(na)$, $g_2(na)$ and $h(ra)$ belong to $\mathcal{D}(\delta)$. Define

$$M = (2\pi)^{-\frac{1}{2}} \, \| \delta_0(a) \| \int_{-\infty}^{\infty} |t\hat{f}(t)|\,dt$$

where $\hat{f}$ is the Fourier transform of f. Then it is finite and

$$\delta_0(a_n) = (2\pi)^{-\frac{1}{2}} \int_{-\infty}^{\infty} \int_0^1 t\hat{f}(t)\, e^{intsa}\delta_0(a)\, d^{int(1-s)a}\, dsdt$$

so that $\| \delta_0(a_n) \| \leq M$. Since $qpe^{ita} = e^{ita}\, pq = pq$ for all $t \in \mathbb{R}$, and $p\delta_0(a)pq = 0$, it follows that $p\delta_0(a_n)pq = 0$. Let $x \in \overline{\{\delta_0(a_n)\}}^{w^*} \subset \mathfrak{A}^{**}$, so that $pxpq = 0$. Since $\delta_0(a_n h(ra)) = 0$ $(n \geq r)$ and $\| a_n \| \longrightarrow 0$, $xh(ra) = 0$. But as $r \to \infty$, $h(ra)$ converges ultrastrongly to $1\text{-}p$, so $x = xp = pxp$. Thus $xq = 0$. Since $\mathcal{J} \cap \mathcal{D}(\delta)$ is dense in $\mathcal{J}$, the unit ball of $\mathcal{J} \cap \mathcal{D}(\delta)$ contains a net $\{y_\iota \mid \iota \in I\}$ converging ultrastrongly to $1\text{-}q$. For any triple $\mu = (m, n, \iota) \in \mathbb{N} \times \mathbb{N} \times I$, let

$$b_\mu = a_n(1-y_\iota) , \quad b_\mu' = \delta_0(a_n)(1-y_\iota) \quad \text{and} \quad b_\mu'' = a_n\delta_0(y_\iota) .$$

Take $\varepsilon > 0$ and let

$$J = \{\mu = (m, n, \iota) \in \mathbb{N} \times \mathbb{N} \times I \;\big|\; \| b_\mu'' \| \leq 2^{-m} , \ \| a_n\delta(y_\iota) \| \leq \varepsilon \}.$$

In the product ordering, J is directed upwards, the nets b_μ and b_μ'' $(\mu \in J)$ are norm convergent to 0, while $\{b_\mu' \mid \mu \in J\}$ has $xq = 0$ as a weak* limit point in $\mathfrak{A}^{**}$. Since $\delta_0(b_\mu) = b_\mu' - b_\mu''$, the point $(0,0)$ belongs to the weak closure of $\{(b_\mu, \delta_0(b_\mu)) \mid \mu \in J\}$ in the Banach space $\mathfrak{A} \oplus \mathfrak{A}$. By the Hahn-Banach separation theorem, there is a net (c_ν) in the convex full of $\{b_\mu \mid \mu \in J\}$ such that $\| c_\nu \| \longrightarrow 0$ and $\| \delta_0(c_\nu) \| \longrightarrow 0$. Since δ is relatively bounded with respect to δ_0 by THEOREM 1, we have $\| \delta(c_\nu) \| \longrightarrow 0$. Now let ϕ be a state annihilating $\mathcal{J}$ with $\phi(a) = 0$. Since ϕ induces a state of $\mathfrak{A}/\mathcal{J}$, the spectral theory of $\pi(a)$ shows that

$$\phi(g_1(na)^2) = \phi(g_2(na)^2) = 0 .$$

Since $a_n = a - n^{-1} g_1(na)g_2(na)$, we have that

$$\delta(a-b_\mu) = n^{-1}\delta(g_1(na))g_2(na) + n^{-1}g_1(na)\delta(g_2(na)) + \delta(a_n)y_\iota + a_n\delta(y_\iota) .$$

Hence

$$|\phi\cdot\delta(a-b_\mu)| = |\phi(a_n\delta(y_\iota))| \le \varepsilon \quad \text{for all} \quad \mu \in J ,$$

so we have $|\phi\cdot\delta(a - c_\nu)| \le \varepsilon$. Letting first $r \longrightarrow \infty$ and then $\varepsilon \longrightarrow 0$, it follows that $\phi \cdot \delta(a) = 0$, which implies the following theorem:

THEOREM 2. (Batty) Let δ_0 be a closed *-derivation of a C*-algebra $\mathfrak{A}$, and let δ be a *-derivation of $\mathfrak{A}$ with $\mathcal{D}(\delta) = \mathcal{D}(\delta_0)$. Let $\mathcal{J}$ be a closed ideal if $\mathfrak{A}$, $a = a^* \in \mathcal{D}(\delta_0)$ and λ be the smallest point of $sp(\pi(a))$ where π is the quotient mapping of $\mathfrak{A}$ onto $\mathfrak{A}/\mathcal{J}$. Suppose $\phi \cdot \delta_0(a) = 0$ whenever ϕ is a state of $\mathfrak{A}$ annihilating $\mathcal{J}$ with $\phi(a) = \lambda$. Then $\phi \cdot \delta(a) = 0$ for all such states ϕ .

In the above theorem suppose δ_0 is (resp. quasi) well-behaved, then the set of δ_0^J-well-behaved operators in $\pi(\mathcal{D}(\delta_0)^s)$ consists of strongly δ^J-well-behaved operators where $\delta_0^J(\pi(a)) = \pi \cdot \delta(a)$ and $\mathcal{D}(\delta_0)^s$ is the self-adjoint part of $\mathcal{D}(\delta_0)$. In fact, δ_0^J-well-behaved operator is strongly δ_0^J-well-behaved . By Theorem 2, $\phi \cdot \delta(a) = 0$ since there is a state ϕ of $\mathfrak{A}$ annihilating $\mathcal{J}$ with $\phi(a) = \|\phi\|$, $\phi \cdot \delta_0(a) = 0$ for any δ_0^J-well-behaved operator $a \in \mathcal{D}(\delta_0)^s$. Therefore δ is (resp. quasi) well-behaved.

COROLLARY 1. Let δ_0, δ be as in THEOREM 2. Suppose δ_0 is (resp. quasi) well-behaved, then δ is (resp. quasi) well-behaved.

COROLLARY 2. If δ_0 is a generator, then for each $r \in \mathbb{R}$ with $|r|$ sufficiently small, $\delta_0 + r\delta$ is also a generator for any *-derivation δ with $\mathcal{D}(\delta) = \mathcal{D}(\delta_0)$.

13.3. QUANTIZATION OF SPACES.

Let $(\mathfrak{A}, G, \alpha)$ be a C*-dynamical system where G is discrete abelian, and let δ_0 be a *-derivation of $\mathfrak{A}$ commuting with α . Then there is a *-derivation $\tilde{\delta}_0$ of the C*-crossed product $\mathfrak{A} \times_\alpha G$ of $\mathfrak{A}$ by α such that

$$\tilde{\delta}_0(x)(g) = \delta_0[x(g)] \quad \text{for} \quad x \in \mathcal{D}(\delta_0) \odot_\alpha G ,$$

the set of all $\mathcal{D}(\delta_0)$-valued functions on G with finite support. Suppose δ_0 is a generator and put $\beta_t = \exp t\delta_0$. Let δ be a *-derivation of $\mathfrak{A} \times_\alpha G$ with $\mathcal{D}(\delta) = \mathcal{D}(\tilde{\delta}_0)$. Then $t \longrightarrow \delta\cdot\tilde{\beta}_t(x)$ is continuous for all $x \in \mathcal{D}(\delta)$ where $\tilde{\beta}$ is a contiuous action of $\mathbb{R}$ on $\mathfrak{A} \times_\alpha G$ such that

$$\tilde{\beta}_t(x)(g) = \beta_t[x(g)] \quad \text{for} \quad x \in L^1(G ; \mathfrak{A}) .$$

Indeed let

$$\mathcal{S} = \{ a \in \mathcal{D}(\delta_0) \mid b \to \delta(ba) \text{ is continuous from } \mathfrak{A}_0 \text{ into } \mathfrak{A} \times_\alpha G \}$$

where $\mathfrak{A}_0$ is as in 13.1. Suppose $(a_n) \subset \mathfrak{A}_0$ converges to 0 in $\mathfrak{A}_0$ and $(\delta(a_n))$ to $x \in \mathfrak{A} \times_\alpha G$. Since

$$\delta(a\lambda(g)b) = \delta(\lambda(g))\alpha_g^{-1}(a)b + \lambda(g)\delta(\alpha_g^{-1}(a)b)$$

and δ_0 commutes with α, one has $x\lambda(g)b = 0$ for all $g \in G$ and $b \in \mathcal{S}$. Then $\Phi(x\lambda(g)) \in L(\mathcal{S})$ where Φ is the conditional expectation of $\mathfrak{A} \times_\alpha G$ onto $\mathfrak{A}$, and $L(\mathcal{S})$ is as in 13.1. By the same way as 13.1, $\Phi(x\lambda(g)) = 0$ for all $g \in G$. Let $x = \Sigma_g x_g \lambda(g)$ be the Fourier expansion of x. Then $x_g = 0$, so $x = 0$. Therefore

$$\| \delta(a) \| \le M(\| a \| + \| \delta_0(a) \|) \quad (a \in \mathfrak{A}_0)$$

for some $M > 0$. Since

$$\delta \circ \tilde{\beta}_t(a\lambda(g)) = \delta\cdot\beta_t(a)\ \lambda(g) + \beta_t(a)\ \delta(\lambda(g)) ,$$

we have the conclusion. Hence there exist derivations $\tilde{\rho}_f$ $(f \in L^1(\mathbb{R}))$ of $\mathfrak{A} \times_\alpha G$ such that $\mathcal{D}(\tilde{\rho}_f) = \mathcal{D}(\delta)$ and

$$\tilde{\rho}_f = \int_{\mathbb{R}} f(t)\ \tilde{\beta}_t \cdot \delta \cdot \tilde{\beta}_{-t}\ dt .$$

Similarly we have derivations $\hat{\rho}_g$ $(g \in G)$ of $\mathfrak{A} \times_\alpha G$ such that $\mathcal{D}(\hat{\rho}_g) = \mathcal{D}(\delta)$ and

$$\hat{\rho}_g = \int_{\hat{G}} \overline{\langle g,p \rangle}\ \hat{\alpha}_p \cdot \delta \cdot \hat{\alpha}_{-p}\ dp$$

where $\hat{\alpha}$ is the dual action of α. Moreover if β is periodic, then $(\hat{\rho}_g)^{\sim}_0 = (\tilde{\rho}_0)^{\wedge}_g$ commutes with $\tilde{\beta}$, and $(\tilde{\rho}_n)^{\wedge}_e = (\hat{\rho}_e)^{\sim}_n$ commute with $\hat{\alpha}$. Especially $(\tilde{\rho}_0)^{\wedge}_e = (\hat{\rho}_e)^{\sim}_0$ commutes with $\hat{\alpha}$ and $\tilde{\beta}$. Now suppose $\mathfrak{A}$ is unital abelian and β is transitive and periodic. Then

$$(\tilde{\rho}_0)^{\wedge}_e = k\tilde{\delta}_0 + \delta_1$$

for some $k \in \mathbb{R}$ and a pregenerator δ_1 of $\mathfrak{A} \times_\alpha G$ such that δ_1 commutes with $\tilde{\delta}_0$ and $\delta_1|_{\mathfrak{A}} = 0$. In fact we denote $(\tilde{\rho}_0)^{\wedge}_e$ by ρ. Since ρ commutes with $\hat{\alpha}$, we known $\rho(a) \in \mathfrak{A}$ for all $a \in \mathcal{D}(\delta_0)$. So if $x_n \in \mathcal{D}(\rho)$ with $x_n \to 0$ and $\rho(x_n) \to y \in \mathfrak{A} \times_\alpha G$, we see

$$\Phi(x_n\lambda(g)^*) \to 0 \quad \text{and} \quad \Phi[(\rho(x_n) - y)\lambda(g)^*] \to 0 \quad \text{for each } g \in G .$$

let $x_n = \sum a_g^{(n)} \lambda(g) \in \mathcal{D}(\rho) = \mathcal{D}(\tilde{\delta}_0)$. Then $a_g^{(n)} \to 0$ and

$$\Phi[\Sigma_k\{\rho(a_k^{(n)})\lambda(k-g) + a_k^{(n)}\rho(\lambda(k))\lambda(g)^* - y_k\lambda(k-g)\}] \to 0$$

where $y = \Sigma_k y_k \lambda(k)$ is the Fourier expansion of y in $\mathfrak{A} \times_\alpha G$ $(y_k \in \mathfrak{A})$. Hence $a_g^{(n)} \to 0$ and $(a_g^{(n)}) \to y_g$ for each $g \in G$. Since $\mathcal{D}(\rho|_{\mathfrak{A}}) = \mathcal{D}(\delta_0)$

it follows from THEOREM 2 that $\rho|_{\mathfrak{A}}$ is closable. So $y_g = 0$ for all $g \in G$. Therefore $y = 0$, which means that ρ is closable. Hence one may assume that it is closed. Let $x \in C^*(G)$ and $(x_\iota) \subset \mathcal{D}(\tilde{\delta}_0)$ with $x_\iota \longrightarrow x$. Put

$$y_\iota = \int_T \tilde{\beta}_t(x_\iota)\, dt \in (\mathfrak{A} \times_\alpha G)^{\tilde{\beta}} = \mathfrak{A}^\beta \times_\alpha G = C^*(G)$$

by duality [7]. Since ρ commutes with $\tilde{\beta}$, and ρ is closed, we have $y_\iota \in \mathcal{D}(\rho)$ and $\rho(y_\iota) \in C^*(G)$. So $\rho|_{C^*(G)}$ is a closed *-derivation of $C^*(G)$. Since $\hat{\alpha}_p \circ \rho \circ \hat{\alpha}_{-p} = \rho$ for $P \in \hat{G}$ and $\mathcal{F} \circ \hat{\alpha}_p \circ \mathcal{F}^{-1} = \tau_p$ on $C(\hat{G})$, we have that $\hat{\rho} = \mathcal{F} \circ \rho \circ \mathcal{F}^{-1}$ commutes with τ on $C(\hat{G})$ where $\mathcal{F}$ is the Fourier isomorphism of $C^*(G)$ onto $C(\hat{G})$, and τ is the shift action of $\hat{G}$ on $C(\hat{G})$. It follows from Goodman-Nakazato [3,5] that there is a one parameter subgroup (p_t) of $\hat{G}$ such that

$$\hat{\rho}(f)(p) = \lim_{t \to 0} t^{-1}(f(p_t p) - f(p)) \quad \text{for all} \quad f \in \mathcal{D}(\hat{\rho}).$$

Since $\langle g, \cdot \rangle \in \mathcal{D}(\hat{\rho})$, one has $\rho(\lambda(g)) = \partial(g)\lambda(g)$ for all $g \in G$ where $\partial(g) = \lim_{t \to 0} (\langle g, p_t \rangle - 1)$. Let $\delta_1(a\lambda(g)) = \partial(g)\, a\lambda(g)$ for all $a \in \mathcal{D}(\delta_0)$ and $g \in G$. Then it is a pregenerator of $\mathfrak{A} \times_\alpha G$ such that $\mathcal{D}(\delta_1) = \mathcal{D}(\tilde{\delta}_0)$, $\delta_1|_{\mathfrak{A}} = 0$ and δ_1 commutes with $\tilde{\delta}_0$. Since ρ is a closed *-derivation of $\mathfrak{A} \times_\alpha G$ and $\rho|_{\mathfrak{A}}$ commutes with β, it follows from Batty [1] that $\rho|_{\mathfrak{A}} = f\, \delta_0$ on $\mathcal{D}(\delta_0)$ for some real valued function f on $sp(\mathfrak{A})$. Since β is transitive, $\rho|_{\mathfrak{A}} = k\delta_0$ for some $k \in \mathbb{R}$. Therefore we have that

$$\rho(a\, \lambda(g)) = k\delta_0(a)\, \lambda(g) + a\partial(g)\, \lambda(g) = (k\tilde{\delta}_0 + \delta_1)(a\lambda(g))$$

for $a \in \mathcal{D}(\delta_0)$ and $g \in G$, which implies the following lemma:

LEMMA 1. Let $(\mathfrak{A}, G, \alpha)$ be a C*-dynamical system where $\mathfrak{A}$ is unital abelian and G is discrete abelian. Let $\beta_t = \exp t\delta_0$ be a transitive action of $\mathbb{T}$ on $\mathfrak{A}$ commuting with α. Suppose δ is a *-derivation of $\mathfrak{A} \times_\alpha G$ with $\mathcal{D}(\delta) = \mathcal{D}(\tilde{\delta}_0)$, then

$$\iint_{\mathbb{T} \times \hat{G}} \hat{\beta}_t \circ \hat{\alpha}_p \circ \delta \circ \hat{\alpha}_{-p} \circ \hat{\beta}_{-t}\, dt = k\tilde{\delta}_0 + \delta_1$$

for some $k \in \mathbb{R}$ and a pregenerator δ_1 of $\mathfrak{A} \times_\alpha G$ such that $\mathcal{D}(\delta_1) = \mathcal{D}(\tilde{\delta}_0)$, $\delta_1|_{\mathfrak{A}} = 0$ and δ_1 commutes with $\tilde{\delta}_0$.

Now let δ be as in the above lemma. Consider $\hat{\rho}_g$, $g \in G$ namely

$$\hat{\rho}_g = \int_{\hat{G}} \overline{\langle g, p \rangle}\, \hat{\alpha}_p \circ \delta \circ \hat{\alpha}_{-p}\, dp .$$

Let $\delta(a) = \Sigma \delta(a)_g\, \lambda(g)$ be tha Fourier expansion of $\delta(a)$. Then $\hat{\rho}_g(a) = \delta(a)_g\, \lambda(g)$. Put $\delta_g(a) = \hat{\rho}_g(a)\, \lambda(g)^*$ for $a \in \mathcal{D}(\delta_0)$ $(g \neq e)$. Then it is linear of $\mathcal{D}(\delta_0)$ into $\mathfrak{A}$ with

$$\delta_g(ab) = \delta_g(a)\ \alpha_g(b) + a\delta_g(b)\ .$$

Suppose there is a unitary $u \in \mathcal{D}(\delta_0)$ such that $1 \notin sp(\alpha_g(u)u^*)$ and $\mathfrak{A} = C^*(u)$. Then

$$\delta_g(u^n) = \sum_{k=0}^{n-1} \alpha_g(u^k)u^{-k}\ \delta_g(u)u^{n-1}\ .$$

Since $1 \notin sp(\alpha_g(u)u^*)$, one has that

$$\sum_{k=0}^{n-1} \alpha_g(u^k)u^{-k} = (\alpha_g(u^n)u^{-n} - 1)(\alpha_g(u)u^* - 1)^{-1}\ .$$

So

$$\delta_g(u^n) = \delta_g(u)(\alpha_g(u - u)^{-1}(\alpha_g - id)(u^n)$$

for all $n \in \mathbb{Z}$ since $\delta_g(1) = 0$. Put $a_g = \delta_g(u)\ (\alpha_g(u) - u)^{-1} \in \mathfrak{A}$. Since $\mathfrak{A} = C^*(u)$, we have that $\delta_g = a_g(\alpha_g\text{-}id)$ on $\mathcal{D}(\delta_0)$.

So

$$\hat{\rho}_g(a) = a_g(\alpha_g\text{-}id)(a)\ \lambda(g) = [a_g\ \lambda(g),\ a] \quad \text{for all} \quad a \in \mathcal{D}(\delta_0)\ .$$

Let $\delta(\lambda(g)) = \sum \delta(\lambda(g))_h \lambda(h)$ be the Fourier expansion of $\delta(\lambda(g))$.

Then $\hat{\rho}_g(\lambda(h)) = \delta(\lambda(h))_{g+h}\lambda(g+h)$.

So

$$\hat{\rho}_g(a\lambda(h)) = \hat{\rho}_g(a)\lambda(h) + a\hat{\rho}_g(\lambda(h)) = [a_g\lambda(g),\ a]\lambda(h) + a\delta(\lambda(k))_{g+h}\lambda(g+h).$$

Now assume that δ commutes with $\tilde{\beta}$. Since α commutes with β and β is ergodic, we have $a_g \in \mathbb{C}$. Then $[a_g\lambda(g),\ a]\lambda(h) = [a_g\lambda(g),\ a\lambda(h)]$.

Since $\hat{\rho}_g - ad(a_g\lambda(g))$ is a derivation on $\mathcal{D}(\tilde{\delta}_0)$, one has that

$$\delta(\lambda(h+k))_{h+k+g}\ u = \delta(\lambda(h))_{h+g}\ \alpha_g(u) + \delta(\lambda(k))_{k+g}\ u$$

for all $h, k \in G$. Since $1 \in \mathcal{D}(\delta)$, one has $\delta(1)_g = 0$. So $\delta(\lambda(h))_{g+h} = 0$ for all $h \in G$, or $\alpha_g(u) = u$. Since $1 \notin sp(\alpha_g(u)u^*)$, we have $\delta(\lambda(h))_{g+h} = 0$ for all $h \in G$. So $\hat{\rho}_g(\lambda(h)) = 0$ for all $h \in G$.

Hence $\hat{\rho}_g = ad(a_g\lambda(g))$ on $\mathcal{D}(\delta)$ $(g \neq e)$. On the other hand

$$\delta = \sum\hat{\rho}_g = \hat{\rho}_e + \sum_{g\neq e}\hat{\rho}_g \quad \text{on } \mathcal{D}(\delta) \text{ by definition.}$$

Since $\hat{\rho}_e$ commutes with $\hat{\alpha}$, $\tilde{\beta}$, it follows from Lemma 1 that

$$\delta = k\tilde{\delta}_0 + \delta_1 + \sum_{g\neq e} ad(a_g\lambda(g)) \quad \text{on } \mathcal{D}(\delta)\ .$$

Let $\delta_F = ad(\sum_{g\in F} a_g\lambda(g))$ for a finite set F of $G\backslash\{e\}$ with $F = -F$.

Since

$$\delta_F \longrightarrow \sum_{g\neq e} \mathrm{ad}(a_g\lambda(g)) = \delta_2$$

pointwisely on $\mathcal{D}(\delta)$, we have that $\delta = k\tilde{\delta}_0 + \delta_1 + \delta_2$ such that k, δ_1 are as in LEMMA 1 and δ_2 is an approximately bounded *-derivation of $\mathfrak{A} \times_\alpha G$ with $\delta_2(\lambda(G)) = 0$. Applying the above argument to $\tilde{\rho}_0$, we deduce the following lemma :

LEMMA 2. Let $(\mathfrak{A}, G, \alpha)$, β, δ_0, and δ be as in LEMMA 1. Suppose there exists a unitary $u \in \mathcal{D}(\delta_0)$ such that $\mathfrak{A} = C^*(u)$ and $1 \notin sp(\alpha_g(u)u^*)$ $(g \neq e)$, then we have that

$$\int_{\mathbb{T}} \tilde{\beta}_t \circ \delta \circ \tilde{\beta}_t \, dt = k\tilde{\delta}_0 + \delta_1 + \delta_2 \quad \text{for some } k \in \mathbb{R} ,$$

a pregenerator δ_1 and an approximately bounded *-derivation δ_2 of $\mathfrak{A} \times_\alpha G$ such that

$$\mathcal{D}(\delta_1) = \mathcal{D}(\delta_2) = \mathcal{D}(\tilde{\delta}_0) , \ \delta_1|_{\mathfrak{A}} = 0 , \ \delta_2(\lambda(G)) = 0$$

and δ_1 commutes with $\tilde{\delta}_0$.

Now suppose α is effective and there is an eigenunitary u for β generating $\mathfrak{A}$. Since β commutes α and β is ergodic, we have $\alpha_g(u)u^* \in \mathbb{C}\,1$. Since $\mathfrak{A} = C^*(u)$ and α is effective, there are $c_g \neq 1$ $(g \neq e)$ with $\alpha_g(u) = c_g u$. Since $\tilde{\beta}_t \cdot \tilde{\rho}_n \cdot \tilde{\beta}_{-t} = e^{int}\tilde{\rho}_n$ for all $n \in \mathbb{Z}$ and $t \in \mathbb{T}$, we see $\tilde{\beta}_t \cdot \tilde{\rho}_n(\lambda(g)) = e^{int}\tilde{\rho}_n(\lambda(g))$. Since $\beta_t(u^n) = e^{itn}u^n$, $\tilde{\rho}_n(\lambda(g)) = u^n b(n,g)$ for some $b(n,g) \in C^*(G)$. Let $\delta(\lambda(g)) = \sum \delta(\lambda(g))_h \lambda(h)$ and $b(n,g) = \sum b(n,g)_h \lambda(h)$ be the expansion of $\delta(\lambda(g))$ and $b(n,g)$. Then

$$\delta(\lambda(g))_h = a(0) + \sum_{n\neq 0} b(n,g)_h \, u^n$$

where $a(0)$ is the 0th-component. Since $\tilde{\rho}_0 = k\tilde{\delta}_0 + \delta_1 + \delta_2$ by LEMMA 2, $\tilde{\rho}_0(\lambda(g)) = \partial(g)\lambda(g)$. Since

$$\int_{\mathbb{T}} \beta_t(\delta(\lambda(g)))dt = \partial(g)1 \ (g \neq h), \ = 0 \ (\text{otherwise}).$$

which is $a(0)$,

$$\delta(\lambda(g)) = \partial(g)\lambda(g) + \sum_{n\neq 0} u^n b(n,g) .$$

By Lemma 1, $(\hat{\rho}_e)^{\sim}_0 = k'\tilde{\delta}_0 + \delta_1'$. Let $\hat{\rho}_e(u) = \sum_n a_n u^n \in \mathfrak{A}$. Since

$$\int_{\mathbb{T}} e^{-it} \beta_t \cdot \hat{\rho}_e(u)dt = k'\delta_0(u) \quad \text{and} \quad \beta_t(u) = e^{it} u ,$$

we have

$$\hat{\rho}_e(u) = k'\delta_0(u) + \sum_{n\neq 1} a_n u^n .$$

Then

$$\hat{\rho}_e(u^n)\lambda(g) = k'\tilde{\delta}_0(u^n\lambda(g)) + \sum_{m\neq 1} na_m u^{m+n-1} \lambda(g) .$$

Consequently we have that

$$\delta(u^n\lambda(g)) = (\Sigma_h\hat{\rho}_h)(u^n)\lambda(g) + u^n(\Sigma_m\tilde{\rho}_m)(\lambda(g)) = (k'\tilde{\delta}_0+\delta_1)(u^n\lambda(g))$$

$$+ \sum_{h\neq e}[a_h\lambda(h), u^n]\lambda(g) + \sum_{m\neq 0}\{u^n\tilde{\rho}_m(\lambda(g)) + na_{m+1}u^{n+m}\lambda(g)\}$$

where

$$\hat{\rho}_k(a) = [a_k\lambda(h), a] \quad \text{for} \quad a \in \mathcal{D}(\delta_0) .$$

Since

$$\delta - k'\tilde{\delta}_0 - \delta_1 - \sum_{h\neq e}\mathrm{ad}(a_h\lambda(h))$$

is a*-derivation,

$$u^n\{\sum_{m\neq 0}\tilde{\rho}_m(\lambda(g)) - \sum_{h\neq e}[a_h\lambda(h), \lambda(g)]\} = u^n x_g\lambda(g) = \delta_x(u^n\lambda(g))$$

for some x in $Z^1_\alpha(G,\mathfrak{A})$, the set of all α-one cocycles of G to $\mathfrak{A}$ vanishing at e. Consequently, we have that

$$\delta(u^n\lambda(g)) = (c\tilde{\delta}_0+\delta_1)(u^n\lambda(g)) + \sum_{h\neq e}[a_n\lambda(h), u^n\lambda(g)] + \delta_x(u^n\lambda(g))$$

for some $c \in \mathbb{R}$, which implies the following theorem:

THEOREM 3. Let $(\mathfrak{A}, G, \alpha)$ be a C*-dynamical system where $\mathfrak{A}$ is unital abelian, G is discrete abelian and α is effective. Let $\beta_t = \exp t\delta_0$ be a periodic action of $\mathbb{R}$ on $\mathfrak{A}$ commuting with α. Suppose there exists an eigen unitary for β generating $\mathfrak{A}$, then any *-derivation δ of $\mathfrak{A} \times_\alpha G$ with $\mathcal{D}(\delta) = \mathcal{D}(\tilde{\delta}_0)$ can be expressed as $\delta = k\tilde{\delta}_0 + \delta_1 + \delta_2 + \delta_x$ for some $k \in \mathbb{R}$, a pregenerator δ_1 and an approximately bounded *-derivation δ_2 and a*-derivation δ_x of $\mathfrak{A} \times_\alpha G$ such that $\mathcal{D}(\delta_1) = \mathcal{D}(\delta_2) = \mathcal{D}(\tilde{\delta}_0)$, $\delta_1|_{\mathfrak{A}} = 0$ and δ_1 commutes with $\tilde{\delta}_0$, where $\delta_x(a\lambda(g)) = ax_g\lambda(g)$ for $x \in Z^1_\alpha(G,\mathfrak{A})$ and $a \in \mathcal{D}(\delta_0)$.

COROLLARY 3. Let $\mathfrak{A}_\theta$ be the irrational rotation algebra. Then there are two non approximately bounded pregenerators δ_1, δ_2 of $\mathfrak{A}_\theta$ such that for any *-derivation δ of $\mathfrak{A}_\theta$ with $\mathcal{D}(\delta) = \mathcal{D}(\delta_1)\cap\mathcal{D}(\delta_2)$, $\delta = k_1\delta_1 + k_2\delta_2 + \delta_3$ for some k_1, $k_2 \in \mathbb{R}$ and approximately bounded *-derivation δ_3 of $\mathfrak{A}_\theta$ with $\mathcal{D}(\delta_3) = \mathcal{D}(\delta)$.

REFERENCES

1. C.J.K. Batty : Derivations on compact spaces, Preprint, 1978.

2. C.J.K. Batty : Small perturbations of C*-dynamical systems, Comm. Math. Phys., 68 (1979), 39-43.

3. F.M. Goodman : Translation invariant closed *-derivations, Preprint, 1980.

4. R. Longo : Automatic relative boundedness of derivations in C*-algebras, J. Func. Anal., 34 (1979), 21-28.

5. H. Nakazato : Closed *-derivations on compact groups, Preprint, 1980.

6. S. Sakai : Derivations in operator algebras, Lecture Notes, 1979.

7. H. Takai : On a duality for crossed products of C*-algebras, J. Func. Anal., 19 (1975), 25-39.

8. H. Takai : On a problem of Sakai in unbounded derivations, Preprint, 1980.

DEPARTMENT OF MATHEMATICS
TOKYO METROPOLITAN UNIVERSITY
TOKYO 158, JAPAN

Proceedings of Symposia in Pure Mathematics
Volume 38 (1982), Part 2

UNBOUNDED *-DERIVATIONS AND INFINITESIMAL GENERATORS ON OPERATOR ALGEBRAS

Ola Bratteli
Palle E. T. Jørgensen

1. INTRODUCTION

One important aspect of the theory of unbounded derivations is related to the existence problem for the dynamics in the C*- or W*-formalism of infinite quantum systems, [10,12,13].

Mathematically, this aspect of the theory centers around the following problems from semi-group theory.

(a) The estimate $\| a + \alpha\delta(a) \| \geq \| a \|$ for $\alpha \in \mathbb{R}$, and elements a in the domain $D(\delta)$ of the given derivation δ

(b) Density of the subspaces $(I \pm \delta)(D(\delta))$

If δ is bounded, then the exponential $e^{t\delta}$ is always well defined. But an unbounded derivation δ is generally not even closable. But if it is, and if $\bar{\delta}$ generates a one-parameter group, then we say that δ is a pre-generator.

We shall adopt the view that for reversible systems the dynamics is given by a strongly continuous one-parameter group of *-automorphisms of a C*-algebra, the observable algebra [10], and for dissipative systems the irreversible dynamics is described by a completely positive semigroup [13]. The differential equation for the dynamics is then formulated in terms of the infinitesimal generator δ. This operator δ is a *-derivation for closed systems (reversible dynamics), and a dispersive operator for open systems (irreversible dynamics).

Definition [11], [26] An operator δ on a C*-algebra $\mathfrak{A}$ is said to be dispersive if $\delta(a^*) = \delta(a)^*$ and

$$\omega(\delta(a)a + a\,\delta(a)) \leq 0$$

for all $a = a^* \in D(\delta)$ and some tangent functional ω at the positive part

---

1980 Mathematics Subject Classification.

0082-0717/80/0000-0506/$04.25

a_+ of a, or equivalently

$$\omega(\delta(a)) \leq 0$$

for all $a = a^* \in D(\delta)$, and corresponding tangent functional ω at a_+.

This notion is strictly stronger than the notion of dissipative or well behaved operator, introduced earlier in this section, and it can be proved, [11], that if δ is a dispersive operator on $\mathfrak{A}$ such that $(I-\alpha\delta)(D(\delta)) = \mathfrak{A}$ for some $\alpha > 0$, then δ generates a positivity preserving semigroup of contractions. Conversely (as pointed out above), such a generator is dispersive.

For a given operator $\delta : D(\delta) \longrightarrow \mathfrak{A}$ defined on a dense subspace $D(\delta)$ in a C*- , or W*-algebra $\mathfrak{A}$, there are two approaches to the problem of constructing the dynamics

(A) Show that δ is closable, and that the closure $\bar{\delta}$ is the generator of a dynamical semigroup (Note that if δ is a *-derivation and $\bar{\delta}$ generates a group, it can easily be seen that this is a group of *-automorphisms)

(B) Show that δ can be extended (possibly beyond its closure) to a generator.

We will concentrate on the first approach in Section 2 and part of Section 3, and thereafter describe the second approach. The first approach has dominated research activity on the subject in the seventies [1,2,10,17,22, 27,30].

2. Approximately invariant subspaces and related topics

In the description of quantum lattice systems and Fermi gases one encounters an increasing sequence $\{\mathfrak{A}_n\}$ of C*-subalgebras of $\mathfrak{A}$ such that $1 \in \mathfrak{A}_n \subseteq D(\delta)$ and $\bigcup_{n=1}^{\infty} \mathfrak{A}_n$ is dense in $\mathfrak{A}$, [33]. We will describe results in a more general framework where $\delta : D(\delta) \longrightarrow \mathfrak{A}$ is merely assumed to be a linear dissipative operator on a Banach space $\mathfrak{A}$, and $\mathfrak{A}_n$ are only assumed to be closed linear subspaces with $\mathfrak{A}_n \subseteq D(\delta)$. If in this situation δ leaves all the subspaces $\mathfrak{A}_n$ globally invariant, $\delta(\mathfrak{A}_n) \subseteq \mathfrak{A}_n$, then $\bar{\delta}$ clearly generates a contraction semigroup leaving all the subspaces globally invariant. We will now consider various perturbations of this situation, in particular cases of approximatively invariant subspaces.
The following theorem of Batty, [2], generalizes earlier results of Kishimoto [22] and Jφrgensen [16].

THEOREM 1 (Batty) Let $\delta : D(\delta) \longrightarrow \mathfrak{A}$ be a dissipative operator on a Banach space $\mathfrak{A}$, and let $\{\mathfrak{A}_n\}$ be an increasing sequence of closed subspaces with $\mathfrak{A}_n \subseteq D(\delta)$ and $\bigcup_n \mathfrak{A}_n$ dense. Let $Q_n : \mathfrak{A} \longrightarrow \mathfrak{A}/\mathfrak{A}_n$ be the quotient mapping and assume that

$$\| Q_n \, \delta \mid \mathfrak{A}_n \| = 0(1)$$

Then $\delta \mid \bigcup_n \mathfrak{A}_n$ is a pregenerator of a contraction semi-group.

In the context of lattice systems the condition of this theorem can be interpreted as an upper bound for the interaction energy across the boundaries of localized regions. The 0(1)-condition is appropriate for one-dimensional systems. More generally, for lattice dimension d the growth of the surface energy will usually be of the form $0(n^{d-1})$. The best result with a growth condition $0(n)$ was obtained by Bratteli and Kishimoto [6], generalizing earlier results by Jφrgensen [16], [17].

THEOREM 2 (Bratteli-Kishimoto) Let $\delta : D(\delta) \longrightarrow \mathfrak{A}$ be a dissipative operator on a Banach space $\mathfrak{A}$, and let $\{\mathfrak{A}_n\}$ be an increasing sequence of closed sub-spaces with $\mathfrak{A}_n \subseteq D(\delta)$ and $\bigcup_n \mathfrak{A}_n$ dense. Assume there exist bounded linear operators $L_{n,m} : \mathfrak{A}_n \longrightarrow \mathfrak{A}_{n+m}$ and constants $M_n, \alpha > 0$ such that

$$\| L \mid \mathfrak{A}_n - L_{n,m} \| \leq M_n \, e^{-\alpha m}$$

for $n = 1,2, \ldots,$ $m = 0,1, \ldots$.

It follows that L is a pregenerator of a contraction semi-group.

The idea of the proof of this theorem is to use a perturbation expansion where the perturbation is not fixed, but depends on the expansion step. This idea was already suggested in [17], where the theorem was proved in the case that $\delta = \delta_1 + \delta_2$, satisfying $\delta_1(\mathfrak{A}_n) \subseteq \mathfrak{A}_n$, $\delta_2(\mathfrak{A}_n) \subseteq \mathfrak{A}_{n+1}$ and $\| \delta_2 \mid \mathfrak{A}_n \| = 0(n)$ for $n = 1,2, \ldots$. Even earlier, Jφrgensen [16], proved a Hilbert space version of THEOREM 2 which allowed even slightly stronger growth, for example $0(n \log n)$, but not $0(n^{1+\varepsilon})$ with $\varepsilon > 0$.

THEOREM 3 (Jφrgensen) Let $H : D(H) \longrightarrow \mathcal{H}$ be a symmetric operator on a Hilbert space $\mathcal{H}$, and let $\{P_n\}$ be an increasing sequence of orthogonal projections on $\mathcal{H}$ converging strongly to 1 such that $P_n \mathcal{H} \subseteq D(H)$. Assume that $H \, P_n \mathcal{H} \subseteq P_{n+1} \mathcal{H}$ and

$$\| H \, P_n - P_n H \, P_n \| \leq a_n$$

for some sequence of positive numbers a_n satisfying

$$\sum_{n=1}^{\infty} a_n^{-1} = \infty$$

It follows that H is essentially self-adjoint on $\bigcup_n P_n \mathcal{H}$.

It was also shown by examples in [16] that the theorem is not true with an $O(n^2)$ bound instead of the present bound. However, Jørgensen [18] later proved that in the case that H is semibounded, the theorem is true with an $O(n^2)$ bound.

THEOREM 4 (Jørgensen) Let $H : D(H) \longrightarrow \mathcal{H}$ be a semibounded symmetric operator on a Hilbert space $\mathcal{H}$, and let $\{P_n\}$ be an increasing sequence of self-adjoint projections on $\mathcal{H}$ converging strongly to 1 such that $P_n \mathcal{H} \subseteqq D(H)$. Assume that $H P_n \mathcal{H} \subseteqq P_{n+1}\mathcal{H}$ and

$$\| H P_n - P_n H P_n \| \le a_n$$

for some sequence of positive numbers on satisfying

$$\sum_{n=1}^{\infty} a_n^{-\frac{1}{2}} = \infty$$

It follows that H is essentially self-adjoint on $\bigcup_n P_n \mathcal{H}$.

In addition to the method of approximately invariant subspaces described above, there are two other general methods to decide when a given derivation (or more general infinitesimal operator) is a pre-generator. They are also useful for applications to quantum lattice systems. The first of these is based an analytic elements, the second an approximate commuting approximants. Each of these three methods has their relative merits and drawbacks, and each cover cases which are not covered by the others. For example, the approximately invariant subspace method gives the best results in dimensions ≤ 2, the analytic element method gives good results in all dimensions for translationally invariant short range interactions, while the approximate commutativity method gives the best results for classical interaction in all dimensions.

The most general analytic element, [19], result appears to be

THEOREM 5 [17] Let $\delta : D(\delta) \longrightarrow \mathfrak{A}$ be a dissipative operator on a Banach space $\mathfrak{A}$, and assume that $D(\delta)$ contains a dense set of analytic elements.

Then $\delta \big| \mathfrak{A}_0$ is a pre-generator.

This criterion is not too useful as it stands, but Robinson [28] and Ruelle [29] has adopted it to quantum lattice system to give an easily applicable criterion. They define a norm on the space of interactions, and finiteness of this norm implies that the associated closed derivation has a dense set of analytic elements, and therefore is a generator. In particular their criterion covers all translationally invariant finite range interactions. The precise statement of this criterion requires details of the notation used to describe quantum lattice systems. Therefore we omit it.

Finally there are the criteria based on approximate commutation. The following result was proved by Bratteli and Kishimoto, [6]. In the case of exact commutation it was proved earlier by Sakai [31]

THEOREM 6 (Bratteli-Kishimoto) Let $\{\delta_n\}$ be a sequence of bounded, dissipative operators on a Banach space $\mathfrak{A}$ with densely defined graph limit δ. Assume that $\| \delta_n \delta_m - \delta_m \delta_n \| \leq M < \infty$ for all n, m, and $\gamma_n = \lim_{m\to\infty} \delta_n \delta_m - \delta_m \delta_n$ exists strongly for all n.

It follows that δ is the generator of a contraction semigroup. If the individual δ_ns are derivations, the semigroup extends to an automorphism group. That δ is the graph limit of δ_n means that the graph of δ on $\mathfrak{A} \times \mathfrak{A}$ is the limit of the graphs of δ_n as $n \to \infty$ in the obvious sense.

Olesen and Pedersen, [25], also proved that the existence of γ_n in THEOREM 6 could be replaced by the condition that

$$\| \delta_n \delta_m - \delta_m \delta_n \| \longrightarrow 0$$

as $n, m \longrightarrow \infty$.

For a more extensive treatment of the application of these results to quantum lattice system, the reader is referred to [10].

3. von Neumann algebras

The extension problem for *-derivations (approach (B)) is paralleled to the extension problem for symmetric operators in Hilbert space, but von Neumann's theory of deficiency spaces, and the Friedrichs-Krein results on semibounded operators, do not carry over to *-derivations. In a spatial representation, there are however interesting applications of the classical theory to the *-derivations [4,5,8,19,20,21]. These spatial representations appear when one has a derivation δ on a C*-algebra $\mathfrak{A}$ and a state ω on $\mathfrak{A}$ such that $\omega\circ\delta$ is "sufficiently bounded", more precisely.

PROPOSITION 7 (Bratteli-Robinson [8]) Let δ be a *-derivation defined on a *-subalgebra $\mathcal{D}$ of the bounded operators on a Hilbert space $\mathcal{H}$. Let $\Omega \in \mathcal{H}$ be a unit vector cyclic for $\mathcal{D}$ in $\mathcal{H}$ and denote the corresponding state by ω, i.e.

$$\omega(A) = (\Omega, A\,\Omega), \qquad A \in \mathcal{D}$$

The following conditions are equivalent

(1) $$|\omega(\delta(A))| \leq L(\omega(A^*A) + \omega(AA^*))^{\frac{1}{2}}$$

for all $A \in \mathcal{D}$ and some $L \geq 0$

(2) There exists a symmetric operator H on $\mathcal{H}$ with the properties

$$D(H) = \mathcal{D}\,\Omega$$

and

$$\delta(A)\psi = iHA\psi - iAH\psi$$

for all $A \in \mathcal{D}$ and $\psi \in D(H)$.

More generally, let $\mathcal{M}$ be a von Neumann algebra an a Hilbert space $\mathcal{H}$, and let $\mathcal{D}$ be a weak *-dense *-subalgebra of $\mathcal{M}$. Let H be a symmetric operator on $\mathcal{H}$, and suppose that the domain $D(H)$ of H is invariant under each operator in $\mathcal{D}$ such that we can form $\mathrm{ad}\,H(A) = HA - AH$ as an operator on $D(H)$. If this operator extends to a bounded operator on $\mathcal{H}$, and this extension is in $\mathcal{M}$, then $\delta = i\,\mathrm{ad}\,H$ is a *-derivation in $\mathcal{M}$ with domain $\mathcal{D}$, and we say that δ is a spatial derivation implemented by H. This definition of δ depends a priori on the choice of $\mathcal{D}$ and $D(H)$. In certain situations this dependence is not very strong, however. The simplest of these is when H is essentially self-adjoint on $D(H)$, i.e.

PROPOSITIONS 8 (Bratteli-Robinson [8])

Let H be a self-adjoint operator on a Hilbert space $\mathcal{H}$ and let

$$\alpha_t(A) = e^{itH} A e^{-itH}$$

be the corresponding one-parameter group of automorphisms of $\mathcal{L}(\mathcal{H})$. Denote by δ the infinitesimal generator by α(δ is defined by the derivative in the weak*-topology on $\mathcal{L}(\mathcal{H})$). For $A \in \mathcal{L}(\mathcal{H})$, the following conditions are equivalent

(1) $A \in D(\delta)$

(2) There exists a core D for H such that the sesquilinear form $\psi,\phi \in D \times D \longrightarrow i(H\psi,A\phi) - i(\psi,AH\phi)$ is bounded

(3) There exists a core D for H such that $AD \subseteq D$ and the mapping

$$\psi \in D \longmapsto iHA\psi - iAH\psi \text{ is bounded}$$

If (2) or (3) are fullfilled, then they are also fullfilled for $D = D(H)$, and the bounded operator associated with the sesquilinear form in (2), as well as the bounded operator defined by (3), both coincides with $\delta(A)$.

If H is semi-bounded, it is well known that the closure of the bilinear form associated to H defines a self-adjoint extension $\hat{H}$ of H, called the Friedrichs extension. It is remarkable that $\hat{H}$ implements the same derivation as H.

THEOREM 9 (Jφrgensen [20]) Let H be a semibounded symmetric operator implementing a *-derivation δ with domain $\mathcal{D}$ in a von Neumann algebra $\mathcal{M}$. Then the domain of the Friedrichs extension $\hat{H}$ of H is invariant under each A in $\mathcal{D}$, and

$$\delta(A)\psi = iHA\psi - iAH\psi$$

for all $A \in \mathcal{D}$ and all vectors $\psi \in D(\hat{H})$.

Thus, one possibility for extending a spatial derivation is extension of the implementing operator H_0. Suppose that H_0 has been extended to a self-adjoint H, and that H implements a *-derivation. Then there still remains the problem of deciding whether the unitary group e^{itH} normalizes the given von Neumann algebra $\mathfrak{M}$. We first describe an abstract approach to this problem.

Let $F(H)$ be the linear subspace of the predual $\mathfrak{M}_*$ spanned by the vector states $(\psi, \cdot\, \psi)$ with $\psi \in D(H)$. Then $F(H)$ is in the domain of the pre-adjoint operator δ_* in $\mathfrak{M}_*$. For $\omega \in \mathfrak{M}_*$, let V_ω denote the partial isometry part in the polar decomposition $\omega = V_\omega|\omega|$

THEOREM 10 (Jørgensen [20]) Let δ be a spatial *-derivation on a von Neumann algebra $\mathfrak{M}$, implemented by a self-adjoint operator H.

The following conditions are equivalent

(1) $e^{itH}\mathfrak{M}e^{-itH} = \mathfrak{M}$ for all $t \in \mathbb{R}$.

(2) The identity

$$\delta_*(\omega)(V_\omega) = 0$$

holds for all $\omega \in F(H)$.

A more concrete approach to this problem, geared to applications to quantum statistical mechanics, has been studied in the papers [4,5,8,9,14] where one assumes the existence of a cyclic vector Ω satisfying $H\Omega = 0$.

If δ is the derivation of $\mathfrak{M}$ implemented by H, the aim of this approach is to replace the condition that δ is a generator (i.e. $(I\pm\delta)(D(\delta)) = \mathfrak{M}$) with the more tractable condition that H is essentially self-adjoint on $D(\delta)\Omega$ (i.e. $(I \pm iH)(D(\delta)\Omega)$ are dense in $\mathcal{H}$). This problem has an easily proved solution in the case that $D(\delta)\Omega$ contains sufficiently many analytic elements for H, more precisely:

PROPOSITION 11 (Bratteli-Herman-Robinson [5]) Let δ be a spatial derivation of a von Neumann algebra $\mathfrak{M}$ implemented by a symmetric operator H and assume that $\mathfrak{M}$ has a cyclic vector Ω such that $H\Omega = 0$. Further, assume there exists a *-subalgebra $\mathcal{D} \subseteq D(\delta)$ such that

(1) $\mathcal{D}$ is weak*-dense in $\mathfrak{M}$

(2) $\delta(\mathcal{D}) \subseteq \mathcal{D}$

(3) $\mathcal{D}\Omega$ consists of analytic elements for H.

It follows that H is essentially self-adjoint and its closure $\bar{H}$ satisfies

$$e^{it\bar{H}}\mathcal{M}e^{-it\bar{H}} = \mathcal{M} \quad \text{for all} \quad t \in \mathbb{R} .$$

In the case where $D(\delta)\Omega$ a priori does not contain analytic vectors for H one can nevertheless prove results in two cases of physical interest. The first case occurs in connection with ground state phenomena. The operator H is interpreted as the Hamiltonian and is positive, and the eigenvector Ω corresponds to the lowest energy state of the system, i.e. the ground state. The second case does not necessarily involve the interpretation of H as a Hamiltonian and it is not longer supposed to be positive. On the other hand the eigenvector Ω is assumed to be both cyclic and, separating for the von Neumann algebra. This situation typically occurs in the description of finite temperature equilibrium states in statistical mechanics.

THEOREM 12 (Bratteli-Robinson [8], [10]) Let $\mathcal{M}$ be a von Neumann algebra on a Hilbert space $\mathcal{H}$ with cyclic vector Ω_1 and δ a spatial derivation of $\mathcal{M}$ implemented by a self-adjoint operator H such that $\Omega \in D(H)$ and $H\Omega = 0$. Assume that δ is defined on the largest domain possible,

$$D(\delta) = \{A,\ A \in \mathcal{M},\ \ i[H,A] \in \mathcal{M}\}$$

and assume that $D(\delta)\Omega$ is a core for H. Further assume that

either $H \geq 0$

or Ω is separating for $\mathcal{M}$.

The following conditions are equivalent

(1) $e^{itH}\mathcal{M}e^{-itH} = \mathcal{M}$, $t \in \mathbb{R}$;

(2) $e^{itH}\mathcal{M}_+\Omega \subseteq \mathcal{M}_+\Omega$, $t \in \mathbb{R}$;

(3) $e^{itH}\mathcal{M}_+\Omega \subseteq \overline{\mathcal{M}_+}\Omega$, $t \in \mathbb{R}$;

(4) $(I \pm iH)^{-1}\mathcal{M}_+\Omega \subseteq D(\delta)_+\Omega$;

(5) $(I \pm iH)^{-1}\mathcal{M}_+\Omega \subseteq \mathcal{M}_+\Omega$;

(6) $(I \pm iH)^{-1}\mathcal{M}_+\Omega \subseteq \overline{\mathcal{M}_+\Omega}$.

(The bar denotes closure)

In the case that $H \geq 0$ one can actually use Borchers' theorem to prove more.

COROLLARY 13 [10] Adopt the general assumptions of Theorem 12 and assume $H \geq 0$.

The following conditions are equivalent

(1) $e^{itH} \mathcal{M} e^{-itH} = \mathcal{M}$, $t \in \mathbb{R}$;

(2) $e^{itH} \in \mathcal{M}$, $t \in \mathbb{R}$;

(3) $e^{itH} A'\Omega = A'\Omega$, $A' \in \mathcal{M}'$, $t \in \mathbb{R}$;

(4) $e^{itH} \mathcal{M}_{s\,a.}\Omega \subseteq \mathcal{M}_{s\,a.}\Omega$

(5) $e^{itH} \overline{\mathcal{M}_{s\,a.}\Omega} \subseteq \overline{\mathcal{M}_{s\,a.}\Omega}$

($\mathcal{M}_{s\,a.}$ denotes the self-adjoint elements in $\mathcal{M}$)

THEOREM 12 is unsatisfactory in the case that Ω is separating because it has too many hypothesis. For example any of the conditions (2)-(6) imply that e^{itH} determines a one-parameter group of Jordan automorphisms of $\mathcal{M}$, and hence this is a one-parameter group of automorphisms and $e^{itH}\mathcal{M} e^{-itH} \subseteq \mathcal{M}$ in the case that $\mathcal{M}$ is a factor, or $\mathcal{M}$ is abelian. On the other hand, the condition that $D(\delta)\Omega$ is a core for H is alone "almost" enough to ensure that $e^{itH}\mathcal{M} e^{-itH} \subseteq \mathcal{M}$ (One can show by counterexamples that neither of these two types of conditions can be removed completely in the general case of a separating and cyclic vector Ω [10],) . The following theorem was first proved for abelian $\mathcal{M}$ by Gallavotti-Pulvirenti [14], then it was generalized to the case where Ω defines a trace state (i.e. $\Delta = 1$) by Bratteli-Robinson [9], and finally the general result was proved by Bratteli-Haagerup [4]

THEOREM 14 (Bratteli-Haagerup-Robinson) Let $\mathcal{M}$ be a von Neumann algebra, on a Hilbert space $\mathcal{H}$, with a cyclic and separating vector Ω. Let H be a self-adjoint operator on $\mathcal{H}$ such that

$$H\Omega = 0 \quad \text{and define}$$

$$\mathcal{D} = \{A \in \mathcal{M},\ \ i[H,A] \in \mathcal{M}\}$$

Let Δ be the modular operator associated to the pair $(\mathcal{M},\Omega)$, and let $\mathcal{H}_{\#}$ be the graph Hilbert space associated to $\Delta^{\frac{1}{2}}$ (i.e. $\mathcal{H}_{\#}$ is the linear space $D(\Delta^{\frac{1}{2}})$ equipped with the inner product

$$(\psi,\ \phi)_{\#} = (\Delta^{\frac{1}{2}}\ \psi,\ \Delta^{\frac{1}{2}}\ \phi) + (\psi,\ \phi))$$

The following conditions are equivalent

(1) $e^{itH}\mathcal{M} e^{-itH} = \mathcal{M}$, $t \in \mathbb{R}$;

(2) (a) $\mathcal{D}\,\Omega$ is a core for H

(b) H and Δ commute strongly, i.e.

$$\Delta^{is}\ H\Delta^{-is} = H \qquad ,\ s \in \mathbb{R}\ ;$$

(3) The restriction of H to $\mathcal{D}\Omega$ is essentially self-adjoint as an operator on $\mathcal{H}_{\#}$.

4. The extension problem

We now consider the extension problem (B) for infinitesimal operators $\delta : D(\delta) \longrightarrow \mathfrak{A}$ with dense domain $D(\delta)$ in a unital C*-algebra. The operator δ is said to be completely dissipative if the induced map, $(a_{ij}) \longrightarrow (\delta(a_{ij}))$, is dissipative in $\mathfrak{A} \otimes M_n$, where M_n denotes the $n \times n$ complex matrix algebra and n ranges over $\mathbb{N}$.

THEOREM 15. (Jφrgensen [21]) Let $\delta : D(\delta) \longrightarrow \mathfrak{A}$ be a linear transformation with dense domain in a nuclear (and unital) C*-algebra $\mathfrak{A}$. Assume that

(i) δ is completely dissipative;

(ii) $1 \in D(\delta)$ and $\delta(1) = 0$;

(iii) $D(\delta)$ is self-adjoint, and $\delta(a^*) = \delta(a)^*$ for all $a \in D(\delta)$.

Then there is an operator system, i.e., a norm-closed, self-adjoint, linear subspace B, $\mathfrak{A} \subset B \subset \mathfrak{A}^{**}$, and a strongly continuous, completely positive, semigroup $\tau_t : B \to B$, $0 \leq t < \infty$, with infinitesimal generator $\tilde{\delta}$ extending δ, i.e., $\delta(a) = \tilde{\delta}(a)$ for all $a \in D(\delta)$.

THEOREM 16 (Jφrgensen [21]) Let $\delta: D(\delta) \longrightarrow \mathfrak{A}$ be a completely dissipative linear operator with dense domain $D(\delta)$ in a unital C*-algebra $\mathfrak{A}$. Suppose $1 \in D(\delta)$ and $\delta(1) = 0$.

Then for all $a \in D(\delta)$ such that $a^*a \in D(\delta)$, we have the inequality

$$\delta(a^*a) \geq \delta(a)^*a + a^*\delta(a)$$

The proof of THEOREM 16 is based on the idea that completely dissipative maps may be regarded as generalized tangents to the completely positive maps. This view leads to applications of Arveson's Hahn-Banach theorem, injectivity, and the Schwarz inequality for completely positive maps.

New research of Walter [34] on semitangents at the identity element in the dual of a locally compact group G of type T has recently been carried out via completely dissipative maps (semiderivations) in $C^*(G)$. The independent work of Walter has revealed that the same abstract tangential approach as described above leads to new insight into the dual object $\hat{G}$. [1(W)]

Suppose finally that δ is a *-derivation. Then it is natural to work with a two-sided condition rather than the one-sided dissipative notion. The concept of a well-behaved derivation (Sakai [30]) has already been introduced

A*-derivation δ is said to be well-behaved if for all positive $a \in D(\delta)$ there is a state ω on $\mathfrak{A}$ such that $\omega(a) = \| a \|$ and $\omega(\delta(a)) = 0$. It can be proved (See for example Sakai [33], Batty [3]) that this is equivalent to each of the conditions

(2) For all $a = a^* \in D(\delta)$ there is a state ω on $\mathfrak{A}$ such that

$$|\omega(a)| = \| a \| \quad \text{and} \quad \omega(\delta(a)) = 0$$

(3) $\pm\,\delta$ are dissipative

(4) $\| a + \alpha\delta(a)\| \geq \| a \|$ for all $\alpha \in \mathbb{R}$ and all $a \in D(\delta)$

More surprising is the following fact.

THEOREM 17 (Jϕrgensen [21]) Let δ be a *-derivation in a C*-algebra, and assume that δ is well behaved. Then each of the operators $\pm\,\delta$ is completely dissipative.

As a COROLLARY of THEOREM 15 and 17 we get

COROLLARY 18. (Jϕrgensen [21]) Let δ be a well behaved *-derivation on a nuclear, unital C*-algebra $\mathfrak{A}$.

Then there is an operator system B, $\mathfrak{A} \subset B \subset \mathfrak{A}^{**}$, $b \in B \Rightarrow b^* \in B$, as in Theorem 15, and a strongly continuous, completely positive, semigroup (τ_t, B) with generator extending δ.

We remark that THEOREM 6, specialized to the C*-algebra of a quantum lattice system, can be cast in a form such that it is a statement about existence of extensions of *-derivations to generators (possibly beyond their closures). Finally, the note of F. Goodman in these proceedings reports on joint work with Jϕrgensen on extendability of *-derivations commuting with a compact action on the C*-algebra in question. [1(G)]

REFERENCES

1. Lecture Notes for the 1980 Summer Institute of Operator Algebras and Applications, Kingston, Ontario. Reference is made to articles of the following authors : F. Goodman (G), M.E. Walter (W).

2. C.J.K. Batty, Dissipative mappings with approximately invariant subspaces, J. Func. Anal. 31 (1979), 336-341.

3. C.J.K. Batty, Small perturbations of C*-dynamical systems, Comm. Math. Phys. 68 (1979), 39-43.

4. O. Bratteli and U. Haagerup, Unbounded derivations and invariant states Comm. Math. Phys. 59 (1978), 79-95.

5. O. Bratteli, R.H. Herman and D.W. Robinson, Quasianalytic vectors and derivations of operator algebras, Math. Scand. 39 (1976), 371-381.

6. O. Bratteli and A. Kishimoto, Generation of semigroups and two-dimensional quantum lattice systems, J. Func. Anal. 35 (1980), 344-368.

7. O. Bratteli and D.W. Robinson, Unbounded derivations of C*-algebras, I Comm. Math. Phys. 42 (1975), 253-268, II Comm. Math. Phys. 46 (1976), 11-30.

8. O. Bratteli and D.W. Robinson, Unbounded derivations of von Neumann algebras, Ann. Inst. H. Poincare, Sec. A, 25 (1976), 139-164.

9. O. Bratteli and D.W. Robinson, Unbounded derivations and invariant trace states, Comm. Math. Phys. 46 (1976), 31-35.

10. O. Bratteli and D.W. Robinson, Operator Algebras and Quantum statistical Mechanics I, Springer Verlag, New York-Heidelberg-Berlin (1979), II to appear in the end of 1980 (Contains additional references).

11. O. Bratteli and D.W. Robinson, Positive C_0-semigroups on C*-algebras, Math. Scand. (to appear).

12. D.E. Evans and J.T. Lewis, Dilations of Irreversible Evolutions in Algebraic Quantum Theory, Dublin Inst. Adv. Studies A 24 (1977).

13. D.E. Evans, A review on semigroups of completely positive maps, in Mathematical Problems in Theoretical Physics, K. Osterwalder Ed., Lecture Notes in Physics 116, Springer Verlag, Berlin-Heidelberg-New York (1980).

14. G. Gallavotti and M. Pulvirenti, Classical KMS condition and Tomita-Takesaki theory, Comm. Math. Phys. 46 (1976), 1-9.

15. F. Goodman and P.E.T. Jϕrgensen, Unbounded derivations commuting with a compact group of automorphisms, work in progress

16. P.E.T. Jϕrgensen, Approximately reducing subspaces for unbounded linear operators, J. Func. Anal. 23 (1976), 392-414.

17. P.E.T. Jϕrgensen, Approximately invariant subspaces for unbounded linear operators II, Math. Ann. 227 (1977), 177-182.

18. P.E.T. Jϕrgensen, Essential self-adjointness of semibounded operators, Math. Ann. 237 (1978), 187-192.

19. P.E.T. Jϕrgensen, On one-parameter groups of automorphisms, and extensions of symmetric operators associated with unbounded derivations in operator algebras, Tohôku Math. J. 30 (1978), 279-305.

20. P.E.T. Jϕrgensen, Commutators of Hamiltonian operators and non-abelian algebras, J. Math. Anal. Appl. 73 (1980), 115-133.

21. P.E.T. Jϕrgensen, The existence problem for dynamics in the C*-formulation of dissipative quantum systems, Preprint, Aarhus Univ. (1980).

22. A. Kishimoto, Dissipations and derivations, Comm. Math. Phys. 47 (1976), 25-32.

23. G. Lumer and R.S. Phillips, Dissipative operators in Banach space, Pac. J. Math. 11 (1961), 679-698.

24. E. Nelson, Analytic vectors, Ann. Math. 70 (1959), 572-615.

25. D. Olesen and G.K. Pedersen, Groups of automorphisms with a spectrum condition and the lifting problem, Comm. Math. Phys. 51 (1976), 85-95.

26. R.S. Phillips, On the generation of semigroups of linear operators, Pac. J. Math. 2 (1952) 343-369.

27. R. Powers and S. Sakai, Unbounded derivations in operator algebras, J. Func. Anal. 19 (1975), 81-95.

28. D.W. Robinson, Statistical mechanics of quantum spin systems, I. Comm. Math. Phys. 6 (1967), 151-160, II. Comm. Math. Phys. 7 (1968),

337-348.

29. D. Ruelle, Statistical Mechanics, Benjamin, New York-Amsterdam (1969).

30. S. Sakai, Recent developments in the theory of unbounded derivations in C*-algebras, in C*-algebras and Applications to Physics, H. Araki and R.V. Kadison eds., Lecture Notes in Mathematics 650, Springer Verlag, Berlin-Heidelberg-New York (1978).

31. S. Sakai, On commutative normal *-derivations, Comm. Math. Phys. 43 (1975), 39-40.

32. S. Sakai, On one-parameter subgroups of *-automorphisms on operator algebras and the corresponding unbounded derivations, Am. J. Math. 98 (1976), 427-440.

33. S. Sakai, The Theory of Unbounded Derivations in C*-algebras, Lecture Notes, Copenhagen University and University of Newcastle upon Tyne (1977).

34. M.E. Walter, Differentiation on the dual of a group: An introduction, Preprint, U. Colorado (1980).

Department of Mathematics
University of Trondheim
Trondheim, Norway

Department of Mathematics
University of Aarhus
Aarhus, Denmark

Proceedings of Symposia in Pure Mathematics
Volume 38 (1982), Part 2

SEMIDERIVATIONS ON GROUP $C^{\dagger}$-ALGEBRAS

an abstract by Martin E. Walter

Consider the following five examples:

1) $L = \frac{d^2}{dx^2}$ on $\mathbb{R}^1$; $L(f\bar{f}) \geq (L\bar{f})f + \bar{f}(Lf)$, pointwise inequality of functions. This is a simple example of a completely dissipative operator from physics. The operator L is a semitangent vector (see definition below) concretely realized by what we shall call a semiderivation (see definition below).

2) The Lévy-Khinchine formula on $\mathbb{R}^n$:

$$-\psi(y) = c + i\ell(y) + q(y) + \int_{\mathbb{R}^n - \{0\}} [1 - \exp(-i(x|y)) - \frac{i(x|y)}{1+\|x\|^2}] \frac{1+\|x\|^2}{\|x\|^2} d\mu(x)$$

where $x,y \in \mathbb{R}^n$, $c \geq 0$, ℓ is a continuous linear form, q is a continuous, nonnegative quadratic form and μ is a non-negative bounded measure on $\mathbb{R}^n - \{0\}$ such that the above integral converges.

3) A function ψ on $\mathbb{R}^1$ that satisfies $\psi(e) \leqq 0$, $\psi(-x) = \overline{\psi(x)}$, and

$$\int_{\mathbb{R}^1} \psi(x) (\frac{d\varphi}{dx})^{\#} * \frac{d\varphi}{dx}(x)dx \geq 0 \quad \text{for all} \quad \varphi \in C_c^{\infty}(\mathbb{R}^1) \quad ,$$

where $\varphi^{\#}(x) = \overline{\varphi(-x)}$, the bar denoting the complex conjugate, $*$ denoting convolution, and φ an infinitely differentiable function with compact support.

4) $H^1(G,H(\pi)) = Z^1(G,H(\pi))/B^1(G,H(\pi))$, the first cohomology group of continuous, unitary representation π of G .

5) The "screw functions" of J. von Neumann and I.J. Schoenberg which allow isometric imbeddings of $\mathbb{R}^n$ into Hilbert space.

The main point of this talk is that these (perhaps) seemingly diverse examples are really examples of the same abstract object; namely, they are all examples of semitangents at 1 on the dual of a group. In this case the dual of locally compact group G is taken to be $P(G)_1$, the collection of all continuous positive definite functions of norm one on G .

DEFINITION: A semitangent vector at 1 to $P(G)_1$ is any continuous complex valued function ψ on G satisfying $\psi(g) = \lim_{j\to\infty} n_j(p_{n_j}(g) - 1)$ for each $g \in G$, where $\{n_j\}$ is some subsequence of the natural numbers converging to

0082-0717/80/0000-0586/$01.75

∞ as $j \to \infty$, and $\{p_{n_j}\} \subset P(G)_1$.

One can prove the following

THEOREM 1. Let ψ be a continuous, complex-valued function on G. Then ψ is a semitangent at 1 to $P(G)_1$, i.e., $\psi \in N_0(G)$, if and only if $\psi(e) = 0$, $\psi(g^{-1}) = \overline{\psi(g)}$ for all $g \in G$, and for each choice of natural number n and each choice of n elements $g_1, g_2, \ldots, g_n$ from G the $n \times n$ matrix $(\psi(g_i^{-1}g_i) - \psi(g_j^{-1}) - \psi(g_i))$ is positive hermitian, i.e.,

$$\sum_{i,j=1}^{n} \{\psi(g_j^{-1}g_i) - \psi(g_j^{-1}) - \psi(g_i)\}\lambda_i\bar{\lambda}_j \geq 0 \tag{1}$$

for any choice of complex numbers $\lambda_1, \lambda_2, \ldots, \lambda_n$.

DEFINITION. A linear operator ∂ defined on a norm dense subspace of $C^*(G)$, denoted $\mathrm{Dom}(\partial)$, with values in $C^*(G)$ is called a semi-derivation on $C^*(G)$ if $x \in \mathrm{Dom}(\partial)$ implies $x^* \in \mathrm{Dom}(\partial)$ and

$$\partial(x^*x) \geq (\partial x^*)x + x^*\partial x \tag{3}$$

whenever $x \in \mathrm{Dom}(\partial)$ and $x^*x \in \mathrm{Dom}(\partial)$. We define a semiderivation on $W^*(G)$ to be a linear operator defined on a σ-weakly dense self-adjoint subspace of $W^*(G)$, denoted $\mathrm{Dom}_\sigma(\partial)$, which satisfies (3) whenever x and x^*x are in $\mathrm{Dom}_\sigma(\partial)$. Note that $\geq$ is the operator order.

PROPOSITION 6. Each semitangent ψ in $N_0(G)$ defines a semiderivation, denoted ∂_ψ, which is a closed operator on $C^*(G)$. A semiderivation, again denoted ∂_ψ, is also defined on $W^*(G)$. If ψ is a tangent (at 1) in $P(G)_1$ then ∂_ψ is a derivation.

We remark that ψ is a tangent (by definition) if ψ and $-\psi$ are semitangents.

Some elementary examples of semitangents can be read off from example 2 on our first page. In particular $ix = \lim_{n\to\infty} n(\exp(ix/n) - 1)$; and the operator which is multiplication by this function is the Fourier transform of d/dx on the dual group of the real line, which is again the real line.

The function $-x^2 = \lim_{n\to\infty} n^2(\cos(x/n) - 1)$ is also in $N_o(R)$, and (as a multiplication operator) it is the Fourier (Plancherel) transform of d^2/dx^2 on R.

One might ask if the third derivative transforms to an element of $N_o(R)$, and the answer is no. The absolute value of a function in $N_o(R)$ can grow no faster than the function x^2. Thus the notion of semidifferentiation is sufficiently restrictive.

Other examples of functions of negative type can be calculated. For example, Haagerup has shown that the function -(word length) on the free group on n generators is of negative type.

In "suitable" co-ordinates the function $-\ln(\cosh t)$ $(\infty - < t < \infty)$ is of negative type on $SL(2,R)$. One might note that all of our examples so far are unbounded. There are groups for which the only functions of negative type are of the form $p - 1$, where p is positive definite of norm one. This class of groups is precisely the class of groups with property (T) , see the talk by C.A. Akemann.

BIBLIOGRAPHY

1. Martin E. Walter, Differentiation of the Dual of a Group: An Introduction, preprint.

2. C.A. Akemann and Martin E. Walter, Unbounded Negative Definite Functions, preprint.

Department of Mathematics
University of Colorado
Boulder, Colorado 80309

Proceedings of Symposia in Pure Mathematics
Volume 38 (1982), Part 2

PERTURBATIONS

R. Herman

We've already seen, in Bratteli's lectures, conditions which will guarantee that a derivation generates a continuous one parameter automorphism group of a C* or W*-algebra. It is then natural to consider perturbation of this "generator" and determine when the perturbation generates an automorphism group. This is one view of perturbations and Longo [8] has a nice result in this direction. The view we propose to employ is that propounded by Buchholtz and Roberts. They posed the problem of finding a characterization of certain "natural" perturbation of generators. Work of the author with Bratteli and Robinson [2] come shortly after [3].

In [2] some cohomological problems arose which implicitly raised question about groups other than the real line. This was the subject of a second paper by the author and Jon Rosenberg [5].

The explicit situation in [3] was to consider the case of two one-parameter *-automorphism groups, α_t, β_t acting on a simple unital C*-algebra $\mathfrak{A}$ or von Neumann algebra $\mathcal{M}$, such that $\| \alpha_t - \beta_t \| \longrightarrow 0$ as $t \longrightarrow 0$. One then wanted imformation as to how the two generators δ_α, δ_β were related. The result of the Buchholtz-Roberts investigation was that $\delta_\alpha = \gamma\circ(\delta_\beta + \delta)\circ\gamma^{-1}$ where δ is a bounded (inner) derivation and γ is an inner automorphism of the algebra, with the implementing unitary having special continuity properties with respect to β_t. This led naturally to the question of what could be said if $\| \alpha_t - \beta_t \|$ was small enough for t small. There the results for the von Neumann algebra situation were remarkably similar to those of [3]. The major difference is in the lack of continuity of the implementing unitary as above. That the two situations were truly distinct is a consequence of example 4.9 of [2], where the following was exhibited: there exist two one parameter automorphism groups of $\mathcal{L}(\mathfrak{H})$, such that $\| \alpha_t - \beta_t \| = \delta$ for all $t \in \mathbb{R}\backslash\{0\}$, and δ is any preassigned number in $[0,2]$. We will now outline how the proof proceeds referring the reader to the original papers for more detail. The results rest heavily on work of Kadison [7], Kadison and Ringrose [6] and Sakai

0082-0717/80/0000-0546/$02.25

[12,13].

A theorem of Kadison and Ringrose [6] tells us that for a C*-algebra, $\mathfrak{A}$, if an automorphism γ is such that $\| \gamma - \iota \| < 2$, then γ lies on a norm continuous one-parameter subgroup of automorphisms of $\mathfrak{A}$. If $\mathfrak{A}$ is simple and unital or a von Neumann algebra, then the derivation which generates the subgroup is inner [7 or 12 and 13] and so γ itself is inner. In fact one can show that log(γ) is a bounded and hence inner [ibid] derivation.

Suppose now that we define $\gamma_t = \alpha_t \circ \beta_t^{-1}$ for $t \in \mathbb{R}$ and that $\| \alpha_t - \beta_t \| < 2$ for t in a neighborhood η of the identity. By the remarks above $\gamma_t = \mathrm{Ad}\, u_t$, for $t \in \eta$, where u_t is a unitary operator in $\mathfrak{A}$. We note that γ_t satisfies

$$\gamma_{ts} = \gamma_t \circ \gamma_s^{\beta_t}$$

where

$$\gamma_s^{\beta_t} = \beta_t \circ \gamma_s \circ \beta_t^{-1} .$$

Since the group generated by η is all of $\mathbb{R}$, the relationship just noted implies that γ_t is inner for all $t \in \mathbb{R}$. At this point we now have that $\alpha_t(x) = u_t \beta_t(x) u_t^*$ for all $t \in \mathbb{R}$ and $x \in \mathfrak{A}$ (M). In order to go futher it is necessary to obtain u_t with additional properties. These are set forth in the following:

Definition. Let G be a topological group and let $\alpha, \beta ; G \longrightarrow \mathrm{Aut}(\mathfrak{A})$ be two representations of G as automorphisms of a unital C*-algebra $\mathfrak{A}$. We say that α and β are *exterior equivalent* if there exists a continuous map $t \longrightarrow u_t$ from G to the unitary group of $\mathfrak{A}$ such that

$$\alpha_t(x) = u_t \, \beta_t(x) u_t^* \quad \text{for } t \in G \text{ and } x \in \mathfrak{A}$$

and

$$u_{ts} = u_t \, \beta_t(u_s) \quad \text{for } t,s \in G,$$

or in short of α and β are related by a "unitary one-cocycle".

We've spoken of continuous maps and for this we need to define the topology on the unitary group of $\mathfrak{A}$. For a unital C*-algebra we mean simply the topology of norm convergence and for a von Neumann algebra, ,M, the strong *-topology. Note that if $\mathfrak{A}$ or M_* is separable we then have Polish groups [2].

Suppose for the moment that we knew α and β to be exterior equivalent. What can one say about the unitary one-cocycle $t \longrightarrow u_t$. Consider only $G = \mathbb{R}$ now. If the map $t \longrightarrow u_t$ were not only continuous but also differentiable then the structure is well known [See for instance, 2] and u_t is given by an expansional as (for $t > 0$)

$$u_t = \sum_{n=0}^{\infty} \int_0^t dt_1 \cdots \int_0^{t_{n-1}} dt_n \, \beta_{t_n}(ih) \cdots \beta_{t_1}(ih)$$

where $ih = \frac{d}{dt} u_t \Big|_{t=0}$. A similar expression holds for $t < 0$. Differentiating the expression $\alpha_t = \mathrm{Ad}\, u_t \circ \beta_t$ then gives $\delta_\alpha = \delta_\beta + \mathrm{ad}(ih)$. However $t \longrightarrow u_t$ is not generally differentiable and so it must be smoothed. The result is the theorem we alluded to in the introduction theorem [then 3.6 of 2]. Let M be a von Neumann algebra with separable pre-dual and α and β two σ-weakly continuous one-parameter groups of *-automorphism of M. If there exists a $\varepsilon > 0$ such that $\| \alpha_t - \beta_t \| < 0.28$ for $0 \le t \le \varepsilon$ then $\delta_\alpha = \gamma \cdot (\delta_\beta + \delta) \circ \gamma^{-1}$.

Here δ is an inner derivation of M and γ is an inner automorphism. The automorphism γ arises precisely because of the smoothing of the cocycle. In fact it disappear if $\| \alpha_t - \beta_t \| = O(t)$ as $t \longrightarrow 0$ [2, Thm 3.1].

Thus for we have assumed that u_t satisfied the conditions of definition. To get to this point there are several difficulties to overcome. One is the continuity of the map $t \longrightarrow u_t$ and the other is that it must obey the cocycle identity. The first is taken care of by applying Borel cross-section theorems [See 2 and 5 for the detailed arguments] so that it is possible to choose the map $t \longrightarrow u_t$ to be Borel. The second problem leads immediately to a cohomological one. It is easy to see that u_{ts} and $u_t\, \beta_t(u_s)$ both implement α_{ts}, so that $z(t,s) = u_{ts}^*\, u_t\, \beta_t(u_s)$ belongs to the center of the algebra. In fact it is a twisted two-cocycle with values in the unitary group of the center. For the case of the real line and the von Neumann algebra situation it turn out that this group always vanishes. (See remark 2.10 of [3]). This isn't quite what was shown in [3] or [5]. Rather vanishing theorems were proved for the relavent portion of H^2 used. The result is that the u_t may be adjusted by unitaries in the center so as to satisfy the cocycle relation, and hence the theorem. As $H^2(G,T) \neq 0$ for many commonly occuring groups, e.g. $\mathbb{R}^n$, $n \geq 2$, the vanishing of the cohomological obstruction could not necessarily be expected. This prompted the investigation [5].

Briefly the cohomological problem indicated above is avoided by directly choosing certain unitary operators so that the relavent two-cocycles is cohomologous to zero. The method used will be described shortly. Other results contained in [5] include, a C*-version of a result of Moore viz. $H^2(G,\mathbb{T}) = 0$ implies $H^2(G,C(X,\mathbb{T})) = 0$ for G a, connected, simply connected, Lie group and X a compact metric space, and a characterization of the connected component (in the topology of pointwise convergence) of the automorphism group of an AF-algebra. As a corollary to the latter we obtain that all representations of a connected group by automorphisms of an AF-Algebra leave any trace invariant.

We will now describe in somewhat more detail the results of [5]. First of all it happens that for many separable C*-algebra the pathological mentioned at the beginning is disappears. Namely we have

Proposition. Let $\mathfrak{A}$ be a unital, separable C*-algebra, all of whose derivations are inner. Let G be a second countable, locally compact group, and let $\alpha,\beta : G \to \mathrm{Aut}(\mathfrak{A})$ be two representations of G as automorphisms of $\mathfrak{A}$. Suppose that for all t in some neighborhood N of the identity e of G, $\| \alpha_t - \beta_t \| < 2$. Then $\| \alpha_t - \beta_t \| \to 0$ as $t \to e$ in G. This having been said we now state

Theorem. Let $\mathfrak{A}$ be a finite, unital, separable, simple C*-algebra and let G be a connected, simply connected Lie group. Suppose that α and β are representations of G as automorphisms of $\mathfrak{A}$ and that β leaves a (normalized continuous) trace τ on $\mathfrak{A}$ invariant. If $\| \alpha_t - \beta_t \| < 2$ for t in some neighborhood of e in G, then α and β are exterior equivalent.

The main trick is, as indicated, a felicitous choice of the unitaries u_t such that $\gamma_t (= \alpha_t \circ \beta_t^{-1}) = \mathrm{Ad}\, u_t$. Note that by Proposition $\gamma_t \to \iota$ as $t \to e$. We can, as above, choose the u_t in a Borel fashion. Note that in our case

$$z(t,s) = u_{ts}^* \, u_t \, \beta_t(u_s) \in \mathbb{T} .$$

Let us write $u_t = e^{ih_t}$ and define $k_t = h_t - \tau(h_t)1$, where τ is the normalized trace in the statement of theorem. Then $\mathrm{Ad}\, e^{ik_t} = \gamma_t$ and $\tau(k_t) = 0$. For t sufficiently near e, the norm of k_t will be small. write $u_{ts} = e^{ik_{ts}}$ and $u_t \, \beta_t(u_s) = e^{ik_t} e^{i\beta_t(k_s)} = e^{iK}$ where $K = k_t + \beta_t(k_s) +$ commutator terms. This last is valid because we can choose k_t and $\beta_t(k_s)$ of sufficiently small norm so as to satisfy the requirement for applying the Campbell-Baker-Hausdorff formula. If in addition $\| K \| < \Pi$ (this we can guarantee as well) then since $\mathrm{Ad}\, e^{ik_{ts}} = \mathrm{Ad}\, e^{iK}$ and $\tau(k_{ts}) = \tau(K) = 0$ we see that

$$u_{ts} = u_t \, \beta_t(u_s)$$

for t,s sufficiently near e. This says that z(t,s) is locally trivial and since G is simply connected it must be trivial. For the complete proof the reader should consult [5].

Related to the question of exterior equivalence for center fixing automorphism groups is the generalization of Moore's result to C*-algebras that we mentioned above. Namely

Theorem. Let G be a connected, simply connected Lie group such that $H^2(G,T) = 0$, Let X be a compact metric space, and let C(X,T) be the group of continuous T-valued functions on X (viewed as a trivial G-module). Then $H^2(G,C(X,T)) = 0$.

Remark $H^2(\mathbb{T},C(S^1,\mathbb{T})) \neq 0$ even though $H^2(\mathbb{T},\mathbb{T}) = 0$, so simple connectivity of G is necessary in the above. Using the above we have [5]

Theorem Let $\mathfrak{A}$ be a separable, unital C*-algebra all of whose derivations are inner. Let G be a connected, simply connected Lie group with $H^2(G,T) = 0$ and let α and β be two representations of G as automorphisms of $\mathfrak{A}$, fixing the center of $\mathfrak{A}$. If $\| \alpha_t - \beta_t \| < 2$ for all t in some neighborhood of e in G, then α is exterior equivalent to β.

In closing we refer the reader as well to the paper of Hansen and Olesen [4] where it is shown that for an abelian, connected group G acting on a factor M an inner perturbation ($\equiv \alpha_t \circ \beta_t^{-1}$ is inner for all t) is given by a quasi-one-cocycle u_t ($u_{t+s} = b(t,s)\, u_t\, \beta_t\, (u_s)$ where $b(t,s)$ is a bihomomorphism from $G \times G \longrightarrow T$).

REFERENCES

1. W. Arveson, An Invitation to C*-algebras, Springer-Verlag, 1976.
2. O. Bratteli, R. Herman and D. Robinson, Perturbations of flows on Banach spaces and operator algebras, Comm. Math. Phys. 59 (1978), 167-196.
3. D. Buchholz and J. Roberts, Bounded perturbations of dynamics, Comm. Math. Phys. 49 (1976), 161-177.
4. F. Hansen and D. Olesen, Perturbations of center fixing dynamical systems, Math. Scand. 41 (1977), 295-307.
5. R. H. Herman and J. Rosenberg, Norm-closed group actions on C*-algebra, submitted for publication.
6. R. Kadison and J. Ringrose, Derivations and automorphisms of operator algebras, Comm. Math. Phys. 4 (1967), 32-63.
7. R. Kadison, Derivations of Operator algebras, Ann. Math. 83 (1966), 280-293.
8. R. Longo, On perturbed derivations
9. C. C. Moore, Group extensions and cohomology for locally compact groups, III, Trans. Amer. Math. Soc. 22 (1976), 1-33,
10. __________, Ibid, IV, Trans. Amer. Math. Soc. 221 (1976), 35-58.
11. K. Parthasarathy, Multipliers on locally compact groups, Springer Lecture Notes in Mathematics 93 (1969).
12. S. Sakai, Derivations of W*-algebras, Ann. of Math. 83 (1966), 287-293.
13. S. Sakai, Derivations of simple C*-algebras, J. Functional Analysis 2 (1968), 202-206.

Department of Mathematics
Pennsylvania State University
University Park, Pa. 16802
U.S.A.

Proceedings of Symposia in Pure Mathematics
Volume 38 (1982), Part 2

GENERATORS OF DYNAMICAL SEMIGROUPS.

David E. Evans.

ABSTRACT. Spatial descriptions of one-parameter semigroups of completely positive maps on C*-algebras are obtained, which are relevant to irreversible Markovian dynamics in quantum systems.

In the C*-algebraic framework of quantum theory, reversible dynamics is described, in the Heisenberg picture, by a strongly continuous one-parameter group of *-automorphisms. One is then interested in deciding when there is a Hamiltonian which implements the time development. Irreversible dynamics for subsystems, represented by a C*-subalgebra B can be considered via a projection or conditional expectation N of A onto B, and then forming the family of maps $\{T_t = N \circ \alpha_t : t \geq 0\}$ on B. If N is a projection of norm one and $N(1_A) = 1_B$, then N is automatically completely positive by a theorem of Tomiyama. Thus one is forced to consider a family $\{T_t : t \geq 0\}$ of completely positive maps of the algebra B. In general, because of memory effects, the family does not satisfy the Markovian law $T_t \circ T_s = T_{t+s}$, $t, s \geq 0$. However in a large number of physical models, the Markov property has been derived rigorously, under weak or singular coupling limits for example (see e.g. [3]). However these procedures can sometimes destroy the complete positivity of the dynamics, as shown in [5], but some clarification of this has been made in [4], where under certain conditions and asymptotic modification, the complete positivity can be retained.

So one can define a dynamical semigroup on a C*-algebra to be strongly continuous one-parameter semigroup of completely positive contractions, and use them as a framework for studying irreversible Markovian dynamics. This definition was introduced by Lindblad [12], motivated by the study of operations which remain positive when interacting with finite quantum systems.

Let $\{e^{tL} : t \geq 0\}$ be a strongly continuous contraction semigroup on a C*-algebra A, which will be a dynamical semigroup if and only if $(\lambda - L)^{-1}$ is completely positive for all large positive real λ. If a derivation δ generates a strongly continuous one-parameter group of *-derivations, then by Hille-Yoshida theory δ^2 generates a strongly continuous contraction

---

1980 Mathematics Subject Classification. 46L

0082-0717/80/0000-0523/$01.75

semigroup. Moreover $(\lambda-\delta)^{-1}$ is completely positive for all large real λ, and hence so is $(\lambda-\delta^2)^{-1} = (\sqrt{\lambda}-\delta)^{-1}(\sqrt{\lambda}+\delta)^{-1}$ for all large positive λ, and so $\{e^{t\delta^2}:t \geq 0\}$ is a dynamical semigroup. Thus derivations and squares of derivations generate dynamical semigroups. Hence if $h, h_\alpha \in A_{s.a.}$ and $\Sigma||h_\alpha||^2 < \infty$, then by the Lie-Trotter product formula

$$L = i\,\mathrm{ad}h + \sum_\alpha (i\,\mathrm{ad}h_\alpha)^2$$
$$= K(\cdot) + (\cdot)K^* + 2\sum_\alpha h_\alpha(\cdot)h_\alpha$$

(where $K = ih - \Sigma h_\alpha^2$) generates a dynamical semigroup. More generally, if $K \in A$ and ψ is a completely positive map on A, then $L = K(\cdot)+(\cdot)K^* + \psi$ generates a dynamical semigroup. In the converse direction, if a (bounded) L generates a dynamical semigroup, then we seek $K \in A$ such that $L-K(\cdot)-(\cdot)K^*$ is completely positive, or equivalently $e^{tL} \geq e^{tK}(\cdot)e^{tK^*}$, $t \geq 0$ by the Lie-Trotter formula. (Here we write $T \geq S$ for maps T,S between C*-algebras if $T-S$ is completely positive).

THEOREM [1]. If $\{e^{tL}:t \geq 0\}$ is a norm continuous dynamical semigroup on a C*-algebra A on a Hilbert space H, then there exists $K \in A''$ s.t. $e^{tL} \geq e^{tK}(\cdot)e^{tK^*}$, $t \geq 0$.

This had been first shown for finite dimensional matrix algebras and hyperfinite von Neumann algebras in [10, 12]. By [13, 8] the above theorem reduces to showing that certain cohomology groups vanish. This was shown, in [1], by methods similar to those employed in [11]. Using the above theorem, the spatial description can be improved in the following sense:

THEOREM [2, 8]. If $\{e^{tL}:t \geq 0\}$ is a norm continuous dynamical semigroup on a C*-algebra A $B(H)$, in its universal representation, then there exists a Hilbert space H_0 and a strongly continuous contraction semigroup G_t on $H \otimes H_0$ such that $e^{tL}(x)\otimes 1 = G_t(x\otimes 1)G_t^*$; $x \in A$, $t \geq 0$.

This result was exploited in [8, 9] to obtain dilations of certain dynamical semigroups on von Neumann algebras to groups of automorphisms.

Some analogous results are known for unbounded generators in particular cases. See for example [6, 7] for spatial descriptions of quasi-free dynamical semigroups on the Fermion algebra, which provide examples of strongly continuous dynamical semigroups which are not norm continuous and not simply of product type.

BIBLIOGRAPHY

1. E. Christensen and D.E. Evans. Cohomology of operator algebras and quantum dynamical semigroups. J. London Math. Soc. (2) 20 (1979), 358-368.

2. E.B. Davies. Some contraction semigroups in quantum probability. Z. Wahrschein. 23 (1972), 261-273.

3. E.B. Davies. Master equations: a survey of rigorous results. Rend. Sem. Mat. Fis. Milano. 47 (1977) 165-173.

4. E.B. Davies. Asymptotic modifications of dynamical semigroups on C*-algebras. Preprint, Oxford 1980.

5. R. Dumcke and H. Spohn. The proper form of the generator in the weak coupling limit. Zeit. für Physik. B34 (1979), 411-412.

6. D.E. Evans. Completely positive quasi-free maps on the CAR algebra. Commun. math. Phys. 70 (1979), 53-68.

7. D.E. Evans. Dissipators for symmetric quasi-free dynamical semigroups on the CAR algebra. J. Functional Anal. (to appear).

8. D.E. Evans and J.T. Lewis. Dilations of dynamical semigroups. Commun. math. Phys. 50 (1976), 219-227.

9. D.E. Evans and J.T. Lewis. Dilations of irreversible evolutions in algebraic quantum theory. Commun. Dubl. Inst. Adv. Studies. Ser. A. 24, 1977.

10. V. Gorini, A. Kossakowski, E.C.G. Sudarshan. Completely positive dynamical semigroups of N-level systems. J. Math. Phys. 5 (1976), 821-825.

11. R.V. Kadison. A note on derivations of operator algebras. Bull. London Math. Soc. 7 (1975), 41-44.

12. G. Lindblad. On the generators of quantum dynamical semigroups. Commun. math. Phys. 48 (1976), 119-130.

13. G. Lindblad. Dissipative operators and cohomology of operator algebras. Letters in Math Phys. 1 (1976), 219-224.

DEPARTMENT OF MATHEMATICS,
UNIVERSITY OF WARWICK,
COVENTRY, CV4 7AL, ENGLAND.

Proceedings of Symposia in Pure Mathematics
Volume 38 (1982), Part 2

THE CHARACTERIZATION OF THE ANALYTIC GENERATOR OF *-AUTOMORPHISM GROUPS

László Zsidó[1]

ABSTRACT. Weak*continuous one-parameter groups of *-automorphisms of W*-algebras are characterized in terms of the positive cone of the domain of their analytic generators.

1. Let M be a W*-algebra and $\alpha : \mathbb{R} \longrightarrow \mathrm{Aut}(M)$ a weak*continuous one-parameter group of *-automorphisms of M. We define the analytic extension α_z of α in $z \in \mathbb{C}$ by

$$(x,y) \in \mathrm{graph}(\alpha_z) \Longleftrightarrow \begin{cases} \mathbb{R} \ni t \longmapsto \alpha_t(x) \in M \text{ has a weak* continuous extension on} \\ \{\zeta \in \mathbb{C} \ ; \ (\mathrm{Im}\zeta)(\mathrm{Im}z) \geq 0, \ |\mathrm{Im}\zeta| \leq |\mathrm{Im}z|\}, \\ \text{which is analytical in the interior} \\ \text{and whose value in } z \text{ is } y \end{cases}$$

Then α_z is a weak*closed linear operator in M and α_{-i} , called the analytic generator of α , determines α uniquely ([1]).

We have

$$(1) \quad x,y \in D_{\alpha_{-i}} \Longrightarrow xy \in D_{\alpha_{-i}} \text{ and } \alpha_{-i}(xy) = \alpha_{-i}(x)\alpha_{-i}(y) ,$$

$$(2) \quad x \in D_{\alpha_{-i}} \Longrightarrow x^* \in D_{(\alpha_{-i})^{-1}} \text{ and } (\alpha_{-i})^{-1}(x^*) = \alpha_{-i}(x)^*,$$

so $D_{\alpha_{-i}}$, endowed with the multiplication induced from M and with the involution $\bullet$ defined by

$$x^{\bullet} = \alpha_{-i}(x)^*$$

is a *-algebra. It turns out that the positive cone V_α of this *-algebra is

---

1980 Mathematics Subject Classification 46L55
[1]Supported by the Deutsche Forschungsgemeinschaft and by the AMS.

0082-0717/80/0000-0591/$01.75

$$\{ x \in D_{\alpha_{-i}} \; ; \; \alpha_{-\frac{i}{2}}(x) \geqslant 0 \}.$$

V_α has the following remarcable factorisation property ([2]) :

(3) For each invertible $x \in M$ there is a unitary $u \in M$ with

$$u\,x \;,\; (u\,x)^{-1} \in V_\alpha \,.$$

Moreover, for given x is the above u uniquely determined.

Finally we note that

(4) $V_\alpha^{o} = \{\varphi \in M_*; \; \varphi(a) \geqslant 0 \text{ for all } a \in V_\alpha\}$ separates the points of M .

2. In November 1977 S.Woronowicz raised me the problem : are the above four conditions (1),(2),(3) and (4) enough to characterize the analytic generator of weak*continuous one-parameter groups of *-automorphisms of W*-algebras ?

The answer I found at the end of 1978 is yes. It follows from the next theorem :

THEOREM. Let M be a W*-algebra, A a subalgebra of M which is a *-algebra with a certain involution $\bullet$, and $1 \in V \subset \{x \in A \; ; \; x = x^{\bullet}\}$ a convex cone with

$$x^{\bullet} V x \subset V \text{ for any } x \in A \,.$$

If for each invertible $x \in M$ there is a unitary $u \in M$ with

$$u^*x \;,\; (u^*x)^{-1} \in V$$

and

$V^{o} = \{\varphi \in M_*; \; \varphi(a) \geqslant 0 \text{ for all } a \in V\}$ separates the points of M

then there exists a weak*continuous one-parameter group α of *-automorphisms of M such that

$$V = V_\alpha \,.$$

3. An application, which actually led S.Woronowicz to the above conjecture, consists in the following proof of the existence of the modular group of faithful normal states :

If M is a W*-algebra, φ a faithful normal state on M , $V = \{a \in M \; ; \; \varphi(a\cdot) \geqslant 0\}$ and A the linear hull of V then the conditions from the above theorem are satisfied and the obtained one-parameter group of *-automorphisms of M satisfies the KMS condition with respect to φ , so it is the modular group of φ .

Actually the above theorem can be considered the characterization of those cones V in a W*-algebra M , for which a "modular group" may be defined, measuring the obstruction from the equality $V = M^+$.

We note also that the proof of the above theorem leads to a criterion for the self-adjointness of the weak operator closure of non-self-adjoint operator algebras.

BIBLIOGRAPHY

1. I.Ciorǎnescu and L.Zsidó, "Analytic generators for one-parameter groups", Tôhoku Math. J., 28 (1976), 327-362.

2. S.Woronowicz, "On the purification of factor states", Comm. Math. Phys., 28 (1972), 221-235.

DEPARTMENT OF MATHEMATICS
UNIVERSITY OF MÜNSTER
ROXELER STR. 64
4400 MÜNSTER, WEST GERMANY

Proceedings of Symposia in Pure Mathematics
Volume **38** (1982), Part 2

ALMOST UNIFORMLY CONTINUOUS DYNAMICAL SYSTEMS

László Zsidó[1]
(report on joint work with G.A.Elliott)

ABSTRACT. One-parameter C*- and W*-dynamical systems, whose analytic generator has positive spectrum, are characterized.

1. We recall, that if (A,α) is a one-parameter C*-dynamical system (resp. (M,α) is a one-parameter W*-dynamical system) then the analytic generator α_{-i} of α is defined by

$$(x,y) \in \mathrm{graph}(\alpha_{-i}) \Longleftrightarrow \begin{cases} R \ni t \longmapsto \alpha_t(x) \text{ has a norm-continuous} \\ \text{(resp. weak*continuous) extension} \\ \text{on } \{\zeta \in C \;;\; -1 \leq \mathrm{Im}\zeta \leq 0\}, \text{ which is} \\ \text{analytical in the interior and} \\ \text{whose value in } -i \text{ is } y \;. \end{cases}$$

The analytic generator α_{-i} is a closed (resp. weak*closed) linear operator and for its spectrum holds either

$$\sigma(\alpha_{-i}) = \mathbf{C}$$

or

$$\sigma(\alpha_{-i}) \subset [0,+\infty)$$

([1]). The case $\sigma(\alpha_{-i}) = \mathbf{C}$ is still possible : this happens, for example, for the W*-dynamical system $(L^{\infty}(R),\tau)$, where

$$\tau_t(f)(s) = f(s-t)$$

([3]).

For uniformly continuous α we have $\sigma(\alpha_{-i}) \subset [0,+\infty)$. Since for commutative dynamical systems also the converse implication holds ([4]), it is reasonable to ask : is every α with $\sigma(\alpha_{-i}) \subset [0,+\infty)$ uniformly continuous ?

1980 Mathematics Subject Classification 46L55
[1]Supported by the Deutsche Forschungsgemeinschaft and by the AMS

2. For a one-parameter C*-dynamical system (A,α) the equivalence

$$(*) \quad \sigma(\alpha_{-i}) \subset [0,+\infty) \Longleftrightarrow \alpha \text{ is uniformly continuous}$$

holds, outside of the above mentioned commutative case, also in the case

A prime

([2]). Similarly, if (M,α) is a one-parameter W*-dynamical system then (*) holds, outside of the case M commutative, also in the case

M factor

([2]).

In general (*) does not hold : if

$$M = \mathrm{Mat}_2(\mathbb{C}) \oplus \mathrm{Mat}_2(\mathbb{C}) \oplus \ldots$$

and

$$\alpha = \alpha^1 \oplus \alpha^2 \oplus \ldots$$

where the α^k 's are arbitrary continuous one-parameter groups of *-automorphisms of $\mathrm{Mat}_2(\mathbb{C})$, then we have

$$\|(1 + \alpha_{-i})^{-1}\| \leq \sqrt{2} \ ;$$

thus $\sigma(\alpha_{-i}) \ni -1$, so

$$\sigma(\alpha_{-i}) \not\subset [0,+\infty)$$

([2]).

However, with an appropriate definition of the "almost uniform continuity" we have always

$$\sigma(\alpha_{-i}) \subset [0,+\infty) \Longleftrightarrow \alpha \text{ is almost uniformly continuous}$$

([2]). More precisely :

Let us define the covering size of a closed set $K \subset \mathbb{R}$ by

$$|K|_{\text{cover}} = \text{card}\,\{k \in Z \ ; \ K \cap [k,k+1] \neq \emptyset\} \ .$$

Then :

THEOREM 1. Let (A,α) be a one-parameter C*-dynamical system. Then

$$\sigma(\alpha_{-i}) \subset [0,+\infty)$$

if and only if there exists a family $\mathcal{Y}$ of prime closed two-sided ideals in A with $\bigcap_{I\in\mathcal{Y}} I = \{0\}$, such that

(i) each $I \in \mathcal{Y}$ is α-invariant and the induced C*-dynamical system $(A/I, \alpha/I)$ is uniformly continuous;

(ii) $$\sup_{I \in \mathcal{Y}} |\sigma(\alpha/I)|_{\text{cover}} < +\infty .$$

Moreover, in this case (i) and (ii) hold for any family $\mathcal{Y}$ of prime closed two-sided ideals in A .

THEOREM 2. Let (M,α) be a one-parameter W*-dynamical system. Then

$$\sigma(\alpha_{-i}) \subset [0,+\infty)$$

if and only if there exists a family $\{p_\iota\}$ of mutually orthogonal central projections with $\sum_\iota p_\iota = 1$ in M, such that

(j) each p_ι is α-invariant and the restriction $\alpha \mid Mp_\iota$ of α to Mp_ι is uniformly continuous ;

(jj) $$\sup_\iota |\sigma(\alpha \mid Mp_\iota)|_{\text{cover}} < +\infty .$$

Actually the suprema in (ii) and in (jj) are

$$\leq 18 \exp(6 \cdot 10^3 \|(1 + \alpha_{-i})^{-1}\| + 4\ 10^8) .$$

We note also, that the above considered "almost uniformly continuous" dynamical systems are stable under perturbations with cocycles $t \longrightarrow u_t$ for which u_{-i} , $(u^{-1})_{-i}$ exist and are close to 1 .

BIBLIOGRAPHY

1. I.Ciŏranescu and L.Zsidó, "Analytic generators for one-parameter groups", Tohoku Math. J., 28 (1976), 327-362.

2. G.A.Elliott and L.Zsidó, "The spectrum of the analytic generator of dynamical systems", to appear.

3. A.Van Daele, "On the spectrum of the analytic generator", Math. Scand., 37 (1975), 307-318.

4. L.Zsidó, "Spectral properties of the analytic generator and singular integrals", Memorie dell'Accademia dei Lincei, to appear.

DEPARTMENT OF MATHEMATICS
UNIVERSITY OF MÜNSTER
ROXELER STR. 64
4400 MÜNSTER, WEST GERMANY

Proceedings of Symposia in Pure Mathematics
Volume 38 (1982), Part 2

ALGEBRAS OF UNBOUNDED OPERATORS

Robert T. Powers[1]

ABSTRACT. The theory of unbounded *-representations of *-algebras, particularly the enveloping algebras of Lie algebras, is discussed. Particular attention is focussed on the (x,p)-algebras associated with the canonical commutation relations, xp - px = iI. It is argued that in order to develop a good theory of unbounded *-representations it will be necessary to find a generalization of the von Neumann uniqueness theorem for representations of the Weyl form of the commutation relations to representations of the (x,p)-algebras. The problems and possibilities of finding such a generalization are discussed. The notions of self-adjointness and complete strong positivity for *-representations are discussed.

There are two basic situations where one encounters algebras of unbounded operators. The first is where one is given a *-algebra A of operators on a Hilbert space and in addition to the algebra A one is given certain unbounded operators $(B_1, B_2, \cdots, B_n)$. For example A may be a von Neumann algebra and the operators $(B_1, B_2, \cdots, B_n)$ may be certain operators affilated with A. Another example could be where A is a C*-algebra with a densely defined derivation δ and H may be an unbounded operator so that $\delta(A) = i[H,A]$ for all A in the domain of δ. Another example is when A is the left Hilbert algebra arising in certain situation where one wishes to consider left multiplication operators which are unbounded. A. Inoue has developed an extensive theory of unbounded Hilbert algebras.

The common feature of all these examples is that while unbounded operators occur they are associated with an algebra of bounded operators. The algebras one forms in these situations are rich in bounded operators.

The second situation one encounters in the study of algebras

1980 Mathematics Subject Classification. Primary 47D40.
1 Supported in part by the National Science Foundation and a Guggenheim Fellowship.

0082-0717/80/0000-0557/$05.50

of unbounded operators is where one is given an abstractly defined *-algebra $\mathfrak{A}$ and one wants to represent this algebra as an algebra of operators on a Hilbert space. For example $\mathfrak{A}$ can be the enveloping algebra of a Lie algebra. In this case the algebras one encounters may have no bounded operators except multiples of the identity. It is this second situation that we wish to discuss.

In this discussion I will concentrate on the (x,p)-algebras associated with the Heisenberg commutation relations. All problems and pathologies that can arise in studying representations of enveloping algebras of Lie algebras seem to occur for (x,p)-algebras. I believe that if a good theory can be developed for (x,p)-algebras it will be relatively easy to extend the theory to enveloping algebras of Lie algebras.

The problem Heisenberg set out to solve was that of finding hermitian matrices x and p satisfying the relation,

$$xp - px = [x,p] = i\hbar I,$$

where $\hbar$ is Planck constant divided by 2π. This constant $\hbar$ is approximately 10^{-27} erg-sec. By rescaling x or p we can set $\hbar$ equal to one. We consider the problem of finding hermitian matrices x and p satisfying,

$$[x,p] = iI \tag{1}$$

If Heisenberg had been a careful mathematician he would have realized that it would not be possible to find such matrices. We have

$$\begin{aligned} e^{itp}xe^{-itp} &= x + it[p,x] + \frac{(it)^2}{2!}[p,[p,x]] + \dots \\ &= x + tI \end{aligned} \tag{2}$$

Then we see that x and x+tI should be unitarily equivalent which is not possible for finite matrices. However, Heisenberg found a solution in the form of infinite matrices,

$$x = \frac{1}{\sqrt{2}}\begin{bmatrix} 0 & \sqrt{1} & 0 & 0 & 0 & \dots \\ \sqrt{1} & 0 & \sqrt{2} & 0 & 0 & \dots \\ 0 & \sqrt{2} & 0 & \sqrt{3} & 0 & \dots \\ 0 & 0 & \sqrt{3} & 0 & \sqrt{4} & \dots \\ 0 & 0 & 0 & \sqrt{4} & 0 & \dots \\ \dots & & & & & \end{bmatrix} \qquad p = \frac{-i}{\sqrt{2}}\begin{bmatrix} 0 & \sqrt{1} & 0 & 0 & 0 & \dots \\ -\sqrt{1} & 0 & \sqrt{2} & 0 & 0 & \dots \\ 0 & -\sqrt{2} & 0 & \sqrt{3} & 0 & \dots \\ 0 & 0 & -\sqrt{3} & 0 & \sqrt{4} & \dots \\ 0 & 0 & 0 & -\sqrt{4} & 0 & \dots \\ \dots & & & & & \end{bmatrix} \tag{3}$$

Later Schrödinger found another solution to Heisenberg's problem. In Schrödinger's formulation of quantum mechanics he introduced a complex wave function $\Psi(x)$ on the real line R. The operator x corresponded to multiplication by x and p corresponded to (-i) times differentiation.

$$(x\Psi)(x) = x\Psi(x) \qquad (p\Psi)(x) = -i\frac{d}{dx}\Psi(x) \tag{4}$$

Schrödinger showed his formulation agreed with Heisenberg's formulation as follows. Let,

$$\Psi_n(x) = \pi^{-\frac{1}{4}}(2^n n!)^{-\frac{1}{2}}(x - \frac{d}{dx})^n e^{-x^2/2}.$$

These functions form an orthonormal basis for $L^2(R)$ and with respect to this orthonormal basis x and p have the same matrix elements as given in equation (3).

This raised the question as to whether there were other ways to realize Heisenberg commutation relations. Herman Weyl in his book The Theory of Groups and Quantum Mechanics pointed out that since equation (2) shows that x and p can not both be bounded the commutation relations, equation (1), are not well defined. Weyl proposed that the question should be formulated in terms of the strongly continuous unitary groups U(s) = exp(isp) and V(t) = exp(itx). Exponentiating both sides of equation (2) we have,

$$\exp(isp)\exp(itx)\exp(-isp) = \exp(it(x+sI))$$

or

$$U(s)V(t)U(s)^{-1} = e^{ist}V(t).$$

Multiplying on the right by U(s) we obtain the Weyl formulation of the commutation relations,

$$U(s)V(t) = e^{ist}V(t)U(s) \tag{5}$$

Von Neumann solved the question posed by Weyl. He showed if $(U(s), V(t);\ s,t \in R)$ are strongly continuous one parameter unitary groups satisfying the Weyl condition, equation (5), on a Hilbert space $\mathcal{H}$ then $\mathcal{H}$ can be decomposed as a direct sum of subspaces $\mathcal{M}_n$ so that $\mathcal{H} = \oplus_{n=1}\mathcal{M}_n$ and $U(s)\mathcal{M}_n \subset \mathcal{M}_n$, $V(t)\mathcal{M}_n \subset \mathcal{M}_n$ for all real s and t. There are unitary operators $\mathcal{T}_n$ from $L^2(R)$ onto $\mathcal{M}_n$ so that

$$(\mathcal{T}_n^{-1}U(s)\mathcal{T}_n f)(x) = (e^{isp}f)(x) = f(x+s)$$
$$(\mathcal{T}_n^{-1}V(t)\mathcal{T}_n f)(x) = e^{itx}f(x).$$

Von Neumann's theorem is summarized by saying that every

representation of the Weyl form of the canonical commutation relations is a direct sum of Schrödinger representations.

Von Neumann proved the analogous theorem for n-degrees of freedom. For n-degrees of freedom the Heisenberg relations are

$$[x_i,x_j] = [p_i,p_j] = 0 \qquad (6)$$
$$[x_i,p_j] = i\,\delta_{ij}I,$$

for i,j = 1,....,n. The Weyl formulation becomes,

$$U(s) = \exp(is\cdot p) \qquad V(t) = \exp(it\cdot x) \qquad (7)$$
$$U(s)V(t) = e^{is\cdot t}\,V(t)U(s)$$

for all $s,t \in R^n$ and $s \to U(s)$, $t \to V(t)$ are strongly continuous unitary group representations of R^n. The von Neumann uniqueness theorem of n-degrees of freedom says that every reresentation of the Weyl form of the commutation relations is a direct sum of representations which are unitarily equivalent to the Schrödinger representation on $L^2(R^n)$ given by

$$(U(s)f)(x) = f(x+s) \qquad (V(t)f)(x) = e^{it\cdot x}f(x) \qquad (8)$$

for $f \in L^2(R^n)$.

My interest in the study of unbounded *-representations of *-algebras arose out of the conviction that there should be a general algebraic version of the von Neumann uniqueness theorem. More precisely let $\mathfrak{A}_1$ be the *-algebra of all polynomials in x and p where x*=x and p*=p and x and p satisfy the Heisenberg commutation relations, equation (1). One can realize $\mathfrak{A}_1$ explicitly as the *-algebra of all differential operators with polynomial coefficients. Let $\mathfrak{A}_n = \mathfrak{A}_1 \otimes \dots\dots \otimes \mathfrak{A}_1$ be the tensor product of $\mathfrak{A}_1$ with itself n-times. One can realize $\mathfrak{A}_n$ as the *-algebra of all partial differential operators on R^n with polynomial coefficients.

The Schrödinger representation of $\mathfrak{A}_n$ is defined as follows. We let $\mathcal{H} = L^2(R^n)$ and let $\mathcal{D}(\pi_0)$ be Schwartz's space $\mathcal{S}(R^n)$ of all infinitely differentiable functions which together with their derivatives decrease at infinity faster than any power of the distance from the origin. We call $\mathcal{D}(\pi_0)$ the domain of π_0. For $f \in \mathcal{D}(\pi_0)$ we define

$$(\pi_0(x_i)f)(x_1,\dots,x_n) = x_i f(x_1,\dots,x_n) \qquad (9)$$
$$(\pi_0(p_i)f)(x_1,\dots,x_n) = -i\frac{\partial}{\partial x_i} f(x_1,\dots,x_n).$$

I believe the following theorem is true.

THEOREM. Every "good" representation of $\mathcal{A}_n$ is a direct sum of representations which are unitarily equivalent to the Schrödinger representation.

The problem, of course, is to define good. At best the definition of good should be a natural condition which easily extends to enveloping algebras of Lie algebras and other algebras. At least the definition of good should be invariant under *-automorphisms. We begin with the definition of *-representations.

DEFINITION 1. A representation π of an algebra $\mathcal{A}$ on a Hilbert space $\mathcal{H}$ consists of a linear manifold $\mathcal{D}(\pi)$, called the domain of π, which is dense in $\mathcal{H}$ and a mapping π of $\mathcal{A}$ to linear transformations of $\mathcal{D}(\pi)$ into itself so that

$$\text{(i) } \pi(\alpha A + B) = \alpha\pi(A) + \pi(B)$$
$$\text{(ii) } \pi(AB) = \pi(A)\pi(B)$$

for all $A, B \in \mathcal{A}$, $\alpha \in \mathbb{C}$.

DEFINITION 2. A representation π of a *-algebra $\mathcal{A}$ is said to be hermitian or a *-representation if in addition to the conditions in definition 1 π satisties

(iii) $(f, \pi(A)g) = (\pi(A^*)f, g)$ for all $A \in \mathcal{A}$ and $f, g \in \mathcal{D}(\pi)$.

It follows that if π is an hermitian representation then $\pi(A)$ is closable since $\pi(A)^* \supset \pi(A^*)$.

DEFINITION 3. If π_1 and π_2 are representations of an algebra $\mathcal{A}$ on a Hilbert space $\mathcal{H}$ we say π_1 is an extension of π_2, denoted $\pi_1 \supset \pi_2$, if $\mathcal{D}(\pi_1) \supset \mathcal{D}(\pi_2)$ and $\pi_1(A) \supset \pi_2(A)$ for all $A \in \mathcal{A}$.

If π is a representation of an algebra $\mathcal{A}$ on a Hilbert space $\mathcal{H}$ with domain $\mathcal{D}(\pi)$, there is a natural induced topology on $\mathcal{D}(\pi)$. This topology is defined as follows. Suppose S is a finite set of elements of $\mathcal{A}$. We define the seminorm $\|\cdot\|_S$

$$\|f\|_S = \sum_{A \in S} \|\pi(A)f\|$$

where $\|\pi(A)f\|$ is the Hilbert space norm of $\pi(A)f$. Note $S \supset S'$ implies $\|f\|_S \geq \|f\|_{S'}$. The induced topology on $\mathcal{D}(\pi)$ is the topology generated by the neighborhoods,

$$N(f; S, \epsilon) = \{g \in \mathcal{D}(\pi); \|f - g\|_S < \epsilon\}.$$

Note $\pi(A)$ is a continuous mapping of $\mathcal{D}(\pi)$ into $\mathcal{D}(\pi)$ in the induced topology for each $A \in \mathfrak{A}$.

DEFINITION 4. Suppose π is a representation of an algebra $\mathfrak{A}$ on a Hilbert space $\mathcal{H}$. We say a set $S \subset \mathcal{D}(\pi)$ is strongly dense in $\mathcal{D}(\pi)$ if S is dense in $\mathcal{D}(\pi)$ in the induced topology.

Just as there is a notion of a closed operator there is a notion of a closed representation.

DEFINITION 5. We say π is a closed representation of $\mathfrak{A}$ if $\mathcal{D}(\pi)$ is complete in the induced topology.

Every hermitian operator is closable. The analogous result holds for *-representations.

LEMMA 6. Suppose π is a *-representation of a *-algebra $\mathfrak{A}$ on a Hilbert space $\mathcal{H}$. Then there is a unique minimal closed extension $\tilde{\pi}$ of π. Furthermore the domain of $\tilde{\pi}$ is given by

$$\mathcal{D}(\tilde{\pi}) = \bigcap_{A \in \mathfrak{A}} \mathcal{D}(\overline{\pi(A)})$$

where $\overline{\pi(A)}$ is the closure of $\pi(A)$. Furthermore, $\tilde{\pi}(A)f = \overline{\pi(A)}f$ for all $A \in \mathfrak{A}$ and $f \in \mathcal{D}(\pi)$.

We refer to [8] for a proof and further details.

One might ask is every closed *-representation of the (x,p)-algebra $\mathfrak{A}_1$ a direct sum of Schrödinger representations. The following example shows the answer is no. Let $\mathcal{H} = L^2(a,b)$ and $\mathcal{D}(\pi)$ be the C^∞-functions which together with all of their derivatives vanish at both end points x=a and x=b. We define

$$(\pi(x)f)(x) = xf(x) \qquad (\pi(p)f)(x) = -if'(x)$$

for all $f \in \mathcal{D}(\pi)$. It is easily seen that π is a closed *-representation of $\mathfrak{A}_1$ but π is not a direct sum of Schrödinger representations since $\pi(x)$ is bounded and for the Schrödinger representation $\pi_0(x)$ is unbounded.

In the late fifties Dixmier [3] published a result that tells us when a *-representation π of $\mathfrak{A}_n$ is a direct sum of Schrödinger representations. His result was an improvement of an earlier result of Rellich.

THEOREM 7. (Dixmier) Suppose π is a *-representation of $\mathfrak{A}_n$. Let $H = \sum_{i=1}^n p_i^2 + x_i^2$. If $\pi(H)$ is essentially self-adjoint then the $\pi(x_i)$ and $\pi(p_i)$ are essentially self-adjoint and the exponentiated groups obtained by exponentiating these operators satisfy the Weyl relations.

Dixmier's theorem was a precersor of a more general theorem of Nelson [7] which states that a representation of the enveloping algebra of a Lie algebra can be exponentiated to give a continuous unitary representation of the Lie group if a certain second order elliptic element Δ is represented by an essentially self-adjoint operator.

Given Dixmier's result it is natural to ask why not define "good" to mean that $\pi(H) = \pi(\Sigma_{i=1}^{n} x_i^2 + p_i^2)$ be essentially self-adjoint. Then by Dixmier's theorem every "good" representation would be a direct sum of Schrödinger representations. The difficulty with this definition of "good" is that it is not invariant under *-automorphisms. From the general algebraic point of view there is no particular reason to single out this element H to be represented by an essentially self-adjoint operator.

The next possibility one might consider is requiring that a "good" representation is one in which $\pi(A)$ is essentially self-adjoint for all $A=A^* \in \mathfrak{A}_n$. This requirement is certainly invariant under *-automophisms, but this requirement has the serious fault of not being valid for the Schrödinger representation. One can compute the deficiency indeces of $\pi_0(xpx)$ and $\pi_0(x^3p + px^3)$ in the Schrödinger representation and one finds the deficiency indeces are (1,1) and (0,2). It follows that requiring hermitian elements of $\mathfrak{A}_n$ be represented by essentially self-adjoint operators is too strong a requirement.

There is a notion of self-adjoint representations of *-algebras that does not require $\pi(A)$ to be essentially self-adjoint for hermitian A. In [8] it was shown.

THEOREM 8. Suppose π is a representation of a *-algebra on a Hilbert space $\mathcal{H}$. Let

$$\mathcal{D}(\pi^*) = \bigcap_{A\in\mathfrak{A}} \mathcal{D}(\pi(A)^*) \qquad \pi^*(A) = \pi(A^*)^*|\mathcal{D}(\pi^*)$$

for all $A \in \mathfrak{A}$. Then π^* is a closed representation of $\mathfrak{A}$ on $\mathcal{H}$ with domain $\mathcal{D}(\pi^*)$. Note $\mathcal{D}(\pi^*)$ can fail to be dense in $\mathcal{H}$. If π is a *-representation then $\pi^* \supset \pi$ with which case $\mathcal{D}(\pi^*)$ contains $\mathcal{D}(\pi)$ and is then dense in $\mathcal{H}$.

We call π^* the hermitian adjoint of π. If π is hermitian π^* may fail to be hermitian just as the adjoint A* of an hermitian operator may fail to be hermitian. Just as for operators, on can show that every hermitian extension of a *-representation is an hermitian restriction of π^*. In fact, if π and π_1 are hermitian and $\pi_1 \supset \pi$ then $\pi^* \supset \pi_1^* \supset \pi_1 \supset \pi$.

For a single hermitian operator A the second adjoint A** is simply the closure $\bar{A}$ of A. For *-representations π the second adjoint π^{**} is a closed hermitian extension of π which is sometimes a proper extension of $\tilde{\pi}$. In his thesis, Scruggs showed that $\pi^{***} = \pi^*$ and hence $\pi^{****} = \pi^{**}$ for all hermitian representations π. Hence, no new representations are obtained after taking two hermitian adjoints.

DEFINITION 9. We say a *-representation π of a *-algebra $\mathfrak{A}$ is self-adjoint if $\pi = \pi^*$. We say π is essentially self-adjoint if the closure $\tilde{\pi}$ is self-adjoint. We say π is algebraically self-adjoint if $\pi^* = \pi^{**}$.

REMARK. If π is a *-representation of a *-algebra $\mathfrak{A}$ and A_i are hermitian elements of $\mathfrak{A}$ so that $\pi(A_i)$ are essentially self-adjoint or i=1,2,... and the subalgebra $\mathfrak{B}$ of $\mathfrak{A}$ generated by the A_i is cofinal in $\mathfrak{A}$ (i.e., for each $A \in \mathfrak{A}$ there is a $H \in \mathfrak{B}$ so that $H \geq A^*A$), then π^{**} is self-adjoint.

The Schrödinger representation of $\mathfrak{A}_n$ is self-adjoint. Also the notion of self-adjointness is invariant under *-automorphisms so "good" representations should be self-adjoint. Self-adjointness is useful in working with the commutant of a *-algebra or studying subrepresentations.

DEFINITION 10. Suppose π is a *-representation of a *-algebra $\mathfrak{A}$ on a Hilbert space $\mathcal{H}$. The commutant of $\pi(\mathfrak{A})$, denoted by $\pi(\mathfrak{A})'$, consists of all bounded operators C on $\mathcal{H}$ so that

$$(f, C\pi(A)g) = (\pi(A^*)f, Cg)$$

for all $f,g \in \mathcal{D}(\pi)$ and $A \in \mathfrak{A}$.

This definition is essentially the definition of the commutant used in the Wightman formulation of quantum field theory. This definition is a weak defintion of the commutant, i.e. this definition gives the biggest commutant. One can show that the commutant $\pi(\mathfrak{A})'$ has the following properties.

(i) $\pi(\mathfrak{A})'$ is a complex linear manifold.
(ii) $\pi(\mathfrak{A})'$ is symmetric, i.e., $C \in \pi(\mathfrak{A})'$ implies $C^* \in \pi(\mathfrak{A})'$
(iii) $\pi(\mathfrak{A})'$ is closed in the weak operator topology.

(iv) $\pi(\mathfrak{A})' = \tilde{\pi}(\mathfrak{A})'$

(v) $\pi^{**}(\mathfrak{A})' = \pi(\mathfrak{A})'$

(vi) If $C \in \pi(\mathfrak{A})'$ then $C\mathcal{D}(\pi) \subset \mathcal{D}(\pi^*)$ and $C\pi(A)f = \pi^*(A)Cf$ for all $A \in \mathfrak{A}$ and $f \in \mathcal{D}(\pi)$.

The commutant $\pi(\mathfrak{A})'$ may fail to be an algebra. This is because the commutant need not map the domain $\mathcal{D}(\pi)$ into itself. By statement (vi) the commutant maps $\mathcal{D}(\pi)$ into $\mathcal{D}(\pi^*)$. Then we have

LEMMA 11. Suppose π is a self-adjoint representation of a *-algebra $\mathfrak{A}$. Then the commutant $\pi(\mathfrak{A})'$ is a von Neumann algebra. Furthermore, for each $C \in \pi(\mathfrak{A})'$ we have $C\mathcal{D}(\pi) \subset \mathcal{D}(\pi)$ and $C\pi(A)f = \pi(A)Cf$ for all $A \in \mathfrak{A}$ and $f \in \mathcal{D}(\pi)$.

We comment next on reducing subspaces. If π is a representation of an algebra $\mathfrak{A}$ on a Hilbert space $\mathcal{H}$ and $\mathfrak{M}$ is a linear manifold contained in $\mathcal{D}(\pi)$ we say $\mathfrak{M}$ reduces π if $\pi(A)\mathfrak{M} \subset \mathfrak{M}$ for all $A \in \mathfrak{A}$. We denote by $\pi|\mathfrak{M}$ the representation π restricted to $\mathfrak{M}$. If π is a *-representation of a *-algebra and $\mathfrak{M}$ is a reducing subspace then $\pi/\mathfrak{M}$ is a *-representation on $\overline{\mathfrak{M}}$.

For a bounded *-representation of a *-algebra $\mathfrak{A}$ for each reducing subspace $\mathfrak{M}$ the hermitian projection onto the closure of $\mathfrak{M}$ is in the commutant $\pi(\mathfrak{A})'$. This is not true in general for unbounded *-representations. We say a reducing subspace is self-adjoint if $\pi|\mathfrak{M}$ is a self-adjoint representation.

THEOREM 12. Suppose π is a self-adjoint representation of a *-algebra $\mathfrak{A}$ on a Hilbert space $\mathcal{H}$. Suppose $E \in \pi(\mathfrak{A})'$ is an hermitian projection. Let $\mathfrak{M} = E\mathcal{D}(\pi)$. Then $\mathfrak{M}$ reduces π and the restriction $\pi/\mathfrak{M}$ is self-adjoint.

Conversely, suppose $\mathfrak{M} \subset \mathcal{D}(\pi)$ is a reducing subspace for π and $\pi/\mathfrak{M}$ is self-adjoint. Then, the hermitian projection E onto the closure of $\mathfrak{M}$ is in the commutant $\pi(\mathfrak{A})'$.

Hence, for self-adjoint representations there is a one-to-one correspondence between hermitian projections in $\pi(\mathfrak{A})'$ and self-adjoint reducing subspaces.

REMARK. For self-adjoint representations π of a *-algebra $\mathfrak{A}$ each closed operator $\overline{\pi(A)}$ and $\pi(A)^*$ is affiliated with the von Neumann algebra $\pi(\mathfrak{A})'' = (\pi(\mathfrak{A})')'$ for each $A \in \mathfrak{A}$.

The question arises as to whether every self-ajoint representation of $\mathfrak{A}_1$ is a direct sum of Schrödinger representations. In the late sixties B. Fugledge [6] constructed a self-adjoint representation of $\mathfrak{A}_1$ which is not a direct sum of Schrodinger representations. Fugledge representation is constructed as follows. Let $\mathcal{H} = L^2(R)$ and $\mathcal{D}(\pi)$ be the span of the functions $x^n \exp(-rx^2 + cx)$ with $n=0,1,2,\ldots$ $r > 0$ and c a complex number. Note the functions in $\mathcal{D}(\pi)$ are entire analytic functions so for $f \in \mathcal{D}(\pi)$ $f(x + iy)$ is well defined for all real x and y. Fugledge's representation is obtained from the relations,

$$(\pi(x)f)(x) = xf(x) + f(x + i(2\pi)^{\frac{1}{2}})$$

$$(\pi(p)f)(x) = -if'(x) + \exp(-(2\pi)^{\frac{1}{2}}x)f(x)$$

This is the Schrodinger representation except the operators A and B have been added to x and p respectively, where

$$(Af)(x) = f(x + i(2\pi)^{\frac{1}{2}}) \qquad (Bf)(x) = \exp(-(2\pi)^{\frac{1}{2}}x)f(x).$$

Note A and B are Fourrier transforms of each other. One can show AB = BA and A commutes with differentiation and B commutes with multiplication by x. From this it follows the operators $\pi(x)$ and $\pi(p)$ defined above satisfy the Heisenberg commutation relations. Fugledge showed that $\pi(x)$ and $\pi(p)$ are essentially self-adjoint on $\mathcal{D}(\pi)$. It then follows from the remark after definition 9 that π^{**} is self-adjoint. Fugledge showed that this representation π^{**} is not a direct sum of Schrodinger representations.

In 1970 Woronowicz [9] showed that order considerations were of importance. A state ω of a *-algebra $\mathfrak{A}$ with unit I is a functional satisfying the conditions,

(i) $\omega(\alpha A + B) = \alpha\omega(A) + \omega(B)$.

(ii) $\omega(A^*A) \geq 0$.

(iii) $\omega(I) = 1$.

for all $A,B \in \mathfrak{A}$ and complex numbers α. Woronowicz considered the quantum moment problem. Given a state ω of the (x,p)-algebra $\mathfrak{A}_1$ the quantum moment problem is to find a positive operator of trace one so that

$$\omega(A) = \mathrm{tr}(\pi_0(A)\rho)$$

for all $A \in \mathfrak{A}_1$, where π_0 is the Schrödinger representation of $\mathfrak{A}_1$ Woronowicz showed that the quantum moment problem can not always be solved. Let $H = \frac{1}{2}(p^2 + x^2 - 1)$. Woronowicz noted that in the Schrödinger representation $\pi_0((H-1)(H-2))$ is positive but $(H-1)(H-2)$ can not be expressed as a sum of elements of the form A^*A. Therefore, there is a state ω of $\mathfrak{A}_1$ so that

$\omega((H-1)(H-2)) < 0$. It follows that for such a state there can be no solution to the quantum moment problem. Woronowicz showed that the quantum moment problem has a solution if and only if ω is strongly positive. We say ω is strongly positive if $\omega(A) \geq 0$ whenever $\pi_0(A) \geq 0$. Woronowicz also showed that if

$$\sum_{n=0}^{\infty} \frac{(\omega(a^n a^{*n}))^{\frac{1}{2}} t^n}{n!} < \infty \quad \text{where } a = \frac{(x+ip)}{\sqrt{2}}$$

for some $t > 0$ then there is a unique ρ so that $\omega(A) = \operatorname{tr}(\pi_0(A)\rho)$

These pathologies associated with representations of $\mathfrak{A}_1$ occur in the study of *-representation of an even more elementary algebra. Let $\mathcal{B}_n$ be the *-algebra of all polynomials in n hermitian generators. Nelson [7] showed there exist self-adjoint representations of $\mathcal{B}_2$ in which the generators A_1 and A_2 are represented by essentially self-adjoint operators $\pi(A_1)$ and $\pi(A_2)$ but the spectral projections of their self-adjoint extensions do not commute. An especially elegant example found by Nelson is as follows. Consider the Riemann surface for the square root function. This is the two sheeted covering of the (x,y)-plane pictured below. Note there is a cut along the negative x-axis. When one crosses the cut one passes from one sheet to the next.

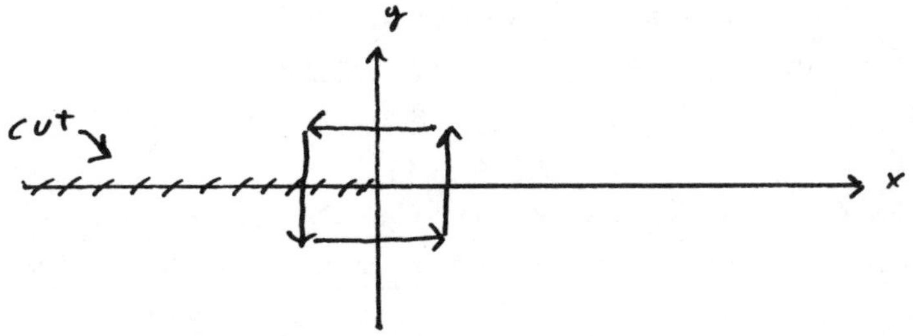

Let $A_1 = -i\partial/\partial x$ and $A_2 = -i\partial/\partial y$. One can show that if one defines these diffenential operators on the C^∞-functions of compact support which vanish in a neighborhood of the origin of the cut plane, then A_1 and A_2 are commuting essentially self-adjoint operators. The exponential of $i\overline{A_1}$ corresponds to translation in the x-direction and the exponential of $i\overline{A_2}$ corresponds to translation in the y-direction. Then one sees

$$\exp(i\overline{A_1})\exp(i\overline{A_2})\exp(-i\overline{A_1})\exp(-i\overline{A_2}) \neq I$$

because as the above diagram indicates such a sequence of translations will cause some points in the plane to be shifted to the corresponding point on the other sheet.

In [8] I showed that there is a connection between this

non-commutativity phenomenon discovered by Nelson and the phenomenon of positive polynomials which are not the sum of squares discovered by Hilbert. Hilbert showed that there exists a sixth order polynomial $P_0(x,y)$ in two real variables x and y such that $P_0(x,y) \geq 0$ for all real x and y and P_0 can not be written as a sum of squares of polynomials. It follows that there exists a state ω on $\mathcal{B}_2$ so that $\omega(P_0(A_1,A_2)) < 0$. Clearly such a state can not correspond to a measure on the plane.

Hilbert only showed the existence of such a polynomial and the proof of its existence is quite involved. More recently explicit polynomials have been found. In a recent paper of Choi and Lam [2] they give such a nice example that I can not refrain from mentioning it. Consider the polynomial

$$P(x,y) = x^4y^2 + y^4 + x^2 - 3x^2y^2$$

Choi and Lam noted that this polynomial is positive for all real x and y since the geometric mean of the first three terms is x^2y^2. Since the arithmetic mean is greater than or equal to the geometric mean the above polynomial is positive. Next suppose $P = \sum_i q_i^2$, where the q_i are at most cubic polynomials. Looking at the expression for P it is apparent that the terms x^3, xy^2, y^3, x^2, y and 1 can not appear in the q_i. Then q_i must be of the form $q_i = a_i x^2y + b_i xy + c_i y^2 + d_i x$. But then the coefficient of x^2y^2 in q_i^2 is positive. Therefore, P can not be written as a sum of squares.

Since the above polynomial is not in the cone spanned by polynomials of the form $\bar{p}p$ there exists a state ω on $\mathcal{B}_2$ so that $\omega(P(A_1,A_2)) < 0$.

DEFINITION 13. We say a state ω of $\mathcal{B}_n$ is strongly positive if

$$\omega(P(A_1,A_2,\ldots,A_n) \geq 0 \quad \text{if } P(x_1,x_2,\ldots,x_n) \geq 0$$

for all real x_i.

A state ω is strongly positive if it is positive on positive polynomials in the A_i. In [8] it was shown that for a self-adjoint representation of $\mathcal{B}_n$ induced from a strongly positive state the operators $\pi(A_i)$ are essentially self-adjoint and their spectral projections do commute. The existence of *-representations in which the spectral projections of the $\pi(A_i)$ fail to commute is connected with the existence of states which are not strongly positive.

In [8] I showed that the self-adjoint representations of the enveloping algebra of a Lie algebra which can be exponentiated to give a strongly continuous unitary representation of the Lie group are those which preserve a certain order structure. The details are as follows.

Let G denote an arbitrary connected simply connected Lie group and $\mathcal{L}(G)$ denote its Lie algebra. Let $\mathcal{O}(G)$ denote the universal enveloping algebra of $\mathcal{L}(G)$. Specifically if $\{X_1, X_2, \ldots, X_n\}$ is a basis for $\mathcal{L}(G)$, then $\mathcal{O}(G)$ consists of all non-commutative polynomials in the X_i modulo the relations

$$X_i X_j - X_j X_i = \sum_{k=1}^{n} C_{ij}^{k} X_k$$

where the C_{ij}^{k} are the structure constants of $\mathcal{L}(G)$ for this basis. We define the *-operation on $\mathcal{L}(G)$ by specifying $X_i^* = -X_i$ for $i=1,\ldots,n$. We have the following theorem.

THEOREM 14. Suppose π is a self-adjoint representation of $\mathcal{O}(G)$. Then the $\pi(iX_k)$ are essentially self-adjoint and exponentiate to give a strongly continuous unitary representation of G if and only if π is completely strongly positive.

Completely strongly positive representations are defined in terms of the universal representation π_0 of $\mathcal{O}(G)$ where π_0 is defined as follows. Let $g \rightarrow U(g)$ be the universal strongly continuous unitary representation of G. This representation is the direct sum of all cyclic representations of G. If $A \in \mathcal{O}(G)$ is an element of the enveloping algebra we denote by d(A) the right invariant differential operator on the group manifold associated with A. The mapping $A \rightarrow d(A)$ is a representation of $\mathcal{O}(G)$ in that $d(\alpha A + BC) = \alpha d(A) + d(B)d(C)$ for all $A,B,C \in \mathcal{O}(G)$ and all complex numbers α. It follows that d is completely specified once it has been specified for all $X \in \mathcal{L}(G)$. Let $X \rightarrow g(X)$ be the exponential map from $\mathcal{L}(G)$ into G. If f is a C^∞-function on the group G we define d(X) on f by the relation

$$(d(X)f)(g) = \lim_{t \to 0} \frac{f(g(tX)g) - f(g)}{t} .$$

Let $\mathcal{D}(\pi_0)$ be the linear manifold of all C^∞-vectors for U(g), i.e. $\mathcal{D}(\pi_0)$ consists of all vectors $f \in \mathcal{H}$ such that the mapping $g \rightarrow U(g)f$ is an infinitely differentiable function of G into $\mathcal{H}$. (We remark that it is enough to know that (f,U(g)f) is a C^∞-function in a neighborhood of the identity of G.) The domain $\mathcal{D}(\pi_0)$ contains the Gårding domain which is well know to be dense in $\mathcal{H}$. One can show π_0 is self-adjoint.

We use π_0 to define an order structure on the space of n x n matrices with entries in $\mathfrak{A}(G)$. We say

$$\{A_{ij}\} \geq 0 \text{ if } \sum_{ij=1}^{n}(f_i, \pi_0(A_{ij})f_j) \geq 0$$

for all $f_i \in \mathcal{D}(\pi_0)$, i=1,...,n. We say a *-representation π is completely strongly positive for all $\{A_{ij}\} \geq 0$ we have

$$\sum_{ij=1}^{n}(f_i, \pi(A_{ij})f_j) \geq 0$$

for all $f_i \in \mathcal{D}(\pi)$. In short, π is completely strongly positive if all matrices with entries in $\mathfrak{A}(G)$ which are positive for the universal representation π_0 are positive for the representation π.

The basic idea of the proof of this result is as follows. We construct a *-algebra $\mathfrak{B}(G)$ which is generated by $\mathfrak{A}(G)$ and the unitary elements U(g) for each $g \in G$, which form a representation of G and act on $\mathfrak{A}(G)$ by *-automorphisms, i.e. $U(g)^*AU(g) = \alpha_g(A)$ for each $A \in \mathfrak{A}(G)$ and $g \to G$ where $g \to \alpha_g$ is a representation of G by *-automorphisms of $\mathfrak{A}(G)$. In essence we have that a self-adjoint representation π of $\mathfrak{A}(G)$ can be exponentiated to give a strongly continuous unitary representation of G if and only if π can be extended from a representation of $\mathfrak{A}(G)$ to a representation of $\mathfrak{B}(G)$. Then the problem is one of extending a representation π of a *-algebra $\mathfrak{A}(G)$ to a larger *-algebra $\mathfrak{B}(G)$.

Every bounded *-representation of a *-algebra can be expressed as a direct sum of cyclic representations each of which can be characterized by a state. In view of this fact, the problem of extending a bounded *-representation of a *-algebra $\mathfrak{A}$ to a larger *-algebra $\mathfrak{B}$ can be expressed in terms of the problem of extending states of a *-algebra $\mathfrak{A}$ to a larger *-algebra $\mathfrak{B}$. This problem can be analysed with the aid of the Hahn-Banach theorem. For unbounded *-representations of *-algebras it is not necessarily true that a *-representation can be expressed as a direct sum of cyclic representations. For this reason in considering the problem of extending unbounded *-representations one needs the more general tool of completely positive maps. Stinespring's result [8] that each unital completely positive map of a C*-algebra generates in a natural way a *-representation of the *-algebra can be generalized to unbounded *-representations of *-algebras.

In [1] Arveson proved an extension theorem for completely positive maps into $\mathcal{B}(\mathcal{H})$. Arveson's theorem can be proved in a more general form which is applicable to unbounded *-representations. Using this generalization of Arveson's theorem one can show that a completely strongly positive self-adjoint representation of $\mathfrak{A}(G)$ can be extended to $\mathcal{B}(G)$ as a completely strongly positive map. Then one can show this extension is in fact a self-adjoint representation of $\mathcal{B}(G)$.

The moral of story is that "good" representations can be characterized by an order structure. The problem is how to characterize this "good" order structure abstractly in terms of the given algebra $\mathfrak{A}(G)$ without reference to the group G.

After trying to prove that every "good" representation of $\mathfrak{A}_n$ is a direct sum of Schrödinger representations for several years I began to have doubts. Let us consider how this theorem could be false. We require the definition of "good" to be invariant under *-automorphisms. Then if this theorem is to have a chance of being true the Schrödinger representation itself must be invariant under *-automorphisms. This raises the problem of characterizing the *-automorphisms of $\mathfrak{A}_n$. For the case of n=1 Dixmier [4] has characterized the automorphisms of $\mathfrak{A}_1$. Consider the *-automorphisms α, β of the form

$$\alpha(x) = x \qquad \alpha(p) = p + tx^n$$

and

$$\beta(x) = x + tp^n \qquad \beta(p) = p.$$

for t real. Dixmier showed that every *-automorphism of $\mathfrak{A}_1$ is composed of a product of *-automorphisms of the above form. The question of whether the Schrödinger representation is invariant under *-automorphisms then hangs on whether or not the Schrödinger representation is invariant under the above automorphisms. The Schrödinger representation is invariant under the above automorphisms. The automorphism α is implemented by the unitary operator given by

$$(Uf)(x) = \exp(it(n+1)^{-1}x^{n+1})f(x)$$

for all $f \in \mathcal{D}(\pi_o)$ and β is implemented the corresponding unitary operator in the Fourrier transformed space. These operators map $\mathcal{D}(\pi_o)$ onto it self, so the Schrödinger representation is invariant under all *-automorphisms.

It follows that for $\mathfrak{A}_1$ some definition of "good" must exist. Whether an attractive definition of "good" can be given remains to be seen. For the case of $\mathfrak{A}_n$ it is not known whether the Schrödinger representation is invariant under *-automorphisms.

During the last few years I have found a number of definitions of "good" which at first looked promising and then later proved to be deficient. Rather than discuss these blind alleys I would like present what I believe is a correct definition of "good".

DEFINITION 15. A *-representation π of a *-algebra $\mathfrak{A}$ is said to be strongly self-adjoint if for every $A \in \mathfrak{A}$ there is an $H = H^* \in \mathfrak{A}$ so that $\pi(H) \geq \pi(A^*A)$ and $\pi(H)$ is essentially self-adjoint.

I believe but have not been able to prove that every strongly self-adjoint representation of $\mathfrak{A}_n$ is a direct sum of Schrödinger representations. I will show that for the case of centrally dominated algebras the notion of strong self-adjoint representations eliminates pathological representations.

A centrally dominated algebra is an algebra such that for every $A \in \mathfrak{A}$ there is an $H = H^*$ in the center of $\mathfrak{A}$ so that $H \geq A^*A$. The enveloping algebra of the Lie algebra of a compact Lie group or the semi-direct product of a compact group with a commutative Lie group is a centrally dominated algebra. The (x,p) algebras, unfortunately, are not centrally dominated. We will need the following lemma due to Nelson [7].

LEMMA 16. Suppose T is a positive essentially self-adjoint operator with dense domain $\mathcal{D}(T)$. Suppose A is an hermitian operator with $\mathcal{D}(A) \supset \mathcal{D}(T)$ and

$$(f,Tf) \geq (Af,Af) \qquad (Tf,Ag) = (Af,Tg)$$

for all $f,g \in \mathcal{D}(T)$. Then A is essentially self-adjoint on $\mathcal{D}(T)$.

Proof: Suppose A and T satisfy the hypothesis of the lemma. Because $T \geq A^*A$ on $\mathcal{D}(T)$ we can extend the above relations to the domain of the closure of T. After making this extension we will be working with $\overline{T}$ which is self-adjoint. To simplify notation let us assume that T is already self-adjoint realizing we can always extend to the domain of $\overline{T}$ if necessary. Then we have T is self-adjoint and $\mathcal{D}(A) \supset \mathcal{D}(T)$. We have for $f,g \in \mathcal{D}(T)$ and real λ,

$$((\lambda T + iI)f,Ag) = (Af,(\lambda T - iI)g)$$

Now suppose that $f \in \mathcal{D}(A^*)$ and $g \in \mathcal{Y}$. We have

$$((\lambda T+iI)^{-1}A^*f,g) = ((\lambda T+iI)(\lambda T+iI)^{-1}f, A(\lambda T-iI)^{-1}g)$$
$$= (A(\lambda T+iI)^{-1}f,g)$$

Since g is arbitrary we have after muliplying by i,

$$A(I - i\lambda T)^{-1}f = (I - i\lambda T)^{-1}A^*f.$$

As $\lambda \to 0$ we have $(I - i\lambda T)^{-1}f \to f$ and $A(I - i\lambda T)^{-1}f \to A^*f$. Hence, the closure of A contains A*. Hence, A is essentially self-adjoint. This completes the proof of the lemma.

THEOREM 17. Suppose π is a strongly self-adjoint representation of a centrally dominated *-algebra $\mathfrak{A}$. Then $\pi(A)$ is essentially self-adjoint for all $A = A^* \in \mathfrak{A}$. Furthermore, if A and B are commuting hermitian elements of $\mathfrak{A}$ then the spectral projections of $\overline{\pi(A)}$ and $\overline{\pi(B)}$ commute.

Proof: Suppose the conditions of the theorem are satisfied. Suppose A = A* is contained in the center of $\mathfrak{A}$. Since π is strongly self-adjoint there is an $H \in \mathfrak{A}$ so that $T = \pi(H)$ is positive, essentially self-adjoint and $T \geq \pi(A^*A)$. Since A is in the center of $\mathfrak{A}$, T and $\pi(A)$ commute so by the previous lemma $\pi(A)$ is essentially self-adjoint. Hence for every hermitian A in the center of $\mathfrak{A}$, $\pi(A)$ is essentially self-adjoint.

Now suppose A is an arbitrary hermitian element of $\mathfrak{A}$. Since $\mathfrak{A}$ is centrally dominated there is an H in the center of $\mathfrak{A}$ so that $H \geq A^*A$. By what we have just seen $\pi(H)$ is essentially self-adjoint and then by the previous lemma $\pi(A)$ is essentially self-adjoint.

Now if A and B are hermitian elements of $\mathfrak{A}$ which commute one can construct the operator $H = \pi(A + iB)$. Since $\pi(A^2+ B^2)$ is essentially self-adjoint it follows that $H^*\overline{H} = \overline{H}H^*$ so that H is normal. It follows that the real and imaginary parts of H, $\overline{\pi(A)}$ and $\overline{\pi(B)}$, are self-adjoint operators with commuting spectral projections. (See exercise 11, page 1259 of [5]). This completes the proof of the theorem.

For centrally dominated *-algebras the notion of strongly self-adjoint representations works well to elininate pathologies. Whether this notion will work for (x,p)-algebras remains to be seen.

BIBLIOGRAPHY

1. W. B. Arveson, "Subalgebras of C*-algebras", Acta. Math. 123(1969), 141-224.

2. M. Choi and T. Lam, "Extremal Positive Semidefinite Forms", Math. Ann. 231(1977), 1-18.

3. J. Dixmier, "Sur la relation i(PQ - QP) = I", Composito Math. 13(1958), 263-269.

4. J. Dixmier, "Sur les algebres de Weyl I", Bull. Soc. Math. France, 96(1968), 209-242.

5. N. Dunford and J. Schwartz, Linear Operators Part II, Interscience, New York, 1963.

6. B. Fuglede, "On the Relation PQ - QP = -iI", Math. Scand. 20(1967), 79-88.

7. E. Nelson, "Analytic Vectors", Ann. of Math. 70(1959), 572-615.

8. R. T. Powers, "Self-adjoint algebras of unbounded operators I, II", Comm. Math. Phys. 21(1971), 85-124; Trans. Amer. Math. Soc. 187(1974), 261-293.

9. S. L. Woronowicz, "The Quantum Moment Problem I, II", Reports on Math. Phys. 1(1970),135-145; Reports on Math. Phys. 1 (1971), 175-183.

DEPARTMENT OF MATHEMATICS
UNIVERSITY OF PENNSYLVANIA
PHILADELPHIA, PA. 19104

Proceedings of Symposia in Pure Mathematics
Volume 38 (1982), Part 2

C^*-ALGEBRAS AND STATISTICAL MECHANICS

N.M. Hugenholtz

Institute for Theoretical Physics of the
University of Groningen, The Netherlands

1. INTRODUCTION

1.1. The KMS condition and some questions. In quantum mechanics systems of a finite or infinite number of degrees of freedom may be described in terms of a C^*-algebra $\mathfrak{A}$ of observables and states of $\mathfrak{A}$. In "good" situations the dynamics is a strongly continuous one-parameter group of automorphisms α_t of $\mathfrak{A}$.

In statistical mechanics one is interested in so-called thermal equilibrium states at different temperatures. Mathematically such states may be defined by means of the KMS condition.

Definition: A state ρ of $\mathfrak{A}$ is said to satisfy the *KMS condition* with respect to a strongly continuous one-parameter group of automorphisms α_t at temperature T if for each $A,B \in \mathfrak{A}$ there is a function f_{AB} of a complex variable z, analytic in the strip $0 < \text{Im}\, z < \beta$, continuous on the boundary, such that for real t

$$f_{AB}(t + i\beta) = \rho(\alpha_t(A)B)$$

and $f_{AB}(t) \quad = \rho(B\, \alpha_t(A))$.

Here $\beta = (kT)^{-1}$, with k the Boltzmann constant.

REMARK: Since α is strongly continuous (in the sense that $t \in \mathbb{R} \to \alpha_t(A)$ is continuous) we don't have to require that f_{AB} is bounded in the strip; this is a consequence.

When in 1967, at the Baton Rouge Conference, Haag, Winnink and I proposed this condition as an equilibrium condition in the C^*-algebraic setting of statistical mechanics there were still many unanswered questions. Were we not talking empty set? In order to conclude that the thermodynamical limit of Gibbs states of finite systems satisfies this condition, we had to make a physically reasonable assumption (in the language of Powers and Sakai we assumed that the

0082-0717/80/0000-0537/$15.75

automorphism group of the dynamics is approximately inner) and we did not know an example of a system where this condition is satisfied, with the exception of non-interacting fermions.

Another problem was immediately apparent. Even if one can prove that an equilibrium state obtained as a thermodynamical limit satisfies the KMS condition, is the condition sufficiently restrictive that it holds only for equilibrium states?

For the case of quantum lattice systems the answers to these questions have been given, partly already in the year following the Baton Rouge Conference, partly four or five years later. A considerable part of this paper is devoted to this development.

Some other questions concerning the KMS condition that have not yet found a satisfactory answer are the following. A state satisfying the KMS condition for some temperature is α_t-invariant. Is the reverse true? The answer is clearly no. But what about the following statement? "The extremal points of the convex set of α_t-invariant states are either ground state or KMS states". The statement is certainly not true for all choices of the dynamics α_t. Can we find conditions on the dynamics such that the above statement is true? The question is motivated by the observation that the stationary (= α_t-invariant) states of actual thermodynamical systems are equilibrium states. This indicates either that in such situations there are no other stationary states than (convex sums of) equilibrium states or that equilibrium states have a stability not possessed by other stationary states.

Is every faithful state of $\mathfrak{A}$ a KMS state for some one-parameter automorphism group? The answer is positive if $\mathfrak{A}$ is a von Neumann algebra but negative in general. Can we find conditions on a state such that it is a KMS state?

The following questions have solely to do with the dynamics. We say that the algebra $\mathfrak{A}$ is asymptotically abelian with respect to the time-evolution if

$$\lim_{|t|\to\infty} [A, \alpha_t(B)] = 0$$

for all $A,B \in \mathfrak{A}$. When does the dynamics have this property?

A question that is related to the physical notion of ergodicity is the following. How many one-parameter automorphism groups of $\mathfrak{A}$ commute with the dynamics α? We call a system ergodic if this set is restricted to the obvious symmetries of the system (such as translations) and α itself.

1.2. <u>What is statistical mechanics?</u> The aim of statistical mechanics is to give the theoretical foundation of thermodynamics. In the most simple situation the system (gas, crystal, magnet) consists of a very large number of identical subsystems (molecules, ions, spins) that interact with one another through forces. The object of the exercise is to explain the thermodynamic properties of such systems from what is known about the interaction. A typical phenomenon in many thermodynamical systems is the occurrence of phase transitions: sudden changes in the properties of the system if external parameters (e.g. the temperature) are changed. Typical examples are condensation, crystallization, and the well-known fact that many metals suddenly become magnetic when the temperature is lowered below a certain critical temperature, the Curie-temperature.

In this article we shall be concerned with a very simple class of models, quantum lattice systems, that are on the one hand simple enough that rigorous statements can be made, on the other hand rich enough to exhibit phase transitions. The discussion will be limited to equilibrium phenomena. As is usual in statistical mechanics, we are really studying infinite systems. Only in the thermodynamic limit where the number of components approaches infinity can we expect the occurrence of phase transitions in the rigorous sense, and will we find discontinuous changes in density, magnetization etc.

1.3. Finite systems. Since the traditional methods of quantum mechanics and statistical mechanics are applicable to finite systems only we shall conclude the introduction with a brief description of a very simple finite quantum system.

Let $\mathcal{H}$ be a n-dimensional Hilbert space and $\mathfrak{A}$ the algebra of $n \times n$ matrices acting on $\mathcal{H}$. The observables of the system are the self-adjoint elements of $\mathfrak{A}$. A s.a. matrix H, the Hamiltonian, determines the time-evolution as follows:

$$\alpha_t(A) = e^{iHt} A e^{-iHt}$$

and vice versa: there is a one-to-one correspondence between s.a. elements of $\mathfrak{A}$ and one-parameter automorphism groups of $\mathfrak{A}$.

The vectors in $\mathcal{H}$ represent the pure states. For $\Phi \in \mathcal{H}$ with $\|\Phi\| = 1$ the state ρ_Φ, defined by

$$\rho_\Phi(A) = (\Phi, A\Phi)$$

is called the *expectation value* of A for the vector Φ. More general states are defined by a density matrix ρ ($\rho \geq 0$ and $\operatorname{Tr} \rho = 1$). In this case the expectation value of A is given by $\operatorname{Tr} \rho A$. The corresponding state (normalized positive linear form) will be written as $\rho(A)$. The pure states clearly correspond to one-dimensional projections. The α_t-invariant states have a density matrix ρ that commutes with H. It is not difficult to see that the extremal α_t-invariant states are pure states corresponding to eigenvectors of H. These states are in general neither ground states nor KMS state.

For a state in thermal equilibrium with a heath bath at temperature T we have

$$\rho = e^{-\beta H} / \mathrm{Tr}\, e^{-\beta H}.$$

We call this the Gibbs state. From now on we shall absorb the factor β in H and write $\rho = \exp(-H) / \mathrm{Tr} \exp(-H)$. If we compare, for arbitrary A and B $\in \mathfrak{A}$, the expressions

$$\rho(\alpha_t(A)B) = (\mathrm{Tr}\, e^{-H})^{-1}\, \mathrm{Tr}(e^{-H} e^{iHt} A\, e^{-iHt} B)$$

and

$$\rho(B\,\alpha_t(A)) = (\mathrm{Tr}\, e^{-H})^{-1}\, \mathrm{Tr}(e^{-H} B\, e^{iHt} A\, e^{-iHt})$$

we see that, due to the invariance of the trace for cyclic permutation,

$$\rho(\alpha_{t-i}(A)B) = \rho(B\,\alpha_t(A)).$$

This means that ρ satisfies the KMS condition with respect to α_t at $\beta = 1$. An easy calculation shows that the Gibbs state is the only state satisfying the KMS condition. If ρ is faithful (i.e. $A \geq 0$ and $\rho(A) = 0$ implies $A = 0$), the density matrix ρ has an inverse. We then define $H = -\log \rho$ and it follows that ρ satisfies the KMS condition with respect to the time-evolution derived from H.

Having thus answered the various questions of section 1.1 for this case, we shall now introduce another equilibrium condition with a more immediate physical meaning. Let ρ be a state and consider

$$S = -\mathrm{Tr}\, \rho \log \rho. \qquad (1.1)$$

As we shall see, this is a measure of the degree of randomness of the system, or, in the language of information theory, S is a measure of our lack of information about the system. S is called the entropy of the state ρ *). If $\{\lambda_1, \lambda_2, \ldots, \lambda_n\}$ is the set of eigenvalues of ρ, then $0 \leq \lambda_i \leq 1$ and $\sum_i \lambda_i = 1$. So, clearly, $S = -\sum_i \lambda_i \log \lambda_i \geq 0$. On the other hand, taking $\rho_2 = n^{-1}\,\mathbb{1}$ in the inequality (A.4) of the Appendix, we obtain

$$\mathrm{Tr}\, \rho \log \rho \geq -\log n,$$

so that

$$0 \leq S \leq \log n. \qquad (1.2)$$

Pure states give maximal information. For such states S = 0. The maximum of S is reached for $\rho = n^{-1}\,\mathbb{1}$. This is the state that gives equal weight to all pure states and contains therefore minimal information. Let us now consider

*) The usual physical definition of entropy is $-k\, \mathrm{Tr}\, \rho \log \rho$, where k is the Boltzmann constant.

the most random state ρ of given average energy $\rho(H)$, i.e., the ρ for which $S(\rho) - \beta\rho(H)$ is maximal. The next theorem says that this state is the Gibbs state with $\rho = \exp(-\beta H) / \mathrm{Tr}\exp(-\beta H)$.

THEOREM 1.1 : *(Variational principle). For* H *self-adjoint and* ρ *a density matrix*

$$F(H) \leq \rho(H) - S(\rho) , \tag{1.3}$$

where

$$F(H) = -\log \mathrm{Tr}\, e^{-H} . \tag{1.4}$$

For given H there is precisely one state, the Gibbs state for which the equality holds.

PROOF : (1.3) follows from (A.4) by choosing $\rho_2 = \exp(-H) / \mathrm{Tr}\exp(-H)$. It is then clear that the equality holds for the Gibbs state, and for this state only. □

The quantity $\rho(H) - S(\rho)$ can be called the *free energy* of the state ρ for Hamiltonian H. The equilibrium state is the state of minimal free energy. F(H) is then the *free energy of the equilibrium state.*

2. QUANTUM LATTICE SYSTEMS

2.1. Definition. Let $\mathbb{Z}^\nu$ be a ν-dimensional lattice. With each lattice point $x \in \mathbb{Z}^\nu$ we associate an n-dimensional Hilbert space $\mathcal{H}_x$. We shall only consider the case n = 2 although the case of arbitrary finite n is hardly more complicated and will be left to the reader.

Let Λ denote a finite subset of $\mathbb{Z}^\nu$, consisting of $|\Lambda|$ points. We define

$$\mathcal{H}(\Lambda) = \prod_{x\in\Lambda} \otimes\, \mathcal{H}_x \quad \text{(tensor product)} .$$

We write $\mathfrak{A}(\Lambda) = B(\mathcal{H}(\Lambda))$ for the set of matrices of rank $2^{|\Lambda|}$ acting on $\mathcal{H}(\Lambda)$. The self-adjoint elements of $\mathfrak{A}(\Lambda)$ are the observables corresponding to the region Λ.

If $\Lambda_1 \subset \Lambda_2$ we have $\mathcal{H}(\Lambda_2) = \mathcal{H}(\Lambda_1) \otimes \mathcal{H}(\Lambda_2/\Lambda_1)$. If we identify $\mathfrak{A}(\Lambda_1)$ with $\mathfrak{A}(\Lambda_1) \otimes \mathbb{1}_{\Lambda_2/\Lambda_1}$, we have the inclusion

$$\mathfrak{A}(\Lambda_1) \subset \mathfrak{A}(\Lambda_2).$$

Similarly, for $\Lambda_1 \cap \Lambda_2 = \emptyset$ the algebras $\mathfrak{A}(\Lambda_1)$ and $\mathfrak{A}(\Lambda_2)$ commute if both are considered as subalgebras of $\mathfrak{A}(\Lambda)$ with $\Lambda_1 \cup \Lambda_2 \subset \Lambda$. The union $\mathfrak{A}_L$ of all $\mathfrak{A}(\Lambda)$ is an algebra, the algebra of local observables. Its completion in the norm $\mathfrak{A}$ is the C^*-algebra of quasi-local observables. It is by definition an U.H.F. algebra.

In order to be able to relate the different points in $\mathbb{Z}^\nu$ we define an isometric mapping $V_{y,x}$ between the Hilbert spaces $\mathcal{H}_x$ and $\mathcal{H}_y$: $\mathcal{H}_y = V_{y,x}\mathcal{H}_x$ with the composition rule $V_{z,x} = V_{z,y} V_{y,x}$. We shall use this to define two important transformation groups as groups of automorphisms of $\mathfrak{A}$.

a. *Gauge transformations* (first kind). Let U be a unitary matrix acting on the Hilbert space $\mathcal{H}_0$ of the point (0,0,...,0). $U_x = V_{x,0}\, U\, V_{0,x}$ is then the corresponding matrix acting on $\mathcal{H}_x$. We construct for each Λ the unitary matrix $U_\Lambda = \prod_{x\in\Lambda} \otimes\, U_x$ and define the map

$$A \in \mathfrak{A}(\Lambda) \to \gamma_U(A) = U_\Lambda\, A\, U_\Lambda^{-1} ,$$

an automorphism of $\mathfrak{A}(\Lambda)$. The extension of γ_U to the whole algebra $\mathfrak{A}$ is the gauge transformation determined by U. It follows immediately from the definition that the gauge transformations form a group.

b. *Translations*. For $a \in \mathbb{Z}^\nu$, the translation τ_a is defined as follows. For $A \in \mathfrak{A}(\Lambda)$

$$\tau_a(A) = V_\Lambda(a)\, A\, V_\Lambda(a)^{-1}$$

where

$$V_\Lambda(a) = \prod_{x\in\Lambda} V_{x+a,x} .$$

We see that τ_a maps $\mathfrak{A}(\Lambda)$ onto $\mathfrak{A}(\Lambda+a)$ $(\Lambda + a = \{x+a,\ x \in \Lambda\})$. The extension of τ_a to $\mathfrak{A}$ is the automorphism of $\mathfrak{A}$ corresponding to a translation over a. The translations form an abelian group.

2.2. Dynamics. As in the finite case, the dynamics or time-evolution shall be a one-parameter group of automorphism $A \to \alpha_t(A)$ of the observable algebra $\mathfrak{A}$. We shall consider only the case that α_t commutes with translations. This implies that α_t cannot be inner since $\mathfrak{A}$ does not contain translation invariant elements except the identity. Hence there is no Hamiltonian and the dynamics of the system must be defined in some other way. The notion that replaces the Hamiltonian is the *potential*. Physically, the potential describes the interaction between the lattice points. Before giving the mathematical definition of a potential, it is useful to consider the following decomposition of elements of $\mathfrak{A}$.

Let τ be the unique tracial state of $\mathfrak{A}$ and $\mathcal{F}$ the Hilbert space obtained as completion of $\mathfrak{A}$ with the scalar product $(A,B) = \tau(A^*B)$. The linear subspaces $\{\mathfrak{A}(X),\ X \subset \mathbb{Z}^\nu,\ |X| < \infty\}$ are not mutually orthogonal, but by means of a unique orthogonalization process we can construct mutually orthogonal subspaces $\{\mathcal{F}(X)\}$, where for each X $\mathcal{F}(X) = \{A \in \mathfrak{A}(X)\colon A \text{ orthogonal to all } \mathfrak{A}(Y) \text{ for } Y \subsetneq X\}$.

It is then easy to prove that, for $|X| \neq 0$,

$$\mathcal{F}(X) = \{A \in \mathfrak{A}(X) : \ Tr^{(Y)} A = 0, Y \subset X\},$$

where $Tr^{(Y)}$ is the trace w.r.t. the Hilbert space $\mathcal{H}(Y)$. Since

$$\mathcal{F} = \sum_X \oplus \mathcal{F}(X),$$

we have for each $A \in \mathfrak{A}$ the unique decomposition

$$A = \sum_X A(X), \tag{2.1}$$

where $A(X)$ has the properties

$$\begin{aligned} &\text{i.} \quad A(X) \in \mathfrak{A}(X) \\ &\text{ii.} \quad Tr^{(Y)} A(X) = 0 \quad \text{for all } Y \subset X, \text{ and } |X| \neq 0. \end{aligned} \tag{2.2}$$

The sum in (2.1) converges in the norm-topology of $\mathcal{F}$ but need not converge in the norm of $\mathfrak{A}$. In any case A is uniquely determined by its components $A(X)$.

DEFINITION 2.1: A translation-invariant *potential* Φ is a function $X \to \Phi(X)$ from the finite subsets of $\mathbb{Z}^\nu$ to the self-adjoint elements of $\mathfrak{A}$ such that

i. $\Phi(X) \in \mathfrak{A}(X)$

ii. $Tr^{(Y)} \Phi(X) = 0, \ Y \subset X$

iii. $\tau_x(\Phi(X)) = \Phi(X + x)$.

REMARK: In anology with (2.1) one would like to write $H = \sum_X \Phi(X)$, where H would be the Hamiltonian. However, due to property iii (translation-invariance) the finite sum diverges even in the $\mathcal{F}$-norm.

We shall not be interested in all such potentials, but only in those where $\|\Phi(X)\|$ decreases sufficiently rapidly with increasing X. We can define a norm

$$\|\Phi\|_1 = \sum_{0 \in X} \frac{\|\Phi(X)\|}{N(X)} \tag{2.3}$$

and the corresponding Banach space

$$\mathcal{B}_1 = \{\Phi : \|\Phi\|_1 < \infty\} .$$

This space is still very large and contains potentials of very long range. It is therefore useful to use other norms. If f is a positive function on the finite subsets of $\mathbb{Z}^\nu$, increasing as a function of $|X|$ or of $d(X)$ (the diameter of X) then

$$\|\Phi\|_f = \sum_{0 \in X} \|\Phi(X)\| f(X)$$

and $\mathcal{B}_f = \{\Phi : \|\Phi\|_f < \infty\}$.

Clearly $\mathcal{B}_f \subset \mathcal{B}_1$.

DEFINITION 2.2: The range $\Delta_\Phi \subset \mathbb{Z}^\nu$ of a potential Φ is

$$\Delta_\Phi = \cup\{X : 0 \in X \text{ and } \Phi(X) \neq 0\}.$$

The linear space of potentials of finite range will be denoted by $\mathcal{B}_0$.

PROPOSITION 2.1: *$\mathcal{B}_0$ is dense in $\mathcal{B}_1$.*

For the proof it is convenient to introduce the following notions. The *distance* $d(x,y)$ of two points x and y is defined as

$$d(x,y) = \Big[\sum_{i=1}^{\nu} (x_i - y_i)^2\Big]^{\frac{1}{2}}.$$

The *diameter* of a set X is $d(X) = \sup\limits_{x,y \in X} d(x,y)$.

PROOF of proposition 2.1: Let $\varepsilon > 0$ and $\Phi \in \mathcal{B}_1$. Define $\Phi_R \in \mathcal{B}_0$ as follows.

$$\Phi_R(X) = \Phi(X) \quad \text{for} \quad d(X) \le 2R$$

$$= 0 \quad \text{for} \quad d(X) \le 2R.$$

Then, if $B(R) = \{x : d(x,0) \le R\}$,

$$\|\Phi - \Phi_R\| = \sum_{X \ni 0 \,:\, X \not\subset B(R)} \frac{\|\Phi_R(X) - \Phi(X)\|}{N(X)} \le \sum_{X \ni 0 \,:\, X \not\subset B(R)} \frac{\|\Phi(X)\|}{N(X)}.$$

Since $\Phi \in \mathcal{B}$, we can choose R such that

$$\sum_{X \ni 0 \,:\, X \not\subset B(R)} \frac{\|\Phi(X)\|}{N(X)} < \varepsilon. \qquad \square$$

Potentials may be used to define the dynamics of the system. Let us call

$$H(\Phi,\Lambda) = \sum_{X \subset \Lambda} \Phi(X)$$

the *Hamiltonaian of the finite sublattice* Λ. If the limit

$$\lim_{\Lambda \to \infty} e^{iH(\Phi,\Lambda)t} A\, e^{-iH(\Phi,\Lambda)t} = \alpha_t^\Phi(A)$$

exists for all $A \in \mathfrak{A}$ and gives rise to a strongly continuous one-parameter automorphism group of $\mathfrak{A}$, we shall call Φ a *dynamical potential* and the group a *dynamical automorphism group* of $\mathfrak{A}$.

Not all elements of $\mathcal{B}_1$ are dynamical but

THEOREM 2.2: *For $f(X) = \exp(|X|)$ the potentials in $\mathcal{B}_f$ are dynamical.*

In order to prove this theorem, choose $A \in \mathfrak{A}(\Lambda_0)$ and take $\Lambda \supset \Lambda_0$. We write

$$\alpha_t^\Lambda(A) = e^{iH(\Phi,\Lambda)t} A\, e^{-iH(\Phi,\Lambda)t}$$

and develop the right-hand side in powers of t,

$$\alpha_t^{\Lambda}(A) = \sum_{n=0}^{\infty} a_n^{\Lambda} t^n ,$$

where

$$a_n^{\Lambda} = \frac{i^n}{n!} [H(\Phi,\Lambda),A]^{(n)} ,$$

and $[B,A]^{(n)}$ stands for the repeated commutator $[B,[B,\dots[B,A]\dots]]$. The following lemma gives an estimate of $[H(\Phi,\Lambda),A]^{(n)}$.

LEMMA 2.2: *If* $\Phi \in \mathcal{B}_1$ *and* $A \in \mathfrak{A}(\Lambda_0)$ *with* $\Lambda_0 \subset \Lambda$ *then*

$$\|[H(\Phi,\Lambda),A]^{(n)}\| \le \|A\| \, e^{|\Lambda_0|} \, n! \, (2\|\Phi\|_f)^n ,$$

with $f(X) = \exp(|X|)$.

PROOF:

$$[H(\Phi,\Lambda),A]^{(n)} = \sum_{X_1\subset\Lambda} \dots \sum_{X_n\subset\Lambda} [\Phi(X_n), \dots [\Phi(X_1),A]\dots] =$$

$$\sum_{k_1,\dots,k_n=0}^{\infty} \; \sum_{\substack{X_1\subset\Lambda : \\ |X_1|=k_1+1}} \dots \sum_{\substack{X_n\subset\Lambda : \\ |X_n|=k_n+1}} [\Phi(X_n), \dots [\Phi(X_1),A]\dots] . \tag{2.5}$$

If $S_i = \Lambda_0 \cup X_1 \cup \dots \cup X_{i-1}$, the summand vanishes unless $X_i \cap S_i \neq \emptyset$ $(i = 1,\dots,n)$.

For $B \in \mathfrak{A}(S)$ we have the inequality

$$\Big\| \sum_{X\subset\Lambda:\, |X|=n} [\Phi(X),B] \Big\| \le \sum_{x\in S} \; \sum_{\substack{x\in X\subset\Lambda : \\ |X|=n}} \|[\Phi(X),B]\| \le 2 \, S \, \|B\| \sum_{0\in X:|X|=n} \|\Phi(X)\| .$$

Using this inequality repeatedly we get

$$\Big\| \sum_{X_1\subset\Lambda:|X_1|=k_1+1} \dots \sum_{X_n\subset\Lambda:|X_n|=k_n+1} \Big[\Phi(X_n), \dots [\Phi(X_1),A]\dots\Big] \Big\|$$

$$\le 2^n \|A\| \prod_{i=1}^{n} \Big[(|\Lambda_0| + k_1 + \dots + k_{i-1}) \sum_{\substack{X_i\ni 0 : \\ |X_i|=k_i+1}} \|\Phi(X_i)\| \Big] . \tag{2.6}$$

Now

$$\frac{1}{n!} \prod_{i=1}^{n} (|\Lambda_0| + k_1 + \dots + k_{i-1}) \le e^{|\Lambda_0| + \sum_i k_i} . \tag{2.7}$$

Substituting (2.6) and (2.7) in (2.5) one gets

$$\|[H(\Phi,\Lambda),A]^{(n)}\| \le n! \; 2^n \|A\| \, e^{|\Lambda_0|} \sum_{k_1,\dots,k_n=0}^{\infty} \prod_{i=1}^{n} \Big[e^{k_i} \sum_{\substack{X_i\ni 0 : \\ |X_i|=k_i+1}} \|\Phi(X_i)\| \Big] \le$$

$$\leq 2^n n!\ \|A\|\ e^{|\Lambda_0|}\ (\|\Phi\|_f)^n,$$

with $f(X) = \exp(|X|)$. □

PROOF of theorem 2.2: Writing $\alpha_t^\Lambda(A) = \sum_{n=0}^{\infty} a_n^\Lambda t^n$, where $A \in \mathfrak{A}(\Lambda_0)$, it follows from lemma 2.3 that $\|a_n^\Lambda\| \leq C\, t_0^{-n}$, with $C = \|A\| \exp(|\Lambda_0|)$ and $t_0^{-1} = 2\|\Phi\|_f$. Hence the power series is majorized in norm by a power series with radius of convergence t_0 and independent of Λ. The next step is to prove the existence of the limit for $\Lambda \to \infty$ of a_n^Λ. Clearly, for Φ of finite range, a_n^Λ is independent of Λ for Λ sufficiently large. Let now Φ' be of finite range and $a_n^{'\Lambda}$ the corresponding coefficient. Using similar estimates as in the proof of lemma 2.3 one proves

$$\|a_n^\Lambda - a_n^{'\Lambda}\| \leq \|A\|\ e^{|\Lambda_0|}\ 2^n \sum_{\ell=0}^{n-1} \|\Phi'\|_f^{\ell}\ \|\Phi\|_f^{n-\ell-1}\ \|\Phi-\Phi'\|_f.$$

This shows that for given n we can choose Φ' such that, independent of Λ, $\|a_n^\Lambda - a_n^{'\Lambda}\| < \frac{1}{2}\varepsilon$, so that the limit

$$\lim_{\Lambda\to\infty} a_n^\Lambda = a_n$$

exists. We now conclude immediately to the existence of

$$\alpha_t(A) = \lim_{\Lambda\to\infty} \alpha_t(A) = \sum_n a_n t^n,$$

for $|t| \leq t_1 < t_0$ and $A \in \mathfrak{A}(\Lambda_0)$. It is easy to see that $\|\alpha_t(A)\| = \|A\|$ for real t, and $\alpha_t(A)\alpha_t(B) = \alpha_t(AB)$, for A and $B \in \mathfrak{A}_L$. By continuity we conclude to the existence of the $\lim \alpha_t^\Lambda(A) = \alpha_t(A)$ for all $A \in \mathfrak{A}$, although in general it is no longer a power series in t. For each t such that $-t_1 \leq t \leq t_1$ α_t is an automorphism of $\mathfrak{A}$.

To extend this result to other values of t we use the group property. For t and t' such that $|t| \leq t_1$, $|t'| \leq t_1$ and $|t+t'| \leq t_1$ we have

$$\alpha_{t+t'}(A) = \alpha_t(\alpha_{t'}(A)).$$

This property can be used to define $\alpha_t(A)$ if t is outside the interval $[-t_1, t_1]$ but $\frac{1}{2}t$ is inside, etc. □

According to Powers and Sakai a strongly continuous one-parameter group of automorphisms α_t of $\mathfrak{A}$ is *approximately inner* if there is a sequence $\{H_n\}$ of self-adjoint elements of $\mathfrak{A}$ such that for each $A \in \mathfrak{A}$

$$\lim_{n\to\infty} e^{iH_n t} A e^{-iH_n t} = \alpha_t(A).$$

Clearly all dynamical groups of automorphisms are approximately inner. It is not known whether the reverse is true. This question is of interest in connection with a conjecture by Powers and Sakai stating that all strongly continuous one-parameter automorphism groups of UHF algebras are approximately inner.

It can be shown that the correspondence between dynamical potentials and dynamical automorphism groups is one-to-one, i.e., different potentials give different groups (see theorem 7.6).

2.3. States. In the case of a finite lattice, the observables are the $2^{|\Lambda|} \times 2^{|\Lambda|}$ matrices on $\mathcal{H}_\Lambda$. The states are then in one-to-one correspondence with density matrices. This will now be generalized to the case of the infinite lattice. We define a *state* to be a normalized positive linear form on $\mathfrak{A}$, or, more explicitly, ρ is a state if

i. $\rho(\lambda_1 A_1 + \lambda_2 A_2) = \lambda_1 \rho(A_1) + \lambda_2 \rho(A_2)$

ii. $\rho(A) \geq 0$ if $A \geq 0$

iii. $\rho(\mathbb{1}) = 1$.

A state ρ is *faithful* if $\rho(A^*A) = 0$ implies $A = 0$. An automorphism α of defines by transposition a linear map $\rho \to \rho \circ \alpha$:

$$(\rho \circ \alpha)(A) = \rho(\alpha(A)).$$

A state ρ is *invariant* under α if $\rho \circ \alpha = \rho$. The invariant states form a convex set.

A state that is invariant under the time evolution (dynamics) is called *stationary*. In equilibrium statistical mechanics one is interested in the set of stationary states or, in particular, the extremal stationary states. A special class of stationary states are the states satisfying the KMS condition. The question whether a given dynamical automorphism group admits KMS states is answered by the following theorem. Suppose Φ is a dynamical potential. Let for each finite Λ ρ_Λ be the Gibbs state of the algebra $\mathfrak{A}(\Lambda)$ with Hamiltonian $H(\Phi,\Lambda)$, and let $\bar{\rho}_\Lambda$ be an extension of ρ_Λ to $\mathfrak{A}$. Since the unit ball of the dual $\mathfrak{A}^*$ of $\mathfrak{A}$ is W*-compact and $\mathfrak{A}$ is separable the countable set $\{\bar{\rho}_\Lambda\}$ contains a convergent subsequence $\{\bar{\rho}_{\Lambda_n}\}$. Let $\rho = \lim \bar{\rho}_{\Lambda_n}$; we say that ρ is a thermodynamical limit of Gibbs states.

THEOREM 2.4: *Each thermodynamical limit of Gibbs states corresponding to a given dynamical potential satisfies the KMS condition with respect to the corresponding dynamical automorphism group.*

PROOF: Let $A, B \in \mathfrak{A}(\Lambda_0)$ and $\Lambda_n \supset \Lambda_0$. We define the entire function f_n by

$f_n(t) = \rho_{\Lambda_n}(B\alpha_t^{\Lambda_n}(A))$. Since ρ_{Λ_n} satisfies the KMS condition w.r.t. $\alpha_t^{\Lambda_n}$ we have $f_n(t+i) = \rho_{\Lambda_n}(\alpha_t^{\Lambda_n}(A)B)$. For z in the strip $0 \leq \mathrm{Im}\, z \leq 1$ $|f_n(z)|$ is majorized by $\|B\| \, \|\alpha_{iy}^{\Lambda_n}(A)\|$, where $y = \mathrm{Im}\, z$, and we may conclude that $|f_n(z)|$ is bounded in the strip $0 \leq \mathrm{Im}\, z \leq 1$. Application of the maximum modulus principle then gives

$$|f_n(z)| \leq \|A\| \, \|B\| .$$

For real t we have

$$|f_n(t) - \rho(B\alpha_t(A))| = |\rho_{\Lambda_n}(B\alpha_t^{\Lambda_n}(A)) - \rho_{\Lambda_n}(B\alpha_t(A)) + \rho_{\Lambda_n}(B\alpha_t(A)) - \rho(B\alpha_t(A))|$$

$$\leq \|B\| \, \|\alpha_t^{\Lambda_n}(A) - \alpha_t(A)\| + |\rho_{\Lambda_n}(B\alpha_t(A)) - \rho(B\alpha_t(A))| .$$

We can make both terms arbitrarily small by choosing n sufficiently large. Hence

$$\lim_{n\to\infty} f_n(t) = \rho(B\alpha_t(A)) .$$

Similarly

$$\lim_{n\to\infty} f_n(t+i) = \rho(\alpha_t(A)B) .$$

We can now conclude that the sequence $\{f_n\}$ converges to a function f, analytic in the interior of the strip $0 < \mathrm{Im}\, z < 1$ and

$$f(t) = \rho(B\alpha_t(A)) \quad ; \quad f(t+i) = \rho(\alpha_t(A)B) .$$

The extension to arbitrary elements A and $B \in \mathfrak{A}$ is straightforward. □

REMARKS:

1. A state satisfying the KMS condition is stationary. Theorem 2.4 provides us therefore with examples of stationary states. It is also easy to prove that a thermodynamical limit of stationary states ρ_Λ (i.e. invariant for α_t^Λ) is stationary (w.r.t. α_t).
2. Even though the dynamical potential Φ is translation-invariant one cannot conclude that ρ is translation-invariant.
3. If we use the inverse temperature β as an explicit parameter, theorem 2.4 implies that there is a KMS state for each $\beta \in \mathbb{R}$.

2.4. Examples. In this section we shall describe some examples that are widely studied because they are on the one hand sufficiently simple that some rigorous statements can be made or even explicit solutions can be given, but on the other hand exhibit non-trivial behaviour as e.g. phase transitions.

The Heisenberg model

The quantum lattice systems we consider in these notes can be interpreted as a system of interacting spins, one in each lattice point. In the case of the Heisenberg model there is only one-point and two-point interaction. The potential Φ can then be written as follows .

$$\Phi(X) = 0 \quad \text{for} \quad |X| > 2 ,$$

$$\Phi(\{x,y\}) = \sum_{i,j=1}^{3} J_{ij}(x,y)\sigma_i(x)\sigma_j(y),$$

$$\Phi(\{x\}) = -\sum_i h_i(x)\sigma_i(x).$$

The 2×2 matrices σ_1, σ_2 and σ_3 used here are the so-called *Pauli matrices*

$$\sigma_1 = \begin{pmatrix} 0 & 1 \\ 1 & 0 \end{pmatrix} \qquad \sigma_2 = \begin{pmatrix} 0 & -i \\ i & 0 \end{pmatrix} \qquad \sigma_3 = \begin{pmatrix} 1 & 0 \\ 0 & -1 \end{pmatrix} .$$

They are hermitian, unitary, traceless and linearly independent matrices satisfying the multiplication rules

$$\sigma_1\sigma_2 = i\sigma_3 \quad \text{and cyclic.}$$

Physically they can be interpreted as operators for the three components of the spin angular momentum.
Translation-invariance requires that $J_{ij}(x,y)$ depends only on x-y and $h_i(x)$ is independent of x. The Hamiltonian for finite Λ is

$$H(\Lambda) = \tfrac{1}{2} \sum_{x,y\in\Lambda} \sum_{i,j=1}^{3} J_{ij}(x,y)\sigma_i(x)\sigma_j(y) - \sum_{x\in\Lambda} \sum_i h_i\sigma_i(x),$$

where we assume that $J_{ij}(x,x) = 0$. The first term on the right-hand side is the proper interaction between the spins, whereas the second term is to be interpreted as the interaction of the spins with an external constant magnetic field $\vec{h}$.

If one requires reflexion symmetry in the x,y and z direction one obtains the *anisotropic Heisenberg model* with Hamiltonian

$$H(\Lambda) = \tfrac{1}{2} \sum_{x,y\in\Lambda} \sum_{i=1}^{3} J_i(x-y)\sigma_i(x)\sigma_i(y) - h \sum_{x\in\Lambda} \sigma_3(x) , \tag{2.8}$$

where the external magnetic field is directed along the z-axis. The special case where $J_i(x) = J(x)$ independent of i is called the *isotropic Heisenberg model*.

From theorem 2.2 we conclude that Φ is dynamic if $\sum_x |J(x)| < \infty$.

If $J_3(x,y) = 0$ in (2.8) one has the XY model.

<u>Ising model</u>

Putting $J_1(x,y) = J_2(x,y) = 0$ in (2.8) one obtains the Hamiltonian of the Ising model

$$H(\Lambda) = \tfrac{1}{2} \sum_{x,y\in\Lambda} J(x-y)\sigma_3(x)\sigma_3(y) - h \sum_{x\in\Lambda} \sigma_3(x).$$

DEFINITION 2.3: The classical algebra $\mathfrak{A}_{C\ell}(\Lambda)$ associated with Λ is the commutative algebra generated by $\{\sigma_3(x), x \in \Lambda\}$. The norm closure $\mathfrak{A}_{C\ell}$ of the union of all $\mathfrak{A}_{C\ell}(\Lambda)$ is the *classical observable algebra*. One observes that $H(\Lambda) \in \mathfrak{A}_{C\ell}(\Lambda)$.

It will be shown that this model is equivalent with the following classical model. One associates with each point $x \in \mathbb{Z}^\nu$ the space $\{1,-1\}_x$ (corresponding with the two eigenvalues of $\sigma_3(x)$), and defines the configuration space K as the direct product

$$K = \prod_{x\in\mathbb{Z}^\nu} \otimes \{1,1\}_x \quad .$$

K is compact in the product topology of the discrete topologies of $\{1,1\}_x$. The algebra of observables is the set C(K) of bounded continuous functions on K. The states are probability measures on K, or, equivalently, states of C(K).

Instead of $\sigma_3(x)$ we now write s(x), an element of C(K), and

$$H(\Lambda) = \tfrac{1}{2} \sum_{x,y\in\Lambda} J(x,y)s(x)s(y) - h \sum_{x\in\Lambda} s(x).$$

If the interaction is not restricted to one- and two-point interaction one writes more generally

$$H(\Lambda) = \sum_{X\subset\Lambda} J(X)s(X) \;,$$

where

$$s(X) = \prod_{x\in X} s(x).$$

In order to prove the equivalence between this classical model and the quantum model we notice first that C(K) is isomorphic with $\mathfrak{A}_{C\ell}$. We then consider for each $x \in \mathbb{Z}^\nu$ the inner automorphism group α_x^θ of $\mathfrak{A}$ defined by

$$\alpha_x^\theta(A) = e^{i\sigma_3(x)\theta} A\, e^{-i\sigma_3(x)\theta} \quad .$$

To understand the physical meaning of this transformation we calculate

$$\alpha_x^\theta(\sigma_i(y)) = \sigma_i(y), \quad y \neq x,$$

$$\alpha_x^\theta(\sigma_3(x)) = \sigma_3(x),$$

$$\alpha_x^\theta(\sigma_1(x)) = \cos\theta\,\sigma_1(x) - \sin\theta\,\sigma_2(x),$$

$$\alpha_x^\theta(\sigma_2(x)) = \sin\theta\,\sigma_1(x) + \cos\theta\,\sigma_2(x).$$

The action of α_x^θ is a rotation of the spin at the lattice point x around the z-axis.

DEFINITION 2.4: A state of $\mathfrak{A}$ is *classical* if it is invariant under α_x^θ for all angles θ and all $x \in \mathbb{Z}^\nu$.

As we shall see there is a one-to-one correspondence between classical states of $\mathfrak{A}$ and states of C(K). In order to prove that we define

$$\eta(x) = \frac{1}{2\pi}\int_0^{2\pi} d\theta\ \alpha_x^\theta.$$

This is a projection such that $\eta(x)(\sigma_3(x)) = \sigma_3(x)$, $\eta(x)(\sigma_1(x)) = \eta(x)(\sigma_2(x)) = 0$. We now define the operator τ on $\mathfrak{A}$ as follows. For $A \in \mathfrak{A}(\Lambda)$ $\tau(A) = \eta(\Lambda)(A)$, where $\eta(\Lambda) = \prod_{x\in\Lambda} \eta(x)$. It is easy to see that τ is a projection operator projecting $\mathfrak{A}$ onto $\mathfrak{A}_{C\ell}$. We now have

THEOREM 2.5: *There is a one-to-one correspondence between states of C(K) and classical states of* $\mathfrak{A}$. *The correspondence is given by* $\tilde{\rho}(A) = \rho(\tau(A))$, *where* ρ *is a state of C(K).*

<u>Classical lattice gas</u>

In this model each lattice site can be either empty or occupied by one particle. So one associates with each point $x \in \mathbb{Z}^\nu$ the space $\{0,1\}_x$. The configuration space K is again the direct product

$$K = \prod_{x\in\mathbb{Z}^\nu} \otimes \{0,1\}_x.$$

A configuration (point in K) may be determined by the values of n(x) for each point $x \in \mathbb{Z}^\nu$ or, equivalently, by the finite or infinite subset $X \subset \mathbb{Z}^\nu$ of occupied sites. The particle number n(x) at the point x is a bounded continuous function on K. A potential Φ associates with every finite X an element $\Phi(X) \in C(K)$ defined as follows

$$\Phi(X) = \phi(X)\,n(X),$$

with $n(X) = \prod_{x\in X} n(x)$. The Hamiltonian for the finite set Λ is

$$H(\Lambda) = \sum_{X\subset\Lambda} \phi(X)n(X).$$

It should be remarked that Φ as defined here does not satisfy the condition (ii) of definition 2.1.

For potentials Φ where $\|\Phi(X)\|$ decreases sufficiently rapidly with increasing d(X) one can prove the equivalence of the classical lattice gas and the Ising model. We shall consider the example of two-particle interaction such that $\sum_y |\phi(x,y)| < \infty$. Since we are dealing with a varying particle number N, we must consider instead of $H(\Lambda)$ the quantity $H(\Lambda) - \mu N(\Lambda)$ occuring in the grand canonical ensemble, where μ is the chemical potential. Making the substitution $n(x) = \frac{1}{2}(s(x)+1)$ we get, if $V = \sum_x \phi(0,x)$,

$$H(\Lambda) - \mu N(\Lambda) = \tfrac{1}{2} \sum_{x,y\in\Lambda} \phi(x,y)n(x)n(y) - \mu \sum_{x\in\Lambda} n(x) = \frac{1}{8} \sum_{x,y\in\Lambda} \phi(x,y)s(x)s(y)$$

$$+ \tfrac{1}{4} \sum_{x\in\Lambda} (\sum_{y\in\Lambda} \phi(x,y))s(x) + \frac{1}{8} \sum_{x,y\in\Lambda} \phi(x,y) - \tfrac{1}{2}\mu \sum_{x\in\Lambda} s(x) - \tfrac{1}{2}\mu\, |\Lambda|$$

$$= \frac{1}{8} \sum_{x,y\in\Lambda} \phi(x,y)s(x)s(y) + (\tfrac{1}{4} V - \tfrac{1}{2}\mu) \sum_{x\in\Lambda} s(x)$$

$$+ (\frac{1}{8} V - \tfrac{1}{2}\mu)|\Lambda| + \text{a surface term.}$$

It is not difficult to show that in the thermodynamical limit the surface term can be neglected. Apart from the constant term, we see that we obtained the Hamiltonian of the Ising model if we make the identifications

$$J(x,y) = \tfrac{1}{4}\phi(x,y) \quad \text{and} \quad h = \tfrac{1}{2}\mu - \tfrac{1}{4}V.$$

This identification of the classical lattice gas with two-particle interaction with the Ising model with two-spin interaction is a simple example of the fact that without condition (ii) in definition 2.1 one may have different potentials giving equivalent theories.

3. CLUSTER PROPERTIES

3.1. Preliminaries. In this chapter we study various cluster properties of states of the quantum lattice algebra $\mathfrak{A}$ defined in chapter 2. Our starting point is the well-known fact that one associates with a state ρ the triple $(\mathcal{H}_\rho, \pi_\rho, \Omega_\rho)$ where π_ρ is a representation of $\mathfrak{A}$ as bounded operators acting on the Hilbert space $\mathcal{H}_\rho$, with cyclic vector Ω_ρ, such that

$$\rho(A) = (\Omega_\rho, \pi_\rho(A)\Omega_\rho).$$

The representation π_ρ is determined uniquely by ρ, up to unitary equivalence.

Let G be a group and $g \to \tau_g$ a homomorphism of G into the automorphisms of $\mathfrak{A}$. By transposition we obtain the action of G on the set of states E

$$(\rho \circ \tau_g)(A) = \rho(\tau_g(A)).$$

If G is a topological group and $\tau_g(A)$ depends continuously on g for all $A \in \mathfrak{A}$, we say that G acts as a *strongly continuous* group of automorphisms.

Let E^G denote the G-invariant states. For $\rho \in E^G$ there is a unique unitary representation $g \to U_\rho(g)$ on $\mathcal{H}_\rho$ such that

$$U_\rho(g)\,\pi_\rho(A)\,U_\rho(g)^{-1} = \pi_\rho(\tau_g(A)) \quad \text{and} \quad U_\rho(g)\Omega_\rho = \Omega_\rho .$$

If the group is strongly continuous, then $U_\rho(g)$ depends continuously on g in the strong operator topology.

For our applications we shall be mainly interested in the following groups.

1. Translations; G is the non-compact discrete abelian group $\mathbb{Z}^\nu$.
2. Time-evolution: $G = \mathbb{R}$.

DEFINITION 3.1: If G acts as a non-compact group of automorphisms of $\mathfrak{A}$, $\mathfrak{A}$ is *asymptotically abelian* with respect to G if

$$\lim_{g\to\infty} [\tau_g(A)B] = 0$$

for all A and $B \in \mathfrak{A}$.

PROPOSITION 3.1: *The quantum lattice algebra* $\mathfrak{A}$ *is asymptotically abelian w.r.t. the translation group.*

Proof: For given $\varepsilon > 0$, $A, B \in \mathfrak{A}$ there exist Λ_1 and Λ_2, $A_1 \in \mathfrak{A}(\Lambda_1)$ and $B_1 \in \mathfrak{A}(\Lambda_2)$ such that $\|A-A_1\| \leq \varepsilon(4\|B\|)^{-1}$ and $\|B-B_1\| \leq \varepsilon(4\|A_1\|)^{-1}$. Hence

$$\|[\tau_x(A),B]\| \leq \|[\tau_x(A_1),B_1]\| + \varepsilon .$$

For x sufficiently large, the commutator on the right-hand side vanishes. □

REMARK: For the time-evolution α_t^Φ derived from a dynamical potential Φ it is not known under what conditions $\mathfrak{A}$ is asymptotically abelian w.r.t. α_t^Φ.

3.2. Asymptotic abelianess. In this section we shall study some consequences of asymptotic abelianess. Although several results we shall derive have more general validity, we shall from now on restrict our attention mainly to the translation group $\mathbb{Z}^\nu$. The set of invariant states will then be denoted by I.

An important tool will be calculating the mean value of an operator-valued or vector-valued function on $\mathbb{Z}^\nu$.

DEFINITION 3.2. An M-net is a net $\{f_\alpha\}$ of functions on $\mathbb{Z}^\nu$ with the properties

i. $f_\alpha(x) \geq 0$ for all $x \in \mathbb{Z}^\nu$ and all α,

ii. $\sum_{x\in\mathbb{Z}^\nu} f_\alpha(x) = 1$ for all α,

iii. $\lim_\alpha \sum_{x\in\mathbb{Z}^\nu} |f_\alpha(x) - f_\alpha(x+y)| = 0$ for all $y \in \mathbb{Z}^\nu$.

An example of an M-net is the following:

$$f_n(x) = n^{-\nu} \quad \text{for } x \text{ in the cube } \{x: 0 \le x_i \le n-1\} ;$$
$$= 0 \quad \text{for all other } x .$$

With a given M-net $\{f_\alpha\}$ one calculates the corresponding mean $\overline{F}^f$ of a function F on $\mathbb{Z}^\nu$ as follows

$$\overline{F}^f = \lim_\alpha \sum_{x\in\mathbb{Z}^\nu} f_\alpha(z)F(z).$$

This mean depends in general on $\{f_\alpha\}$. We say F has a mean $\overline{F}$ if $\overline{F}^f$ is the same for all $\{f_\alpha\}$.

THEOREM 3.2. (Mean ergodic theorem). *Let* $\{f_\alpha\}$ *be an M-net,* $x \to U(x)$ *a unitary representation of the group* $\mathbb{Z}^\nu$ *on* $\mathcal{H}$, *and* P *the projection on the subspace of* $\mathcal{H}$ *spanned by the invariant vectors. Then*

$$\lim_{\alpha\to\infty} \sum_x f_\alpha(x)U(x) = P,$$

in the strong operator topology.

PROOF: We must prove that for all $\psi \in \mathcal{H}$

$$\lim_\alpha \|[\sum_x f_\alpha(x)U(x) - P]\psi\| = 0.$$

It is sufficient to prove this for all ψ for which $P\psi = 0$, i.e. for all ψ in the kernel of P. This kernel is the orthogonal complement of $P\mathcal{H} = \bigcap_x\{\Phi \in \mathcal{H}: (U(x)-1)\Phi = 0\} = \bigcap_x\{(U(x)^*-1)\Phi, \Phi \in \mathcal{H}\}^\perp = [\bigcup_x\{(U(x)^*-1)\Phi, \Phi \in \mathcal{H}\}]^\perp$. Hence the kernel of P is $\bigcup_x\{(U(x)^*-1)\Phi\}$. We can find a vector $\psi' = \sum_{i=1}^n \lambda_i(U(x_i)^*-1)\Phi_i$ such that $\|\psi-\psi'\| < \frac{1}{2}\varepsilon$. We first prove that

$$\lim_\alpha \| [\sum_x f_\alpha(x)U(x) - P]\psi'\| = 0$$

by proving it for each term in ψ' separately.

$$(\sum_x f_\alpha(x)U(x)-P)(U(y)-1)\Phi = \sum_x(f_\alpha(x-y)-f_\alpha(x))U(x)\Phi - P(U(y)-1)\Phi =$$
$$= \sum_x(f_\alpha(x-y)-f_\alpha(x))U(x)\Phi.$$

Hence

$$\lim_\alpha \|(\sum_x f_\alpha(x)U(x)-P)(U(y)-1)\Phi\| \le \lim_\alpha \sum_x |f_\alpha(x-y)-f_\alpha(x)| \|\Phi\| = 0.$$

We can then conclude that

$$\lim_\alpha \|(\sum_x f_\alpha(x)U(x)-P)\psi\| \le \varepsilon.$$ □

For the proof of theorem 3.2 no use was made of the fact that $\mathfrak{A}$ is asymptotically abelian. We need this property however for the following theorems. Let $\rho \in E^G$ and $(\mathcal{H}_\rho, \pi_\rho, \Omega_\rho, U_\rho(g))$ the corresponding Hilbert space, representation, cyclic vector and unitary representation of G. The projection on the invariant vectors for $U_\rho(g)$ will be denoted by P_ρ.

DEFINITION 3.3: $\mathfrak{A}$ is G-abelian if for all $\rho \in E^G$ the von Neumann algebra generated by $P_\rho \pi_\rho(\mathfrak{A})P_\rho$ is abelian.

THEOREM 3.3: *$\mathfrak{A}$ is G-abelian if $\mathfrak{A}$ is asymptotically abelian w.r.t. the non-compact group G.*

PROOF: We shall give the proof for the special case of the quantum lattice algebra and the translation group. We must prove that $[P_\rho \pi_\rho(A)P_\rho, P_\rho\pi_\rho(B)P_\rho]$ vanishes for all $A,B \in \mathfrak{A}$. It is clearly sufficient to prove that $(\psi,(\pi_\rho(A)P_\rho \pi_\rho(B) - \pi_\rho(B)P_\rho\pi_\rho(A))\psi) = 0$ for all $\psi \in P_\rho\mathcal{H}$. Let $\{f_\alpha\}$ be an M-net. Theorem (3.2) implies that

$$(\psi,(\pi_\rho(A)P_\rho \pi_\rho(B) - \pi_\rho(B)P_\rho \pi_\rho(A))\psi) =$$

$$= \lim_\alpha \sum_x f_\alpha(x)(\psi,(\pi_\rho(A)U(x)\pi_\rho(B) - \pi_\rho(B)U(x)^*\pi_\rho(A))\psi) =$$

$$= \lim_\alpha \sum_x f_\alpha(x)(\psi,\pi_\rho([A,\tau_x(B)])\psi).$$

Since $\mathfrak{A}$ is asymptotically abelian the last factor vanishes for $x\to\infty$. This implies that the mean vanishes. □

LEMMA 3.4: *Let $\mathcal{R}\subset B(\mathcal{H})$ be a von Neumann algebra and P a projection operator in $\mathcal{R}$. Denote by $\mathcal{R}_P$ the restriction of $P\mathcal{R}P$ to $P\mathcal{H}$ and by $(\mathcal{R}')_P$ the restriction of $\mathcal{R}'$ to $P\mathcal{H}$. Then $\mathcal{R}_P$ and $(\mathcal{R}')_P$ are von Neumann algebras and*

$$(\mathcal{R}_P)' = (\mathcal{R}')_P .$$

PROOF: $\mathcal{R}'$ commutes with $P\mathcal{R}P$ and the same holds for the restrictions to $P\mathcal{H}$. So $\mathcal{R}_P$ and $(\mathcal{R}')_P$ commute. Let $T \in B(P\mathcal{H})$ and assume that T commutes with $(\mathcal{R}')_P$. Let $S \in B(\mathcal{H})$ be defined by TP. It follows that $S \in \mathcal{R}$. Since $T = S_P$ we conclude that $T \in \mathcal{R}_P$. We have thus proved that $\mathcal{R}_P = ((\mathcal{R}')_P)'$. Since it is evident that $(\mathcal{R}')_P$ is a von Neumann algebra, we have $(\mathcal{R}_P)' = (\mathcal{R}')_P$. □

THEOREM 3.5: *Let $\mathfrak{A}$ be the quantum lattice algebra, $\rho \in I$ and P_ρ the projection on the invariant vectors, then*

a) $[\pi_\rho(\mathfrak{A}) \cup \{P_\rho\}]' = [\pi_\rho(\mathfrak{A}) \cup \{U(\mathbb{Z}^\nu)\}]'$,

b) $P_\rho[\pi_\rho(\mathfrak{A}) \cup \{P_\rho\}]' = P_\rho[P_\rho \pi_\rho(\mathfrak{A})P_\rho]' = P_\rho[P_\rho \pi_\rho(\mathfrak{A})P_\rho]''$,

c) The mapping $B \in [\pi_\rho(\mathfrak{A}) \cup \{U(\mathbb{Z}^\nu)\}]' \to P_\rho B$ *is a *-isomorphism of* $[\pi_\rho(\mathfrak{A}) \cup \{U(\mathbb{Z}^\nu)\}]'$ *onto* $P_\rho[\pi_\rho(\mathfrak{A}) \cup \{U(\mathbb{Z}^\nu)\}]'$,

d) The set $P_\rho\pi_\rho(\mathfrak{A})P_\rho$ *is strongly dense in* $[P_\rho\pi_\rho(\mathfrak{A})P_\rho]''$.

PROOF: a) $P_\rho \in [U(\mathbb{Z}^\nu)]''$; hence $[\pi_\rho(\mathfrak{A}) \cup \{U(\mathbb{Z}^\nu)\}]' \subset [\pi_\rho(\mathfrak{A}) \cup \{P_\rho\}]'$. On the other hand, let $C \in [\pi_\rho(\mathfrak{A}) \cup \{P_\rho\}]'$; then $CU(x)\pi_\rho(A)\Omega_\rho = C\pi_\rho(\tau_x(A))\Omega_\rho = \pi_\rho(\tau_x(A))C\Omega_\rho = U(x)\pi_\rho(A)U(x)^{-1}C\Omega_\rho = U(x)\pi_\rho(A)C\Omega_\rho = U(x)C\pi_\rho(A)\Omega_\rho$.

Hence C commutes with U(x).

b) If $P \in \mathcal{R}$, then $(\mathcal{R}_P)' = ((P\mathcal{R}P)')_P$ so that $(\mathcal{R}')_P = ((P\mathcal{R}P)')_P$. Now both $\mathcal{R}'$ and $(P\mathcal{R}P)'$ commute with P, and consequently lemma 3.4 can be rewritten as $P\mathcal{R}' = P(P\mathcal{R}P)'$. We now take $\mathcal{R} = [\pi_\rho(\mathcal{O}\!\mathcal{L}) \cup \{P_\rho\}]''$ and observe that $P_\rho[\pi_\rho(\mathcal{O}\!\mathcal{L}) \cup \{P_\rho\}]''P_\rho = [P_\rho \pi_\rho(\mathcal{O}\!\mathcal{L}) P_\rho]''$. We find $P_\rho[\pi_\rho(\mathcal{O}\!\mathcal{L}) \cup \{P_\rho\}]' = P_\rho[P_\rho\pi_\rho(\mathcal{O}\!\mathcal{L}) P_\rho]'$. For the second equation we notice that the abelian algebra $[P_\rho \pi_\rho(\mathcal{O}\!\mathcal{L}) P_\rho]''$ restricted to $P_\rho\mathcal{H}_\rho$ is abelian and has a cyclic vector Ω_ρ. Hence, this algebra is maximal abelian and thus equal to its commutant.

c) If $A \in [\pi_\rho(\mathcal{O}\!\mathcal{L}) \cup \{P_\rho\}]'$ then A commutes with P_ρ and thus the mapping $A \to P_\rho A$ is a *-homomorphism. On the other hand, suppose $P_\rho A = 0$ then $P_\rho A \Omega_\rho = 0$ or $A\Omega_\rho = 0$. Since Ω_ρ is cyclic for $\pi_\rho(\mathcal{O}\!\mathcal{L})$ it is separating for $\pi_\rho(\mathcal{O}\!\mathcal{L})'$ and $A = 0$.

d) Operators of the form $P_\rho\pi_\rho(A_1)P_\rho\pi_\rho(A_2)P_\rho \dots \pi_\rho(A_n)P_\rho$ form a total set for $[P_\rho\pi_\rho(\mathcal{O}\!\mathcal{L})P_\rho]''$. We write, using theorem 3.2, $P_\rho\pi_\rho(A_1)P_\rho \dots \pi_\rho(A_n)P_\rho =$

$$\lim_{\{\alpha\}} \sum_{\{x\}} f_{\alpha_1}(x_1) \dots f_{\alpha_{n-1}}(x_{n-1})\, P_\rho\, \pi_\rho(A_1)\, U(x_1)\, \pi_\rho(A_2) \dots U(x_{n-1})\, \pi_\rho(A_n) P_\rho =$$

$$\lim_{\{\alpha\}} \sum_{\{x\}} f_{\alpha_1}(x_1) \dots f_{\alpha_{n-1}}(x_{n-1})\, P_\rho\, \pi_\rho(A_1\, \tau_{x_1}(A_2)\, \tau_{x_1+x_2}(A_2) \dots$$

$\tau_{x_1+\dots+x_{n-1}}(A_n))P_\rho$, which is in the strong closure of $P_\rho\pi_\rho(\mathcal{O}\!\mathcal{L})P_\rho$. □

REMARK: The isomorphism of $[\pi_\rho(\mathcal{O}\!\mathcal{L}) \cup U(\mathbb{Z}^\nu)]'$ and $P_\rho[P_\rho\pi_\rho(\mathcal{O}\!\mathcal{L})P_\rho]''$ means that the invariant elements in the commutant of $\pi_\rho(\mathcal{O}\!\mathcal{L})$ form an abelian von Neumann algebra.

The set I of translation-invariant states is a compact convex set. The states in $\mathcal{E}(I)$, the extremal points of I are sometimes called *ergodic states*. This set is not empty. Indeed, every $\rho \in I$ can be decomposed uniquely in terms of extremal invariant states. Furthermore, as will be seen later, states representing a single thermodynamical phase are extremal invariant.

The following theorem deals with states of the quantum lattice algebra. As in the previous theorem, explicit use is made of $\mathbb{Z}^\nu$-abelianess.

THEOREM 3.6: *For* $\rho \in I$ *the following conditions are equivalent.*

a. $\rho \in \mathcal{E}(I)$,

b. $[\pi_\rho(\mathcal{O}\!\mathcal{L}) \cup U(\mathbb{Z}^\nu)]''$ *is irreducible in* $\mathcal{H}_\rho$,

c. P_ρ *is one-dimensional,*

d. ρ *is weakly clustering, i.e., for all* $A, B \in \mathcal{O}\!\mathcal{L}$ *and all M-nets* $\{f_\alpha\}$

$$\lim_\alpha \sum_x f_\alpha(x)\, \rho(A\, \tau_x(B)) = \rho(A)\rho(B),$$

e. for all $A \in \mathcal{O}\!\mathcal{L}$ *and for all M-nets* $\{f_\alpha\}$

$$\lim_\alpha \sum_x f_\alpha(x)\ \pi_\rho(\tau_x(A)) = \rho(A)\, \mathbb{1},$$

where the limit is the strong operator limit.

PROOF: a $\Leftrightarrow$ b. The existence of a state ρ_1 such that $\rho - \lambda\rho_1 \geq 0$ for some $\lambda \in (0,1)$ is equivalent with the existence of a positive operator in $\pi_\rho(\mathfrak{A})'$ that commutes with $U(\mathbb{Z}^\nu)$.

c $\Rightarrow$ b. Let $B \in [\pi_\rho(\mathfrak{A}) \cup U(\mathbb{Z}^\nu)]'$; then $B\,\pi_\rho(A)\Omega_\rho = \pi_\rho(A)B\,\Omega_\rho = \pi_\rho(A)B\,P_\rho\Omega_\rho = \pi_\rho(A)P_\rho B\,\Omega_\rho = (\Omega_\rho, B\,\Omega_\rho)\ \pi_\rho(A)\Omega_\rho$; hence $B = (\Omega_\rho, B\,\Omega_\rho)\,\mathbb{1}$.

b $\Rightarrow$ c. $P_\rho[\pi_\rho(\mathfrak{A}) \cup U(\mathbb{Z}^\nu)]' = P_\rho[P_\rho\,\pi_\rho(\mathfrak{A})P_\rho]''$ by theorem 3.5. The restriction of this last algebra to $P_\rho\mathcal{H}_\rho$ has Ω_ρ as a cyclic vector. From b) follows that this algebra consists of multiples of P_ρ. Hence $P_\rho\mathcal{H}_\rho = [\Omega_\rho]$.

e $\Rightarrow$ d is trivial.

d $\Rightarrow$ c. $0 = \lim_\alpha \sum_x f_\alpha(x)\ (\rho(A^*\tau_x(B)) - \rho(A^*)\rho(B)) = \lim_\alpha \sum_x f_\alpha(x)\ (\pi_\rho(A)\Omega_\rho, (U(x) - P_{\Omega_\rho})\ \pi_\rho(B)\Omega_\rho) = (\pi_\rho(A)\Omega_\rho, (P_\rho - P_{\Omega_\rho})\ \pi_\rho(B)\Omega_\rho)$.

c $\Rightarrow$ e. Since $\|\sum_x f_\alpha(x)\ \pi_\rho(\tau_x(A))\| \leq \|A\|$ it is sufficient to prove that for a dense set of vectors $\Phi \in \mathcal{H}_\rho$

$$\lim_\alpha \sum_x f_\alpha(x)\ \pi_\rho(\tau_x(A))\Phi = \rho(A)\Phi.$$

We choose $\Phi = \pi_\rho(B)\Omega_\rho$. Then $\lim_\alpha \sum_x f_\alpha(x)\ \pi_\rho(\tau_x(A))\pi_\rho(B)\Omega_\rho = \lim_\alpha \sum_x f_\alpha(x)\ \pi_\rho(B)\ \pi_\rho(\tau_x(A))\Omega_\rho = \lim_\alpha \sum_x f_\alpha(x)\ \pi_\rho(B)\ U(x)\ \pi_\rho(A)\Omega_\rho = \pi_\rho(B)\ P_\rho\ \pi_\rho(A)\Omega_\rho = (\Omega_\rho, \pi_\rho(A)\Omega_\rho)\ \pi_\rho(B)\Omega_\rho = \rho(A)\ \pi_\rho(B)\Omega_\rho$, where we used asymptotic abelianess for the first, theorem 3.2 for the third and c) for the fourth equality. □

REMARK: According to this theorem, extremal invariant states are characterized by the property that there are no non-trivial translation-invariant elements in the commutant or by the uniqueness of Ω_ρ as invariant vector. The fact that for extremal invariant states the space average of an observable (in the strong operator topology) is a multiple of the identity is an important tool for finding explicit solutions for a certain class of models (molecular field models).

A stronger property than weakly clustering is the following.

DEFINITION 3.4: A state $\rho \in I$ is *strongly clustering* if for all $A, B \in \mathfrak{A}$

$$\lim_{|x|\to\infty} \rho(A\,\tau_x(B)) = \rho(A)\rho(B).$$

If ρ is *primary* (i.e., $\pi_\rho(\mathfrak{A})''$ is a factor) it satisfies the strong clustering property. The next theorem gives an even stronger result.

THEOREM 3.7: *The state ρ is primary if and only if, for each $A \in \mathfrak{A}$ and $\varepsilon > 0$, there is a finite sublattice Λ such that*

$$|\rho(AB) - \rho(A)\rho(B)| \leq \varepsilon\ \|B\|$$

for all $B \in \mathfrak{A}(\Lambda^c)$.

For the proof we need the following lemma.

LEMMA 3.8: *For Λ finite and ρ a state of $\mathfrak{A}$*

$$\pi_\rho(\mathfrak{A}(\Lambda^c))'' = \pi_\rho(\mathfrak{A}(\Lambda))' \cap \pi_\rho(\mathfrak{A})''.$$

This result is a form of duality. All elements of the von Neumann algebra $\pi_\rho(\mathfrak{A})''$ that commute with the local algebra $\pi_\rho(\mathfrak{A}(\Lambda))$ are in $\pi_\rho(\mathfrak{A}(\Lambda^c))''$.

PROOF of lemma 3.8: Suppose $X \in \pi_\rho(\mathfrak{A}(\Lambda))' \cap \pi_\rho(\mathfrak{A})''$. We shall prove that $X \in \pi_\rho(\mathfrak{A}(\Lambda^c))''$. We must, therefore, show that for $\varepsilon > 0$ and a finite set of vectors $\{f_k, g_k \in \mathcal{H}_\rho ; k = 1,\dots,m\}$ there is a $B \in \mathfrak{A}(\Lambda^c)$ such that

$$|(f_k, (X - \pi_\rho(B))g_k)| < \varepsilon, \quad k = 1,2,\dots,m.$$

Since $X \in \pi_\rho(\mathfrak{A})''$ there is a $C \in \mathfrak{A}$ such that

$$|(\pi_\rho(e_{1r})f_k, (X - \pi_\rho(C))\, \pi_\rho(e_{1r})g_k)| < \frac{\varepsilon}{n},$$

for $k = 1,2,\dots,m$ and $r = 1,2,\dots,n = 2^{|\Lambda|}$. The $\{e_{ij}\}$ are the matrix units of $\mathfrak{A}(\Lambda)$. Using the fact that X commutes with these matrix units we get

$$|(f_k, (\pi_\rho(e_{rr})X - \pi_\rho(e_{r1}\, C\, e_{1r}))g_k)| \le \frac{\varepsilon}{n}$$

or, after summation over r,

$$|(f_k, (X - \pi_\rho(B))g_k)| \le \varepsilon,$$

where $B = \sum_{r=1}^{n} e_{r1}\, C\, e_{1r} \in \mathfrak{A}(\Lambda^c)$. □

PROOF of theorem 3.7: Assume the clustering property and let $C \in \mathfrak{Z} = \pi_\rho(\mathfrak{A})' \cap \pi_\rho(\mathfrak{A})''$. For each Λ, we have $C \in \pi_\rho(\mathfrak{A}(\Lambda))' \cap \pi_\rho(\mathfrak{A})''$ or, by lemma 3.8, $C \in \pi_\rho(\mathfrak{A}(\Lambda^c))''$. From our assumption we have, for all $B \in \mathfrak{A}(\Lambda^c)$, $(\Omega_\rho, \pi_\rho(A)\, \pi_\rho(B)\Omega_\rho) - (\Omega_\rho, \pi_\rho(A)\Omega_\rho)\, (\Omega_\rho, \pi_\rho(B)\Omega_\rho) \le \varepsilon \|B\| = \varepsilon \|\pi_\rho(B)\|$. Since $C \in \pi_\rho(\mathfrak{A}(\Lambda^c))''$ we get $(\Omega_\rho, \pi_\rho(A)C\,\Omega_\rho) - (\Omega_\rho, \pi_\rho(A)\Omega_\rho)\, (\Omega_\rho, C\,\Omega_\rho) \le \varepsilon\|C\|$. Since the left-hand side is independent of ε we conclude that $(\Omega_\rho, \pi_\rho(A)\ (C - (\Omega_\rho, C\,\Omega_\rho)\,\mathbb{1})\Omega_\rho) = 0$, and since this holds for all $A \in \mathfrak{A}$ we have $C = (\Omega_\rho, C\,\Omega_\rho)\,\mathbb{1}$. The proof of the reverse statement follows from the following theorem. □

THEOREM 3.9: *If ρ is primary, then for each $A \in \mathfrak{A}$ and $\varepsilon > 0$ there is a finite sublattice Λ such that*

$$|\rho(AB) - \rho(A)\rho(B)| \le \varepsilon\, [\rho(B^*B) + \rho(BB^*)]^{\frac{1}{2}}$$

for all $B \in \mathfrak{A}(\Lambda^c)$.

For the proof we make use of

LEMMA 3.10: *Let $\mathcal{R}$ be a factor generated by an increasing sequence $\{\mathfrak{A}_n\}$ of subalgebras and let, for each n, w_n be a s.a. element of $\mathfrak{A}_n' \cap \mathcal{R}$. If there is a subsequence $\{w_{n'}\}$ such that, for some Φ and $\psi \in \mathcal{H}$,*

$$\Phi = \text{weak} \lim_{n' \to \infty} w_{n'} \psi$$

then there is a complex number λ *such that*

$$\Phi = \lambda\psi.$$

PROOF: Suppose that $A \in [\mathfrak{A}_n \cup \mathcal{R}']''$, then $(A\psi, \Phi) = \lim_{n'\to\infty} (A\psi, w_{n'}\psi) = \lim_{n'\to\infty} (Aw_{n'}\psi, \psi) = (A\Phi, \psi)$. Since $\mathcal{R}$ is a factor $B(\mathcal{H}) = [\mathcal{R} \cup \mathcal{R}']''$, and hence $\bigcup_n [\mathfrak{A}_n \cup \mathcal{R}']''$ is dense in $B(\mathcal{H})$. Consequently, the equality $(A\psi, \Phi) = (A\Phi, \psi)$ holds for all A in a dense subset of $B(\mathcal{H})$ and hence for all $A \in B(\mathcal{H})$. Choose A such that $A\chi = (\Phi, \chi)\psi$. Then $(\Phi, \psi)(\psi, \Phi) = (\Phi, \Phi)(\psi, \psi)$, which implies that $\Phi = \lambda\psi$. □

PROOF of theorem 3.9: Suppose that $\{\Lambda_n\}$ is an increasing sequence of finite sublattices such that, for each finite Λ, there is an n such that $\Lambda \subset \Lambda_n$. Let $A \in \mathfrak{A}$. Suppose the theorem is false, then there is an $\varepsilon > 0$ such that for every n there is a $B_n \in \mathfrak{A}(\Lambda_n^C)$ such that $|\rho(A B_n) - \rho(A)\rho(B_n)| > \varepsilon[\rho(B_n^* B_n) + \rho(B_n B_n^*)]^{\frac{1}{2}}$. It is clear that $B_n \neq 0$. If

$$C_n = B_n \, \varepsilon^{-1} [\rho(B_n^* B_n) + \rho(B_n B_n^*)]^{-\frac{1}{2}}$$

we have

$$|\rho(A C_n) - \rho(A)\rho(C_n)| > 1,$$

and

$$\rho(C_n^* C_n) + \rho(C_n C_n^*) = \varepsilon^{-2}.$$

The last equation implies that $\{C_n \Omega_\rho\}$ and $\{C_n^* \Omega_\rho\}$ are bounded sequences. Using weak compactness we conclude that there is a subsequence $\{n'\}$ such that

$\Phi_1 = w \lim_{n'\to\infty} C_{n'} \Omega_\rho$ and $\Phi_2 = w \lim_{n'\to\infty} C_{n'}^* \Omega_\rho$

and hence

$$\Phi_1 + \Phi_2 = w \lim_{n'\to\infty} (C_{n'} + C_{n'}^*)\Omega_\rho,$$

$$i(\Phi_1 - \Phi_2) = w \lim_{n'\to\infty} i(C_{n'} + C_{n'}^*)\Omega_\rho,$$

We conclude from lemma 3.10 that $\Phi_1 = \lambda \Omega_\rho$ and $\Phi_2 = \mu \Omega_\rho$. Using this we get

$$\lim_{n'\to\infty} (\rho(A C_{n'}) - \rho(A)\rho(C_{n'})) = 0,$$

which contradicts our earlier result that

$$|\rho(A C_n) - \rho(A)\rho(C_n)| > 1.$$

□

We now come to another clustering property that is stronger than the "only if" part of theorem 3.7 but does not contain theorem 3.9. To define this property we need the notion of product state with respect to two disjoint regions.

PROPOSITION 3.11: *Let* S_1 *and* S_2 *be two finite or infinite disjoint regions in* $\mathbb{Z}^\nu$. *Let* ρ_1 *be a state of* $\mathfrak{A}(S_1)$ *and* ρ_2 *a state of* $\mathfrak{A}(S_2)$. *Then there is one and only one state* ρ *of* $\mathfrak{A}(S_1 \cup S_2)$ *such that for* $A_1 \in \mathfrak{A}(S_1)$ *and* $A_2 \in \mathfrak{A}(S_2)$ $\rho(A_1A_2) = \rho_1(A_1)\rho_2(A_2)$.

PROOF: $\mathfrak{A}(S_1 \cup S_2)$ is the norm closure of

$$\bigcup_{\Lambda_1 \subset S_1} \bigcup_{\Lambda_2 \subset S_2} \mathfrak{A}(\Lambda_1 \cup \Lambda_2),$$

where Λ_1 and Λ_2 are finite. Let $A \in \mathfrak{A}(\Lambda_1 \cup \Lambda_2)$, and let ρ_{Λ_1} and ρ_{Λ_2} be the density matrices in $\mathfrak{A}(\Lambda_1)$ and $\mathfrak{A}(\Lambda_2)$ corresponding to the restrictions of ρ_1 to $\mathfrak{A}(\Lambda_2)$ and ρ_2 to $\mathfrak{A}(\Lambda_2)$ respectively. Then $\rho_{\Lambda_1} \otimes \rho_{\Lambda_2}$ is a density matrix in $\mathfrak{A}(\Lambda_1 \cup \Lambda_2)$. We define $\rho(A) = \mathrm{Tr}(\rho_{\Lambda_1} \otimes \rho_{\Lambda_2})A$. If $A_1 \in \mathfrak{A}(\Lambda_1)$ and $A_2 \in \mathfrak{A}(\Lambda_2)$ we get $\rho(A_1A_2) = \rho_1(A_1)\rho_2(A_2)$. It is immediately seen that ρ is defined on $\bigcup \mathfrak{A}(\Lambda_1 \cup \Lambda_2)$ and can be extended to the norm closure, and that for $A = A_1A_2$, with $A_1 \in \mathfrak{A}(S_1)$ and $A_2 \in \mathfrak{A}(S_2)$ $\rho(A_1A_2) = \rho_1(A_1)\rho_2(A_2)$. The uniqueness of ρ follows from the fact that the set $\{A_1A_2; A_1 \in \mathfrak{A}(S_1), A_2 \in \mathfrak{A}(S_2)\}$ is a total set for $\mathfrak{A}(S_1 \cup S_2)$. □

DEFINITION 3.5: The state ρ obtained in proposition 3.11 is called the *product state* of ρ_1 and ρ_2. If ρ is a state of $\mathfrak{A}$ and S_1, S_2 two disjoint finite or infinite regions in $\mathbb{Z}^\nu$, the state $\rho^\times$ of $\mathfrak{A}(S_1 \cup S_2)$ obtained as product state of $\rho|\mathfrak{A}(S_1)$ and $\rho|\mathfrak{A}(S_2)$ is called a *factorization* of ρ w.r.t. the regions S_1 and S_2.

THEOREM 3.12: *If* $\Lambda \subset \mathbb{Z}^\nu$ *is finite and* $\{\Lambda_n\}$ *an increasing sequence of finite subsets of* $\mathbb{Z}^\nu$ *such that any finite subset of* $\mathbb{Z}^\nu$ *is contained in* Λ_n *for some* n, *and if* $\mathfrak{A}_n = \mathfrak{A}(\Lambda \cup \Lambda_n^c)$; *then*

$$\lim_{n\to\infty} \|(\rho - \rho^\times)|\mathfrak{A}_n\| = 0,$$

if ρ *is primary. Here* $\rho^\times$ *is the factorization of* ρ *with respect to* Λ *and* Λ^c.

We shall first prove a technical lemma.

LEMMA 3.13: *If* $\{\mathcal{R}_n\}$ *is a decreasing sequence of von Neumann algebras on* $\mathcal{H}$, *and* ω *a normal linear functional on* $\mathcal{R} = \bigcup_n \mathcal{R}_n$ *such that* $\omega|\bigcap_n \mathcal{R}_n = 0$; *then*

$$\lim_{n\to\infty} \|\omega|\mathcal{R}_n\| = 0.$$

PROOF: If ω is a normal linear functional on a von Neumann algebra $\mathcal{R}$, there is always an element $A \in \mathcal{R}_1$ (the unit sphere) such that $\omega(A) = \|\omega\|$. Hence we have for each n an operator $A_n \in (\mathcal{R}_n)_1$ such that $\omega(A_n) = \|\omega|\mathcal{R}_n\|$. Let A be a weak limit point of $\{A_n\}$. Clearly $A \in \bigcap_n \mathcal{R}_n$, and thus $\omega(A) = 0$. But $\omega(A) = \lim_{n'} \omega(A_{n'}) =$

$\lim_{n'} \|\omega|\mathcal{R}_{n'}\|$. Since the sequence $\{\|\omega|\mathcal{R}_n\|\}$ is monotonically decreasing we conclude that $\lim_n \|\omega|\mathcal{R}_n\| = 0$. □

PROOF of theorem 3.12: Let $\mathcal{R}_n = \pi_\rho(\mathfrak{A}(\Lambda \cup \Lambda_n^c))''$; then $\mathcal{R}_{n+1} \subset \mathcal{R}_n$ and $\mathcal{R} = \bigcup_n \mathcal{R}_n \subset \pi_\rho(\mathfrak{A})''$. The extensions $\tilde{\rho}$ and $\tilde{\rho}^\times$ of ρ and $\rho^\times$ to $\pi_\rho(\mathfrak{A})''$ are normal, and so are their restrictions to $\mathcal{R}$. One has

$$\bigcap_n \mathcal{R}_n = \bigcap_n [\pi_\rho(\mathfrak{A}(\Lambda)) \cup \pi_\rho(\mathfrak{A}(\Lambda_n^c))]'' = [\pi_\rho(\mathfrak{A}(\Lambda)) \cup \bigcap_n \pi_\rho(\mathfrak{A}(\Lambda_n^c))]''.$$

From lemma 3.8 we conclude that $\bigcap_n \pi_\rho(\mathfrak{A}(\Lambda_n^c))'' = \mathfrak{Z}$. Since ρ is primary the center $\mathfrak{Z}$ is trivial and $\bigcap_n \mathcal{R}_n = \pi_\rho(\mathfrak{A}(\Lambda))$. The state $\tilde{\rho}$ and $\tilde{\rho}^\times$ coincide on $\pi_\rho(\mathfrak{A}(\Lambda))$. According to lemma 3.13 we have

$$\lim_{n\to\infty} \|(\tilde{\rho} - \tilde{\rho}^\times)\, \mathcal{R}_n\| = 0$$

and hence

$$\lim_{n\to\infty} \|(\rho - \rho^\times)|\, \mathfrak{A}(\Lambda \cup \Lambda_n^c)\| = 0.$$ □

We now collect the results of the theorems 3.7, 3.9 and 3.12 in the following theorem.

THEOREM 3.14: *For a state* ρ *on the quantum lattice algebra* $\mathfrak{A}$ *the following conditions are equivalent.*

a) ρ *is primary.*

b) For $A \in \mathfrak{A}$ *and* $\varepsilon > 0$ *there is a finite sublattice* Λ *such that*

$$|\rho(AB) - \rho(A)\rho(B)| \le \varepsilon \|B\|,$$

for all $B \in \mathfrak{A}(\Lambda^c)$.

c) For $A \in \mathfrak{A}$ *and* $\varepsilon > 0$ *there is a finite sublattice* Λ *such that*

$$|\rho(AB) - \rho(A)\rho(B)| \le \varepsilon[\rho(B^*B) + \rho(BB^*)]^{\frac{1}{2}},$$

for all $B \in \mathfrak{A}(\Lambda^c)$.

d) For finite Λ, *and* $\{\Lambda_n\}$ *an increasing sequence of finite sets*

$$\lim_{n\to\infty} \|(\rho - \rho^\times)|\, \mathfrak{A}(\Lambda \cup \Lambda_n^c)\| = 0,$$

where $\rho^\times$ *is the factorization of* ρ *with respect to the regions* Λ *and* Λ^c.

4. THERMODYNAMICAL FUNCTIONS

4.1. The mean free energy. The thermodynamical functions free energy, (internal) energy and entropy, as defined in 1.3, do not exist for infinite systems. They are so-called *extensive* quantities, i.e. for large Λ they are asymptotically proportional to $|\Lambda|$. For infinite systems one considers instead the mean value of these quantities per lattice point. We start with the mean free energy, which for a finite lattice is defined by

$$\frac{F(\Phi,\Lambda)}{|\Lambda|} = \frac{1}{|\Lambda|} \log \operatorname{Tr} e^{-H(\Phi,\Lambda)} ,$$

and we shall prove that the limit $\Lambda \to \infty$ exists and derive properties of the mean free energy f as function of Φ.

The proof will be divided into a number of lemmas.

LEMMA 4.1: $\|H(\Phi,\Lambda)\| \leq |\Lambda| \, \|\Phi\|_1$.

PROOF: $\|H(\Phi,\Lambda)\| \leq \sum_{X\subset\Lambda} \|\Phi(X)\| \leq \sum_{x\in\Lambda} \sum_{X\ni x} \frac{\|\Phi(X)\|}{|X|} = |\Lambda| \, \|\Phi\|_1$. □

PROPOSITION 4.2: For $\Phi,\Psi \in \mathcal{B}_1$,

$$|F(\Phi,\Lambda) - F(\Psi,\Lambda)| \leq \|\Phi - \Psi\|_1 \, |\Lambda| .$$

PROOF: Immediate consequence of the inequality (A.1) of the appendix and lemma 4.1. □

COROLLARY 4.3: *For* $\Phi \in \mathcal{B}_1$ *one has*

$$\left| \, |\Lambda|^{-1} F(\Phi,\Lambda) + \log 2 \, \right| \leq \|\Phi\|_1 .$$

PROOF: Take $\Psi = 0$ in proposition 4.2 and notice that $F(0,\Lambda) = -|\Lambda| \log 2$. □

As a consequence of corollary 4.3 we note that $|\Lambda|^{-1} F(\Phi,\Lambda)$ is bounded uniformly in Λ.

In the remainder of this section we shall use the following notation. If Λ_1 and Λ_2 are finite and disjoint and $\Phi \in \mathcal{B}_0$, then

$$N(\Lambda_1,\Lambda_2,\Phi) = |\cup\{X \subset \mathbb{Z}^\nu : X \cap \Lambda_1 \neq \emptyset, \; X \cap \Lambda_2 \neq \emptyset \;\text{ and }\; \Phi(X) \neq 0\}| .$$

LEMMA 4.4: *For* $\Phi \in \mathcal{B}_0$ *and* $\Lambda_1 \cap \Lambda_2 = \emptyset$

$$|F(\Phi,\Lambda_1 \cup \Lambda_2) - F(\Phi,\Lambda_1) - F(\Phi,\Lambda_2)| \leq N(\Lambda_1,\Lambda_2,\Phi) \, \|\Phi\|_1 .$$

PROOF:

$$H(\Phi,\Lambda_1 \cup \Lambda_2) - H(\Phi,\Lambda_1) - H(\Phi,\Lambda_2) = {\sum_X}' \Phi(X),$$

where the summation extends over all X such that $X \cap \Lambda_1 \neq \emptyset$ and $X \cap \Lambda_2 \neq \emptyset$ and $\Phi(X) \neq 0$. Hence

$$\|H(\Phi,\Lambda_1 \cup \Lambda_2) - H(\Phi,\Lambda_1) - H(\Phi,\Lambda_2)\| \leq {\sum_x}' \sum_{X\ni x} \frac{\|\Phi(X)\|}{|X|} ,$$

where is summed over all points in $\cup\{X \subset \mathbb{Z}^\nu : X \cap \Lambda_1 \neq \emptyset, \; X \cap \Lambda_2 \neq \emptyset$ and $\Phi(X) \neq 0\}$. Using the above notation we can write

$$\|H(\Phi,\Lambda_1 \cup \Lambda_2) - H(\Phi,\Lambda_1) - H(\Phi,\Lambda_2)\| \leq N(\Lambda_1,\Lambda_2,\Phi) \, \|\Phi\|_1 .$$

On the other hand

$$F(\Phi,\Lambda_1\cup\Lambda_2)-F(\Phi,\Lambda_1)-F(\Phi,\Lambda_2)=$$

$$-\log \mathrm{Tr}^{(\Lambda_1\cup\Lambda_2)} e^{-H(\Phi,\Lambda_1\cup\Lambda_2)}+\log \mathrm{Tr}^{(\Lambda_1)} e^{-H(\Phi,\Lambda_1)}+\log \mathrm{Tr}^{(\Lambda_2)} e^{-H(\Phi,\Lambda_2)}=$$

$$-\log \mathrm{Tr}^{(\Lambda_1\cup\Lambda_2)} e^{-H(\Phi,\Lambda_1\cup\Lambda_2)}+\log \mathrm{Tr}^{(\Lambda_1\cup\Lambda_2)} e^{-[H(\Phi,\Lambda_1)+H(\Phi,\Lambda_2)]}.$$

The proof now follows immediately from (A.1). □

LEMMA 4.5: *If* $\Phi\in\mathcal{B}_0$, $a\in\mathbb{N}$ *and* $K_i(a)$, $i=1,2,\dots,n$, *are* n *non-overlapping cubes, each containing* a^ν *points, then*

$$\lim_{a\to\infty}\left[(n\,a^\nu)^{-1}F(\Phi,\bigcup_i K_i(a))-a^{-\nu}F(\Phi,K(a))\right]=0,$$

uniformly in n.

PROOF: Using lemma 4.4. we can write

$$\left|F(\Phi,\bigcup_{i=1}^{n}K_i(a))-F(\Phi,\bigcup_{i=1}^{n-1}K_i(a))-F(\Phi,K(a))\right|\leq N(K(a),K(a)^C,\Phi)\,\|\Phi\|_1.$$

Repeating this inequality for $n-1$, $n-2$, etc., we get

$$\left|F(\Phi,\bigcup_{i=1}^{n}K_i(a))-n\,F(\Phi,K(a))\right|\leq n\,N(K(a),K(a)^C,\Phi)\,\|\Phi\|_1,$$

or

$$\left|(n\,a^\nu)^{-1}F(\Phi,\bigcup_{i=1}^{n}K_i(a))-a^{-\nu}F(\Phi,K(a))\right|\leq a^{-\nu}N(K(a),K(a)^C,\Phi)\,\|\Phi\|_1.$$

For $a\to\infty$ the right-hand side approaches zero, which concludes the proof. □

COROLLARY 4.6: *For* $\Phi\in\mathcal{B}_0$ *the limit*

$$\lim_{a\to\infty}a^{-\nu}F(\Phi,K(a))=f(\Phi)$$

exists.

PROOF: For $\varepsilon>0$ there exists $a_0\in\mathbb{N}$, independent of n, such that

$$\left|(n\,a)^{-\nu}F(\Phi,K(n\,a))-a^{-\nu}F(\Phi,K(a))\right|<\tfrac{1}{2}\varepsilon,$$

for all $a\geq a_0$. But then

$$\left|a_1^{-\nu}F(\Phi,K(a_1))-a_2^{-\nu}F(\Phi,K(a_2)\right|\leq\left|(a_1a_2)^{-\nu}F(\Phi,K(a_1a_2))-a_1^{-\nu}F(\Phi,K(a_1)\right|+$$

$$\left|(a_1a_2)^{-\nu}F(\Phi,K(a_1a_2))-a_2^{-\nu}F(\Phi,K(a_2))\right|\leq\varepsilon$$

for $a_1\geq a_0$ and $a_2\geq a_0$. □

So far we have shown that the limit of $|\Lambda|^{-1}F(\Phi,\Lambda)$ exists for cubes. We shall now give the prove for more general shapes of Λ.

DEFINITION 4.1: Let the infinite lattice $\mathbb{Z}^\nu$ be divided into ν-dimensional cubes of length $a \in \mathbb{N}$. For a given finite $\Lambda \subset \mathbb{Z}^\nu$, let Λ_a^- be the union of all cubes contained in Λ and Λ_a^+ the union of all cubes with non-vanishing intersection with Λ. Clearly $|\Lambda_a^-| \leq |\Lambda| \leq |\Lambda_a^+|$. We say that a sequence $\{\Lambda^n\} \to \infty$ *in the sense of Van Hove* if for all $a \in \mathbb{N}$, $|\Lambda^n| \to \infty$ and $|\Lambda_a^{n+}| / |\Lambda_a^{n-}| \to 1$.

LEMMA 4.7: *For* $\Phi \in \mathcal{B}_0$ *the limit*

$$\lim_{n\to\infty} |\Lambda_n|^{-1} F(\Phi,\Lambda_n) = f(\Phi),$$

if $\{\Lambda_n\} \to \infty$ *in the sense of Van Hove.*

PROOF: By lemma 4.4 we have

$$|F(\Phi,\Lambda) - F(\Phi,\Lambda_a^-) - F(\Phi,\Lambda/\Lambda_a^-)| \leq N(\Lambda_a^-,\Lambda/\Lambda_a^-,\Phi)\, \|\Phi\|_1 .$$

This inequality can be rearranged as follows.

$$\left| |\Lambda|^{-1} F(\Phi,\Lambda) - a^{-\nu} F(\Phi,K(a)) \right| \leq \frac{|\Lambda/\Lambda_a^-|}{|\Lambda|} |\Lambda/\Lambda_a^-|^{-1} \quad |F(\Phi,\Lambda/\Lambda_a^-)| +$$
$$\frac{|\Lambda/\Lambda_a^-|}{|\Lambda|} |\Lambda_a^-|^{-1} \; |F(\Phi,\Lambda_a^-)| + \left| |\Lambda_a^-|^{-1} F(\Phi,\Lambda_a^-) - a^{-\nu} F(\Phi,K(a)) \right| +$$
$$\frac{N(\Lambda_a^-,\Lambda/\Lambda_a^-,\Phi)}{|\Lambda|} \|\Phi\|_1 .$$

We first choose a so large that

$$\left| |\Lambda_a^-|^{-1} F(\Phi,\Lambda_a^-) - a^{-\nu} F(\Phi,K(a)) \right| < \tfrac{1}{4}\varepsilon .$$

By then choosing Λ sufficiently large (in the sense of Van Hove) we can make the other terms each $< \frac{1}{4}\varepsilon$. For the last term we see that as follows.

$$N(\Lambda_a^-,\Lambda/\Lambda_a^-,\Phi) \leq N(\Lambda_a^-,\Lambda_a^+/\Lambda_a^-,\Phi) \leq a^{-\nu}(|\Lambda_a^+| - |\Lambda_a^-|)\, N(K(a),K(a)^C,\Phi).$$

Hence

$$\frac{N(\Lambda_a^-,\Lambda/\Lambda_a^-,\Phi)}{|\Lambda|} \leq \frac{|\Lambda_a^+| - |\Lambda_a^-|}{|\Lambda|} \; \frac{N(K(a),K(a)^C,\Phi)}{a^\nu} .$$

The first factor vanishes for $\Lambda \to \infty$, the second for $a \to \infty$. We conclude that for a and Λ sufficiently large

$$\left| |\Lambda|^{-1} F(\Phi,\Lambda) - a^{-\nu} F(\Phi,K(a)) \right| < \varepsilon \quad .$$

□

We can now prove the main theorem.

THEOREM 4.8: *For* $\Phi \in \mathcal{B}_1$ *the thermodynamical limit*

$$\lim_{n\to\infty} |\Lambda_n|^{-1} F(\Phi,\Lambda_n) = f(\Phi)$$

exists if $\{\Lambda_n\} \to \infty$ *in the sense of Van Hove;* f *is a concave function with the property*

$$|f(\Phi) - f(\Psi)| \leq \|\Phi - \Psi\|_1 .$$

PROOF: The extension of the result of lemma 4.7 to $\mathcal{B}_1$ is a consequence of lemma 4.2. The same is true for the inequality. That f is concave follows if we prove that for finite Λ $F(\Phi,\Lambda)$ has that property. This is a direct consequence of (A.6) (appendix). □

4.2. The mean entropy. Let $\rho \in I$. The restriction of ρ to $\mathfrak{A}(\Lambda)$, where Λ is a finite subset of $\mathbb{Z}^\nu$, defines a density matrix ρ_Λ. We now define the *local entropy*

$$S(\rho,\Lambda) = -\operatorname{Tr}\rho_\Lambda \log \rho_\Lambda.$$

PROPOSITION 4.9: *The local entropy* $S(\rho,\Lambda)$ *has the following properties.*

i. $0 \le S(\rho,\Lambda) \le |\Lambda| \log 2$;

ii. Subadditivity: for $\Lambda_1 \cap \Lambda_2 = \emptyset$, $S(\rho,\Lambda_1 \cup \Lambda_2) \le S(\rho,\Lambda_1) + S(\rho,\Lambda_2)$;

iii. For $\Lambda \subset \Lambda'$, $S(\rho,\Lambda') - S(\rho,\Lambda) \le (|\Lambda'| - |\Lambda|) \log 2$.

PROOF: i. is the inequality (1.2). To prove iii. we write $\Lambda' = \Lambda \cup \Lambda'/\Lambda$; from ii. $S(\rho,\Lambda') \le S(\rho,\Lambda) + S(\rho,\Lambda'/\Lambda)$ and from i. $S(\rho,\Lambda'/\Lambda) \le (|\Lambda'| - |\Lambda|) \log 2$. There remains to prove ii. We notice that for $\Lambda_1 \cap \Lambda_2 = \emptyset$ we have

$$\mathcal{H}_{\Lambda_1\cup\Lambda_2} = \mathcal{H}_{\Lambda_1} \otimes \mathcal{H}_{\Lambda_2}, \quad \rho_{\Lambda_1} = \operatorname{Tr}^{\mathcal{H}_{\Lambda_2}} \rho_{\Lambda_1\cup\Lambda_2} \quad \text{and} \quad \rho_{\Lambda_2} = \operatorname{Tr}^{\mathcal{H}_{\Lambda_1}} \rho_{\Lambda_1\cup\Lambda_1}.$$

We use the inequality (A.4) and substitute $\rho_1 = \rho_{\Lambda_1\cup\Lambda_2}$, $\rho_2 = \rho_{\Lambda_1} \otimes \rho_{\Lambda_2}$. We notice that

$$\log(\rho_{\Lambda_1} \otimes \rho_{\Lambda_2}) = \log \rho_{\Lambda_1} \otimes \mathbb{1} + \mathbb{1} \otimes \log \rho_{\Lambda_2},$$

and

$$\operatorname{Tr}(\rho_{\Lambda_1} \otimes \rho_{\Lambda_2}) = 1,$$

and obtain

$$\operatorname{Tr}(\rho_{\Lambda_1\cup\Lambda_2} \log \rho_{\Lambda_1\cup\Lambda_2}) - \operatorname{Tr}(\rho_{\Lambda_1\cup\Lambda_2}(\log\rho_{\Lambda_1} \otimes \mathbb{1})) - \operatorname{Tr}(\rho_{\Lambda_1\cup\Lambda_2}(\mathbb{1} \otimes \log\rho_{\Lambda_2})) \ge 0$$

from which ii. follows. □

We shall use proposition 4.10 to prove the existence of the mean entropy.

PROPOSITION 4.10: *If* $\Lambda \to F(\Lambda)$ *is a positive function on the finite subsets of* $\mathbb{Z}^\nu$ *such that*

a. $0 \le F(\Lambda) \le |\Lambda| F$, $F > 0$,

b. For $\Lambda_1 \cup \Lambda_2 = \emptyset$, $F(\Lambda_1 \cup \Lambda_2) \le F(\Lambda_1) + F(\Lambda_2)$,

c. $F(\Lambda + x) = F(\Lambda)$, *for all* $x \in \mathbb{Z}^\nu$.

Then

$$\lim_{\Lambda\to\infty} \frac{F(\Lambda)}{|\Lambda|} = f,$$

for any Van Hove sequence with the additional property that $|\Lambda|/|K(\Lambda)| \ge \delta > 0$, *where* $K(\Lambda)$ *is the smallest cube containing* Λ.

REMARK: The condition $|\Lambda|/K(\Lambda) \geq \delta > 0$ excludes the possibility that Λ expands in one direction faster than in another and becomes increasingly flat.

PROOF of proposition 4.10: If $\{K_i(a)\}$ is a set of non-overlapping cubes of side a, then, by the subadditivity,

$$F(\bigcup_i K_i(a)) \leq \sum_i F(K_i(a)),$$

or

$$\frac{F(\bigcup_i K_i(a))}{|\bigcup_i K_i(a)|} \leq \frac{F(K(a))}{|K(a)|} .$$

Consequently, the sequence $\{F(K(2^n))/|K(2^n)|\}$ is monotonically decreasing, so that

$$f = \lim_{n\to\infty} \frac{F(K(2^n))}{|K(2^n)|} = \inf_n \frac{F(K(2^n))}{|K(2^n)|} .$$

We write $K_n = K(2^n)$ and divide $\mathbb{Z}^\nu$ into cubes of sides 2^n. Let Λ_n^- be the union of all such cubes contained in Λ and Λ_n^+ the union of the smallest covering of Λ, so that $\Lambda_n^- \subset \Lambda \subset \Lambda_n^+$. Then

$$F(\Lambda) \leq F(\Lambda_n^-) + F(\Lambda/\Lambda_n^-),$$

or

$$\frac{F(\Lambda)}{|\Lambda|} \leq \frac{|\Lambda_n^-|}{|\Lambda|} \frac{F(\Lambda_n^-)}{|\Lambda_n^-|} + \frac{|\Lambda| - |\Lambda_n^-|}{|\Lambda|} \frac{F(\Lambda/\Lambda_n^-)}{|\Lambda/\Lambda_n^-|} ,$$

and thus

$$\frac{F(\Lambda)}{|\Lambda|} \leq \frac{F(K_n)}{|K_n|} + \frac{|\Lambda| - |\Lambda_n^-|}{|\Lambda|} F . \tag{4.1}$$

For given $\varepsilon > 0$ and sufficiently large n

$$\frac{F(K_n)}{|K_n|} < f + \varepsilon.$$

Choosing now Λ sufficiently large, we have, using the Van Hove condition,

$$\frac{|\Lambda| - |\Lambda_n^-|}{|\Lambda|} F \leq \frac{|\Lambda_n^+| - |\Lambda_n^-|}{|\Lambda|} F \leq \varepsilon ,$$

so that

$$\frac{F(\Lambda)}{|\Lambda|} < f + 2\varepsilon. \tag{4.2}$$

We obtain another inequality by enclosing Λ in the smallest cube K_m $(m > n)$. Then

$$\frac{F(K_m)}{|K_m|} \le \frac{|\Lambda|}{|K_m|}\,\frac{F(\Lambda)}{|\Lambda|} + \frac{|K_m| - |\Lambda|}{|K_m|}\,\frac{F(K_m/\Lambda)}{|K_m/\Lambda|} .$$

Using now (4.1) for Λ replaced by K_m/Λ, we get

$$\frac{F(K_m/\Lambda)}{|K_m/\Lambda|} \le \frac{F(K_n)}{|K_n|} + \frac{|\Lambda_n^+| - |\Lambda|}{|K_m| - |\Lambda|}\,F;$$

substitution of this in the previous inequality yields

$$\frac{F(K_m)}{|K_m|} \le \frac{|\Lambda|}{|K_m|}\,\frac{F(\Lambda)}{|\Lambda|} + \frac{|K_m| - |\Lambda|}{|K_m|}\,\frac{F(K_n)}{|K_n|} + \frac{|\Lambda_n^+| - |\Lambda|}{|\Lambda|}\,\frac{|\Lambda|}{|K_m|}\,F.$$

Hence

$$f - \varepsilon \le \frac{|\Lambda|}{|K_m|}\,\frac{F(\Lambda)}{|\Lambda|} + \left(1 - \frac{|\Lambda|}{|K_m|}\right)(f + \varepsilon) + \frac{|\Lambda|}{|K_m|}\,\varepsilon ,$$

or

$$f \le \frac{F(\Lambda)}{|\Lambda|} + 2\,\frac{|K_m|}{|\Lambda|}\,\varepsilon .$$

Now, from our condition on the limit $\Lambda \to \infty$, we have

$$\frac{|K_m|}{|\Lambda|} \le 2^\nu\,\frac{|K(\Lambda)|}{|\Lambda|} \le 2^\nu\,\delta^{-1}$$

and

$$f \le \frac{F(\Lambda)}{|\Lambda|} + 2^{\nu+1}\,\delta^{-1}\,\varepsilon. \tag{4.3}$$

Combining (4.2) and (4.3) we complete the proof. □

THEOREM 4.11: *For* $\rho \in I$, *the mean entropy per lattice site*

$$s(\rho) = \lim_{\Lambda\to\infty} \frac{S(\rho,\Lambda)}{|\Lambda|}$$

exists, provided $\Lambda \to \infty$ *in the sense of Van Hove, and has the following properties.*

i. $0 \le s(\rho) \le \log 2$.

ii. The function $\rho \to s(\rho)$ *is affine upper semicontinuous on* I.

PROOF: The theorem as stated is slightly more general than what we shall prove. We shall assume that $\Lambda \to \infty$ such that $|\Lambda|/|K(\Lambda)| \ge \delta > 0$. We can then apply proposition 4.10 to conclude that $s(\rho)$ exists. Property i. follows from property i. of proposition 8.3. To prove ii., we notice that for $A \in \mathfrak{A}(\Lambda)$ the function $\rho \in E \to \rho(A) = \mathrm{Tr}^{(\Lambda)}\rho_\Lambda A$ is continuous if E is equipped with the W^*-topology. Choosing for A matrix units we conclude that all matrix elements of ρ_Λ are continuous and so is $S(\rho,\Lambda) = -\mathrm{Tr}(\rho_\Lambda \log \rho_\Lambda)$. Since, from proposition 4.10, $s(\rho)$ equals the infimum of a sequence of continuous functions $s(\rho)$ is upper semicontinuous. There remains to prove that s is affine. We first notice that if $\rho = \lambda_1\rho_1 + \lambda_2\rho_2$, with $0 \le \lambda_i \le 1$, and $\lambda_1 + \lambda_2 = 1$, then also $\rho_\Lambda = \lambda_1\rho_{1\Lambda} + \lambda_2\rho_{2\Lambda}$.

Using (A7) we have

$$-\lambda_1 \mathrm{Tr}(\rho_{1\Lambda} \log \rho_{1\Lambda}) - \lambda_2 \mathrm{Tr}(\rho_{2\Lambda} \log \rho_{2\Lambda})$$

$$\leq - \mathrm{Tr}((\lambda_1 \rho_{1\Lambda} + \lambda_2 \rho_{2\Lambda}) \log (\lambda_1 \rho_{1\Lambda} + \lambda_2 \rho_{2\Lambda}))$$

$$= - \lambda_1 \mathrm{Tr}(\rho_{1\Lambda} \log (\lambda_1 \rho_{1\Lambda} + \lambda_2 \rho_{2\Lambda})) - \lambda_2 \mathrm{Tr}(\rho_{2\Lambda} \log (\lambda_1 \rho_{1\Lambda} + \lambda_2 \rho_{2\Lambda}))$$

$$\leq - \lambda_1 \mathrm{Tr}(\rho_{1\Lambda} \log(\lambda_1 \rho_{1\Lambda})) - \lambda_2 \mathrm{Tr}(\rho_{2\Lambda} \log (\lambda_2 \rho_{2\Lambda}))$$

$$= - \lambda_1 \mathrm{Tr}(\rho_{1\Lambda} \log \rho_{1\Lambda}) - \lambda_2 \mathrm{Tr}(\rho_{2\Lambda} \log \rho_{2\Lambda}) - \lambda_1 \log \lambda_1 - \lambda_2 \log \lambda_2 .$$

For the last inequality we used (A.8). We now have

$$\lambda_1 S(\rho_1,\Lambda) + \lambda_2 S(\rho_2,\Lambda) \leq S(\lambda_1\rho_1 + \lambda_2\rho_2,\Lambda) \leq \lambda_1 S(\rho_1,\Lambda) + \lambda_2 S(\rho_2,\Lambda) + \log 2 ,$$

and thus

$$s(\lambda_1\rho_1 + \lambda_2\rho_2) = \lambda_1 s(\rho_1) + \lambda_2 s(\rho_2). \qquad \square$$

REMARK: In order to prove the existence of the limit $s(\rho) = |\Lambda|^{-1} S(\rho,\Lambda)$ for $\Lambda \to \infty$ in the sense of Van Hove, one makes use of the strong subadditivity of the entropy: $S(\rho,\Lambda_1 \cup \Lambda_2) - S(\rho,\Lambda_1) - S(\rho,\Lambda_2) + S(\rho,\Lambda_1 \cap \Lambda_2) \leq 0$.

The results of proposition 4.10 and theorem 4.11 can be generalized to periodic states.

PROPOSITION 4.10$'$: *The conclusion of proposition 4.10 remains valid if condition c) is replaced by c$'$.* $F(\Lambda + px) = F(\Lambda)$, *for some* $p \in \mathbb{N}$ *and all* $x \in \mathbb{Z}^\nu$.

PROOF: The only change necessary in the proof of proposition 4.10 is the replacement of $K_n = K(2^n)$ by $K_n' = K(2^n p)$. $\square$

THEOREM 4.11$'$: *For a state* ρ *that is periodic with period* p, *i.e.,* ρ *is invariant under the translations* τ_{px}, *with* $p \in \mathbb{N}$, *the mean entropy*

$$s(\rho) = \lim_{\Lambda\to\infty} \frac{S(\rho,\Lambda)}{|\Lambda|}$$

exists, provided $\Lambda \to \infty$ *in the sense of Van Hove, and has the following properties.*

i. $0 \leq s(\rho) \leq \log 2$

ii. The function $\rho \to s(\rho)$ *is affine upper semicontinuous on the convex set of states with period* p.

iii. $s(\rho \circ \tau_x) = s(\rho)$ *for all* $x \in \mathbb{Z}^\nu$.

PROOF: We consider again the slightly less general case where $|\Lambda|/|K(\Lambda)| \geq \delta > 0$. It is then an immediate consequence of propostion 4.10$'$. $\square$

4.3 The mean energy. The existence of the mean energy is assured by the following theorem.

THEOREM 4.12: *For* $\rho \in I$ *and* $\Phi \in \mathcal{B}_1$

$$\lim_{\Lambda\to\infty} |\Lambda|^{-1} \rho(H(\Phi,\Lambda)) = \rho(A_\Phi),$$

if $\Lambda \to \infty$ *in the sense of Van Hove. Here*

$$A_\Phi = \sum_{X\ni 0} \frac{\Phi(X)}{|X|} .$$

PROOF: We have

$$\frac{\rho(H(\Phi,\Lambda))}{|\Lambda|} = \frac{1}{|\Lambda|} \sum_{X\subset\Lambda} \rho(\Phi(X)) = \frac{1}{|\Lambda|} \sum_{x\in\Lambda} \sum_{X:x\in X\subset\Lambda} \frac{\rho(\Phi(X))}{|X|}$$

and

$$\rho(A_\Phi) = \frac{1}{|\Lambda|} \sum_{x\in\Lambda} \sum_{X:x\in X} \frac{\rho(\Phi(X))}{|X|} .$$

Hence

$$\left|\frac{\rho(H(\Phi,\Lambda))}{|\Lambda|} - \rho(A_\Phi)\right| \le \frac{1}{|\Lambda|} \sum_{x\in\Lambda} \sum_{X:x\in X\not\subset\Lambda} \frac{|\rho(\Phi(X))|}{|X|} .$$

For given $\varepsilon > 0$, there exists a subset Λ_0 containing 0 such that

$$\sum_{X:0\in X\not\subset\Lambda_0} \frac{\|\Phi(X)\|}{|X|} < \varepsilon .$$

Let $\Lambda_x = \{\Lambda_0 + x\}$ and $\Lambda_1 \subset \Lambda$ the largest subset of Λ such that, for $x \in \Lambda_1$, $\Lambda_x \subset \Lambda$. Now

$$\left|\frac{\rho(H(\Phi,\Lambda))}{|\Lambda|} - \rho(A_\Phi)\right|$$

$$\le \frac{1}{|\Lambda|} \sum_{x\in\Lambda_1} \sum_{X:x\in X\not\subset\Lambda_x} \frac{\|\Phi(X)\|}{|X|} + \frac{1}{|\Lambda|} \sum_{x\in\Lambda/\Lambda_1} \sum_{x\in X} \frac{\|\Phi(X)\|}{|X|}$$

$$\le \frac{|\Lambda_1|}{|\Lambda|}\varepsilon + \frac{|\Lambda| - |\Lambda_1|}{|\Lambda|} \|\Phi\|_1 .$$

If $\Lambda \to \infty$ in the sense of Van Hove, the second term will be $<\varepsilon$ for Λ sufficiently large. □

5. THE VARIATIONAL PRINCIPLE AND TANGENT PLANES

5.1. The variational principle. Having defined the mean free energy, the mean entropy and the mean energy for a state $\rho \in I$, we shall formulate the variational principle for states of the infinite lattice. We start with

PROPOSITION 5.1: *For* $\rho \in I$ *and* $\Phi \in \mathcal{B}_1$,

$$f(\Phi) \le \rho(A_\Phi) - s(\rho).$$

PROOF: This is an immediate consequence of theorem 1.1 and chapter 4. □

To show that there are states for which the equality holds, we proceed as follows. Let $\varepsilon > 0$ be given. We shall construct a state $\rho' \in I$, such that

$$f(\Phi) > \rho'(A_\Phi) - s(\rho') - \varepsilon.$$

Since for finite Λ, the Gibbs state satisfies the variational equality, we let ρ' look as much as possible like a Gibbs state. Let $K(n) = \{x : 0 \leq x_i < n, i = 1,2,\ldots,\nu\}$ and for $a \in \mathbb{Z}^\nu$ $K_a = K(n) + na$. The cubes $\{K_a ; a \in \mathbb{Z}^\nu\}$ are disjoint and cover the entire lattice. For each $a \in \mathbb{Z}^\nu$, let

$$\rho_a = \frac{e^{-H(\Phi,K_a)}}{\operatorname{Tr} e^{-H(\Phi,K_a)}}$$

be the density matrix of the Gibbs state in K_a. We now define the state $\tilde{\rho}_n$ of $\mathfrak{A}$ as the product state of all $\rho_a, a \in \mathbb{Z}^\nu$. Let $A \in \mathfrak{A}(\Lambda)$ and let Λ be covered by $\{K_{a_i}, i = 1,2,\ldots,p\}$, then

$$\tilde{\rho}_n(A) = \operatorname{Tr} A \rho_{a_1} \otimes \rho_{a_2} \otimes \ldots \otimes \rho_{a_p},$$

where the trace is taken in the Hilbert space $\mathcal{H}(\bigcup_i K_{a_i})$. By taking the mean of $\tilde{\rho}_n \circ \tau_x$ over all $x \in K(n)$ we get the translation invariant state

$$\rho_n = n^{-\nu} \sum_{x \in K(n)} \tilde{\rho}_n \circ \tau_x .$$

We shall calculate the mean entropy and the mean energy of this state. Since s is affine we get, using theorem 4.11' property iii.,

$$s(\rho_n) = |K(n)|^{-1} \sum_{x \in K(n)} s(\tilde{\rho}_n \circ \tau_x) = s(\tilde{\rho}_n) = \lim_{m \to \infty} \frac{S(\tilde{\rho}_n, K(mn))}{|K(mn)|} .$$

We can write $K(mn) = \bigcup_{i=1}^{p} K_{a_i}$, with $p = m^\nu$. Then

$$S(\tilde{\rho}_n, K(mn)) = -\operatorname{Tr}[(\rho_{a_1} \otimes \rho_{a_2} \otimes \ldots \otimes \rho_{a_p}) \log (\rho_{a_1} \otimes \ldots . \otimes \rho_{a_p})]$$

$$= -m^\nu \operatorname{Tr}(\rho_{K(n)} \log \rho_{K(n)}).$$

We conclude that

$$s(\rho_n) = -|K(n)|^{-1} \operatorname{Tr}(\rho_{K(n)} \log \rho_{K(n)}). \tag{5.1}$$

We estimate the mean energy as follows.

$$\rho_n(A_\Phi) = |K(n)|^{-1} \sum_{x \in K(n)} \tilde{\rho}_n(\tau_x(A_\Phi))$$

$$= |K(n)|^{-1} \sum_{x\in K(n)} \sum_{X\ni x} \frac{\tilde{\rho}_n(\Phi(X))}{|X|}$$

$$= |K(n)|^{-1} \sum_{x\in K(n)} \sum_{X:x\in X\subset K(n)} \frac{\tilde{\rho}_n(\Phi(X))}{|X|} + |K(n)|^{-1} \sum_{x\in K(n)} \sum_{X:x\in X\not\subset K(n)} \frac{\tilde{\rho}_n(\Phi(X))}{|X|} .$$

The first term at the right-hand side is $|K(n)|^{-1}\tilde{\rho}_n(H(\Phi,K(n)))$, and thus

$$|\rho_n(A_\Phi) - |K(n)|^{-1}\tilde{\rho}_n(H(\Phi,K(n)))| \leq |K(n)|^{-1} \sum_{x\in K(n)} \sum_{X:x\in X\not\subset K(n)} \frac{\|\Phi(X)\|}{X} .$$

Let $\varepsilon > 0$ be given; there exists $m \in \mathbb{N}$ such that

$$\sum_{0\in X:d(X)>m} \frac{\|\Phi(X)\|}{|X|} \leq \frac{\varepsilon}{2} .$$

If we define

$$K(n,m) = \{x \in K(n) : d(x,K(n)^C) > m\},$$

we can choose n so large that

$$(|K(n)| - |K(n,m)|)/|K(n)| \leq \tfrac{1}{2}\varepsilon \|\Phi\|_1^{-1}.$$

Then

$$|\rho_n(A_\Phi) - |K(n)|^{-1} \tilde{\rho}_n(H(\Phi,K(n)))|$$

$$\leq |K(n)|^{-1} \sum_{x\in K(n,m)} \sum_{X:x\in X\not\subset K(n)} \frac{\|\Phi(X)\|}{X}$$

$$+ |K(n)|^{-1} \sum_{x\in K(n)/K(n,m)} \sum_{X:x\in X\not\subset K(n)} \frac{\|\Phi(X)\|}{X}$$

$$\leq \frac{|K(n,m)|}{|K(n)|} \frac{\varepsilon}{2} + \frac{|K(n)| - |K(n,m)|}{|K(n)|} \|\Phi\|_1 \leq \varepsilon.$$

Since the restriction of $\tilde{\rho}_n$ to $\mathfrak{A}(K(n))$ is the Gibbs state $\rho_{K(n)}$, we have, for n sufficiently large

$$|\rho_n(A_\Phi) - |K(n)|^{-1} \rho_{K(n)}(H(\Phi,K(n)))| \leq \varepsilon. \tag{5.2}$$

Finally, for n sufficiently large,

$$|f(\Phi) - |K(n)|^{-1} F(\Phi,K(n))| \leq \varepsilon . \tag{5.3}$$

Since $\rho_{K(n)}$ is the Gibbs state for the finite lattice K(n) we have the equality

$$F(\Phi,K(n)) = \rho_{K(n)}(H(\Phi,K(n)) - S(\rho_{K(n)},K(n)).$$

Combining this with (5.1), (5.2) and (5.3) we obtain

$$f(\Phi) \geq \rho_n(A_\Phi) - s(\rho_n) - 2\varepsilon.$$

We can already conclude that

$$f(\Phi) = \inf_{\rho \in I} (\rho(A_\Phi) - s(\rho)).$$

Because the function $\rho \to \rho(A_\Phi) - s(\rho)$ is affine lower semicontinuous the infimum is reached in at least one point. We proved

THEOREM 5.2: *For* $\Phi \in \mathcal{B}_1$ *we have*

$$f(\Phi) = \min_{\rho \in I} (\rho(A_\Phi) - s(\rho)).$$

The set $I^\Phi \subset I$ *of states for which the minimum is reached is a non-empty closed convex set.*

DEFINITION 5.1: The states $\rho \in I^\Phi$ are the *equilibrium states* corresponding to the potential Φ.

REMARK: It is easy to extend the variational principle to a larger class of states than I, e.g., states with a given period. In that case one can define equilibrium states that are not translation invariant.

A thermodynamical limit of Gibbs states is not always an equilibrium state in the above sense. Let for finite $\Lambda \subset \mathbb{Z}^\nu$ ρ^Φ_Λ be the corresponding Gibbs state. If there is a unique thermodynamical limit ρ^Φ if $\Lambda \to \infty$ in such a way that any finite Λ_0 is eventually contained in Λ, then it is easy to see that this limit is translation invariant. Since for each Λ

$$F(\Phi,\Lambda) = \rho^\Phi_\Lambda(H(\Phi,\Lambda)) - S(\rho^\Phi_\Lambda,\Lambda)$$

we have, in the limit $\Lambda \to \infty$,

$$f(\Phi) \geq \rho^\Phi(A_\Phi) - s(\rho^\Phi),$$

where we used the upper semicontinuity of s. This, however, implies

$$f(\Phi) = \rho^\Phi(A_\Phi) - s(\rho^\Phi).$$

PROPOSITION 5.3: *If the Gibbs states* $\{\rho^\Phi_\Lambda\}$ *have a unique limit*

$$\rho^\Phi = \lim_\Lambda \rho^\Phi_\Lambda ,$$

then $\rho^\Phi \in I^\Phi$.

If the thermodynamical limit is not unique then, due to compactness, there is always a subsequence $\{\Lambda_n\}$ for which the limit of $\rho^\Phi_{\Lambda_n}$ exists. In that case this limit need not be translation invariant.

The case that I^Φ consists of more than one state is of much interest. For such Φ there is a *first order phase transition*. The extremal points of I^Φ represent the pure phases. These are also extremal translation invariant, by the next theorem.

THEOREM 5.4: *If* $\rho \in \mathcal{E}(I^\Phi)$ *then* $\rho \in \mathcal{E}(I)$.

PROOF: Suppose $\rho \notin \mathcal{E}(I)$, then there exist ρ_1 and $\rho_2 \in I$, $\rho_1 \neq \rho_2$ and $0 < \lambda < 1$ such that $\rho = \lambda\rho_1 + (1-\lambda)\rho_2$. Since $\rho \in \mathcal{E}(I^\Phi)$ ρ_1 and ρ_2 cannot both be in I^Φ. Suppose that $\rho_1 \notin I^\Phi$, then $f(\Phi) = \rho(A_\Phi) - s(\rho) = \lambda(\rho_1(A_\Phi) - s(\rho_1)) + (1-\lambda)(\rho_2(A_\Phi) - s(\rho_2)) > \lambda f(\Phi) + (1-\lambda)f(\Phi) = f(\Phi)$. We have a contradiction. □

5.2. Tangent planes. Since the mean free energy f is a concave function on $\mathcal{B}_1$ we shall give the definition of the tangent plane to the graph of a concave function.

DEFINITION 5.2: If f is a real concave function on a Banach space X, then a bounded linear function α on X is a *tangent plane* to the graph of f at the point x if for all $y \in X$

$$f(x+y) - f(x) \leq \alpha(y).$$

In this section we shall establish a one-to-one correspondence between equilibrium states and tangent planes to the graph of the mean free energy. If $\rho \in I^\Phi$, then $f(\Phi) = \rho(A_\Phi) - s(\rho)$, and $f(\Phi+\Psi) \leq \rho(A_\Phi) + \rho(A_\Psi) - s(\rho)$ or $f(\Phi+\Psi) - f(\Phi) \leq \rho(A_\Psi)$. Now $|\rho(A_\Psi)| \leq \|\Psi\|_1$ and we conclude that $\Psi \to \rho(A_\Psi)$ is a tangent plane to the graph of f at the point Φ. We notice further that different states of I^Φ define different tangent planes at Φ: suppose that $\rho_1(A_\Psi) = \rho_2(A_\Psi)$ for all $\Psi \in \mathcal{B}_1$, then $\rho_1 = \rho_2$ since the set $\{\tau_x(A_\Psi), x \in \mathbb{Z}^\nu, \Psi \in \mathcal{B}_1\}$ is dense in $\mathfrak{A}$.

We shall now proceed to the reverse problem. If α is a tangent plane at Φ, is there a state $\rho \in I^\Phi$, such that $\alpha(\Psi) = \rho(A_\Psi)$ for all $\Psi \in \mathcal{B}_1$?

DEFINITION 5.3: We say that f is *differentiable* at the point Φ if there is a unique tangent plane Df_Φ at Φ. We write $\mathcal{D}$ for the set of points where f is differentiable.

That $\mathcal{D}$ contains "almost all" points of $\mathcal{B}_1$ is expressed by the following lemma, for the proof of which we refer to Dunford and Schwartz, theorem V,9.8.

LEMMA 5.5: *A convex function on a Banach space is differentiable at a residual set (i.e. the complement of a set of the first category).*

Let now $\Phi \in \mathcal{D}$. Then I^Φ consists of only one state ρ^Φ, and, if α^Φ is the tangent plane at Φ, $\alpha^\Phi(\Psi) = \rho^\Phi(A_\Psi)$. The one-to-one correspondence between tangent plane at Φ and states of I^Φ has now been established for all $\Phi \in \mathcal{D}$. For the general case we need the following lemma on residual sets.

LEMMA 5.6: *Let X be a Banach space and $Y \subset X$ a residual set; then there exists a residual set $X_1 \subset X$ such that for each $x \in X_1$ $x\lambda \in Y$ for λ in a residual set in $\mathbb{R}$.*

PROOF: We assume first that Y is a dense open set. Let $\{a_i, i = 1,2,\dots\}$ be dense in $\mathbb{R}$ and let $X_i = \{x \in X :$ there is $\lambda \in \mathbb{R}$ with $|\lambda - a_i| < \frac{1}{i}$ such that $\lambda x \in Y\}$. We see immediately that $X_i = \bigcup_{\lambda \neq 0: |\lambda - a_i| < \frac{1}{i}} \lambda^{-1} Y$; hence X_i is a dense open set. The intersection $X_1 = \bigcap_i X_i$ is a residual set. Now for $x \in X_1$ there is for each i a λ_i: $|a_i - \lambda_i| < \frac{1}{i}$ such that $\lambda_i x \in Y$. The set $\{\lambda_i\}$ is dense in $\mathbb{R}$. Hence the intersection of the line $\{\lambda x, \lambda \in \mathbb{R}\}$ with Y contains a dense set of points. But for each point $\lambda_i x$ there is a neighborhood also contained in Y. Hence the intersection of $\{\lambda x, \lambda \in \mathbb{R}\}$ with Y is a dense open set and $\lambda x \in Y$ for λ in a dense open set $S_x \subset \mathbb{R}$.

If, more generally, Y is a residual set, then there is a countable set of dense open sets Y_n, such that $\bigcap_n Y_n \subset Y$. For each Y_n there is a X_1^n. Let $x \in X_1 = \bigcap_n X_1^n$; then for each n $\lambda x \in Y_n$ for λ in a dense open set S_x^n of $\mathbb{R}$. Hence $\lambda x \in Y$ for λ in the residual set $\bigcap_n S_x^n$. □

We shall use this lemma for the proof of an important theorem on tangent planes.

THEOREM 5.7: *If* f *is a convex function on a separable Banach space* X *and* $\eta \in X^*$ *a tangent plane to the graph of* f *at the point* x_0, *then* η *is contained in the* W^**-closed convex hull of the set* $\mathcal{Z}$ *of tangent planes at* x_0 *defined by* $\mathcal{Z} = \{\zeta \in X^* : \exists$ *sequence* $x_n \to x_0$ *such that* f *is differentiable at each* x_n *and* $Df_{x_n} \to \zeta\}$.

PROOF: Without loss of generality, we may assume that $x_0 = 0$. We first consider the special case $X = \mathbb{R}$. If $\{x_n\}$ is a sequence of points where f is differentiable which converges to zero from the right, then $f'(x_n) \to f_R'(0)$, where $f_R'(0)$ is the right derivative of f at 0. Similarly, if $\{y_n\}$ is a sequence of points where f is differentiable which converges to zero from the left then $f'(y_n) \to f_L'(0)$. In this case $\mathcal{Z}$ consists of the right derivative and the left derivative. Every tangent in $x = 0$ is a convex sum of these.

In the general case, assume that η is not contained in the W^*-closed convex hull of $\mathcal{Z}$. Then there exists $x \in X$ and a real number m such that $\eta(x) > m$ and $\zeta(x) < m$ for all $\zeta \in \mathcal{Z}$. Since for some neighborhood of 0 we have $|f(x) - f(0)| \leq M\|x\|$, the tangent planes at 0 are bounded in norm by M. We can therefore find $\varepsilon > 0$, such that for $\|x - x'\| < \varepsilon$ $\eta(x') > m$ and $\zeta(x') < m$. Using lemma 5.6 we can choose x' such that f is differentiable at the points $\lambda x'$ with λ in a residual set in $\mathbb{R}$. Considering $\eta(\lambda x')$ and $f(\lambda x')$ as functions of λ, we see that η is a tangent at f in $\lambda = 0$. If $\{\lambda_n\}$ is a decreasing sequence of points in the residual set such that f is differentiable in $\lambda_n x'$ and if $\lambda_n \to 0$ then there is a subsequence $\{\lambda_{n'}\}$ such that $Df_{\lambda_{n'}x'}$ converges to an element ζ' of $\mathcal{Z}$. Clearly $\zeta'(\lambda x')$ as function of λ is the right derivative of $f(\lambda x')$. Hence $\eta(\lambda x') \leq \zeta(\lambda x')$ and thus $\eta(x') < m$. Contradiction. □

We shall apply this theorem to the case of f as a function on $\mathcal{B}_1$. We assume that $\Phi \notin \mathcal{D}$. Let $\zeta \in \mathcal{Z}$; then there exists a sequence $\{\Phi_n \in \mathcal{D}\}$, $\Phi_n \to \Phi$, such that $Df_{\Phi_n} \to \zeta$. Since $\Phi_n \in \mathcal{D}$ there is a state ρ_n on $\mathfrak{A}$ such that $Df_{\Phi_n}(\Psi) = \rho_n(A_\Psi)$ and

$$\lim_{n\to\infty} \rho_n(A_\Psi) = \zeta(\Psi)$$

for all $\Psi \in \mathcal{B}_1$. This implies that the sequence $\{\rho_n\}$ converges to a state ρ and $\rho(A_\Psi) = \zeta(\Psi)$. We shall prove that $\rho \in I^\Phi$. Since $\rho_n \in I^{\Phi_n}$ we have $f(\Phi_n) = \rho_n(A_{\Phi_n}) - s(\rho_n)$, and thus

$$|f(\Phi) - \rho(A_\Phi) + s(\rho)| \leq |f(\Phi) - f(\Phi_n)| + |\rho(A_\Phi) - \rho_n(A_\Phi)| +$$

$$|\rho_n(A_\Phi) - \rho_n(A_{\Phi_n})| + |s(\rho) - s(\rho_n)|.$$

For n sufficiently large, all terms at the righthand side become arbitrarily small. This shows that for each $\zeta \in \mathcal{Z}$ there is a $\rho \in I^\Phi$ such that $\rho(A_\Psi) = \zeta(\Psi)$. If α is a tangent plane at the point Φ, α is in the W^*-closure of the convex hull of $\mathcal{Z}$, by theorem 5.7. But then there is a state ρ_α in the convex hull of the states corresponding to the elements of $\mathcal{Z}$, such that $\rho_\alpha(A_\Psi) = \alpha(\Psi)$, and $\rho_\alpha \in I^\Phi$. We have now proved

THEOREM 5.8: *There is a one-to-one correspondence between the tangent planes to the graph of* f *at* Φ, *and the equilibrium states for* Φ. *The correspondence is given by*

$$\alpha(\Psi) = \rho(A_\Psi),$$

where α *is a tangent plane at* Φ *and* ρ *is the corresponding element of* I^Φ.

6. EQUIVALENCE OF THE VARIATIONAL PRINCIPLE AND THE KMS CONDITION

If we want to compare the variational principle with the KMS condition we shall have to restrict our attention to such potentials for which the KMS condition can be defined. We, therefore, choose $f(X) = \exp(|X|)$ and consider the corresponding Banach space $\mathcal{B}_f$ as defined in section 2.2. By theorem 2.2, the potentials in $\mathcal{B}_f$ are dynamical, in the sense that they define a strongly continuous one-parameter group of automorphisms of $\mathfrak{A}$. It is important to notice that the results of the previous chapter, in particular theorems 5.2 and 5.8 remain valid in the smaller space $\mathcal{B}_f$. We shall write $\mathcal{D}_f = \mathcal{D} \cap \mathcal{B}_f$ for the set of potentials in $\mathcal{B}_f$ with a unique tangent plane.

The set of states satisfying the KMS condition for $\beta = 1$ with respect to the group of automorphisms α_t^Φ, for given $\Phi \in \mathcal{B}_f$, we shall denote by K^Φ. In this chapter we shall show that $K^\Phi \cap I = I^\Phi$.

6.1. Equilibrium states satisfy the KMS condition. We shall first consider the case that $\Phi \in \mathcal{D}_f$.

THEOREM 6.1: *If* $\Phi \in \mathcal{D}_f$ *and if* ρ_Φ *is the corresponding equilibrium state, then* ρ_Φ *satisfies the KMS condition w.r.t.* α_t^Φ.

PROOF: If $\{\rho_n\}$ is the sequence of states in I constructed in section 5.1, we have

$$\rho_\Phi = \lim_{n\to\infty} \rho_n .$$

The proof that $\rho_\Phi \in K^\Phi$ is similar to but slightly more complicated than the proof of theorem 2.4, stating that the thermodynamical limit of Gibbs states satisfies the KMS condition. The complication is that ρ_n is not itself a Gibbs state but a space average over Gibbs states.

Let $A, B \in \mathfrak{A}(\Lambda_0)$ and $\varepsilon > 0$. We can find $\Lambda_1 \supset \Lambda_0$ such that for all $\Lambda \supset \Lambda_1$

$$\| \alpha_t^\Lambda(A) - \alpha_t^\Phi(A) \| < \varepsilon / \|B\| .$$

Then also, for all $\Lambda \supset \Lambda_1 + x$,

$$\| \alpha_t^\Lambda(\tau_x(A)) - \alpha_t^\Phi(\tau_x(A)) \| < \varepsilon / \|B\| .$$

Using the same notation as in section 5.1 we define, for sufficiently large n,

$$F_n(z) = |K(n)|^{-1} \sum_{x \in K(n):\Lambda_0 + x \subset K(n)} \rho_{K(n)}(\tau_x(B)\, \alpha_z^{K(n)}(\tau_x(A))),$$

where $\rho_{K(n)}$ is the Gibbs state of the region K(n), and satisfies the KMS condition w.r.t. $\alpha_t^{K(n)}$. For $0 \le \operatorname{Im} z \le 1$,

$$|F_n(z)| \le \|A\| \, \|B\| .$$

For real t,

$$|F_n(t) - \rho_n(B\alpha_t^\Phi(A))| \le |K(n)|^{-1} \sum_{\substack{x\in K(n):\\ \Lambda_1 + x \subset K(n)}} | \tilde{\rho}_n(\tau_x(B\alpha_t^\Phi(A)))$$

$$- \tilde{\rho}_n(\tau_x(B)\,\alpha_t^{K(n)}(\tau_x(A)))| + |K(n)|^{-1} \sum_{\substack{x\in K(n):\\ \Lambda_1 + x \not\subset K(n)}} | \tilde{\rho}_n(\tau_x(B\alpha_t^\Phi(A)))|$$

$$+ |K(n)|^{-1} \sum_{\substack{x\in K(n):\\ \Lambda_1 + x \not\subset K(n),\, \Lambda_0 + x \subset K(n)}} | \tilde{\rho}_n(\tau_x(B)\, \alpha_t^{\Lambda_n}(A)))| .$$

The first term on the righthand side is $<\varepsilon$. The other terms are surface terms and can be made arbitrarily small for n sufficiently large. We have shown that

$$\lim_{n\to\infty} F_n(t) = \rho_\Phi(B\alpha_t^\Phi(A)).$$

A similar calculation yields

$$\lim_{n\to\infty} F_n(t+i) = \rho_\Phi(\alpha_t^\Phi(A)B).$$

We conclude that the sequence $\{F_n\}$ of entire functions converges to a function F, analytic in the strip $0 < \operatorname{Im} z < 1$, continuous on the boundary and such that, for real t, $F(t) = \rho_\Phi(B\alpha_t^\Phi(A))$ and $F(t+i) = \rho_\Phi(\alpha_t^\Phi(A)B)$. The extension of this result to arbitrary A and $B \in \mathfrak{A}$ is straightforward. □

We shall now generalize this result to arbitrary $\Phi \in \mathcal{B}_f$. We have seen previously (in the proof of theorem 5.8) that any $\rho \in I^\Phi$ is in the w^*-closed convex hull of a set of states $\mathcal{X}^\Phi \subset I^\Phi$ where, for $\rho' \in \mathcal{X}^\Phi$, $\rho' = \lim_n \rho_{\Phi_n}$, with $\Phi_n \in \mathcal{D}_f$, $\Phi_n \to \Phi$ and ρ_{Φ_n} the unique equilibrium state for Φ_n. We shall prove that the states in $\mathcal{X}^\Phi$ satisfy the KMS condition w.r.t. α_t^Φ. It is clearly sufficient to prove that

$$\lim_n \rho_{\Phi_n}(\alpha_t^{\Phi_n}(A)B) = \rho'(\alpha_t^\Phi(A)B)$$

and

$$\lim_n \rho_{\Phi_n}(B\,\alpha_t^{\Phi_n}(A)) = \rho'(\alpha_t^\Phi(A)B).$$

This is a direct consequence of $\rho' = \lim_n \rho_{\Phi_n}$ and the fact that

$$\lim_n \|\alpha_t^{\Phi_n}(A) - \alpha_t^\Phi(A)\| = 0,$$

as follows immediately from the proof of theorem 2.2. Since K^Φ is a w^*-closed convex set any state in the w^*-closed convex hull of $\mathcal{X}^\Phi$ satisfies the KMS condition. We have proved

THEOREM 6.2: *Every equilibrium state* $\rho \in I^\Phi$, *for* $\Phi \in \mathcal{B}_f$, *satisfies the KMS condition w.r.t.* α_t^Φ.

So far, we have proved $I^\Phi \subset K^\Phi$. The reverse statement $K^\Phi \cap I \subset I^\Phi$ is more complicated. The remaining sections of this chapter are devoted to this problem.

6.2. <u>The relative hamiltonian</u>. Before introducing the Gibbs condition we shall define what one means by the perturbed dynamics α_t^h obtained from the dynamics α_t by the effect of a relative hamiltonian $h = h^* \in \mathfrak{A}$.

THEOREM 6.3: *Let* $\mathfrak{A}$ *be a* C^**-algebra with a strongly continuous one parameter group of automorphisms* α_t *with generator* δ; *let* $h \in \mathfrak{A}$. *Then the derivation* δ^h, *defined for all* A *in the domain of* δ, *by*

$$\delta^h(A) = \delta(A) + i[h,A], \tag{6.1}$$

is the generator of a strongly continuous one parameter group of automorphisms α_t^h, *given by*

$$\alpha_t^h(A) = U_t^h \alpha_t(A) U_t^{h-1}, \tag{6.2}$$

where U_t^h *is a continuous family of unitary elements of* $\mathfrak{A}$ *and can be written as*

$$U_t^h = \sum_{n=0}^{\infty} i^n \int_0^t dt_1 \int_0^{t_1} dt_2 \cdots \int_0^{t_{n-1}} dt_n\, \alpha_{t_n}(h) \cdots \alpha_{t_2}(h)\, \alpha_{t_1}(h).$$

PROOF: As the series converges absolutely for all t, we have $U_t^h \in \mathfrak{A}$. It satisfies the differential equation

$$\frac{d}{dt} U_t^h = i\, U_t^h \alpha_t(h). \tag{6.3}$$

For A in the domain of δ, (6.2) can be differentiated at $t=0$, yielding (6.1). For α_t^h to be a group, U_t^h must satisfy the cocycle equation $U_{t_1+t_2}^h = U_{t_1}^h \alpha_{t_1}(U_{t_2}^h)$. Equation (6.3) is precisely the differential form of the cocycle equation. □

DEFINITION 6.1: If α_t is the dynamics of a system, then α_t^h will be called the *perturbed dynamics*, due to the perturbation or *relative hamiltonian* h.

Suppose that ρ satisfies the KMS condition with respect to α_t. Is there a corresponding KMS state with respect to α_t^h? In order to answer this question we first consider the simple case of the algebra $\mathfrak{A}$ of $n \times n$ matrices. Let the dynamics be given by the s.a. matrix H and the perturbation by h. The corresponding Gibbs states are ρ and ρ^h with density matrices

$$\rho = \frac{e^{-H}}{\mathrm{Tr}(e^{-H})} \quad \text{and} \quad \rho^h = \frac{e^{-(H+h)}}{\mathrm{Tr}(e^{-(H+h)})}.$$

With the scalar product $(A,B) = \mathrm{Tr}\, A^*B$ the set $\mathfrak{A}$ becomes a n^2-dimensional Hilbert space which we denote by $\mathcal{K}$. The algebra can be represented as matrices acting on $\mathcal{K}$ by left multiplication: $\pi(A)B = AB$. Consider now the vector $\rho^{\frac{1}{2}} \in \mathcal{K}$. Then $(\rho^{\frac{1}{2}}, \pi(A)\rho^{\frac{1}{2}}) = \mathrm{Tr}\,\rho A = \rho(A)$. Since $\rho^{\frac{1}{2}}$ is cyclic for $\pi(\mathfrak{A})$ the triple $\{\mathcal{K}, \pi, \rho^{\frac{1}{2}}\}$ is the GNS representation corresponding to the state ρ. Similarly $(\rho^h)^{\frac{1}{2}}$ is the cyclic vector for the "perturbed" state ρ^h. Writing $\rho^{\frac{1}{2}} = \Omega_\rho$, the vector corresponding to the state ρ^h is, except for a normalization constant,

$$\Omega_{\rho^h} = e^{-\frac{1}{2}(H+h)}\, e^{\frac{1}{2}H}\, \Omega_\rho .$$

The operator connecting these two vectors can be rewritten as follows. Let $T(s) = \exp(-s(H+h))\exp(sH)$. We see immediately that

$$\frac{dT}{ds} = -T(s)\, e^{-sH} h\, e^{sH} = -T(s)\, \alpha_{is}(h).$$

This allows us to write T(s) as a series

$$T(s) = 1 - \int_0^s ds_1\, \alpha_{is_1}(h) + \int_0^s ds_1 \int_0^{s_1} ds_2\, \alpha_{is_2}(h)\alpha_{is_1}(h) + \dots .$$

Putting $s = \frac{1}{2}$, we get

$$\Omega_{\rho^h} = \Big[\sum_{n=0}^{\infty} (-1)^n \int_0^{\frac{1}{2}} ds_1 \int_0^{s_1} ds_2 \cdots \int_0^{s_{n-1}} ds_n\, \alpha_{is_n}(h)\, \alpha_{is_{n-1}}(h) \cdots \alpha_{is_1}(h)\Big]\, \Omega_\rho . \tag{6.3}$$

Each term in the series is well-defined and the series converges.

We now proceed to the general case of a C^*-algebra with dynamics α_t, a relative hamiltonian h, and a perturbed dynamics α_t^h. As we shall see, the connection between the unperturbed KMS state and the perturbed one is again given by formula (6.3), where Ω_ρ is the cyclic vector determined by the unperturbed KMS state ρ and the vector Ω_{ρ^h} determines the perturbed equilibrium state ρ^h. The situation is, however, much more complicated, as $\alpha_t(h)$ does not exist, in general, for complex t. The correct interpretation of (6.3) is then as follows. The vector-valued function

$$\alpha_{t_n}(h)\,\alpha_{t_{n-1}}(h)\ldots\alpha_{t_1}(h)\,\Omega_\rho$$

of the real variables $t_1,t_2,\ldots,t_n$ can be extended to an analytic function of $z_1,z_2,\ldots,z_n$ in a cone, which includes the region of integration.

THEOREM 6.4: *Assume that* $\mathfrak{A}$ *is a* C^**-algebra with a strongly continuous group of automorphisms* α_t *and that* ρ *satisfies the KMS condition with respect to* α_t. *Let* $h=h^*\in\mathfrak{A}$ *and let* α_t^h *be the corresponding perturbed automorphisms. Then the vector state* ρ^h *corresponding to the vector* Ω_{ρ^h} *given by formula (6.3) satisfies the KMS condition with respect to* α_t^h.

This theorem is a corollary of a theorem from the theory of modular automorphisms of von Neumann algebras. Before stating that theorem we shall discuss briefly the connection between the KMS condition and the modular theory of von Neumann algebras.

If the state ρ of $\mathfrak{A}$ satisfies the KMS condition, then there exists the quadruple $\{\mathcal{H}_\rho,\pi_\rho,\Omega_\rho,U_t^\rho\}$. The state ρ can be extended to a normal faithful state $\tilde\rho$ of the von Neumann algebra $\pi_\rho(\mathfrak{A})''$ by the definition $\tilde\rho(x)=(\Omega_\rho,x\,\Omega_\rho)$, $x\in\pi_\rho(\mathfrak{A})''$. Similarly, α_t can be extended to an automorphism $\tilde\alpha_t$ of $\pi_\rho(\mathfrak{A})''$ by $\tilde\alpha_t(x)=U_t^\rho\,x\,U_t^{\rho-1}$. It is not difficult to show that $\tilde\rho$ satisfies the KMS condition w.r.t. $\tilde\alpha_t$. The vector Ω_ρ is cyclic and separating. Except for a sign, $\tilde\alpha_t$ equals the modular automorphism group $\sigma_t^{\tilde\rho}$. The relation is

$$\sigma_t^{\tilde\rho}=\tilde\alpha_{-t}\ .$$

THEOREM 6.5: *If* $\mathcal{R}$ *is a von Neumann algebra with cyclic and separating vector* Ω *and corresponding modular automorphisms* σ_t *and if* $h=h^*\in\mathcal{R}$, *then*

$$\Omega^h=\sum_{n=0}^{\infty}\int_0^{\frac12}ds_1\int_0^{s_1}ds_2\ldots\int_0^{s_{n-1}}ds_n\,\sigma_{-is_n}(h)\,\sigma_{-is_{n-1}}(h)\ldots\sigma_{-is_1}(h)\,\Omega$$

is well-defined and converges absolutely. Ω^h is cyclic and separating for $\mathcal{R}$. The corresponding modular automorphisms σ_t^h satisfy, for $x\in\mathcal{R}$,

$$\sigma_t^h(x)=V_t^h\,\sigma_t(x)\,V_t^{h-1}\ ,$$

where

$$V_t^h=\sum_{n=0}^{\infty}(-i)^n\int_0^{t}dt_1\ldots\int_0^{t_{n-1}}dt_n\,\sigma_{t_n}(h)\ldots\sigma_{t_1}(h).$$

For the proof we refer to the references given at the end of this paper.

6.3. The Gibbs condition. The *interaction energy* $W_\Lambda(\Phi)$ between Λ and Λ^c for a given potential Φ is defined by

$$W_\Lambda(\Phi) = \sum_{\substack{X:X\cap\Lambda\neq\emptyset \\ X\cap\Lambda^c\neq\emptyset}} \Phi(X) \quad .$$

$W_\Lambda(\Phi)$ is finite for all potentials such that $\sum_{0\in X} \|\Phi(X)\| < \infty$.

DEFINITION 6.2: A state ρ satisfies the *Gibbs condition* if

i. The cyclic vector Ω_ρ is separating for $\pi_\rho(\mathfrak{A})''$;

ii. For each Λ the perturbed state ρ^h, with $h = -W_\Lambda(\Phi)$, is a product state of the form $\rho_\Lambda \otimes \phi_{\Lambda^c}$, where ρ_Λ is the Gibbs state for the region Λ and ϕ_{Λ^c} is a state of $\mathfrak{A}(\Lambda^c)$.

THEOREM 6.6: *For* $\Phi \in \mathcal{B}_f$, *with* $f(X) = \exp(|X|)$, *the Gibbs-condition and the KMS condition are equivalent.*

PROOF: Suppose ρ satisfies the KMS condition. By theorem 6.4, the perturbed state ρ^h, with $h = h^* \in \mathfrak{A}$, satisfies the KMS condition with respect to the perturbed dynamics α_t^h, where $\alpha_t^h(A) = U_t^h \alpha_t(A) U_t^{h-1}$ and

$$U_t^h = \sum_{n=0}^{\infty} i^n \int_0^t dt_1 \int_0^{t_1} dt_2 \ldots \int_0^{t_{n-1}} dt_n \, \alpha_{t_n}(h) \ldots \alpha_{t_1}(h).$$

From theorem 2.2 we know that, for $A \in \mathfrak{A}$,

$$\alpha_t(A) = \lim_{\Lambda\to\infty} e^{iH(\Phi,\Lambda)t} A \, e^{-iH(\Phi,\Lambda)t} \quad .$$

Hence, as can be checked easily,

$$U_t^h = \lim_{\Lambda\to\infty} \sum_{n=0}^{\infty} i^n \int_0^t dt_1 \ldots \int_0^{t_{n-1}} e^{iH(\Phi,\Lambda)t_n} h \, e^{-iH(\Phi,\Lambda)(t_n - t_{n-1})} h \ldots h \, e^{-iH(\Phi,\Lambda)t_1}$$

$$= \lim_{\Lambda\to\infty} e^{i(H(\Phi,\Lambda)+h)t} e^{-iH(\Phi,\Lambda)t} \quad .$$

If we now take $h = -W_\Lambda(\Phi)$ and assume that $\Lambda' \supset \Lambda$ we get

$$U_t^h = \lim_{\Lambda'\to\infty} e^{i(H(\Phi,\Lambda)+H(\Phi,\Lambda'/\Lambda)t} e^{-iH(\Phi,\Lambda')t}$$

and from this

$$\alpha_t^h(A) = \lim_{\Lambda'\to\infty} e^{i(H(\Phi,\Lambda)+H(\Phi,\Lambda'/\Lambda)t} A \, e^{-i(H(\Phi,\Lambda)+H(\Phi,\Lambda'/\Lambda)t}$$

$$= e^{iH(\Phi,\Lambda)t} \alpha_t^{\Lambda^c}(A) \, e^{-iH(\Phi,\Lambda)t} \quad .$$

Consider now $\rho^h(A_1A_2)$, with $A_1 \in \mathfrak{A}(\Lambda)$ and $A_2 \in \mathfrak{A}(\Lambda^c)$. For fixed A_2 $\rho^h(A_1A_2)$ is a linear form on $\mathfrak{A}(\Lambda)$ satisfying the KMS condition corresponding to the hamiltonian $H(\Phi,\Lambda)$. The only linear form having that property is the Gibbs state ρ_Λ. We conclude that

$$\rho^h(A_1A_2) = \rho_\Lambda(A_1)\cdot\phi(A_2),$$

so that ρ^h satisfies the Gibbs condition. To prove the reverse statement, let ρ satisfy the Gibbs condition. Then ρ^h, for $h = -W_\Lambda(\Phi)$, is the product state $\rho_\Lambda \otimes \phi$. Let σ_t^ρ and $\sigma_t^{\rho^h}$ be the modular automorphisms of $\pi\rho(\mathfrak{A})''$ corresponding to ρ and ρ^h respectively, with generators δ^ρ and δ^{ρ^h}. Then for $A \in \mathfrak{A}(\Lambda)$

$$\sigma_t^{\rho^h}(A) = e^{-iH(\Phi,\Lambda)t} A\, e^{iH(\Phi,\Lambda)t},$$

or

$$\delta^{\rho^h}(A) = -i[H(\Phi,\Lambda),A].$$

Since, on the other hand,

$$\sigma_t^\rho(A) = V_t^{h-1}\,\sigma_t^{\rho^h}(A)V_t^h,$$

we have

$$\delta^\rho(A) = \delta^{\rho^h}(A) + i[h,A].$$

Hence, for $A \in \mathfrak{A}_L$ and Λ sufficiently large

$$\delta^\rho(A) = -i[H(\Phi,\Lambda) + W_\Lambda(\Phi),A].$$

If δ^Φ is the generator of α_t^Φ, we have for $A \in \mathfrak{A}_L$

$$\delta^\Phi(A) = -i\lim_{\Lambda\to\infty}[H(\Phi,\Lambda),A],$$

or, since for $A \in \mathfrak{A}_L$

$$\lim_{\Lambda\to\infty}[W_\Lambda(\Phi),A] = 0,$$

$$\delta^\Phi(A) = -i\lim_{\Lambda\to\infty}[H(\Phi,A) + W_\Lambda(\Phi),A] = \delta^\rho(A),$$

for all $A \in \mathfrak{A}_L$. Since $\mathfrak{A}_L$ is a core of δ^Φ, we conclude that $\sigma_t^\rho(A) = \alpha_t^\Phi(A)$. □

6.4. Equivalence of the Gibbs condition and the variational principle. We already know that the variational principle implies the Gibbs condition (theorem 6.2 and 6.6). In this section we shall discuss the reverse.

THEOREM 6.7: *For translation invariant states and $\Phi \in \mathcal{B}_f$, with $f(X) = \exp(|X|)$ the Gibbs condition implies the variational principle.*

PROOF: We shall sketch the proof for the special case of finite range potentials. Assume $\rho \in I$ satisfies the Gibbs condition for $\Phi \in \mathcal{B}_0$. Let Λ and $\Lambda_0 \supset \Lambda$ be finite subsets of $\mathbb{Z}^\nu$ such that $W_\Lambda(\Phi) \in \mathfrak{A}(\Lambda_0)$. We now define: $\rho(\Lambda_0)$ is the density matrix corresponding to the restriction of ρ to $\mathfrak{A}(\Lambda_0)$; $\rho(\Lambda_0,h)$ is the density matrix of the state of $\mathfrak{A}(\Lambda_0)$ obtained from $\rho(\Lambda_0)$ by perturbing with $h \in \mathfrak{A}(\Lambda_0)$; $\rho(\Lambda_0)^{(\Lambda)}$ and $\rho(\Lambda_0,h)^{(\Lambda)}$ are their restrictions to $\mathfrak{A}(\Lambda)$, i.e., the partial traces with respect to $\mathcal{H}_{\Lambda_0/\Lambda}$. Since ρ is faithful, $\rho(\Lambda_0) > 0$. If we write $A(\Lambda_0) = -\log(\rho(\Lambda_0))$, we have

$$\rho(\Lambda_0) = e^{-A(\Lambda_0)}, \quad \rho(\Lambda_0,h) = \frac{e^{-(A(\Lambda_0)+h)}}{\mathrm{Tr}\, e^{-(A(\Lambda_0)+h)}}$$

and thus

$$\log\rho(\Lambda_0) - \log\rho(\Lambda_0,h) = h + \log(\mathrm{Tr}\, e^{-(A(\Lambda_0)+h)}) . \tag{6.4}$$

We now use the inequality (A.9) for $\rho(\Lambda_0)$, $\rho(\Lambda_0,h)$ and their restrictions to $\mathfrak{A}(\Lambda)$.

$$0 \leq \mathrm{Tr}^{(\Lambda)}(\rho(\Lambda_0)^{(\Lambda)} \log\rho(\Lambda_0)^{(\Lambda)}) - \mathrm{Tr}^{(\Lambda)}(\rho(\Lambda_0)^{(\Lambda)} \log\rho(\Lambda_0,h)^{(\Lambda)})$$

$$\leq \mathrm{Tr}^{(\Lambda_0)}(\rho(\Lambda_0)\log\rho(\Lambda_0)) - \mathrm{Tr}^{(\Lambda_0)}(\rho(\Lambda_0)\log\rho(\Lambda_0,h))$$

$$= \mathrm{Tr}^{(\Lambda_0)}(\rho(\Lambda_0)h) + \log \mathrm{Tr}^{(\Lambda_0)} e^{-(A(\Lambda_0)+h)} ,$$

where we used (6.4). By the Golden-Thompson inequality (A.10) we have

$$\log \mathrm{Tr}^{(\Lambda_0)} e^{-(A(\Lambda_0)+h)} \leq \log \mathrm{Tr}^{(\Lambda_0)}(e^{-A(\Lambda_0)} e^{-h}) \leq \log \| e^{-h} \| \leq \| h \| .$$

So we have

$$0 \leq \mathrm{Tr}^{(\Lambda)}(\rho(\Lambda_0)^{(\Lambda)} \log\rho(\Lambda_0)^{(\Lambda)}) - \mathrm{Tr}^{(\Lambda)}(\rho(\Lambda_0)^{(\Lambda)} \log\rho(\Lambda_0,h)^{(\Lambda)}) \leq 2\| h \| .$$

We now take the limit $\Lambda_0 \to \infty$. We see without difficulty that

$$\lim_{\Lambda_0\to\infty} - \mathrm{Tr}^{(\Lambda)}(\rho(\Lambda_0)^{(\Lambda)} \log\rho(\Lambda_0)^{(\Lambda)}) = S(\rho,\Lambda).$$

The second term is less trivial. Intuitively one would expect that

$$\lim_{\Lambda_0\to\infty} \rho(\Lambda_0,h) = \rho^h,$$

and this is indeed the case. The proof is rather involved and will be omitted. We have now the inequality

$$0 \leq -S(\rho,\Lambda) - \mathrm{Tr}^{(\Lambda)}(\rho^{(\Lambda)} \log \rho^{h(\Lambda)}) \leq 2\| h \| .$$

We take $h = -W_\Lambda(\Phi)$, and since ρ satisfies the Gibbs condition, we know that $\rho^{h(\Lambda)} = \exp(-H(\Phi,\Lambda))/\mathrm{Tr}\exp(-H(\Phi,\Lambda))$. This gives

$$0 \leq -S(\rho,\Lambda) + \rho(H(\Phi,\Lambda)) - F(\Phi,\Lambda) \leq 2\| W_\Lambda(\Phi) \| .$$

If we divide this by $|\Lambda|$ and take the limit $\Lambda \to \infty$ we obtain

$$f(\Phi) = \rho(A_\Phi) - S(\rho),$$

since $\lim |\Lambda|^{-1} \| W_\Lambda(\Phi) \| = 0$. □

6.5. Other equilibrium conditions. Various other equilibrium conditions have been proposed that are either equivalent with the KMS condition or imply this condition. The discussion will be brief. A more detailed account can be found in the papers by D. Kastler and G.L. Sewell in this volume.

We have seen in chapter 2 that the KMS condition is intimately related to the Gibbs condition. Since the temperature occurs explicitly in these conditions, they are truly thermodynamical conditions. Nevertheless, it is possible to derive the KMS condition, using only purely mechanical notions. The key condition is stability for local perturbations, which is defined as follows. In section 6.2 we introduced the perturbed dynamics α_t^h resulting from a perturbation or relative hamiltonian $h \in \mathcal{O}\!\mathcal{l}$. A stationary state ρ is called *stable for local perturbations* if there exists, for each $h = h^* \in \mathcal{O}\!\mathcal{l}$, a state ρ_h, stationary for the perturbed dynamics α_t^h, such that

i. $\displaystyle W^*\text{-}\lim_{h\to 0} \rho_h = \rho$

ii. $\displaystyle W^*\text{-}\lim_{t\to\pm\infty} \rho_h \circ \alpha_t = \rho.$

It can be shown that a stationary state ρ, that is stable for local perturbations and, in addition, satisfies a certain clustering property, is either a KMS state for some temperature or the spectrum of the generator H_ρ of U_t^ρ is one-sided. In the latter case ρ is ground state or ceiling state.

This result goes a long way toward answering one of the basic questions mentioned in the introduction. Sufficiently clustering stationary states are either thermodynamical equilibrium states, or they are unstable in the above sense.

In contrast to this purely mechanical condition, the following conditions are derived from thermodynamical considerations. The first of these is local thermodynamic stability (LTS), a local version of the variational principle, and is formulated in terms of conditional entropy and conditional free energy.

DEFINITION 6.3: The *conditional entropy* $\tilde{S}_\Lambda(\rho)$ of the state ρ relative to the region Λ is

$$\tilde{S}_\Lambda(\rho) = \lim_{\Lambda'\to\infty} [S(\rho,\Lambda') - S(\rho,\Lambda'/\Lambda)].$$

The *conditional free energy* is

$$\tilde{F}_\Lambda(\Phi,\rho) = \rho(H(\Phi,\Lambda) + W_\Lambda(\Phi)) - \tilde{S}_\Lambda(\rho).$$

DEFINITION 6.4: A state ρ of $\mathcal{O}\!\mathcal{l}$ satisfies the *LTS condition* if for each finite $\Lambda \subset \mathbb{Z}^\nu$ and for each state ψ of $\mathcal{O}\!\mathcal{l}$, for which $\psi|\mathcal{O}\!\mathcal{l}(\Lambda^c) = \rho|\mathcal{O}\!\mathcal{l}(\Lambda^c)$,

$$\tilde{F}_\Lambda(\Phi,\rho) \leq \tilde{F}_\Lambda(\Phi,\psi).$$

It has been shown that for potentials in $\mathcal{B}_f$, where $f(X) = \exp(|X|)$, the LTS condition and the KMS condition are equivalent.

Another condition, similar in spirit to the LTS condition, is that of passivity. It is a condition on the amount of energy transmitted to the system by a time-dependent local perturbation h_t. Suppose $h_t = 0$ except in the finite interval $[0,T]$ and let $L^h(\rho)$ be the work done by the external perturbation in that interval if the system was originally in the state ρ.

DEFINITION 6.5: The state ρ is *passive* if $L^h(\rho) \geq 0$ for all differentiable perturbations h_t that vanish outside $[0,T]$. This condition is clearly related to the second law of thermodynamics. Concerning passive states, there is the following result. If ρ is either a ground state, or a KMS state with positive β, then ρ is passive. On the other hand, a state $\rho \in \mathcal{E}(I)$ is either a ground state or a KMS state with positive β, if ρ is passive.

Apart from the conditions LTS and passivity that have a rather direct physical meaning, there are certain correlation inequalities for equilibrium states, that have been shown to imply equilibrium.

7. POTENTIAL SPACES

7.1. Potentials and observables. As long as we are interested in translation invariant states only it is useful to consider equivalence classes of observables that have the same expectation value for all states in I.

DEFINITION 7.1: $\mathcal{M}$ is the closed linear subspace of $\mathfrak{A}$ spanned by the set

$$\{\tau_x(A) - A \,;\, x \in \mathbb{Z}^\nu \,,\, A \in \mathfrak{A}\} \;.$$

One proves without difficulty that $\mathcal{M}$ is identical with the set $\{B \in \mathfrak{A} : \rho(B) = 0$ for all $\rho \in I\}$ and with the set $\{B \in \mathfrak{A} : \lim_{\Lambda\to\infty} |\Lambda|^{-1} \sum_{x\in\Lambda} \tau_x(B) = 0\}$.

Let X be the quotient space $\mathfrak{A}/\mathcal{M}$. For $x \in X$, the norm is defined by

$$\|x\| = \inf_{A\in x} \|A\| \,.$$

With this norm X is a Banach space. Concerning this norm we have the following

THEOREM 7.1: *For* $x \in X$, *we have*

$$\|x\| = \lim_{\Lambda\to\infty} |\Lambda|^{-1} \Big\| \sum_{x\in\Lambda} \tau_x(A)\Big\| \,,$$

for all $A \in x$.

PROOF: The uniqueness is proved as follows. If A_1 and $A_2 \in x$, then $A_1 - A_2 \in \mathcal{M}$; hence $\lim_{\Lambda\to\infty} |\Lambda|^{-1} \sum_{x\in\Lambda} \tau_x(A_1 - A_2) = 0$. To prove the existence of the limit, we choose $A \in \mathfrak{A}$ and define $F(\Lambda) = \|\sum_{x\in\Lambda} \tau_x(A)\|$. The function $\Lambda \to F(\Lambda)$ satisfies the conditions of proposition 4.10, and the limit of $|\Lambda|^{-1} F(\Lambda)$ exists for $\Lambda \to \infty$. The proof that this limit is indeed $\|x\|$, follows from the fact that, for $A \in \mathfrak{A}$, $|\Lambda|^{-1} \|\sum_{x\in\Lambda} \tau_x(A)\| \leq \|A\|$. □

It is easy to see that there is an isometry between the dual X^* of the Banach space X and the translation invariant elements of $\mathfrak{A}^*$. If Q is the canonical map of $\mathfrak{A}$ onto X, and $\tilde{\omega} \in X^*$ we have $\tilde{\omega}(QA) = \omega(A)$.

We shall now show that the Banach space $\mathcal{B}_1$ of potentials can be mapped isometrically on a subset of the selfadjoint elements of X. As we saw in section 2.2 there is for each $A \in \mathfrak{A}$ a function $X \to A(X) \in \mathfrak{A}(X)$ with the property that for each X (finite) the partial traces $\mathrm{Tr}^{(Y)} A(X) = 0$ for $Y \subset X$. We can define another norm

$$\|A\|_1 = \sum_X \|A(X)\|,$$

so that the set

$$\mathfrak{A}_1 = \{A \in \mathfrak{A} : \|A\|_1 < \infty\}$$

is a Banach space. The quotient space $X_1 = \mathfrak{A}_1/\mathcal{M}$ is a Banach space with its norm derived from the norm of $\mathfrak{A}_1$:

$$\|x\|_1 = \inf_{A \in x} \|A\|_1.$$

If $A \in \mathfrak{A}_1$ we define another element $\bar{A} \in \mathfrak{A}_1$ as follows.

$$\bar{A} = A(\emptyset) + \sum_{X \ni 0} \frac{\bar{A}(X)}{|X|},$$

where, for $X \neq \emptyset$, $\bar{A}(X) = \sum_x \tau_{-x}(A(X+x))$. It is easy to see that $\bar{A} = A \pmod{\mathcal{M}}$ and that

$$\begin{aligned}
&\text{i.} && \bar{A}(X) \in \mathfrak{A}(X);\\
&\text{ii.} && \mathrm{Tr}^{(Y)} \bar{A}(X) = 0, Y \subset X;\\
&\text{iii.} && \tau_x(\bar{A}(X)) = \bar{A}(X+x);\\
&\text{iv.} && \sum_{0 \in X} \frac{\|\bar{A}(X)\|}{|X|} < \infty.
\end{aligned} \tag{7.1}$$

Furthermore, $|\bar{A}|_1 = |A(\emptyset)| + \sum_{0 \in X} \frac{\|\bar{A}(X)\|}{|X|}$.

THEOREM 7.2: *For* $x \in X_1$, *we have*

$$\|x\|_1 = \|\bar{A}\|_1,$$

for all $A \in x$.

PROOF: The uniqueness follows from the fact that if $A_1 = A_2 \pmod{\mathcal{M}}$ then $\bar{A}_1 = \bar{A}_2$. For this it is sufficient to prove that for $C \in \mathcal{M}$ $\bar{C} = 0$. $\mathcal{M}$ is generated by elements of the form $C = \tau_x(D) - D$, for $D \in \mathfrak{A}$. $C(X) = \tau_x(D(X-x)) - D(X)$, and $\bar{C}(X) = \sum_y \tau_{-y} C(X+y) = \sum_y \tau_{x-y}(D(X-x+y)) - \sum_y \tau_{-y}(D(X+y)) = 0$. Furthermore,

$$\|\bar{A}\|_1 = |A(\emptyset)| + \sum_{X \ni 0} \frac{\|\bar{A}(X)\|}{|X|} \leq |A(\emptyset)| + \sum_{X \ni 0} \frac{1}{|X|} \sum_x \|A(X+x)\| = |A(\emptyset)|$$

$$+ \sum_x \sum_{X \ni x} \frac{\|A(X)\|}{|X|} = |A(\emptyset)| + \sum_{X \neq \emptyset} \|A(X)\| = \|A\|_1. \qquad \square$$

The properties (7.1) are precisely the same as the properties of $\Phi(X)$ in the definition 2.1 of potentials, given in section 2.2. We shall now define a one-to-one isometric mapping of $\mathcal{B}_1$ onto a subspace of X_1. For $\Phi \in \mathcal{B}_1$, $QA_\Phi \in X_1$, and $\|\Phi\|_1 = \|QA_\Phi\|_1$. Since for each $X \neq \emptyset$, $\tau(\Phi(X)) = 0$, we have $\tilde{\tau}(QA_\Phi) = \tau(A_\Phi) = 0$, where τ is the tracial state. Conversely, if $x = x^* \in X_1$, and $\tilde{\tau}(x) = 0$, we have for each $A \in x : \tau(A) = A(\emptyset) = 0$. We define the potential Φ_x as follows. $\Phi_x(X) = \bar{A}(X)$, for $A \in x$. Again we have $\|\Phi_x\|_1 = \|x\|_1$.

Having made the identification of potentials in $\mathcal{B}_1$ and traceless self-adjoint elements of X_1, we can give a meaningful extension of $\mathcal{B}_1$ to $\mathcal{B}$, where $\mathcal{B}$ can be identified with the traceless self-adjoint elements of X. We shall show that the variational principle is still valid in the space $\mathcal{B}$. We have, for $\Phi \in \mathcal{B}_1$

$$f(\Phi) = \min_{\rho \in I} (\rho(A_\Phi) - s(\rho)).$$

$\rho(A_\Phi)$ is defined for $\Phi \in \mathcal{B}$. Also $f(\Phi)$ can be extended to $\mathcal{B}$. Instead of proposition 4.2 one proves the stronger result

$$|\Lambda|^{-1} \, |F(\Phi_1,\Lambda) - F(\Phi_2,\Lambda)| \leq \|\Phi_1 - \Phi_2\|.$$

Hence, for $\Phi_1, \Phi_2 \in \mathcal{B}_1$, we have

$$|f(\Phi_1) - f(\Phi_2)| \leq \|\Phi_1 - \Phi_2\|.$$

This shows that f can be extended to $\mathcal{B}$ by continuity. We have proved an extension of theorem 5.2 to $\mathcal{B}$.

THEOREM 7.3: *For $\Phi \in \mathcal{B}$ we have*

$$f(\Phi) = \min_{\rho \in I} (\rho(A_\Phi) - s(\rho)).$$

The set I^Φ of states for which the minimum is reached is a non-empty closed convex set.

The identification of the potential space $\mathcal{B}$ with the space of equivalence classes of traceless observables allows a simple formulation of the problem mentioned at the end of section 1.1. Let $\Phi \in \mathcal{B}$ be a dynamical potential. The dynamics α_t^Φ commutes with translations which makes $\mathcal{M}$ stable under α_t^Φ. Let $\tilde{\alpha}_t^\Phi$ be the induced mapping of X onto itself.

DEFINITION 7.2: The *commutant of* Φ in $\mathcal{B}$ is the set of elements of $\mathcal{B}$ invariant under $\tilde{\alpha}_t^\Phi$.

Problem: Find the commutant of a given dynamical potential.

7.2. Physically admissible potentials. We have seen before that not all potentials in $\mathcal{B}$ are dynamical in the sense that they give rise to a dynamics of the system. We shall now present other arguments that $\mathcal{B}$ is too large.

We quote the following theorem on convex functions on a Banach space.

THEOREM 7.4: *If* P *is a continuous convex function on a Banach space* X, *then the tangent functionals to the graph of* P *are norm-dense in the set of* P-*bounded linear functionals.*

A linear functional ω *is* P-*bounded if there is a* $C \in \mathbb{R}$ *such that* $\omega(x) \leq P(x) + C$ *for all* $x \in X$.

We apply this to the concave function f on $\mathcal{B}$. Due to the variational inequality $f(x) \leq \tilde{\rho}(x) - s(\rho)$, for all $x \in \mathcal{B}$, $\tilde{\rho}$ is f-bounded.

THEOREM 7.5: *The set* $\bigcup_{\Phi \in \mathcal{B}} I^{\Phi}$ *is norm dense in the set* I.

As we shall see, this theorem has far-reaching consequences. As mentioned in chapter 3, a state $\rho \in I$ can be decomposed uniquely in extremal invariant states, i.e., there is a unique probability measure μ_ρ concentrated on $\mathcal{E}(I)$ and such that

$$\rho = \int \omega \, d\mu_\rho(\omega).$$

It can be shown that for two states ρ_1 and $\rho_2 \in I$ $\|\rho_1 - \rho_2\| = \|\mu_{\rho_1} - \mu_{\rho_2}\|$. Consider now the following situation. Let Φ_1 and Φ_2 be two different elements of $\mathcal{B}$ with a unique tangent plane. The corresponding equilibrium states ρ^{Φ_1} and ρ^{Φ_2} are in $\mathcal{E}(I)$. Let $\rho = \frac{1}{2}\rho^{\Phi_1} + \frac{1}{2}\rho^{\Phi_2}$. According to theorem 7.5 there is a potential Φ and a state $\tilde{\rho} \in I^{\Phi}$, such that $\|\rho - \tilde{\rho}\| < \varepsilon$. If $\mu_{\tilde{\rho}}$ is the decomposition of $\tilde{\rho}$ in ergodic states, we know from theorem 5.4 that $\mu_{\tilde{\rho}}$ is concentrated on $\mathcal{E}(I^{\Phi})$. Since $\|\mu_{\tilde{\rho}} - \mu_{\tilde{\rho}}\| < \varepsilon$, we conclude that for ε small enough, the states ρ^{Φ_1} and ρ^{Φ_2} will occur in the decomposition of $\tilde{\rho}$. This means that $\rho^{\Phi_1} \in \mathcal{E}(I^{\Phi})$ and the same for ρ^{Φ_2}.

We have here an example of states each of which is an equilibrium state for two different potentials. Since we can identify equilibrium states with tangent planes, we see that there are tangent planes that have two and therefore infinitely many tangent points.

DEFINITION 7.3: A concave function f on a Banach space X is *strictly concave* if for $x_1 \neq x_2$ and $\lambda \in (0,1)$,

$$f(\lambda x_1 + (1-\lambda)x_2) > \lambda f(x_1) + (1-\lambda) f(x_2).$$

As we have seen by the previous example the mean free energy f is not strictly concave on the Banach space $\mathcal{B}$. The fact that a state can be equilibrium for different potentials is physically undesirable and a consequence of allowing to large a space of potentials, as is borne out by the next theorem.

THEOREM 7.6: *The mean free energy* f *is strictly concave on* $\mathcal{B}_f$, *with* $f(X) = \exp(|X|)$.

PROOF: Let ρ be a tangent plane to f at the point $\Phi \in \mathcal{B}_f$. Then ρ satisfies the KMS condition w.r.t. α_t^{Φ} (theorem 6.2). Suppose there is another point $\Phi' \in \mathcal{B}_f$ such that ρ is tangent plane at Φ'. Then ρ is also a KMS state w.r.t.

$\alpha_t^{\Phi'}$. But the uniqueness of the modular group implies $\alpha_t^{\Phi} = \alpha_t^{\Phi'}$. It remains to show that then $\Phi = \Phi'$. For $A \in \mathfrak{A}(\Lambda_0)$, we have

$$\frac{d}{dt} \alpha_t^{\Phi}(A) \Big|_{t=0} = i \sum_X [\Phi(X), A].$$

We have completed the proof if we show that $\sum_X [\Phi(X), A] = 0$ for all $A \in \mathfrak{A}_L$ implies $\Phi = 0$. Taking partial traces of $\sum_X [\Phi(X), A] = 0$ with respect to regions outside Λ we obtain $\sum_{X \subset \Lambda} [\Phi(X), A] = 0$. Since this is true for all $A \in \mathfrak{A}(\Lambda)$ we get $\sum_{X \subset \Lambda} \Phi(X) = 0$, from which we conclude that $\Phi(X) = 0$ for $X \subset \Lambda$. □

There is an interesting application of this theorem to the quantum (or classical) lattice gas. For such models the potential consists of two terms $\Phi - \mu \Phi_1$, where $\Phi(X) = 0$ for $|X| < 2$ and Φ_1 is a one-point potential $\Phi_1(\{x\}) = \frac{1}{2}\sigma_3(x)$. We assume furthermore that $\Phi \in \mathcal{B}_f$ and that Φ and Φ_1 commute in the sense of definition 7.2. The pressure of the gas is then $p(\mu) = -f(\Phi - \mu \Phi_1)$, and the particle density $n = \frac{dp}{d\mu}$. Eliminating μ, one finds p as function of n. It is an easy consequence of theorem 7.6 that p is a continuous function of n. If for some value of μ p is not differentiable, the graph of p as function of n has a horizontal piece.

Let us recall that the space $\mathcal{B}_f$ discussed so far is the set of potentials such that $\|\Phi\|_f = \sum_{0 \in X} \|\Phi(X)\| f(X) < \infty$, where $f(X) = \exp(|X|)$. The function f has the effect of suppressing n-point interactions for large n. The decrease of $\Phi(X)$ as function of the distance between points may however be very slow. For two-point interactions the only condition is absolute summability:

$$\sum_x \|\Phi(\{x,0\})\| < \infty. \tag{7.2}$$

Recent studies indicate that restrictions on the decrease of $\Phi(X)$ with increasing $d(X)$ are necessary for a satisfactory behaviour of f in more phase regions. As expressed in lemma 5.5 the points in $\mathcal{B}_f$ for which f is not differentiable form a set of the first category (a countable union of nowhere dense sets). This is in agreement with an empirical rule, the so-called *Gibbs phase rule*, stating that points of coexistence of n phases are located on smooth hypersurfaces of codimension n, and that for Φ on such a hypersurface, there is a neighborhood of Φ such that f varies smoothly (analytically) and the number n of coexisting phases is constant if the potential moves in this neighborhood along the hypersurface of coexistence.

It has been shown that the spaces $\mathcal{B}$ and even $\mathcal{B}_f$, with $f(X) = \exp(|X|)$, do not meet these requirements. In $\mathcal{B}$ f is nowhere Fréchet differentiable and hence nowhere analytic, a result that is hardly surprising since $\mathcal{B}$ is too large anyhow. But even for the much smaller space of two-point potentials satisfying the condition (7.2) the outcome is negative. Explicit examples are known, where for a two-point potential Φ two phases are in equilibrium and f is not analytic in Φ if the potential moves along the coexistence surface.

The problem of finding a function g on the finite subsets of $\mathbb{Z}^\nu$ such that in the Banach space $\mathcal{B}_g$ the Gibbs phase rules apply, is as yet unsolved. The above investigations indicate that if such a space exists, g(X) must increase as a function of d(X).

APPENDIX. INEQUALITIES FOR FINITE MATRICES

A.1. *For* A *and* B *hermitian matrices,*

$$|\log \operatorname{Tr} e^A - \log \operatorname{Tr} e^B| \leq \|A - B\| .$$

PROOF: Let $A(x) = B + x(A - B)$; then $A(1) = A$ and $A(0) = B$. We have

$$\log \operatorname{Tr} e^A - \log \operatorname{Tr} e^B = \int_0^1 dx \frac{d}{dx} (\log \operatorname{Tr} e^{A(x)}) = \int_0^1 dx \frac{\operatorname{Tr} ((A - B) e^{A(x)})}{\operatorname{Tr} e^{A(x)}} ,$$

and hence

$$|\log \operatorname{Tr} e^A - \log \operatorname{Tr} e^B| \leq \|A - B\| . \qquad \square$$

A.2. *If* A *and* B *are hermitian matrices and* f *is real, convex on* $\mathbb{R}$,

$$\operatorname{Tr}(f(A) - f(B) - (A - B)f'(B)) \geq 0.$$

If f *is strictly convex, the equality implies* $A = B$.

PROOF: We notice first that the convexity of f does not imply that $A \to f(A)$ is convex. Let $\{\phi_i; a_i\}$ and $\{\psi_i; b_i\}$ be the set of eigenvectors and eigenvalues of A and B respectively. Then

$$(\phi_i, [f(A) - f(B) - (A - B)f'(B)]\phi_i) = (\phi_i, [f(a_i) - f(B) - (a_i - B)f'(B)]\phi_i)$$

$$= \sum_j |(\phi_i, \psi_j)|^2 [f(a_i) - f(b_j) - (a_i - b_j)f'(b_j)] \geq 0. \qquad \square$$

A.3. *For* A *and* B *positive matrices,*

$$\operatorname{Tr} A \log A - \operatorname{Tr} A \log B \geq \operatorname{Tr}(A - B) .$$

Equality implies $A = B$.

PROOF: Substitute $f(t) = t \log t$ in A.2. □

A.4. *For* ρ_1 *and* ρ_2 *density matrices,*

$$\operatorname{Tr} \rho_1 \log \rho_1 - \operatorname{Tr} \rho_1 \log \rho_2 \geq 0.$$

Equality implies $\rho_1 = \rho_2$.

PROOF: Immediate consequence of A.3. □

A.5. *(Peierls's inequality). For* A *a hermitian matrix and* $\{\phi_i\}$ *a, not necessarily complete, orthonormal set of vectors,*

$$\sum_i e^{(\phi_i, A\phi_i)} \leq \operatorname{Tr} e^A .$$

PROOF: It is sufficient to assume that $\{\phi_i\}$ is complete. Let $B = \sum_i (\phi_i, A\phi_i) P_{\phi_i}$, where P_{ϕ_i} is the projection operator on ϕ_i. We now apply A.2 for $f(t) = e^t$, and notice that

$$\operatorname{Tr}(A - B) f'(B) = \sum_i (\phi_i, (A - B) e^B \phi_i) =$$

$$= \sum_i (\phi_i, (A - (\phi_i, A\phi_i)) e^{(\phi_i, A\phi_i)} \phi_i) = 0. \qquad \square$$

A.6. *The function* $A \to \log \operatorname{Tr} A$ *from the hermitian matrices to* $\mathbb{R}$ *is an increasing convex function.*

PROOF: Let $0 \leq \lambda \leq 1$. Then, by A.5,

$$\operatorname{Tr} e^{A + (1-\lambda)B} = \sup_{\{\phi_i\}} \sum_i e^{\lambda(\phi_i, A\phi_i) + (1-\lambda)(\phi_i, B\phi_i)}$$

$$= \sup_{\{\phi_i\}} \sum_i (e^{(\phi_i, A\phi_i)})^{\lambda} (e^{(\phi_i, B\phi_i)})^{1-\lambda} ,$$

where the supremum is taken over all bases $\{\phi_i\}$. By Hölder's inequality

$$\sum_i (e^{(\phi_i, A\phi_i)})^{\lambda} (e^{(\phi_i, B\phi_i)})^{1-\lambda} \leq (\sum_i e^{(\phi_i, A\phi_i)})^{\lambda} (\sum_i e^{(\phi_i, B\phi_i)})^{1-\lambda} ,$$

so that

$$\operatorname{Tr} e^{A + (1-\lambda)B} \leq \sup_{\{\phi_i\}} (\sum_i e^{(\phi_i, A\phi_i)})^{\lambda} (\sum_i e^{(\phi_i, B\phi_i)})^{1-\lambda}$$

$$\leq (\sup_{\{\phi_i\}} \sum_i e^{(\phi_i, A\phi_i)})^{\lambda} (\sup_{\{\phi_i\}} \sum_i e^{(\phi_i, B\phi_i)})^{1-\lambda} = (\operatorname{Tr} e^A)^{\lambda} (\operatorname{tr} e^B)^{1-\lambda} .$$

Since $\log t$ for $t > 0$ is increasing, we conclude that

$$\log \operatorname{Tr} e^{\lambda A + (1-\lambda)B} \leq \lambda \log \operatorname{Tr} e^A + (1-\lambda) \log \operatorname{Tr} e^B .$$

That $\operatorname{Tr} e^A$ is increasing is proved as follows. Let $A \leq B$, then

$$\operatorname{Tr} e^A = \sup_{\{\phi_i\}} \sum_i e^{(\phi_i, A\phi_i)} \leq \sup_{\{\phi_i\}} \sum_i e^{(\phi_i, B\phi_i)} = \operatorname{Tr} e^B . \qquad \square$$

A.7. *The function* $\rho \to \operatorname{Tr}(\rho \log \rho)$ *is convex on the set of density matrices.*

PROOF: $\mathrm{Tr}((\lambda\rho_1 + (1-\lambda)\rho_2)\log(\lambda\rho_1 + (1-\lambda)\rho_2)) = \lambda\,\mathrm{Tr}(\rho_1 \log(\lambda\rho_1 + (1-\lambda)\rho_2)) + (1-\lambda)\,\mathrm{Tr}(\rho_2 \log(\lambda\rho_1 + (1-\lambda)\rho_2))$. By A.4 we have $\mathrm{Tr}(\rho_1 \log(\lambda\rho_1 + (1-\lambda)\rho_2)) \leq \mathrm{Tr}\,\rho_1 \log \rho_1$ and $\mathrm{Tr}(\rho_2 \log(\lambda\rho_1 + (1-\lambda)\rho_2)) \leq \mathrm{Tr}\,\rho_2 \log \rho_2$. Combining these we get the desired inequality. □

A.8. *If the matrices* A *and* B *satisfy* $0 < A \leq B$ *then* $\log A \leq \log B$.

PROOF: We shall prove first that $A^{-1} \geq B^{-1}$. Let $C = B^{-\frac{1}{2}} A B^{-\frac{1}{2}}$ then $C > 0$ and $\|C\| \leq 1$. Hence $\mathbb{1} \leq C^{-1} = B^{\frac{1}{2}} A^{-1} B^{\frac{1}{2}}$ or $B^{-1} \leq A^{-1}$. Using this, we find

$$\log A - \log B = -\int_0^\infty dx((A + x\mathbb{1})^{-1} - (B + x\mathbb{1})^{-1}) \geq 0. \qquad \square$$

A.9. *If* ρ_1 *and* ρ_2 *are two density matrices, the quantity*

$$S(\rho_1|\rho_2) = \mathrm{Tr}\,(\rho_1 \log \rho_1 - \rho_1 \log \rho_2)$$

is called the relative entropy. From A.4 we conclude that $S(\rho_1|\rho_2) \geq 0$.

If the matrix algebra $\mathfrak{A} = \mathfrak{A}_n \otimes \mathfrak{A}_m$, *with* $\mathfrak{A}_n(\mathfrak{A}_m)$ *the algebra of* $n \times n$ $(m \times m)$ *matrices and if for* $A \in \mathfrak{A}$ $A^{(n)}$ *denotes the partial trace of* A *with respect to* $\mathfrak{A}_m$:

$$A^{(n)} = \mathrm{Tr}^{(m)} A \in \mathfrak{A}_n,$$

then

$$0 \leq S(\rho_1^{(n)}|\rho_2^{(n)}) \leq S(\rho_1|\rho_2).$$

PROOF: See references [31,49].

A.10. *For* A *and* B *hermitian matrices*

$$\mathrm{Tr}\, e^{A+B} \leq \mathrm{Tr}(e^A e^B).$$

PROOF: See Golden [18] or Thompson [48].

NOTES AND REFERENCES

Additional information on the subject matter of this paper may be found in Ruelle [42] and the second volume of Bratteli and Robinson [10]. Another interesting general reference for lattice systems is Israel [26], which contains an extensive discussion of some of the questions raised in chapter 7. For classical systems, see also Ruelle [45].

Some classical papers that have had a profound influence on the development of C^*-algebraic methods to physics are Haag [20], Haag and Kastler [23], Araki and Woods [9].

Chapter 1. The KMS condition as a boundary condition for Green functions in statistical mechanics was introduced by Kubo, and subsequently used by Martin and Schwinger as a systematic way of calculating thermodynamic functions of equilibrium systems. As an equilibrium condition in the framework of C^*-algebras it was first discussed by Haag, Hugenholtz and Winnink [21].

Chapter 2. Theorem 2.2 is a slight generalization by Ruelle [42] of a result by Robinson [37]. The definition of potentials in this chapter is due to Griffiths and Ruelle [19] (see also Roos [39]). Approximately inner automorphism groups were introduced in [34].

Chapter 3. The notion of asymptotic abelianess and the theory of ergodic states are mainly due to Doplicher, Kastler, and Robinson [12,27] and to Ruelle [40]. Theorem 3.7 on the clustering of primary states was proven by Powers [33] and the somewhat stronger result in theorem 3.9 by Araki and Kishimoto [7]. The notion of a product state and proposition 3.11 are due to Powers [32]. Theorem 3.12 is an adaptation to quantum lattice systems of a theorem by Haag, Kadison, and Kastler [22].

Chapter 4. The theory of the variational principle for quantum lattice systems was developed by Robinson, Lanford e.a. as a quantum analogue of work on classical lattice systems by Ruelle, Robinson, Gallavotti and Miracle-Sole [17,41,38]. The content of this chapter is mainly due to Robinson [36] and to Lanford and Robinson [28]. For strong subadditivity of the entropy and the proof of the shape independence of the mean entropy see the papers by Lieb and Ruskai [30] and by Robinson and Ruelle [38].

Chapter 5. This chapter is based on the papers [36,29] by Robinson and Lanford.

Chapter 6. Section 6.1 is due to Lanford and Robinson [37,29]. The theory of the relative Hamiltonian, the Gibbs condition and its equivalence with KMS and the variational principle is due to Araki [1,2,3,4,5] and Araki and Ion [6]. The work on stability for local perturbation is due to Haag, Kastler and Trych-Pohlmeyer [24]. The local thermodynamic stability condition was introduced for classical systems by Sewell [46] and for quantum systems by Araki and Sewell [8,47]. Passivity was proposed by Pusz and Woronowicz [35]. For a discussion of various correlation inequalities see papers by Fannes and Verbeure [14,15,16].

Chapter 7. Section 7.2 is partly based on a paper by Israel [25]. Theorem 7.6 on the strict concavity of the mean free energy is due to Griffiths and Ruelle [19]. For a discussion of the Gibbs phase rule see Ruelle [43,44]. Daniëls and van Enter [11] and van Enter [13].

Acknowledgement. This paper is partly based on lectures given by the author while visiting the University of Pennsylvania during the academic year 1979-1980. The author wishes to express his sincere thanks to R.V. Kadison for helpful discussions and for his kind hospitality at the department of mathematics. Financial support by the National Science Foundation is also gratefully acknowledged.

BIBLIOGRAPHY

1. H. Araki, "Expansionals in Banach algebras", Ann. Sci. École Norm. Sup. 6 (1973), 67.

2. H. Araki, "Relative Hamiltonian for faithful normal states of von Neumann algebras", Publ. RIMS Kyoto Univ. 9 (1973), 165.

3. H. Araki, "Positive cone, Radon-Nikodym theorems, relative Hamiltonian and the Gibbs condition in statistical mechanics. An application of the Tomita-Takesaki theory", Proceedings of the International School of Physics "Enrico Fermi", Course LX (1973), North-Holland, Amsterdam, 1976, 64-100.

4. H. Araki, "On the equivalence of the KMS condition and the variational principle of quantum lattice systems", Commun. Math. Phys. 38 (1974), 1-10.

5. H. Araki, "On KMS states of a C^*-dynamical system", Proceedings second Japan-USA seminar, H. Araki and R.V. Kadison Ed., Springer, N.Y., 1978, 66-84.

6. H. Araki, and P.D.F. Ion, "On the equivalence of KMS and Gibbs conditions for states of quantum lattice systems", Commun. Math. Phys. 35 (1974), 1-12.

7. H. Araki, and A. Kishimoto, "On clustering property", RIMS - 198 (1976).

8. H. Araki, and G.L. Sewell, "KMS conditions and local thermodynamic stability of quantum lattice systems", Commun. Math. Phys. 52 (1977), 103-109.

9. H. Araki, and E.J. Woods, "Representations of the canonical commutation relations describing a non-relativistic free boson gas", J. Math. Phys. 4 (1963), 637-662.

10. O. Bratteli, and D.W. Robinson, Operator algebras and quantum statistical mechanics II, Springer, New York, 1979.

11. H.A.M. Daniëls, and A.C.D. van Enter, "Differentiability properties of the pressure in lattice systems", Commun. Math. Phys. 71 (1980), 65-76.

12. S. Doplicher, D. Kastler, and D.W. Robinson, "Covariance algebras in field theory and statistical mechanics", Commun. Math. Phys. 3 (1966), 1-28.

13. A.C.D. van Enter, "A note on the stability of phase diagrams in lattice systems", to appear in Commun. Math. Phys.

14. M. Fannes, and A. Verbeure, "Correlation inequalities and equilibrium states", Commun. Math. Phys. 55 (1977), 125-131.

15. M. Fannes, and A. Verbeure, "Correlation inequaltities and equilibrium states II", Commun. Math. Phys. 57 (1977), 165-172.

16. M. Fannes, and A. Verbeure, "Global thermodynamical stability and correlation inequalities", J. Math. Phys. 19 (1978), 558-570.

17. G. Gallavotti, and S. Miracle-Sole, "Statistical mechanics of lattice systems", Commun. Math. Phys. 5 (1967), 317-323.

18. S. Golden, "Lower bounds for the Helmholtz function", Phys. Rev. 137B (1965), 1127-1128.

19. R.B. Griffiths, and D. Ruelle, "Strict convexity ("continuity") of the pressure in lattice systems", Commun. Math. Phys. 23 (1971), 169-175.

20. R. Haag, "On quantum field theories", Dan. Mat. Fys. Medd. 29 (1955), no. 12.

21. R. Haag, N.M. Hugenholtz, and M. Winnink, "On the equilibrium states in quantum statistical mechanics", Commun. Math. Phys. 5 (1967), 215-236.

22. R. Haag, R.V. Kadison, and D. Kastler, "Asymptotic orbits in a free Fermi gas", Commun. Math. Phys. 33 (1973), 1-22.

23. R. Haag, and D. Kastler, "An algebraic approach to quantum field theory", J. Math. Phys. 5 (1964), 848-861.

24. R. Haag, D. Kastler, and E.B. Trych-Pohlmeyer, "Stability and equilibrium states", Commun. Math. Phys. 38 (1974), 173-193.

25. R.B. Israel, "Existence of phase transitions for long-range interactions", Commun. Math. Phys. 43 (1975), 59-68.

26. R.B. Israel, Convexity in the theory of lattice gases, Princeton Univ. Press, Princeton, 1979.

27. D. Kastler, and D.W. Robinson, "Invariant states in statistical mechanics", Commun. Math. Phys. 3 (1966), 151-180.

28. O.E. Lanford III, and D.W. Robinson, "Mean entropy of states in quantum statistical mechanics", J. Math. Phys. 9 (1968), 1120-1125.

29. O.E. Lanford III, and D.W. Robinson, "Statistical mechanics of quantum spin systems. III", Commun. Math. Phys. 9 (1968), 327-338.

30. E.H. Lieb, and M.B. Ruskai, "Proof of the strong subadditivity of quantum-mechanical entropy", J. Math. Phys. 14 (1973), 1938-1941.

31. G. Lindblad, "Entropy, information and quantum measurements", Commun. Math. Phys. 33 (1973), 305-322.

32. R.T. Powers, "Representations of the canonical anticommutation relations", Thesis Princeton Univ. (1967).

33. R.T. Powers, "Representations of uniformly hyperfinite algebras and their associated rings", Ann. Math. 86 (1967), 138-171.

34. R.T. Powers, and S. Sakai, "Existence of ground states and KMS states for approximately inner dynamics", Commun. Math. Phys. 39 (1975), 273-288.

35. W. Pusz, and S.L. Woronowicz, "Passive states and KMS states for general quantum systems", Commun. Math. Phys. 58 (1978), 273-290.

36. D.W. Robinson, "Statistical mechanics of quantum spin systems", Commun. Math. Phys. 6 (1967), 151-160.

37. D.W. Robinson, "Statistic mechanics of quantum spin systems. II", Commun. Math. Phys. 7 (1968), 337-348.

38. D.W. Robinson, and D. Ruelle, "Mean entropy of states in classical statistical mechanics", Commun. Math. Phys. 5 (1967), 288-300.

39. H. Roos, "Strict convexity of the pressure: a note on a paper of R.B. Griffiths and D. Ruelle", Commun. Math. Phys. 36 (1974), 263-276.

40. D. Ruelle, "States of physical systems", Commun. Math. Phys. 3 (1966), 133-150.

41. D. Ruelle, "A variational formulation of equilibrium statistical mechanics and the Gibbs phase rule", Commun. Math. Phys. 5 (1967), 324-329.

42. D. Ruelle, Statistical mechanics, W.A. Benjamin, New York, 1969.

43. D. Ruelle, "On manifolds of phase coexistence", Theoret. Math. Phys. 30 (1977), 24-29.

44. D. Ruelle, "A heuristic theory of phase transitions", Commun. Math. Phys. 53 (1977), 195-208.

45. D. Ruelle, Thermodynamical formalism, Addison-Wesley, London, 1978.

46. G.L. Sewell, "Equilibrium and meta-stable states of classical systems", Ann. Physics 97 (1976), 55-79.

47. G.L. Sewell, "KMS conditions and local thermodynamic stability of quantum lattice systems. II", Commun. Math. Phys. 55 (1977), 53-62.

48. C. Thompson, "Inequality with applications in statistical mechanics", J. Math. Phys. 6 (1965), 1812-1813.

49. H. Umegaki, Kodai Math. Sem. Rep. 14 (1962), 59-85.

Institute of Theoretical Physics of the
University, Groningen, Netherlands.

Proceedings of Symposia in Pure Mathematics
Volume 38 (1982), Part 2

DOES ERGODICITY PLUS LOCALITY IMPLY THE GIBBS STRUCTURE?

Daniel Kastler

ABSTRACT. One first reviews the main results obtained thus far in the algebraic statistical mechanics of closed systems. The dynamically stable "pure phases" of an asymptotically abelian C^*-system are either ground states or Gibbs (i.e. KMS) states for a mixture of time and gauge, the concept of temperature evolving from stability, that of chemical potential from extending stable states from the observable algebra to the field algebra (or from testing their behaviour under localized morphisms). Future prospects: deeper explanation of the KMS structure from locality and a "quantum ergodicity" of the dynamics; and study of the relation of the degree of (global) stability versus fall off of correlations (clustering).

1. INTRODUCTION. Mathematical physics pursues two types of aims: on the one hand the construction of specified models for specified physical systems; on the other the explanation of generic features of whole classes of physical systems from general properties common to all systems of the class. Operator algebras can be applied to either of these pursuits. And, at the moment, there are indications that we might stand at the beginning of a constructive era, with parallelism in mathematics and physics. This talk however is devoted to past work of the generic type, centering around the question: why is it that for widely different physical systems one observes a common "Gibbs structure" of their equilibrium states? What are the common properties of these systems which are responsible for the occurrence of the Gibbs structure? Since, as we shall see, the proper mathematical expression of the Gibbs structure is the KMS property, this problem lies within a canonical mathematical frame, indeed within the context of a mainstream of mathematical progress in the past decade in operator algebras.

Before going into our question, I want to remind you of the general frame of algebraic quantum field theory in the relativistic case. The latter consists of the following general assumptions:

(i) to each region (= bounded open set) $\mathcal{O} \subset R^4$ there is a local C^* algebra $\mathfrak{A}_{\mathcal{O}}$ with embeddings $\mathfrak{A}_{\mathcal{O}_1} \hookrightarrow \mathfrak{A}_{\mathcal{O}_2}$ corresponding coherently to the

0082-0717/80/0000-0540/$06.75

embeddings $\mathcal{O}_1 \subset \mathcal{O}_2$. The C^* - inductive limit $\mathfrak{A} = C^*\lim_{\rightarrow} \mathfrak{A}_{\mathcal{O}}$ is called the quasilocal algebra.

(ii) for regions $\mathcal{O}_1$ and $\mathcal{O}_2$ lying spatially to each other the subalgebras $\mathfrak{A}_{\mathcal{O}_1}$ and $\mathfrak{A}_{\mathcal{O}_2}$ of $\mathfrak{A}$ commute (locality property)

(iii) The Poincare group $\mathcal{P}$ acts continuously and covariantly on $\mathfrak{A}$ (i.e. one has a group homomorphism $\Lambda \in \mathcal{P} \rightarrow \alpha_\Lambda \in \mathrm{Aut}\,\mathfrak{A}$ with $\Lambda \rightarrow \alpha_\Lambda(A)$ norm continuous for every $A \in \mathfrak{A}$ and $\mathfrak{A}_{\alpha_\Lambda(\mathcal{O})} = \alpha_\Lambda(\mathfrak{A}_{\mathcal{O}})$ for all $\Lambda \in \mathcal{P}$ and all regions $\mathcal{O}$).

In fact the axiomatic frame needed in this talk is merely the existence of a representation $t \in \mathbb{R} \rightarrow \alpha_t \in \mathrm{Aut}\,\mathfrak{A}$ of the translations in time satisfying the assumption of asymptotic abelianness, i.e. the requirement

$$[A,\alpha_t(B)] \xrightarrow[t=\infty]{} 0 \tag{1}$$

for an appropriate topology. Physically, it is clear that (i) should proceed from (ii) and (iii) via the fact that time evolution produces a spreading in space. However the detailed mechanism of this is not clear and we will here take the asymptotic abelian property (i) as a basic assumption.

Let us mention briefly the general philosophy of algebraic field theory: we consider the physical laws to be expressed by the local and the automorphic structure of $\mathfrak{A}$; whilst the states of $\mathfrak{A}$ (along with the representations they generate) entail the description of the physical situations of the system. It is worthwhile in this connection to note the following alternative: a simple separable C^* - algebra $\mathfrak{A}$, is either the compacts on separable Hilbert space (with its unique representation)[1], or an antiliminar algebra with a maze of inequivalent representations of all types I, II and III. Since the C^* - algebras of physically interesting systems are presumably simple[2], we observe here the preexistence, in mathematics, of the distinction encountered in quantum physics between ordinary quantum mechanics on the one side (systems with finitely many degrees of freedom with uniqueness of representation, cf. the Mackey-Stone-Von Neumann uniqueness theorem), and field theory on the other (infinite quantum systems with a continuum of inequivalent representations reflecting the enormous variety of diverse physical situations of the system).

In this talk we deal with the C^* - algebra $\mathfrak{A}$ of an infinite system and focus on the states of $\mathfrak{A}$ which are both simplest and most basic physically, namely the time-invariant or stationary states. Our frame is then a C^* - system $(\mathfrak{A},\mathbb{R},\alpha)$ with $\mathbb{R}$ the group of time translations (generally a C^* - system $(\mathfrak{A},G,\alpha)$ is the triple of a C^* - algebra $\mathfrak{A}$, a locally

---

[1] up to quasiequivalence.

[2] indeed, there are no exact "quotient structures" of physics.

compact group G and a representation α of G as automorphisms of $\mathfrak{A}$, with $g \in G \to \alpha_g(A)$ continuous in norm for each $A \in \mathfrak{A}$).

Our first task consists in drawing the consequences of asymptotic abelianness for the system's stationary states

2. INVARIANT STATES OF ASYMPTOTICALLY ABELIAN C^* - SYSTEMS.

DEFINITION 1. Let ω be an α-invariant state of the C^*-system $(\mathfrak{A},\mathbb{R},\alpha)$: ω is clustering whenever

$$(2) \qquad \omega(A\alpha_t(B)) - \omega(A)\omega(B) \underset{t=\infty}{\longrightarrow} 0, \ A,B \in \mathfrak{A};$$

and weekly clustering whenever

$$(3) \qquad \frac{1}{2T}\int_{-T}^{+T} \omega(A\alpha_t(B))\,dt - \omega(A)\omega(B) \underset{T=\infty}{\longrightarrow} 0, \ A,B \in \mathfrak{A}.$$

THEOREM 1.[1)] Let ω be an α-invariant state of the C^*-system $(\mathfrak{A},\mathbb{R},\alpha)$ with GNS construction $(\pi_\omega, U_\omega, \mathfrak{H}_\omega, \xi_\omega)$. Assume asymptotic abelianness in the sense

$$(4) \qquad \|\pi_\omega\{[A,\alpha_t(B)]C\}\ \xi_\omega\| \underset{|t|=\infty}{\longrightarrow} 0, \ A,B,C \in \mathfrak{A},$$

where $[\ ,\ ]$ denotes a commutator; and set

$$(5) \qquad \begin{cases} F_{AB}(t) = \omega(B\alpha_t(A)) \\ G_{AB}(t) = \omega(\alpha_t(A)B), \end{cases} \quad A,B \in \mathfrak{A},\ t \in \mathbb{R}.$$

The following conditions (i), (ii), (iii), (iv) are equivalent

(i) ξ_ω spans the U_ω-invariant subspace of $\mathfrak{H}_\omega$

(ii) ω is weakly clustering

(iii) ω is an extremal element of the convex set of α-invariant states of $\mathfrak{A}$

(iv) ω generates an irreducible representation of $\mathfrak{A} \times_\alpha \mathbb{R}$ (in other terms $\pi_\omega(\mathfrak{A})$ and $U_\omega(\mathbb{R})$ taken together are irreducible).

Moreover one has the following "spectral alternative": denoting by $S = \mathrm{Sp}U_\omega$ the spectrum of the representation U_ω:

(a) either S is one-sided (i.e. $S \wedge (-S) = \emptyset$); in that case ω is a pure state.

(b) or $S = \lambda\,\mathbb{Z}$ for some $\lambda \in \mathbb{R}$; in that case $\pi_\omega(\mathfrak{A})$ is abelian.

(c) or $S = \mathbb{R}$; in that case the sets $\mathfrak{F} = \{F_{AB}; A,B \in \mathfrak{A}\}$ and $\mathfrak{G} = \{G_{AB}; A,B \in \mathfrak{A}\}$ are both weak * total in $B(\mathbb{R}) = \mathfrak{M}_1(\hat{\mathbb{R}})$.

If ω is assumed clustering, case (b) is excluded.

---

1) Parts (i) (ii) (iii) and (a) (b) (c) of this theorem both hold under more general conditions.

The first half of this theorem characterizes the extremal α- invariant states (the "pure phases" of the physical systems) as the states whose correlations vanish with time in the mean (weak clustering). The second half establishes in fact a dichotomy, case (b) being excluded in physics by our belief that the (non abelian) C^*-algebra $\mathfrak{A}$ is simple: the statement becomes thus that the energy spectrum is either one-sided, as is the case for ground (zero-temperature) states; or that it covers the whole reals, as is the case for temperature states. In this sense the distinction between zero and finite temperature to some extent already arises from asymptotic abelianness. At this point, however, the concept of temperature is not yet evolved: our next step (section 2 below) will consist in developing this concept. For this however we need first to state the KMS property.

3. THE KMS PROPERTY AS THE NOVEL FORMULATION OF THE GIBBS ANSATZ. The "Gibbs' Ansatz", base of the traditional quantum statistical mechanics, is a successful recipe (due to Gibbs and Boltzmann) for obtaining the equilibrium state $\omega_{\beta,\mu}$ to the temperature $T = (k\beta)^{-1}$ and chemical potential μ : The instruction reads :

$$(6)\qquad \omega_{\beta,\mu}(A) = \frac{\mathrm{Tr}\{e^{-\beta(H-\mu N)}A\}}{\mathrm{Tr}\{e^{-\beta(H-\mu N)}\}}\,, \qquad A \in \mathfrak{A}.$$

Here H is the hamiltonian, and N the particle-number operator of the quantum system under consideration. The formula as it stands does not yet specify equilibrium thermodynamics for the following reason: for (6) to make sense the operator $e^{-\beta(H-\mu N)}$ must be trace class, whence the need to use box quantization (we have to make the spectrum of $H - \mu N$ artificially discrete). Since, however, the system in a box is in essence antithermodynamical (for instance it does not fulfill return to equilibrium) one has to supplement Ansatz (6) with the prescription of "thermodynamical limit" : in fact the state $\omega_{\beta,\mu}$ is defined as the limit of (6) for a box tending to ∞ (with β and μ kept constant). Though numerically invaluable Ansatz (6) is thus inadequate for foundational purposes. We shall replace it by a statement entailing the same information but valid for "infinite systems" (without recourse to the thermodynamic limit): the KMS (Kubo-Martin-Schwinger) principle which we now proceed to state.

DEFINITION 2. Let ω be a σ-invariant[1)] state of the C^*-system $(\mathfrak{A},\mathbb{R},\sigma)$ and let the corresponding "two point functions" F_{AB}, G_{AB}, $A,B \in \mathfrak{A}$ be as in (5) with α replaced by σ. We say that ω is β-KMS for σ, $\beta \in \mathbb{R}$, whenever, to all $A,B \in \mathfrak{A}$, there is a function u_{AB} of the complex variable

---

[1)] σ-invariance could be omitted from this definition, since it follows from (7) below.

continuous in the strip $0 \leq \mathrm{Im} z \leq \beta$, holomorphic in its interior, such that

$$F_{AB}(t) = u_{AB}(\beta i + t),\ G_{AB} = u_{AB}(t),\ t \in \mathbb{R}. \tag{7}$$

Alternative formulations of this condition are as follows: instead of (7) we could require

$$\hat{F}_{AB}(p) = e^{\beta p}\hat{G}_{AB}(p),\quad A,B \in \mathfrak{A}, \tag{8}$$

(understood as a relation between distributions)[1]. We could instead also require that

$$\omega(B\sigma_{\beta i}(A)) = \omega(AB),\quad A,B \in \mathfrak{A}_{an} \tag{9}$$

where $\mathfrak{A}_{an}$ is the set of σ-analytic elements of $\mathfrak{A}$. (8) is just the Fourier translation of (7), which in turn results from (9) replacing there A by $\sigma_t(A)$ and using the density of $\mathfrak{A}_{an}$ in $\mathfrak{A}$.

We are now able to state the <u>KMS principle</u>. Let $\mathfrak{A}$ be the given algebra (of fields) acted upon by the commuting groups $\mathbb{R}$ and $\mathcal{G}$ of time translations, resp. <u>gauge of the first kind</u> (i.e. we have a C^*-system $(\mathfrak{A}, \mathbb{R} \times G, \alpha \times \gamma)$, $\mathbb{R} \times G$ the direct product of the reals and the compact <u>gauge group</u> G)[2]. Equilibrium states ω to the (<u>inverse</u>) <u>temperature</u> β and the (<u>generalized) chemical potential</u> $\mu, \mu \in \mathfrak{G}$, the Lie algebra of G, fulfill the β-KMS property for σ, with

$$\sigma_t = \alpha_t \gamma_{\xi_t},\quad t \in \mathbb{R}. \tag{10}$$

where $t \to \xi_t$ denotes the one-parameter subgroup of G generated by μ.

In the usual case (one species of particles) the gauge group G reduces to the one dimensional torus T^1 with Lie algebra $\mathbb{R}$, thus the chemical potential is a real number. In that case we have $\xi_t = \mu t$, thus

$$\sigma_t = \alpha_t \gamma_{\mu t},\quad t \in \mathbb{R}, \tag{11}$$

and it is readily checked that the Gibbs Ansatz (6) implies the KMS-principle: indeed the time translation - resp. gauge-automorphism groups are then defined by their infinitesimal generators H resp - N[3]:

$$\alpha_t(A) = e^{iHt}Ae^{-Ht} \tag{12}$$

$$\gamma_t(A) = e^{-iNt}Ae^{iNt} \tag{13}$$

$$t \in \mathbb{R},\ a \in \mathfrak{A},$$

---

[1] in fact between measures since ω is then automatically α-invariant.

[2] we state the "KMS principle" in the generalized frame of an arbitary compact gauge group. The usual case corresponds to $G = T^1$ (see below).

[3] We remind that (6) is stated for a concrete C^*-algebra $\mathfrak{A}$.

therefore (11) means that

$$\sigma_t(A) = e^{i(H - \mu N)t} A e^{-i(H - \mu N)t} \tag{14}$$

so that (9) immediately follows from (6) using commutativity under the trace. The fact that the KMS principle contains all the information carried by the Gibbs Ansatz conversely follows from the known form of the (then inner) modular automorphisms for finite Von Neumann algebras. Gibbs' Ansatz and KMS principle are thus equivalent whenever the former can be formulated. The latter however applies without the restrictive condition that the Von Neumann algebra[1] be finite-and this is exactly what we need in order to encompass infinite quantum systems[2].

Having now adequately described the structure to be explained from first principles, we proceed to describing the explanation. This will be effected in two steps, which will respectively lead us to the concept of temperature and to that of (a generalized) chemical potential. These steps will correspond to the splitting of the KMS principle into the two following parts I and II. Denote by $\mathfrak{A}^G$ (the <u>algebra of observables</u>)[3] the gauge invariant part of the field algebra $\mathfrak{A}$:

$$\mathfrak{A}^G = \{A \in \mathfrak{A};\ \gamma_g(A) = A \quad \text{for all} \quad g \in \mathcal{G}\}. \tag{15}$$

and observe that owing to the commutativity of time and gauge, $\mathfrak{A}^G$ is globally invariant for α. The above KMS principle then splits into:

I. Equilibrium states in restriction to $\mathfrak{A}^{\mathcal{G}}$ are states of the C^*-system $(\mathfrak{A}^{\mathcal{G}}, R, \alpha)$ which fulfill the β-KMS condition for the time translation automorphism group α.

II. Equilibrium states on $\mathfrak{A}$ fulfill the β-KMS condition for a 1-parameter group σ of the type (10) (mixture of time and gauge)

(It should be clear that the KMS principle reduces to Principle I on $\mathfrak{A}^G$ on which the gauge group acts trivially and hence σ reduces to α)

4. THE NOTION OF TEMPERATURE AS INFERRED FROM THE DYNAMICAL STABILITY OF THE EQUILIBRIUM STATES.

One of the ways in which Principle I above can be drawn from first principles[4] is a derivation from a postulated dynamical stability of the equilibrium states. Specifically we have:

---

1) weak closure of the representation of $\mathfrak{A}$ generated by ω.

2) for which the Von Neumann algebra is a type III factor (see section 6 below).

3) The non gauge-invariant "fields" in $\mathfrak{A}$ are not observable (for instance, in the usual case of $\mathcal{G} = T^1$, their phase is inobservable).

4) For other derivations of the KMS-property from postulated properties of the equilibrium states, see the report of G.L. Sewell in these Proceedings.

THEOREM 2. Let $(\mathfrak{A}^G,\mathbb{R},\alpha)$ be a C^*-system for which we assume asymptotic abelianness in the following sense: there is a dense, α-invariant * subalgebra $\mathfrak{A}_{loc}$ of $\mathfrak{A}^G$ such that[1])

(16) $$\{t \to \|[A,\alpha_t(B)]\|\} \in L^1(\mathbb{R}), \quad A,B \in \mathfrak{A}_{loc}$$

Further let ω be an α-invariant state of $\mathfrak{A}^G$ and assume that

(i) ω is hyperclustering of order 4 on $\mathfrak{A}_{loc}$

(ii) ω is stable under inner (= local) perturbations of the dynamics.

Then if ω is not a trace state, it is either a ground state or a β-KMS state for α, with β a real number.

We need to define the specifications (i) and (ii) in the above statement.

DEFINITION 3. An α-invariant state ω of a C^*-system $(\mathfrak{A}^G,\mathbb{R},\alpha)$ is <u>hyperclustering of order</u> p whenever all its truncated expectation values $\omega_p^T(\alpha_{t_1}(A_1),\dots,\alpha_{t_p}(A_p))$ up to order p are L^1 in all differences of their arguments.[2]) (This concept is a strengthening of the clustering property (2), in that it requires L^1-type decrease and extends to correlations of order 4.)

For our next definition we need to recall the notion of <u>inner perturbation</u> α^h <u>by</u> $h = h^* \in \mathfrak{A}$ <u>of a 1-parameter automorphism group</u> α <u>of a C^*-algebra</u> $\mathfrak{A}$: the perturbed 1-parameter group α^h can either be defined by the differential equation

(16) $$i\,\frac{d}{dt}\Big|_{t=0}\alpha_t^h(A) = i\,\frac{d}{dt}\Big|_{t=0}\alpha^t(A) + [h,A], \qquad A \in \mathfrak{A}_{diff}$$

(required for all α-differentiable elements A in $\mathfrak{A}$; $[h,A]$ denotes a commutator); or globally as

(17) $$\alpha_t^h(A) = P_t^h\alpha_t(A)P_t^{h^*}, \qquad A \in \mathfrak{A},$$

where the function $P^h\colon \mathbb{R} \to \mathfrak{A}$ is specified by

(18) $$i\,\frac{dP_t^h}{dt} = P_t^h\,\alpha_t(h), \quad P_o^h = \mathbb{1}\,.$$

One checks that P_t^h is a unitary 1-cocycle in the sense that

(19) $$\begin{cases} P_{s+t}^h = P_s^h\,\alpha_s(P_t^h) \\ P_t^{h*} = (P_t^h)^{-1} = \alpha_t(P_{-t}^h) \end{cases}, \quad s,\ t \in \mathbb{R},$$

a property which ensures that (17) defines an automorphism group α^h of $\mathfrak{A}$.

---

1) $\mathfrak{A}_{loc}$ is obtained as the algebra of strictly local elements.

2) For the definition of the latter see e.g. [47]. For p = 2 $\omega_p^T(\alpha_{t_1}(A),\alpha_{t_2}(B))$ equals the l.h.s. of (2) where $t = t_1 - t_2$.

We shall apply the notion of inner perturbations α^h to the observable algebra $\mathfrak{A}^G$ with its dynamical group α: such a perturbation than represents a local alteration of the dynamics since it corresponds (as shown by (16)) to "adding $h = h^* \in \mathfrak{A}$ as an increment to the hamiltonian", h being is a local quantity according to our interpretation of $\mathfrak{A}^G$ ($\mathfrak{A}^G$ is considered as representing physically the "(quasi) local observables" cf. Introduction (i)). We can now explain what we mean by local stability of a stationary state

DEFINITION 4. Let ω be an α-invariant state of the algebra $\mathfrak{A}^G$ of quasilocal observables. We say that ω is stable under local (or inner) perturbation of the dynamics whenever there is a map $h \to \omega^h$ from a neighborhood of zero in the self-adjoint part of $\mathfrak{A}^G$ into the latter's state space such that

(i) $\omega^h \circ \alpha_t^h = \omega^h$

(ii) $\omega^{\lambda h}(A) \xrightarrow[\lambda=0]{} \omega(A), \quad A \in \mathfrak{A}^G$

(iii) $\omega^h(\alpha_t(A)) \xrightarrow[t=\pm\infty]{} \omega(A), \quad A \in \mathfrak{A}^G.$

The physical interpretation of these requirements is clear: (i) means that the "perturbed state" a^h is stationary for the perturbed dynamics, (ii) requires ω^h to lie near ω for small h, (iii) is a condition of "return to equilibrium" (after removal of the perturbation). All of these conditions are physically cogent for an equilibrium state ω (except perhaps condition (iii) for $t \to -\infty$ which is a sort of "time reversal" requirements). Theorem 2 then appears as offering a basic physical motivation for Principle I above. Another physical motivation, much less demanding w.r.t. clustering, is afforded by

THEOREM 3. The conclusion of Theorem 2 remains valid for $\mathfrak{A}^G$ simple nonabelian if assumptions (i) resp. (ii) are replaced by

(i a) ω is weakly clustering (i.e. extremal invariant cf. Theorem 1 (ii), (iii))

(ii a) ω is coexistent in the sense of

DEFINITION 5. Let $(\mathfrak{A}, \mathbb{R}, \alpha)$ be an asyptotically abelian system and let ω be an α-invariant state of $\mathfrak{A}$: ω is coexistent whenever for any other asymptotically abelian system $(\mathfrak{A}', \mathbb{R}, \alpha')$ and α'-invariant state ω' of $\mathfrak{A}'$ the product-state $\omega \otimes \omega'$ of the product-system $(\mathfrak{A} \otimes \mathfrak{A}', \mathbb{R}, \alpha \otimes \alpha')$ is stable for inner perturbations of the product-system dynamics.

While the stability encountered in Theorem 2 is of the type "contamination by an impurity", that of Theorem 3 refers to stability against local perturbations arising from neighbouring systems.

We now turn to our explanation of the concept of chemical potential (Principle II above). It is gratifying that (granted one has explained the KMS structure for ω on $\mathfrak{A}^G$) one does not need for this any additional physical input.

5. THE NOTION OF (GENERALIZED) CHEMICAL POTENTIAL AS DERIVED FROM EXTENDING STATES FROM $\mathfrak{A}^G$ TO $\mathfrak{A}$.

THEOREM 4. Let $\{\mathfrak{A}, \mathbb{R} \times G, \alpha \times \gamma\}$ be a C^*-system, with G a compact group. Assume asymptotic abelianness in the sense that

$$\|[A,\alpha_t(B)]\| \xrightarrow[|t|=\infty]{} 0, \quad A,B \in \mathfrak{A}. \tag{20}$$

where [ ] denotes a commutator[1]. And denote by $\mathfrak{A}^G$ the fixed-point subalgebra of $\mathfrak{A}$ under the action γ of G. Then

(i) each extremal α-invariant state ω_o of $\mathfrak{A}^G$ possesses an extremal α-invariant extension ω to $\mathfrak{A}$. Moreover any other such extension is of the form $\omega \circ \gamma_g$ for some $g \in G$.

(ii) Assume now the state ω_o of $\mathfrak{A}^G$ to be extremal β-KMS for α [2] and faithful in the sense that $\omega_o(A^*A) = o$, $A \in \mathfrak{A}$, entails $A = 0$ [3]. Each of the extremal α-invariant extensions ω of ω_o to $\mathfrak{A}$ is then β-KMS for a one-parameter group of $\mathfrak{A}$ of the type $t \to \alpha_t\gamma_{\xi_t}$ with $t \to \xi_t$ a continuous one-parameter subgroup of the center of the stabilizer G_ω of ω in G:

$$G_\omega = \{g \in G;\ \omega \circ \gamma_g = \omega\} \tag{21}$$

REMARK 1. In the usual case $G = T^1$, G_ω is either the whole T^1 or discrete, in which case it cannot accommodate a one-parameter subgroup: thus the chemical potential vanishes whenever the extension ω is not gauge-invariant.

REMARK 2. If ω_o is not faithful (possible if $\mathfrak{A}^G$ is not simple as e.g. the gauge-invariant part of the CAR algebra) there is a possibility that ω be "vacuum like" in certain directions of gauge space cf. [43].

REMARK 3. The β-KMS property of the extremal invariant state ω_o causes the latter to be primary, and thus clustering (primary states of asymptotically invariant systems are clustering). However, Theorem 4

---

[1] We state this Theorem in the version corresponding to Bose fields. For Fermi fields one would have to assume $\{\mathfrak{A},\mathbb{R},\alpha\}$ to be asymptotically Grassmannian cf. [43].

[2] One shows that extremal invariant β-KMS states are extremal β-KMS. Thus this assumption for the state ω_o corresponds to the output of Theorems 2 and 3 above.

[3] Observe that faithfulness of ω_o is automatic if the algebra $\mathfrak{A}^G$ is simple, since the KMS nature of ω_o causes the kernel of π_{ω_o} to coincide with the left kernel of ω_o.

(ii) requires only that the extension ω be extremal invariant, i.e. weakly clustering (cf. Theorem 1, (ii), (iii)). It actually happens that the extension ω be only weakly clustering, thus nonprimary. In this case its central measure, which is extremal α-invariant [1)] is a transitive ergodic measure due to the compactness of G. The corresponding primary central components are still β-KMS for $t \to \alpha_t \gamma_{\xi_t}$ (central decomposition commutes with the KMS structure) however they are no longer α-invariant. Such a "breaking of time invariance", with periodicity of the extension ω w.r.t. time, is observed in nature (Josephson effect) and in models (Einstein-condensed thermodynamic limit of the free Bose gas under certain boundary conditions [57].

REMARK 4. Theorem 4 assumes invariance and asymptotic abelianness w.r.t. time. One might be interested in situations (e.g. spin systems) where similar assumption held for the space translations (cf. [44]).

Theorem 4 assumes the existence of a field algebra $\mathfrak{A}$ with an action of a gauge group yielding the observable algebra $\mathfrak{A}^G$ as its fixed point algebra. However, fundamentally, $\mathfrak{A}^G$ is the object of primary physical interest and is correspondingly the basic mathematical entity from which the field algebra $\mathfrak{A}$ can be constructed, cf. the algebraic theory of superselection sectors). One thus may wish to develop the concept of the chemical potential uing $\mathfrak{A}^G$ alone without recourse to the field algebra $\mathfrak{A}$. This is indeed possible if one uses the relativistic local structure of $\mathfrak{A}^G$:

THEOREM 5. Let $\mathfrak{A}_{ob}$ be a quasi-local algebra of observables for which we assume the usual relativistic structure . If ζ is a localized automorphism of $\mathfrak{A}$ carrying charge n, and v_t is the continuous unitary $\mathfrak{A}$-valued cocycle relating ζ to its time-translation conjugate:

(22) $$\zeta^{-1}\alpha_t \circ \zeta = \mathrm{Ad}\, v_t \circ \alpha_t, \qquad t \in \mathbb{R},$$

we have, for a state ω_o of $\mathfrak{A}_{ob}$ fulfilling the conditions of Theorem 4 (ii):

(a) the state $\omega_o \circ \zeta$ is quasi-equivalent to ω_o

(b) the Radon-Nikodym derivative of $\omega_o \circ \zeta$ w.r.t. ω_o (in the sense of Connes [33]) relates as follows to v_t:

(23) $$(D(\omega_o \circ \zeta) : D\omega_o)_t = e^{in\beta(\mu+c)t}\pi_\omega(v_{-\beta t})$$

where c is a real constant independent of ω_o.

We could have anticipated from Theorem 4 that the localized morphisms

---

1) for the dual action of α on the state space (due to extremal invariance of ω).

2) We have no reason to note the latter $\mathfrak{A}^G$ since it is not defined as the fixpoints of a field algebra under gauge.

ζ should play a role in the way the chemical potential is "seen" by the observable algebra. These localized morphisms are indeed the "algebraic aspect" of the fields, the latter being obtained from implementing the former by operators (in a larger Hilbert space). We presented above the theory of the chemical potential as an explanation of part II of a "KMS principle" furnished by the historical theory. But in fact, after one has derived the KMS structure of the equilibrium states[1] from their dynamical stability, thus revealing the temperature as a parameter labelling these states, the problem at hand is to seek a complete classification of those states. Now Theorem 4 reveals the existence of another classification parameter, the chemical potential μ, which appears there as showing up by looking at the extension of those states to the field algebra (note that the uniqueness result Theorem 4 (i) implies that μ pertains to the state in restriction to $\mathfrak{A}^G$). Since μ is an attribute of the state on $\mathfrak{A}^G$, and the true algebraic respondent of the fields are the localized morphisms ζ of $\mathfrak{A}^G$, its is not surprising that one should be able to detect μ by looking at the way in which the state ω behaves with respect to ζ: this is indeed what is shown in Theorem 5. Concerning assertion (a) there, a physical comment is in order: it is not surprising that for a temperature-(and thus density-) state, $\omega_o \circ \zeta$ should be quasi-equivalent to ω_o. Indeed we know from the theory of superselection sectors that ζ "adds a finite change" to the state ω_o. Now, whilst this operation performed on the vacuum state leads to a radically different (= disjoint) "superselection sector", it will only slightly affect a density state possessing already an infinite "total charge".[2]

6. TYPES OF THE VON NEUMANN FACTORS GENERATED BY EQUILIBRIUM STATES.

Returning to the spectral alternative (a), (c) in Theorem 1 (remember that case (b) is excluded by the unwanted abelianness of $\pi_\omega(\mathfrak{A}))$[3] we now have, postulating stability of the state ω[4], the added structure that ω is β-KMS in case (b).

Case (a), corresponding to ground states[5], gives rise to type I factors, since Borcher's Theorem then implies that ω is pure, as we already mentioned in Theorem 1 (a).

---

1) considered as states of the physical meaningful observable algebra $\mathfrak{A}_o = \mathfrak{A}^G$.

2) S. Doplicher had realized long before the appearance of Theorem 5 that the quasiequivalence $\omega_o \circ \zeta \sim \zeta$ was a basic property of density states, which he proposed to take as a starting point for their investigation. Unfortunately the Connes cocyles did not exist at the time!

3) and in fact thus absent, $\mathfrak{A}$ being presumably simple

4) against "contamination" (Theorem 2) or "coexistence" (Theorem 3). In the first case we need also a strenghtening of clustering.

5) or "ceiling states" since our theory does not yield the sign of temperature.

Case (b) corresponds to an equilibrium state ω with a finite temperature if we exclude trace states[1]: in that case ω generates a <u>type III factor</u> as shown by the following single argument[2]: extremal invariance of ω implies that U_ω acts ergodically on the commutant $\pi_\omega(\mathfrak{A}^G)$: but the KMS condition turns this (via the modular involution) into ergodicity of the action of U_ω or $\mathfrak{M} = \pi_\omega(\mathfrak{A}^G)$. This implies in turn that Connes' invariant $\Gamma(\mathfrak{M})$ coincides with the Arveson spectrum $Sp(\alpha)$, which is identical with $SpU_\omega = R$ since ξ_ω separates $\mathfrak{M}$. Thus $\Gamma(\mathfrak{M}) = R$ and $\mathfrak{M}$ is a factor of type III_1 ($\mathfrak{M}$ was a factor in the first place since ω is a primary being extremal α-invariant KMS). One has in fact the additional feature that $\mathfrak{M}$ fulfills property Λ'_λ or Araki for all λ [52].

7. IS THERE A BETTER THEORY BASED ON A QUANTAL ERGODICITY MAKING CLUSTERING STATIONARY STATES AUTOMATICALLY INVARIANT? The theory proposed above consists of two technically separated pieces (sections 4 and 5 above, temperature, resp. chemical potential). Now the explanation of the chemical potential seems optimal: one does not hope to do better than deriving the situation from merely extending states from $\mathfrak{A}^G$ to $\mathfrak{A}$ (or looking at their behaviour under localized morphisms of $\mathfrak{A}^G$) without the need of any additional input. The first step however (temperature), whilst giving a satisfactory explanation of the previously mysterious (if mathematically canonical) KMS condition, does not seem to us to offer the deepest possible explanation. Indeed, whilst local dynamical stability of the equilibrium states is beyond any doubt a physically necessary requirement (and thus a correct assumption!), one would prefer to be able to <u>prove</u> this feature than have to assume it. Actually, common experience seems to show that most stationary pure phases are stable - if this was not the case we would not be here discussing this problem!

Apart from this physical drawback one could also criticize the theory of Section 4 from an esthetical standpoint: this theory is based on a property of the <u>dynamics</u> (asymptotic abelianness, a locality condition in time) and a property of the <u>states</u> (stability against local dynamical perturbations). It would seem more satisfactory to draw the properties of the equilibrium states (defined as the pure phases, i.e. the extremal invariant states) from <u>properties of the dynamics alone</u>. We believe in fact that this should be possible, if we revitalize (in a quantum context) the historically ambient

---

[1] The latter could in fact exist in a physical system corresponding to the limiting case of infinite temperature. Such states would generate a factor of <u>type II</u>, and are possibly of great physical interest (chaotic state with prevalence of kinetic energy - at the beginning of the universe?)

[2] We recall that $(\pi_\omega, U_\omega, \mathfrak{H}_\omega, \xi_\omega)$ denotes the GNS construction of ω.

idea of ergodicity[1]. Indeed we believe that locality (e.g. asymptotic abelianness in time) plus quantum ergodicity (e.g. unsplittability of the quantum system) should yield essentially that all (sufficiently clustering) pure phases are Gibbs states if not ground states (= KMS for finite or zero temperature). In this connection the following heuristic argument is tempting: consider a fluid contained in a vessel and partition by thought the container in two macroscopic parts 1 and 2 (e.g. on both sides of a plane cutting though the vessel). Now, due to locality, the total system is very nearly a dynamical tensor product of system 1 and system 2 (a molecule in 1 interacts appreciably with a molecule in 2 only if both molecules lie very near the frontier, so that the interaction energy between 1 and 2 is negligeably small compared with the total energy).

On the other hand, we cannot have this type of tensorization in two subsystems due to the pervading nature of the interaction (ergodicity of the dynamics, unsplittability of the quantum system): if smoke is introduced in system 1, it will eventually diffuse everywhere!

Here two principles are at work, locality and ergodicity, of which one implies almost tensorization and the other strictly forbids it: thus it is tempting to think that both principles taken together entail a great amount of structure and we believe that they should in fact imply the Gibbs structure.

Returning to stability in fact corroborates this hope. If we had indeed that the Gibbs structure followed from locality plus ergodicity (since we know that KMS is essentially equivalent to stability, in the presence of locality)[2] we would have that ergodicity gives us stability free of charge: but this seems quite natural physically. Indeed ergodicity means that each part of the system is exposed to the influence of "the outside": but "being swept over by outside influence" should strongly act against the possibility of local "peaking up" of the situation, and thus work towards stability.

Whilst these different aspects seem to fall into place philosophically, this offers us no indications of a strategy of proof. It is not entirely clear in the first place how ergodicity should be phrased in the algebraic frame. The most natural requirement from the physical point of view would be the unsplittability of the quantal system in two subsystems e.g., technically, the prohibition of a (non-trivial) tensorization $\mathfrak{A} = \mathfrak{A}_1 \otimes \mathfrak{A}_2$

---

[1] of the dynamics not of the states! The term: ergodic state is quite misleading in this context, as was the fact that Liouville measure is ergodic (on energy surfaces): a fact which should in our opinion be interpreted as an ergodicity of the dynamics (here fortuitously of the Liouville "state")

[2] We proved in section 4 that stability entails KMS in the presence of asymptotic abelianness and we know [29] that generally KMS implies stability.

of the algebra, accompanied by a tensorization $\alpha = \alpha_1 \otimes \alpha_1$ of the invariance group.[1) 2)] Another possible formulation of quantal ergodicity would be a requirement of "smallness" of the commutant of the time translations in Aut $\mathfrak{A}$ (few "constants of the motion"). Note that such a requirement works in the same direction as "unsplittability", since splitting in subsystems 1 and 2 clearly tends to increase the commutant of time (different speeds in parts 1 and 2, possible existence of a "flip").

We end this speculative section mentioning some criteria for the KMS-property which might indicate a possible strategy of proof.

THEOREM 6. Let ω (with notation as in Theorem 1) be an α-invariant state of the C^*-system $(\mathfrak{A},G,\alpha)$, with G abelian and $\mathrm{Sp}U_\omega = \hat{G}$. Assume asymptotic abelianness in the sense of equation (4) for α (or more generally for an action τ of an amemable group H on $\mathfrak{A}$ which commutes with and leaves ω invariant α. Denote by $\mathfrak{G}$ either $L^\infty(G)$ topologized by $\sigma(L^\infty(G),L'(G))$ or $B(G)$ topologized by $\sigma(B(G), C_\infty(\hat{G}))$.[3)] If there is a closable linear operator T_o in $\mathfrak{G}$ with domain including $\{F_{AB};A,B \in \mathfrak{A}\}$ such that $T_oF_{AB} = G_{AB}, A,B \in \mathfrak{A}$, then ω is KMS for some continuous one-parameter subgroup of G.

This conclusion is maintained if ω is assumed τ-invariant and τ-clustering (instead of merely weekly τ clustering) and the linearity assumption of T_o is dropped.

THEOREM 7. Let again ω be an α-invariant state of $(\mathfrak{A},G,\alpha)$ with G abelian and $\mathfrak{A}$ unital. Assume $\mathrm{Sp}U_\omega = \hat{G}$; and τ such that all its truncated expectation values of order $\leqq 4$ are bounded by a $C_\infty(G)$ function in all differences of their arguments. If there is a closable linear operator T_o on $C(G)$ such that $T_o(F_{AB} - \omega(A)\omega(B)) = G_{AB} - \omega(A)\omega(B), A,B \in \mathfrak{A}$, ω is KMS for some continuous one-parameter subgroup of G.

From these theorems we learn that we could derive KMS (= the Gibbs structure) for an asymptotically abelian system if each (sufficiently) clustering invariant state gave rise to a closable map T_o turning each F_{AB} into the corresponding G_{AB}. Could it be that the existence of such a map follows from ergodicity (as a "tightness" condition)?

8. A COMMENT ON STABILITY VERSUS CLUSTERING.

Even though dynamical stability might not be the best motivation for the Gibbs structure, it is doubtlessly a main physical feature of equilibrium states. Now we can

---

1) The relevant invariance group should at least contain the 4-translations, as shown by the example of a equilibrium state in translatory motion (KMS not for time, but for a one-parameter subgroup of the 4-translations).

2) Basic physical laws are often of the nature of "prohibitions".

3) $B(G)$ denotes the linear space of positive type function on G, $C_\infty(G)$ (with its sup norm) the continuous functions on G vanishing at ∞.

consider dynamical stability of a much stronger (= global) type than the local stability found above to be a characteristic of KMS states. And experience shows that stability tends to decrease with increasing range of correlations (decreasing clustering) of the state at hand. In that respect the following fact is of interest: let ω be a β-KMS state of a C^*-system $(\mathfrak{A},\mathbb{R},\alpha)$ and let $h = h^* \in \mathfrak{A}$ be analytic for α. For λ sufficiently small there is a unique β-KMS state $\omega^{\lambda h}$ for the perturbed automorphism $\alpha^{\lambda h}$ (cf. equations (16) and following); moreover $\omega^{\lambda h}$ is given in terms of the truncated expectation values ω_n^T of ω by the following expansion:

$$\omega^{\lambda h}(A) = \omega(A) + \sum_{n=1}^{\infty} (-i\lambda)^n \cdot \int_{0<s_1<s_2<\cdots<\beta} ds_1 \ldots ds_n \omega_n^T(A,\alpha_{is_1}(h),\ldots,\alpha_{is_n}(h)) \tag{24}$$

Now suppose that we want to investigate stability of ω under global perturbations of the type $\alpha^{\bar{h}}$ with

$$\bar{h} = \int f(\vec{x})\alpha_{\vec{x}}(h)d^3x \ , \tag{25}$$

with f some continuous bounded function, e.g.

$$\bar{h} = \int \alpha_{\vec{x}}(h)d^3x \ , \tag{26}$$

(here $\alpha_{\vec{x}}$, $\vec{x} \in \mathbb{R}^3$ denotes a representation of the spatial translations as automorphisms of $\mathfrak{A}$ assumed to exist). Now although (25) and (26) have only a formal meaning (they obviously do not make sense as elements of $\mathfrak{A}$!), one can nevertheless show the existence of a bonafide perturbed dynamical automorphism group $\alpha^{\bar{h}}$ [1)]. Therefore it makes sense to try and find a corresponding perturbed state $\omega^{\lambda\bar{h}}$ β-KMS for $\alpha^{\bar{h}}$ by plugging h into (24) as replacing the local h. The integral in the r.h.s. of (24) is then replaced by

$$\int ds_1 \ldots ds_n d^3x_1 \ldots d^3x_n \omega_n^T(A,\alpha_{is_1+\vec{x}_1}(h),\ldots,\alpha_{is_n+\vec{x}_n}(h)), \tag{27}$$

and clearly, the last expansion has an increased chance to make sense for rapid fall-off of the truncated function ω^T, i.e. for strong clustering of ω. In particular if we have the property referred to as <u>integrability</u> (and proved to hold at high temperature in certain models):

---

1) This is not astonishing in view of equation (16), since for A local most of the integration (25) is wasted in the commutator $[\bar{h},A]$ because of locality.

$$(28) \qquad \int d^3x_1 \ldots d^3x_n |\omega_n^T(A,\alpha_{\vec{x}_1}(h),\ldots,\alpha_{\vec{x}_n}(h)| \leqq n!\, C^n$$

for some positive constant C^n, we see, boldly replacing in (27) the integration w.r.t. the variable s_i by the integration volume $\beta^n/n!$ [1)] that the r.h.s. of (27) gets majorized by

$$(29) \qquad \frac{\beta^n}{n!}\, n!\, C^n = (\beta C)^n$$

whence convergence of the expansion of $\omega^{\lambda\bar{h}}$ within the radius $1/\beta C$. Now suppose that the state ω is less clustering than this, but enough for each individual term of the expansion (27) of $\omega^{\lambda\bar{h}}$ to make sense (this would e.g. be the case if we had a majorization of the type (28) with $n!$ replaced by $(n!)^2$). We could than try summation methods to evaluate this expansion, for instance, Borel summation. In this context we find the following phenomenon interesting in connection with phase transition: a theorem in Borel summability theory says that if a series $a_n\lambda^n$ is Borel summable for a value λ_o of the variable, the Borel sum is analytic within the circle with diameter $(0,\lambda_o)$. But it can occur that this holds for, say, $\lambda_o = 1$ and $\lambda_0=-1$ with different analytic functions on $(0,1)$ resp. $(-1,0)$. This looks very much like what happens in a phase transition case! It would be nice to know more about the mutual relationship between the degree of clustering and the degree of stability against global dynamical perturbations -and this in conjunction with the phase transition situation. One could for instance hope that for a fluid, one would have analyticity of the expansion of $\omega^{\lambda\bar{h}}$ away from phase transition lines (for a crystal one would expect this to hold only for perturbations by an "isotropic" h - averaged over rotational degrees of freedom)[2)]. Such questions are tightly related with the problem of symmetry breaking, for which a theory of KMS based on dynamical ergodicity would be likely to shed some light.

9. INDICATIONS ON PROOFS.

THEOREM 1. The implications (i) $\Leftrightarrow$ (ii) $\Rightarrow$ (iii) $\Leftrightarrow$ (iv) hold for any C^*-system, asymptotic abeliannes is needed only for reversing the middle one. Since

$$(30) \qquad \omega(A\alpha_t(B)) - \omega(A)\omega(B) = (\pi(A)^*\xi_\omega | (U_t - |\xi_\omega)(\xi_\omega|)\,|\pi(B)\xi_\omega)$$

(i) $\Leftrightarrow$ (ii) is immediate from the mean ergodic theorem. (iii) $\Leftrightarrow$ (iv) is obtained by a straight forward generalization of the argument showing

---

[1)] This bold replacement turns out to be rigourously true in the classical case.

[2)] It is provocative that assuming this, one seems to be able to derive the observed types of symmetry breaking (crystallographic groups ... etc.).

equivalence of purity of ω and irreducibility of $\pi_\omega(\mathfrak{A})$. (i) $\Rightarrow$ (iii): we must show that given $\varphi,\psi \in \mathfrak{H}_\omega$ there is $X \in (U_\omega(\mathbb{R}) \cup \pi_\omega(\mathfrak{A}))''$ with $X\varphi$ arbitarly close to ψ. Now, by the cyclicity of ξ_ω, there is $A \in \mathfrak{A}$ with $(\xi_\omega, \pi(A)\varphi) \neq 0$, since the projection E_o on the U_ω-invariant vectors of $\mathfrak{H}_\omega$ (i.e. on ξ_ω by (i)) yields a non vanishing multiple $E_o\pi(A)\varphi$ of ξ_ω. Now $E_o \in U_\omega(\mathbb{R})''$ by the mean ergodic theorem. Cyclicity of ξ_ω then yields the desired $X = \pi(B)E_o\pi(A)$, $B \in \mathfrak{A}$.

The converse direction (iii) $\Rightarrow$ (i) is afforded by <u>G-abelianness</u>, i.e. the fact that $E_o\pi_\omega(\mathfrak{A})E_o$ is abelian, the latter resulting from asymptotic abelianness since

(31) $$0 = \lim_{T=\infty} \int_{-T}^{+T} E_o\pi_\omega([A,\alpha_t(B)])E_o dt = [E_o\pi_\omega(A)E_o, E_o\pi_\omega(E_o)]$$

according to the mean ergodic theorem. Now if $(U_\omega(\mathbb{R}) \cup \pi_\omega(\mathfrak{A}))''$ is irreducible, so is $E_o(U_\omega(\mathbb{R}) \cup \pi_\omega(\mathfrak{A}))E_o = E_o\pi_\omega(\mathfrak{A})E_o$. The fact that the latter is also abelian forces E_o to have rank 1.

We now sketch the proof of the spectral alternative (a), (c), assuming ω clustering. It is easy to see that $\lambda \in \mathrm{Sp}U_\omega$ iff for each open set $\mathcal{V}$ containing λ there is $A \in \mathfrak{A}$ with Arveson spectrum $\mathrm{Sp}^\alpha(A)$ within $\mathcal{V}$ and $\pi_\omega(A)\xi_\omega \neq 0$. We now prove that $\mathrm{Sp}U_\omega$ is additive: $\lambda_1,\lambda_2 \in \mathrm{Sp}U_\omega \Rightarrow \lambda_1 + \lambda_2 \in \mathrm{Sp}U_\omega$. Let $\mathcal{V}$ be open with $\mathcal{V} \ni \lambda_1 + \lambda_2$; choose $\mathcal{V}_1,\mathcal{V}_2$ open with $\mathcal{V}_i \ni \lambda_i, \mathcal{V}_1 + \mathcal{V}_2 \subset \mathcal{V}$; and A_1,A_2 with $\mathrm{Sp}^\alpha(A_i) \subset \mathcal{V}_i$ and $\pi_\omega(A_i)\xi_\omega \neq 0$. We have $\mathrm{Sp}^\alpha(A_1\alpha_t(A_2)) \subset \mathcal{V}_1 + \mathcal{V}_2 \subset \mathcal{V}$ for all $t \in \mathbb{R}$. And, by clustering, for t big enough $\pi_\omega(A_1\alpha_t(A_2)) \neq 0$; we proved the additivity of $\mathrm{Sp}U_\omega$. Now the spectral alternative follows readily: if (a) does not hold there are $\alpha,\beta > 0$ with $\alpha, -\beta \in \mathrm{Sp}U_\omega$. By additivity $n\alpha - m\beta \in \mathrm{Sp}U_\omega$ for all integers m,n. Now if α/β is irrational $n\alpha - m\beta$ will come arbitrarily near to each $\lambda \in \mathbb{R}$, then $\mathrm{Sp}U_\omega = \mathbb{R}$ since closed. But the mishap that α/β be rational can be mended by slightly perturbing β, using the easy fact that asymptotic abelianness and clustering imply

(32) $$\omega(A\alpha_t(B)C) \xrightarrow[|t|=\infty]{} \omega(AC)\omega(B)$$

which precludes the existence of isolated points of $\mathrm{Sp}U_\omega$. We proved the spectral alternative. Now in case (a) π_ω has to be irreducible because $(U_\omega(\mathbb{R}) \cup \pi_\omega(\mathfrak{A}))''$ is irreducible by (iii), and we know (Borchers Theorem)[1)] that positivity of $\mathrm{Sp}U_\omega$ allows to find, to each $t \in \mathbb{R}$, a unitary $V(t)$ with $\mathrm{Ad}V_\omega(t) = \mathrm{Ad}U_\omega(t)$ and $U_\omega^*(t)V(t) \in \pi(\mathfrak{A})''$. Finally the weak * totality of $\mathfrak{F}$ or $\mathfrak{G}$ in $B(\mathbb{R})$ stems from the fact that $\mathrm{Sp}U_\omega = \mathbb{R}$ by Fell's result

---

1) Result obtained using derivation theory.

relating faithfulness of representations (fullness of spectrum for abelian groups) with weak * density.[1)]

THEOREM 2. The first step in the proof consists in showing that stability of ω against local perturbations implies that

$$(33) \qquad \int_{-\infty}^{+\infty} [F_{AB}(t) - G_{AB}(t)]dt = 0, \quad A,B \in \mathfrak{A}_{loc} .$$

The trick leading to this is the following: from (16) we have, taking the derivative at zero of both sides of Definition 4(i) applied to $B \in \mathfrak{A}_{diff}$:

$$(34) \qquad \omega^{(h)}\left(\frac{d}{dt}\bigg|_{t=0} \alpha_t(B)\right) = i\omega^{(h)}[h,B]$$

Making here

$$(35) \qquad B = \int_S^T \alpha_t(A)dt, \quad A \in \mathfrak{A}_{loc}, \ S,T \in \mathbb{R},$$

for which $\frac{d}{dt}\Big|_{t=0} \alpha_t(B) = \alpha_T(A) - \alpha_S(A)$, we get

$$(36) \qquad \omega^{(h)}(\alpha_T(A) - \alpha_S(A)) = i\omega^{(h)}([h, \int_S^T \alpha_t(A)dt])$$

We now take the limits $T \to +\infty$, $S \to -\infty$. In view of (ii) in Definition 4, we get

$$(37) \qquad \omega^{(h)}(\int_{-\infty}^{+\infty} [h,\alpha_t(A)]dt) = 0$$

We need only, to obtain (33), to replace $\omega^{(h)}$ by ω in the latter equation: this is done by making there $h \to \lambda h$, and going to zero with λ, using (ii) in Definition 4. For this it is essential that the integrand in (37) be within $\mathfrak{A}$, which is guaranteed by assumption (16). We obtained (33) up to notation.

The next step consists in deriving from (33) the "two-fold equation"

$$(38) \qquad \int_{-\infty}^{+\infty} [F_{A_1B_1}(t)F_{A_2B_2}(t) - G_{A_1B_1}(t)G_{A_2B_2}(t)]dt = 0, \ A_i,B \in \mathfrak{A}_{loc}$$

This is obtained by substituting $A = A_1\alpha_u(A_2)$, $B = B_1\alpha_u(B_2)$ in (33) and taking the limit $u \to \alpha$ which leads to the splitting

$$(39) \qquad F_{A_1\alpha_u(A_2),B_1\alpha_u(B_2)} \xrightarrow[u=\infty]{} F_{A_1B_1}F_{A_2B_2},$$

[1)] An analogous result holds for weak * totality in $L^\infty(\mathbb{R})$ but requires for its proof a group-duality result.

and analogously for the functions G, in virtue of asymptotic abelianness and clustering. Here we must use Lebesgue dominated convergence, which motivates the hyperclustering assumption in Theorem 4.

To draw the conclusion from the two-fold relation (39) we now notice that replacing there A_2 by $\alpha_s(A_2)$ the latter reads, up to notation

(40)
$$\mathfrak{F}_{A_1B_1} * \mathfrak{G}_{A_2B_2} = \mathfrak{G}_{A_1B_1} * \mathfrak{F}_{A_2B_2}$$

or in terms of Fourier transforms

(41)
$$\hat{\mathfrak{F}}_{A_1B_1}\hat{\mathfrak{G}}_{A_2B_2} = \hat{\mathfrak{G}}_{A_1B_1}\hat{\mathfrak{F}}_{A_2B_2}$$

Now either $\mathrm{Sp}\,U_\omega$ is one-sided or $\mathrm{Sp}\,U_\omega = R$. In the latter case division of (42) by $\hat{\mathfrak{G}}_{A_2B_2}$ [1] is allowable and, since the choices of A_1B_1 and A_2B_2 are independant, we infer from (41) the existence of a function Φ such that

(42)
$$\hat{F}_{AB} = \Phi\,\hat{G}_{AB}, \qquad A,B \in \mathfrak{A}_{loc}$$

The fact that Φ must be a real character of $\mathbb{R}$ can now be seen either by formal manipulation or using Theorem 6.

THEOREM 4 is to a large extent an item in the duality theory of compact groups. Recall the following classical facts: For a compact group G there is a bijection between the closed subgroups K of G and the globally right-translation invariant C^*-subalgebras $\mathfrak{a}$ of $\mathcal{C}(G)$[2]. We have

(43)
$$\begin{cases} K \to \mathfrak{a} = \mathcal{C}(K\backslash G) = \{f \in \mathcal{C}(G);\ \lambda(k)f = f\ \forall\ k \in K\} \\ \mathfrak{a} \to K = \{k \in G;\ f(kg) = f(g)\,\forall\ f \in \mathfrak{a} \text{ and } g \in G\} \end{cases}$$

Further, with K_i, $i = 1,2$, closed subgroups of G, each C^*-isomorphism of $\mathcal{C}(K_1\backslash G)$ onto $\mathcal{C}(K_2\backslash G)$ commuting with the right translations $\rho(g)$, $g \in G$, is a left translation $\lambda(h)$ for some $h \in G$ such that $K_2 = hK_1h^{-1}$.

These facts easily follow from the Stone-Weierstrass theorem and from the fact that isomorphisms between abelian C^*-algebras are transposed of homeomorphisms between their spectra.

The key to the uniqueness of extension is now: For a weakly clustering α-invariant state ω of the C^*-system $\{\mathfrak{A},R,\alpha\}$, the norm closure $\overline{\mathcal{C}^1_\omega(G)}$ in $\mathcal{C}(G)$ of the set $\mathcal{C}^1_\omega(G) = \{\varphi^{(A)}: g \in G \to \omega(\gamma_g(A)); A \in \mathfrak{A}\}$ coincides with

---

[1] Appropriately chosen. This point is delicate and was inadequately treated in the first papers on the subject. A clarification is offered at the end of [55].

[2] $\mathcal{C}(G)$ denotes the C^*-algebra of continuous functions on G with its supnorm.

$C(G_\omega\backslash G)$ [1].

Indeed $\overline{C^1_\omega(G)}$ is globally invariant under $\rho(g)$ since $\rho(g)\varphi^{(A)} = \varphi^{(\gamma_g(A))}$; self adjoint since $\overline{\varphi^{(A)}} = \varphi^{(A^*)}$; and multiplicative since, by clustering

$$(44)\qquad \frac{1}{2T}\int_{-T}^{+T} \omega(\gamma_g(A)\alpha_t\gamma_g(B))dt \underset{T=\infty}{\longrightarrow} \varphi(\gamma_g(A))\cdot\varphi(\gamma_g(B))$$

(the functions at the l.h.s. build an equicontinous set). Let now ω_1 and ω_2 be two extremal α-invariant extensions of ω_o to $\mathfrak{A}$, with $K_i = K_{\omega_i}$. We have that

$$(45)\qquad a(A,B,t) = \int \gamma_g(A^*\alpha_t(B))dg \in \mathfrak{A}$$

Thus

$$(46)\qquad \frac{1}{2T}\int_{-T}^{+T} \omega_i(a(A,B,t)dt \underset{T=\infty}{\longrightarrow} \int \varphi_i^{(A^*)}(g)\varphi_i^{(B)}(g)dg$$

thus there is a linear isometry $V: L^2(K_1\backslash G) \to L^2(K_2\backslash G)$ with $V\varphi_1^{(A)} = \varphi_2^{(A)}$, $A \in \mathfrak{A}$. Further, by (44) and equicontinuity V is multiplicative, thus, if $T_i(f)\xi = f\xi, f \in C(K_i\backslash G), \xi \in L^2(K_1\backslash G)$ we have $T_2(Vf) = VT_1(f)V^*$, $f \in C(K_1\backslash G)$. From this follows that V is an isomorphism of $C(K_1\backslash G)$ onto $C(K_2\backslash G)$, which commutes obviously with the $\rho(g), g \in G$. From the facts recalled above we then have a $g \in G$ with $\varphi_2^{(A)} = \lambda(g)\varphi_1^{(A)}, A \in \mathfrak{A}$ in other terms $\varphi_2 = \varphi_1 \circ \gamma_g$.

(ii): the proof is based on a density theorem for the "two point function" $\varphi^{(A,B)}(g) = \omega(A\gamma_g(B))$ analogous to the density of the "one point functions" mentioned before (also due to clustering), plus an argument of a group duality flavour.

THEOREMS 6 AND 7. The proofs are technical and based on group duality results. Instead of describing them let us give here a rough heuristic argument giving an intuitive picture of what happens in the case $G = \mathbb{R}$. In the alternative $\mathrm{Sp}U_\omega = \mathbb{R}$ we have density of both sets $\mathfrak{F} = \{F_{AB}; A,B \in \mathfrak{A}\}$ and $\mathfrak{G} = \{G_{AB}; A,B \in \mathfrak{A}\}$ in "all functions on the reals", thus we can roughly assume that the map T acts on arbitrary functions. Now it is clear that T commutes with translating a function by $t, t \in \mathbb{R}$, (such translations being obtained for F_{AB} or G_{AB} by the replacement $A \to \alpha_t(A)$). This implies formally that T acts as a multiplication by a function Φ on Fourier transforms. Remains to see that Φ is a real character. Now, using the trick (39), and applying clustering, we see that T commutes with pointwise multiplication in t-space. Thus Φ must commute with convolutions in Fourier

---

[1] We recall that G_ω is the stabilizer of ω in G.

space, and is consequently a character - real as one sees by looking at F_{AA^*} and G_{AA^*}.

10. BIBLIOGRAPHICAL INDICATIONS. Algebraic field theory and KMS condition: [1] [3] [6] [7] [19] [40] [47] [59]. Theorem 1: [10] [11] [14] [15] [16] [18] [23] [26] [42] [55]. Theorem 2 and 3: [34] [40] [29] [30] [31] [45] [54] [55]. Theorems 4 and 5: [43] [44] [47] [56]. Section 6: [6] [20] [21] [33] [40] [50] [52]. Theorems 6 and 7: [55].

ACKNOWLEGMENTS. The author is indebted to Professors T. Gamelin, A. Hales and M. Takesaki for the warm hospitality and stimulating atmosphere of the Department of Mathematics-UCLA-where these notes were prepared. Thanks are due to Keren Evans for her typing of the manuscript.

BIBLIOGRAPHY

1. I.E. SEGAL, Postulates for General Quantum Mechanics, Ann. Math. 48, (1947), 930.

2. R. KUBO, Statistical-Mechanical Theory of Irreversible Processes. I - General Theory and Simple Applications to Magnetic and Conduction Problems, J. Phys. Soc. Japan, 12 (1957), 570.

3. R. HAAG, Discussion des "axiomes" et des propriétés asymptotiques d'une théorie des champs locale avec particules composées, Colloques Intern. du CNRS LXXV Lille 1957 (1959)

4. P.C. MARTIN and J. SCHWINGER, Theory of Many Particles Systems, I, Phys. Rev., 115 (1959), 1342.

5. A.S. WIGHTMANN, Proc. Intern. Congress of Mathematicians, August 1962.

6. H. ARAKI and E.J. WOODS, Representations of the Canonical Commutation Relations describing on a non Relativistic Infinite Free Bose Gas, J. Math. Phys. 4 (1963), 637.

7. R. HAAG and D. KASTLER, An Algebraic Approach to Quantum Field Theory, J. Math. Phys., 5 (1964), 848.

8. H.J. BORCHERS, Energy and Momentum as Observables in Quantum Field Theory, Commun. Math. Phys 2 (1966) 49.

9. R.V. KADISON, Derivations of Operator Algebras, Ann. Math., 83 (1966), 280.

10. D. RUELLE, States of Physical Systems, Commun. Math. Phys. 3, (1966) 133.

11. S. DOPLICHER, D. KASTLER and D.W. ROBINSON, Covariance Algebras in Field Theory and Statistical Mechanics, Commun. Math. Phys. 3 (1966), 1.

12. I. KOVACS and J. SZÜCS, Ergodic type Theorem in von Neumann Algebras, Acta Sc. Math. 27 (1966), 233.

13. G.F. DELL'ANTONIO, On some Groups of Automorphisms of physical Observables, Commun. Math. Phys., 2 (1966), 384.

14. D. KASTLER and D.W. ROBINSON, Invariant States in Statistical Mechanics, Commun. Math. Phys. 3 (1966) 151.

15. S. DOPLICHER, R.V. KADISON, D. KASTLER and D.W. ROBINSON, Asymptotically Abelian Systems, Commun. Math. Phys. 61 (1967), 101.

16. D.W. ROBINSON and D. RUELLE, Extremal invariant States. Ann. Inst. H. Poincaré 6 (1967) 299.

17. M. TOMITA, Standard Forms and of Von Neumann Algebras, Vth Functional Analysis Symposium of the Math. Soc. of Japan Sendai (1967).

18. E. STÖRMER, Large Groups of Automorphisms of C^*-algebras, Commun. Math. Phys. 6 (1967) 101.

19. R. HAAG, N. HUGENROLTZ and M. WINNINK, On the Equilibrium States in Quantum Statistical Mechanics, Commun. Math. Phys. 5 (1967), 215.

20. N. HUGENHOLTZ, On the Factor Type of Equilibrium States in Quantum Statistical Mechanics Commun. Math. Phys., 6 (1967) 189.

21. E. STÖRMER, Types of von Neumann Algebras associated with extremal invariant States, Commun. Math. Phys. 6 (1967), 194.

22. S. DOPLICHER, R. HAAG, J.E. ROBERTS, Fields, Observables and Gauge Transformation I and II. Commun. Math. Phys., 13, 1 (1969) and 15, 173 (1969).

23. S. DOPLICHER, D. KASTLER and E. STÖRMER, Invariant States and Asymptotic Abelianness, J. Funct. Anal. 3 (1969), 411.

24. R.V. KADISON, Some analytic Methods in the Theory of Operator Algebras, Lectures in Modern Analysis and Applications, Springer Lecture Notes in Math. 140 (1970) 8.

25. M. TAKESAKI, Tomita's Theory of Modular Hilbert Algebras and its Applications, Springer Lecture Notes in Mathematics, 128 (1970).

26. D. KASTLER, G. LOUPIAS, L. MICHEL, and M. MEBKHOUT, Central Decomposition of invariant States, Commun. Math. Phys. 27 (1972), 195.

27. H. ARAKI, Remarks on Spectra of modular Operators of von Neumann Algebras, Commun. Math. Phys., 28 (1972), 267.

28. E. STÖRMER, Spectra of States and Asymptotically Abelian C^*-algebras, Commun. Math. Phys. 28 (1972), 279.

29. H. ARAKI, Relative Hamiltonian for faithful normal States of a von Neumann Algebra, Pub. Res. Inst. Math. Sci. Kyoto University 9 (1973), 165.

30. H. ARAKI, Expansional in Banach Algebras Ann. Scient. Ecole Norm. Sup. 6 (1973), 1.

31. D.W. ROBINSON, Return to Equilibrium, Commun. Math. Phys. 31 (1973) 171.

32. H.J. BORCHERS, Uber Ableitungen von C^*-Algebren, Nach. Gött. Akad., 2 (1973).

33. A. CONNES, Une classification des facteurs de type III, Ann. Scient. Ecole Norm. Sup. 6 (1973), 133.

34. R. HAAG, D. KASTLER and E.B. TRYCH-POHLMEYER, Stability and Equilibrium States, Commun. Math. Phys. 38 (1974) 173.

35. E.B. TRYCH-POHLMEYER, Ph.d. Thesis Hamburg (1974).

36. W. ARVESON, On Groups of Automorphism of Operator Algebras, J. Funct. Anal., 15 (1974), 217.

37. S. DOPLICHER, R. HAAG, and J.E. ROBERTS, Local Observables and Particle Statistics I and II. Commun. Math. Phys. 23, 199 (1971) and 35, 49 (1974).

38. J. ROBERTS, Spontaneously broken Gauge Symmetries and Superselection Rules. School of Mathematical Physics Camerino (Italy). October 1974.

Marseille Preprint 74/p.665.

39. M. AIZENMANN, G. GALLAVOTI, S. GOLDSTEIN and J.L. LEBOWITZ, On the Stability of Equilibrium States of classical Systems, Comm. Math. Phys. 48 (1976).

40. D. KASTLER, Equilibrium States of Matter and Operator Algebras, Symposia Mathematica XX (1976), 49.

41. O. BRATTELI, and D. KASTLER, Relaxing the clustering condition in the derivation of the KMS condition, Commun. Math. Phys. 46 (1976) 37.

42. R. HERRMANN, and D. KASTLER, Energy Spectrum of Extremal Invariant States Commun. Math. Phys. 56 (1977), 87.

43. H. ARAKI, R. HAAG, D. KASTLER and M. TAKESAKI, Extension of States and Chemical Potential. Commun. Math. Phys. 53 (1977), 97.

44. H. ARAKI, A. KISHIMOTO, Symmetry and Equilibrium States, Commun. Math. Phys. 52 (1977), 211.

45. F. HOEKMAN, Ph.d. Thesis, Groningen.

46. R.V. KADISON, Unbounded Similarity, Algebres d'Opérateurs et leurs Applications en Physique mathématique, Coll. Intern, CNRS 274 (1979), 133.

47. D. KASTLER, Foundations of Equilibrium Statistical Mechanics ibid 20.

48. M. PULVIRENTI, Stability, KMS and Self-adjointness in the Liouville Operator in classical Systems ibid. 301.

49. M. TAKESAKI, Fourier Analysis on compact Automorphism Groups (An Application of the Tannaka Duality Theorem) ibid. 361.

50. D. TESTARD, Some Properties of the Representation of the Quasilocal Observables in Statistical Mechanics and Quantum Field Theory, ibid. 409.

51. E.B. TRYCH-POHLMEYER, The Stability Properties of Equilibrium States, ibid. 437.

52. D. TESTARD, Asymptotic Ratio Set ov von Neumann Algebras generated by Temperature States in Statistical Mechanics. Rep. Math. Phys. 12 (1977) 115.

53. R. HAAG, and E. TRYCH-POHLMEYER, Stability Properties of Equilibrium States Commun. Math. Phys. 56 (1977) 213.

54. O. BRATTELI, A. KISHIMOTO, and R. ROBINSON, Stability Properties and the KMS Condition, Commun. Math. Phys. 61 (1978), 209.

55. D. KASTLER and M. TAKESAKI, Group Duality and the Kubo-Martin-Schwinger Condition, Commun. Math. Phys. 70 (1979), 193.

56. M. MEBKHOUT, Algebraic Theory of the Chemical Potential in the Case of the Usual Gauge Group, Annals. of Physics, 123, 2 (1979), 317.

57. O. BRATTELI and D.W. ROBINSON, Operator Algebras and Quantum Statistical Mechanics I, Springer-Verlag, New York Heidelberg Berlin 1973 and Vol. II to come.

58. D. KASTLER, Algebraic Statistical Mechanics of closed Systems. A. Visconti Colloquium Volume CNRS (1980).

DEPARTMENT OF MATHEMATICS
UNIVERSITY OF CALIFORNIA, LOS ANGELES
LOS ANGELES, CALIFORNIA 90024

PHYSIQUE THEORIQUE BP907
UER LUMINY
MARSEILLE 13288 CEDEX 2
FRANCE

Proceedings of Symposia in Pure Mathematics
Volume 38 (1982), Part 2

STABILITY IN STATISTICAL MECHANICS

Geoffrey L. Sewell

The purpose of this talk will be to discuss the characterisations of equilibrium and metastable states in terms of a variety of stability conditions. It will emerge that finer gradations of stability than those given by the KMS conditions and by the minimisation of the free energy density of a system are needed for the purpose of describing certain physically interesting states. A more detailed review of the subject may be found in Ref. [1].

1. THE MODEL. We shall be concerned with an infinitely extended assembly of particles in the space X, which we take, for simplicity, to be the lattice $\mathbb{Z}^d$, though some of our treatment can be carried through for the case where X is $\mathbb{R}^d$. We denote by L the family $\{\Lambda\}$ of finite point subsets of X.

In a standard way [2,3], we take the observables of the system to be the self-adjoint elements of a quasi-local C*-algebra $\mathfrak{A}$. Thus, $\mathfrak{A}$ is constructed as the norm completion of $\mathfrak{A}_L := \bigcup_{\Lambda \in L} \mathfrak{A}(\Lambda)$, where $\mathfrak{A}(\Lambda)$, the C*-algebra of observables for the region Λ, is isotonic w.r.t. Λ. It will be assumed that the local algebras $\{\mathfrak{A}(\Lambda)\}$ are finite type-I factors, and therefore $\mathfrak{A}$ is UHF. Space-translations are represented by a homomorphism σ of the additive group X into Aut $\mathfrak{A}$, such that $\sigma(x)\mathfrak{A}(\Lambda) = \mathfrak{A}(\Lambda+x)$. The states of the system are taken to be the set Ω of positive normalised linear functionals on $\mathfrak{A}$. In particular, the central tracial state is denoted by τ. Thus, the restriction of $\omega(\in \Omega)$ to a local algebra $\mathfrak{A}(\Lambda)$ corresponds to an element ρ_Λ^ω of $\mathfrak{A}(\Lambda)$, defined by the formula [4]

$$\omega(A) = \tau(\rho_\Lambda^\omega A) \quad \forall\ A \in \mathfrak{A}(\Lambda) \tag{1}$$

We introduce the equivalence relation $\sim$ in Ω, such that $\omega' \sim \omega$ if, for some $\Lambda_0 \in L$, $\rho_\Lambda^{\omega'} = \rho_\Lambda^\omega$ whenever $\Lambda \cap \Lambda_0 = \emptyset$. The set of translationally invariant elements of Ω will be denoted by Ω_σ.

The interactions in the system are taken to correspond to a family of local Hamiltonians $\{H_\Lambda = H_\Lambda^* \in \mathfrak{A}(\Lambda) | \Lambda \in L\}$,

0082-0717/80/0000-0572/$02.75

satisfying the translational invariance condition that $\sigma(x)H_\Lambda = H_{\Lambda+x}$. The free energy of the region Λ, induced by the state ω at temperature β^{-1} is then [4]

$$F_\Lambda(\omega) = \omega(H_\Lambda) + \beta^{-1}\tau(\rho_\Lambda^\omega \ \ell n \ \rho_\Lambda^\omega) \tag{2}$$

We define the <u>free energy density functional</u> $f : \Omega_\sigma \to \mathbb{R}$ by the formula

$$f(\omega) = \lim_{\Lambda\uparrow} |\Lambda|^{-1} F_\Lambda(\omega) \ , \tag{3}$$

the existence of this limit being guaranteed by the subadditivity of entropy [5], provided that the interactions are suitably tempered [2]. Let $\Omega^{(o)}$ be the set of states ω such that, for all $\omega' \sim \omega$, $\exists \lim_{\Lambda\uparrow} (\omega'(H_\Lambda) - \omega(H_\Lambda)) \in \mathbb{R} \cup \{\infty\}$. We define the <u>incremental free energy</u> $\Delta F(\omega|\omega')$, due to a transition from $\omega(\in \Omega^{(o)})$ to $\omega'(\sim\omega)$ (local modification of state!) by the formula

$$\Delta F(\omega|\omega') = \lim_{\Lambda\uparrow} (F_\Lambda(\omega') - F_\Lambda(\omega)), \tag{4}$$

the existence of this limit also being ensured by the subadditivity of entropy.

The <u>dynamics</u> of the system corresponds in simple cases, i.e., when the forces are of suitably short range, to a one-parameter group $\alpha(\mathbb{R})$ of automorphisms of $\mathfrak{A}$, whose generator δ is determined by the formula [6]

$$\delta(A) = \text{norm} - \lim_{\Lambda\uparrow} i\,[H_\Lambda, A] \quad \forall\ A \in \mathfrak{A}_L \tag{5}$$

In such cases, we have a <u>C*-dynamical system</u> $(\mathfrak{A}, \Omega, \alpha)$. In more complex cases, where the limit in (5) does not exist, the dynamics becomes essentially state- (or representation-) dependent [7]: thus, in the normal folium Ω_π of a representation π of $\mathfrak{A}$, it may correspond to a one-parameter group $\alpha_\pi(\mathbb{R})$ of automorphisms of $\pi(\mathfrak{A})''$, whose generator δ_π is given by the formula

$$\delta_\pi \pi(A) = s - \lim_{\Lambda\uparrow} i\pi([H_\Lambda, A]) \quad \forall\ A \in \mathfrak{A}_L \tag{6}$$

In this case, we have a <u>W*-dynamical system</u> $(\pi(\mathfrak{A})'', \Omega_\pi, \alpha_\pi)$.

Let ω be a state on $\mathfrak{A}$, π its GNS representation and $\hat{\omega}_\pi$ its canonical extension to $\pi(\mathfrak{A})''$. Then ω is a <u>KMS state</u> for inverse temperature β^{-1} if (1) $\hat{\omega}_\pi$ is faithful, and (2) its modular automorphism group $\hat{\alpha}_\pi(\mathbb{R})$ satisfies the condition $\hat{\alpha}_\pi(t) \circ \pi = \pi \circ \alpha(\beta^{-1}t)$, in the C*-case, and $\hat{\alpha}_\pi(t) = \alpha_\pi(\beta^{-1}t)$ in the W*-case.

The KMS conditions are also equivalent to a variety of correlation inequalities [8], e.g., to the following ones, which arise [9,10] in connection with the thermodynamical stability conditions specified below.

$$\Phi(\omega(A^*A),\omega(AA^*)) \leq \begin{cases} -i\omega(A^*\delta A) \ \forall \ A \in D(\delta) & (C^*\text{-case}) \\ -i\hat{\omega}_\pi(\pi(A^*)\delta_\pi\pi(A)) \ \forall \ \pi(A) \in D(\delta_\pi) & (W^*\text{-case}) \end{cases} \tag{7}$$

$$\text{where } \Phi(u,v) = \begin{cases} u \ln u - u \ln v & \text{for } u,v > 0;\ u+v>0 \\ 0 & \text{for } u = v = 0 \end{cases} \tag{8}$$

2. STABILITY CONDITIONS. We shall now define certain conditions of stability of states against thermodynamical and dynamical perturbations, and relate them to those of KMS.

A state ω is said to be locally thermodynamically stable (LTS) [11] if it belongs to the class $\Omega^{(o)}$ and if

$$\Delta F(\omega|\omega') \geq 0 \ \forall \ \omega' \sim \omega . \tag{9}$$

A translationally invariant state $\omega(\in \Omega_\sigma)$ is said to be globally thermodynamically stable (GTS) [11] if it minimises the free energy density functional f: note here that f is lower semicontinuous w.r.t. the w*-topology of Ω, and thus attains its infimum.

Next we formulate two types of stability against local perturbations of the forces of a C*-dynamical system $(\mathcal{A},\Omega,\sigma)$: W*-systems may be treated analogously, with minor technical modifications [12]. For $h = h^* \in \mathcal{A}$ and $\lambda \in \mathbb{R}$, we define (cf. [13]) $\alpha^{\lambda h}(\mathbb{R})$ to be the one-parameter group of automorphisms of $\mathcal{A}$, whose generator is

$$\delta^{(\lambda h)} = \delta + i\lambda[h,\ ] \tag{10}$$

We say that a state ω of the system is locally dynamically stable (LDS) [13] if (1) it is stationary w.r.t. $\alpha(\mathbb{R})$ and (2) for all self-adjoint elements h of $\mathcal{A}$, there exists a state $\omega^{(\lambda h)}$ that is stationary w.r.t. $\alpha^{(\lambda h)}(\mathbb{R})$ and tends (w*) to ω as $\lambda \to 0$.

Finally, let $t \to h_t$ be a continuously differentiable function from $\mathbb{R}$ into the self-adjoint elements of $\mathcal{A}$, with compact support in $\mathbb{R}_+$, and let $\{\alpha_1^{(h)}(t) | t \in \mathbb{R}\}$ be the one-parameter family of automorphisms of $\mathcal{A}$ governing the dynamics of the system in the presence of the perturbative Hamiltonian h_t, i.e.

$$\alpha_1^{(h)}(t)A = A + \int_0^t dt'\alpha_1^{(h)}(t')(\delta(A) + i[h_{t'},A]) \ \forall A \in D(\delta) \tag{11}$$

We say that $\omega(\in \Omega)$ is passive [14] if (1) it is stationary w.r.t. $\alpha(\mathbb{R})$

and (2) the state $\omega_t^{(h)} := \omega \circ \alpha_1^{(h)}(t)$ satisfies the following condition, which signifies that energy cannot be extracted from the system by means of a temporary local perturbation:-

$$\int dt\ \omega_t^{(h)}\left(\frac{dh_t}{dt}\right) \geq 0. \tag{12}$$

The passivity condition is thus an extrapolation of Kelvin's principle to even the microscopic level.

The stability conditions we have described are related to those of KMS as follows.

(I) For systems with finite range forces, LTS <=> KMS [9,11] and GTS <=> KMS + translational invariance [4,10].

(II) For systems with suitable long range forces, GTS => KMS, LTS, but, in some cases, there are translationally invariant KMS (or LTS) states that are not GTS [1,15].

(III) LDS + supplementary conditions (clustering, asymptotic abelianness) => KMS => LDS [13].

(IV) Passivity + supplementary conditions (ergodicity) => KMS => Passivity [14].

These results may be summarised by saying that

(1) the KMS conditions are equivalent, modulo certain supplementary ones, to the various <u>local</u> stability conditions (LTS, LDS, Passivity) considered here; and

(2) the GTS condition is equivalent to KMS + translational invariance for systems with finite range forces only: for systems with suitable long range forces it is strictly stronger than the latter combination of conditions.

3. EQUILIBRIUM AND METASTABLE STATES. We designate the equilibrium states to be those that are stable against all dynamical and thermodynamical perturbations. Thus, in view of the results of the previous Section, we propose the following characterisation of these states.

(E) <u>The equilibrium states are those that satisfy both the KMS and GTS conditions.</u>

Note that the condition becomes redundant here only if the forces are of short range (cf. results (I), (II) above).

As regards metastable states*, these are the ones that, from an empirical standpoint, simulate those of equilibrium except that they do not minimise the free energy of the system. Thus, they have 'very long'

---

* Examples of these states are abundant, e.g., supercooled or superheated phases of matter, certain allotropes of some materials (even diamond!), supercurrent carrying states of superconductors.

lifetimes in the face of local mechanical and thermal disturbances and enjoy 'good' thermodynamical properties. For reasons that should presently become clear, we propose a statistical mechanical characterisation of these states that divides them into two classes, that we term the ideal and the normal ones.

Thus, we say that ω is an ideal metastable state if

(IM1) it is not GTS;

(IM2) it is KMS; and

(IM3) it minimises the restriction of f, the free energy density functional, to some convex subspace Ω_M of Ω, that is stable under space translations.

Clearly, (IM1) is essential for any characterisation of metastability. Further [1], (IM2) guarantees that, apart from the stability properties already discussed, ω remains stationary, i.e., it has an infinite lifetime, in the face of thermal perturbations satisfying detailed balance, while (IM3) ensures its good thermodynamic properties. Thus, we contend that the conditions (IM), if fulfilled, would account for the main properties of metastable states. However, it follows from the results (I) - (IV) of Section 2 that (IM1,2) are incompatible in systems with short range forces, and therefore the only systems that might support (IM) states are ones with suitable long range interactions. Concrete examples of systems that have been shown to support them are the following.

(1) Lattice systems with many-body body forces, leading to the formation of 'clusters', which possess a 'superheated liquid phase' [15].

(2) The BCS, and similar mean field theoretical models, which support a supercooled normal phase below the transition temperature [16].

(3) (Probably!) A non-relativistic assembly of gravitating fermions, which appears likely to have a supercooled gaseous-type phase [17].

In view of the limited range of applicability of the IM conditions, we introduce a second class of states, that we designate the normally metastable ones, by replacing (IM2) by a condition of lower stability. Specifically, we characterise these states by normal metastability conditions (M1-3), where (M1) $\equiv$ (IM1), (M3) $\equiv$ (IM3) and (M2) is as follows.

(M2) There exists a family Γ of one-parameter semigroups $\gamma(\mathbb{R}_+)$ of completely positive transformations of $\mathcal{A}$ (C*-case), governing the dynamics of the system when coupled to suitable Markovian thermal reservoirs* at temperature β^{-1}, a subset $\mathcal{A}_0$ of $\mathcal{A}$, and a

* Thus, $\gamma(\mathbb{R}_+)$ may be characterised by the condition of detailed balance [18].

'small' parameter λ on which $\omega, \{H_\Lambda\}$ and Ω_M depend such that, if $\gamma(\mathbb{R}_+) \in \Gamma$, then

$$|\omega(\gamma_t A) - \omega(A)| < \psi(t/\tau(\gamma))\,||A|| \ \forall \ A \in \mathcal{Q}_0,\ t \in \mathbb{R}_+,$$

where the function $\psi: \mathbb{R}_+ \to \mathbb{R}_+$ contains no explicit λ-dependence,

$$\lim_{s \to 0} \psi(s) = 0$$

and $$\lim_{\lambda \to 0} \tau(\lambda) = \infty.$$

Thus, (M2) is an asymptotic stability condition, which requires that the restriction of to a chosen subset $\mathcal{Q}_0$ of observables should have an infinite lifetime, in the limit $\lambda \to 0$, in the presence of a suitable class of interactions (represented by Γ) with its thermal environment. The essential Physical idea behind this condition is that the escape of the state from Ω_M be impeded by free energy barriers which thus lead to an infinite lifetime for $\omega_{|\mathcal{Q}_0}$ in the limit $\lambda \to 0$: this is a less restrictive version of a similar idea, used in a classical context in Ref. [19] .

An example that has been proved to satisfy our conditions for normal metastability is provided by a state of an Ising ferromagnet, below the transition temperature, with polarization opposed to some weak external magnetic field [20] : here the lifetime of this state as restricted to the polarisation density tends to infinity as the external field tends to zero, in the presence of a suitably wide class of thermal environments.

We conclude this section with a brief discussion of an example of metastability that raises intriguing questions about the stability conditions we have formulated. The example is that of a supercurrent-carrying state of a superconductor. From an empirical standpoint, this enjoys perfect thermodynamical properties and appears to have an infinite lifetime in the face of prevailing laboratory perturbations [21] . One might imagine therefore that the state satisfies our ideal metastability conditions. However, we have shown that this cannot be the case by establishing [1, Section 5] that models satisfying the requirements of local gauge invariance and local charge conservation cannot support translationally-invariant, current-carrying KMS states. This was proved by showing that translationally invariant current-carrying states of such models must violate the inequality (7) for certain observables. Evidently, this result means that even the extraordinarily high stability of the supercurrent-carrying states is less than that demanded by the KMS conditions.

4. CONCLUDING REMARKS. The picture we have arrived at is one in which equilibrium and metastable states are classified in terms of their stability properties. Thus, the highest grade of stability is GTS, the next is KMS, and then there are lower grades, not yet systemmatically formulated, that suffice to provide the appearance, from an empirical standpoint, of perfect stability, as in the case of the supercurrent-carrying state discussed in Section 3.

REFERENCES

1. G. L. Sewell, Physics Reports 57, 307 (1980).

2. D. Ruelle, Statistical Mechanics (Benjamin, N.Y., 1969).

3. G. G. Emch, Algebraic Methods in Statistical Mechanics and Quantum Field Theory (Wiley-Interscience, New York, London, 1972).

4. H. Araki, Commun. Math. Phys. 38, 1 (1974).

5. E. H. Lieb and M. B. Ruskai, J. Math. Phys. 14, 1938 (1973).

6. R. F. Streater, Commun. Math. Phys. 6, 233 (1967);
D. W. Robinson, Commun. Math. Phys. 6, 151 (1967).

7. D. A. Dubin and G. L. Sewell, J. Math. Phys. 11, 2990 (1970).
M. Winnink, pp. 311-333 of 'Statistical Mechanics and Field Theory', Ed. R. N. Sen and C. Weil, Israel University Press, Jerusalem-London, 1973.
G. L. Sewell, Commun. Math. Phys. 33, 43 (1973)
O. Bratteli and D. W. Robinson: Commun. Math. Phys. 50, 135 (1976)

8. M. Fannes and A. Verbeure, Commun. Math. Phys. 55, 125 (1977);
Commun. Math. Phys. 57, 165 (1977).

9. G. L. Sewell, Commun. Math. Phys. 55, 53 (1977).

10. M. Fannes and A. Verbeure, J. Math. Phys. 19, 558 (1978).

11. H. Araki and G. L. Sewell, Commun. Math. Phys. 52, 103 (1977).

12. P. Torres, Thesis, University of London, 1976.

13. R. Haag, D. Kastler and E. Trych-Pohlmeyer, Commun. Math. Phys. 38, 173 (1974).

14. W. Pusz and S. L. Woronowicz, Commun. Math. Phys. 58, 273 (1978)

15. G. L. Sewell, Commun. Math. Phys. 55, 63 (1977).

16. G. G. Emch and H. J. F. Knops, J. Math. Phys. 11, 3008 (1970).

17. H. Narnhofer and G. L. Sewell, Commun. Math. Phys. 71, 1 (1980).

18. A. Kossakowski, A. Frigerio, V. Gorini and M. Verri, Commun. Math. Phys. 57, 97 (1977).

19. J. L. Lebowitz and O. Penrose, J. Stat. Phys. 3, 211 (1971).

20. P. Vanheuverzwijn, J. Math. Phys. 20, 2667 (1979).

21. F. London, Superfluids, Vol. 1 (J. Wiley and Sons, London, 1950).

PHYSICS DEPARTMENT, QUEEN MARY COLLEGE, UNIVERSITY OF LONDON, LONDON E1 4NS

Proceedings of Symposia in Pure Mathematics
Volume 38 (1982), Part 2

PHASE TRANSITIONS

Ola Bratteli

In the C*-algebraic approach to quantum statistical mechanics of closed systems the KMS (Kubo-Martin-Schwinger) condition is often used as a characterization of equilibrium. One reason for this is that infinite volume Gibbs states are KMS states. Recently this characterization has been justified in several ways.

1. For quantum lattice systems the KMS condition is equivalent to a local maximum entropy principle, and for translationally invariant states the condition is equivalent to a global maximum entropy principle (See G.L. Sewell's talk, or [4]).

2. Under a strong ergodicity hypothesis on the dynamics (L^1-asymptotic abelianess) the KMS condition is equivalent to a certain stability condition under perturbation of the dynamics (See D. Kastler's talk, or [4]).

3. Under relatively general circumstances the KMS condition is equivalent to the principle that the system cannot perform work under a temporary change in the dynamics. This is closely related to the second law of thermodynamics, [6], [4].

Let $\mathfrak{A}$ be a C*-algebra with identity I, and let $t \in \mathbb{R} \longrightarrow \gamma_t$ be a strongly continuous one-parameter group of *-automorphisms of $\mathfrak{A}$, with generator $\delta : \gamma_t = \exp\{t\delta\}$. Let $E_{\mathfrak{A}}$ be the state-space of $\mathfrak{A}$.

If $\omega \in E_{\mathfrak{A}}$ and $\beta \in R$, then ω is called a (γ,β)-KMS state if

$$\omega(A\gamma_{i\beta}(B)) = \omega(BA)$$

for all $A, B \in \mathfrak{A}$ such that B is entire analytic for γ. The state ω is called a $(\gamma,+\infty)$-KMS state if it is a ground state, i.e.

$$-i\omega(A^*\delta(A)) \geq 0 \quad \text{for all} \quad A \in D(\delta) \ ,$$

and ω is called a $(\gamma,-\infty)$-KMS state if it is a ceiling state, i.e.

$$-i\omega(A^*\delta(A)) \leq 0 \quad \text{for all} \quad A \in D(\delta) \ .$$

The parameter β is interpreted as the inverse temperature of the state.

1980 Mathematics Subject Classification.

0082-0717/80/0000-0505/$01.75

Let K_β be the set of (γ,β)-KMS states. The set K_β is a compact, convex subset of $E_{\mathfrak{A}}$ for each $\beta \in [-\infty,+\infty]$. K_β is a simplex if $\beta \in \langle -\infty,+\infty\rangle$, and $K_{\pm\infty}$ are faces in $E_{\mathfrak{A}}$. On the other hand, it happens only under very special circumstances that K_β is a face for $|\beta| < +\infty$, or that $K_{\pm\infty}$ are simplexes.

We next consider the structure of the map $\beta \longrightarrow K_\beta$. If ω_α is a weak*-convergent net in $E_{\mathfrak{A}}$ such that $\omega_\alpha \in K_{\beta_\alpha}$ for each α, and $\beta_\alpha \in [-\infty,+\infty]$ converges to some $\beta \in [-\infty,+\infty]$, then $\lim_\alpha \omega_\alpha \in K_\beta$. It follows that

$$\mathcal{K} = \{(\beta,\omega),\ \omega \in K_\beta\ ,\ \ \beta \in [-\infty,+\infty]\}$$

is a closed subset of the topological product $[-\infty,+\infty] \times E_{\mathfrak{A}}$. If we define a continuous map $\Pi : \mathcal{K} \longrightarrow [-\infty,+\infty]$ by $\Pi(\beta,\omega) = \beta$, then $(\mathcal{K},\Pi)$ has a natural structure as a bundle of compact convex sets. This bundle is called the temparature state space of the C*-dynamical system. The restriction $(\mathcal{K}_0,\Pi)$ to $\langle -\infty,+\infty\rangle \subset [-\infty,+\infty]$ is called the restricted temperature state space, and $(\mathcal{K}_0,\Pi)$ is a simplex bundle over $\mathbb{R}$, i.e.

1. $\mathcal{K}_0$ is a Hausdorff topological space such that $\Pi^{-1}(F)$ is a compact subset of $\mathcal{K}_0$ for each compact $F \subseteq \mathbb{R}$.

2. $\Pi : \mathcal{K}_0 \longrightarrow \mathbb{R}$ is continuous

3. $K_\beta = \Pi^{-1}(\beta)$ is a simplex for all $\beta \in \mathbb{R}$ (The case $K_\beta = \phi$ is allowed)

4. $A(\mathcal{K})$ is full

Here $A(\mathcal{K})$ is the set of continuous real functions on $\mathcal{K}_0$ which are affine on each fibre K_β, and that $A(\mathcal{K})$ is full means that each continuous real affine function on a fibre K_β has an extension to $A(\mathcal{K}_0)$.

We say that the system represented by $(\mathfrak{A},\gamma)$ has a (first order) phase transition at inverse temperature β if there exists more than one (γ,β)-KMS state, i.e. if K_β consists of more than one point. The following result shows that in the general frame-work of C*-dynamical systems every conceivable structure of phase transitions can occur, at least in the separable case, and even within the class of simple C*-algebras.

THEOREM [2], [3] Let $(\mathcal{K},\Pi)$ be any metrizable simplex bundle over $\mathbb{R}$. It follows that there exists a C*-dynamical system $(\mathfrak{A},\gamma)$ such that the restricted temperature state space of $(\mathfrak{A},\gamma)$ is isomorphic to $(\mathcal{K},\Pi)$. $\mathfrak{A}$ can be taken to be simple, separable and nuclear. γ can be taken to be periodic with specified period 2Π.

In particular, if F is any closed subset of $[-\infty,+\infty]$ one can find a simple unital C*-algebra $\mathfrak{A}$, and a one-parameter group γ of *-automorphisms of $\mathfrak{A}$ such that there exists (γ,β)-KMS states if and only if $\beta \in F$. The system can be chosen such that these KMS states are unique, [1]. In the general case of the theorem one can choose the C*-dynamical system $(\mathfrak{A},\gamma)$ such that the set $K_{+\infty}/K_{-\infty}$ of ground/ceiling states is the state space of an arbitrary simple separable unital AF-algebra, and if one remove the periodicity requirement on γ, $K_{+\infty}/K_{-\infty}$ can be taken to be arbritory faces of the state spaces of such algebras. For a more complete review of these results, see [5].

In the case of UHF algebras $\mathfrak{A}$, and dynamics γ on $\mathfrak{A}$ defined by potentials, there exist several results on absence and presence of first order phase transitions. Some of these are reviewed in Sakais section of this article. For a comprehensive review of results of this nature and further references the reader is referred to [4].

REFERENCES

1. O. Bratteli, G.A. Elliott and R.H. Herman, On the possible temperatures of a dynamical system, Comm. Math. Phys. 74 (1980), 281-295.

2. O. Bratteli, G.A. Elliott and A. Kishimoto, The temperature state space of a C*-dynamical system, I, Yokohama Univ. J., to appear.

3. O. Bratteli, G.A. Elliott and A. Kishimoto, The temperature state space of a C*-dynamical system, II, in Preparation

4. O. Bratteli and D.W. Robinson, Operator Algebras and Quantum Statistical Mechanics, I, Springer Verlag, New York-Heidelberg-Berlin (1979), II, to appear the end of 1980.

5. O. Bratteli, On temperature states and phase transitions in C*-dynamical systems, to appear in the Proceedings of the International Conference on Operator Algebras and Group Representations, Neptun, Romania, 1-13 September 1980.

6. W. Pusz and S.L. Wororowicz, Passive States and KMS states for general quantum systems, Comm. Math. Phys. 58 (1978), 273-290.

Department of Mathematics
University of Trondheim
Trondheim, Norway

Proceedings of Symposia in Pure Mathematics
Volume 38 (1982), Part 2

NATURE OF BOSE CONDENSATION

A. Verbeure

ABSTRACT. After the work of Araki and Woods [1] on the infinite free Bose gas, we discuss two other solvable models. We report on the nature of the equilibrium states for the so-called imperfect Bose gas in $\nu \geqslant 3$ dimensions [2] and for the one dimensional Bose in a dilated external field recently studied by van den Berg[3].

I. IMPERFECT BOSE GAS.

Let ν be the dimension, Λ any centered cube in $\mathbb{R}^\nu$.

For ϕ $L^2(\Lambda, dx)$,

$a(\phi)$ and $a^+(\phi)$ are the Fock creation and annihilation operators.

The model is given by the following local Hamiltonians:

$$H_\Lambda = T_\Lambda - \mu_\Lambda N_\Lambda + \frac{\lambda}{2} \frac{N_\Lambda^2}{V(\Lambda)} \quad ; \lambda > 0 \tag{1}$$

where T_Λ is the kinetic energy, N_Λ the number of particles operator, μ_Λ is the chemical potential fixed by the condition

$\omega_{\beta,\Lambda}(N_\Lambda) = \rho\, V(\Lambda)$ for some $\rho > 0$ in $\mathbb{R}$, where $\omega_{\beta,\Lambda}$ is the Gibbs state for (1).

We have the following:

THEOREM [2]

If $\nu \geqslant 3$, then

1. Let $\phi_o^\Lambda(x) = \frac{1}{V(\Lambda)^{1/2}}$ for $x \in \Lambda$

$= 0$ for $x \notin \Lambda$

then

$$\rho_o \equiv \lim_{\Lambda\to\infty} \frac{\omega_{\beta,\Lambda}(a^+(\phi_o^\Lambda)\ a(\phi_o^\Lambda)\ )}{V(\Lambda)}$$

0082-0717/80/0000-0585/$02.25

$$\geq \rho - \left(\frac{1}{2\pi}\right) \int_{\mathbb{R}^\nu} \frac{dk}{e^{\beta\frac{k^2}{2}} - 1}$$

2. $W^\star - \lim_\Lambda \omega_{\beta,\Lambda}$ exists, denote it by the state ω_β on the CCR-algebra.
It satisfies the following continuity property:

$$f \in \mathcal{D}(\mathbb{R}^\nu) \to \omega_\beta(W(f)) \qquad \text{continuous,}$$

$W(f) = \exp i(a^+(f) + a(f))$
$f \in H =$ closure of $\mathcal{D}(\mathbb{R}^\nu)$ with respect to

$$|||f|||^2 = \rho_o |\tilde{f}(0)|^2 + \int dk \frac{|\tilde{f}(k)|^2}{e^{\beta\frac{k^2}{2}} - 1}$$

3. Let π_β be the G.N.S. representation of ω_β, then

$$\text{str.}\lim_\Lambda \frac{\pi_\beta(N_\Lambda)}{V(\Lambda)} = \rho\, 1$$

$$\text{str.}\lim_\Lambda \frac{a_\beta^+(\phi_o^\Lambda)\, a_\beta(\phi_o^\Lambda)}{V(\Lambda)} = \rho_o\, 1$$

4. Let $\phi_o = \left(\left(\frac{1}{2\pi}\right)^{\nu/2}, 0\right) \in H$, then, if $\rho_o \neq 0$;

$U_o \equiv \dfrac{\bar{a}_\beta(\phi_o)}{\rho_o^{1/2}}$ is a unitary element of the center of the von Neumann algebra $M = \pi_\beta(\text{CCR})''$

Let $U_o = \int_0^1 e^{i\,2\pi\alpha} E(d\alpha)$ be the spectral resolution of ω_β, then

$$\omega_\beta(\cdot) = \int_0^1 \omega_\beta(E(d\alpha)\,.)$$

is the central decomposition of ω_β.

REMARKS.

Point 1. Yields the existence of a Bose condensation in the zero mode.

Point 2. shows the richness of the von Neumann algebra M in the sense that there exists many fields below the critical temperature.

Point 3. can be interpreted as a form of equivalence of ensembles

Point 4. is a decomposition of the limit state into facto states below the critical temperature.

II. ONE DIMENSIONAL BOSE GAS

None of the statements in I are valid. However with external fields van den Berg [3] shows that condensation is possible by arguing on the free energy. The question here is how do the states look on the CCR for this model?

We define the model as follows:

$$\alpha_t^L\, a(f) = a(e^{-it\, h_L}\, f)$$

$$h_L = -\frac{1}{2}\frac{\partial^2}{\partial x^2} - \mu_L + V(\tfrac{x}{L}) \quad \text{on} \quad \mathcal{L}^2(\Lambda)$$

and $\Lambda = [-L,L]$.

Conditions on the potential V:

(a) $|V(x) - V(y)| \leqslant K|x-y|^{\alpha}$, $\alpha > \frac{1}{2}$, K IR

(b) $\overline{\lim_n}\; n^{\delta} \int_{-1}^{1} dq\; e^{-\beta n\, V(q)} < \infty \quad ; \delta > \frac{1}{2}$

Consider the following set of functions

$$\{ f_{k,q}^L \;|\, k \in \mathbb{R} \;,\; q \in [-1,1] \}$$

where

$$f_{k,q}^L(x) = e^{ikx} \left(\frac{L^{-\varepsilon}}{2}\right)^{1/2} e^{-\frac{L^{-\varepsilon}}{2}|x-Lq|} \quad \text{if } x \in [-L,L] \text{ for some } \varepsilon : \; 0 < \varepsilon < 1$$

$$= 0 \quad \text{if } x \notin [-L,L]$$

THEOREM [4]

1. If $k \neq 0$ or $q \neq q_m$ where $V(q_m) = \min_q V(q)$ and V satisfies condition (a) then:

$$\lim_{L} \omega_{\beta,L} (a^+(f^L_{k,q})q(f^L_{k,q}))$$

$$= \{\exp \beta(\frac{k^2}{2} + V(q) - V(q_m)) - 1\}^{-1}$$

$$\lim_{L} \omega_{\beta,L}(a^{\pm}(f^L_{k,q})a(f^L_{k',q'})) = 0 \quad \text{if } q \neq q'$$

2. If V satisfies conditions (a) and (b) then

$$\underline{\lim}_{L\to\infty} \sum_{q_m} \frac{\omega_{\beta,L} (a^+(f^L_{o,q_m})a(f^L_{o,q_m}))}{2L}$$

$$\geq \rho - \frac{1}{2\pi} \int_{\mathbb{R}} dk \int_{-1}^{1} dq \frac{1}{\exp \beta(\frac{k^2}{2} + V(q) - V(q_m))^{-1}}$$

3. $W^\star - \lim_{L\to\infty} \omega_{\beta,L}$ exists, and its explicit from can be written down.

REMARKS

Point 1. The functions $f^L_{k,q}$ are something like apprimate coherent states and we prove that they "diagonalize" the Hamiltonian in the limit $L \to \infty$. The second property shows the product property of the state for different values of q.
Point 2. yields the existence of condensation and the particles condensate in the minima of the potential with zero momentum. The external potential forces in a way the particles to condensate in these minima.

REFERENCES

1 H. Araki, J. Woods; J. Math. Phys. 4, 637 (1963).

2 M. Fannes, A. Verbeure; J. Math. Phys. 21, 1809 (1980).

3 M. van den Berg; Physics Letters , to appear.

4 J. Messer, A.Verbeure; in preparation.

INSTITUUT VOOR THEORETISCHE NATUURKUNDE
UNIVERSITEIT LEUVEN
Celestijnenlaan 200 D
B- 3030 Leuven, Belgium

Proceedings of Symposia in Pure Mathematics
Volume **38** (1982), Part 2

PERTURBATIONS OF FREE EVOLUTIONS AND POISSON MEASURES

Ph. COMBE*, R. HOEGH-KROHN**, R. RODRIGUEZ*,
M. SIRUGUE***, M. SIRUGUE-COLLIN****

ABSTRACT. We analyse the structure of the unitary cocycles between two one parameter groups of $*$-automorphisms of the generalized canonical commutation C^*-algebra in terms of a stochastic process, whose paths have a clear interpretation in terms of classical mechanics.

The C^*-algebras of generalized canonical commutation relations (G.C.C.R.) $\Delta\overline{(G,\zeta)}$ have one parameter groups of $*$-automorphisms $\hat{\alpha}_t^o$ which correspond to one parameter groups of automorphisms α_t^o of G. These automorphism groups generalize free evolution. In this report we prove that within a representation of $\Delta\overline{(G,\zeta)}$ where $\hat{\alpha}_t^o$ is unitarily implemented, we can perturb the corresponding unitary group by gentle perturbations. The corresponding unitary cocycle is analysed in terms of well defined Feynman path integral associated to a generalized Poisson process, whose paths are the classical paths associated with α_t^o, except for a finite number of times. Moreover the measure defining the Feynman integral is the Poisson measure associated with the perturbation.

Of central importance in what follows are the C^*-algebras $\Delta\overline{(G,\zeta)}$ of generalized canonical commutation relations, which can be briefly described as follows [1] : let G be a locally compact abelian group and ζ a multiplier on G, namely a function from $G \times G$ to the torus T, such that $\zeta(e,g) = 1, \forall g \in G$ and

$$\zeta(g_1,g_2)\zeta(g_1g_2,g_3) = \zeta(g_1,g_2g_3)\zeta(g_2,g_3)$$

One has the following proposition :

PROPOSITION. If ζ is non degenerated in the sense that $\zeta(g,g')\overline{\zeta(g',g)} = 1$, $\forall g' \in G$, implies $g = e$ (the identity of G), then there exists only one C^*-algebra $\Delta\overline{(G,\zeta)}$ such that its $*$-representations are in one to one correspondence with the unitary projective representations of G with multiplier ζ. Moreover $\Delta\overline{(G,\zeta)}$ is generated by the unitaries W_g, $g \in G$ such that $W_g^* = W_{g^{-1}}$ and $W_g W_{g'} = \zeta(g,g')W_{gg'}$.

0082-0717/80/0000-0514/$02.00

On these algebras one can define the generalized free evolutions : let $t \to \alpha_t^o$ be a one-parameter group of automorphisms of G, such that $\zeta(\alpha_t^o g, \alpha_t^o g') = \zeta(g,g')$, then $\hat{\alpha}_t^o(W_g) = \lambda(t,g) W_{\alpha_t^o g}$, where λ is a one cocycle in T, extends to a one-parameter group of *-automorphisms of $\overline{\Delta(G,\zeta)}$.

Now we shall describe the Poisson Processes which are associated with such a generalized free evolution, [2]. The underlying probability space can be realized as follows : let G be an abelian locally compact group and $T > o$, then we consider the space $\Omega = \bigsqcup_{n \geq o} \Omega_n$ where

$$\Omega_n = \left\{ \omega_n = (n, t_1, \ldots t_n, g_1 \ldots g_n) \; ; \; o < t_1 < t_2 \ldots \; t_n < T, \; g_i \in G \right\}$$

moreover Ω_o is just a point ω_o.

F_T is the smallest σ-algebra on Ω which contains the sets $\mathcal{V}^{(o)} = \{\omega_o\}$ and

$$\mathcal{V}^{(n)}_{a_i \mathcal{B}_i} = \left\{ (n, t_i, g_i) \; : \; t_i \in a_i \, , \, g_i \in \mathcal{B}_i \right\}$$

where the a_i's are disjoint ordered Lebesgue measurable subsets of $[0,T]$ and the $\mathcal{B}_i$'s are Borel subsets of G. The following proposition can be found, e.g. in [3]

PROPOSITION. Given a (positive) bounded measure μ on G there exists a (positive) bounded measure P_μ on Ω which extends the additive function on the $\mathcal{V}^{(n)}$ defined by

$$P_\mu(\mathcal{V}^{(o)}) = 1 \quad \text{and} \quad P(\mathcal{V}^{(n)}_{a_i, \mathcal{B}_i}) = \prod_{i=1}^{n} |a_i| \mu(\mathcal{B}_i)$$

where $|a_i|$ is the Lebesgue measure of a_i.

(Ω, F_T, P_μ) is a probability space (up to the normalization of P_μ).

If F is an integrable function on Ω then

$$\int_\Omega P_\mu(d\omega) F(\omega) = \sum_{n \geq o} \int_o^T dt_n \ldots \int_o^{t_2} dt_1 \int_G \ldots \int_G d\mu(g_1) \ldots d\mu(g_n) \; F(n, t_i g_i)$$

Then we define the generalized Poisson process X associated with a one-parameter group α_t^o of automorphisms of G by

$$X_\tau(n, t_i g_i) = e \quad \text{if} \quad \tau \in \left]t_n, T\right]$$

$$X_\tau(n, t_i g_i) = \alpha^o_{-t_i}(g_i^{-1}) \ldots \alpha^o_{-t_n}(g_n^{-1}) \qquad \tau \in \left]t_{i-1}, t_i\right]$$

$$X_\tau(n, t_i g_i) = \alpha^o_{-t_1}(g_1^{-1}) \ldots \alpha^o_{-t_n}(g_n^{-1}) \qquad \tau \in \left[0, t_1\right]$$

Furthermore

$$\Delta X_\tau(n, t_i g_i) = \lim_{\epsilon \downarrow o} \; X_{\tau - \epsilon}(\omega) \, X_{\tau + \epsilon}(\omega)^{-1}$$

and N is the usual Poisson Process $N(n, t_i g_i) = \sum_{i=1}^{n} \theta_+(\tau - t_i)$.

Then we can state the main theorem : [4]

THEOREM. Let π be a measurable representation of $\overline{\Delta(G,\zeta)}$ in the Hilbert space $\mathcal{H}$. Let $\hat{\alpha}_t^o$ be a generalized free evolution which is unitarily implemented in this representation by a strongly continuous group of unitaries $t \to \exp(i \, t \, H_o)$. If moreover V is an operator on $\mathcal{H}$ of the form

$$V = \int_G d\mu(g)\ \pi(W_g)$$

where μ is a bounded measure on G, then

$$(\Phi | e^{iTH_o} e^{-iT(H_o+V)} \Psi)$$
$$= \int_\Omega P_\mu(d\omega) \left\{ (-i)^{N_o}\ \overline{Z(X^{-1}, \Delta X)}\ F_{\Phi\Psi}(X_o) \right\} (\omega)$$

where the functions

$$\omega = (n, t_i g_i) \in \Omega \rightarrow Z(X^{-1}, \Delta X)(\omega) = \prod_{i=1}^{n} \zeta(X_{t_i}(\omega)^{-1}, \Delta X_{t_i}(\omega))$$

and

$$\omega = (n, t_i g_i) \in \Omega \rightarrow F_{\Phi\Psi}(\omega) = (\Phi\pi(W_{X_o^{-1}(\omega)})\Psi)$$

are measurable bounded functions on Ω.

We want to mention that this general result has a special application to quantum field theory and more precisely to the Sine-Gordon model. Let

$$V_k^\Lambda = \lambda \int_\Lambda dx\ :\cos(\phi_k(x,o)):$$

be the interaction of the Sine-Gordon model where Λ is a finite box, ϕ_k is the field with ultraviolet cut off. Then there exists a natural generalized Poisson process X_{xt} on a probability space Ω which is of the type previously described, but with basis $\Lambda \times \mathbb{Z}_2$ and a Poisson measure which corresponds to the measure μ on that basis which is the tensor product of the Lebesgue measure on Λ by the invariant measure on $\mathbb{Z}_2$:

$$X_{xt}(\omega = (n, t_i, x_i, \epsilon_i)) = \sum_{i=1}^{n} \epsilon_i \int_R d\xi\, \Delta_F(x - x_i - \xi, t - t_i)\, \psi_k(\xi)$$

where $\psi_k(\xi)$ is an ultraviolet cut off function, and Δ_F the Feynman propagator.

One can analyse the structure of the transition matrix in the Fock vacuum Ψ according to the formula

$$(\Psi | \exp(iH_oT)\ \exp(-iT(H_o + V_\Lambda^k))\Psi) = \int_\Omega P_{\Lambda T}(d\omega)\ F_\Lambda^k(\omega)$$

where $F_\Lambda^k(\omega = (n, t_i, x_i, \epsilon_i))$

$$= \left(\frac{i\lambda}{2}\right)^n \exp\left(-\frac{\alpha^2}{4} \sum_{i \neq j} \epsilon_i \epsilon_j \iint d\xi\, d\zeta\, \psi_k(x_i - \xi)\, \Delta_F(\xi - \zeta\,;\ t_i - t_j)\, \psi_k(x_j - \zeta)\right)$$

BIBLIOGRAPHY

[1] MANUCEAU, J., SIRUGUE, M., TESTARD, D., VERBEURE, A., Commun.Math. Phys., 32 (1973), 231.

[2] COMBE, Ph., HOEGH-KROHN, R., RODRIGUEZ, R., SIRUGUE, M., SIRUGUE-COLLIN, M., Poisson Processes on Groups and Feynman Path Integrals, to appear in Commun.Math.Phys.

[3] MASLOV, V.P., CHEBOTAREV, A.M., in Proceedings of the Colloquium on Feynman Path Integral, Marseille 1978, Springer Lecture Notes in Physics, n° 106 (1979).

[4] COMBE, Ph., HOEGH-KROHN, R., RODRIGUEZ, R., SIRUGUE, M., SIRUGUE-COLLIN, M., Feynman Path Integral and Poisson Process with Piecewise Classical Path (in preparation).

* Centre de Physique Théorique, CNRS, MARSEILLE, and Université d'Aix-Marseille II, Centre de Luminy

** Universitet i OSLO

*** Centre de Physique Théorique, CNRS, MARSEILLE

**** Centre de Physique Théorique, CNRS, Marseille, and Université de Provence

Current Address : Centre de Physique Théorique
CNRS - Luminy - Case 907
F-13288 MARSEILLE CEDEX 2 (FRANCE)

Proceedings of Symposia in Pure Mathematics
Volume 38 (1982), Part 2

STRUCTURAL QUESTIONS IN QUANTUM FIELD THEORY

Rudolf Haag

ABSTRACT. The structural assumptions of local, relativistic quantum physics are described and their relation to observable and unobservable fields, charge quantum numbers and particle aspects is sketched.

1. MECHANICS VERSUS FIELD THEORY.

Before the advent of quantum theory there were two basic conceptual schemes in terms of which physical phenomena were ordered. The first was Newtonian mechanics. Here the physical system is a collection of indestructible particles. The "state" of the system is given by the position of each particle in space as a function of time (orbits of the particles in space-time). The laws which determine the possible orbits are expressed by means of the concept of forces which depend on the instantaneous configuration of the particles in space ("action at a distance").

The other scheme (Faraday-Maxwell-Einstein) was based on the concept of fields and the principle of locality ("Nahwirkungsprinzip"). Here the physical system is space-time; the state is given by various field quantities F_i which are continuous functions over space-time. The laws are hyperbolic partial differential equations for the functions $F_i(x)$ guaranteeing that the maximal velocity for the propagation of any effect is the velocity of light.

COMMENT. The special theory of relativity does not only require invariance under Lorentz-transformations. One can construct Lorentz invariant mechanics with an action at a disstance (ugly though that is). Einstein's causality principle which sharpens the principle of locality is an independent, essential ingredient. It assigns to each space-time point x a region which may be influenced by the situation at x. It is the set of points in positive time-like directions from x i.e. the interior of the closed

0082-0717/80/0000-0533/$03.25

positive light cone. We may call it the "causal shadow" of x. Two points x, y which lie "space-like" to each other so that neither is in the causal shadow of the other can have no causal connection.

2. QUANTUM MECHANICS AND QUANTUM FIELD THEORY.

In Quantum Physics the quantities which were used to describe the state of a system in the classical theory have no definite numerical values.They become "observables" which are mathematically represented by self-adjoint elements of a noncommutative $*$-algebra. Corresponding to the two basic classical schemes one has quantum mechanics and quantum field theory. The latter has been developed up to the level in which it includes the principles of special relativity. No satisfactory fusion between the concepts of general relativity and quantum physics exists.

a) LINEAR THEORY. For a classical, relativistic field theory with linear field equations a corresponding quantum field theory can be defined with any desired amount of mathematical rigour and conceptual conciseness. The amazing result is that it describes the same physical phenomena as the quantum mechanics of a system of arbitrarily many indistinguishable particles which do not interact. So in some sense the fundamentally different concepts of mechanics and field theory appear to be brought into harmony by the quantum theory. However one should keep in mind that in the special case of linear theories the Nahwirkungsprinzip looses its bite since the forces between the particles are zero.

b) NON LINEAR THEORY. Consider the classical field equation

$$\frac{\partial^2 \phi}{\partial t^2} - \Delta \phi + g\,\phi^3 = 0 \quad . \tag{1}$$

If the constant g is positive then there exist global solutions and the generic behavior for large positive or negative times is a spreading out over large parts of space and consequently a decrease of the magnitude of ϕ at every point. Thus ϕ^3 becomes ultimately negligeable compared to ϕ and the solution of (1) is asymptotic at $t \to \pm\infty$ to solutions of the linear field equation

$$\frac{\partial^2 \phi}{\partial t^2} - \Delta \phi = 0 \tag{2}$$

One may say that $\phi(x)$ interpolates between two solutions of (2) $\phi^{in}(x)$ resp. $\phi^{out}(x)$, to which it is asymptotic at $t \to -\infty$ resp. $t \to +\infty$.

For the corresponding quantum theory one expects an analoguous behavior: the quantum field ϕ should interpolate between two "free" fields (Fields obeying the linear equation) ϕ^{in}, ϕ^{out} which in turn may be related to the description of freely moving incoming resp. outgoing particles, the difference between ϕ^{in} and ϕ^{out} representing the cumulative effect of the interaction during finite times. Thus, if one can compute the relation between ϕ^{in} and ϕ^{out} on has all the information about possible collision processes, one can determine the scattering matrix S and collision cross sections.

The actual formulation of a quantum theory corresponding to the field equation (1) is not a simple task and has not been achieved in a fashion which could convince a mathematician of the existence of such a theory. One encounters first of all the problem that the field at a point cannot be a bona fide observable. It is a singular object like Dirac's δ-function. Therefore any nonlinear function of the field at a point can at best be defined in terms of very careful limiting processes. This can indeed be done by the famous renormalization procedure if one treats the problem in the frame work of perturbation expansions (power series in the coupling constant g) where one can also exhibit the existence of the asymptotic fields ϕ^{in}, ϕ^{out} and compute the S-matrix to any finite order in g. It is however not known, whether this really defines a theory.

The one-to-one connection between fields and particle types which exists in linear quantum field theories and (probably) in a simple non linear model like the one described above should not seduce us to believe that the concepts of "field" and "particle" are essentially synonymous in quantum physics. In my opinion there is a clear hierarchy: The principle of locality is the essential ingredient of relativistic quantum physics. The fields may be regarded as a convenient vehicle in terms of which this principle can be formulated (see below). The particles are the stable excitations which the theory allows, they are the roof, not the foundation of the building. There are models of relativistic, local field theories which do not have any stable excitations at all (no particles) and others in which the number of different types of particles is much larger than the number of fundamental fields and where non of these particle types is related to the fundamental fields in the manner described above. Such a behavior is expected e.g. in "Quantum Chromodynamics" which is the most hopeful candidate for a theory of strong interactions at present time.

3. LOCAL, RELATIVISTIC QUANTUM PHYSICS.

In discussing the general structure of a theory which incorporates the principles of quantum physics and special relativity it is better not to begin by trying to define a concept of quantum fields. Experience with Quantum Electrodynamics - which is the best understood realistic model of a quantum field theory - shows that an axiomatization of the basic fields leads to a complicated not very natural looking set of assumptions. There are technicalities which result from the fact that the fields at a point are singular objects and field averages are still represented by unbounded operators. More serious, however, is the fact that the basic fields used in Quantum Electrodynamics are unobservable and as a consequence they are endowed with properties which have to be postulated ad hoc; these properties emerged in the long historical development of the theory by trial and error and can not be understood as first principles. We shall therefore formulate local, relativistic quantum physics in a way in which fields are not the primary objects.

How can the principle of strict locality be incorporated into quantum physics? It asserts that for any chosen region $\mathcal{O}$ of space-time there are observables which can be measured precisely inside that region and, in fact, without loss of generality we can limit our attention to the bounded observables. In orthodox quantum physics the set of bounded observables corresponds precisely to the set of self adjoint elements of a C^* -algebra. Hence we are led to the following picture:

i) There is a correspondence between open regions of Minkowski space M (= space-time in special relativity) and C^* -subalgebras of a C^* -algebra $\mathfrak{A}$

$$\mathcal{O} \rightarrow \mathfrak{A}(\mathcal{O}) \; ; \quad \mathcal{O} \subset M \, , \; \mathfrak{A}(\mathcal{O}) \subset \mathfrak{A} \, . \tag{3}$$

The physical interpretation of the scheme is tied to this correspondence (3): the self adjoint elements of $\mathfrak{A}(\mathcal{O})$ represent precisely the bounded observables of the region $\mathcal{O}$. No other physical interpretation is available nor is any more needed. $\mathfrak{A}(\mathcal{O})$ is called the algebra of local observables in $\mathcal{O}$. $\mathfrak{A}$ is the algebra of all quasilocal observables; it is the C^*-algebra generated by all $\mathfrak{A}(\mathcal{O})$. We shall refer to (3) as a "local net of C^*-algebras" and assert that if the net (3) is given then the physical theory is completely fixed. Of course the net has to have various properties in order to describe a reasonable physical theory. According to the principles of special relativity one should have

ii) Einstein causality. If $\mathcal{O}_1$ and $\mathcal{O}_2$ are regions which lie spacelike to each other then $\mathfrak{A}(\mathcal{O}_1)$ should commute with $\mathfrak{A}(\mathcal{O}_2)$. (A measurement in $\mathcal{O}_1$ cannot have any effect on a measurement in $\mathcal{O}_2$.)

iii) Existence of a dynamical law with hyperbolic propagation. If $\mathcal{O}_1$ is in the causal shadow of $\mathcal{O}$ then

$$\mathfrak{A}(\mathcal{O}_1) \subset \mathfrak{A}(\mathcal{O})$$

iv) Poincaré invariance. The Poincaré group $\mathcal{P}$ (translations and Lorentz-transformations in M) is represented by automorphisms acting on $\mathfrak{A}$. $g \in \mathcal{P} \rightarrow \alpha_g \in \operatorname{Aut} \mathcal{M}$. Moreover

$$\alpha_g\left(\mathfrak{A}(\mathcal{O})\right) = \mathfrak{A}(g\mathcal{O}). \tag{4}$$

There is one further essential requirement, the stability. In its simplest form it asserts that there should be a ground state, the physical vacuum. A state is mathematically described as a positive linear form over $\mathfrak{A}$ with norm 1. Given a state ω the Gelfand-Naimark-Segal construction gives a representation π_ω of $\mathfrak{A}$ by bounded operators acting on a Hilbert space. If the state is invariant under the (dual action of the) automorphisms α_g then we get also a representation of the Poincaré group by unitary operators in this Hilbert space. If this representation is strongly continous for the subgroup of translations we may talk about the energy-momentum spectrum i.e. the spectrum of the self adjoint operators which represent the generators of the translation group. The invariant state ω_0 is represented by a vector Ω_0 in the Hilbert space which is an eigenvector of the energy-momentum operators to the eigenvalue o. If the energy spectrum is contained in the positive semi-axis then we may call ω_0 the ground state (in the associated GNS-representation). Thus we could demand

v) Stability. There exists a unique state ω_0 , the physical vacuum, which is Poincaré invariant and is the ground state in the associated GNS-representation.

Further requirements on the net may be considered e.g.

vi) Additivity.

$$\mathfrak{A}(\cup \mathcal{O}_i) = \vee \mathfrak{A}(\mathcal{O}_i) \tag{5}$$

where the right hand side denotes the C^*-algebra generated by all the $\mathfrak{A}(\mathcal{O}_i)$.

This brings the theory in closer connection to field theory since it implies that the algebra $\mathfrak{A}$ is generated from the algebras of arbitrarily small cells. The observalbe fields may then be considered as the germs of the net; a generating subset of these germs ("quantum fields") provides a coordinatization of the net.

4. SUPERSELECTION SECTORS. CHARGE QUANTUM NUMBERS. UNOBSERVABLE FIELDS. STATISTICS.

We have mentioned the vacuum state ω_0 and the representation π_0 of $\mathfrak{A}$ and U_0 of $\mathcal{P}$ which results from this state by the GNS construction. In general there will be other (inequivalent) representations of $\mathfrak{A}$ which are of interest. As "representations of interest" we may consider e.g. all irreducible representations in which the Poincaré group is implemented by unitaries in a strongly continous fashion and the energy spectrum is bounded below. Or we may consider all representations which contain vector states describing "local excitations of the vacuum" i.e. states which coincide with the vacuum for observations at space-like large distances from the origin. Both criteria single out essentially the same family of representations. Each representation (up to equivalence) in this family is called a superselection sector. The labels which distinguish the different sectors may be interpreted in their physical significance as charge quantum numbers.

I shall no enter into a discussion of the superselection structure of a local, relativistic quantum theory but refer to the talk of K. Fredenhagen at this conference. Therefore a few remarks on this topic may suffice here. The existence of non trivial charge quantum numbers is related to the existence of morphisms of the algebra $\mathfrak{A}$ which are not inner and have certain localization properties. There is a composition law of charges and to each charge there is a unique conjugate charge. Also associated with each charge is a "statistics type" which comes into play when one considers the multiple composition of the charge with itself (physically leading to states with n-fold charge) and which corresponds to the distinction of Bose and Fermi statistics (or para-Bose, para-Fermi). It is gratifying to see how these well known possibilities arise as a natural consequence from the principles of local quantum physics.

This brings us immediately to the significance of unobservable fields. If one wants to describe all the states in all superselection sectors together one might take the direct sum of all the representations with different charge quantum numbers. The observables will not connect the different sectors. Nevertheless it may be convenient to introduce operators leading from one sector to a different one (operators which change the charge). These are, however, not observables.

5. PARTICLES.

In order to search for particles we must have detectors. So we shall discuss first which elements of the algebra $\mathfrak{A}$ can be regarded as the mathematical counterparts of a physical detector with essential radius r, centerd at the origin of the coordinate system. Let $\mathcal{O}_r$ denote the cylinder in space-time, whose basis is a 3-dimensional ball of radius r, whose height a small time interval τ and whose center is the origin. Then consider an element $C \in \mathfrak{A}$ such that

$$\text{i)} \quad C = C^2 = C^* \qquad (6)$$

$$\text{ii)} \quad \omega_0(C) = 0 \qquad (7)$$

$$\text{iii)} \quad \| C - \mathfrak{A}(\mathcal{O}_r) \| < \varepsilon, \qquad (8)$$

in other words a projector with vanishing vacuum expectation value which can be approximated up to accuracy ε in norm by elements from $\mathfrak{A}(\mathcal{O}_r)$. C corresponds to an observable which is essentially localized in the specified region, which gives no signal in the vacuum state and gives a signal for some states which are local excitations in the neighborhood of the origin. It is a detector of something in this region. The fact that we do not know what it detects is not disconcerting. This problem is precisely the same for the experimentalist who has to use monitoring experiments and accumulated experience to find out the sensitivity of a particular detector.

Having the mathematical counterparts of single detectors we can construct coincident arrangements. Thus

$$D = \alpha_{x_1}(C_1)\,\alpha_{x_2}(C_2)\,\alpha_{x_3}(C_3)$$

with $x_i = \underset{\sim}{x}_i, t$ (equal time coordinates)

and $|\underset{\sim}{x}_i - \underset{\sim}{x}_0'| \gg r$

describes a 3-fold coincidence arrangement at time t i.e. 3 detectors placed at the space positions $\underset{\sim}{x}_i$, far separated from each other relative to their size. If ω is a state for which $\omega(D) \approx 1$ then it describes a situation in which a simultaneous signal in each of the three detectors is produced. Such a signal cannot be produced by a single particle. It needs at least 3 particles.

After these remarks it is evident how to define a single particle state. It is a state which can at no time produce a signal in a 2-fold coincident arrangement but does produce signals in suitably chosen single detectors. We may call such a state a permanently singly localized state. As a criterion we can take (with $x_i = \underset{\sim}{x}_i, t$)

$$\int_{|\underset{\sim}{x}_1 - \underset{\sim}{x}_2| > d} \omega(\alpha_{x_1}(C_1)\,\alpha_{x_2}(C_2))\, d^3x < \varepsilon(d) \quad \text{uniformly in t} \tag{9}$$

where $\varepsilon(d) \to \infty$ as $d \to \infty$ for any detector C. In addition we need of course, $\|\omega - \omega_0\| = 2$ which excludes that ω has any vacuum component. Actually it would suffice to require (9) in the limit $t \to \infty$. The essential point is that the number of simultaneous localization centers at $t \to \infty$ corresponds to the number of outgoing particles in the state.

Now there arise the questions

i) When does the theory have any states which are asymptotically singly localized?

ii) Do such states have sharp mass values i.e. do they correspond to particles in the usual sense?

iii) Can all states in the vacuum representation be decomposed according to asymptotic particle number?

We know that ii) and iii) will not be true in general when there are excitations of arbitrarily small energy (infra red problem). In the simplest situation, when one has an energy gap between the vacuum and all other states and when there are no charge quantum numbers one can intuitively see which properties of $\mathfrak{A}$ and α_g are needed in order to have an affirmation answer to i), ii), iii), though even in that case one is still far from a rigorous proof.

II. Institut für Theoretische Physik
Luruper Chaussee 149
2000 Hamburg 50
Germany

Proceedings of Symposia in Pure Mathematics
Volume 38 (1982), Part 2

NEW LIGHT ON THE MATHEMATICAL STRUCTURE OF ALGEBRAIC FIELD THEORY

JOHN E. ROBERTS

ABSTRACT. The problem of superselection sectors in algebraic field theory is reexamined so as to draw parallels with some standard mathematical questions. It is shown how sectors correspond to certain Hermitian bimodules over a sheaf of von Neumann algebras on Minkowski space for the topology generated by the spacelike complements of compact sets. The composition of sectors corresponds to the tensor product of bimodules. Thus, the theory of superselection sectors may, in spirit, be regarded as the study of Hermitian vector bundles relative to a structural sheaf characteristic of the underlying physical theory.

0. INTRODUCTION. Algebraic field theory is designed as a conceptual scheme for elementary particle physics in the sense that quantum mechanics provides one with a conceptual scheme for atomic physics. The germinating ideas can be found in the work of Haag in the late fifties [1] and the mathematical contours of the subject became established in the early sixties in papers of Araki [2,3], Borchers [4,5], Haag and Kastler [6].

The title, tenor and context of this talk make it imperative for me to emphasise that it is the development of elementary particle physics rather than of operator algebras which ultimately determines the success or failure of algebraic field theory. To illustrate this, I single out three developments in elementary particle physics against which one might measure the past and future achievements of algebraic field theory.

a) Symmetries. Exact and approximate symmetries have been a dominating feature of elementary particle physics from the outset. Exact symmetries manifest themselves in exact conservation laws, like the conservation of electric charge and of baryon number. Approximate symmetries manifest themselves in approximate conservation laws, like the conservation of isospin and of strangeness, and in the appearance of multiplets of particles having approximately the same mass, such as proton and neutron or the three π-mesons.

b) Gauge Theories. Quantum electrodynamics is the outstanding example of a field theory and yields amazingly accurate quantitative predictions. It

---

1980 Mathematics Subject Classification. 46L60, 81E05, 46M20.

differs from other early field theoretical models in being a gauge theory, i.e. it has a local gauge symmetry

$$\psi(x) \to e^{i\theta(x)}\,\psi(x); \qquad A^{\mu}(x) \to A^{\mu}(x) + \partial^{\mu}\theta(x).$$

Physicists were therefore naturally led to examine other gauge theories. In the sixties, these 'Yang-Mills' theories seemed to be just one more playground of theoretical physics. However, by the early seventies two gauge theories had emerged, the Salam-Weinberg theory and quantum chromodynamics, which describe a unified theory of weak and electromagnetic interactions and a theory of strong interactions respectively. From the mid-seventies, in the light of experimental evidence, physicists have begun to believe that these theories represent a significant advance in understanding these fundamental interactions.

c) Solitons and Instantons. This intriguing development started with the observation in the last century of solitary waves on a canal propagating without change of shape and evolved into the study of exact solutions of non-linear classical field equations, such as the Sine-Gordon equation and the Korteweg-de Vries equation. The relevant aspects of solitons are those which transcend these completely integrable systems and are linked to topological invariants for the set of solutions of the field equations or, to use the physicists' language, to topological quantum numbers. This feature is shared by the instantons which arose as special self-dual solutions of the classical Euclidean Yang-Mills field equations. There has been an intense and fruitful collaboration between mathematicians and physicists in recent years on instantons and solitons in classical field theories. In quantum field theory, physicists have been left to their own devices: one knows that, for all their mathematical similarity, the physical significance of solitons and instantons is quite different, but their role in quantum field theory is still not really understood, neither physically nor mathematically. In elementary particle physics, this subject is still the exclusive preserve of the theoretician; there is no experimental evidence for solitons or instantons. However, one good reason for including this topic is that examples in classical field theory suggest that solitons and instantons are likely to appear in conjunction with gauge theories.

Judged against this background, algebraic field theory has had only limited success. The pattern and structural importance of symmetries can be well understood on the basis of superselection theory. But even this success is tempered by the failure to extend the analysis to gauge theories. Algebraic field theory has not yet made important contributions to understanding either gauge theories, solitons or instantons although there are some promising starting points for such contributions. All the evidence suggests that some form of non-commutative differential geometry based on operator algebras must describe gauge theories and their associated topological quantum numbers. This fits well into the current pattern of research in operator algebras and may prove a fruitful field of collaboration between mathematicians and physicists in the next decade. In the hope of stimulating such a development, I have decided to take a fresh look at the relevant mathematical aspects of algebraic field theory to bring out parallel developments in mathematics. The more geometric interpretation of superselection structure that emerges has the right qualitative innovatory features. Although a genuinely radical break with the traditions of quantum field theory may eventually prove necessary, this approach has the merit of providing some immediately accessible research projects that can steer future developments.

In the written version of this talk, I have tried to present a coherent account of the results to the point where they allow one to reinterpret the composition of sectors as the tensor product of bimodules. None of the techniques involved is particularly difficult or original but many disparate results have had to be tailored to the demands of this project. Section 1 introduces algebraic field theory and superselection structure. Section 2 is devoted to the underlying operator-algebraic structures; it treats Hermitian modules in relation to W*-categories and representations of operator algebras. Section 3 shifts to W*-systems. Although it concludes by characterizing the Morita equivalence of W*-systems, its main purpose is to illustrate the role of 1-cohomology in a group-theoretical context before it appears in section 4 in analysing the locality condition imposed on superselection sectors. Section 5 provides a little background in sheaf theory so that Hermitian modules over a sheaf of von Neumann algebras can be studied in section 6. The analogy between Hermitian vector bundles and superselection sectors can then be drawn in section 7 although the real applications of Hermitian modules to superselection structure can first be given in section 9 after Hermitian bimodules have been studied in section 8.

1. ALGEBRAIC FIELD THEORY.

Algebraic field theory starts by considering the von Neumann algebra $\mathfrak{A}(U)$ generated by the observables which can be measured within the bounded open set U of space-time. I shall suppose:

a) Isotony. $\mathfrak{A}(U_1) \subset \mathfrak{A}(U_2)$, $U_1 \subset U_2$.

b) Local Commutativity. $A_1A_2 = A_2A_1$, if $A_i \in \mathfrak{A}(U_i)$, $i = 1,2$, and $U_1 \subset U_2'$, where U' denotes the spacelike complement of U.

c) Additivity. $\mathfrak{A}(\bigcup_i U_i) = \bigvee_i \mathfrak{A}(U_i)$, i.e. each $\mathfrak{A}(U)$ is generated as a von Neumann algebra by the subalgebras associated with any open covering.

These properties may be summed up by saying that $\mathfrak{A}$ is an *additive local net* of von Neumann algebras over the set of bounded open sets in Minkowski

space ordered under inclusion. I shall suppose furthermore that
d) Covariance. There is a σ-continuous action $L \to \alpha_L$ of the Poincaré group $\mathcal{P}$ by automorphisms of $\mathfrak{A}$ satisfying

$$\alpha_L(\mathfrak{A}(U)) = \mathfrak{A}(LU), \quad L \in \mathcal{P}.$$

e) Vacuum. There is a pure locally normal state ω_o, the vacuum state, with $\omega_o \circ \alpha_L = \omega_o$, $L \in \mathcal{P}$.
f) Positivity of the Energy. Let $\{\pi_o, \mathcal{U}_o\}$ be the covariant GNS-representation, the vacuum representation, with cyclic vector Ω, the vacuum vector, associated with $\{\omega_o, \alpha\}$. Let $\mathcal{U}_o(a) = e^{iP.a}$ for $a \in \mathbb{R}^4$, the translation subgroup of $\mathcal{P}$. Then the support of the total energy-momentum 4-vector P is contained in the forward light cone, i.e. $P^o \geq 0$ and $P.P \geq 0$.

If you compare this brief list of assumptions with other lists in the literature, you will notice that there has been relatively little fluctuation. It should be complemented by some form of time-slice axiom, a substitute for equations of motion, which should, one day, play a role in algebraic field theory. The combination of local commutativity and positivity of the energy is very restrictive and all deep results in the theory make essential use of both properties. In fact, the resulting structure is uncomfortably tight in that there are no known realizations whose physical content is not trivial. This is probably the main barrier to effective progress and there is a special branch of field theory, constructive field theory, where much ingenuity and many man-hours have been devoted to this problem.

The discovery of superselection sectors [7] has evolved into one of the key problems of algebraic field theory: determine the structure of the set of representations which are relevant to elementary particle physics. A superselection sector is just the set of pure states affiliated with an irreducible representation of this set. There is no definitive characterization of relevant representations covering all theories one would envisage. However, there are two types of conditions which have hitherto been employed successfully. One may demand covariance conditions; for example, that there is a continuous unitary representation $\mathcal{U}_\pi$ of the covering group $\tilde{\mathcal{P}}$ of the Poincaré group with spectrum in the forward light cone and satisfying

$$\mathcal{U}_\pi(L)\, \pi(A) = \pi(\alpha_L A)\, \mathcal{U}_\pi(L), \quad A \in \mathfrak{A},\ L \in \tilde{\mathcal{P}}.$$

Or one may demand locality conditions; for example, that there is a closed double cone $\mathcal{O}$ such that

$$\pi|\,\mathcal{O}' + a \simeq \pi_o|\,\mathcal{O}' + a, \quad a \in \mathbb{R}^4. \qquad \text{(L)}$$

The notation adopted here is that $\mathfrak{A}|V$ denotes the subnet $U \to \mathfrak{A}(U)$ with $U \subset V$ and $\pi|V$ denotes the restriction of π to this subnet.

These two types of conditions are not independent; in specific cases, they may determine the same class of representations on separable Hilbert spaces. For a review of the current status of the analysis of these conditions, I refer you to the contribution of Buchholz and Fredenhagen in these Proceedings. Instead, I want to show how, with hindsight, you can go about analysing the locality condition (L), a task originally undertaken in [8,9], so as to gain fresh insight into the mathematical structures involved.

2. REPRESENTATION AND MODULES.

Two simple exercises will help to put the analysis of superselection structure into its proper mathematical context. Let π_0 be a faithful normal representation of a von Neumann algebra M. Given two normal representations π and π' of M, let

$$(\pi,\pi') = \{t : H_\pi \to H_{\pi'} : t\pi(m) = \pi'(m)t,\ m \in M\}$$

denote the set of intertwiners from π to π'. They are the set of arrows from π to π' in the W*-category [10] of normal representations of M which will be denoted by Rep M. Let $M' = (\pi_0,\pi_0)$ denote the commutant of M in the representation π_0. If $s,t \in (\pi_0,\pi)$, $m \in M'$ then $tm \in (\pi_0,\pi)$ and $s^*t \in M'$. Thus (π_0,π) is a right M'-module equipped with an M'-valued inner product. The first exercise will be to identify Rep M as a category of M'-modules.

Let A be a C*-algebra, then a right Hermitian A-module X is a right Banach A-module with an A-valued inner product $\langle\ ,\ \rangle$, conjugate-linear in the first variable and linear in the second, such that, for all $x,y \in X$ and $a \in A$,

a) $\langle x,x\rangle \geq 0$, c) $\langle x,ya\rangle = \langle x,y\rangle a$,

b) $\langle x,y\rangle^* = \langle y,x\rangle$, d) $\| x \|^2 = \| \langle x,x\rangle \|$.

Such modules have been treated by Paschke [11], Rieffel [12] and Kasparov [13] under the names right Hilbert module, rigged space and Hilbert C*-module respectively. We consider these modules as the objects of a category hmod$A$. An arrow T from X to Y in hmod$A$ is a linear map from X to Y such that for some $K \geq 0$

$$\langle Tx,Tx\rangle \leq K\langle x,x\rangle, \qquad x \in X,$$
$$T(xa) = (Tx)a, \qquad x \in X,\ a \in A.$$

The least value of $\sqrt{K}$ will be denoted $\| T \|$.

The arrows of hmod$A$ are just the bounded A-module homomorphisms and $\| T \| = \sup_{\| x \| \leq 1} \| Tx \|$. More surprisingly, the second condition is superfluous [11; Thm.2.8], but this result will not be used here. The Banach space of arrows from X to Y in hmod$A$ will be denoted (X,Y).

If T is a map from X to Y and we can find a map T^* from Y to X with

$$\langle Tx,y\rangle = \langle x,T^*y\rangle, \quad x\in X, \; y\in Y$$

then automatically $T\in (X,Y)$ and $T^*\in(Y,X)$. The arrows with adjoints form a proper subcategory of hmodA which is a C*-category in the sense of [10].

We regard A itself as a right Hermitian A-module under right multiplication with an inner product defined by $\langle a,b\rangle = a^*b$. If $x\in X$, then setting $T_x a = xa$ defines an element T_x of (A,X) with adjoint and $\|T_x\| = \|x\|$. A is a generator in hmodA, i.e. given $T\in (X,Y)$ with $T\neq 0$, there is an $S\in (A,X)$ with $TS\neq 0$. If A has an identity, the map $T\to T(1)$ is an isomorphism of (A,X) with X and, if we identify (A,A) and A using this isomorphism, $T\to T(1)$ is a unitary equivalence of Hermitian A-modules.

There is a Gelfand-Naimark-Segal construction associated with hmodA. If ω is a positive linear functional on A, let $F_\omega(X)$ denote the Hilbert space associated with the semidefinite scalar product $\omega(\langle x,x'\rangle)$ on X. $T\in(X,Y)$ gives rise to a bounded linear map $F_\omega(T) : F_\omega(X) \to F_\omega(Y)$, since $\omega(\langle Tx,Tx\rangle) \le \|T\|^2\omega(\langle x,x\rangle)$, and $F_\omega(T^*) = F_\omega(T)^*$ if T has an adjoint. F_ω is a contractive linear functor from hmodA to the category H of Hilbert spaces. Taking the direct sum $F = \oplus_\omega F_\omega$ over all, or some full subset of, states of A we get an isometric embedding of hmodA into H. This is only a trivial variation on known results, [12; Lemma 2.4] and [11; Prop. 2.6], allowing one to reduce questions about Hermitian A-modules to questions about operators on Hilbert spaces.

A right Hermitian A-module X is said to be self-dual if every element of (X,A) has an adjoint. It then follows that every bounded A-module homomorphism from X to an Hermitian A-module has an adjoint [11; Prop. 3.4]. The full subcategory of self-dual modules is thus a C*-category which will be denoted HmodA. If A has an identity, every finitely generated projective A-module is self-dual. We shall be principally interested in modules over a von Neumann algebra and, as Paschke showed [11; § 3], the theory of self-dual modules is then particularly tractable. Before reformulating and summarizing his main results in Theorem 2.2, we prove an important lemma whose basic ingredients can be found in [14; Thm. 6.5].

LEMMA 2.1. Let A, B, C be objects of a W*-category A and consider (A,B) and (A,C) as right Hermitian (A,A)-modules. Then, if $T : (A,B) \to (A,C)$ is a bounded (A,A)-module homomorphism, there is an $x\in (B,C)$ such that

$$T(b) = xb, \quad b\in (A,B).$$

PROOF. Let J be the directed set of finite subsets of (A,B). If $J\in J$, then by taking polar decompositions, we can easily construct [14; Lemma 6.7] a finite set $\{u_i\}_{i\in I_J}$ of partial isometries of (A,B) with $u_i^*u_j = 0$ if $i\neq j$ and

$\sum_{i\in I_J} u_i u_i^* b = b$, $b \in J$. Now set $x_J = \sum_{i\in I_J} T(u_i)u_i^*$ then $x_J b = T(e_J b)$, $b\in(A,B)$, where $e_J = \sum_{i\in I_J} u_i u_i^*$. Thus $\|x_J\| = \sup\{\|x_J b\| : b\in(A,B),\ \|b\| \le 1\} \le \|T\|$.

Hence the net $\{x_J\}_{J\in\mathcal{J}}$ has a cluster point in the σ-topology, i.e. the weak topology of the dual system $((A,B), (A,B)_*)$, with $xb = T(b)$ as required.

If we require that $xe = x$, where $e \in (B,B)$ is the left support of (A,B), then x is uniquely determined.

THEOREM 2.2. If M is a von Neumann algebra, then HmodM is a W*-category and a reflective subcategory of hmodM.

PROOF. Consider the embedding $F = \oplus_\omega F_\omega$ of hmodM in H, where the direct sum is taken over the normal states of M. We identify arrows of hmodM with their images under F, then, since (M,X) and X are canonically isomorphic, a bounded linear operator S from $F(X)$ to $F(Y)$ is in (X,Y) if and only if $S(M,X)\subset(M,Y)$. Thus (X,M) is σ-closed and, if X is self-dual, $(M,X) = (X,M)^*$ is σ-closed. Furthermore, the arrows of hmodM with adjoints between objects X with (M,X) σ-closed form a W*-category A. Applying Lemma 2.1 with $A = M$, we deduce that A is a full subcategory and hence that $(X,M) = (M,X)^*$ for all objects of A. Thus A = HmodM. Now, let X be an object of hmodM and let $R(X)$ denote the σ-closure of (M,X). The inner product and action of M on $(M,X)\simeq X$ extend by σ-continuity to $R(X)$ making it an Hermitian M-module. Let $U \in (X,R(X))$ denote the arrow defined by $Ux = T_x$. U, identified with $F(U)$, is a unitary from $F(X)$ to $F(R(X))$ and $UR(X) = (M,R(X))$ so $R(X)$ is self-dual. If Y is self-dual and $S\in(X,Y)$ then SU^* is the unique arrow T from $R(X)$ to Y with $TU = S$. Thus HmodM is a reflective subcategory of hmodM (see e.g. [15; IV.3]).

The intimate relation between W*-categories and self-dual modules over von Neumann algebras indicated by Lemma 2.1 can be made more precise: let A be a W*-category and A, B, C objects of A. Let $F_A(B)$ denote (A,B) considered as a right Hermitian (A,A)-module. If $b \in (B,C)$, let $F_A(b) : F_A(B) \to F_A(C)$ be defined by $F_A(b)a = ba$, $a \in (A,B)$.

Then:

THEOREM 2.3. $F_A : A \to \mathrm{Hmod}(A,A)$ is a full normal *-functor. If A is a generator of A, then F_A is faithful and if, furthermore, A has direct sums and sufficient subobjects, then F_A is an equivalence of W*-categories.

PROOF. The previous lemma shows, in particular, that $F_A(B)$ is a self-dual module and also that F_A, which is obviously a $*$-functor, is full. Since the kernel of F_A is closed in the σ-topology, F_A is normal. If A is a generator, the left support of each (A,B) is 1_B so that F_A is faithful. Now $F_A(A) = (A,A)$ is a generator of Hmod(A,A). Hence by [10; Prop. 7.3c], if A has direct sums and sufficient subobjects, any object of Hmod(A,A) is unitarily equivalent to an object in the image of F_A, so that F_A is an equivalence of W*-categories.

The W*-category HmodM has direct sums and sufficient subobjects and a generator M. Thus, by [10; Prop. 7.3f], any self-dual Hermitian M-module is equivalent to a module of the form $e\,H \otimes M$, where H is a Hilbert space and e is a projection from $\mathcal{B}(H) \otimes M = (H \otimes M, H \otimes M)$. As a corollary of Theorem 2.3, we see that any self-dual module has a unique predual [10; Cor. 2.8], that an Hermitian M-module X is self-dual if and only if it has a predual and (X,X) is a von Neumann algebra, and that a σ-closed submodule of a self-dual module is self-dual.

If we apply Theorem 2.3 to RepM and recall that a faithful normal representation is a generator of RepM [14; Prop. 1.3], we get, as a solution to the exercise,

COROLLARY 2.4. Let π_o be a faithful normal representation of a von Neumann algebra M, then $F_{\pi_o} : \mathrm{RepM} \to \mathrm{Hmod}(\pi_o, \pi_o)$ is an equivalence of W*-categories.

This result should obviously be seen in relation to the Morita equivalence of von Neumann algebras [14; Thm. 8.15]. We return to this point in the next section.

Instead, we add a few remarks, needed later, on the category Hmod of self-dual Hermitian modules. An object of Hmod is a pair {M,X}, where M is a von Neumann algebra and X an object of HmodM. An arrow in Hmod from {M,X} to {N,Y} is a pair {ϕ,T}, where $\phi : M \to N$ is an arrow in the category of von Neumann algebras, i.e. a normal identity-preserving *-homomorphism, and T is a linear map from X to Y such that for some $K \geq 0$

$$\langle Tx, Tx\rangle \leqslant K\phi\langle x,x\rangle\,, \qquad x \in X$$
$$T(xm) = Tx\phi m, \qquad x \in X,\ m \in M.$$

The least value of $\sqrt{K}$ will be denoted $\|T\|$.

If we define the s-topology on a self-dual module X over a von Neumann algebra M by the seminorms $x \to \sqrt{\omega\langle x,x\rangle}$, with ω a normal state of M, then T is obviously s-continuous. Hence, by [10; Cor. 2.9], T is also σ-continuous.

We have an obvious forgetful functor G from Hmod to the category of von Neumann algebras and this has a left adjoint F defined by $F(M) = \{M, O_M\}$, where O_M denotes the zero module over M. HmodM is a subcategory of Hmod; it is the fibre of G over M.

Hmod has a terminal object, the zero module over the zero von Neumann algebra. If we set $\prod_{i \in I} \{M_i, X_i\} = \{\prod_{i \in I} M_i, \prod_{i \in I} X_i\}$, where the products are defined in the sense of Banach spaces and the algebraic structure is defined coordinatewise, we get an object of Hmod but it is not a product in the categorical sense. This defect of Hmod is shared by the category of Banach spaces and can be remedied similarly, namely, by passing to the subcategory defined by $\|T\| \leq 1$. However, we wish to have a category with equalizers too and define Imod to be the subcategory of Hmod obtained by requiring T to be ϕ-isometric,

$$\langle Tx,Tx\rangle = \phi\langle x,x\rangle, \qquad x\in X.$$

The forgetful functor from Imod to the category of von Neumann algebras still has a left adjoint given by the same formula. Equalizers can be defined in the obvious manner: if $\{\phi_1,T_1\}$ and $\{\phi_2,T_2\}$ are arrows from $\{M,X\}$ to $\{N,Y\}$ in Imod, we let $R = \{m\in M : \phi_1 m = \phi_2 m\}$ and $W = \{x\in X : T_1x = T_2x\}$ then the injection $\{R,W\} \to \{M,X\}$ is the equalizer of $\{\phi_1,T_1\}$ and $\{\phi_2,T_2\}$. The only point which is not immediately obvious is that W is self-dual as an Hermitian module over the von Neumann algebra R. However, since T_1 and T_2 are σ-continuous, W is σ-closed and self-duality follows from

LEMMA 2.5. Let A be a W*-category with objects A,B . Let R be a von Neumann subalgebra of (A,A) and W a σ-closed subspace of (A,B) such that, if $w\in W$ and $r\in R$, $w^*w\in R$ and $wr\in W$. Then W is self-dual as an Hermitian R-module.

PROOF. Let $N = \{b\in(B,B) : bW\subset W$ and $b^*W\subset W\}$,then, since $b \to ba$ and $b \to b^*$ are σ-continuous in a W*-category, N is a von Neumann algebra. The arrows in R, W, W* and N together from a W*-category, so, by Lemma 2.1, W is self-dual.

Now a category with products and equalizers is actually complete (see e.g. [16; I 2.3] or [15; V Cor. 2]) so we can sum up the discussion in the preceding paragraph in

PROPOSITION 2.6. The category Imod is complete.

Evidently, if we dropped the self-duality condition in the definitions of Hmod and Imod, the analogous results are even easier to prove and, in this case, we might as well work with Hermitian modules over general C*-algebras.

3. COVARIANT REPRESENTATIONS AND MODULES. Let $\{M,\alpha\}$ be a W*-system, i.e. a von Neumann algebra carrying a σ-continuous action α of a locally compact group G. The second exercise is to start from a faithful, normal, covariant representation of $\{M,\alpha\}$ and to identify the category of normal covariant representations as a suitable category of modules. This will involve establishing the analogues of the results in the previous section in the presence of a group action and we shall be guided by the idea of treating $\{M,\alpha\}$ as a "von Neumann algebra in the category of G-sets".

A Hilbert space in this category is a Hilbert space H_σ with a continuous unitary representation σ of G on H_σ. The W*-category of such Hilbert spaces will be denoted $H(G)$. As a W*-category $H(G)$ is equivalent to H, the W*-category of Hilbert spaces, but we shall regard $H(G)$ as a W*-category in the category of G-sets since it carries a natural action of G, defined on objects by

$gH_\sigma = H_\sigma$ and on the arrows by

$$x \in (H_\sigma, H_\tau) \to gx = \tau(g)x\sigma(g)^*.$$

We refer to a W*-category carrying a σ-continuous action of a locally compact group G leaving the objects fixed as a W*-G-category. If A and B are two such categories, then $\Gamma(A)$ denotes the W*-category of fixed points of A under the action of G and (A,B) the W*-G-category whose objects are normal $*$-functors from A to B commuting with the action of G and whose arrows are the bounded natural transformations carrying the action induced by B.

With these conventions, we can define the W*-G-category Rep$\{M,\alpha\}$ of normal covariant representations by treating $\{M,\alpha\}$ as a W*-G-category with a single object and setting

$$\mathrm{Rep}\{M,\alpha\} = (\{M,\alpha\}, H(G)).$$

It should be noted that the arrows of Rep$\{M,\alpha\}$ just intertwine the associated representations of the algebra M whereas the arrows of ΓRep$\{M,\alpha\}$ also intertwine the associated representations of G.

We get a natural category of $\{M,\alpha\}$-modules by reinterpreting the definition of HmodM in the category of G-sets. Thus a self-dual Hermitian $\{M,\alpha\}$-module is a pair $\{X,\beta\}$ where X is a self-dual Hermitian M-module and β is a σ-continuous action of G on X such that, if $g \in G$, $x,y \in X$ and $m \in M$,

$$\text{e) } \langle\beta_g x, \beta_g y\rangle = \alpha_g\langle x,y\rangle, \qquad \text{f) } \beta_g(xm) = \beta_g(x)\,\alpha_g(m).$$

An arrow $T: \{X,\beta\} \to \{Y,\gamma\}$ in the W*-G-category Hmod$\{M,\alpha\}$ is simply a $T \in (X,Y)$ and $gT = \gamma_g T \beta_g^{-1}$. Thus Hmod$\{M,\alpha\}$ is equipped with a full and faithful normal *-functor Hmod$\{M,\alpha\} \to$ HmodM forgetting the action of G.

Hermitian $\{M,\alpha\}$-modules are familiar under another guise. If $\{X,\beta\}$ is such a module then, as we have seen, there is a Hilbert space H and an isometry $w \in (X, H\otimes M)$ in HmodM. Now $H \otimes M$ becomes an $\{M,\alpha\}$-module with the action $\iota \otimes \alpha$ so we can regard w as an isometry in Hmod$\{M,\alpha\}$. If we set $z_g = w(gw)^*$, $g \in G$ then

$$z_{gg'} = z_g\, g(z_{g'}), \quad g,g' \in G, \quad z_g^* = g(z_{g^{-1}}), \quad g \in G.$$

Thus z is a 1-cocycle with values in $\{B(H) \otimes M, \iota \otimes \alpha\}$ in the sense of Connes and Takesaki [17; Def. III 1.1].

The corresponding 1-cohomology is an obstruction to $\{X,\beta\}$ being equivalent to an $\{M,\alpha\}$-submodule of $\{H\otimes M, \iota\otimes\alpha\}$. This point of view, however, does not do full justice to the non-Abelian 1-cohomology of a group and it is better to follow Serre [18] and notice that the 1-cocycle z tells one how to define an action β^z of G on the M-module $eH\otimes M$, where $e = ww^*$, so as to make it equivalent to $\{X,\beta\}$, namely, by setting

$$\beta_g^z n = z_g\, \iota\otimes\alpha_g(n), \qquad n \in eH\otimes M.$$

This means that the 1-cohomology classifies the self-dual Hermitian $\{M,\alpha\}$-modules. To make this more precise, we define a W*-category $\tilde{Z}^1(G,\{M,\alpha\})$ whose objects are 1-cocycles with values in $\{\mathcal{B}(H)\otimes M,\iota\otimes\alpha\}$ for some Hilbert space H. Given two such cocycles z and z', $t : z \to z'$ in $\tilde{Z}^1(G,\{M,\alpha\})$ if $t\in\mathcal{B}(H,H')\otimes M$ with $t = te_z = e_{z'}t$, where $e_z = z_g z_g^*$, and

$$tz_g = z'_g\ \iota\otimes\alpha_g(t), \quad g\in G.$$

The algebraic operations and the norm on the arrows are defined in the obvious manner.

THEOREM 3.1. There is an equivalence $F : \tilde{Z}^1(G,\{M,\alpha\})\to\Gamma\mathrm{Hmod}\{M,\alpha\}$ defined by $F(z) = \{e_zH\otimes M,\beta^z\}$ for each object z of $\tilde{Z}^1(G,\{M,\alpha\})$ and, if $t : z \to z'$, then $F(t)$ is t considered as an object of $(e_zH\otimes M,e_{z'}H'\otimes M)$.

PROOF. A routine computation shows that β^z is a σ-continuous action of G on $e_zH\otimes M$ satisfying e) and f) above. If $t : z \to z'$ then $F(t)\beta_g^z = \beta_g^{z'}F(t)$, $g\in G$, so that F is a faithful *-functor. An arrow from $F(z)$ to $F(z')$ in $\Gamma\mathrm{Hmod}\{M,\alpha\}$ corresponds to a $t\in\mathcal{B}(H,H')\otimes M$ with $t = te_z = e_{z'}t$ intertwining the actions of G and is thus of the form $F(t)$ for some $t : z \to z'$. It remains to show that any object $\{X,\beta\}$ of $\Gamma\mathrm{Hmod}\{M,\alpha\}$ is equivalent to an object in the image of F. However, as remarked above, picking an isometry $w : \{X,\beta\} \to \{H\otimes M,\iota\otimes\alpha\}$ in $\mathrm{Hmod}\{M,\alpha\}$ and setting $z_g = w(gw)^*$ turns w into a unitary equivalence from $\{X,\beta\}$ to $\{eH\otimes M,\beta^z\}$ in $\Gamma\mathrm{Hmod}\{M,\alpha\}$.

We note in passing that outer equivalent actions give rise to isomorphic W*-G-categories of modules. Since if a is a unitary 1-cocycle with values in $\{M,\alpha\}$, the normal *-functor from $\mathrm{Hmod}\{M,\alpha\}$ to $\mathrm{Hmod}\{M,{}^a\alpha\}$ which maps $\{X,\beta\}$ to $\{X,{}^a\beta\}$, where

$${}^a\beta_g(x) = \beta_g(x)\, a_g^*, \quad g\in G,$$

and is the identity on arrows commutes with the action of G and has an inverse defined analogously starting from the unitary 1-cocycle $g \to a_g^*$ with values in $\{M,{}^a\alpha\}$.

If A is an object of a W*-G-category $\mathcal{A}$, then (A,A) is to be interpreted as a von Neumann algebra carrying an action of G. We now have the analogue of Theorem 2.3 since the functor F_A commutes with the action of G.

THEOREM 3.2. Let $\mathcal{A}$ be a W*-G-category and A an object of $\mathcal{A}$, then $F_A : \mathcal{A} \to \mathrm{Hmod}(A,A)$ is a full normal *-functor. If A is a generator of $\mathcal{A}$, then F_A is faithful and if, furthermore, $\mathcal{A}$ has direct sums and sufficient subobjects, then F_A is an equivalence of W*-G-categories.

Now, a normal covariant representation of $\{M,\alpha\}$ is a generator of $\mathrm{Rep}\{M,\alpha\}$ if and only if the underlying representation of M is faithful.

Hence we get, as a solution to the exercise,

COROLLARY 3.3. Let $\{\pi,\sigma\}$ be a covariant normal representation of a W*-system $\{M,\alpha\}$ with π faithful, then

$$F_{\{\pi,\sigma\}} : \mathrm{Rep}\{M,\alpha\} \to \mathrm{Hmod}(\{\pi,\sigma\},\{\pi,\sigma\})$$

is an equivalence of W*-G-categories.

This result can be looked at more symmetrically: we say that two W*-systems $\{M,\alpha\}$ and $\{N,\beta\}$ over a group G are *in opposition* if there are normal covariant representations $\{\pi,\sigma\}$ of $\{M,\alpha\}$ and $\{\pi',\sigma\}$ of $\{N,\beta\}$ with π and π' faithful and $\pi(M)' = \pi'(N)$. We now have

THEOREM 3.4. Two W*-systems over G, $\{M,\alpha\}$ and $\{N,\beta\}$ are in opposition if and only if $\mathrm{Rep}\{M,\alpha\}$ and $\mathrm{Hmod}\{N,\beta\}$ are equivalent as W*-G-categories.

PROOF. Let $F : \mathrm{Hmod}\{N,\beta\} \to \mathrm{Rep}\{M,\alpha\}$ be an equivalence of W*-G-categories, then, if A denotes $\{N,\beta\}$ looked at as an object of $\mathrm{Hmod}\{N,\beta\}$, (A,A) is isomorphic to $\{N,\beta\}$. Hence $\{N,\beta\}$ is isomorphic to $(F(A), F(A))$. But, since A is a generator of $\mathrm{Hmod}\{N,\beta\}$, $F(A)$ is a generator of $\mathrm{Rep}\{M,\alpha\}$ so that the underlying representation of M is faithful. Hence $\{M,\alpha\}$ and $\{N,\beta\}$ are in opposition. The converse follows from Corollary 3.3.

The next result characterizes the Morita equivalence of W*-systems.

THEOREM 3.5. The following conditions on two W*-systems over G, $\{M_1,\alpha_1\}$ and $\{M_2,\alpha_2\}$, are equivalent.

a) There is a W*-system $\{N,\beta\}$ in opposition to both $\{M_1,\alpha_1\}$ and $\{M_2,\alpha_2\}$.

b) $\mathrm{Rep}\{M_1,\alpha_1\}$ and $\mathrm{Rep}\{M_2,\alpha_2\}$ are equivalent W*-G-categories.

c) $\mathrm{Hmod}\{M_1,\alpha_1\}$ and $\mathrm{Hmod}\{M_2,\alpha_2\}$ are equivalent W*-G-categories.

d) There is a W*-G-category A with generators A_1 and A_2 with (A_i,A_i) isomorphic to $\{M_i,\alpha_i\}$, $i = 1,2$.

e) There is a W*-system $\{P,\gamma\}$ and invariant projections e_1, e_2 with $e_1 + e_2 = 1$ and central support 1 such that the reduced system $\{P_{e_i},\gamma_{e_i}\}$ is isomorphic to $\{M_i,\alpha_i\}$, $i = 1,2$.

PROOF. It follows from Theorem 3.4 that a) implies b) and c).
If $F : \mathrm{Rep}\{M_1,\alpha_1\} \to \mathrm{Rep}\{M_2,\alpha_2\}$ is an equivalence of W*-G-categories and A is a generator of $\mathrm{Rep}\{M_1,\alpha_1\}$, then $F(A)$ is a generator of $\mathrm{Rep}\{M_2,\alpha_2\}$ and we can take $\{N,\beta\}$ to be a W*-system isomorphic to (A,A) and hence to $(F(A), F(A))$ to see that b) implies a). A similar argument taking $A = \mathrm{Hmod}\{M_2,\alpha_2\}$ shows that c) implies d). d) implies e) since we can take $\{P,\gamma\}$ to be the 2×2-matrix W*-system $M(A_1,A_2)$. It remains to show that e) implies a). Let $\{\pi,\sigma\}$ be a normal covariant representation of $\{P,\gamma\}$ with π faithful and consider the reduced representation $\{\pi_{e_i},\sigma_{e_i}\}$ of $\{P_{e_i},\gamma_{e_i}\}$. π_{e_i} is faithful and, since e_i has central support 1 in P, $\{\pi_{e_i}(P_{e_i})', \mathrm{ad}\sigma_{e_i}\}$ is isomorphic to $\{\pi(P)', \mathrm{ad}\sigma\}$. Hence $\{\pi(P)', \mathrm{ad}\sigma\}$ is in opposition to $\{M_i,\alpha_i\}$, $i = 1,2$,

completing the proof.

In condition d), (A_2,A_1) is a 'self-dual $\{M_1,\alpha_1\}$-$\{M_2,\alpha_2\}$-equivalence bimodule' and, if this notion is formalized in analogy with [14; Def. 7.5], the existence of such a bimodule is trivially equivalent to condition d). Condition e) yields the analogue of [19; Thm. 1.1]. The notion of W*-systems in opposition defines an involution on the set of Morita equivalence classes of W*-systems over G.

4. ČECH COHOMOLOGY. The analysis of the locality condition (L) will reveal a close analogy with the above analysis of covariant normal representations. (L) asserts that there are unitaries ψ_a with

$$\psi_a\pi(A) = \pi_0(A)\psi_a, \qquad A\in \mathfrak{A}|\mathcal{O}'+a$$

Writing $\mathcal{S}(V)$ for $[\pi_0|V]'$ and setting $z_{aa'} = \psi_a\psi^*_{a'}$ gives a unitary $z_{aa'}\in\mathcal{S}((\mathcal{O}'+a)_\cap(\mathcal{O}'+a'))$ satisfying the 1-cocycle identity

$$z_{aa'}z_{a'a''} = z_{aa''}.$$

Thus we have a unitary 1-cocycle z in Čech cohomology with respect to the open covering

$$\mathcal{U}(\mathcal{O}) = \{\mathcal{O}'+a : a\in\mathbb{R}^4\}$$

taking values in the presheaf $\mathcal{S}$ of von Neumann algebras over Minkowski space. Now given such a 1-cocycle z, set

$$\pi_z(A) = z_{oa}\pi_0(A)z^*_{oa}, \qquad A\in\mathfrak{A}|\mathcal{O}'+a$$

If $A\in\mathfrak{A}|(\mathcal{O}'+a)_\cap(\mathcal{O}'+a')$ then

$$z_{oa}\pi_0(A)z^*_{oa} = z_{oa}z_{aa'}\pi_0(A)z^*_{aa'}z^*_{oa} = z_{oa'}\pi_0(A)z^*_{oa'}.$$

Thus π_z is well defined on $\mathfrak{A}$ and is manifestly a representation satisfying (L). Furthermore, if $z_{aa'} = \psi_a\psi^*_{a'}$, as above, then the unitary ψ_0 intertwines π and π_z.

This is the basic point needed to establish that the 1-cohomology classifies the representations satisfying (L). To clear up the remaining details, we make the 1-cocycles into the objects of a W*-category $Z^1(\mathcal{U}(\mathcal{O}),\mathcal{S})$ by taking an arrow from z to z' in $Z^1(\mathcal{U}(\mathcal{O}),\mathcal{S})$ to be a function $a\to t_a\in\mathcal{S}(\mathcal{O}'+a)$ such that

$$t_az_{aa'} = z'_{aa'}t_{a'}, \qquad a,a'\in\mathbb{R}^4.$$

Let Rep$\{\mathcal{U}(\mathcal{O});\mathfrak{A}\}$ denote the W*-category of representations satisfying (L) and their intertwining operators. Define $F : Z^1(\mathcal{U}(\mathcal{O}),\mathcal{S}) \to$ Rep$\{\mathcal{U}(\mathcal{O});\mathfrak{A}\}$ by $F(z) = \pi_z$ and $F(t) = t_0$ then F is obviously a faithful normal *-functor. It is full since $s : \pi_z\to\pi_{z'}$ implies $z'^*_{oa}sz_{oa} \in \mathcal{S}(\mathcal{O}'+a)$ so that we need only set $t_a = z'^*_{oa} s z_{oa}$ to get an arrow from z to z' with $F(t) = s$. We have already shown that any object π of Rep$\{\mathcal{U}(\mathcal{O});\mathfrak{A}\}$ is unitarily equivalent to some π_z so that F is an equivalence of W*-categories.

The conclusion is valid for a more general class of coverings.

THEOREM 4.1. If any bounded subset of $\mathbb{R}^4$ is contained in some set of an open covering $\mathcal{U}$, then Rep$\{\mathcal{U};\mathfrak{A}\}$ and $Z^1(\mathcal{U},\mathcal{S})$ are equivalent W*-categories.

In view of the result of the preceding section, it is natural to ask whether $Z^1(\mathcal{U},\mathcal{S})$ is not equivalent to some category of $\mathcal{S}$-modules. Results of this nature can be derived but only after invoking the additivity of the observable net $\mathfrak{A}$. Additivity for $\mathfrak{A}$ implies that $\mathcal{S}$ satisfies the following intersection property:

$$\mathcal{S}(\bigcup_i V_i) = \bigcap_i \mathcal{S}(V_i).$$

This property is reminiscent of sheaf theory and we establish the connection in the following section. The reader unfamiliar with sheaf theory will probably find that the results and proofs expressed in the mathematically more natural language of sheaf theory become conceptually simpler in terms of the intersection property. To date, it suffies for the applications to algebraic field theory so the reader,interested in reaching these applications quickly, might do well by studying carefully the setting of section 9 and proving for himself the appropriate specialization of the relevant intermediary results.

5. SHEAVES. There is a well established notion of a sheaf over a topological space X with values in a category $\mathcal{C}$ (see e.g. [16], [20], [21]). A presheaf $\mathcal{S}$ with values in $\mathcal{C}$ is a contravariant functor from the category of open sets of X to $\mathcal{C}$ and $\mathcal{S}$ is a sheaf if, for each object c of $\mathcal{C}$, the presheaf $V \to \mathcal{C}(c,\mathcal{S}(V))$ is a sheaf of sets. Here $\mathcal{C}(c,\mathcal{S}(V))$ denotes the set of arrows in $\mathcal{C}$ from c to $\mathcal{S}(V)$.

For a sheaf $\mathcal{S}$ of von Neumann algebras we must, in particular, have $\mathcal{S}(\emptyset) = 0$, so $V \to \mathcal{S}(V) = [\pi_0|V]'$ is certainly not a sheaf. However, we can modify $\mathcal{S}$ to get a sheaf $\tilde{\mathcal{S}}$ simply by setting

$$\tilde{\mathcal{S}}(V) = \prod_j \mathcal{S}(V_j),$$

where the product is taken over the connected components V_j of V and the restriction mapping is the obvious one. Thus $\tilde{\mathcal{S}}(V) = \mathcal{S}(V)$, whenever V is a non-empty connected open set. Precisely the same construction is needed to pass from a constant presheaf to a "constant" sheaf. An alternative way of looking at the constant sheaf $\tilde{M}$ modelled on a von Neumann algebra M is to regard $\tilde{M}(V)$ as the von Neumann algebra of locally constant, bounded functions from V to M. Similarly, we can define $\tilde{\mathcal{S}}(V)$ to be the von Neumann algebra of locally constant, bounded functions t from V to $B(H_{\pi_0})$ satisfying

$$t(x)\,\pi_0(A) = \pi_0(A)\,t(x), \quad A \in \mathfrak{A}(U)$$

for each connected open set U with $x \in U \subset V$.

Since the intersection of any pair of elements from $\mathcal{U}(\mathcal{O})$ is non-empty and connected, $Z^1(\mathcal{U}(\mathcal{O}),S) = Z^1(\mathcal{U}(\mathcal{O}),\mathfrak{Z})$. In the next section, we shall show how $Z^1(\mathcal{U},\mathfrak{Z})$ is equivalent to a category of $\mathfrak{Z}$-modules for an arbitrary open covering $\mathcal{U}$.

The remainder of the present section is devoted to a few elementary constructions of sheaf theory in the context of sheaves of Banach spaces, C*-algebras and von Neumann algebras. We define these to be sheaves with values in the category of Banach spaces and contractive linear mappings, C*-algebras and *-homomorphisms and von Neumann algebras and normal, identity-preserving *-homomorphisms, respectively. Each of these categories has products and this allows an alternative criterion for a presheaf to be a sheaf which is sometimes used as a definition.

PROPOSITION 5.1. A presheaf S with values in a category C with products is a sheaf if and only if for every open cover $U = \bigcup_i U_i$

$$S(U) \overset{\alpha}{\to} \prod_i S(U_i) \underset{\gamma}{\overset{\beta}{\rightrightarrows}} \prod_{j,k} S(U_j \cap U_k)$$

is exact.

Here the i-th coordinate of α is the restriction map from $S(U)$ to $S(U_i)$, the (j,k)-th coordinate of β and γ is the projection onto $S(U_j)$ or $S(U_k)$ respectively followed by restriction to $S(U_j \cap U_k)$, and exactness means that α is the equalizer of β and γ.

Here, rather than a proof, is the elementary form of the statement specialized to the categories in question: given $s_i \in S(U_i)$ with $\sup_i \| s_i \| < +\infty$ and $s_j = s_k$ on $U_j \cap U_k$[1], there is a unique $s \in S(U)$ with $s = s_i$ on U_i. Furthermore $\| s \| = \sup_i \| s_i \|$. For C*-algebras or von Neumann algebras, this last condition is evidently superfluous.

If U is an open subset of X and S a presheaf over X, then $S|U$ denotes the presheaf over U obtained by restricting S to the open subsets of U. If S is a sheaf then so is $S|U$. If S_1 and S_2 are two presheaves over X with values in C, then $\mathrm{Hom}(S_1,S_2)$ denotes the set of natural transformations from S_1 to S_2. $\mathit{Hom}(S_1,S_2)$ denotes the presheaf $U \to \mathrm{Hom}(S_1|U,S_2|U)$ with the obvious restriction mappings. If S_1 and S_2 are sheaves then so is $\mathit{Hom}(S_1,S_2)$.

---

[1] The reader is invited to convince himself that this phrase is a harmless substitute for explicit restriction mappings.

As usual, Γ denotes the functor from the category of presheaves (or sheaves) over X with values in C to C itself defined by $\Gamma(S) = S(X)$ and $\Gamma(t) = t_X$ for $t \in \mathrm{Hom}(S_1,S_2)$.

If C is the category of Banach spaces and contractive linear mappings, it is convenient to modify the above definitions slightly by allowing the natural transformations to take values in the category B of Banach spaces and bounded linear mappings. In this case, $\mathrm{Hom}(S_1,S_2)$ will denote the Banach space of bounded natural transformations with the norm $\|t\| = \sup_U \|t_U\|$. $Hom(S_1,S_2)$ now becomes a presheaf or sheaf of Banach spaces and Γ takes values in B.

A similar convention will be used for sheaves of dual Banach spaces, i.e. sheaves with values in the category of dual Banach spaces (with specified predual) and contractive σ-continuous linear maps.

A theory of sheaves of Banach spaces has been gradually emerging over the past few years. The results are not used here but the reader interested in such questions as the relation to Banach bundles and continuous fields of Banach spaces might consult [22, 23, 24].

6. HERMITIAN S-MODULES. Returning to the context of algebraic field theory and the sheaf $\mathfrak{Z}$ of von Neumann algebras defined at the beginning of the last section, let us see how $\mathfrak{Z}$-modules arise. If π is a locally normal representation of the net $\mathfrak{A}$ and V is an open set, we define $M_\pi(V)$ to be the von Neumann algebra of locally constant bounded functions t from V to $B(H_{\pi_0},H_\pi)$ satisfying

$$t(x)\,\pi_0(A) = \pi(A)\,t(x), \qquad A \in \mathfrak{A}(U)$$

for each connected open set U with $x \in U \subset V$. M_π has an obvious structure of right $\mathfrak{Z}$-module and M_{π_0} is just $\mathfrak{Z}$ itself but now regarded as a right $\mathfrak{Z}$-module. Furthermore, π satisfies the locality condition (L) if and only if there is a closed double cone $\mathcal{O}$ such that

$$M_\pi|\mathcal{O}'+a \simeq M_{\pi_0}|\mathcal{O}'+a, \qquad a \in \mathbb{R}^4. \qquad (L')$$

In this section, we shall give a precise definition of the class of $\mathfrak{Z}$-modules involved and show that they are classified by the 1-cohomology relative to the covering $U(\mathcal{O})$ with values in $\mathfrak{Z}$. This then yields equivalences of W*-categories

$$\mathrm{Hmod}(U(\mathcal{O}),\mathfrak{Z}) \simeq Z^1(U(\mathcal{O}),\mathfrak{Z}) \simeq \mathrm{Rep}\{U(\mathcal{O});\mathfrak{A}\}$$

which are the analogues of the results on covariant representations in section 3.

Let S be a sheaf of von Neumann algebra then M will be called a self-dual right Hermitian S-module, or simply an Hermitian S-module, if the pair

$\{M,S\}$ is a sheaf in the category Imod of section 2. M_π is an Hermitian $\mathfrak{S}$-module for each locally normal representation π. The discussion of products and equalizers in Imod shows that if $\{M,S\}$ is a sheaf in Imod, S is a sheaf of von Neumann algebras and M a sheaf of dual Banach spaces.

If M and N are S-modules, Hom(M,N) refers to the underlying sheaves of dual Banach spaces whilst $T \in \mathrm{Hom}_S(M,N)$ means that $T_U : M(U) \to N(U)$ in Hmod$S(U)$ for each open set U and $U \to T_U$ and $U \to T_U{}^*$ are in Hom(M,N) and Hom(N,M) respectively. If $T \in \mathrm{Hom}_S(M,N)$, then $\prod_U T_U$ is an arrow from $\prod_U M(U)$ to $\prod_U N(U)$ in the W*-category $\prod_U \mathrm{Hmod}S(U)$. In this way, we easily see that the category of Hermitian S-modules is a W*-subcategory of $\prod_U \mathrm{Hmod}S(U)$ which will be denoted Hmod$S$. Hmod$S$ has direct sums and sufficient subobjects. The other definitions of the last section now apply analogously; thus $M|U$ denotes the $S|U$-module got by restricting to the open set U, $Hom_S(M,N)$ is now a sheaf of dual Banach spaces, $Hom_S(M,M)$ a sheaf of von Neumann algebras and Γ a normal *-functor from Hmod$S$ to Hmod$S(X)$.

If S itself is considered in a natural way as an object of Hmod$S$, then $T \to T_X(1_X)$ is an isomorphism of $\mathrm{Hom}_S(S,S)$ and $\Gamma(S) = S(X)$. Thus $Hom_S(S,S)$ and S are canonically isomorphic sheaves of von Neumann algebras.

We now analyse Hermitian S-modules that are locally unitarily equivalent using a routine sheaf-theoretical argument (cf. [25]). Let M and N be Hermitian S-modules and let $R = Hom_S(N,N)$. Suppose $\mathcal{U} = \{U_i\}_{i\in I}$ is an open covering and $\psi_i : M|U_i \to N|U_i$ unitary S-module equivalences, $i \in I$. Define $z_{ij} = \psi_i\psi_j{}^*$ on $U_i \cap U_j$, then $z_{ij} \in R(U_i \cap U_j)$ and $z_{ij}z_{jk} = z_{ik}$ on $U_i \cap U_j \cap U_k$. In other words, z is a unitary 1-cocycle with respect to the covering $\mathcal{U}$ with values in R.

We now show how, conversely, such a unitary 1-cocycle can be used to define an Hermitian S-module M_z. Let $M_z(V)$ be the subspace of $\prod_i N(V \cap U_i)$ of elements $N = (N_i)$ such that

$$z_{ij,V\cap U_j} N_j = N_i \qquad \text{on } V \cap U_i \cap U_j.$$

If $U \subset V$, we have an obvious restriction mapping from $M_z(V)$ to $M_z(U)$. The action of $S(V)$ on $M_z(V)$ is defined by

$$(NS)_i = N_i S \quad \text{on } V \cap U_i, \qquad S \in S(V),$$

and the $S(V)$-valued inner product by

$$\langle N,N'\rangle = \langle N_i,N_i'\rangle \quad \text{on } V \cap U_i.$$

The inner product is well defined since $\|\langle N_i,N_i'\rangle\| \le \|N\|\,\|N'\|$ and the unitarity of z_{ij} implies $\langle N_i,N_i'\rangle = \langle N_j,N_j'\rangle$ on $V \cap U_i \cap U_j$.

PROPOSITION 6.1. M_z is an Hermitian S-module. If $\phi_{i,V}(N) = N_i$ for $V \subset U_i$ and $N \in M_z(V)$ then ϕ_i is a unitary of $Hom_S(M_z,N)(U_i)$ with $\phi_i\phi_j^* = z_{ij}$ on $U_i \cap U_j$. If z arises as above from the Hermitian S-module M, there is a unitary χ of $Hom_S(M_z,M)(X)$ with $\chi = \psi_i^*\phi_i$ on U_i.

PROOF. $\phi_{i,V}$ is obviously an isometry in Hmod$S(V)$ from $M_z(V)$ to $N(V)$ and is natural in V. Computing adjoints we get $(\phi_{k,V}^*C)_i = z_{ik,V}C$ on $V \cap U_i$, $C \in N(V)$. Thus $\phi_i\phi_j^* = z_{ij}$ on $U_i \cap U_j$ and ϕ_i is a unitary equivalence of $M_z|U_i$ and $N|U_i$. Now $V \to \prod_i N(V \cap U_i)$ is a sheaf of Banach spaces and furthermore if $N = (N_i)$ with $N_i \in N(V \cap U_i)$ and $z_{ij,V \cap U_j}N_j = N_i$ on $V_l \cap U_i \cap U_j$, $l \in L$ and $V = \bigcup_{l \in L} V_l$,

then $N \in M_z(V)$ so M_z is a sheaf of Banach spaces. Since $M_z|U_i$ are Hermitian $S|U_i$-modules, it follows easily from Proposition 5.1. that M_z is an Hermitian S-module. Now suppose z arises from M, then $\phi_i\phi_j^* = z_{ij} = \psi_i\psi_j^*$ on $U_i \cap U_j$. Hence $\psi_i^*\phi_i = \psi_j^*\phi_j$ on $U_i \cap U_j$ and, since $Hom_S(M_z,M)$ is a sheaf of Banach spaces, there is a unitary $\chi : M_z \to M$ with $\chi = \psi_i^*\phi_i$ on U_i.

This shows that the Hermitian S-modules equivalent to N on restriction to the sets of an open covering are classified by a 1-cohomology. To make this more precise, we define a W*-category $Z^1(U,R)$ whose objects are the unitary 1-cocycles with respect to the covering U with values in R. An arrow $t : z \to z'$ in $Z^1(U,R)$ is an element $t \in \prod_i R(U_i)$ such that $t_iz_{ij} = z'_{ij}t_j$ on $U_i \cap U_j$. $Z^1(U,R)$ is a W*-category with the structure inherited from the von Neumann algebra $\prod_i R(U_i)$. We let Hmod(U,N) denote the full subcategory of Hmod$S$ whose objects satisfy $M|U_i \simeq N|U_i, i \in I$.

THEOREM 6.2. Let S be a sheaf of von Neumann algebras, N an S-module, $R = Hom_S(N,N)$ and $U = \{U_i\}_{i \in I}$ an open covering. There is an equivalence of W*-categories $F : Z^1(U,R) \to$ Hmod(U,N) where $F(t) : M_z \to M_{z'}$ is defined by $F(t) = \phi_i'^*t_i\phi_i$ on U_i, in the above notation, for $t : z \to z'$.

PROOF. $\phi_i'^*t_i\phi_i = \phi_i'^*t_iz_{ij}\phi_j = \phi_i'^*z'_{ij}t_j\phi_j = \phi_j'^*t_j\phi_j$ on $U_i \cap U_j$. Hence $F(t) : M_z \to M_{z'}$ as required and, since ϕ_i is unitary, F is evidently a faithful *-functor. If $T : M_z \to M_{z'}$, define $t_i = \phi_i'T\phi_i^*$ on U_i, then $t_iz_{ij} = \phi_i'T\phi_j^* = z'_{ij}t_j$ on $U_i \cap U_j$ so $t : z \to z'$ and F is full. Since every object of Hmod(U,N) is equivalent to an object of the form M_z by Proposition 6.1., F is an equivalence.

7. HERMITIAN VECTOR BUNDLES. The results of the previous section allow us to locate the superselection structure of algebraic field theory within the mainstream of mathematics. Probably the first place one meets 1-cocycles with respect to a covering is in the theory of fibre bundles.

If X is a topological space and V an open set of X, let $C(V)$ denote the C*-algebra of bounded, continuous functions on V. Then $V \to C(V)$ is a sheaf of C*-algebras. If E is a finite-dimensional Hermitian vector bundle over X and $E(V)$ denotes the continuous sections s of $E|V$ with $\sup_{x \in V}(s(x),s(x)) < +\infty$, then $V \to E(V)$ is in a natural way an Hermitian C-module.

Now let $\mathcal{U} = \{U_i\}_{i \in I}$ be an open covering of X such that $E|U_i$ is trivial. If E is n-dimensional, the transition functions z_{ij} are unitaries of $M_n \otimes C(U_i \cap U_j)$ and give rise to a unitary 1-cocycle z which is an object of $Z^1(\mathcal{U}, M_n \otimes C)$. We can equally well regard z as arising as in the previous section by implementing the equivalences $E|U_i \simeq M_n \otimes C|U_i, i \in I$. Furthermore, as we saw there, the C-module E can be constructed directly up to unitary equivalence from the 1-cocycle z without recourse to the bundle E.

Thus the superselection sectors are analogues of Hermitian vector bundles where the sheaf C has been replaced by the sheaf $\mathfrak{S}$ of von Neumann algebras constructed in section 5. The restriction to 1-dimensional 'bundles' implied by the locality condition (L') is only apparent, since $M_n \otimes \mathfrak{S}|\mathcal{O}'+a \simeq \mathfrak{S}|\mathcal{O}'+a$ (cf. [8; Lemma 2.5]). In fact, the class of $\mathfrak{S}$-modules satisfying (L') is closed under subobjects and countable direct sums.

On the other hand, (L') does single out coverings of the form $\mathcal{U}(\mathcal{O})$ and a different special class appears in the work of Buchholz and Fredenhagen [26]. It is not yet clear what the natural class of coverings is. Mathematically, one might envisage using all translation covariant coverings, thereby regarding Minkowski space as a topological space acted on by the translation group. Both locality and covariance conditions can also be treated simultaneously combining the techniques of sections 3 and 6; the corresponding 1-cocycles formed part of the original analysis of superselection structure (cf. [9; § II]).

It appears that sheaf theory will prove to be a natural mathematical framework for field theory. Adopting this standpoint does not mean renouncing the net of observables since there are often natural nets associated with sheaves. It does require a certain change in thinking patterns on localization in field theory, an effort rewarded by strategic gains. The sheaf $\mathfrak{S}$ is not the only basic sheaf in algebraic field theory and another sheaf seems to play the role of the 'sheaf of observables'. The concept of a topological space equipped with a sheaf of rings is of sufficient importance in mathematics to have been given the name 'ringed space'. Such sheaves endow topological spaces with additional structure; the sheaf of differentiable functions reflects a differentiable structure, the sheaf of analytic functions an analytic structure. In this spirit, we may regard the sheaf of observables as endowing Minkowski space with a non-commutative mathematical structure which reflects the presence of quantum fields.

8. HERMITIAN BIMODULES. Up till now, only the uninteresting aspects of superselection structure have been touched on. The subject begins to acquire physical interest when one can form 'tensor products' of the relevant representations of the observable net corresponding to the addition of charges. Mathematically, this possibility comes as a complete surprise. The module standpoint has the merit of making it clear where one should look: to take tensor products of modules over a non-commutative ring, one needs a bimodule structure.

Before coming to sheaves, there are a few algebraic preliminaries to be discussed. If A is a C*-algebra, an Hermitian A-bimodule is an object X of hmod$A$ equipped with a *-homomorphism $\rho : A \to (X,X)$. We write ax in place of $\rho(a)x$ unless the action needs to be specified explicitly. The Hermitian A-bimodules are the objects of a category hbimod$A$, where $T : X \to Y$ if T is an arrow of hmod$A$ commuting with the left action of A. A itself will be regarded in the obvious fashion as an object of hbimod$A$.

The tensor product $X \underset{A}{\otimes} Y$ of two Hermitian A-bimodules X and Y has the expected properties: given $x \in X$ and $y \in Y$ there is an $x \otimes y \in X \underset{A}{\otimes} Y$. Finite sums of such elements are norm dense in $X \underset{A}{\otimes} Y$. The left and right actions of A on these elements are given by

$$a(x \otimes y) = ax \otimes y, \quad (x \otimes y)a = x \otimes ya.$$
$$\langle x \otimes y,\, x \otimes y \rangle = \langle y, \langle x,x \rangle y \rangle.$$

If $S : X \to X'$ and $T : Y \to Y'$ in hbimod$A$, there is an arrow $S \otimes T : X \underset{A}{\otimes} Y \to X' \underset{A}{\otimes} Y'$ with $(S \otimes T)(x \otimes y) = Sx \otimes Ty$.
If S and T have adjoints, so does $S \otimes T$ and $(S \otimes T)^* = S^* \otimes T^*$. The existence of $X \underset{A}{\otimes} Y$ with these properties hinges on the estimate

$$0 \leq \sum_{i,j=1}^{m} \langle Ty_i, \langle Sx_i, Sx_j \rangle Ty_j \rangle \leq \| S \|^2 \| T \|^2 \sum_{i,j=1}^{m} \langle y_i, \langle x_i, x_j \rangle y_j \rangle$$

valid for $T : Y \to Y'$ in hbimod$A$, $S : X \to X'$ in hmod$A$ and $x_i \in X$, $y_i \in Y$, $i = 1,2,\ldots m$. This estimate is easily proved by expressing it in terms of the composition of operator-valued matrices.

There is a coherent unitary equivalence expressing the associativity of the tensor product. $X \underset{A}{\otimes} A$ is unitarily equivalent to X. $A \underset{A}{\otimes} X$ is unitarily equivalent to X whenever AX is norm dense in X; in this case, X is an Hermitian A-rigged A-module in the sense of [14; Def. 3.5]. The corresponding full subcategory of hbimodA is then a monoidal category (see e.g. [15; VII], [27]).

If M is a von Neumann algebra, a self-dual Hermitian M-bimodule is an object X of HmodM equipped with an identity-preserving, normal *-homomorphism $\rho : M \to (X,X)$. The corresponding full subcategory of hbimodM will be denoted HbimodM. Tensor products can be formed within HbimodM : $X \underset{M}{\otimes} Y$ can be characterized as above, replacing A by M, except that finite sums $\sum_{i=1}^{n} x_i \otimes y_i$ are only required to be σ-dense in $X \underset{M}{\otimes} Y$. In fact, since HmodM is a reflective subcategory of hmodM by Theorem 2.2 , the tensor product in HbimodM can be constructed from that in hbimodM by passing to the associated self-dual module. HbimodM is a monoidal W*-category, cf. [28; § 3].

Of particular interest here are the Hermitian M-bimodules, $G(\rho)$ say, arising when ρ is an endomorphism of $M = (M,M)$. ρ will be considered as an object in the strict monoidal W*-category EndM [28; Ex. 3.7]. An arrow from ρ to ρ' in EndM is an $r \in M$ with $r\rho(m) = \rho'(m)r$, $m \in M$. If $s : \sigma \to \sigma'$ in EndM then $r \times s : \rho\times\sigma \to \rho'\times\sigma'$ is just $r\rho(s)$ intertwining $\rho\sigma$ and $\rho'\sigma'$. A simple computation shows that EndM completely determines these Hermitian M-bimodules:

PROPOSITION 8.1. Given $r : \rho \to \rho'$ in EndM, let $G(r)m = rm$, $m \in M$, then G : EndM $\to$ HbimodM is a full and faithful *-functor. Furthermore there is a unitary $\tilde{G}_{\rho,\sigma} : G(\rho) \otimes G(\sigma) \to G(\sigma\times\rho)$ in HbimodM with $\tilde{G}_{\rho,\sigma}(m \otimes m') = \sigma(m)m'$, $m,m' \in M$. If $s : \sigma \to \sigma'$ then $\tilde{G}_{\rho',\sigma'}\, G(r) \otimes G(s) = G(s\times r)\, \tilde{G}_{\rho,\sigma}$.

Apart from reversing the order in the tensor product, $(G,\tilde{G},1_M)$ is just a monoidal functor in the sense of [27]. From now on, all modules will be understood to be self-dual.

If S is a sheaf of von Neumann algebras, a (self-dual) Hermitian S-bimodule is a pair σM consisting of an Hermitian S-module M and a $\sigma \in \mathrm{Hom}(S, Hom_S(M,M))$ defining the left action of S on M. It is an object in a W*-category Hbimod$S$, where an arrow from σM to τN is a $T \in \mathrm{Hom}_S(M,N)$ intertwining σ and τ, i.e. such that for any open set V

$$T\sigma_V(S) = \tau_V(S)T \text{ on } V, \quad S \in S(V).$$

If $U = \{U_i\}_{i \in I}$ is an open covering, then Hbimod(U,N) denotes the full subcategory of Hbimod$S$ whose objects are of the form σM with $M \in \mathrm{Hmod}(U,N)$. There is now an analogue of Theorem 6.2 classifying Hbimod(U,N) by a 1-cohomology. This will be described here in the interesting special case, relevant to algebraic field theory, that N is just S considered as an Hermitian S-module, where the coefficients for the 1-cohomology are particularly simple.

Let End$S$ denote the W*-category whose objects ρ are elements of Hom(S,S) where an arrow from ρ to ρ' is a $t \in S(X)$ such that, for each open set V and

$s \in S(V)$, $t\rho_V(s) = \rho'_V(s)t$ on V. The coefficient object for the 1-cohomology is the sheaf $EndS$ of W*-categories defined by setting $EndS(V) = EndS|V$ with the obvious restriction mappings. A unitary 1-cocycle with respect to U taking values in $EndS$ is a pair (ρ,z) where ρ_i is an object of $EndS(U_i), i \in I$. z_{ij} is a unitary from ρ_j to ρ_i on $U_i \cap U_j$ in $EndS(U_i \cap U_j)$, $i,j \in I$ and $z_{ij}z_{jk} = z_{ik}$ on $U_i \cap U_j \cap U_k$, $i,j,k \in I$. These cocycles are the objects of a W*-category $Z^1(U,EndS)$ where $t : (\rho,z) \to (\rho',z')$ if $t_i : \rho_i \to \rho'_i$ in $EndS(U_i)$ and $t_i z_{ij} = z'_{ij}t_j$ on $U_i \cap U_j$.

Now, given an object σM of Hbimod(U,S), pick unitaries $\psi_i \in Hom_S(M,S)(U_i)$ and set $z_{ij} = \psi_i\psi_j{}^*$ on $U_i \cap U_j$. In what follows, we identify the canonically isomorphic sheaves $Hom_S(S,S)$ and S so that $z_{ij} \in S(U_i \cap U_j)$ and $z_{ij}z_{jk} = z_{ik}$ on $U_i \cap U_j \cap U_k$. Now define $\rho_i \in EndS(U_i)$ by setting, for $V \subset U_i$ and $s \in S(V)$,

$$\rho_{i,V}(s) = \psi_i\sigma_V(s)\psi_i{}^* \text{ on } V.$$

If $V \subset U_i \cap U_j$, we have $z_{ij}\rho_{j,V}(s) = \psi_i\sigma_V(s)\psi_j{}^* = \rho_{i,V}(s)z_{ij}$ on V, so that (ρ,z) is an object of $Z^1(U,EndS)$.

Conversely, if $\zeta = (\rho,z)$ is an object of $Z^1(U,EndS)$, we define an Hermitian S-bimodule $F(\zeta)$ by equipping the Hermitian S-module M_z of the previous section with a left action τ of S defined by

$$\tau_V(s) = \phi_i{}^*\rho_{i,V\cap U_i}(s|_{V\cap U_i})\phi_i \text{ on } V \cap U_i, \quad s \in S(V),$$

where ϕ_i is as in Proposition 6.1. Since $\phi_i\phi_j{}^* = z_{ij}$ on $U_i \cap U_j$ and $z_{ij} : \rho_j \to \rho_i$ on $U_i \cap U_j$, this does define $\sigma \in \mathrm{Hom}(S,Hom_S(M_z,M_z))$.

If $\zeta = (\rho,z)$ arises as above from σM and χ is as in Proposition 6.1, then χ intertwines the actions τ and σ since

$$\chi\tau_V(s) = \psi_i{}^*\rho_{i,V\cap U}(s|_{V\cap U_i})\phi_i = \sigma_V(s)\chi \text{ on } V \cap U_i.$$

Hence τM_z and σM are unitarily equivalent Hermitian S-bimodules. Similar simple local calculations and Theorem 6.2 now yield:

THEOREM 8.2. Let S be a sheaf of von Neumann algebras, $U = \{U_i\}_{i \in I}$ an open covering, then $\zeta \to F(\zeta)$ extends to a unitary equivalence $F : Z^1(U,EndS) \to \mathrm{Hbimod}(U,S)$ defined by $F(t) = \phi'_i{}^* t_i \phi_i$ on U_i for $t : \zeta \to \zeta'$.

The naive definition of the tensor product of two Hermitian S-bimodules σM and τN cannot be expected to work in general: if $\pi P(V) = \sigma M(V) \otimes_{S(V)} \tau N(V)$ with the obvious restriction mappings and an action π given by

$$\pi_V(s)_U\,(M \otimes N) = \sigma(s|_U)M \otimes N, \quad M \in M(U), N \in N(U),\ U \subset V,$$

then $\{P,S\}$ is, a priori, only a presheaf in Imod$S$. The correct definition would be got by passing to the associated sheaf, but I do know whether this is possible in full generality in this specific context. For the present purposes, the following argument suffices. If σM and τN are unitarily equivalent to Hermitian S-bimodules ρS and $\rho' S$, where $\rho,\rho' \in \mathrm{End}S$, then πP will be unitarily equivalent to $\rho'\rho S$ (see Proposition 8.1) and hence an Hermitian S-bimodule. So if σM and τN are objects of Hbimod$(\mathcal{U},S)$, $P|U_i \simeq S|U_i, i \in I$. But the sheaf $\{\tilde{P},S\}$ associated with $\{P,S\}$ can then be constructed trivially by taking $\tilde{P}(V)$ to be the subspace of $\prod_{i\in I} P(V \cap U_i)$ of elements $(P_i)_{i\in I}$ with $P_i = P_j$ on $U_i \cap U_j$ with the obvious restriction mappings. The arrow $P \to (P|_{V\cap U_i})_{i\in I}$, $P \in P(V)$ from P to $\tilde{P}$ is a universal arrow from P to Hermitian S-modules. This implies, in particular, that the action extends to give an Hermitian S-bimodule $\pi\tilde{P}$. Hence we can define $\sigma M \underset{S}{\otimes} \tau N = \pi\tilde{P}$ within Hbimod$(\mathcal{U},S)$ and it is routine to check that the obvious definitions make Hbimod$(\mathcal{U},S)$ into a monoidal W*-category.

On the other hand, each $\mathit{EndS}(V)$ is a strict monoidal W*-category in the same way as EndM is and the monoidal structure is compatible with the restriction mappings. Thus $Z^1(\mathcal{U},\mathit{EndS})$ is itself a strict monoidal W*-category with the induced operations, e.g. $(\rho',z')\times(\rho,z) = (\rho'\rho,z'\times z)$, where $(z'\times z)_{ij} = z'_{ij} \times z_{ij} = z'_{ij}\rho'_{j,U_i\cap U_j}(z_{ij})$. The relation between the monoidal structures is described by

THEOREM 8.3. If ζ and ζ' are objects of $Z^1(\mathcal{U},\mathit{EndS})$ there is a unitary $\tilde{F}_{\zeta,\zeta'} : F(\zeta)\otimes F(\zeta') \to F(\zeta'\times\zeta)$ natural in ζ and ζ', i.e. given $t : \zeta_1 \to \zeta_2$ and $t' : \zeta'_1 \to \zeta'_2$,

$$F(t'\times t)\,\tilde{F}_{\zeta_1,\zeta'_1} = \tilde{F}_{\zeta_2,\zeta'_2}\,F(t)\otimes F(t').$$

PROOF. Adapting the definition of $\tilde{G}$ in Proposition 8.1, we have a unitary equivalence

$$\tilde{G}_{\rho_i,\rho'_i} : \rho_i S|U_i \otimes \rho'_i S|U_i \to \rho'_i\rho_i S|U_i$$

natural in ρ_i and ρ'_i. Let $\zeta^o = \zeta' \times \zeta$ and set

$$\tilde{F}_{\zeta,\zeta'} = \phi_i^{o*}\tilde{G}_{\rho_i,\rho'_i}\,\phi_i \otimes \phi'_i \quad \text{on } U_i.$$

A local computation shows that $\tilde{F}_{\zeta,\zeta'}$ is a well defined unitary satisfying the required identity.

9. COMPOSITION OF SECTORS. We now describe a new approach to superselection sectors based on the concept of Hermitian module and begin by defining a coarser topology on Minkowski space that will eventually allow us to interpret the composition of sectors as the tensor product of bimodules.

The spacelike complements of compact sets from a base for a topology τ on Minkowski space. Of course, τ is not Hausdorff although it is T_1. For the present purpose, τ has the agreeable property that the intersection of two non-empty τ-open sets is non-empty. This allows one to define a sheaf S_τ for the τ-topology by setting $S_\tau(V) = [\pi_0|V]'$ if V is a non-empty τ-open set and $S_\tau(\emptyset) = 0$. If $\emptyset \neq U \subset V$, the restriction mapping $S_\tau(V) \to S_\tau(U)$ is just the inclusion mapping.

Thus S_τ is no more complicated than the nets of algebraic field theory and the conceptual simplifications this allows are worth noting explicitly. S_τ can be regarded as a *-subalgebra of $B(H_{\pi_0})$ which is the union of distinguished von Neumann subalgebras $S_\tau(V)$, for each non-empty τ-open set V, with $S_\tau(V) \subset S_\tau(U)$ whenever $U \subset V$, and satisfies the intersection property $S_\tau(\bigcup_i V_i) = \bigcap_i S_\tau(V_i)$. Similarly, we can regard an Hermitian S_τ-module M as the union of Hermitian $S_\tau(V)$-modules $M(V)$ with the analogous properties. Furthermore $\psi \in Hom_{S_\tau}(M,M')(V)$ can be regarded as a bounded S_τ-module homomorphism from M to M' such that $\psi(M(U)) \subset M'(U)$ for $U \subset V$.

If M is an object of $\mathrm{Hmod}(\mathcal{U}(\mathcal{O}),S_\tau)$, i.e. an Hermitian S_τ-module with $M|\mathcal{O}'+a \simeq S_\tau|\mathcal{O}'+a$, $a \in \mathbb{R}^4$, then, by picking an associated 1-cocycle z and applying the construction of section 4, we can define a representation π of $\mathfrak{A}$ satisfying (L) such that M is unitarily equivalent to the Hermitian S_τ-module M_π where $M_\pi(V) = \{t \in B(H_{\pi_0},H_\pi) : t\pi_0(A) = \pi(A)t,\ A \in \mathfrak{A}|V\}$. There is a more canonical construction which will be sketched here as it has the conceptual merit of separating the concrete realization of M in terms of operators on Hilbert space from the construction of the representation π. It uses a generalized Gelfand-Naimark-Segal construction as in the concrete realization of hmodA and HmodM in section 2. Let $H(M)$ denote the Hilbert space constructed from the scalar product $(\Omega,\langle M,M'\rangle\Omega)$ on M. A $\psi \in Hom_{S_\tau}(M,M')(V)$ can be regarded as a bounded operator from $H(M)$ to $H(M')$ and, since M is canonically isomorphic to $Hom_{S_\tau}(S_\tau,M)$, the elements of M can themselves be regarded as bounded operators from $H_{\pi_0} = H(S_\tau)$ to $H(M)$.

There is now a unique representation π of $\mathfrak{A}$ on $H(M)$ such that $M_\pi = Hom_{S_\tau}(S_\tau,M)$. If $\mathfrak{A}$ satisfies *essential duality*, i.e. if the net $\mathfrak{A}^d$ over the set K of closed double cones defined by $\mathfrak{A}^d(\mathcal{O}) = S_\tau(\mathcal{O}')$ is local, this representation has a canonical extension to $\mathfrak{A}^d$ [29; Thm. 4.1].

This construction will now be reinterpreted as saying that every object of $\mathrm{Hmod}(U(\mathcal{O}),S_\tau)$ is in a canonical way an Hermitian S_τ-bimodule. The hypothesis of essential duality has been shown by Bisognano and Wichmann [30] to be derivable from plausible assumptions on a Wightman field theory. It implies that S_τ has certain local commutativity properties: to express them, call two τ-open sets V_1 and V_2 *amply spacelike* if there are spacelike double cones $\mathcal{O}_1$ and $\mathcal{O}_2$, with $\mathcal{O}_1' \subset V_1$ and $\mathcal{O}_2' \subset V_2$. As a rough guide to intuition, this means that one can set up arbitrarily large laboratories in each of V_1 and V_2 which are mutually spacelike; it does not, of course, mean that V_1 and V_2 are themselves spacelike. S_τ will be said to be *local* if $S_\tau(V_1)$ and $S_\tau(V_2)$ commute whenever V_1 and V_2 are amply spacelike; this is obviously equivalent to essential duality for $\mathfrak{A}$.

An Hermitian S_τ-bimodule πM will be said to be *local* if $\psi\pi(S) = S\psi$, whenever $S \in S_\tau(V_1)$, $\psi \in Hom_{S_\tau}(M,S_\tau)(V_2)$ and V_1 and V_2 are amply spacelike. S_τ is local if and only if it is local as an Hermitian S_τ-bimodule.

LEMMA 9.1. If S_τ is local and z is an object of $Z^1(U(\mathcal{O}),S_\tau)$ then $z_{aa'} \in S_\tau(\mathcal{O}_1')'$ for a, $a' \in (\mathcal{O}+\mathcal{O}_1)'$.

PROOF. If there exists $\mathcal{O}_2 \in K$ with $\mathcal{O}_1 \subset \mathcal{O}_2' \subset (\mathcal{O}'+a) \cap (\mathcal{O}'+a')$, then $z_{aa'} \in S_\tau(\mathcal{O}_2') \subset S_\tau(\mathcal{O}_1')'$ since S_τ is local. Hence $(\mathcal{O}+\mathcal{O}_1)'$ can be covered with open sets U_i such that $z_{aa'} \in S_\tau(\mathcal{O}_1')'$ if a, $a' \in U_i$. Since $(\mathcal{O}+\mathcal{O}_1)'$ is connected, the cocycle identity implies that $z_{aa'} \in S_\tau(\mathcal{O}_1')'$ for all $a,a' \in (\mathcal{O}+\mathcal{O}_1)'$.

The Lemma depends critically on the connectivity of the spacelike complement of double cones. It is not valid in a two dimensional space-time where there can be solitonic sectors satisfying (L).

THEOREM 9.2. If S_τ is local, every object M of $\mathrm{Hmod}(U(\mathcal{O}),S_\tau)$ has a unique structure of local Hermitian S_τ-bimodule. $\mathrm{Hmod}(U(\mathcal{O}),S_\tau)$ can then be identified with a full monoidal subcategory of $\mathrm{Hbimod}(U(\mathcal{O}),S_\tau)$.

PROOF. Let $\psi_a \in Hom_{S_\tau}(M,S_\tau)(\mathcal{O}'+a)$ be unitary and suppose that πM is a local S_τ-bimodule, then π is unique since

$$\pi(S) = \psi_a^* S\psi_a,\quad S \in S_\tau(V) \text{ if } V \text{ and } \mathcal{O}'+a \text{ are amply spacelike.} \qquad (*)$$

If V and $\mathcal{O}'+a_i$ are amply spacelike, $i = 1,2$, there exists $\mathcal{O}_i \in K$ with $\mathcal{O}_i' \subset V$ and $a_i \in (\mathcal{O}+\mathcal{O}_i)'$. Picking $a \in (\mathcal{O}+\mathcal{O}_1)' \cap (\mathcal{O}+\mathcal{O}_2)'$ and setting $z_{aa'} = \psi_a\psi_{a'}^*$, we see, by Lemma 9.2, that $\psi_{a_1}^* S\psi_{a_1} = \psi_a^* S\psi_a = \psi_{a_2}^* S\psi_{a_2}$, $S \in S_\tau(V)$. Thus $\pi(S)$ is well defined by (*) and $\pi(S) \in Hom_{S_\tau}(M,M)(V \cap (\mathcal{O}'+a))$. Any τ-open set V can be covered by amply spacelike $\mathcal{O}'+a$, so $\pi(S) \in Hom_{S_\tau}(M,M)(V)$ by the intersection property. π is clearly a left action of S_τ on M. If $\pi'M'$ is constructed similarly and $\phi \in Hom_{S_\tau}(M,M')(V_1)$ with V_1 and V amply spacelike, pick a with V and $V_1 \cap (\mathcal{O}'+a)$ amply spacelike, then $\phi\pi(S) = \phi\psi_a^* S\psi_a = \psi_a'^* S\psi_a' = \pi'(S)\phi$,

$S \in \mathcal{S}_\tau(V)$, since $\psi'_a \phi \psi_a^* \in \mathcal{S}_\tau(V_1 \cap (\mathcal{O}'+a))$ and $\mathcal{S}_\tau$ is local. Taking $M' = \mathcal{S}_\tau$, this shows that πM is a local bimodule and taking $V_1 = \mathbb{R}^4$, which is amply spacelike to any non-empty τ-open set V, we see that $\mathrm{Hmod}(\mathcal{U}(\mathcal{O}),\mathcal{S}_\tau)$ can be identified with a full subcategory of $\mathrm{Hbimod}(\mathcal{U}(\mathcal{O}),\mathcal{S}_\tau)$. It remains to show that it is closed under tensor products. Let $\pi''M'' = \pi M \otimes \pi' M'$; specializing the definition to the present situation, we have $\pi''M''(V) = \bigcap_a \pi M(V_a) \underset{\mathcal{S}_\tau(V_a)}{\otimes} \pi' M'(V_a)$, where $V_a = V \cap (\mathcal{O}'+a)$. There is a unitary $\psi''_a \in Hom_{\mathcal{S}_\tau}(M'',\mathcal{S}_\tau)(\mathcal{O}'+a)$ with $\psi''_a(M \otimes M') = \psi'_a(\pi'\psi_a(M)M')$, $M \in M$, $M' \in M'$.

Hence if V and $\mathcal{O}'+a$ are amply spacelike and $S \in \mathcal{S}_\tau(V)$,

$$\psi''^*_a S \psi''_a (M \otimes M') = \psi_a^*(1) \otimes \psi'^*_a S \psi'_a(\pi'\psi_a(M)M') = \psi_a^*(1) \otimes \pi'(S\psi_a(M))M'$$

$= \psi_a^* S \psi_a M \otimes M' = \pi(S) M \otimes M'$. Thus $\psi''^*_a S \psi''_a = \pi(S) \otimes 1 = \pi''(S)$, $S \in \mathcal{S}_\tau(V)$, so that $\pi''M''$ is a local bimodule as required.

In [8], $\mathfrak{A}$ was assumed to satisfy duality and a composition law up to unitary equivalence for representations satisfying (L) was defined by composing localized morphisms. This is equivalent to forming the tensor product of the associated $\mathcal{S}_\tau$-bimodules as may be verified either by direct computation or by invoking Theorem 8.3. Having established that the composition of sectors corresponds to the tensor product of bimodules, it is clear that the other important constructions of [8,9] answer natural questions on the structure of local bimodules. The permutation symmetry tells one that the tensor product of local bimodules is commutative, up to natural unitary equivalence. The existence of conjugate sectors shows that a local bimodule actually has a second $\mathcal{S}_\tau$-valued inner product : it is a bi-Hermitian module so that the left and right actions enter in a completely symmetric way. Interchanging these actions then corresponds to passing to the conjugate sector.

The weaker localization condition employed by Buchholz and Fredenhagen [26] calls for a different topology τ_S on Minkowski space. A non-empty open set of Minkowski space is τ_S-open if it contains some translate of the spacelike cone S. The significance of these various topologies is not yet clear. It might be interesting to investigate why topologies appear which single out a particular direction in Minkowski space.

The more interesting potential field of applications concerns gauge theories, solitons and instantons. Indeed, the point of view here should prove effective in all problems where the cohomology [29,31] of nets of observables or fields plays a role as it emphasises the objects classified by the cohomology rather than the cohomology itself. In particular, by interpreting the 2-cohomology in this way, one should discover the correct way of handling the connection and curvature of a gauge theory in quantum field theory. To draw a parallel with the 1-cohomology, the S_τ-bimodules appear nowhere in conventional treatments of field theory, but they do appear implicitly in the field bundle of [9]. The field bundle proved a very effective substitute for conventional fields, for example, in allowing one to construct collision states.

BIBLIOGRAPHY

1. R. Haag, Discussion des "axiomes" et des propriétés asymptotiques d'une théorie des champs locale avec particules composées, Colloques Internationaux du CNRS LXXV Lille 1957, CNRS, Paris, 1959.

2. H. Araki, Einführung in die Axiomatische Quantenfeldtheorie, Lecture Notes, Zürich 1961-62.

3. H. Araki, Von Neumann algebras of local observables for the free scalar field, J. Math. Phys., 5(1964), 1-13.

4. H.J. Borchers, The vacuum state in Quantum Field Theory II, Commun. Math. Phys., 1(1965), 57-79.

5. H.J. Borchers, Local rings and the connection of spin with statistics, Commun. Math. Phys., 1(1965), 281-307.

6. R. Haag, D. Kastler, An algebraic approach to Quantum Field Theory, J. Math. Phys., 5(1964), 848-861.

7. G.C. Wick, A.S. Wightman, E.P. Wigner, The intrinsic parity of elementary particles,Phys. Rev., 88(1952), 101-105.

8. S. Doplicher, R. Haag, J.E. Roberts, Local observables and particle statistics I, Commun. Math. Phys., 23(1971), 199-230.

9. S. Doplicher, R. Haag, J.E. Roberts, Local observables and particle statistics II, Commun. Math. Phys., 35(1974), 49-85.

10. P. Ghez, R. Lima, J.E. Roberts, W*-categories, Submitted to the J. Operator Theory.

11. W.L. Paschke, Inner product modules over B*-algebras, Trans. Amer. Math. Soc., 182(1973), 443-468.

12. M.A. Rieffel, Induced representations of C*-algebras, Advan. Math., 13(1974), 176-257.

13. G.G. Kasparov, Hilbert C*-modules: theorems of Stinespring and Voiculescu, to appear in J. Operator Theory.

14. M.A. Rieffel, Morita equivalence for C*-algebras and W*-algebras, J. Pure Appl. Alg., 5(1974), 51-96.

15. S. Mac Lane, Categories for the working mathematician, Springer, New York, Heidelberg, Berlin, 1971.

16. M. Artin, A. Grothendieck, J.L. Verdier, Théorie des topos et cohomologie étale des schémas, Lecture Notes in Mathematics 269, Springer, Berlin, Heidelberg, New York, 1972.

17. A. Connes, M.A. Takesaki, The flow of weights on factors of type III, Tôhoku Math. J., 29(1977), 473-575.

18. J.P. Serre, Cohomologie Galoisienne, Lecture Notes in Mathematics 5, Springer, Berlin, Heidelberg, New York, 1965.

19. L.G. Brown, P. Green, M.A. Rieffel, Stable isomorphism and strong Morita equivalence of C*-algebras, Pacific J. Math., 71(1977), 349-363.

20. J.W. Gray, Sheaves with values in a category, Topology, 3(1965), 1-18.

21. B. Mitchell, Theory of categories, Academic Press, New York, London, 1965.

22. K.H. Hofmann, Bundles and sheaves are equivalent in the category of Banach spaces, Lecture Notes in Mathematics 575, 53-69, Springer, Berlin, Heidelberg, New York, 1977.

23. B. Banaschewski, Sheaves of Banach spaces, Quaest. Math., 2(1977), 1-22.

24. C.J. Mulvey, Banach sheaves, J. Pure Appl. Alg., 17(1980), 69-84.

25. J. Frenkel, Cohomologie non-Abélienne et espaces fibrés, Bull. Soc. Math. France, 85(1957), 135-220.

26. D. Buchholz, K. Fredenhagen, Locality and structure of particle states in Relativistic Quantum Theory, to appear.

27. S. Eilenberg, G.M. Kelly, Closed Categories, Proceedings of the Conference on Categorical Algebra, La Jolla 1965, Springer, Berlin, Heidelberg, New York 1966.

28. J.E. Roberts, Cross products of von Neumann algebras by group duals, Symposia Mathematica, 20(1976), 335-363.

29. J.E. Roberts, Net cohomology and its applications to Field Theory, In : Quantum Fields-Algebras, Processes, ed. L. Streit, 239-268, Springer, Wien, New York, 1980.

30. J.J. Bisognano, E.H. Wichmann, On the duality condition for quantum fields, J. Math. Phys., 17(1976), 303-321.

31. J.E. Roberts, A survey of local cohomology, Lecture Notes in Physics 80, 81-93, Springer, Berlin, Heidelberg, New York 1978.

FACHBEREICH PHYSIK
UNIVERSITÄT OSNABRÜCK
D-4500 OSNABRÜCK
FEDERAL REPUBLIC OF GERMANY

Proceedings of Symposia in Pure Mathematics
Volume 38 (1982), Part 2

ALGEBRAIC AND MODULAR STRUCTURE OF VON NEUMANN ALGEBRAS OF PHYSICS

Roberto Longo

1. INTRODUCTION. Since the appearance of the algebraic approach to Quantum Physics, several authors, as we shall see, have studied the structure of the von Neumann algebras involved, both in Quantum Field Theory and in Statistical Mechanics. Sometimes they used ad hoc computations, sometimes they developed more general arguments.

The results in this direction became more precise by the Tomita-Takesaki theory. As a matter of fact it is now possible to easily derive several of the new and old results by general theorems about W^*-dynamical systems. The aim of this paper is to collect some of these theorems. The exposition will not follow the chronological development; instead we prefer to state the theorems in an abstract form, so far as it is possible, which clarifies their role and their nature. A few new arguments may occur. Furthermore we include the description of the modular objects of certain von Neumann algebras of local observables.

2. MODULAR INVARIANTS AND W^*-DYNAMICAL SYSTEMS. Let (A, G, α) be a C^*-dynamical system with A unital and let $\omega \in A^*$ be a α-invariant state. Denote by (π, ξ, H) the GNS triple associated to ω. By a standard procedure, the unitary representation U of G on H defined by $U(g)\ \pi(A)\xi = \pi(\alpha_g(A))$, $A \in A$, $g \in G$, determines a covariant representation of A, that is $\pi(\alpha_g(A)) = U(g)\pi(A)U(g)^*$. One thus obtains a W^*-dynamical system (R, G, α) where $R = \pi(A)''$ and (by an improper notation) $\alpha_g(A) = U(g)\pi(A)U(g)$, $A \in R$.

Subject classification 46L55, 46L60.

0082-0717/80/0000-0548/$05.00

Now inequivalent representations of A may have a different physics significance (see e.g. [20, 55]) and one is naturally led to determine algebraic invariants for R, assuming (A, G, α) to satisfy properties of physics interest, in order to distinguish among representations. We begin by listing some of those assumptions, see e.g. [1-6]. In the following M denotes the Godement (or any) mean on G.

1C. *Weakly clustering*:

$$M_g\,[\omega(\alpha g(A)B) - \omega(A)\,\omega(B)] = 0, \qquad A, B \in A.$$

1W. ξ *is the unique* U-*invariant vector* (up to scalars).

2C. *Weakly mixing*:

$$M_g\,|\omega(\alpha_g(A)A^*) - |\omega(A)|^2| = 0, \qquad A \in A .$$

2W. *The trivial subrepresentation on* $\mathbb{C}\xi$ *is the only finite-dimensional subrepresentation of* U.

3C. *Mixing (or clustering)*: As $g \to \infty$

$$\omega(\alpha_g(A)B) - \omega(A)\,\omega(B) \to 0, \qquad A, B \in A.$$

3W. $U(g) \to [\mathbb{C}\xi]$ *weakly as* $g \to \infty$.

4C. $\omega([\alpha_g(A), B]^* [\alpha_g(A), B]) \to 0, \qquad A, B \in A.$

4W. *For every* $x \in [R'\xi]$ *and* $A,B \in \pi(A)$

$$[\alpha_g(A), B]x \to 0 \text{ strongly.} \tag{1}$$

Note that if ξ is separating for R, this entails the strong asymptotic abelianness property, namely the commutator in (1) converges for every $A, B \in R$.

5. *Norm asymptotic abelianness:*

$$\|\,[\alpha_g(A), B]\,\| \to 0 \qquad A, B \in A.$$

By considering different kinds of convergence in 4 and 5 one obtains different notions. Assume for definiteness the usual convergence as $g \to \infty$.

Then $i{+}1 \Rightarrow i$ $(i = 1,2,4)$ and $4 \Rightarrow 3$ if R is a factor. To simplify the exposition we shall make the following assumption throughout this section only.

ASSUMPTION 1. ξ is separating for R.

Properties iC and iW are equivalent $(i = 1\text{-}4)$ and are formulated in terms of A and R respectively. We shall henceforth focus on the W*-dynamical system (R, G, α) and derive some consequences. Notations will be as above.

1. The following theorem was proved by Hugenholtz [7] in a restricted from, and then by Størmer [8] in this generality, see [9].

THEOREM 1. *If the weakly clustering property holds, then R is of type III or ω is a trace.*

Note that, by assumption 1, weakly clustering is equivalent to ergodicity for α, namely $R^{\alpha} = \mathbb{C}I$, where R^{α} denotes the algebra of the α-fixed elements.

2. The hypotheses in the Hugenholtz-Størmer theorem, are not sufficient to determine the Connes [10] invariants $T(R)$ and $S(R)$, so we examine further assumptions. Denote by $Z(R)$ the center of R.

DEFINITION 1 (R, G, α) satisfies *condition M* if, for any given unitary $V \in R$, $\alpha_g(V)^*V \in Z(R)$, all $g \in G \Rightarrow V \in Z(R)$.

Condition M means that α admits no generalized eigenoperator with eigenvalue in $Z(R)$. It is easily seen [11] that this follows by the strong asymptotic abelianness (1), however the interest of condition M relays on the fact that, if R is a factor, this condition follows by the weakly mixing property 2, and it is equivalent to it for example if G is abelian.

Let's denote by spec(ω) the spectrum of the modular operator relative to ξ. In [12] spec(ω) was defined without explicit reference to the modular objects, thus it may be defined in terms of A by using the Kaplansky density theorem, namely spec(ω) is equal to the set of all $\lambda \geq 0$ such that, for any given $\varepsilon > 0$, there exists a bounded sequence (net in the non-separable case) $A_n \in A$ with $\omega(A_n^* A_n) > 1$ such that

$$\overline{\lim_n} \, |\omega(A_n B) - \lambda\omega(B A_n)|^2 \leq \varepsilon \; \omega(B B^*), \qquad B \in A .$$

Denote by $\perp$ the annihilator with respect to the usual duality $((\lambda\ t) \to \lambda^{it})$ between $\mathbb{R}$ and $\mathbb{R}^+$.

THEOREM 2. *Assume that* (R, G, α) *satisfies condition* M. *Then* $\mathrm{spec}(\omega)\setminus\{0\}^{\perp} = T(R)$ *and, if* α *is ergodic,* $T(R)^{\perp} = \mathrm{spec}(\omega)\setminus\{0\}$. *Hence* $\mathrm{spec}(\omega)\setminus\{0\}$ *is a closed subgroup of* $\mathbb{R}^{+}$ *which is an algebraic invariant for* R .

PROOF. To prove the first equality we have to show that $T(R)$ is the kernel of the map $t \in \mathbb{R} \to \sigma_t^{\omega} \in \mathrm{Aut}(R)$, where σ^{ω} is the modular group with respect to ξ. Indeed let $t \in T(R)$ and $V \in R$ a unitary implementing σ_t^{ω}. Since ω is α-invariant, σ^{ω} and α commute, therefore

$$VAV^{*} = \sigma_t^{\omega}(A) = \alpha_g \circ \alpha_t^{\omega} \circ \alpha_g^{-1}\ (A)$$

$$= \alpha_g(V\, \alpha_g^{-1}(A)\, V^{*}) = \alpha_g(V)\, A\, \alpha_g(V)^{*}, \qquad A \in R$$

that is V and $\alpha_g(V)$ implements the same automorphism of R, hence $\alpha_g(V)V^{*} \in Z(R)$, $g \in G$. By condition M, $V \in Z(R)$, thus $\sigma_t^{\omega}(A) = VAV^{*} = A$, $A \in R$. If α is ergodic, then a standard argument shows that $\mathrm{spec}(\omega)\setminus\{0\}$ is a group (see e.g. [11]), thus $\mathrm{spec}(\omega)\setminus\{0\} = (\mathrm{spec}(\omega)\setminus\{0\})^{\perp\perp} = T(R)^{\perp}$. ■

Note that $T(R)^{\perp} = S(R)$ if R is a factor of type $\neq \mathrm{III}_0$ [10], while for a general von Neumann algebra R one has $T(R) = \cap\, \mathrm{spec}(\varphi)\setminus\{0\}$, where $\varphi \in R_{*}$ ranges among faithful states with multiplicative $\mathrm{spec}(\varphi)$. We refer to [11] for the following examples.

EXAMPLES. For every infinite cyclic subgroup Λ of $\mathbb{R}^{+}$, there exists a W^{*}-dynamical system (R, G, α) based on a III_0-factor R such that α is weakly mixing with respect to a faithful α-invariant state $\omega \in R_{*}$ with $\mathrm{spec}(\omega) = \Lambda^{-}$.

3. No property of the state, namely mixing type conditions or stronger properties as Lebesgue spectrum (i.e. the spectral measure of U is absolutely continuous with respect to the Haar measure), are known to suffice for the equality $\mathrm{spec}(\omega) = S(R)$ to hold, even assuming $\mathrm{spec}(\omega) = \mathbb{R}^{+}$ and R a factor.

4-5. In [12] Størmer proved that properties 4 and 3 together implies $\mathrm{spec}(\omega) = S'(R)$ where $S'(R) = \cap\, S(ERE)$, E ranging among non-zero projection of R. Araki [13] then proved that the strong asymptotic abelianness assumption (1) alone entails $\mathrm{spec}(\omega) = S(R)$; see [9]. However the theory of Connes on the essential spectrum Γ [10] enables one to easily derive these results by more general facts about commuting pairs of W^{*}-

-dynamical systems [14,15] as follows.

Let (R, H, β) be a W^*-dynamical system with H abelian, and let $\Gamma_B(\beta) = \cap\, sp(\beta^E)$ where E varies among projections E of R^β with central cover $C(E) = I$ and $\beta^E = \beta|ERE$. Thus $\Gamma_B(\beta) = \Gamma(\beta)$ if R is a factor, and the same argument as for Γ shows that $\Gamma_B(\sigma^\varphi) = S(R)$, for any faithful state $\varphi \in R_*$. Denote by $R_\varphi = R^{\sigma^\varphi}$ the centralizer of φ.

THEOREM 3. *Let* (R, G, α) *and* (R, H, β) *be* W^**-dynamical systems with* G *amenable and* H *abelian. Suppose that* α *and* β *commute and* $\omega \in R_*$ *is a faithful* α*-invariant state. If*

$$M_g\, \omega([\alpha_g(A), B]^* [\alpha_g(A),B]) = 0 \tag{2}$$

for all $A \in R$ *and* $B \in Z(R_\omega)$, *then*

$$sp(\beta) = \Gamma_B(\beta)$$

PROOF. We have to show that if the spectral subspace $R(V, \beta) \neq \{0\}$ for some open $V \subset \hat{H}$ then $ERE(V, \beta^E) \neq \{0\}$ for every projection $E \in R^\beta$ with $c(E) = I$. Let $0 \neq A \in R(V, \beta)$; since α commutes with β, we have $E\alpha_g(A)E \in ERE(V, \beta^E)$, $g \in G$. We shall show that $E\alpha_g(A^*)E\, \alpha_g(A)E = 0$, all $g \in G$, implies $A = 0$, and we are done. Now by (2) $E\alpha_g(A^*)E\, \alpha_g(A)E$ and $E\alpha_g(A^*A)E$ are asymptotically equal (i.e. their difference strongly tends to zero in mean) therefore $E\epsilon_\alpha(A^*A) = E\epsilon_\alpha(A^*A)E = M_g(E\,\alpha_g(A^*A)E) = M_g(E\,\alpha_g(A^*)E\,\alpha_g(A)E) = 0$, where ϵ_α is the canonical conditional expectation of R onto $R^\alpha \subset Z(R)$. Since $c(E) = I$, we then have $\epsilon_\alpha(A^*A) = 0$, thus $A = 0$ because ϵ_α is faithful. ∎

The above theorem is suitable for being translated in the C^*-algebra setting [15]. It soon entails the following corollary which provides an unified statement for the results of Størmer and Araki and the theorem of Connes to the effect that $spec(\omega) = S(R)$ if R_ω is a factor [10].

COROLLARY 1. *Let* (R, G, α) *be as above and assume that (2) holds for all* $A \in R$, $B \in Z(R_\omega)$, *then* $spec(\omega) = S(R)$.

3. SPECTRUM CONDITION AND SEMIGROUPS OF ENDOMORPHISMS. We consider now another physics property rich of consequences: the spectrum condition. We

start by giving an abstract definition.

DEFINITION 2. Let U be a unitary representation of a locally compact abelian group G. We say that the spectrum of U, $sp(U)$, is *asymmetric* if $sp(U) \cap sp(U)^{-1} \subset \{e\}$ where e is the identity of $\hat{G}$.

The above definition contains the standard spectrum condition ($G = \mathbb{R}^n$, $sp(U) \subset V_+$). We refer to [51] for an algebraic formulation of this property. Notice that definition 2 doesn't refer to any notion of order on G, as other generalizations do. Furthermore, as pointed out to the author by M. Takesaki, it may be defined and used even if G is non-commutative: in such a case $sp(U)$ asymmetric means that if V and its complex conjugate $\bar{V}$ are representations of G, both quasi-contained in U, then V is trivial. Much of the following can be carried on in this non-commutative context, but we don't pursue here in this direction.

The following lemma contains the well-known vacuum version of Borchers theorem [16]; a proof in the one-parameter case appeared in [17].

LEMMA 1. *Let* (R, G, α) *be a* W^**-dynamical system with* G *abelian,* α *implemented by a unitary group* U, *and* ξ *a* U*-invariant cyclic vector for* R. *If* $sp(U)$ *is asymmetric then* $U(G) \subset R$.

PROOF. Let $\beta_g(A) = U(g)AU(g)^*$, $A \in R'$, $g \in G$; we have to show that β is trivial. Let $E = [R'\xi]$ and consider the W^*-dynamical system $(R'E, G, \beta^E)$ obtained reducing β to $R'E$. Note that β^E is implemented by $U^E = U|E(H)$. Since ξ is U^E-invariant, cyclic and separating for $R'E$, it follows easily (see [2]) that $sp(U^E) = sp(\beta^E)$. Now $sp(U^E)$ is asymmetric and $sp(\beta^E) = sp(\beta^E)^{-1}$, therefore $sp(\beta^E) = \{e\}$ and β^E is trivial. Since $c(E) = I$, $\beta_g(A)E = \beta^E{}_g(AE) = AE$, $A \in R'$, entails $\beta_g(A) = A$, which concludes the proof. ■

REMARK 1. Assuming G connected, the above argument can be easily continued to show that α is inner, only assuming that $sp(U) \cap sp(U)^{-1}$ is compact. See also [18] for related arguments.

We shall see that, for the applications in Quantum Field theory, one needs to deal with a generalization of a W^*-dynamical system where the automorphism group is replaced by a semigroup of endomorphisms. Given a locally compact group G, we say here that a semigroup $S \subset G$ is *total* in G if $G = S \cup S^{-1}$ as setwise union. By extending a result of Driessler [19] (see

(b) below) one has the following [11];

THEOREM 4. *Let R be a Von Neumann algebra on a Hilbert space H, U a unitary representation of a locally compact abelian group* G, $\xi \in H$ *a cyclic vector for* R *which is unique* U*-invariant and* $S = \{g \in G, U(g) R U(g)^* \subset R\}$. *Further assume that* S *is total in* G *and* $sp(U)$ *is asymmetric. Then either* $S \neq G$ *and R is a* III_1*-factor or* $S = G$ *and* $R = B(H)$.

PROOF. Let's assume ξ to be separating, instead that cyclic, for R. We shall prove that R is a III_1-factor or $R = \mathbb{C}I$; the theorem will then easily follow by considering the action of S^{-1} on R' implemented by U, and by applying lemma 1 to check that $G = S$ implies $R = B(H)$ (notice that $[\mathbb{C}\xi] \in U(G)'')$. Let ω be the faithful positive functional $\omega(T) = (T\xi,\xi)$, $T \in R$, and let $A = A^* \in R_\omega$. For every $g \in S$, $U(g)A\,U(g)^* \in R$, therefore

$$F(g) \equiv (AU(g)A\xi,\xi) = (AU(g)A\,U(g)^*\xi,\xi)$$
$$= (U(g)\,AU(g)^*\,A\xi,\xi) = (A\,U(g^{-1})\,A\xi,\xi)$$

that is $F(g) = F(g^{-1}), g \in S$, hence $F(g) = F(g^{-1})$ for all $g \in G$ because S is total in G. Now F is a positive-definite function, therefore it is the Fourier transform of a measure $\hat{F}$ on $\hat{G}$, and $\hat{F}(p) = \hat{F}(p^{-1})$ Since the support of $\hat{F}$ is contained in $sp(U)$ and $sp(U)$ is asymmetric, $\hat{F}$ must be concentrated in $\{e\}$ thus F is costant. In particular $F(g) = F(e)$ or

$$(U(g)A\xi,\ A\xi) = (A\xi,A\xi) = \|A\xi\| = \|U(g)A\xi\|\ \|A\xi\|.$$

By the limit case of the Schwartz inequality we have $U(g)A\xi = A\xi$, $g \in G$, hence $A\xi \in \mathbb{C}\xi$ because ξ is unique U-invariant, thus $A \in \mathbb{C}I$ since ξ is separating. We have then proved that $R_\omega = \mathbb{C}I$. Since $Z(R) \subset R_\omega$, R is a factor. Furthermore it is easily seen that the triviality of R_ω entails that R is a III_1-factor or $R = \mathbb{C}I$ (see e.g. [11]) and we are done. ■

In algebraic Quantum Field Theory one considers a net of von Neumann algebras $R(O)$ indexed by regions O of the Minkowsky space $\mathbb{R}^4$ [20,21]. Translations by vectors $x \in \mathbb{R}^4$ act covariantly, namely $R(O + x) = U(x)\,R(O)U(x)^*$, where U is a four-parameter unitary group with $sp(U) \subset V_+$ (the forward light cone) and leaving uniquely invariant a vector $\xi \in H$ called the vacuum.

By the Reeh-Schlieder, theorem ξ is cyclic for $R(O)$ if O has non-empty interior, and it is separating separating if O', the space-like complement of O, has non-empty interior, since by locality $R(O)' \supset R(O')$.

We collect now some old and more recent results that may be derived by theorem 4 (see also [19]) with $G = \mathbb{R}$ and U a one-parameter time-like translation unitary group, say for definiteness $U(t) \equiv U((t,t,o,o))$, so that $sp(U) \subset \mathbb{R}^+$ and ξ is unique U-invariant.

(a) *Irriducibility of the algebra of all local observables* (Ruelle [22]): $R(\mathbb{R}^4) = B(H)$. In fact $R(O) = B(H)$ if O is a light-like (or time-like) tube.

(b) $R(W)$ *is a factor of type* III_1 if W is a wedge shaped region (defined by $x_1 > |x_0|$). This result is due to Bisognano and Wichmann and Kastler [23,24] in a Wightman theory (see theorem 7 below) and to Driessler [19] if there is a mass gap in the spectrum of the translation unitary group (or if there is a cluster property).

(c) *Either* $R(V_+)$ *is a type* III_1 *factor or* $R(V_+) = B(H)$. By a result of Buccholz [25], $R(V_+)$ is a III_1-factor for example in the free massless field (see theorem 8 below), while $R(V_+) = B(H)$ if there is a mass gap in the spectrum, by a result of Sadowski and Woronowicz [28] (see theorem 5 below).

(d) $R(O)$ *is properly infinite if* O *has non-empty interior* (Kadison [26]). In fact we may assume O bounded and $O \subset W$, thus $R(O) \subset R(W)$ hence $R(O)' \supset R(W)'$; since $R(W)'$ is an infinite factor, $R(O)'$ must be properly infinite and the same is true for $R(O)$ because these algebras are in a standard form. The original argument of Kadison was to prove that two finite von Neumann algebra $M \subset N$ cannot have the same cyclic separating vector.

(e) *Statistical indipendence*: if $O_1, O_2 \subset \mathbb{R}^4$ are non-empty open convex sets and $O_1 \subset O_2'$ then $AB = 0$, $A \in R(O_1)$, $B \in R(O_2)$ entails $A = 0$ or $B = 0$ (Borchers [45]), see also [54]). In fact we can embed O_1 in a wedge region W_1 and O_2 in W_1', hence $R(O_1)$ is contained in a factor $M = R(W_1)$ and $R(O_2) \subset R(W_1') \subset M'$, hence the result follows by the Murray-von Neumann lemma to the effect that the algebra generated by a factor M and M' is naturally isomorphic to the algebraic tensor product of M with M'.

If O_1 is a double cone and has positive distance with $O_2 = O_2''$, it is reasonable to further expect that the von Neumann algebra generated by $R(O_1)$ and $R(O_2)$ is naturally isomorphic to $R(O_1) \otimes R(O_2)$; this result is true in a free field by Buchholz theorem 11, but cannot be valid in general if O_1 and O_2 have zero distance [36].

(f) $R(O) = \mathbb{C}I$ *if* O *contains only one point* (Wightman [27]). Notice that, for any $O_o \subset \mathbb{R}^4$, $R(O_o)$ is defined as the intersection of all $R(O)$ with $O \supset O_o$ and O open. We show that $R(O) = \mathbb{C}I$ if O is the edge of W; this follow because by translation covariance $R(\bar{W}) = R(W)$, hence $R(O)$ is contained in the center of $R(W)$ which is trivial.

We prove now the result of Sadowski and Woronowicz to the effect that $R(V_+) = B(H)$ in the situation stated in (c) above.

THEOREM 5. *With the same hypothesis of theorem 4, assume further that* $G = \mathbb{R}$, $S \supset \mathbb{R}^+$ *and* $sp(U) \subsetneq \mathbb{R}^+$. *Then* $R = B(H)$.

PROOF. Let $A \in R$, $B \in R'$; then $G(t) \equiv ([U(t)AU(-t),B]\xi,\xi) = 0$, $t \in \mathbb{R}^+$, thus $\hat{G}$ is the distributional boundary value of an analytic function in a half plane. On the other hand $G(t) = (AU(-t)B\xi,\xi) - (BU(t)A\xi,\xi)$, hence $\operatorname{supp} \hat{G} \subset sp(U) \cup - sp(U)$ and $\hat{G}$ vanishes on a non-empty open set. By the edge of the wedge theorem $\hat{G} = 0$, thus $G = 0$, that is to say $([U(t)AU(-t),B]\xi,\xi) = 0$ for all $t \in \mathbb{R}$. By theorem 4, the union of $U(t)RU(-t)$, $t \in \mathbb{R}$, is irreducible, therefore $([T,B]\xi,\xi) = 0$ for all $T \in B(H)$, which easily entails $B\xi \in \mathbb{C}\xi$, hence $B \in \mathbb{C}I$ since ξ is separating for R'. ∎

We now state a result of Driessler [29] useful in a dilatation invariant theory. Its proof is similar to that of theorem 13.

THEOREM 6. *Let* $M \subset R \subset B(H)$ *be von Neumann algebras and* α_n *a sequence of automorphisms of* $B(H)$ *such that* $\alpha_n(M) \subset M$, $n \in \mathbb{N}$, $\alpha_n(A) \to \omega(A)$ *weakly and* $[\alpha_n(A),B] \to 0$ *strongly for every* $A \in M$, $B \in R$, *where* ω *is a normal state of* $B(H)$. *Then* R *is of type III or* ω *is a trace on* M.

REMARK 2. Let $(R, \mathbb{Z}, \alpha)$ be a W^*-dynamical system and $\omega = (\cdot\, \xi, \xi)$ an α-invariant state. Suppose that $M \subset R$ is a von Neumann algebra such that $\alpha_n(M) \subset M$, $n \in \mathbb{N}$, and let A be the union of $\alpha_n(M)' \cap R$, $n \in \mathbb{N}$. If α is ergodic and ξ is cyclic and separating for A and R then α is strongly asymptotically abelian on R (expand $\| [\alpha_n(A),B]\xi\|^2$ and approximate $B\xi$ by $C\xi$, $C \in A$; cf [29]).

4. MODULAR STRUCTURE OF SOME $R(O)$. We now describe the identification of the modular objects for some algebras of local observables. Let Λ be the one-parameter group of pure Lorentz transformations of $\mathbb{R}^4$ with respect to the x_1-axis, thus

$$\Lambda(t) = \begin{bmatrix} ch(t) & sh(t) & 0 & 0 \\ sh(t) & ch(t) & 0 & 0 \\ 0 & 0 & 1 & 0 \\ 0 & 0 & 0 & 1 \end{bmatrix}$$

and let U be the unitary covariant representation of the Poincaré group, hence $U(\Lambda(t))$ is a one-parameter unitary group such that $U(\Lambda(t))\ R(W)\ U(\Lambda(t))^* = R(\Lambda(t)W) = R(W)$, where W is the previously considered wedge region.

Denote by Δ_O and J_O the modular operator and the modular conjugation of $R(O)$ with respect to the vacuum ξ (provided it is cyclic and separating). Assuming the $R(O)$'s are generated by a (for example scalar) Wightman field, which is essentially selfadjoint and local in the needed way, Bisognano and Wichmann [23] estabilished the following.

THEOREM 7. *With the above notations*, $\Delta_W^{it} = U(\Lambda(2\pi t))$ *and* $J_W = U(R_1)\Theta$, *where* R_1 *is the rotation of* π *above the* x_1*-axis and* Θ *is the* PCT *operator. In particular*

$$R(W)' = J_W\ R(W)\ J_W = R(W')$$

namely duality holds for W.

The intuitive idea in theorem 7 relies on the formal identity $U(\Lambda(-i\pi))A\xi = U(R_1)\ \Theta\ A^*\xi$; the proof requires a careful use of the analytic continuation procedure by means of the analytic properties of the Wightman functions. See [52] for a shorter proof and [53] for further developments.

Let consider now the free massless field ϕ. This field is dilatation invariant, namely there exist a vacuum-fixing unitary group D such that $D(t)\ R(O)\ D(t) = R(e^{-t}\ O)$. Let Z be the unitary second quantazation of

$-I$ $(Z\,\phi(f)\,Z^* = -\phi(f),\ Z\xi = \xi)$. By using an argument similar to that one of Bisognano and Wichmann, Buchholz [25] proved the following.

THEOREM 8. *Relatively to the free massless field one has* $\Delta_{V_+}^{it} = D(t)$ *and* $J_{V_+} = Z\Theta$. *In particular*

$$R(V_+)' = J_{V_+} R(V_+) J_{V_+} = R(V_-)$$

(*time-like duality holds for* V_+)

Let ρ be the relativistic ray inversion map in $\mathbb{R}^4$, $\rho(x) = \frac{x}{x^2}$, $x^2 = x_o^2 - x_1^2 - x_2^2 - x_3^2 \neq 0$, and v the one-parameter group of generalized linear fractional transformations of $\mathbb{R}^4$ defined by

$$v(t)\,(x_o,x_1,0,0) = (x_o(t),x_1(t),0,0)$$

$$x_i(t) = \tfrac{1}{2}\left\{\frac{(x_o+x_1+1) + e^{-t}(x_o+x_1-1)}{(x_o+x_1+1) - e^{-t}(x_o+x_1-1)} + (-1)^i\,\frac{(x_o-x_1+1) + e^{-t}(x_o-x_1-1)}{(x_o-x_1+1) - e^{-t}(x_o-x_1-1)}\right\}$$

$(i=0,1)$ and the requirement to commute with space-rotations. Neither ρ or $v(t)$ are everywhere defined, but $\mathcal{O}_1$ is in the domain of $v(t)$, $t \in \mathbb{R}$, where $\mathcal{O}_1$ is the normalized double cone (centered at 0, ray = 1). It has been verified in [30] that ρ and $v(t)$ induce second quantization unitaries $\Gamma(U_\rho)$ and $\Gamma(V(t))$ such that the following holds.

THEOREM 9. *In the free massless field one has* $\Delta_{\mathcal{O}_1}^{it} = \Gamma(V(2\pi t))$ *and* $J_{\mathcal{O}_1} = T\,\Theta\,\Gamma\,(U_\rho)$, *where* T *is the time reflection operator. In particular one has*

$$R(\mathcal{O}_1)' = R(\mathcal{O}_1') = R(\mathcal{O}_1^t)^{(*)}$$

where $\mathcal{O}_1^t = \{x \in \mathbb{R}^4,\ (x-y)^2 > 0, y \in \mathcal{O}_1\}$, *thus space-like and time-like duality holds for* $\mathcal{O}_1$.

Theorem 9 may be proved by using a geometrical idea which, up to the author's knowledge, is due to V. Glaser: one can use ρ to transform a double cone onto a wedge region. In fact, by suitably composing ρ with translations, one can transform $\mathcal{O}_1$ onto W or onto V_+. As a consequence $R(\mathcal{O}_1)$, $R(W)$ and

(*)With a suitable definition for the local algebras.

$\mathcal{R}(V_+)$ are isomorphic and theorems 7,8,9 are equivalents in the free massless field.

The Haag space-like duality property [20] was first proved in [32] for free fields. See also [33,34,35] for further approachs.

5. FURTHER COMMENTS ON $\mathcal{R}(\mathcal{O}_1)$. In general one is not able to compute the type of $\mathcal{R}(\mathcal{O}_1)$, neither to check if it is a factor. The only general result in this direction is that one of Kadison ((d) of section 3) and the result of Borchers [45] to the effect that every non-zero projection of $\mathcal{R}(\mathcal{O}_1)$ is equivalent to I in any $\mathcal{R}(\mathcal{O})$ with $\bar{\mathcal{O}}_1$ contained in the interior of $\mathcal{O}$. Furthermore Roberts [36] showed that in a dilatation invariant theory $\mathcal{R}(\mathcal{O}_1)$ cannot be a type I factor, in fact $\mathcal{R}(\mathcal{O}_1)$ is type III by theorem 6. On the other hand in a free field one may use direct computations [37,38,39] and obtain the following result [39].

THEOREM 10. *In the free field of mass* $m \geq 0$ $\mathcal{R}(\mathcal{O}_1)$ *is the unique ITPFI factor of type* III_1.

Alternatively one may obtain theorem 10 in the following way. First one notices that it is sufficient to consider the m=0 case only, since the $\mathcal{R}(\mathcal{O}_1)$'s are isomorphic for different m [40]. As we previously explained $\mathcal{R}(\mathcal{O}_1)$, $\mathcal{R}(W)$ and $\mathcal{R}(V_+)$ are isomorphic if m=0. Therefore $\mathcal{R}(\mathcal{O}_1)$ is a III_1-factor by (b) or (c) of section 3. The hyperfiniteness of $\mathcal{R}(\mathcal{O}_1)$ may be deduced by the following theorem of Buchholz [41].

THEOREM 11. *In a free field there exists a type I factor* M *such that* $\mathcal{R}(\alpha\mathcal{O}_1) \subset M \subset \mathcal{R}(\beta\mathcal{O}_1)$ *for any given* $\beta > \alpha > 0$.

Finally Testard [42,46] obtained the following result (in a Wightman frame so that theorem 7 holds) and then deduced the IPTFI property by an unpublished result of Connes.

THEOREM 12. *The asymtotic ratio set of* $\mathcal{R}(W)$ *is equal to* $\mathbb{R}^+$.

Actually Testard proved a more general theorem and also applied it to III_1-factors in Statistical Mechanics [2].

We refer to [21,27,31,43,44,45] for results about the relative structure of the $\mathcal{R}(\mathcal{O})$'s, for example to the effect that two $\mathcal{R}(\mathcal{O})$'s coincide or differ for different $\mathcal{O}$'s.

6. NON-VACUUM THEORY. Few results of the type here described have been obtained without the tool of an invariant normal state. Willig [47] and M.S. Glaser [48] proved that type I_∞ and type II_∞ factors respectively cannot carry a strongly asymtotically abelian action. We conclude these notes by showing their result. However we remark that the strong asymptotic abeliannes property can be quite difficult to check not having a faithful normal invariant state. The argument in the following proof is partly taken from [29]. See [49,50] for examples of asymptotic abelian actions on type II, or type III factors.

THEOREM 13. *Let* R *be a von Neuman algebra and* α_n *a sequence of unital endomorphisms of* R *such that* $[\alpha_n(A),B] \to 0$ *strongly for all* $A, B \in R$. *Then* R *is the direct sum of a finite and a type* III *von Neumann algebra.*

PROOF. Let G be the maximal finite central projection of R and $F = I-G$. Then $\alpha_n(RF) \subset RF$, since otherwise RG should contain the properly infinite Von Neumann algebra $\alpha_n(RF)G$, therefore we may assume R to be properly infinite.

Assume that R is not type III and let τ be a normal tracial state on a reduced von Neumann algebra ERE by a non-zero projection $E \in R$. Define $\omega(A) = \mathrm{LIM}\ \tau(E\,\alpha_n(A)E)$, $A \in R$, where LIM denotes the limit along a free ultrafilter. Then ω is a (not necessarily normal) state of R; we shall show that ω is a trace, which is a contradiction. Indeed we have $\omega(AB) = \mathrm{LIM}\ \tau(E\,\alpha_n(AB)E) = \mathrm{LIM}\ \tau(E\,\alpha_n(A)E\,\alpha_n(B)E) = \mathrm{LIM}\ \tau(E\alpha_n(B)E\alpha_n(A)E) =$
$= \mathrm{LIM}\ \tau(E\alpha_n(BA)E) = \omega(BA)$. ∎

BIBLIOGRAPHY

1. D. Ruelle, Statistical Mechanics, Benjamin, Reading (1969).

2. D. Kastler, Equilibrium States of Matter and Operator Algebras, in Symposia Math. XX, 49, Academic Press (1976).

3. S. Doplicher, D. Kastler, D.W. Robinson, Covariance Algebras in Field Theory and Statistical Mechanics, Commun.Math.Phys. 3 (1966), 1.

4. S. Doplicher, R.V. Kadison, D. Kastler, D.W. Robinson, Asymptotically Abelian Systems, Commun.Math.Phys. 6 (1967), 101.

5. S. Doplicher, D. Kastler, Ergodic States in a Now-commutative Ergodic Theory, Commun.Math.Phys. 7 (1968), 1.

6. S. Doplicher, D. Kastler, E. Størmer, Invariant States and Asymtotic Abelianness, J. Functional Anal. 3 (1969), 419.

7. N.M. Hugenholtz, On the Factor Type of Equilibrium States in Quantum Statistical Mechanics, Commun.Math.Phys. 6 (1967), 189.

8. E. Størmer, Types of von Neumann Algebras Associated with Extremal Invariant States, Commun. Math. Phys. 6 (1967), 194.

9. E. Størmer, Contribution to these Proceedings and references therein.

10. A Connes, Une classification des Facteurs de type III, Ann. Ec. Norm. Sup. 6 (1973), 133.

11. R. Longo, Notes on Algebraic Invariants for Non-Commutative Dynamical Systems, Commun. Math. Phys. 69 (1979), 195.

12. E. Størmer, Spectra of States and Asymptotically Abelian C^*-Algebras, Commun.Math.Phys. 28 (1972), 279 (Corrections, ibidem 33 (1974), 341).

13. H. Araki, Remarks on Spectra of Modular Operators of Von Neumann Algebras, Commun. Math. Phys. 28 (1972), 267.

14. R.H. Herman, Spectra of Automorphism Group of Operator Algebras, Duke Math. J. 41 (1974), 667.

15. R.H. Herman, R. Longo, A Note on the Γ-Spectrum of an Automorphism Group, Duke Math. J. 47 (1980), 27.

16. H.J. Borchers, Energy and Momentum as Observables in Quantum Field Theory, Commun.Math.Phys. 2 (1966), 49.

17. R.V. Kadison, Some Analytic Methods in the Theory of Operator Algebras, in Lecture Notes in Math., Springer, 140 (1969), 8.

18. A Kishimoto, Extremal Invariant States with a Spectrum Condition and Borchers' Theorem, Letters in Math.Phys. 3 (1979), 193.

19. W. Driessler, Comments on Light-like Translations and Applications in Relativistic Quantum Field Theory, Commun. Math. Phys. 44 (1975), 133.

20. R. Haag, D. Kastler, An Algebraic Approach to Quantum Field Theory, J. Math. Phys. 5 (1964), 848.

21. N.N. Bogoliubov, A.A. Logunov, I.T. Todorov, Introduction to Axiomatic Quantum Field Theory, Benjamin, Reading (1975).

22. D. Ruelle, On the Asymptotic Condition in Quantum Field Theory, Helv. Phys. Acta 35 (1962), 147.

23. J.J. Bisognano, E.H. Wichmann, On the Duality for a Hermitian Scalar Field, J. Math. Phys. 16 (1975), 985.

24. D. Kastler, E.H. Wichmann, unpublished.

25. D. Buchholz, On the Structure of Local Quantum Fields with non--trivial Interaction, preprint.

26. R.V. Kadison, Remarks on the type of von Neumann Algebras of Local Observables in Quantum Field Theory, J. Math. Phys. 4 (1963), 1511.

27. A.S. Wightman, La Théorie Quantique Locale et la Théorie Quantique de Champs, Ann. Inst. H. Poincaré, 1 (1964), 403.

28. P. Sadowski, S.L. Woronowicz, Total Sets in Quantum Field Theory, Rep. Math. Phys. 2 (1971), 113.

29. W. Driessler, On the type of Local Algebras in Quantum Field Theory, Commun. Math. Phys. 53 (1977), 295.

30. P. Hislop, R. Longo, preprint.

31. H. Araki, Von Neumann Algebras of Local Observables for the Scalar Field, J. Math. Phys. 5 (1964), 1.

32. H. Araki, A Lattice of von Neumann Algebras Associated with Quantum Theory of Free Boson Field, J. Math. Phys. 4 (1963), 1343.

33. J.P. Eckmann, K. Osterwalder, An Application of Tomita's Theory of Modular Hilbert Algebras: Duality for Free Bose Fields, J. Functional Anal. 13 (1973), 1.

34. M.A. Rieffel, A Commutation Theorem and Duality for Free Bose Fields, Commun. Math. Phys. 39 (1974), 153.

35. P. Leyland, J.E. Roberts, D. Testard, Duality for Quantum Free Fields, Marseille CNRS preprint.

36. J.E. Roberts, Some Applications of Dilatation Invariance to Structural Question in the Theory of Local Observables, Commun. Math.Phys. 37 (1974), 273.

37. H. Araki, Type of von Neumann Algebras Associated to the Free Scalar Field, Progr. Theoret. Phys. 32 (1964), 956.

38. G.F. Dell'Antonio, Structure of the Algebras of Some Free Systems, Commun. Math. Phys. 9 (1968) 51.

39. H. Araki, T. Woods, A Classification of Factors, Publ. R.I.M.S. Kyoto Univ. 4 (1968), 51.

40. J.P. Eckmann, J. Frölich, Unitary Equivalence of Local Algebras in the Quasi free Representation, Ann. Inst. H. Poincaré A, 20, n. 2 (1974), 201.

41. D. Buchholz, Product States for Local Algebras, Commun. Math.Phys. 36 (1974), 287.

42. D. Testard, Asymptotic Ratio Set of Von Neumann Algebras Generated by Temperature States in Statistical Mechanics, Rep. Math. Phys. 12 (1977), 115.

43. H.J. Borchers, Über die Vollständigkeit Lorentz invarianter Felder in Einer Zeitartigen Röhre, Nuovo Cimento 19 (1961), 787.

44. H. Araki, A Generalization of Borchers' Theorem, Helv. Acta Phys. 36 (1963), 132.

45. H.J. Borchers, A Remark on a Theorem of B. Misra, Commun. Math. Phys. 4 (1967) 315.

46. D. Testard, Some Properties of the Representation of the Quasi-local Observables in Statistical Mechanics and Quantum Field Theory, in Colloques Internationaux du CNRS, n° 274, Editions du CNRS, Marseille (1979), 409.

47. P. Willig, B(H) is Very Non-Commutative, Proc. AMS 24 (1970), 204.

48. M.S. Glaser, Asymptotic Abelianness of Infinite Factors, Trans. AMS 178 (1973), 41.

49. S. Sakai, C^*-Algebras and W^*-Algebras, Springer-Verlag (1971).

50. A. Connes, E.J. Woods, Existence des Facteurs Infinis Asymptotiquement Abéliens, CR Acad. Sc. Paris 279 (1974), 189.

51. S. Doplicher, An Algebraic Spectrum Condition, Commun. Math.Phys. 1 (1965), 1.

52. C. Rigotti, On the Essential Duality Condition for Hermitian Scalar Field, in Colloques Internationaux du CNRS, N° 274, Editions du CNRS, Marseille (1979), 307.

53. C. Rigotti, Remarks on the Modular Operators and Local Observables, Commun. Math. Phys. 61 (1978), 267.

54. G. Simonelli, Un Teorema di Borchers e sue Applicazioni, Tesi di Laurea n. 19672, Università di Roma (1973).

55. R.T. Powers, UHF Algebras and their Applications to répresentations of the Anticommutations Relations, in Cargese Lectures in Physics, 4 (1970), 137.

ISTITUTO MATEMATICO "G. CASTELNUOVO"
UNIVERSITA' DI ROMA
ROME, ITALY

Proceedings of Symposia in Pure Mathematics
Volume **38** (1982), Part 2

LOCAL NETS OF C^*-ALGEBRAS AND THE STRUCTURE OF ELEMENTARY PARTICLES

Detlev Buchholz and Klaus Fredenhagen

ABSTRACT: The net-structure of a C^*-algebra $\mathfrak{A}$ of local observables implies that representations describing massive particles are normal relative to a vacuum representation on certain subalgebras of $\mathfrak{A}$ which are assigned to particular regions of space-time. Physically, this means that particles can be localized in the causal complement of such regions. Using this fact, one can analyse the representations containing multiparticle states and derive many qualitative features of elementary particle physics.

Quantum Field Theory can be described by a net of C^*-algebras,

$$\mathcal{O} \longrightarrow \mathfrak{A}(\mathcal{O})$$

where $\mathcal{O}$ runs over the bounded regions of Minkowski space and where $\mathfrak{A}(\mathcal{O})$ is interpreted as the algebra generated by observables which can be measured within the region $\mathcal{O}$. The algebra $\mathfrak{A}$ of all local observables is defined as the C^*-inductive limit of this net,

$$\mathfrak{A} = \overset{C^*}{\bigcup_{\mathcal{O}}} \mathfrak{A}(\mathcal{O})$$

and for unbounded regions S one sets

$$\mathfrak{A}(S) := \overset{C^*}{\bigcup_{\mathcal{O} \subset S}} \mathfrak{A}(\mathcal{O})$$

This net is subject to the following conditions:

(i) Locality: If the region $\mathcal{O}_1$ is contained in the space-like complement $\mathcal{O}_2'$ of the region $\mathcal{O}_2$, then the associated algebras $\mathfrak{A}(\mathcal{O}_1)$ and $\mathfrak{A}(\mathcal{O}_2)$ commute.

(ii) Covariance: The group of translations in Minkowski space is represented by automorphisms of

$$x \longrightarrow \alpha_x \in \mathrm{Aut}(\mathfrak{A})$$

0082-0717/80/0000-0508/$04.00

such that $\alpha_x(\mathfrak{A}(\mathcal{O})) \subset \mathfrak{A}(\mathcal{O} + x)$ holds for any x.

According to Haag and Kastler [1], one knows that in principle any physical information of the theory is contained in the "local net" $\mathcal{O} \longrightarrow \mathfrak{A}(\mathcal{O})$. But it appears to be difficult to extract the relevant information in practice. This paper is devoted to the problem of explaining some qualitative features of elementary particle physics in this mathematical setting.

On the experimental side, physicists prepare states and do measurements. Being interested in elementary particle physics, they try to prepare a restricted class of states, as for example the vacuum, one-particle states and states which look like multi-particle states for asymptotic times.

Mathematically, this set of "particle states" is described by vector states in suitable representations of the algebra of observables $\mathfrak{A}$. In general, the pure particle states are not associated to a single irreducible representation of $\mathfrak{A}$, but fall into different sectors which are distinguished by so-called superselection rules (charges). Here a sector is defined as the set of vector states in one irreducible representation of $\mathfrak{A}$. Examples of superselection rules are the distinction between half-integer and integer spin, the distinction between Fermi- and Bose-statistics, the electric charge and the so-called colour charge. There also exist several superselection rules which hold only approximately, as e.g. isospin, strangeness etc.

On physical grounds, this set of superselection sectors is expected to have the following structure:

(i) There is a composition law of sectors, i.e. if the representations π_1 and π_2 describe particle states, then there also exists a composed representation π such that the equivalence class $[\pi]$ of π depends only on the equivalence classes of π_1 and π_2.

$$[\pi_1]\,[\pi_2] \longrightarrow [\pi_1]\circ[\pi_2] = [\pi] \qquad \text{(commutative)}$$

This composition law corresponds to the additivity of charges and should therefore have the following properties:

a) there is a representation π_0 induced by the vacuum, and

$$[\pi]\circ[\pi_0] = [\pi]$$

because the vacuum carries no charge.

b) to any π there is a conjugate representation $\bar{\pi}$, representing anti-particles, such that any $\pi_1 \in [\pi]\circ[\bar{\pi}]$ contains a subrepresentation equivalent to π_0.

This property is actually the basis for the construction of states belonging to π, starting from states in the vacuum sector. The idea is as follows: given a state in the vacuum sector which can be interpreted as being composed of states from the sectors of π and $\bar{\pi}$, one gets a state from the sector of π simply by shifting the part corresponding to the conjugate representation $\bar{\pi}$ far away.

(ii) any sector has a well defined type of statistics.

(iii) all particle states have a certain stability property.

As a first step in deriving these features of elementary particle physics from properties of $\mathfrak{A}$ one has to select the representations of $\mathfrak{A}$ which describe particle states. Several selection criteria have been proposed. The oldest and most general criterion is due to Borchers [2] who studied representations $(\pi, \mathcal{H}_\pi)$ of $\mathfrak{A}$ with the following properties:

(B) (i) Covariance: There exists a strongly continuous unitary representation $x \longrightarrow U_\pi(x)$ of the translation group in $\mathcal{H}_\pi$ such that for any $A \in \mathfrak{A}$ and any translation x

$$U_\pi(x)\,\pi(A)\,U_\pi(-x) = \pi(\alpha_x(A)).$$

(ii) Positivity of energy: The spectrum of the unitary group of translations, Sp $\mathcal{U}_\pi$, is contained in the forward light cone $\overline{V_+} = \{p \in \mathbb{R}^4 \mid p_0 \geq 0,\ p^2 \geq 0\}$.

Criterion (B) is a form of the stability property. It expresses the fact that one can extract only a finite amount of energy from elementary particle systems. Any representation occuring in elementary particle physics should satisfy this criterion. However, a complete analysis of the structure of the class of all representations $\pi \in (B)$ is difficult and has not yet been carried out. Amongst the few results which have been obtained, we mention the following ones:

There is a choice of the representation U_π such that

(i) $U_\pi(x) \in \pi(\mathfrak{A})''$ (Borchers [3]). So any subrepresentation of π satisfies criterion (B) too.

(ii) The lower boundary of $\mathrm{Sp}\,\mathcal{U}_\pi$ is Lorentz invariant [4] . This holds although Lorentz covariance of the representation has not been assumed.

Among the unsolved problems, we mention the question of whether there is always a vacuum representation $\pi_0 \in (B)$ such that the states which are associated to a representation $\pi \in (B)$ are in some sense localized excitations of this vacuum.

If a theory describes massless particles, one may doubt that the condition (B) on the energy imposes severe restrictions on the localization properties of states. However, in theories where only massive particles occur, one may expect that only well-localized states can have finite energy. Such localization properties of states are of importance for the determination of the structure of sectors. This has been demonstrated first by Borchers in a special case [5] and was then discussed in general by Doplicher, Haag and Roberts [6] . Doplicher, Haag and Roberts proposed the following selection criterion for the representations $(\pi, \mathcal{H}_\pi)$ of interest:

(DHR) (i) There is a vacuum representation $\pi_0 \in (B)$.

(ii) π is covariant, and for some bounded region $\mathcal{O}$,

$$\pi|\mathfrak{A}(\mathcal{O}') \simeq \pi_0|\mathfrak{A}(\mathcal{O}')$$

This criterion happened to be very successful as may be seen from the following list of results which have been obtained for representations $\pi \in$ (DHR):

(i) There is a composition law of representations.

(ii) There is an intrinsic notion of statistics.

In the case of finite statistics:

(iii) Any representation $\pi \in$ (DHR) with finite statistics satisfies (B).

(iv) To any representation $\pi \in$ (DHR) with finite statistics there exists a conjugate representation $\bar{\pi}$, any particle possesses an antiparticle, and scattering states can be constructed.

So for these representations the structural properties expected on physical grounds can be derived from the properties of $\mathfrak{A}$. Yet unfortunately, there exist important examples of representations which violate the (DHR)-criterion and to which this analysis therefore does not apply. These are representations which describe charged states in gauge theories and representations describing solitons in two-dimensional models.

For this reason we have tried to find a selection criterion which is more general than (DHR) and which covers at least all models describing only massive particles. To this end we have, as a first step, analyzed the properties of the fundamental representations describing massive one-particle states.

DEFINITION: Let $\pi \in (B)$ be a factorial representation of $\mathfrak{A}$. π is called a one-particle representation if $(o < m < M)$

$$\mathrm{Sp}\, \mathcal{U}_\pi = \left\{ p \in \mathbb{R}^4 \,|\, p_0 \geq o,\ p^2 = m^2 \quad \text{or} \quad p^2 \geq M^2 \right\}$$

This condition excludes the occurence of massless particles in the theory. Furthermore, since $0 \notin \mathrm{Sp}\, \mathcal{U}_\pi$, there exists no vacuum state in the sector of π. As a consequence of the isolation of the mass shell $p^2 = m^2$ in $\mathrm{Sp}\, \mathcal{U}_\pi$, operations with sufficiently small energy - momentum transfer transform one-particle states into one-particle states. There are well-localized operators in $\mathcal{B}(\mathcal{H}_\pi)$ which have this property, e.g.

$$\pi(A)(f) = \int d^4x\, f(x)\, \pi(\alpha_x(A))$$

with $A \in \mathfrak{A}(\mathcal{O})$ for some bounded region $\mathcal{O}$ and f the Fourier transform of an infinitely differentiable function with compact support. Such operators are almost local in the following sense: an operator $A \in \mathcal{B}(\mathcal{H}_\pi)$ is called almost local if for any neighbourhood $\mathcal{O}$ of the origin in Minkowski space and any $\lambda > o$, there is an operator $A_\lambda \in \pi(\mathfrak{A}(\lambda \cdot \mathcal{O}))''$ such that

$$\lim_{\lambda \to \infty} \lambda^n \|A - A_\lambda\| = o$$

for all $n \in \mathbb{N}$.

Crucial for the analysis of one-particle states is the following result [7] which shows that energy and momentum of a particle are essentially localized in a finitely extended region:

LEMMA. Let E be the spectral projection which belongs to some sufficiently small nonvoid open set on the one-particle mass shell in $\mathrm{Sp}\, \mathcal{U}_\pi$. Then there exist almost local operators

$B, B_\mu \in \mathcal{B}(\mathcal{H}_\pi)$ such that

$$\begin{array}{ll} \text{(i)} & P_\mu BE = B_\mu E \\ \text{(ii)} & B_\mu^* BE = B^* B_\mu E \\ \text{(iii)} & BE \neq 0 \end{array}$$

where P_μ, $\mu = 0,..3$ are the generators of U_π.

An immediate consequence of this lemma is the existence of a vacuum state ω_0. In fact, let $\Phi \in \mathcal{H}_\pi$ such that $\|BE\Phi\| = 1$, and let φ be the state on $\mathfrak{A}$, induced by $BE\Phi$. Then for any $A \in \mathfrak{A}$

$$\varphi \circ \alpha_x (A) \longrightarrow \omega_0(A)$$

if x tends to infinity in any spacelike direction, and this convergence holds uniformly on local algebras. If x tends to infinity in a fixed spacelike direction b, say, then this uniformity even holds for the algebra $\mathfrak{A}(S')$ of the spacelike complement of any cone S pointing in this direction. These spacelike cones are given by

$$S = a + \bigcup_{\lambda > 0} \lambda \mathcal{O}$$

where a is the apex of the cone and $\mathcal{O}$ is some closed spacelike double cone containing b in its interior.

By methods due to Borchers [10], the above lemma leads to the following result:

THEOREM 1: <u>Let π be a one-particle representation. Then there exists a unique vacuum representation $\pi_0 \in (B)$ such that</u>

$$\pi \mid \mathfrak{A}(S') \simeq \pi_0 \mid \mathfrak{A}(S')$$

<u>for any pointed spacelike cone S.</u>

This property of one-particle representations is weaker than the (DHR)-criterion, but it should be stressed that it is a consequence of the postulated structure of Sp $\mathcal{U}_\pi$. We therefore propose the following selection criterion for particle representations:

(S) (i) There is a vacuum representation $\pi_0 \in (B)$.

(ii) π is covariant, and for any spacelike cone S

$$\pi \mid \mathfrak{A}(S') \simeq \pi_0 \mid \mathfrak{A}(S') \quad .$$

This criterion does not exclude charged sectors in gauge theories a priori, the question of the existence of such sectors in the absence of massless particles being part of the confinement problem. At present it is unknown whether there exist representations $\pi \in (S)$ which violate (DHR), but there seems to be no reason forbidding the occurence of such representations in general. We found it therefore worth-wile to extend the (DHR)-analysis to representations $\pi \in (S)$.

The first and crucial step in this analysis is to find a way of composing the representations $\pi \in (S)$. If for example a particle is described in a representation π_1 and another particle in a representation π_2 , one would also like to have a representation containing states with both particles.

For representations fulfilling criterion (DHR) the construction goes as follows:

a) Since π is covariant, one has for any translation x

$$\pi \mid \mathfrak{A}(\mathcal{O}'+x) \simeq \pi_0 \mid \mathfrak{A}(\mathcal{O}'+x)$$

So for all $A \in \mathfrak{A}$

$$\|\pi(A)\| = \|\pi_0(A)\| \quad .$$

Hence π can be considered as a representation of $\pi_0(\mathfrak{A})$, and one may identify $\mathfrak{A}$ and $\pi_0(\mathfrak{A})$ by dropping the symbol π_0 .

b) Using the unitary equivalence of $\pi \mid \mathfrak{A}(\mathcal{O}_0')$ and $\pi_0 \mid \mathfrak{A}(\mathcal{O}_0')$, one finds a representation $\rho \simeq \pi$ in the vacuum Hilbert space $\mathcal{H}_0$ with

$$\rho(A) = A$$

for all $A \in \mathfrak{A}(\mathcal{O}_0')$. One refers to this property by saying that ρ is localized in $\mathcal{O}_0$. Of course, ρ is locally normal.

c) Since ρ is localized in some region $\mathcal{O}_0$, one has for $\mathcal{O} \subset \mathcal{O}_0$

$$\rho(\mathfrak{A}(\mathcal{O})) \subset \mathfrak{A}(\mathcal{O}'))' \quad . \qquad (*)$$

At this point, it is necessary to add an assumption which relates the commutant of $\mathfrak{A}(\mathcal{O}')$ to the algebra $\mathfrak{A}$, namely

$$\mathfrak{A}(\mathcal{O}')' = \mathfrak{A}(\mathcal{O})'' =: \mathcal{B}(\mathcal{O})$$

for all double cones $\mathcal{O}$. This property is called duality and can be derived in field theoretic models [9] . The net $\mathcal{O} \to \mathcal{B}(\mathcal{O})$ is local and covariant, and ρ can be uniquely extended to a locally normal morphism $\hat{\rho}$ of

$$\mathcal{B} = \overline{\bigcup_{\mathcal{O}} \mathcal{B}(\mathcal{O})}^{C^*}$$

From (*) one conclues

$$\hat{\rho}\,(\mathcal{B}) \subset \mathcal{B}$$

This leads to the following natural way of composing representations:

$$[\pi_1] \circ [\pi_2] := [\hat{\rho}_1\, \rho_2] \quad , \quad \rho_i \simeq \pi_i \quad , \; i = 1,2.$$

$\hat{\rho}_1 \rho_2$ satisfies (DHR) and its equivalence class is independent of $\rho_i \simeq \pi_i$, i = 1,2.

We now want to apply these ideas to the construction of composite representations in the case of criterion (S) :

a') Again $\|\pi(A)\| = \|\pi_0(A)\|$, so we may drop the symbol π_0 .

b') We pick a representation $\rho \simeq \pi$, $\rho : \mathfrak{A} \to \mathcal{B}(\mathcal{H}_0)$ with

$$\rho(A) = A$$

for all $A \in \mathfrak{A}(S_0')$. It turns out that ρ is weakly continuous on $\mathfrak{A}(S')$ for any spacelike cone S .

c') If S' contains the localization cone S_0 of ρ , we have

$$\rho(\mathfrak{A}(S')) \subset \mathfrak{A}(S)'$$

As in the case of criterion (DHR), we assume duality in the form

$$\mathfrak{A}(S)' = \mathfrak{A}(S')''$$

for any spacelike cone S . But nevertheless there arises a difficulty in composing the representations $\pi \in (S)$. Since the set of spacelike cones is not directed, it is not clear how to extend ρ to the algebra which is generated by all algebras $\mathfrak{A}(S')''$, where S runs over all spacelike cones.

We solve this problem in the following pedestrian way: We fix an auxiliary spacelike cone S_a in the spacelike complement of the localization region S_0 of ρ . Since the set

$$(S_a' + x)_{x \in \mathbb{R}^4}$$

is directed, we may define the algebra

$$\mathcal{B}^{S_a} = \overset{C^*}{\bigcup_x} \; (S_a' + x)''$$

We note:

(i) ρ has a unique extension ρ^{S_a} to $\mathcal{B}^{S_a}$ which is normal on $\mathfrak{A}(S_a' + x)''$ for any x .

(ii)

$$\rho^{S_a}(\mathcal{B}^{S_a}) \subset \mathcal{B}^{S_a}$$

We now introduce the set of representations

$$\Delta_{S_o} := \{ \varrho : \mathfrak{A} \longrightarrow \mathcal{B}(\mathcal{H}_o),\ \varrho \in (S) \mid \varrho \text{ localized in } S_o \}$$

and define the composition of representations as follows:

DEFINITION:

$$[\pi_1] \circ [\pi_2] := [\varrho_1 \overset{S_a}{} \varrho_2] \quad , \Delta_{S_o} \ni \varrho_i \simeq \pi_i, i=1,2, \quad S_a \subset S_o'$$

With this definition we have the following theorem which shows in particular that this composition of representations does not depend on the auxiliary cone S_a .

THEOREM 2: (i) $\varrho_1 \overset{S_a}{} \varrho_2$ <u>does not depend on the choice of S_a</u> (<u>within the above restrictions</u>).

(ii) $\varrho_1 \overset{S_a}{} \varrho_2 \in \Delta$

(iii) $[\varrho_1 \overset{S_a}{} \varrho_2]$ <u>does neither depend on S_o nor on the choice of</u> $\varrho_i \simeq \pi_i$, <u>i = 1,2</u>.

Having constructed the composite sectors, we can now establish a notion of statistics for representations obeying criterion (S). This notion is nearly identical to the one used in the case of (DHR)-representations.

Let $\varrho \in \Delta_{S_o}$, $\varrho_i \simeq \varrho$, $\varrho_i \in \Delta_{S_i}$, i = 1, ..n, where all cones S_o, S_1, .. S_n are localized in the spacelike complement of S_a , and S_1, .. S_n are mutually spacelike. Then the product of morphisms $\varrho_i^{S_a}$, i = 1,..,n, does not depend on the order of factors:

$$\prod_{i=1}^{n} \varrho_i^{S_a} = \prod_{i=1}^{n} \varrho_{p^{-1}(i)}^{S_a} \qquad \text{for any permutation } p \in P_n$$

Let ω_o denote the vacuum state and let $\Psi_1, \ldots, \Psi_n \in \mathcal{H}_o$ be vectors which represent the states $\omega_o \circ \varrho_1, \ldots, \omega_o \circ \varrho_n$ in the representation ϱ , respectively, i.e.

$$(\Psi_i, \varrho(A)\Psi_i) = \omega_o \circ \varrho_i(A)$$

for all $A \in \mathfrak{A}$, i = 1,...,n. The state $\omega_o \circ \varrho_1^{S_a} \cdots \varrho_n$ can then be considered as the product of the states $\omega_o \circ \varrho_i$, i=1,..n, and this product is commutative. There is also a corresponding product of vectors

$$(\Psi_1, \ldots, \Psi_n) \longrightarrow \Psi_1 \times \ldots \times \Psi_n$$

where $\Psi_1 \times \ldots \times \Psi_n$ represents the state $\omega_o \circ \varrho_1^{S_a} \cdots \varrho_n^{S_a}$ in the representation $(\varrho^{S_a})^n$, but this product of vectors need not be commutative. It turns out that in 3 (or more) spatial dimensions (in the DHR-case in 2 (or more) spatial dimensions) there exists a unitary representation ε_ρ^n of the permutation group P_n such that

$$\varepsilon_\rho^n(p)(\Psi_1 \times .. \times \Psi_n) = \Psi_{p^{-1}(1)} \times .. \times \Psi_{p^{-1}(n)} .$$

Since the corresponding states are independent of the order of factors, the permutation operators $\varepsilon_\rho^n(p)$ have to commute with all abservables in the representation $(\rho_{S_a})^n$. In addition we note that ε_ρ^n is independent of the choice of S_a in the spacelike complement of the localization cone S_o of ρ, and the equivalence class of ε_ρ^n depends only on the equivalence class of ρ.

The classification of the representations ε_ρ^n of the permutation group P_n can be carried through as in the case of (DHR). We briefly review the results.

If ρ is irreducible, the equivalence classes of ε_ρ^n for all n are determined by a statistics parameter λ_ρ with possible values $0, \pm\frac{1}{d}, d \in \mathbb{N}$. If $\lambda_\rho = \frac{1}{d}$, all irreducible representations of P_n corresponding to Young Tableaux with at most d rows occur as subrepresentations (Para-Bose-Statistics). If $\lambda_\rho = -\frac{1}{d}$, all irreducible representations of P_n corresponding to Young Tableaux with at most d columns occur as subrepresentations (Para-Fermi-Statistics). If $\lambda_\rho = 0$, all irreducible representations of P_n occur (This is the so-called infinite statistics case). Any subrepresentation occurs with the same multiplicity, and only the mentioned representations can occur.

In general, one does not know whether ρ is a direct sum of irreducible representations. But it has been shown [6] that any ρ can be decomposed into a direct sum of irreducible representations with finite statistics and a representation with infinite statistics. Since only in the case of finite statistics finer structural properties can be obtained, one would like to exclude the pathological possibility of infinite statistics.

As known from the work of Doplicher, Haag and Roberts, the finiteness of statistics of π is equivalent to the existence of a conjugate sector $\bar{\pi}$. But applying the methods used for the construction of the vacuum sector π_o from the one-particle representation π, one can construct such a conjugate representation $\bar{\pi}$ directly [8] :

Let $V: \mathcal{H}_\pi \longrightarrow \mathcal{H}_o$ be a unitary intertwiner between the restrictions of π and π_o to $\mathfrak{A}(S')$ where S is some spacelike cone. Then the states ψ_x in the vacuum sector, induced by the vectors $VU_\pi(x)\, BE\,\phi \in \mathcal{H}_o$, B, E, ϕ as defined above, look on $\mathfrak{A}(S')$ like the one-particle states $\varphi \circ \alpha_x$ which approximate the vacuum. Using the lemma, one sees that ψ_x approximates some state ψ if x becomes spacelike to $\mathcal{O} + S'$ for any bounded region $\mathcal{O}$. It turns out that the associated representation π_ψ of $\mathfrak{A}$ satisfies criterion (S) and is conjugate to π in the sense explained above. π_ψ also fulfils the spectrum condition (criterion (B)). This fact can be used to show that π has subrepresentations with finite statistics. Using the fact that representations with finite statistics are finite direct sums of irreducible representations with finite statistics [6], and keeping in mind that π is factorial, we get the following theorem:

THEOREM 3: <u>Let π be a one-particle representation. Then π is a multiple of an irreducible one-particle representation with finite statistics</u>.

The mentioned results provide good reasons to expect that any representation which has to be considered in particle physics is a direct sum of irreducible representations $\pi \in (S)$ with finite statistics. Actually, for these representations structural properties can be derived which show that the class (S) of representations indeed describe elementary particle physics. We list the relevant results: Let $\pi \in (S)$ be irreducible with finite statistics. Then

(i) π has positive energy, i.e. $\pi \in (B)$.

(ii) There exists a conjugate representation $\bar{\pi} \in (S)$. $\bar{\pi}$ is irreducible and has the same statistics parameter as π. The equivalence class of $\bar{\pi}$ is uniquely fixed.

(iii) If π is a one-particle representation, so is $\bar{\pi}$, and they have the same mass spectrum (Existence of antiparticles).

(iv) Given one-particle states $\Psi_1, \ldots, \Psi_n$, there are incoming and outgoing scattering states

$$(\Psi_1 \times \ldots \times \Psi_n)_{\substack{in \\ out}}$$

inducing states in (S).

This analysis is more or less the same as in the (DHR)-case. One problem in the (S)-case comes from the fact that it is not possible to find local fields which generate the charged sectors from the vacuum. In the (S)-case the interpolating fields are localized only in spacelike cones. So in scattering theory one has to choose these cones to be suitably related to the momentum support of the wave functions for the scattering states. Also the analytic properties of the scattering matrix in momentum space change, but the related problems have not yet been completely examined.

BIBLIOGRAPHY

1. R.Haag, D.Kastler, "An algebraic approach to field theory" Journ.math.Phys. 5 (1964), 848-861.

2. H.-J.Borchers, "On the vacuum state in quantum field theory II", Commun.math.Phys. 1 (1965), 57-59.

3. H.-J.Borchers, "Energy and momentum as observables in quantum field theory", Commun.math.Phys. 2 (1966), 49-54.

4. H.-J.Borchers, D.Buchholz, to be published.

5. H.-J.Borchers, "Local rings and the connection of spin with statistics", Commun.math.Phys. 1 (1965), 281-307.

6. S.Doplicher, R.Haag, J.Roberts, "Local observables and particle statistics", Commun.math.Phys. 23 (1971), 199-230 and Commun.math.Phys. 35 (1974), 49-85.

7. D.Buchholz, K.Fredenhagen, "Locality and the structure of particle states in relativistic quantum field theory", to be published.

8. K.Fredenhagen, "On the existence of antiparticles", Commun.math.Phys. (to appear).

9. J.J.Bisognano, E.H.Wichmann, "On the duality condition for a hermitean scalar field", Journ.math.Phys.16 (1975), 985-1007.

10. H.-J.Borchers, "A remark on a theorem of Misra", Commun. math.Phys. 4 (1967), 315-323.

II. INSTITUT FÜR
THEORETISCHE PHYSIK
UNIVERSITÄT HAMBURG
D-2000 HAMBURG 50

FAKULTÄT FÜR PHYSIK
UNIVERSITÄT FREIBURG
D-7800 FREIBURG i.BR.

Proceedings of Symposia in Pure Mathematics
Volume 38 (1982), Part 2

QUASI-PERIODIC HAMILTONIANS. A MATHEMATICAL APPROACH

J. BELLISSARD* and D. TESTARD**

A quasi-periodic hamiltonian on $L^2(\mathbb{R}^d)$ is a self-adjoint operator H such that the function $(z \in \mathbb{C} - \mathbb{R})$

$$s \in \mathbb{R}^d \longrightarrow U(s)\,(H-z)^{-1}\,U(s)^{-1} \in \mathcal{B}(L^2(\mathbb{R}^d))$$

is continuous and almost periodic [1], where $U(s)$ is the translation by s acting as a unitary operator in $L^2(\mathbb{R}^d)$. The general theory of almost periodic function allows us to find an abelian compact metrizable group, X , a continuous homomorphism $\gamma: \mathbb{R}^d \to X$ with a dense image and a norm-resolvent continuous family $(H_x)_{x \in X}$ of self-adjoint operators such that

1) $$U(s)\,H_x\,U(s)^{-1} = H_{x+\gamma(s)}$$

2) $$H_{x=0} = H$$

If $K = \operatorname{Ker}\gamma$ has exactly d generators we get a periodic hamiltonian and $X = \mathbb{R}^d/K$. K is a lattice, and its annihilator $K^\perp \sim \hat{X}$ is called the reciprocal lattice, whereas $\hat{\mathbb{R}}^d/K^\perp$ is called the "Brillouin zone".

In the other case, we get a "virtual subgroup" K given by the action $\gamma: \mathbb{R}^d \to X$ in the sense of Mackey [2] . The annihilator $K^\perp$ of K is the set of characters of $\mathbb{R}^d$ which can be continued as continuous functions on X . It follows that $K^\perp$ is isomorphic to $\hat{X}$ as group, but it is oftenly far from closed in $\mathbb{R}^d$. Thus the "Brillouin zone" $\hat{\mathbb{R}}^d/K^\perp$ is a tiny space and we need the non commutative integration theory to treat this case [3].

Let Γ be the groupoid of the action $\mathbb{R}^d \times_\gamma X$ and $\hat{\Gamma}$ its dual, given by the dual action $\hat{\gamma}$ of $\hat{X}$ on $\hat{\mathbb{R}}^d$. The C^*-algebra of Γ [4] can be represented by a set of random operators $(A_x)_{x \in X} = A$ on $L^2(\mathbb{R}^d)$, where

$$U(s)\,A_x\,U(s)^{-1} = A_{x+\gamma(s)}$$

If a(x,s) is the kernel of A we get :

$$A_x\psi(s) = \int_{\mathbb{R}^d} a(x-\gamma(s),\, s'-s)\,\psi(s')\,d^ds'$$

In this representation, an almomst periodic operator is affiliated to $C^*(\Gamma)$.

0082-0717/80/0000-0503/$02.00

We define the trace on $C^*(\Gamma)$ by :

$$tr(A) = \int_{x\in X} dx\ a(x,0)$$

the same construction holds for $\hat{\Gamma}$ and gives a trace $\tilde{tr}$.

The Fourier transform of A is the element of $C^*(\hat{\Gamma})$ such that $(n \in \hat{X},\ k \in \hat{\mathbb{R}}^d)$

$$(\mathcal{F}A)(k,n) = \int_{X\times\mathbb{R}^d} a(x,s)e^{2i\pi(<k,s> - <n,x>)}\,dx\,ds.$$

where we have used an additive notation for characters.

PROPOSITION [9] :

1) The Fourier transform is an isomorphism from $C^*(\Gamma)$ to $C^*(\hat{\Gamma})$ such that $A \in C^*(\Gamma)$

$$\tilde{tr}(\mathcal{F}A) = tr(A)$$

2) If H is a quasi-periodic hamiltonian and $(\tilde{H}_k)_{k\in\hat{\mathbb{R}}^d}$ its Fourier transform then :

$$H = \int^{\oplus}_{\hat{\mathbb{R}}^d/\hat{X}} d\Lambda(k)\ \tilde{H}_k$$

The last integral has to be understood on the singular sense [3].

Now the main result is given by :

THEOREM [9] :

Let I be a Borel subset of $\mathbb{R}$ and H be a quasi-periodic hamiltonian.

i) The spectrum of H_x is independent of $x \in X$.

ii) $\mathrm{Sp}\,H = \bigcup_{k\in\hat{\mathbb{R}}^d} \mathrm{Sp}\,\tilde{H}_k$. If γ is injective, then $\mathrm{Sp}\,\tilde{H}_k$ is independent of $k \in \mathbb{R}^d$.

iii) If $\mathrm{Sp}\,H_k \cap I$ is pure point, then $\mathrm{Sp}\,H \cap I$ is purely continuous.

CONJECTURE [9]:

iii) can be replaced by :

$\mathrm{Sp}\,\tilde{H}_k \cap I$ is pure point (resp. singular continuous, absolutely continuous) if and only if $\mathrm{Sp}\,H \cap I$ is absolutely continuous (resp. singular continuous, pure point).

EXAMPLE 1

$$H = -\frac{d^2}{ds^2} + V \qquad \text{on } L^2(\mathbb{R}),\ \text{with} \qquad (\omega \in \mathbb{R}^\nu)$$

$$V(s) = \sum_{n\in\mathbb{Z}^\nu} v_n\, e^{2i\pi <n,\omega>s} = \hat{V}(\omega.s)$$

We assume that for some ε, c_1, c_2, $|<n,\omega>| \geq c_1/_{|n|^{\nu+\varepsilon}}$, $n \in \mathbb{Z}^\nu/\{0\}$ and $|v_n| \leq c_2\, e^{-r\|n\|}$ $(r > 0)$

Then $X = \mathbb{T}^\nu$, γ describes the Kronecker flow along ω, the reciprocal lattice is $<\omega, \mathbb{Z}^\nu>$ and the Brillouin zone $\hat{\mathbb{R}}/<\omega,\mathbb{Z}^\nu>$ is a tiny space. We find $\tilde{H}_k = (k - i<\omega,\frac{\partial}{\partial y}>)^2 + \hat{V}(y)$ on $L^2(\mathbb{T}^\nu)$.

THEOREM (Dinaburg-Sinaï, Rüssman [5,6])

There is $E_0 > 0$ such that the spectrum of $\tilde{H}_k$ is pure point providing $E > E_0$, and $|k - 2\pi\langle n,\omega\rangle| \geq C_3 |n|^{-\nu-\varepsilon}$ $n \in \mathbb{Z}^\nu/\{0\}$

Then there is an analytic function $a_k(1/\sqrt{E})$ in $E > E_0$, such that the eigenvalues of $\tilde{H}_k$ are given by

$$E - a_k(1/\sqrt{E}) = k - 2\pi\langle n,\omega\rangle \quad n \in \mathbb{Z}^\nu$$

EXAMPLE 2 : "The tight binding approximation" [7,8]

$$\psi \in \ell^2(\mathbb{Z}) \qquad (H^\lambda_\alpha \psi)(n) = \psi(n+1) + \psi(n-1) + 2\lambda \cos 2\pi(n\theta + \alpha)\psi(n)$$

with θ irrational. Then $H_{\alpha=0}$ is quasi-periodic, $X = \mathbb{T}$, and Γ is its own dual. We get $(\beta \in \mathbb{T})$

$$\tilde{H}^\lambda_\beta = \lambda H^{1/\lambda}_\beta \qquad \text{(Aubry-André's duality)}$$

THEOREM (Aubry-André, see [7])

If $\lambda > 1$, H^λ_0 has a pure point spectrum.

If $\lambda < 1$, H^λ_0 has an absolutely continuous spectrum.

If $\lambda = 1$, H^λ_0 has a singular continuous spectrum.

EXAMPLE 3 : "Quasi-periodic Kronig-Penney Model"

On $L^2(\mathbb{R})$:

$$H = -\frac{d^2}{dt^2} + \sum_{n=-\infty}^{+\infty} \lambda_n \delta(t-t_n)$$

where there are smooth functions λ and f on $\mathbb{T}$ such that

$$\lambda_n = \lambda(n\theta),\ t_{n+1} - t_n = f(n\theta)$$

(θ is irrational).

THEOREM [9]

If $|\theta - p/q| \geq C/q^{2+\varepsilon}$, $\forall p, q \in N$, there are constants E_0, E_1 such that in $[0,E_0]$ and $[E,\infty]$ the spectrum of H is pure point, provided

$$|\sin(k - 2\pi n\theta)| > \frac{C}{n^{1+\varepsilon}} \qquad \forall n \in \mathbb{Z}$$

In this case there is an analytic function $a_k(\lambda, \sqrt{E})$ both in λ and $\sqrt{E}$ convergent in $E > E_1$, or $E_0 < E$, such that the eigenvalues are solutions of

$$\cos(k - 2\pi n\theta) = \cos\sqrt{E} + a_k(\lambda, \sqrt{E})$$

ACKNOWLEDGEMENTS

The authors are indebted to Prof. Kuiper for the hospitality at I.H.E.S.. They thank D. Kastler for convincing them to learn this theory. They are indebted to A. Connes for suggesting the leading ideas of this work. One of us (J.B.) thanks S. Aubry for fruitful discussions.

BIBLIOGRAPHY

1 See for instance J. DIXMIER, "Les C^*-algèbres et leurs représentations", p. 296, Gauthier-Villars, Paris, 2nd. Ed. (1969).

2 G. MACKEY, "Ergodic Theory and Virtual Groups", Math. Ann. 166, 187-207 (1966).

3 A. CONNES, "Sur la théorie non-commutative de l'intégration", in Lecture Notes in Mathematics, n° 725, P. de la Harpe Ed., Springer, Berlin, Heidelberg, New York (1979).

4 J. RENAULT, "A Groupoïd Approach to C^*-Algebras, in Lecture Notes in Mathematics, n° 793, Springer, Berlin, Heidelberg, New York (1980).

5 E.I. DINABURG and Ya. G. SINAI, "The One-Dimensional Schrödinger Equat-ion with a Quasi-Periodic Potential", Funct. Anal. Appl. 9, 279-285 (1975).

6 H. RUSSMANN, "On the One-Dimensional Schrödinger Equation with a Quasi-Periodic Potential", Preprint Mainz, Germany (1979).

7 S. AUBRY and G. ANDRE, "Analyticity Breaking and Anderson Localization in Incommensurate Lattices", Proceedings of the VIII International Colloquium on Group Theoretical Methods in Physics, Kiryat Anavim, (1979).

8 M. Ya. AZBELL, Phys. Rev. Lett. 43, 1954 (1979).

9 J. BELLISSARD and D. TESTARD, work in preparation.

* Université de Provence, Marseille and Centre de Physique Théorique, CNRS, Marseille

** Centre Universitaire d'Avignon and Centre de Physique Théorique, CNRS, Marseille

CNRS - LUMINY - Case 907
CENTRE DE PHYSIQUE THEORIQUE
F-13288 MARSEILLE CEDEX 2 (FRANCE)

Proceedings of Symposia in Pure Mathematics
Volume 38 (1982), Part 2

POSITIVE LINEAR MAPS

Man-Duen Choi[†]

1. INTRODUCTION

In this expository article, we are concerned with some basic aspects of positive linear maps on C^*-algebras. Since the structure of positive linear maps is drastically non-trivial even for the finite-dimensional case, it could be appropriate to regard any general C^*-algebra in the context as the $n \times n$ matrix-algebra. Readers are also referred to [1; 7; 17] for some recent research and references on the subject.

Throughout this article, C^*-algebras possess a unit I and are written in German type $\mathfrak{A}$, $\mathfrak{B}$. M_n denotes the algebra of all $n \times n$ complex matrices. $M_n(\mathfrak{A})$ is the algebra of $n \times n$ matrices over $\mathfrak{A}$. We write $\mathfrak{A}^h = \{$all hermitian elements in $\mathfrak{A}\}$, $\mathfrak{A}^+ = \{$all positive elements in $\mathfrak{A}\}$ — in particular, $M_n^+ = \{$all $n \times n$ positive semi-definite hermitian matrices$\}$. A linear map $\Phi: \mathfrak{A} \to \mathfrak{B}$ is said to be *unital* if $\Phi(I) = I$; Φ is said to be *positive [*resp. *hermitian-preserving]* if $\Phi(\mathfrak{A}^+) \subseteq \mathfrak{B}^+$ [resp. $\Phi(\mathfrak{A}^h) \subseteq \mathfrak{B}^h$]. For each matrix $A = [\alpha_{jk}]_{j,k}$, we write A^t for the transpose matrix $[\alpha_{kj}]_{j,k}$.

§2 is devoted to thorough examinations of the "global" structure of positive linear maps. For the sake of simplicity, the investigation is restricted to the finite-dimensional case. Foremost, we call attention to the subtlty and the complexity of the partial ordering in M_n after tensor product. This in turn shows the discrimination, in a precise way, between the class of positive linear maps and the proper subclass of completely positive linear maps. Moreover, in exactly the same manner as positive real polynomials need not be sums of squares, there exist positive linear maps that are not "decomposable".

In §3, we present some elegant features of a single positive linear map, by exhibiting several assorted inequalities of the Schwarz type. Finally, we conclude with a conjecture pertaining to tracial linear maps.

1980 Mathematics Subject Classification. 46L05, 15-02.
[†]Partially supported by NSERC of Canada.

2. POSITIVE LINEAR MAPS BETWEEN MATRIX ALGEBRAS

2A. THE ORDER STRUCTURE OF MATRIX ALGEBRAS

For any sets $S \subseteq M_n$ and $T \subseteq M_m$, we write $S \otimes T = \{\Sigma S_j \otimes T_j : S_j \in S, T_j \in T\} \subseteq M_n \otimes M_m$. The algebraic tensor product $M_n \otimes M_m$ can be regarded as $M_n(M_m)$, and thus be identifiable with M_{nm}. Namely, if $\{E_{jk}\}_{j,k}$ and $\{F_{pq}\}_{p,q}$ are the matrix units of M_n and M_m respectively, then $\{E_{jk} \otimes F_{pq}\}_{j,k,p,q}$ will form the matrix units of $M_n \otimes M_m$. While M_{nm} inherits most algebraic features of M_n and M_m, it seems peculiar that M_{nm}^+ is not fully compatible with M_n^+ and M_m^+. In fact, the tensor product structure induces the following relations:

$$\begin{array}{ccccc} M_n^h \otimes M_m^h & = & (M_n \otimes M_m)^h & \underset{\text{def}}{\simeq} & M_{nm}^h \\ \uparrow & & \uparrow & & \uparrow \\ M_n^+ \otimes M_m^+ & \subseteq & (M_n \otimes M_m)^+ & \underset{\text{def}}{\simeq} & M_{nm}^+ \end{array}$$

NOTES. (i) It is clear that $M_n^h \otimes M_m^h \subseteq (M_n \otimes M_m)^h$. Since both sides are *real* linear spaces of same dimension, it follows that $M_n^h \otimes M_m^h = (M_n \otimes M_m)^h$. Alternatively, we may pursue the standard computation

$$U = \Sigma A_j \otimes B_j \in (M_n \otimes M_m)^h$$

$$\Rightarrow U = \tfrac{1}{2}(U+U^*) = \Sigma\left[\frac{A_j+A_j^*}{2} \otimes \frac{B_j+B_j^*}{2} - \frac{A_j-A_j^*}{2i} \otimes \frac{B_j-B_j^*}{2i}\right]$$

$$\in M_n^h \otimes M_m^h .$$

(ii) The inclusion $M_n^+ \otimes M_m^+ \subseteq (M_n \otimes M_m)^+$ is always *proper* except if $n = 1$ or $m = 1$. For example, let $U = \Sigma_{j=1}^2 \Sigma_{k=1}^2 E_{jk} \otimes F_{jk}$. Then $\frac{1}{2}U$ is a rank-1 projection, thus $U \in (M_n \otimes M_m)^+$. On the other hand, by the extremality, any rank-1 element in $M_n^+ \otimes M_m^+$ can be written as $A \otimes B$ with $A \in M_n^+$, $B \in M_m^+$. Obviously, U cannot be written as $A \otimes B$, so $U \notin M_n^+ \otimes M_m^+$.

(iii) Define a linear map $\Theta : M_n \otimes M_m \to M_n \otimes M_m$ by

$$\Theta(\Sigma A_j \otimes B_j) = \Sigma A_j^t \otimes B_j .$$

It is clear that

$$M_n^+ \otimes M_m^+ = \Theta(M_n^+ \otimes M_m^+) \subseteq \Theta\big((M_n \otimes M_m)^+\big).$$

Thus we have

$$M_n^+ \otimes M_m^+ \subseteq (M_n \otimes M_m)^+ \cap \Theta\big((M_n \otimes M_m)^+\big).$$

It is natural to ask whether the preceding inclusion $\subseteq$ can be replaced by identity $=$. More explicitly, we ask

QUESTION. *Suppose* $B_{jk} \in M_m$ $(1 \le j \le n,\ 1 \le k \le n)$. *If both* $[B_{jk}]_{j,k}$ *and* $[B_{kj}]_{j,k} \in (M_n \otimes M_m)^+$, *does it follow that* $[B_{jk}]_{j,k} \in M_n^+ \otimes N_m^+$?

The underlying structure theory of this question is rather deep even if $n, m \le 3$. Woronowicz [19] has shown that the answer is yes if $(n,m) = (2,3)$, but no if $(n,m) = (2,4)$. In Note (iv) below, there is a concrete example showing that the answer is no again if $(n,m) = (3,3)$.

(iv) To pursue our further investigation, we ask

QUESTION. *If* $U \in (M_n \otimes M_m)^+$ *and* $\Theta(U) = U$ [namely, if $B_{jk} = B_{kj} \in M_m$ $(1 \le j \le n,\ 1 \le k \le n)$ and $U = [B_{jk}]_{j,k} \in (M_n \otimes M_m)^+ = M_n(M_m)^+$], *does it follow that* $U \in M_n^+ \otimes M_m^+$?

If $n = 2$, the answer is yes (see [5, Theorem 7; 19, Theorem 5.3]. This means

$$\begin{bmatrix} A & B \\ B & C \end{bmatrix} \in (M_2 \otimes M_m)^+ \Rightarrow \begin{bmatrix} A & B \\ B & C \end{bmatrix} \in M_2^+ \otimes M_m^+ .$$

In $n = 3$, the answer is no as shown by a counter-example: Let

$$U = [B_{jk}]_{j,k=1}^{3} = \left[\begin{array}{ccc|ccc|ccc} 1 & 0 & 0 & 0 & 1 & 0 & 0 & 0 & 1 \\ 0 & 2 & 0 & 1 & 0 & 0 & 0 & 0 & 0 \\ 0 & 0 & \frac{1}{2} & 0 & 0 & 0 & 1 & 0 & 0 \\ \hline 0 & 1 & 0 & \frac{1}{2} & 0 & 0 & 0 & 0 & 0 \\ 1 & 0 & 0 & 0 & 1 & 0 & 0 & 0 & 1 \\ 0 & 0 & 0 & 0 & 0 & 2 & 0 & 1 & 0 \\ \hline 0 & 0 & 1 & 0 & 0 & 0 & 2 & 0 & 0 \\ 0 & 0 & 0 & 0 & 0 & 1 & 0 & \frac{1}{2} & 0 \\ 1 & 0 & 0 & 0 & 1 & 0 & 0 & 0 & 1 \end{array}\right] \in (M_3 \otimes M_3)^+ .$$

It can be checked by direct computation $U \notin M_3^+ \otimes M_3^+$.

2B. POSITIVE LINEAR MAPS VS. COMPLETELY POSITIVE LINEAR MAPS

The global structure of positive linear maps is very complicated. For the sake of contrast, we also look into the structure of completely positive linear maps. By one of several equivalent definitions (cf. [5]), we say that a linear map $\Phi : M_n \to M_m$ is *completely positive* if there exist $n \times m$ matrices V_j such that $\Phi(A) = \Sigma V_j^* A V_j$ for all $A \in M_n$.

Let $L(M_n, M_m) = \{$all linear maps: $M_n \to M_m\}$. There is a natural linear isomorphism between $L(M_n, M_m)$ and $M_n \otimes M_m \simeq M_{nm}$, assigning each linear map $\Phi : M_n \to M_m$ to a big matrix $[\Phi(E_{jk})]_{j,k=1}^n \in M_n(M_m)$. Moreover, M_p is identifiable with {linear functionals on M_p} since each $A \in M_p$ induces a linear functional ρ_A by $\rho_A(X) = \mathrm{trace}(AX)$. Henceforth, we get a chart showing natural correspondences among different classes (see [5, pp. 286-287]).

$L(M_n, M_m)$	$\longleftrightarrow$	$M_n \otimes M_m$	$\longleftrightarrow$	linear functionals on $M_n \otimes M_m$
$\hookuparrow$		$\hookuparrow$		$\hookuparrow$
hermitian-preserving linear maps	$\longleftrightarrow$	$(M_n \otimes M_m)^h$	$\longleftrightarrow$	linear functionals assuming real values on $(M_n \otimes M_m)^h$
$\hookuparrow$		$\hookuparrow$		$\hookuparrow$
positive linear maps	$\longleftrightarrow$	?	$\longleftrightarrow$	linear functionals assuming positive values on $M_n^+ \otimes M_m^+$
$\hookuparrow$		$\hookuparrow$		$\hookuparrow$
completely positive linear maps	$\longleftrightarrow$	$(M_n \otimes M_m)^+$	$\longleftrightarrow$	linear functionals assuming positive values on $(M_n \otimes M_m)^+$

As revealed above, the class of positive linear maps remains obstinate in structure theory. It is no wonder in the recent research, completely positive linear maps, rather than general positive linear maps, have become the natural morphisms in C^*-algebra theory.

2C. DECOMPOSABLE POSITIVE LINEAR MAPS

We say that a positive linear map $\Phi : M_n \to M_m$ is *decomposable* if there exist $n \times m$ matrices V_j and W_k such that

$$\Phi(A) = \Sigma V_j^* A V_j + \Sigma W_k^* A^t W_k$$

for all $A \in M_n$. [Equivalently, we may define a decomposable positive linear map as a compression of a Jordan homomorphism (see e.g., [16, p. 267; 18]).]

There arises a natural question: *Must every positive linear map be decomposable?* This is intimately related to a more general question originated from a classical result of Hilbert [11]: *Which positive semi-definite real polynomials can be written as sums of squares?* To look into the structure, we examine the following cognate statements.

(I) Every positive linear map is decomposable.

(II) Let F be a bihermitian form $F(\mu\,;\lambda) = \Sigma\alpha_{pqjk}\bar{\mu}_p\mu_q\bar{\lambda}_j\lambda_k$ with complex indeterminates $\mu = (\mu_1, \ldots, \mu_m)$, $\lambda = (\lambda_1, \ldots, \lambda_n)$. If $F(\mu\,;\lambda) \geq 0$ for all μ, λ, then there exist bilinear forms $g_i(\mu\,;\lambda) = \Sigma\beta^{(i)}_{pj}\mu_p\lambda_j$, and *dual*-bilinear forms $h_i(\mu\,;\lambda) = \Sigma\gamma^{(i)}_{pj}\bar{\mu}_p\lambda_j$ such that $F = \Sigma(\bar{g}_i g_i + \bar{h}_i h_i)$.

(III) Every positive semi-definite real biquadratic form is sum of squares of bilinear forms; namely, if $F(r\,;s) = \Sigma\alpha_{pqjk}r_p r_q s_j s_k$ $(p \leq q,\ j \leq k)$ with *real* indeterminates $r = (r_1, \ldots, r_m)$, $s = (s_1, \ldots, s_n)$, and if $F(r\,;s) \geq 0$ for all r, s, then there exist real bilinear forms $f_i(r\,;s)$ such that $F = \Sigma f_i^2$.

(IV) Every positive semi-definite real polynomial is sum of squares of polynomials; namely, if $F(x_1, \ldots, x_n) \in \mathbb{R}[x_1, \ldots, x_n]$ is ≥ 0 for all real x_j, then there exist real polynomials f_i such that $F = \Sigma f_i^2$.

By a routine proof (the main ingredients of the proof can be found in [5,6]), we can verify the implications (IV) => (III) => (II) <=> (I). Thus, a single concrete counter-example for (I), will disprove all statements (I) - (IV). Notably, Hilbert has already shown, in principle, that there exist counter-examples for Statement (IV), but his method is very complicated and does not lend itself to a really practical construction (however, see [9] for research on this topic). Moreover, a *faulty* proof of Statement (III) has also appeared in the literature of circuit theory, leading to misinformation on network structure.

The promised counter-example ([7, Appendix B]) is a linear map $\Phi: M_3 \to M_3$ such that

$$\Phi\left(\begin{bmatrix}\alpha_{11} & \alpha_{12} & \alpha_{13}\\ \alpha_{21} & \alpha_{22} & \alpha_{23}\\ \alpha_{31} & \alpha_{32} & \alpha_{33}\end{bmatrix}\right) = \begin{bmatrix}\alpha_{11} & -\alpha_{12} & -\alpha_{13}\\ -\alpha_{21} & \alpha_{22} & -\alpha_{23}\\ -\alpha_{31} & -\alpha_{32} & \alpha_{33}\end{bmatrix} + \begin{bmatrix}\alpha_{33} & 0 & 0\\ 0 & \alpha_{11} & 0\\ 0 & 0 & \alpha_{22}\end{bmatrix}$$

This map Φ appears to be the simplest example of a positive linear map that is not decomposable (see also [6, Theorem 2], [19], [20]).

3. INEQUALITIES

3A. SCHWARZ INEQUALITIES

It has become an extremely difficult problem to classify all positive linear maps. Nevertheless, there are still some elegant structure theorems revealing the subtlty of the order structure of C^*-algebras. The following is the pioneering result.

THEOREM (Kadison [12]). *Suppose* $\Phi : \mathfrak{A} \to \mathfrak{B}$ *is a unital positive linear map. Then* $\Phi(A^2) \geq \Phi(A)^2$ *for all* $A \in \mathfrak{A}^h$.

REMARK. This inequality provides some non-trivial information about positive linear maps. Indeed, the naïve definition says that $\Phi(A^2) \geq 0$ for all $A \in \mathfrak{A}^h$, while the Theorem above says that $\Phi(A^2) \geq \Phi(A)^2 \geq 0$. On the other hand, although $\Phi(A^4) \geq 0$ for all $A \in \mathfrak{A}^h$, it will be false to assert $\Phi(A^4) \geq \Phi(A)^4$ in general. The following is another pertinent result:

THEOREM ([4, Corollary 2.3]). *Suppose* $\Phi : \mathfrak{A} \to \mathfrak{B}$ *is a unital positive linear map. Then* $\Phi(A^{-1}) \geq \Phi(A)^{-1}$ *for all positive invertible* $A \in \mathfrak{A}$.

REMARK. The proof of all cognate theorems runs as follows: For each $A \in \mathfrak{A}^h$, $C^*(A)$ is a commutative C^*-algebra. Thus $\Phi|_{C^*(A)}$ is completely positive and has a tractable structure.

3B. ASSORTED INEQUALITIES

The theorem below states a delicate inequality covering several known results about positive linear maps.

THEOREM. *Suppose* $\Phi : \mathfrak{A} \to \mathfrak{B}$ *is a unital positive linear map. Suppose* $T \geq A^*A$ *and* $TA = AT$. *Then* $\Phi(T) \geq \Phi(A^*)\Phi(A)$, $\Phi(T) \geq \Phi(A)\Phi(A^*)$.

NOTES. (i) Let $A = A^*$, $T = A^2$. We get Kadison's inequality $\Phi(A^2) \geq \Phi(A)^2$.

(ii) Let $T = I$. Then this theorem reduces to a well-known fact discovered by Russo and Dye [14, Corollary 1]: Every unital positive linear map is contractive.

(iii) To prove this theorem, we may make use of the fact: $M_2(\mathfrak{A})$, rather than $\mathfrak{A}$ itself, has sufficiently many commutative $*$-subalgebras.

Ando [1] has indicated that each positive linear map possesses a sort of 2-positive effect. The following is a result along this line.

THEOREM [7, Proposition 4.3]. *Suppose* $\Phi : \mathfrak{A} \to \mathfrak{B}$ *is a positive linear map with* $\Phi(I)$ *being invertible. Then* $\Phi(ST^{-1}S) \geq \Phi(S)\Phi(T)^{-1}\Phi(S)$ *for all* $S \in \mathfrak{A}^h$ *and all positive invertible* $T \in \mathfrak{A}$.

NOTE. Suppose Φ is unital in the Theorem above. Then the special case $T = I$ leads to Kadison's inequality $\Phi(S^2) \geq \Phi(S^2)$, while another special case $S = I$ leads to the inequality $\Phi(T^{-1}) \geq \Phi(T)^{-1}$.

3C. SOME OPEN QUESTIONS

As an easy consequence of Kadison's inequality, each unital positive linear map $\Phi : \mathfrak{A} \to \mathcal{L}$ satisfies the inequality

$$\Phi(A^*A) + \Phi(AA^*) \geq \Phi(A^*)\Phi(A) + \Phi(A)\Phi(A^*)$$

for all $A \in \mathfrak{A}$ (see [16, Lemma 7.3]). It may be interesting to know when we can split the inequality. Namely, we ask

QUESTION. *Suppose* $\Phi : \mathfrak{A} \to \mathcal{L}$ *is a unital positive linear map. Under what extra judicious condition can we conclude* $\Phi(A^*A) \geq \Phi(A^*)\Phi(A)$?

NOTES. (i) In case Φ is completely positive (or 2-positive [4, Corollary 2.8]), then $\Phi(A^*A) \geq \Phi(A^*)\Phi(A)$ for all $A \in \mathfrak{A}$.

(ii) It is known [7, Proposition 3.6] that if A_0 is subnormal, then $\Phi(A_0^*A_0) \geq \Phi(A_0^*)\Phi(A_0)$ and $\Phi(A_0A_0^*) \geq \Phi(A_0)\Phi(A_0^*)$.

(iii) It may be plausible to conjecture: If A_0 is an operator such that $\Phi(A_0^*A_0) = \Phi(A_0A_0^*)$, then it follows that $\Phi(A_0^*A_0) \geq \Phi(A_0^*)\Phi(A_0)$. This turns out to be false as shown below. Define $\Phi : M_2 \to M_2$ by $\Phi(A) = \frac{1}{2}(A + A^t)$. If $A_0 = A_0^t$, then the equality $\Phi(A_0^*A_0) = \Phi(A_0A_0^*)$ is automatic, while the inequality $\Phi(A_0^*A_0) \geq \Phi(A_0^*)\Phi(A_0)$ is equivalent to A_0 being normal. Thus any non-normal A_0 of the form $\begin{bmatrix} \alpha & \beta \\ \beta & \gamma \end{bmatrix}$ with $\alpha, \beta, \gamma \in \mathbb{C}$ will serve as a counter-example.

We say that a linear map $\Phi : \mathfrak{A} \to \mathcal{L}$ is *tracial* if Φ is positive and $\Phi(A^*A) = \Phi(AA^*)$ for all $A \in \mathfrak{A}$. It is evident that tracial linear maps are the natural generalizations of tracial states. And, any structure theorems about the tracial linear maps would contribute to the understanding of "finite" C^*-algebras. The following question is of particular interest:

QUESTION. *Must every tracial linear map be completely positive?*

In particular, we propose a

CONJECTURE. *If* $\Phi : \mathfrak{A} \to \mathcal{L}$ *is a unital tracial linear map, then* $\Phi(A^*A) \geq \Phi(A^*)\Phi(A)$ *for all* $A \in \mathfrak{A}$.

REMARK. Of course, the conjecture is true if $\mathcal{L} = \mathbb{C}$. The *crucial* case is $\mathcal{L} = M_2$. Indeed, if the Conjecture is true for $\mathcal{L} = M_2$, then the Conjecture will remain true for a general $\mathcal{L}$. Furthermore, note that the Conjecture is meaningful only when $\mathfrak{A}$ is an infinite-dimensional finite C^*-algebra.

BIBLIOGRAPHY

1. T. Ando, "Concavity of certain maps among positive definite matrices and its application to Hadamard products", *Linear Alg. Appl.* 26 (1979), 203-241.

2. W.B. Arveson, "Subalgebras of C*-algebras, I", *Acta Math.* 123 (1969), 141-224; "II", *Acta Math.* 128 (1972), 271-308.

3. M.D. Choi, "Positive linear maps on C*-algebras", *Can. J. Math.* 24 (1972), 520-529.

4. M.D. Choi, "A Schwarz inequality for positive linear maps on C*-algebras", *Ill. J. Math.* 18 (1974), 565-574.

5. M.D. Choi, "Completely positive linear maps on complex matrices", *Linear Alg. Appl.* 10 (1975), 285-290.

6. M.D. Choi, "Positive semidefinite biquadratic forms", *Linear Alg. Appl.* 12 (1975), 95-100.

7. M.D. Choi, "Some assorted inequalities for positive linear maps on C*-algebras", *J. Operator Theory*, to appear.

8. M.D. Choi and E.G. Effres, "Injectivity and operator spaces", *J. Func. Anal.* 24 (1977), 156-209.

9. M.D. Choi and T.Y. Lam, "Extremal positive semidefinite forms", *Math. Ann.* 231 (1977), 1-18.

10. Ch. Davis, "A Schwarz inequality for convex operator functions", *Proc. Amer. Math. Soc.* 8 (1957), 42-44.

11. D. Hilbert, "Über die Darstellung definiter Formen als Summe von Formenquadraten", *Math. Ann.* 32 (1888), 342-350.

12. R.V. Kadison, "A generalized Schwarz inequality and algebraic invariants for C*-algebras", *Ann. Math.* 56 (1952), 494-503.

13. E.H. Lieb and M.B. Ruskai, "Some operator inequalities of the Schwarz type", *Adv. in Math.* 12 (1974), 269-273.

14. B. Russo and H.A. Dye, "A note on unitary operators in C*-algebras", *Duke Math. J.* 33 (1966), 413-416.

15. W.F. Stinespring, "Positive functions on C*-algebras", *Proc. Amer. Math. Soc.* 6 (1955), 211-216.

16. E. Størmer, "Positive linear maps of operator algebras", *Acta Math.* 110 (1963), 233-278.

17. E. Størmer, "Positive linear maps of C*-algebras", *Lecture Notes in Physics*, Vol. 29, pp. 85-106, Springer-Verlag, 1974.

18. E. Størmer, "Decomposition of positive projections on C*-algebras", to appear.

19. S.L. Woronowicz, "Positive maps of low dimensional matrix algebras", *Rep. Math. Phys.* 10 (1976), 165-183.

20. S.L. Woronowicz, "Nonextendible positive maps", *Comm. Math. Phys.* 51 (1976), 243-282.

DEPARTMENT OF MATHEMATICS
UNIVERSITY OF TORONTO
TORONTO, CANADA M5S 1A1

Proceedings of Symposia in Pure Mathematics
Volume **38** (1982), Part 2

REPRESENTATIONS OF INSEPARABLE C*-ALGEBRAS

Simon Wassermann

1. In 1961 Glimm proved his celebrated result [2, Theorem 1] that a separable C*-algebra which is not postliminal has a type II factor representation. Although Sakai [3] subsequently showed that any non-postliminal C*-algebra has a type III factor representation, it is still not known whether an arbitrary inseparable non-postliminal C*-algebra need have type II representations. Recent attempts to settle this question have led to positive results in some interesting special cases.

2. Let $M = M_1 \oplus M_2 \oplus \ldots$, the C*-algebra of bounded sequences (x_n), with $x_n \in M_n = M_n(\mathbb{C})$, with pointwise algebraic operations and the supremum norm. If $\mathcal{U}$ is a free ultrafilter on $\mathbb{N}$, a trace state on M is given by

$$f_{\mathcal{U}}((x_n)) = \lim_{\mathcal{U}} tr_n(x_n).$$

Wright [5] showed that $f_{\mathcal{U}}$ is a type II_1 factor state.

If M is embedded as a C*-subalgebra in a C*-algebra A, a norm-one projection $\theta: A \to M$ is defined as follows. Let e_r be the identity projection of M_r. Then

$$e_r A e_r \cong R_r \otimes M_r,$$

where M_r is the relative commutant of M_r in $e_r A e_r$.

If f_r is a state on R_r, $f_r \otimes id_{M_r}: e_r A e_r \to M_r$ is a projection of norm one, and θ is given by

$$\theta(x) = ((f_r \otimes id_{M_r})(e_r x e_r)) \qquad (x \in A).$$

PROBLEM 1. Is $f_{\mathcal{U}} \circ \theta$ a type II state when the f_r are all pure?

There are partial answers to this question in two directions:

0082-0717/80/0000-0587/$01.50

(a) Let $H = H_1 \oplus H_2 \oplus \ldots$, where H_n is an n-dimensional Hilbert space. Then $M \cong L(H_1) \oplus L(H_2) \oplus \ldots \subseteq L(H)$, and with $A = L(H)$, $R_r = \mathbb{C}$. Also $\theta(x) = (e_r x e_r)$, where e_r is the projection onto H_r; in this case θ is unique.

THEOREM (Anderson, [1]). $f_{\mathcal{U}} \circ \theta$ is a type II factor state on L(H).

(b) Let $A_1, A_2, \ldots$ be non-zero C*-algebras, and let $f_n \in P(A_n)$. If the A_n are unital and $A = \oplus (A_n \otimes M_n)$, $M \cong \oplus (\mathbb{C}.1_n \otimes M_n) \subseteq A$ and

$$\theta((x_n)) = (f_n \otimes id_n)(x_n) \quad ((x_n) \in A)$$

defines a norm-one projection $A \to M$.

THEOREM (Wassermann, [4]). $f_{\mathcal{U}} \circ \theta$ is a type II factor state of A.

The proof uses the known special case $A = M$.

COROLLARY. If R is a properly infinite von-Neumann algebra with infinite dimensional centre, R has a type II_∞ factor representation.

EXAMPLE. $R = N \bar{\otimes} \ell^\infty(\mathbb{Z})$, where N is a type III factor.

PROBLEM 2. Does every type III factor have a type II factor representation?

BIBLIOGRAPHY

1. J. Anderson, "Extreme points in sets of positive linear maps on $\mathcal{B}(\mathcal{H})$", J. Functional Analysis, 31 (1979), 195-217.

2. J. Glimm, "Type I C*-algebras", Ann. Math., 73 (1961), 572-612.

3. S. Sakai, "On a characterization of type I C*-algebras", Bull. Amer. Math. Soc., 72 (1966), 508-512.

4. S. Wassermann, "Type II representations of inseparable C*-algebras", Bull. London Math. Soc., to appear.

5. F.B. Wright, "A reduction theory for algebras of finite type", Ann. Math., 60 (1954), 560-570.

DEPARTMENT OF MATHEMATICS
UNIVERSITY OF GLASGOW
GLASGOW, SCOTLAND G12 8QW

Proceedings of Symposia in Pure Mathematics
Volume 38 (1982), Part 2

C*-ALGEBRAS WITH A COUNTABLE DUAL

Horst Behncke

This work was partly done in collaboration with G. Elliott. We give a complete classification of all separable C*-algebras A satisfying

(1) $\hat{A}$, the dual of A, is countable and

(2) The order topology on $\hat{A}$ agrees with the Jacobson topology.

Any separable C*-algebra A satisfying (1) is a postliminary AF algebra, which possesses a composition series (J_ρ) $o \leq \rho \leq \alpha$ with

i) $J_o = (o)$, $J_\alpha = A$, α a countable ordinal

ii) $J_{\rho+1}/J_\rho$ is the largest dual ideal in A/J_ρ

iii) $J_\rho = \overline{\bigcup_{\sigma<\rho} J_\sigma}$ if ρ is a limit ordinal

In particular Prim $\hat{A} = \hat{A}$ and $\hat{A}$ can be considered a partially ordered set via inclusion. (2) implies that any decreasing chain in $\hat{A}$ is finite.

By induction on α one shows that the dimension group $G(A)$ of any C*-algebra A with (1) and (2) is the lexicographic product of $\mathbb{Z}$ indexed by $\hat{A}$

(3) $G(A) = \bigoplus_{x \in \hat{A}} \mathbb{Z}$ with lexicographic order

0082-0717/80/0000-0501/$01.75

Let $U(A) = \{ x \in \hat{A} \mid A/x \text{ is unital} \}$

Then $U(A)^c$, the complement of $U(A)$ is open and the dimension range $D(A)$ of A satisfies

$$D(A) = D(A) \oplus \bigoplus_{x \notin U(A)} \mathbb{Z}$$

With this one can determine a function $d: \hat{A} \to \overline{\mathbb{N}} = \{\alpha, 0, 1, 2, \ldots\}$ with

$d(x) > 0$ if x is maximal and

$d(x) = 0$ whenever either $\{\gamma \mid \gamma > x\}$ contains a chain without a maximal element or $\sum_{y>x} d(y) = \alpha$

Such a function will be called a defector. Then

Lemma: $\gamma \in D(A)$ if $\gamma \in G(A)_+$ and

i) $\gamma(x) \geq 0$ is arbitrary if $x \notin U(A)$

ii) $x \in U(A)$ then $\gamma(x) \leq d(x)$ unless $\gamma(y) < d(y)$ for some $y>x$

For this reason we write $D(d)$ instead of $D(A)$

Conversely if P is a countable partially ordered set and d a defector on P, one can construct explicitly a C*-algebra $A(P,d)$ such that

$A(P,d)^\wedge = P$ and

$D(A(P,d)) = D(d)$

This and [2, theorem 4.3] shows

Theorem: For any separable C*-algebra A with (1) and (2) there exists a defector d on $\hat{A}$ such that $A \approx A(P,d)$

Moreover one can easily describe the isomorphism classes of the model algebras $A(P,d)$ in terms of d in this case.

These results generalize those of the author and Leptin [1]

References

[1] H. Behncke, H. Leptin, Classification of C*-algebras with a finite dual. J. Functional Analysis 16 1974, 241-257

[2] G. Elliott, On the classification of inductive limits of sequences of semisimple finite dimensional algebras. J. Algebra 38 1976 29-44.

Proceedings of Symposia in Pure Mathematics
Volume 38 (1982), Part 2

ON THE EXTENSIONS OF C*-ALGEBRAS RELATIVE TO FACTORS OF TYPE II_∞

Hideo Takemoto

Let M be an infinite semifinite factor with separable predual and denote by I the norm closure of the ideal of elements of finite rank. Let A be a separable C*-algebra. In [2], Brown, Douglas and Fillmore introduced the semigroup Ext(A,I) of extensions of I by A, this is, sub-C*-algebras $I \subset B \subset M$ with B/I equal to A, with an appropriate addition and equivalence relation. Here we shall consider what they called strong equivalence, with respect to which B_1 is equivalent to B_2 if there is an inner automorphism of M taking B_1 and inducing the identity on A in M/I.

In this talk, we make the technical assumption that M is an approximately finite-dimensional factor of type II_∞ and a C*-algebra A is an approximately finite-dimensional C*-algebra. Then, we can show that Ext(A,I) is a group and is in fact computed to consist of just one element.

At first, we shall show that any two trivial extensions of I by A are equivalent. This fact can be show via a generalization of Voiculescu's result (see [4] and also [1]) in the case of type I.

LEMMA. Suppose that $A \cap I = \{0\}$. Let π be a *-isomorphism of A into M such that $\pi(1) = 1$ and $\pi(A) \cap I = \{0\}$. Then, there exists a unitary operator u in M such that

$$x - u\pi(x)u^* \in I$$

for every $x \in A$.

By using the above lemma and more considerations, we can show

1980 Mathematics Subject Classification. 46L99.

0082-0717/80/0000-0581/$01.75

that any two trivial extensions are equivalent.

THEOREM I. Let σ and τ be two trivial extensions of I by A, then σ and τ are equivalent.

Next, we will show that every extension of I by A is trivial. If A is a UHF-algebra or each column of A gives the multiplicities of one minimal direct summand of one finite-dimensional algebra in the diagram in the various minimal direct summands of the next algebra, then we can give a direct proof. In a general case, we will show this fact by using the K-theory.

Let $I \to B \to A$ be an extension of I by A in M. Then, we will show the following lemma.

LEMMA. The induced sequence $K_0I \to K_0B \to K_0A$ splits as a sequence of preordered abelian groups. (The preorder under consideration is of course that determined by the semigroup of equivalence classes of projections in each of the algebras I, B and A, and in matrix algebras over them.)

By the above lemma and more consicerations, we can show that every extension of I by A is trivial.

THEOREM II. Every extension of I by A in M is trivial.

By Theorem I and II, we get the semigroup of extensions of I by A consist of only one element.

COROLLARY. The semigroup Ext(A,I) of extensions of I by A in M consist of only zero element.

BIBLIOGRAPHY

1. W.Arveson, Notes on extensions of C*-algebras, Duke Math. Journ., 144(1977), 329 - 355.

2. L.Brown, R.Douglas and P.Fillmore, Unitary equivalence module compact operators and extensions of C*-algebras, Springer-Verlag, Lecture Note, 345(1973), 58 - 128.

3. G.A.Elliott and H.Takemoto, On the extensions of C*-alge-

bras relative to factors of type II_∞, to appear.

4. D.Voiculescu, A non-commutative Weyl-von Neumann theorem, Rev. Roum. Pures Appl., 21(1976), 97 - 113.

DEPARTMENT OF MATHEMATICS
COLLEGE OF GENERAL EDUCATION
TOHOKU UNIVERSITY
SENDAI, JAPAN

Proceedings of Symposia in Pure Mathematics
Volume 38 (1982), Part 2

A STRUCTURE THEORY IN THE REGULAR MONOTONE COMPLETION OF C*-ALGEBRAS

kazuyuki Saitô[1]

A C*-algebra has a nice property as an ordered vector space, however, it has no order completeness property. So I would like to give here an order completion of a C*-algebra and discuss about structures of its completion.

A regular monotone completion of a unital C*-algebra A is an ordered pair (C,γ) where C is a monotone complete C*-algebra (automatically an AW*-algebra) and γ is an unital *-monomorphism which is continuous in the sense that $\gamma(x_\alpha) \uparrow \gamma(x)$ in C_h (the hermitian part of C) whenever $x_\alpha \uparrow x$ in A_h (the hermitian part of A) with the following properties:

(1) C_h is the smallest monotone closed subset of C_h which contains $\gamma(A_h)$,

(2) $\gamma(A_h)$ is order dense in C_h, that is, whenever x is an element in C_h, then x is the least upper bound in C_h of the set $\{\gamma(a);\ a \in A_h$ and $\gamma(a) \leqq x\}$.

The concept of the regular monotone completion of a unital C*-algebra was established by Maitland Wright (for separable case) and by Hamana (for general case), to provide a method to construct non W*, AW*-algebras from any given C*-algebras.

Then it is natural to ask how properties of a given C*-algebra A are reflected in properties of the regular monotone completion $\overline{A}$ of A and vice versa.

1. A CONSTRUCTION OF $\overline{A}$ FOR A ([1],[2] and [6]). Let A be a unital C*-algebra, then A_h is an order unit vector space (with respect to the natural order). Let V be the unique Dedekind completion of A_h (for every x in V, $x = \text{Sup}\{a;\ a \in A_h$ and $a \leqq x\}$ in the bounded complete vector lattice V). Let $\overline{A_h}$ be the monotone closure of A_h in V, then we can get that $\overline{A_h}$ is isometrically order isomorphic to the hermitian part of the C*-algebra $\overline{A}$ constructed as a monotone closure of

---

1980 Mathematics Subject Classification. 46L05, 46A40, 06F20.

[1]Supported partially by the National Science Foundation.

A in the injective envelope I(A) of A as a C*-algebra where I(A) has the following properties ([1]):

(a) I(A) is an injective C*-algebra (automatically monotone complete),

(b) the identity mapping on I(A) is a unique completely positive linear mapping of I(A) into itself which fixes each element of A.

Thus, we can get that $(\overline{A},i)$ is a regular monotone completion of A (For a separable A, Maitland Wright proved the existence of $\overline{A}$ by a different method).

2. THE UNIQUENESS OF $\overline{A}$. Let (C_1,α_1) and (C_2,α_2) be regular monotone completions of a unital C*-algebra A, then there is a unique *-isomorphism β from C_1 onto C_2 such that the diagram

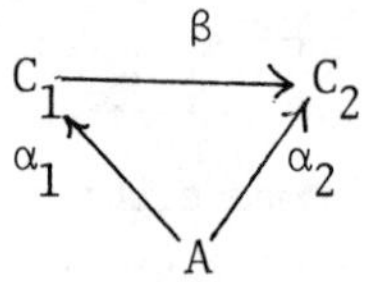

commutes.

Key point of the proof is to construct an order isomorphism from C_1 onto C_2 with $\beta\alpha_1 = \alpha_2$ and by the order density of $\alpha_j(A_h)$ in $(C_j)_h$ (j = 1,2), one can show that β is a *-isomorphism (see also [2] and [6]).

From now on, to simplify the notations, we shall denote the unique regular monotone completion of A by $(\overline{A},i)$. If A is non unital, then we consider $\overline{A_1}$ (where $A_1 = C^*(A,1)$) and we use the notation $\overline{A}$ instead of $\overline{A_1}$ for convenience sake.

3. A STRUCTURE THEORY OF $\overline{A}$. The following theorem plays a fundamental role in our considerations.

THEOREM ([5]). Let A be a unital C*-algebra and I be an ideal of A (throughout this note, the term " ideal " is understood to a closed two-sided ideal). Let $\{e_\alpha\}$ be an increasing approximate for I, then the supremum z of $\{e_\alpha\}$ in $\overline{A}$ is a unique central projection such that there is a *-isomorphism β from $\overline{I}$ onto $\overline{A}z$ such that the diagram

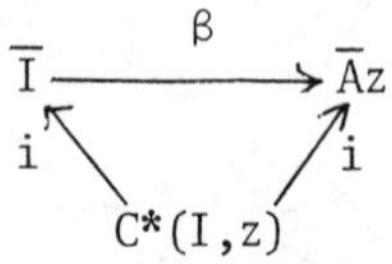

commutes.

COROLLARY (see also [1], [6] and [5]). $\overline{A}$ is a factor if and only if A is prime.

COROLLARY ([5]). Let A be a C*-algebra and M(A) be the multiplier algebra of A, then $\overline{M(A)}$ and $\overline{A}$ are isomorphic and the injective image of M(A) in $\overline{A}$ is the idealizer of A in $\overline{A}$.

THEOREM ([4], [5]). Suppose that A is a C*-algebra, then A is GCR if and only if $\overline{A/I}$ is of type I (as an AW*-algebra [3]) for every ideal I of A. A is NGCR if and only if $\overline{A}$ does not have any type I direct summand. Moreover, if A is separable and NGCR, then $\overline{A}$ has no W*-direct summand and has no non trivial separable representations (in particular, if A is primitive, separable and NGCR, then $\overline{A}$ is a σ-finite, non W*, type III AW*-factor).

Let $C_b(\mathrm{Prim}(A))$ be the C*-algebra of all bounded complex continuous functions on the Primitive ideal space of A, then by Dauns-Hofmann's theorem, the center $Z(M(A))$ of M(A) is *-isomorphic with $C_b(\mathrm{Prim}(A))$ via the canonical *-isomorphism ϕ.

THEOREM ([5]). Suppose that A is GCR and separable, then $Z(\overline{A})$ (the center of $\overline{A}$) is *-isomorphic with the regular monotone completion of $Z(M(A))$, more precisely to say, there is a *-isomorphism $\overline{\phi}$ from $Z(\overline{A})$ onto $\overline{C_b(\mathrm{Prim}(A))}$ such that the diagram

$$\begin{array}{ccc} Z(\overline{A}) & \xrightarrow{\ \overline{\phi}\ } & \overline{C_b(\mathrm{Prim}(A))} \\ i \uparrow & & \uparrow i \\ Z(M(A)) & \xrightarrow{\ \phi\ } & C_b(\mathrm{Prim}(A)) \end{array}$$

commutes.

If A is a C*-subalgebra of a C*-algebra B, then can we conclude that $\overline{A}$ is embedded as a C*-subalgebra of $\overline{B}$? This is false as the following example shows: Let A be a UHF-algebra (infinite dimensional) acting on a separable Hilbert space H such that $A \cap C(H) = \{0\}$ (where $C(H)$ is the C*-algebra of all compact operators on H), then $B = A + C(H)$ is a C*-algebra such that $\overline{B} = B(H)$. If $\overline{A} \subset B(H)$ as a C*-subalgebra, then $\overline{A}$ has a non trivial separable representation and this is a contradiction by the above theorem.

BIBLIOGRAPHY

1. M.Hamana, " Injective envelopes of C*-algebras ", J.Math.Soc.Japan, 31(1979), 181-197.

2. M.Hamana, " Regular embedding of C*-algebras in monotone complete C*-algebras ", To appear in J.Math.Soc.Japan.

3. I.Kaplansky, " Projections in Banach algebras ", Ann. of Math., 53(1951), 235-249.

4. K.Saitô, " AW*-algebras with monotone convergence property and examples by Takenouchi and Dyer ", Tôhoku Math.J., 31(1979), 31-40.

5. K.Saitô, " A structure theory in the regular σ-completion of a C*-algebra " To appear in J.London Math.Soc..

6. J.D.Maitland Wright, " Regular σ-completions of C*-algebras ", J. London Math.Soc., 12(1976), 299-309.

Mathematical Institute
Tôhoku University
Sendai, Japan 980

Proceedings of Symposia in Pure Mathematics
Volume 38 (1982), Part 2

COMMUTING PAIRS OF C*ALGEBRAS IN THE CALKIN ALGEBRA

Kenneth R. Davidson

Let H be a separable Hilbert space, and let π be the canonical quotient map of $B(H)$ onto the Calkin algebra. Suppose A and B are a commuting pair of abelian C*subalgebras of the Calkin algebra. We consider the problem of finding a commuting pair of C*subalgebras $\mathfrak{A}$ and $\mathfrak{B}$ of $B(H)$ such that $\pi\mathfrak{A} = A$ and $\pi\mathfrak{B} = B$. In this note, we restrict our attention to the singly generated case $A = C^*(n)$ and $B = C^*(m)$, where n and m are commuting normal elements of the Calkin algebra.

It is possible that $\mathfrak{A}$ and $\mathfrak{B}$ may contain non-zero compact operators. For example, if U is the bilateral shift on $L^2(T)$ and P is the orthogonal projection onto $H^2(T)$, $A = C^*(\pi U)$ and $B = C^*(\pi P)$, then A and B cannot be lifted to a commuting pair of abelian C*algebras. However, let $U_1 = U-(I-P)UP$. Then $C^*(U_1)$ and $C^*(P)$ commute, and $C^*(U_1)$ contains all compact operators which commute with P.

The $C^*(n,m)$ is isomorphic to $C(X)$ where X is the joint spectrum $\sigma(n,m)$ in C^2. This determines an element $\tau(n,m)$ in $\mathrm{Ext}(X)$.[1]

<u>Proposition 1</u>. The lifting problem depends only on $\tau(n,m)$. The set $L(X)$ of τ in $\mathrm{Ext}(X)$ which lift is a semigroup. □

Let p_1 and p_2 be the coordinate projections in C^2. Let $X_1 = p_1X = \sigma(n)$ and $X_2 = p_2(X) = \sigma(m)$.

<u>Theorem 2</u>. If $X = X_1 \times X_2 \subseteq C^2$ and τ belongs to $\mathrm{Ext}(X)$, then τ belongs to $L(X)$ if and only if (§): For each λ in $C\backslash X_1$, there is a finite set $\{b_1,\ldots,b_n\}$ in X_2 such that for every idempotent p in $C(X_2)$ with $p(b_i) = 0$, $i = 1,\ldots,n$, the Fredholm index $\mathrm{ind}\,\tau((z_1-\lambda)p(z_2)+1-p(z_2)) = 0$; and the corresponding condition with X_1 and X_2 interchanged holds. □

Suppose X is a subset of $T \times R$ corresponding to a unitary u and

0082-0717/80/0000-0519/$01.50

self adjoint element a. Let $B = \{b \in R: T\times\{b\} \subseteq X\}$ and $Y = T \times B$. Let $i: Y \longrightarrow X$ be the inclusion map.

Theorem 3. $L(X) = i_*L(Y)$.

Theorem 4. Suppose $X_2 = p_2X$ is totally disconnected, then τ in $\mathrm{Ext}(X)$ belongs to $L(X)$ if and only if (§): For every x in X_2 and λ in C such that (λ,x) is not in X, there is a clopen neighbourhood U of x_2 such that every idempotent p in $C(X_2)$ with support $(p) \subseteq U\backslash\{x\}$, the Fredholm index $\mathrm{ind}\,\tau((z_1-\lambda)p(z_2)+1-p(z_2)) = 0$. □

If $X \subseteq C^2$, $\mathrm{Ext}(X) \cong sH_1(X) \oplus sH_3(X)$ where SH_i is reduced Steenrod homology.[2]

Theorem 5. $L(X) \subseteq sH_1(X)$. □

Corollary 6. If for every x in X_1, $\mathrm{Ext}(p_1^{-1}(2) \cap X) = 0$ and for every x in X_2, $\mathrm{Ext}(p_2^{-1}(2) \cap X) = 0$, then $L(X) = 0$. □

[1] L.G. Brown, R.G. Douglas, and P.A. Fillmore, Extensions of C*algebras, operators with compact self commutators, and K-homology, B.A.M.S. 79 (1973), 973-978.

[2] J. Kaminker and C. Schochet, K-theory and Steenrod homology: applications to Brown-Douglas-Fillmore theory of operator algebras, T.A.M.S. 227 (1977), 63-107.

University of Waterloo
Waterloo, Ontario
Canada N2L 3G1

Proceedings of Symposia in Pure Mathematics
Volume 38 (1982), Part 2

C^*-ALGEBRAS OF MULTIVARIABLE WIENER-HOPF OPERATORS

Paul S. Muhly[1]
and
Jean N. Renault

ABSTRACT. The C^*-algebra $\mathfrak{W}$ generated by the Wiener-Hopf operators over a cone in $\mathbb{R}^n$ is shown to be the image of a groupoid C^*-algebra under a faithful representation. When the cone is either polyhedral or homogeneous and self-dual, the representation may be used to exhibit a composition series for $\mathfrak{W}$ with very explicit subquotients. This yields, in particular, the spectrum of $\mathfrak{W}$ and the topology on it.

Let P be a closed cone in $\mathbb{R}^n$ and assume that P is the closure of its interior. For $f \in L^1(\mathbb{R}^n)$ define $W(f)$ on $L^2(P)$ by the formula $W(f)\xi(x) = \int_P f(x-t)\xi(t)dt$ and let $\mathfrak{W}$ be the C^*-algebra generated by the $W(f)$ as f runs over $L^1(\mathbb{R}^n)$. We use the theory of groupoid C^*-algebras [5] to describe $\mathfrak{W}$. The groupoid $\mathfrak{G}$ associated with P is obtained as follows. Let Y be the maximal ideal space of the C^*-algebra generated by the functions $f * 1_P$ where f runs over $L^1(\mathbb{R}^n)$ and for $(y,t) \in Y \times \mathbb{R}^n$ we define $y+t \in Y$ by the formula $(f * 1_P)^\wedge(y+t) = (f_t * 1_P)^\wedge(y)$ where the hat denotes the Gelfand transform and $f_t(s) = f(s-t)$. Then the map $(y,t) \longrightarrow y+t$ converts $(Y,\mathbb{R}^n)$ into a locally compact transformation group. Let X denote the closure of $\{y_0 + t \mid t \in P\}$ in Y where y_0 corresponds to evaluation at 0 in $\mathbb{R}^n$. Then X is compact in Y and our groupoid is the reduction of the transformation group groupoid $(Y,\mathbb{R}^n)$ to X. Thus as a set $\mathfrak{G} = \{(x,t) \in X \times \mathbb{R}^n \mid x+t \in X\}$. The unit space of $\mathfrak{G}$ may be identified with X and $\mathfrak{G}$ has a natural left Haar system $\{\lambda^x\}_{x\in X}$ where $\lambda^x = \delta_x \times \lambda$ and λ is Lebesgue measure on $\mathbb{R}^n$. We denote the C^*-algebra of this groupoid (with Haar system) by $C^*(\mathfrak{G},\lambda)$.

THEOREM 1. For f in $C_c(\mathfrak{G})$ define $\widetilde{W}(f)$ on $L^2(P)$ by the formula $\widetilde{W}(f)\xi(t) = \int f(y_0+t,s)\xi(t+s)1_P(t+s)d\lambda(s)$, $\xi \in L^2(P)$. Then $\widetilde{W}$ extends to a C^*-isomorphism from $C^*(\mathfrak{G},\lambda)$ onto $\mathfrak{W}$.

1980 Mathematics Subject Classification. 46L99.
[1]Supported in part by the National Science Foundation.

The virtue of this theorem is that it allows us to apply the orbital analysis of $\mathfrak{G}$, as in [5], to identify the ideal structure of $\mathfrak{W}$. In general, $\mathfrak{G}$ may be very complicated. However, when P is either polyhedral or homogeneous and self-dual, $\mathfrak{G}$ is fairly tractable. We concentrate on the latter type of cone here. It is well known [4] that such a cone may be represented as the set of squares in a formally real, finite dimensional, Jordan algebra $\mathfrak{A}$. We assume for the sake of discussion that $\mathfrak{A}$ is simple. We denote the idempotents in $\mathfrak{A}$ by $\mathbb{I}$ and the quadratic representation of $\mathfrak{A}$ by U. If $e \in \mathbb{I}$, $e^{\perp}$ stands for $\mathbb{1} - e$ where $\mathbb{1}$ is the identity. It turns out that as a set Y may be taken to be $\{(e,a) \in \mathbb{I} \times \mathfrak{A} \mid a \in U(e^{\perp})\mathfrak{A}\}$ and that $\mathfrak{A}$ acts on Y according to the formula $(e,a) + b = (e, a + U(e^{\perp})b)$. The space X, then, is simply $\{(e,a) \in Y \mid a \in P\}$. Thus, the orbit structure on X is quite clear. Unfortunately, the topology on Y is a bit difficult to describe. Let $\mathbb{I}_k$ denote the idempotents in $\mathfrak{A}$ of degree k, i.e., $e \in \mathbb{I}_k$ if and only if e is the sum of k orthogonal minimal idempotents, and let $Z_k = \{(e,c) \in \mathbb{I}_k \times \mathfrak{A} \mid U(e^{\perp})c = 0\}$. The assumption that $\mathfrak{A}$ is simple implies that $\mathbb{I}_k$ is a homogeneous space and so carries a natural topology; Z_k is a vector bundle over $\mathbb{I}_k$ with the obvious topology.

THEOREM 2. Suppose that $\mathfrak{A}$ is of degree n (i.e., suppose that $\mathbb{1}$ has degree n), then $\mathfrak{W}$ admits a composition series

$$\{0\} = I_{-1} \subseteq I_0 \subseteq I_1 \subseteq \cdots \subseteq I_n = \mathfrak{W}$$

such that for $k = 0,1,\cdots,n-1$, I_k/I_{k-1} is isomorphic to $C_0(Z_k) \otimes \mathcal{K}$ where $\mathcal{K}$ is the C^*-algebra of compact operators on an infinite dimensional separable Hilbert space, and I_n/I_{n-1} is isomorphic to $C_0(\mathfrak{A})$.

COROLLARIES. i) $\mathfrak{W}$ is postliminal and solvable in the sense of Dynin [3].

ii) I_0 is the algebra of compact operators on $L^2(P)$.

iii) I_{n-1} is the commutator ideal of $\mathfrak{W}$.

iv) As a set, the spectrum of $\mathfrak{W}$ is $\bigcup_{k=0}^{n} Z_k$ and a basis for the topology on the spectrum is given by the sets

$$\mathcal{U} \cup Z_{k-1} \cup Z_{k-2} \cup \cdots \cup Z_0$$

where $\mathcal{U}$ is open in Z_k, $k = 0,1,\cdots,n$.

When P is the forward light cone in $\mathbb{R}^n$, i.e., when $P = \{(x_1,\cdots,x_n) \mid x_n \geq 0,\ x_n^2 \geq \sum_{k=1}^{n-1} x_k^2\}$, then using the Fourier and Cayley transforms it is possible to see that these results complement the recent work of Berger, Coburn, and Koranyi [1]. A similar analysis of $\mathfrak{W}$ may be made when P is polyhedral. The results of such an analysis generalize some of the work of Douglas and Howe [2] and of Dynin [3].

BIBLIOGRAPHY

1. C. Berger, L. Coburn, and A. Koranyi, Opérateurs de Wiener-Hopf sur les spheres de Lie, C. R. Acad. Sci. Paris 290(1980), Série A, 989-991.

2. R.G. Douglas and R. Howe, On the C^*-algebra of Toeplitz operators on the quarter plane, Trans. Amer. Math. Soc. 158(1971), 203-217.

3. A. Dynin, Inversion problem for singular integral operators: C^*-approach, Proc. Nat. Acad. Sci. 75(1978), 4668-4670.

4. M. Koecher, Jordan algebras and their applications, mimeographed notes, University of Minnesota, 1962.

5. J. Renault, A Groupoid Approach to C^*-algebras, Lecture Notes in Mathematics, 793, Springer-Verlag, New York, 1980.

THE UNIVERSITY OF IOWA
IOWA CITY, IOWA 52242

TULANE UNIVERSITY
NEW ORLEANS, LOUISIANA 07118

Proceedings of Symposia in Pure Mathematics
Volume 38 (1982), Part 2

THE CONCEPT OF BIBOUNDED OPERATORS ON W^*-ALGEBRAS

Burkhard Kümmerer

ABSTRACT. Given a probability measure space (X,Σ,μ) then we have the canonical embeddings

$$L^\infty(X,\Sigma,\mu) \underset{j_1}{\hookrightarrow} L^2(X,\Sigma,\mu) \underset{j_2}{\hookrightarrow} L^1(X,\Sigma,\mu)$$

where j_2 is the preadjoint of j_1. We study a non-commutative generalization of this situation replacing $L^\infty(X,\Sigma,\mu)$ by a σ-finite W^*-algebra $\mathfrak{A}$, $L^1(X,\Sigma,\mu)$ by its predual $\mathfrak{A}_*$, and $L^2(X,\Sigma,\mu)$ by the standard Hilbert space associated to $\mathfrak{A}$. The injections depend on the choice of a faithful normal state φ on $\mathfrak{A}$. We call a weak* continuous bounded linear operator T on $\mathfrak{A}$ φ-bibounded if its preadjoint T_* leaves $j_2 \circ j_1(\mathfrak{A})$ invariant in $\mathfrak{A}_*$. For these operators we prove an analogue of the little Riesz convexity theorem and give some applications to spectral and ergodic theory.

This note is based on Groh-Kümmerer [3] to which we refer for details and proofs. The commutative situation is treated in Schaefer [5].

1. THE CANONICAL INJECTIONS. We consider a σ-finite W^*-algebra $\mathfrak{A}$ with fixed faithful normal state φ. By the work of Araki [1] and Connes [2] we associate with $\mathfrak{A}$ the standard Hilbert space $(\mathcal{H},\mathcal{P},J)$, where the Hilbert space $\mathcal{H}$ is ordered by the self dual cone $\mathcal{P}$ and J is the modular involution. We regard $\mathcal{H}$ as the Hilbert space arising from the GNS construction for $\mathfrak{A}$ with respect to φ with cyclic separating vector $\xi_\varphi \in \mathcal{P}$. The corresponding modular operator is denoted by Δ_φ. In an arbitrary ordered vector space $(E,\leq)$ we define for $x,y \in E$, $x \leq y$ the order intervall $[x,y] := \{z \in E : x \leq z \leq y\}$.

We define the map $j_1 : \mathfrak{A} \longrightarrow \mathcal{H} : x \longmapsto \Delta_\varphi^{1/4} x \xi_\varphi$ and list some properties:

1980 Mathematics Subject Classification. 46 L 10, 47 B 55.

0082-0717/80/0000-0542/$02.00

PROPOSITION. 1. j_1 is a norm contractive injection with dense image.
2. j_1 is weak*-weak continuous and we can define its preadjoint $j_2 := (j_1)_*: \mathcal{H} \longrightarrow \mathfrak{A}_*$ and $j := j_2 \circ j_1 : \mathfrak{A} \longrightarrow \mathfrak{A}_*$. j_2 and j_1 are norm contractive injections.
3. The restriction of j_1 (j_2) to the order intervall $[-\mathbb{1}, \mathbb{1}]$ $([-\xi_\varphi, \xi_\varphi])$ is an order isomorphism onto the order intervall $[-\xi_\varphi, \xi_\varphi]$ $([-\varphi, \varphi])$.
4. The restriction of j_2 to the order intervall $[-\xi_\varphi, \xi_\varphi]$ is a homeomorphism for the weak (norm) topologies on $\mathcal{H}$ and $\mathfrak{A}_*$.

2. φ-BIBOUNDED OPERATORS.

DEFINITION. Let $\mathcal{L}(E)$ denote the algebra of all bounded linear operators on a Banach space E. We call $T \in \mathcal{L}(\mathfrak{A})$ <u>φ-bibounded</u> if T is weak* continuous and its preadjoint T_* satisfies $T_*(j(\mathfrak{A})) \subseteq j(\mathfrak{A})$.

EXAMPLES. 1. If $T \in \mathcal{L}(\mathfrak{A})$ is weak* continuous and if $T_*([-\varphi, \varphi]) \subseteq [-\varphi, \varphi]$, then T is φ-bibounded.
2. Let $\mathfrak{A}_\varphi := \{x \in \mathfrak{A} : \varphi(xy) = \varphi(yx) \text{ for all } y \in \mathfrak{A}\}$ be the centralizer of $\mathfrak{A}$ with respect to φ. Then for $a \in \mathfrak{A}_\varphi$ the operators L_a and R_a defined by $L_a: x \longmapsto ax$ and $R_a: x \longmapsto xa$ $(x \in \mathfrak{A})$ are φ-bibounded.

We list some properties of φ-bibounded operators:

PROPOSITION. 1. If T is φ-bibounded then $j^{-1} \circ T_*^* \circ j$ defines a φ-bibounded operator $T^+ \in \mathcal{L}(\mathfrak{A})$.
2. The set $\mathcal{L}_{bi}(\mathfrak{A}) := \{T \in \mathcal{L}(\mathfrak{A}) : T \text{ is } \varphi\text{-bibounded}\}$ endowed with the norm $\|T\|_{bi} := \max\{\|T\|, \|T^+\|\}$ is a Banach algebra with involution $T \longmapsto T^+$.
3. If T is φ-bibounded then the map $j_1(x) \longmapsto j_1(T(x))$ $(x \in \mathfrak{A})$ extends to a bounded operator $T_\varphi \in \mathcal{L}(\mathcal{H})$. (This may be considered as a noncommutative version of the little Riesz convexity theorem.)

3. APPLICATIONS TO SPECTRAL THEORY.

THEOREM. Suppose $T \in \mathcal{L}_{bi}(\mathfrak{A})$ then
1. $\sigma(T_\varphi) \subseteq \sigma(T) \cup \overline{\sigma(T^+)} = \sigma_{bi}(T)$ where $\sigma_{bi}(T)$ denotes the spectrum of T in $\mathcal{L}_{bi}(\mathfrak{A})$.

2. If $\sigma(T)$ and $\sigma(T^+)$ have topological dimension zero then $\sigma(T_\varphi) = \sigma(T) = \overline{\sigma(T^+)} = \sigma_{bi}(T)$.

COROLLARY. Consider $T \in \mathcal{L}_{bi}(\mathfrak{A})$ such that $T \cdot T^+ = T^+ \cdot T$.
1. If $\|T\| = 1$, $T(\mathbb{1}) = \mathbb{1}$ and $\sigma(T) \subseteq \Gamma$, $\sigma(T^+) \subseteq \Gamma$, Γ being the unit circle, then T is a Jordan *-isomorphism.
2. If $\sigma(T) = \{1\}$ then $T = T^+ = id_{\mathfrak{A}}$.

4. APPLICATIONS TO ERGODIC THEORY.

DEFINITION. 1. A semigroup $S \subseteq \mathcal{L}(E)$, E any Banach space, is called <u>mean ergodic</u> if there exists a projection $P \in \mathcal{L}(E)$ with $P \cdot T = T \cdot P = P$ for all $T \in S$ and $P \in \overline{coS}$, $\overline{coS}$ being the closure of the convex hull coS of S in the strong operator topology. P is uniquely determined and projects onto the fixed space of S . We call P the <u>mean ergodic projection</u> corresponding to S .
2. A semigroup $S \subseteq \mathcal{L}(\mathfrak{A})$ is called <u>weak* mean ergodic</u> if S consists of weak* continuous operators and if the preadjoint semigroup $S_* := \{T_* : T \in S\}$ is mean ergodic in $\mathcal{L}(\mathfrak{A}_*)$. In this case we denote by P_* the corresponding mean ergodic projection in $\mathcal{L}(\mathfrak{A}_*)$.
3. $T \in \mathcal{L}_{bi}(\mathfrak{A})$ is called a <u>φ-bicontraction</u> if $\|T\|_{bi} \leq 1$.

THEOREM. 1. Any semigroup $S \subseteq \mathcal{L}(\mathfrak{A})$ consisting of φ-bicontractions is weak* mean ergodic.
2. If $S \subseteq \mathcal{L}(\mathfrak{A})$ is an amenable semigroup of φ-bibounded operators such that S and $S^+ := \{T^+ : T \in S\}$ are bounded then S is weak* mean ergodic.

For more information about weak* mean ergodic semigroups on W^*-algebras we refer to Kümmerer-Nagel [4] .

Let $T \in \mathcal{L}(\mathfrak{A})$ be a φ-bicontraction. By the above theorem the semigroup $\{T_*^n : n \in \mathbb{N}\}$ is mean ergodic with mean ergodic projection $P_* = \lim_n \frac{1}{N}\sum_{i=0}^{N-1} T_*^i$ and we can prove the following noncommutative version of a theorem of Akcoglu-Sucheston:

THEOREM. For a φ-bicontraction T the following are equivalent:
1. $\lim_n T_*^n = P_*$ in the weak operator topology on $\mathcal{L}(\mathfrak{A}_*)$.
2. $\lim_M \frac{1}{M}\sum_{i=0}^{M} T_*^{n_i} = P_*$ in the strong operator topology for each infinite subsequence $(n_i)_{i\in\mathbb{N}}$ of $\mathbb{N}$.

BIBLIOGRAPHY

1. H. Araki, "Some properties of the modular conjugation operator of von Neumann algebras and a non-commutative Radon-Nikodym theorem with a chain rule", Pacific J. Math. 50 (1974), 309-354.

2. A. Connes, "Charactérisation des espaces vectoriels ordonnés sous-jacent aux algèbres de von Neumann", Ann. Inst. Fourier, Grenoble 24 (1974), 121-155.

3. U. Groh, B. Kümmerer, "Bibounded Operators on W^*-algebras", Preprint, Tübingen 1980 .

4. B. Kümmerer, R. Nagel, "Mean ergodic semigroups on W^*-algebras", Acta Sci. Math. Szeged 41 (1979), 151-159 .

5. H.H. Schaefer, Banach Lattices and Positive Operators, Springer Verlag, Berlin - Heidelberg- New York, 1974.

MATHEMATISCHE FAKULTÄT
UNIVERSITÄT TÜBINGEN
AUF DER MORGENSTELLE 10
7400 TÜBINGEN
GERMANY

Proceedings of Symposia in Pure Mathematics
Volume 38 (1982), Part 2

Contractive Projections on C^*-algebras

Y. Friedman and B. Russo

Let A be a C^*-algebra and let P be a contractive projection on A , i.e., a linear map $P:A \to A$ with $P^2 = P$ and $\|P\| = 1$. The range $B = P(A)$ of P is then a closed linear subspace of A .

In the case when the range $B = P(A)$ of P is assumed to be a C^*-subalgebra of A , it is known that P is positive and satisfies the conditional expectation property (Tomiyama [7]). In the other direction, if P is assumed completely positive and unital (therefore contractive) then its range B has the structure of a C^*-algebra (though not necessarily a C^*-subalgebra of A , Choi-Effros [3]). This result has been extended as follows: if A is a JC algebra and P is a positive unital projection on A , then P(A) is also a JC algebra (not necessarily a JC-subalgebra of A , Effros-Størmer [4]).

Our work is concerned with describing the algebraic properties of an arbitrary contractive projection P and its range P(A) . The first phase of this project is the memoir [1] of Arazy and Friedman in which a complete description is given of all contractive projections (and their ranges) on the algebra C_∞ of all compact operators on a separable complex Hilbert space. Their results show that the range of a contractive projection on C_∞ is a direct sum of Jordan algebras and Lie algebras of operators. Since the work of Effros and Størmer explains the appearance of Jordan algebras, it remained to explain, in the setting of a general C^*-algebra, the appearance of the Lie algebras in the work of Arazy and Friedman.

The evidence we have obtained so far indicates that the range P(A) should have an algebraic structure which includes, to some extent, that of Jordan algebras and Lie algebras, e.g., Jordan triple system (Jacobson [6]). This has been confirmed in the case of a commutative C^*-algebra in our paper [5], the

0082-0717/80/0000-0525/$02.00

results of which will be described below.

While the role of C^*-algebras and Jordan algebras in mathematical physics (quantum statistical mechanics) is well known, it is only recently that ternary algebras, in particular the type that occur as the range of a contractive projection on a C^*-algebra, have been used to construct Lie algebras and Lie superalgebras (Bars-Günaydin [2]). The latter play important roles in mathematical physics, e.g., in gauge theories of elementary particles. The fact that ternary algebras appear naturally as the range of contractive projections suggests a connection between two different models of mathematical physics.

Our first theorem is the basic structure theorem for contractive projections on commutative C^*-algebras. To facilitate its statement we need some notation. If $\{\mu_i\}_{i\in I}$ is a family of Borel measures on a locally compact space K with polar decompositions $\mu_i = \varphi_i \cdot |\mu_i|$ and mutually disjoint supports, then for $f \in C_0(K)$ we shall let Qf be the function which is equal to $\langle f,\mu_i\rangle\bar{\varphi}_i$ on $\operatorname{supp}\mu_i$. A priori Qf is just a numerical function on $S \equiv \bigcup_{i\in I} \operatorname{supp}\mu_i$ which is defined just locally almost everywhere.

THEOREM 1. <u>A linear map</u> $P:C_0(K) \to C_0(K)$ <u>is a contractive projection if and only if there exist norm one Borel measures</u> $\{\mu_i\}_{i\in I}$ <u>with mutually disjoint supports such that</u> Qf <u>is continuous on</u> S <u>and</u> $P = EQ$ <u>where</u> E <u>is a linear isometric operator from</u> $Q(C_0(K)$ <u>to</u> $C_0(K)$ <u>such that</u> Eg <u>extends</u> g <u>from</u> S <u>to</u> K <u>for all</u> $g \in Q(C_0(K))$.

We call $\{\mu_i\}_{i\in I}$ the <u>atoms</u> of P and $Q:C_0(K) \to C_b(S)$ the <u>essential part</u> of P. Theorem 1 has many applications due to the simple form of Q and that fact that $\bar{S}$ is a (Shilov) boundary for the range of P. For example it implies that a bicontractive projection (i.e., $\|I-P\| = 1$ also) on $C_0(K)$ has the form $Pf = \frac{1}{2}(f+\lambda\cdot f\circ\sigma)$ where σ is an involutive homeomorphism of K and $\lambda:K \to \mathbf{C}$ satisfies $\lambda\circ\sigma = \bar{\lambda}$, $|\lambda| = 1$. Another consequence is that the only contractive projection on $C[0,1]$ fixing the three functions $1,x,x^2$ is the identity.

In order to state Theorem 2 we need a definition. A C^*-<u>ternary algebra</u> is a Banach space X with a tri-linear map $[\cdot,\cdot,\cdot]:X\text{x}X\text{x}X \to X$ satisfying $[[a,b,c],d,e] = [a,[d,c,b],e] = [a,b,[c,d,e]]$, $\|[a,b,c]\| \le \|a\|\|b\|\|c\|$, $\|[a,a,a]\| = \|a\|^3$. A Gelfand Naimark type representation theorem holds for such objects (Zettl [8]).

THEOREM 2. <u>Let</u> P <u>be a contractive projection on</u> $C_0(K)$. <u>Then the range</u> $P(A)$ <u>is a</u> C^*-<u>ternary algebra under the triple product</u> $(f,g,h) \to [f,g,h] \equiv P(f\bar{g}h)$.

COROLLARY. $P(A)|\bar{S}$ <u>is a ternary subalgebra of</u> $C_0(\bar{S})$.

Here are two examples illustrating Theorems 1 and 2.

<u>Example 1</u>. On $C[0,1]$ let $Pf(t) = f(0)(1-t) + f(1)t$. The range of P consists of the linear functions, the atoms are δ_0, δ_1 and $Qf = f(0)\chi_{\{0\}} + f(1)\chi_{\{1\}}$.

<u>Example 2</u>. On $C_0(\mathbb{R})$ let $Pf(x) = \frac{1}{2}(f(x) - f(-x))$. The range of P consists of the odd functions in $C_0(\mathbb{R})$, the atoms are $\frac{1}{2}(\delta_t - \delta_{-t})$ for $t \in (0,\infty)$ and $Qf = Pf|(\mathbb{R}-\{0\})$.

Here are two non-commutative examples.

<u>Example 3</u>. On $M_n(\mathbb{R})$ let $Px = \frac{1}{2}(x-x^T)$. The range of P consists of the real anti-symmetric matrices so is closed under $(x,y,z) \to xy^Tz + zy^Tx$.

<u>Example 4</u>. On $M_4(\mathbb{C})$ let

$$s_1 = \left(\begin{array}{cc|c} \begin{matrix} 0 & 1 \\ -1 & 0 \end{matrix} & & O \\ \hline & O & O \end{array}\right) \quad s_2 = \left(\begin{array}{c|c} O & \begin{matrix} 1 & 0 \\ 0 & 0 \end{matrix} \\ \hline \begin{matrix} -1 & 0 \\ 0 & 0 \end{matrix} & O \end{array}\right) \quad s_3 = \left(\begin{array}{c|c} O & \begin{matrix} 0 & 1 \\ 0 & 0 \end{matrix} \\ \hline \begin{matrix} 0 & 0 \\ -1 & 0 \end{matrix} & O \end{array}\right)$$

$$t_1 = \left(\begin{array}{c|c} O & O \\ \hline O & \begin{matrix} 0 & 1 \\ -1 & 0 \end{matrix} \end{array}\right) \quad t_2 = \left(\begin{array}{c|c} O & \begin{matrix} 0 & 0 \\ 0 & -1 \end{matrix} \\ \hline \begin{matrix} 0 & 0 \\ 0 & 1 \end{matrix} & O \end{array}\right) \quad t_3 = \left(\begin{array}{c|c} O & \begin{matrix} 0 & 0 \\ 1 & 0 \end{matrix} \\ \hline \begin{matrix} 0 & -1 \\ 0 & 0 \end{matrix} & O \end{array}\right)$$

and for $(a_{ij}) \in A$ let

$$P(a_{ij}) = \frac{a_{12}-a_{21}}{2}s_1 + \frac{a_{34}-a_{43}}{2}t_1 + \frac{a_{13}-a_{31}}{2}s_2$$

$$+ \frac{a_{42}-a_{24}}{2}t_2 + \frac{a_{14}-a_{41}+a_{23}-a_{32}}{4}(s_3+t_3) .$$

Then the range of P is closed under the operation $(a,b,c) \to ab^*c + cb^*a$.

Examples 3 and 4 and Theorem 2 suggest a strong connection between contractive projections on C^*-algebras and Jordan triple systems.

REFERENCES

1. J. Arazy and Y. Friedman, Contractive projections in C_1 and C_∞ , Memoirs A.M.S., 13 (1978), No. 200.

2. I. Bars and M. Gunaydin, Construction of Lie algebras

and Lie superalgebras from ternary algebras, J. Math. Phys., 20 (1979), 1977-1993.

3. M. D. Choi and E. Effros, Injectivity and operator spaces, J. Funcl. Anal., 24 (1974), 156-209.

4. E. Effros and E. Størmer, Positive projections and Jordan structure in operator algebras, Math. Scand., 45 (1979), 127-138.

5. Y. Friedman and B. Russo, Contractive projections on $C_0(K)$, to appear

6. N. Jacobson, Structure and representations of Jordan algebras, A.M.S. Collog. Publications, Providence, 1968.

7. Y. Tomiyama, On the projection of norm one in W^*-algebras, Proc. Japan Acad., 33 (1957), 608-612.

8. H. Zettl, Charakterisierung ternärer Operatorenringe, doctorral dissertation, University Erlangen-Nürnberg, 1979.

University of California
Irvine,CA 92717

Proceedings of Symposia in Pure Mathematics
Volume **38** (1982), Part 2

PROBLEMS ON JOINT QUASITRIANGULARITY FOR N-TUPLES OF ESSENTIALLY COMMUTING, ESSENTIALLY NORMAL OPERATORS

Norberto Salinas

An n-tuple $\underline{t} = (T_1,..,T_n)$ of operators on a (separable, infinite dimensional) Hilbert space $\mathcal{H}$ is said to be jointly quasitriangular, i.e. $\underline{t} \in QT_n$, if there exists an increasing $\{P_m\}$ of finite rank projections on $\mathcal{H}$ such that $P_m \underset{s}{\rightarrow} 1$ and $\lim_{m \to \infty} \|(1-P_m)T_k P_m\| = 0$, $1 \leq k \leq n$. (cf. [7] and [12]).

The following theorem is the celebrated spectral chacterization of single quasitriangular operators obtained in [1, Theorem 5.4].

THEOREM. A.F.V. . An operator T on $\mathcal{H}$ is non-quasitriangular if and only if there exists λ in $\mathbb{C}$ such that $T-\lambda$ is semi-Fredholm and $\operatorname{ind}(T-\lambda) < 0$.

REMARK. The if part of the above result was first obtained in [6,Theorem 2.4]. On the other hand, if T is essentially normal (i.e. T^*T-TT^* is compact), then the only if part follows from the celebrated result of [3,Theorem 11.1] which states that the essential spectrum and the indices outside the essential spectrum constitute a complete set of unitary invariants, up to compact perturbations, for the class of essentially normal operators. Thus, given an essentially normal operator T on $\mathcal{H}$, with non negative indices outside its essential spectrum $\sigma_e(T)$, one first construct a quasitriangular essentially normal operator S such that $\sigma_e(S) = \sigma_e(T)$, and $\operatorname{ind}(S-\lambda) = \operatorname{ind}(T-\lambda)$, for every $\lambda \notin \sigma_e(S)$, and one then applies the above mentioned result of [3].

The canonical example of a non-quasitriangular essentially normal operator is the unilateral shift. The following result shows that for n-tuples $\underline{t} = (T_1,..,T_n)$ of essentially normal operators that essentially commute (i.e. $T_jT_k - T_kT_j$ is compact, for $1 \leq j , k \leq n$) the situation is quite different.

THEOREM 1. Let $\underline{t}$ be an n-tuple of essentially commuting essentially normal operators with $n > 1$, and let $X = \sigma_e(\underline{t})$. If X is connected and there exists a contractible compact $Y \subseteq \mathbb{C}^n$ such that $\partial Y \subseteq X \subseteq Y$, then $\underline{t} \in QT_n$.

AMS(MOS) Subject Classification (1980): 46L05,47A66,47B35,47C15,47D25.

0082-0717/80/0000-0570/$01.75

The main ingredients in the proof of the above theorem is Hartog's theorem (a classical result in the theory of functions of several complex variables) and a modification of an argument due to Voiculescu [12].

The following results indicates that for $n > 1$ the essential spectrum $\sigma_e(\underline{t})$ and the indices outside $\sigma_e(\underline{t})$ are not enough to characterize joint-quasitriangular n-tuples of essentially commuting essentially normal operators.

THEOREM 2. Let $\Omega \subseteq \mathbb{C}^n$, $n > 1$ be an open contractible strongly pseudo-convex domain with smooth boundary and $\underline{t}_\Omega = (T_{z_1}, \ldots, T_{z_n})$ be the n-tuple of Toeplitz operators associated with the coordinate maps on Ω [11]. Then for every $\phi : \{1,\ldots,n\} \to \{1,*\}$, we have

$$\underline{t}_\Omega^\phi = T_{z_1}^{\phi(1)}, \ldots, T_{z_n}^{\phi(n)} \in QT_n \quad .$$

In particular, for $n = 2$ we have: if $0 \in \Omega$, then $\underline{t}_\Omega^\phi$ is a Fredholm pair and there exist ϕ' and ϕ'' such that

$$\operatorname{ind} \underline{t}^{\phi'} = \operatorname{ind} \begin{pmatrix} T_{z_1}^{\phi'(1)} & T_{z_2}^{\phi'(2)} \\ -T_{z_2}^{*\phi'(2)} & T_{z_1}^{*\phi'(1)} \end{pmatrix} < 0 ,$$

and $\operatorname{ind} \underline{t}^{\phi''} > 0$ (see [5] for the definition of Fredholm pairs and their indices).

We are indebted to Ron Douglas who explained to us how extension theory for C*-algebras could be used to attack the classification problem for jointly quasitriangular n-tuples of essentially commuting, essentially normal operators.

DEFINITION. Let $X \subseteq \mathbb{C}^n$ be compact, $\tau : C(X) \to \mathcal{Q}(\mathcal{H})$ be a unital *-monomorphism. We say that τ is quasitriangular, i.e. $[\tau] \in \operatorname{Ext}_{qt}(X)$, if there exists $\underline{t} \in QT_n$ such that $\tau z_k = \pi T_k$, $1 \leq k \leq n$, where $\pi : \mathcal{L}(\mathcal{H}) \to \mathcal{Q}(\mathcal{H})$ is the Calkin map, and $z_k : X \to \mathbb{C}$ is the k-th coordinate map.

REMARK. (a) It can be seen that $[\tau] \in \operatorname{Ext}_{qt}(X)$ if and only if $\pi^{-1} \tau P(X)$ is a quasitriangular algebra [2,§2], where $P(X)$ is the closure in $C(X)$ of the analytic polynomials.

(b) We recall [4] that there exists a natural map $\gamma_\infty : \operatorname{Ext}(X) \to \operatorname{Hom}(K^1(X),Z)$ induced by Index Theory, i.e. if (f_{ij}) is an invertible matrix in $C(X) \otimes M_m$, then $(\gamma_\infty[\tau])[(f_{ij})] = \operatorname{ind}((\tau f_{ij}))$.

DEFINITION. We define $K_p^1(X)$ to be the semigroup of equivalence classes in $K^1(X)$ containing elements of $P(X) \otimes M_m$ for some $m = 1,2,\ldots$.

THEOREM 3. If $[\tau] \in \operatorname{Ext}_{qt}(X)$, then $\gamma_\infty[\tau]$ is non-negative on $K_p^1(X)$.

QUESTIONS. (I) When does the converse of Theorem 3 hold? (II) When is $\operatorname{Ext}(X)$ an ordered group with $\operatorname{Ext}_{qt}(X)$its positive cone? (III) When is $K^1(X)$ an ordered group with $K_p^1(X)$ its positive cone?

REMARK. (a) Partial results concerning an affirmative answer of question (I) were obtained in [8] and [10]. (b) Given a unital separable C*-algebra A and a normed closed subalgebra B of A one can define $Ext_{qt}(A;B)$ and $K_1(A;B)$ in an analogous way to the definition of $Ext_{qt}(X)$ and $K_p^1(X)$. If B is a C*-subalgebra of A, then $[\tau] \varepsilon Ext_{qt}(A;B)$ if and only if $\pi^{-1}\tau B$ is quasidiagonal [9]. If A is quasidiagonal, then $Ext_{qt}(A;B)$ is a closed subsemigroup of $Ext(A)$ containing the neutral element, and a result similar to Theorem 3 holds. One can ask questions similar to (I), (II) and (III) in this more general context. (c) A more detailed exposition of the results presented in this note will appear in [10].

BIBLIOGRAPHY

1. C. Apostol, C. Foias and D. Voiculescu, "Some results on non-quasi-triangular operators, IV", Rev. Roum. Math. Pures Appl., 18(1973), 1473-1494.

2. W. Arveson, "Interpolation problems in nest algebras", J. Func. Anal., 20, No. 3, (1975), 208-233.

3. L. Brown, R. Douglas and P. Fillmore, "Unitary equivalence modulo the compact operators and extensions of C*-algebras". Proceedings of a conference on Operator Theory, Lecture Note in Mathematics, No. 345, Springer-Verlag, New York, (1973), 58-128.

4. L. Brown, R. Douglas and P. Fillmore, "Extensions of C*-algebras and K-Homology", Ann. of Math. (2), 105(1977), 265-324.

5. R. Curto, "Fredholm and invertible n-tuples of operators. The Deformation Problem", to appear in Trans. Amer. Math. Soc.

6. R. Douglas and C. Pearcy, "A note on quasitriangular operators", Duke Math. J. 37 (1970), 177-188.

7. P. Halmos, "Quasitriangular operators", Acta Sci. Math. (Szeged) 29 (1968), 283-293.

8. G. Kaplan, "Joint quasitriangularity of 2-tuples of essentially normal, essentially commuting operators on infinite dimensional Hilbert spaces", Dissertation, SUNI at Stony Brook, (1979).

9. N. Salinas, "Homotopy invariance of Ext(A)", Duke Math. J., 44, No.4 (1977), 777-794.

10. N. Salinas, "Quasitriangular extensions of C*-algebras and problems on joint quasitriangularity of operators", preprint.

11. V. Venugopalkrisma, "Fredholm operators associated with strongly pseudoconvex domains", J. Funct. Anal. 9 (1972), 349-373.

12. D. Voiculescu, "Some extensions of quasitriangularity II, Rev. Roum. Math. Pures Appl., 18(1973), 1439-1459.

DEPARTMENT OF MATHEMATICS
THE UNIVERSITY OF KANSAS
LAWRENCE, KANSAS 66045

Proceedings of Symposia in Pure Mathematics
Volume 38 (1982), Part 2

SPECTRAL PERMANENCE FOR JOINT SPECTRA

Raul E. Curto

Let A be a C*-subalgebra of a unital C*-algebra B and $1 \in A$. It is well known that for an element $a \in A$, $\sigma_A(a)=\sigma_B(a)$. In this note we generalize this result to n-variables. Let $a=(a_1,\ldots,a_n)$ be a commuting n-tuple of elements of A. Consider $E(A,a)=\{E_k^n(A),d_k\}_{k\in\mathbb{Z}}$ and $E(B,a)=\{E_k^n(B),d_k\}_{k\in\mathbb{Z}}$, the Koszul complexes associated with a, where each a_i is regarded as a left multiplication on A or B, respectively. For instance, if n=2,

$$E(A,a):\ 0 \longrightarrow A \xrightarrow{d_2} A\oplus A \xrightarrow{d_1} A \longrightarrow 0\ ,$$

where $d_2(b)=(-a_2b)\oplus(a_1b)$ and $d_1(b_1\oplus b_2)=a_1b_1+a_2b_2$; $b,b_1,b_2 \in A$. Let Sp(a,A) and Sp(a,B) denote the Taylor spectra of a on A or B, respectively, i.e., $Sp(a,A)=\{\lambda\in\mathbb{C}^n\colon E(A,a-\lambda)$ is not exact$\}$ and similarly for Sp(a,B) (see [1] and [3] for definitions).

THEOREM 1. $Sp(a,A)=Sp(a,B)$.

THEOREM 2. $0\notin Sp(a,A)$ if and only if

$$\hat{a}=\begin{pmatrix} d_1 & 0 & \\ d_2^* & d_3 & \\ 0 & d_4^* & \ddots \end{pmatrix} \in M_{2n-1}(A)$$

is invertible, where d_i is the ith boundary map in the Koszul complex for a $(i=1,\ldots,n)$.

EXAMPLE. (n=2)

$$0 \longrightarrow A \xrightarrow{\begin{pmatrix}-a_2\\ a_1\end{pmatrix}} A\oplus A \xrightarrow{(a_1\ a_2)} A \longrightarrow 0 \quad \text{is exact} \quad \text{iff} \quad \begin{pmatrix} a_1 & a_2 \\ -a_2^* & a_1^* \end{pmatrix} \text{ is invertible.}$$

COROLLARY 1. If $A\subset L(H)$, the algebra of bounded operators on a Hilbert space H, then $Sp(a,A)=Sp(a,H)$.

AMS(MOS) Subject Classifications(1980): 47C15,46L05,46L10,47A62.

More generally, let $\sigma_{\pi,k}$ and $\sigma_{\delta,k}$ be the joint spectra considered by Słodkowski [2], that is, for instance,

$$\sigma_{\delta,k}(a,A)=\{\lambda\in\mathbb{C}^n : E(A,a-\lambda) \text{ is not exact at some stage } i,\ 0\leq i\leq k\}.$$

THEOREM 3. $\sigma_{\pi,k}(a,A)=\sigma_{\pi,k}(a,B)$ and $\sigma_{\delta,k}(a,A)=\sigma_{\delta,k}(a,B)$ (all k)

COROLLARY 2. If $A\subset L(H)$, then

$$\sigma_{\pi,k}(a,A)=\sigma_{\pi,k}(a,H) \text{ and } \sigma_{\delta,k}(a,A)=\sigma_{\delta,k}(a,H) \quad (\text{all } k).$$

COROLLARY 3. Let C be a unital C*-algebra and $\phi : A\to C$ be a *-homomorphism. Then

$$\sigma_{\pi,k}(\phi(a),C)\subset\sigma_{\pi,k}(a,A) \text{ and } \sigma_{\delta,k}(\phi(a),C)\subset\sigma_{\delta,k}(a,A)$$

for all k. If ϕ is injective, there is equality in both cases.

The results above are based on the following generalization of [1, Proposition 3.4]. Let $0\leq i_0,i_1,\ldots,i_{k+1}\in\mathbf{Z}$, $A_j=A\otimes\mathbb{C}^{i_j}$ and

$$(\mathbf{A}):\quad A_{k+1}\xrightarrow{a_{k+1}}A_k\xrightarrow{a_k}\cdots\to A_j\xrightarrow{a_j}A_{j-1}\to\cdots\to A_1\xrightarrow{a_1}A_0\to 0\ ,$$

where the a_j's are matrices over A of the right size and $a_ja_{j+1}=0$. Let $a_j^*=(a_{ml}^*)$ if $a_j=(a_{lm})$. Then $(\mathbf{A})$ is exact at $j=0,1,\ldots,k$ iff $a_j^*a_j+a_{j+1}a_{j+1}^*\in M_{i_j}(A)^{-1}$ for all $j=0,1,\ldots,k$.

Details of this work will appear elsewhere.

REFERENCES

1. R. E. Curto, Fredholm and invertible n-tuples of operators. The deformation problem, Transactions AMS, to appear.

2. Z. Słodkowski, An infinite family of joint spectra, Studia Mathematica, 61(1977), 239-255.

3. J. L. Taylor, A joint spectrum for several commuting operators, Journal of Functional Analysis, 6, 2(1970), 172-191.

DEPARTMENT OF MATHEMATICS
THE UNIVERSITY OF KANSAS
LAWRENCE, KANSAS 66045

Proceedings of Symposia in Pure Mathematics
Volume 38 (1982), Part 2

UNBOUNDED NEGATIVE DEFINITE FUNCTIONS AND PROPERTY T FOR LOCALLY COMPACT GROUPS

Charles A. Akemann (joint work with Martin Walter)

A complex function f on a locally compact group G is called negative definite if, for all n-tuples $x_1,\ldots,x_n$ in G, the $n \times n$ matrix

$$\left(f(x_i) + \overline{f(x_j)} - f(x_j^{-1}x_i) \right)_{i,j=1}^{n}$$

is positive semi-definite. The group G has property T if the trivial representation is isolated in the dual space of G. We shall be primarily interested in non-compact groups since our results are trivial in the compact case. An example of such a group is $SL(3,R)$. For a further discussion see [D.A. Kazhdan, Connection of the dual space of a group with the structure of its closed subgroups, Functional Analysis and its Applications, 1(1) 1967, 63-65]. Our principal theorem states that G has property T iff there are no continuous unbounded negative definite functions on G. As a Corollary we can easily see why a discrete group with property T cannot be inner amenable. For more applications of the "property T" concept, see the papers by Alain Connes in this volume. The paper summarized above will appear in the Canadian Journal of Mathematics.

University of California, Santa Barbara

0082-0717/80/0000-0495/$01.25

ABCDEFGHIJ–CM–898765432